Advances in Filtration and Separation Technology

VOLUME 2

Filtration and Separation in Environmental Control Technology

Advances in Filtration and Separation Technology

VOLUME 2

Filtration and Separation in Environmental Control Technology

American Filtration Society

Gulf Publishing Company
Houston, London, Paris, Zurich, Tokyo

Advances in Filtration and Separation Technology

VOLUME 2

Filtration and Separation in Environmental Control Technology

Library of Congress Catalog Card Number: 90-082432

ISBN 0-87201-096-1

The information herein was presented at the second regional meeting of the American Filtration Society, March 19–24, 1990, Washington, D.C.

Contents

THEORY, EXPERIMENT AND MECHANISM OF LIQUID FILTRATION—I

PHARMACEUTICALS

AUTOMOTIVE FILTRATION

ELECTROSTATICS

THEORY, EXPERIMENT AND MECHANISM OF LIQUID FILTRATION—II

FILTRATION TESTING—II

FILTER MATERIALS TECHNOLOGY

ENERGY GENERATION—III FLUID/SOLID SEPARATION IN COAL PROCESSING

METAL WORKING

PARTICLE SEPARATION IN MINERAL PROCESSING—II

PLANT SCALE COMPARISON OF
MULTI-MEDIA AND DUAL MEDIA FILTERS

Mr. Elliott Barnett
Martin Marietta Energy Systems, Inc.
Post Office Box 2009
Oak Ridge, Tennessee 37831 (615) 574-9595
Dr. Bruce Robinson, University of Tennessee
Mr. Wayne Loveday, Knoxville Utilities Board
Mr. Joe Snyder, Knoxville Utilities Board

The objectives of this work were to compare the filtration performance of a multi-media filter versus a dual media filter at the Mark B. Whitaker Filtration Plant in Knoxville, Tennessee. The main considerations and parameters monitored during the filter study included effluent turbidity, headloss difference (i.e. energy costs), initial costs (media and installation), each filter's ability to remove particles, percentage of time each filter stayed in service, amount of backwash water used, ability to meet potential future effluent requirements for turbidity and aluminum, resistance to flow surges, and amount of media loss. During the entire study the filtration rates were varied, but both filters were maintained at approximately the same filtration rate, and subjected to identical settled water conditions. Both the dual and multi-media filters produced an excellent quality drinking water. Neither filter media type had any difficulty producing an effluent of less than 0.5 NTU's. However, the multi-media filter was more successful at attaining a 0.1 NTU potential future effluent standard with the pretreatment used. While cost analysis favored the dual media, the multi-media filter produced a higher quality drinking water than the dual media filter during the nine months of the study.

In April of 1988 two filters at the Mark B. Whitaker (MBW) Water Treatment Plant in Knoxville, TN were reconstructed. One filter was reconstructed as a dual media filter and the other as a multi media filter. The dual media filter consisted of 20 inches of anthracite coal (effective size 0.98 mm and uniformity coefficient 1.37), 10 inches of filter sand (effective size 0.52 mm and uniformity coefficient 1.30), and 12 inches of support gravel. The multi media filter consisted of 18 inches of anthracite coal (effective size 1.07 mm and uniformity coefficient 1.63), 9 inches of filter sand (effective size 0.53 mm and uniformity coefficient 1.30), 3 inches of fine garnet (effective size 0.28 and uniformity coefficient 1.61), 3 inches of large garnet (effective size 1.60 and uniformity coefficient 1.19), and 9 inches of support gravel. Each filter had a surface area of approximately 850 ft^2. Both filters had identical surface washing devices, backwash troughs and underdrain systems.

The filter comparison study began in May of 1988 and ended in February of 1989. During the entire test the filtration rates were varied, but both filters were maintained at approximately the same filtration rate, and received identical settled water. The average settled water conditions for the entire study were an influent turbidity of 4.24 NTU's, a pH of 6.95, a Temperature of 19.39 C^o, an aluminum concentration of 852.8 ppb, an alkalinity of 57.21 ppm, and a hardness of 86.54 ppm.

The main objectives of the filter comparison study were to determine which filter type would perform the best and which was the most cost effective. The main considerations were:

1. Effluent turbidity
2. Head loss difference (i.e. energy costs)
3. Initial cost of each media and installation costs
4. Each filter's ability to remove particles
5. The percentage of time each filter stayed in service
6. The amount of backwash water used by each filter
7. Each filter's ability to meet potential future effluent requirements for turbidity and aluminum
8. Each filter's ability to resist flow surges
9. Media loss

Beginning in July 1988 after two months of "shakedown" each filter was operated at a nominal filtration rate of 3, 4, 5, and 6 gpm/ft^2 in an incremental manner. In other words, the filters were operated at 3 gpm/ft^2 for a long enough period of time to obtain a sufficient number of runs in order to make a reasonable comparison at that rate, and then the filtration rate was increased to 4 gpm/ft^2 and so on up to 6 gpm/ft^2. After both filters had proceeded through all four filtration rates they were then subjected to maximum rates by opening the effluent valve of the filters all the way and allowing each filter to be operated at whatever maximum rate it was hydraulically capable of. This average maximum rate was 6.50 gpm/ft^2 for the dual media and 6.57 gpm/ft^2 for the mixed media. Beginning in November 1988 the filters were run through the same series of filtration rates in order to obtain data at different settled water conditions, i.e. summer versus winter conditions.

<u>Effluent Turbidity</u>

Both the dual and multi media filters produced an excellent quality drinking water throughout the entire nine months of the filter study. The present standard for effluent turbidity is 1.0 NTU and neither filter exceeded this criteria during the entire filter study. The average effluent turbidity for either filter never exceeded 0.20 NTU's at any filtration rate. This shows how well both filters performed at producing high quality drinking water.

The percentage of the filter run that the effluent turbidity was at or below 0.5 NTU was observed due to the likelihood that the new effluent criteria for drinking water will soon be this low, and because 0.5 NTU was used as the maximum turbidity in determining how long a filter to waste procedure would occur during each filter run. Both the multi and dual media filters produced an effluent that stayed at or below 0.5 NTU greater than 95 percent of the time, at each filtration rate.

Due to the likelihood that the effluent turbidity standard for drinking water will become 0.1 NTU during the useful life of the filters, each filter's ability to meet this potential limit was examined. The effluent turbidity of the multi media filter stayed at or below 0.1 NTU for a greater percentage of it's filter run than the dual media effluent did, at all filtration rates except 3 gpm/ft^2 during the second rotation of filtration rates. The multi media filter obtained an effluent turbidity of 0.1 NTU or less an average of 76 percent of it's filter runs, whereas the dual media filter averaged 40 percent of it's filter runs. It is very likely that either filter could achieve 0.1 NTU with proper pretreatment (e.g. polymers). It is speculated that the dual media may require higher polymer dosages than the mixed media, at least partially offsetting the capital cost advantages of the dual media.

During the second round of filtration rates the effluent turbidities of both filters increased slightly. This is believed to have been caused by the decrease of approximately 10 C^o in the influent water temperature.

<u>Head Loss Difference</u>

During the normal operation of water filtration the filter accumulates particles that are removed from the water. After a while the head loss builds up in the filter to a point where the filter must be backwashed. Since the multi media filter has two extra layers of media, the fine and large garnet, the initial head loss through the layers of filter media is greater than through the dual media filter. The initial head loss through each filter was dependent on the filtration rate at which it was operated at. The average rate of head loss increase in both filters was approximately the same, 0.15 ft/hr.

The only time at which head loss difference translated into a cost difference at the MBW Water Plant was when the plant operated at above 45 MGD. Below 45 MGD, water flowed from the sedimentation/flocculation basins to the filters by gravity flow. At production rates greater than 45 MGD transfer pumps were required to pump the additional water to the top of the filters. A calculation of the amount of energy required by each filter estimated that the multi media would cost $385.25 per year more to operate because of the slightly higher head loss.

<u>Initial Costs</u>

The capital costs, i.e. media and installation costs, of the multi media filter were approximately $6000 more than the dual media filter.

Particle Analysis

A particle analysis was performed with a HIAC particle counter on the influent and effluent water of both filters to determine the percent removal of particles and to make a comparison to see which filter was more effective at removing particles. In order to make this test as congruent as possible both filters were operated at the same filtration rate, washed one right after the other, and water samples were taken at the same time. Therefore, both filter runs occurred simultaneously and both filters were subjected to identical influent water conditions. The chosen nominal flowrate was 6 MGD, since the plant had ten filters and desired to be able to operate at 60 MGD. The results of the particle analysis showed excellent particle removal by both of the filters. The average percent particle removal was 99.8 percent for both the dual and multi media filters.

Backwash Water Requirements

As was expected, on the average the multi media filter required a slightly longer backwash, 12 minutes versus 10 minutes for the dual media, as judged by the operators to produce a clean filter. This amount of time out of service translates into time when water is not being produced. Both filters were in service for more than 99 percent of the time during the filter study. However, the dual media filter was in service slightly longer than the multi media filter. The multi media filter also required a small amount more of wash water than the dual media. Assuming a water production cost of $0.25/1000 gallons the multi media costs for backwash water would have been between $430.00 and $1730.00 per year more than the dual media, depending on the filtration rate maintained. Both filters would require roughly the same filter-to-waste at the beginning of a run, approximately 0.4 to 0.5 percent of the run on average.

Aluminum Removal

Aluminum has been proposed to be regulated in drinking water although the Environmental Protection Agency (EPA) has dropped it from it's list of contaminants to be regulated at present. Europe does have such a standard, and if a U.S. standard is adapted it will probably be set at 50 ppb.

Both filters removed approximately the same amount of aluminum throughout the entire filter study. The percentage difference in aluminum removal between the two filters was never greater than 5 percent. On the average, neither filter removed aluminum to a level as low as 50 ppb.

Flow Surges

Both the dual and the multi media filters were subjected to rapid 33 percent flowrate increases in order to determine whether corresponding responses could be seen in their effluent turbidities. A 33 percent flowrate increase was chosen to simulate a four filter declining rate filtration plant. When one of the four filters was being washed the other three filters would instantly experience a 33 percent flowrate increase. Both times the experiment was performed no significant increase in effluent turbidity was noticed when the flowrate was rapidly increased. There was no significant difference in the response of either filter.

Media Loss

Two core samples were taken in each filter to determine if any media was being lost during normal backwashing of the filters. There was no detectable media loss in either filter.

Conclusions

Although the initial and operational costs of the multi media filter exceeded the dual media, the multi media produced a higher quality drinking water. The multi media filter produced an effluent with a turbidity consistently less than the dual media filter and is therefore considered to have produced a better quality drinking water.

The Effect of Temperature Induced Changes in the Carbonate Buffer System on Adsorption/Destabilization Flocculation of Kaolinite With Alum or Iron

Dr. Adrian T. Hanson
Civil Engineering
New Mexico State University
Las Cruces, New Mexico
and
Dr. J. L. Cleasby
Civil Engineering
Iowa State University
Ames, Iowa

Introduction Water utilities treating surface waters in the temperate region face some unique challenges; the water which they treat may be 5 °C or colder for 5 months of the year. Velz, in 1934, performed research into the effects of temperature on flocculation with metal salts. Velz noted that the flocculation process was not temperature sensitive if the appropriate pH was maintained. Camp, in 1940, following Velz's lead, did further work on cold temperature effects on floc formation using iron salts to form iron hydroxide floc in distilled water. Camp reported a significant increase in the optimum flocculation pH with a drop in temperature from 28 °C to 4 °C. A phenomenon noted earlier by Velz. Morris and Knocke (1984) presented data on the effects of temperature on the coagulation of clay using iron and alum. This work appeared to contradict the findings of both Velz and Camp. However, Work by Hanson (1989), and Hanson and Cleasby (1989) unified the findings of Velz, Camp, and Morris and Knocke, by pointing out the relationship between pOH and flocculation with metal salts.

The carbonate buffer system, which is the dominant buffer systems in natural waters, controls the pH response of natural waters. As the temperature of the water being treated changes, both the equilibrium concentrations of the various species and the reaction kinetics may change. One must ask if the temperature range experienced in nature will cause changes in the carbonate system chemistry which are important to the flocculation process. It has been well established that flocculation with metal salts is sensitive to temperature induced changes in the pH/pOH relationship. It is possible that changes in the buffer system will effect changes in the pH response which are also have a significant impact on the flocculation process significant.

The carbonate system consists of 4 major species; CO_3^{--}, HCO_3^-, CO_2, and H_2CO_3. In the range of pH values of interest the CO_3^{--} is negligible, so we will only consider the partitioning of the others. Figures 1 and 2, calculated from Stumm and Morgan (1982), indicate that, in the pH range of interest, a change in temperature from 20 to 5 °C does not significantly effect the equilibrium speciation or the buffer capacity of the carbonate system.

The other potential concern is reaction kinetics. When alum ($Al_2(SO_4)_3*18H_2O$) is added to water the following overall reaction takes place.

$$Al_2(SO_4)_3*18H_2O + 3Ca(HCO_3)_2 \dashrightarrow 2Al(OH)_3 + 3CaSO_4 + 18H_2O + 6CO_2$$

This is actually a composite reaction which gives a simple description of a more complex series of reactions. It is suggested that the actual sequence of events is as follows. The alum is added and quickly forms $Al(OH)_3$ precipitate. This leaves an excess of H^+ ions in solution, because each of the aluminum has tied up three hydroxides. The carbonic acid-bicarbonate reaction:

$$H_2CO_3 \longleftrightarrow H^+ + HCO_3^-$$

is quick, on the order of 0.05 to 0.10 seconds at 25 °C (Stumm and Morgan, 1981). However, the carbonic acid-carbon dioxide reaction:

$$H_2CO_3 \longleftrightarrow CO_2 + H_2O$$

is slow, on the order of 24 to 40 seconds at 25 °C (Stumm and Morgan, 1981).

If we assume a temperature correction of the form $K = Ae^{(-Ea/RT)}$, then $K_1/K_2 = 2.5$ for a temperature change from 25 °C to 5 °C. The estimated reaction times at the two temperatures are shown in Table 1.

Table 1. Reaction Kinetics for Carbonate System at two Temperatures

Reaction	Reaction Time (Seconds)	
	25 °C	5 °C
H_2CO_3/HCO_3^-	0.01-0.05	0.25-0.13
H_2CO_3/CO_2	25-40	62.5-100

Table 2 contains approximate reaction times for the reactions between the various alum products and the colloid surfaces present in water. The relative time scales indicate that the kinetics of the carbon dioxide/carbonic acid reaction could be important in the particle destabilization process.

Table 2. Reaction type versus time scale (Amirtharajah, 1987)

Reaction	Time Scale (seconds)
Al (III) monomer adsorption	< 0.1
Al (III) polymer formation and adsorption	0.1 to 1
Formation of sweep floc aluminum hydroxide precipitate	1 to 7

<u>Experimental</u> The objective of this work is to demonstrate what impact, if any, changing temperature has on the carbonate buffer system, and what impact those changes have on the flocculation process. The experimental conditions are shown in Table 3. The effects of temperature were measured by changing the system chemistry and measuring the impact of these changes on the flocculation process. This required the performance and monitoring of flocculation experiments in a controlled temperature setting. The controlled

temperature conditions were achieved using a walk-in constant temperature room. The temperature in the room was monitored and controlled using a personal computer (PC) based data acquisition and control system. The flocculation work was performed in a bench scale batch reactor similar to the reactor used by Argaman and Kaufman (1968). The turbine impeller, shown in Figure 3, was used. The following process control parameters were monitored:

- o reactor pH
- o reactor temperature
- o impeller rpm
- o energy input

A dispersed Kaolinite clay was used as the primary particle system. The suspension dilution water was Ames, IA tap water buffered with 100 mg/L $NaHCO_3$ (Table 4).

The impact of the system chemistry on the flocculation process was measured by monitoring the reduction in primary particles over time. The particle size distribution analysis were determined using automatic image analysis (Lemont OASYS), coupled directly to a microscope.

<u>Results</u> Figure 4 shows the change in the pH of the suspension being flocculated following the addition of the alum coagulant. There are a number of features of interest in this figure.

First, note that the final equilibrium pH of the 5 °C suspension is approximately 0.5 pH units higher than the final equilibrium pH of the 20 °C suspension. This difference in pH represents the shift needed to maintain a constant pOH. The rate at which the 5 °C system recovers from the initial pH depression is exactly the same at a pH of 6.8 as it is at a pH of 7.4. Thus, the pH recovery rate shown in the figure is not a sensitive function of pH, but is linked to temperature.

Second, the pH adjustment for experiments carried out a 20 °C, was performed during the rapid mix process. The solution pH was lowered to approximately 7.5 to 7.2 prior to rapid mixing. The alum was injected at the beginning of rapid mixing, and the pH was fine tuned to pH = 6.8 during rapid mixing. At 5 °C fine tuning the pH during the rapid mix was not possible. The addition of the alum suppressed the pH and the pH did not recover until the rapid mix was over.

Third, Comparing data for 20 and 5 °C at a pH of 6.8, one sees that the rate at which the pH drops and lowest pH reached is similar for both temperatures. These two things give an indication of the rapidity and extent of the aluminum hydroxide reaction. It appears that, within the resolution limits of this data, the alum reaction which caused the depression was not significantly retarded by the 15 degree change in temperature. Recalling Table 2, the reactions of importance in adsorption destabilization flocculation occur in a time frame measured in 0.1 seconds to 2 seconds. From this one might expect that the extended duration of the pH depression at 5 °C will not alter the destabilization of the clay particles. Based on this it is suggested that one might not expect the changes in the carbonate buffer system kinetics do not seem to inhibit the alum reaction measurably, and one must look else where to explain any measured effect of temperature on flocculation. Figure 5 indicates the impact of temperature and pH on alum flocculation, and Figure 6 indicates the impact of these variables on iron flocculation. From Figures 5 and 6 one can conclude that, if the pOH is held constant, the changes noted in the pH response does not alter the flocculation efficiency of metal salts in the adsorption/destabilization flocculation region.

The pOH at the beginning of the rapid mixing process seems to be critical at low temperatures. However, the pH, and thus the pOH, does not appear nearly as critical during rapid mixing. It seems probable that the pOH at the time the aluminum is injected determines the adsorption rate of the aluminum hydroxide polymers from solution onto the solid surfaces, and it determines what aluminum hydroxide polymers will be formed. The drop in pH caused by the aluminum reaction would enhance the adsorption process rather than hinder it. The lower pH may cause the particles to be restabilized with a positive charge, and very little flocculation will occur until the pH stabilizes. This might partially explain the lag noted in the constant pOH, low temperature flocculation test, in comparison to the 20 °C test. One also notes that, although the flocculation efficiency for the 5 °C experiment reaches a value equivalent to the removal efficiency measured at 20 °C, there is a time lag prior to flocculation starting. One must wonder if this time lag in the onset of measurable flocculation is related to the slow pH recovery at low temperature.

This leads one to ask if the slow pH recovery can be eliminated, thus improving the low temperature flocculation efficiency. To answer this question one must determine the exact mechanism causing the extended duration of the pH depression at low temperature. The evidence presented indicates that the carbonate system chemistry is the cause. However, there are a two other potential causes which need to be addressed prior to fully embracing the kinetics argument, these are: a slow response from the pH probe, Poor gas transfer due to reduced mixing efficiency.

The pH probe response time was tested using the 4, 7, and 10 buffers at both high and low temperature, and the probe was eliminated as a potential source of the slow pH recovery.

Figures 7 and 8 illustrate the impact of mixing intensity and time on the pH recovery rate. The low energy condition represents a G of 450 sec^{-1} (250 rpm), and the high energy condition represents a G of 1020 sec^{-1} (500 rpm). If the diffusion of CO_2 from the bulk liquid to the atmosphere is limiting the pH recovery of the system, increased mixing intensity should affect the recovery time. Figure 7 shows that increasing the mixing intensity from G=450 sac^{-1} to 1000 sec^{-1} had no impact on the recovery rate. The rapid mix phase was 60 seconds in duration, and Figure 7 indicates that the system pH did not reach equilibrium for approximately 100-120 seconds. Figure 8 represents the impact of extending the high energy rapid mixing to 180 seconds. It is seen that the duration of the rapid mix phase was also not significant.

Having considered the probe and the mixing regime, the obvious conclusion is that the buffer system chemistry caused the extended pH depression. This possibility was suggested by the literature presented earlier. Figure 3 indicates how well the experimental data agrees with the theory presented. At 20 °C, it took about 15 seconds for the pH to recover about 1/2 way out of its depression. At that point acid was added to adjust the pH to 6.8. However, it appears that it would have achieved equilibrium in about 30-40 seconds. We also see that the recovery, at 5 °C, takes about 120 seconds at both pH values. It appears that the slow pH recovery agrees very well with what carbonate chemistry would lead us to expect, and the slow recovery is indeed due to the slow equilibration time of the H_2CO_3/CO_2 partitioning reaction.

It is also interesting to note that the magnitude of the depression is about the same for both the 5 and 20 °C samples near a pH of 7, and is considerably larger for the test run at pH of 7.5. This demonstrates two things. First, the change in the magnitude of the depression with pH is a nice demonstration of the change in the carbonate system buffer capacity with pH. Figure 2 shows the buffer intensity decreasing rapidly between pH=6.2 and pH=8.0. Thus, one would anticipate that the magnitude of the depression would be larger at a pH of 7.4 than at a pH of 6.8. Second, the similarity in the magnitude of the depression between the 20 °C and the 5 °C demonstrates that the carbonate chemistry in this pH range is insensitive to temperature changes of this magnitude. Theory would lead us to expect this, as is shown in the curves in Figure 1 in the Literature Review.

Conclusions The following conclusions were drawn from this work:

1. The extended pH depression noted during rapid mixing with metal salts is a function of the carbonate buffer systems partioning kinetics (carbonic acid/carbon dioxide).

2. The extended pH depression is not a function of mixing intensity, mixing duration, or pH probe response time.

3. The extended pH depression does not appear to have an adverse impact on the ultimate flocculation efficiency with metal salts at low temperature in the adsorption/destabilization region, but it may be partially responsible for the lag in the onset of flocculation measured at low temperature.

Acknowledgements: This work was funded in part by the National Science Foundation through award number CES8613737, and by the Engineering Research Institute at Iowa State University. The assistance of Dave Scheoller of the Civil Engineering Analytical lab in installing and operating the computer based data acquisition and process control is gratefully acknowledged.

Bibliography

Argaman, Y., and W. J. Kaufman, Turbulence in Orthokinetic Flocculation, Sanitary Engineering Research Laboratory, College of Engineering and School of Public Health, University of California - Berkley, SERL Report No. 68-5, July 1968.

Camp, Thomas R., D. A. Root, and B. V. Bhoota, Effects of Temperature on the Rate of Floc Formation, JAWWA, Vol. 32, No. 11, Nov. 1940, pp. 1913 - 1927.

Hanson, Adrian, The Effect of Water Temperature and Reactor Geometry on Turbulent Flocculation. Doctoral Dissertation, Iowa State University, 1989.

Hanson, A. T., and J. L. Cleasby, The Effects of Temperature on Turbulent Flocculation: Fluid Dynamics and Chemistry. AWWA Annual Conference Proceedings, Los Angeles, California, AWWA, 1989.

Morris J. K., and W. R. Knocke, Temperature Effects on the Use of Metal-Ion Coagulants in Water Treatment, JAWWA, Vol. 76, No. 3 March 1984, pp. 74 - 79.

Stumm, W., and J. J. Morgan. Aquatic Chemistry. 2nd edition. John Wiley and Sons, New York, 1981.

Velz, C. J., Influence of Temperature on Coagulation, Civil Engineering, Vol. 4, No. 7, 1934, pp 345 - 349.

Table 3. Summary of Experimental Conditions used in Flocculation of Kaolinite in a Turbulent Flow Field

Primary Particles: Kaolin (Kentucky Ball Clay) 1.88 μm dia.
Particle Concentration: 25 mg/L, 24 NTU, 5.8×10^6 particles/mL
Dilution Water: Ames, IA tap water buffered with 100 mg/L $Na(HCO_3)$
Coagulant: Alum as $Al_2(SO_4)_3*18\ H_2O$; 10 mg/mL aged at room temperature overnight.
Coagulant Dose: 5 mg/L as $Al_2(SO_4)_3*18\ H_2O$ Base pH: 6.8 at 20 °C

Coagulant: Ferric Sulfate as $Fe_2(SO_4)_3*6\ H_2O$; 10 mg/mL aged at room temperature overnight.
Coagulant Dose: 4 mg/L as $Fe_2(SO_4)_3*6\ H_2O$ Base pH: 5.5 at 20 °C

Rapid Mixing: 60 seconds

Flocculation: 30 minutes @ 20 °C; 45 minutes @ 5 °C

<u>Turbulent Parameters</u>

Temperature (°C)	Turbine RPM	Energy / Unit Mass (cm^2/sec^3)	G-RMS Vel. Grad. (sec^{-1})	Kolmog. Micro. Scale (μm)	Turbulent Parameter Held Const.
20	30	4.89	22	214	Baseline
20	60	37.60	60	129	Baseline
5	30	4.89	18	291	Const. ϵ
5	60	37.6	50	175	Const. ϵ

Table 4. Dilution water chemical analysis summary

Parameter	Units	Dilution Water Batch 1	2	3	4	5	6
HCO3-	mg/L	114	114	116.6	119	114	139
CO3--	mg/L	3.5	7.4	6.3	11.8	16.4	---
PH	-log[H+]	8.54	8.8	8.75	8.89	8.95	8.3
NO3+NO2-N	mg/L as N	0.61	0.45	0.78	0.63	0.33	0.2
TOTAL-P	mg/L as PO4	0.097	0.32	0.1	0.075	0.15	0.14
SULFATE	mg/L as SO4	129	121	119	121	114	111
CHLORIDE	mg/L	44.4	31	29.6	31.2	33.4	32
Na	mg/L	48.5	47	47.7	46.2	44.7	52.1
K	mg/L	2.43	2.65	2.73	2.56	2.48	2.56
Ca	mg/L	56.9	50.4	62.4	60.9	58.4	59.5
Mg	mg/L	10.5	11	6.11	7.94	9.88	4.56
Mn	mg/L	0.0018	0.004	0.012	0.004	0.004	0.001
Fe	mg/L	0.017	0.083	0.132	0.054	0.056	0.015
Al	mg/L		0.03	0.06	0.03	0.038	0.02
NPOC	mg/L	1.54			0.6	1.5	1.88

Batch	1	2	3	4	5	6
Date	11/04/87;	12/02/87;	01/28/88;	02/26/88;	5/27/88;	8/18/88

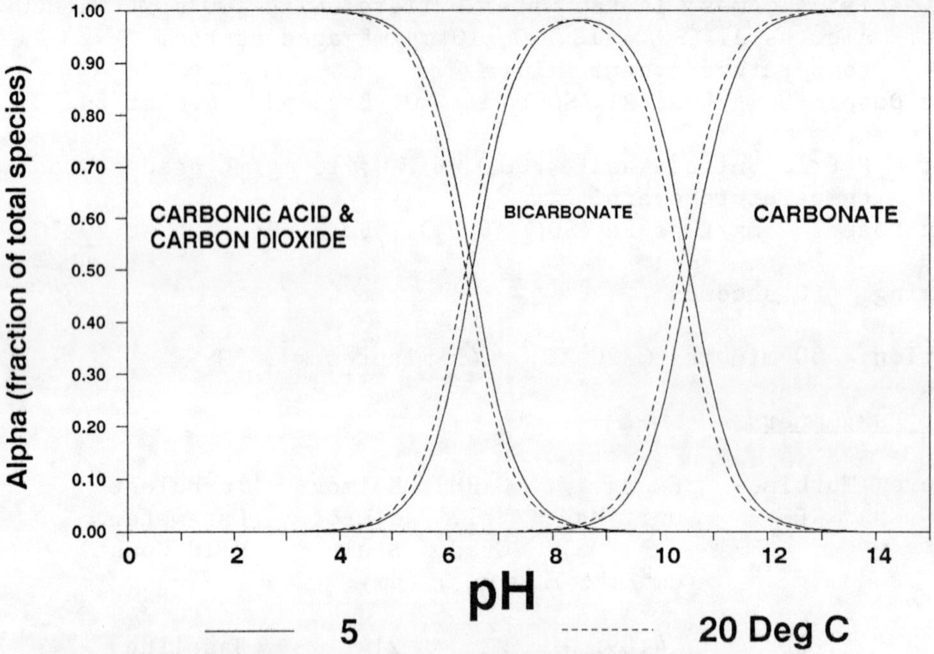

Figure 1. The effect of temperature on the speciation of the carbonate system (Stumm and Morgan, 1981)

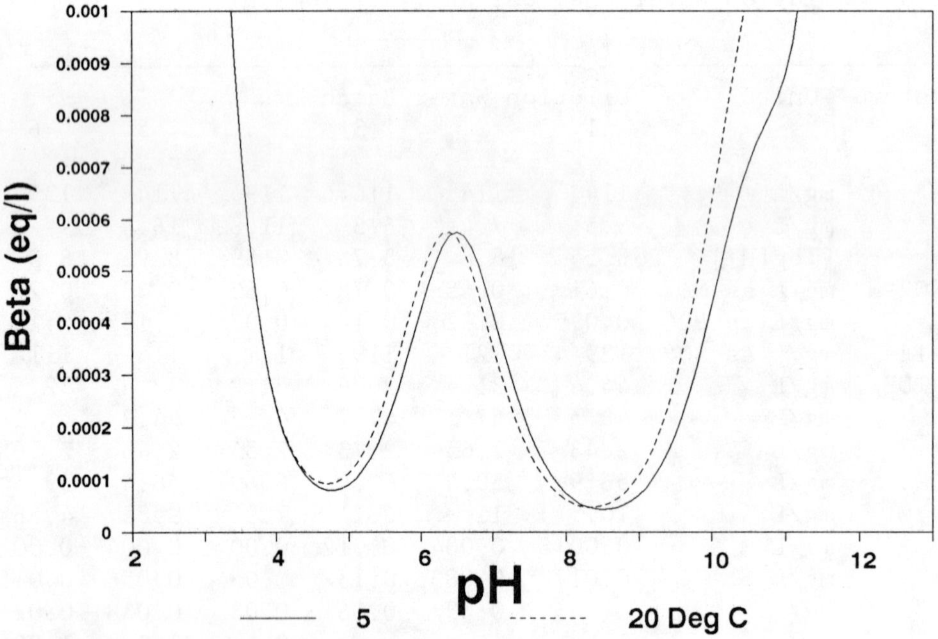

Figure 2. The effect of temperature on the buffer intensity of the carbonate system (Stumm and Morgan, 1981)

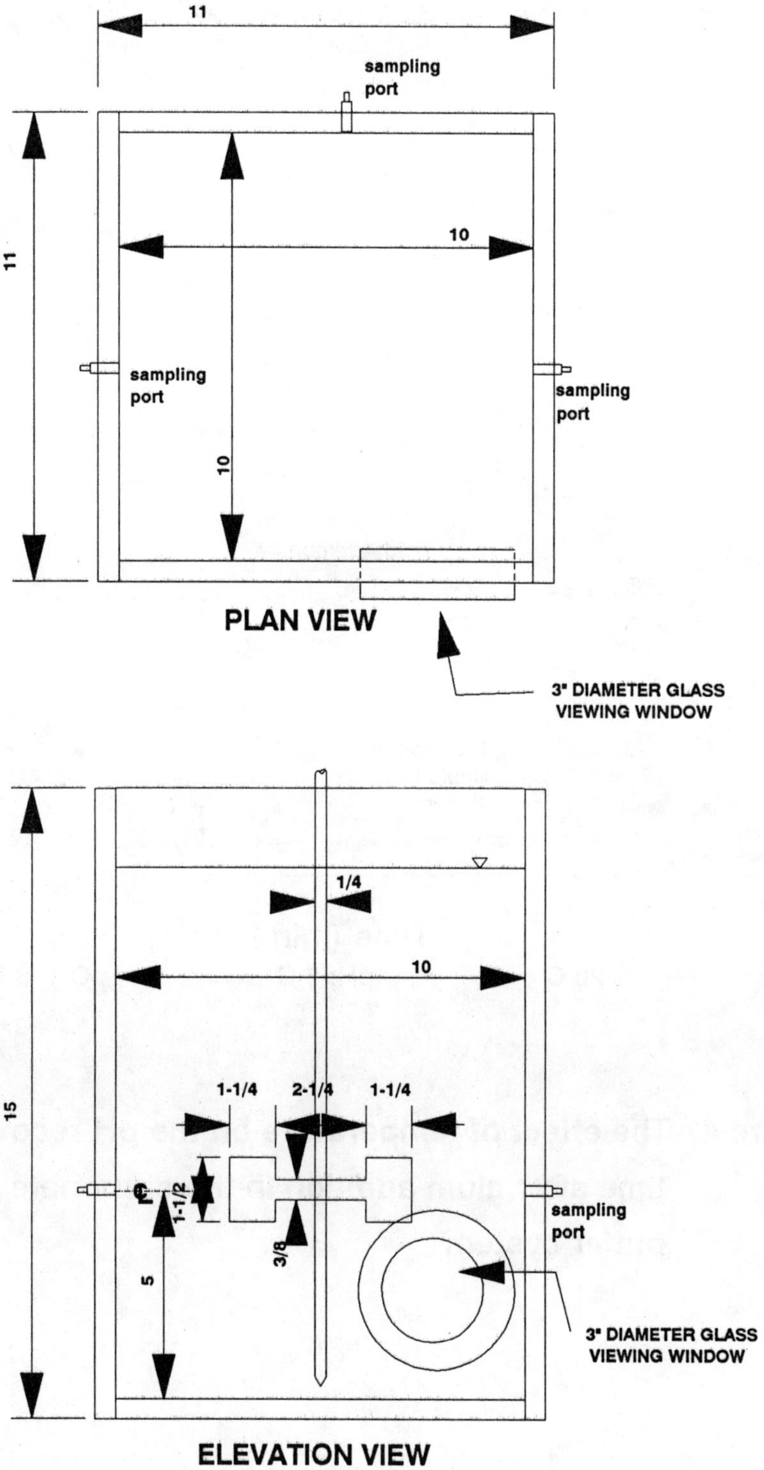

Figure 3. Schematic of the batch reactor (Argaman & Kaufman, 1968)

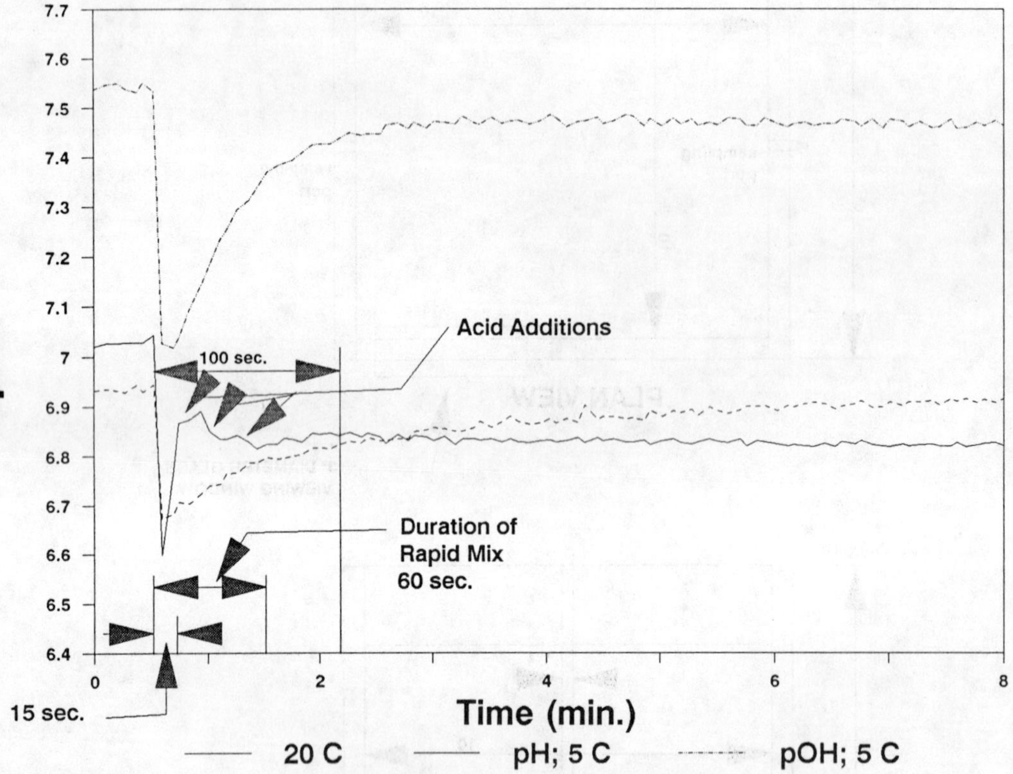

Figure 4. The effect of temperature on the pH recovery time after alum addition in the carbonate buffer system

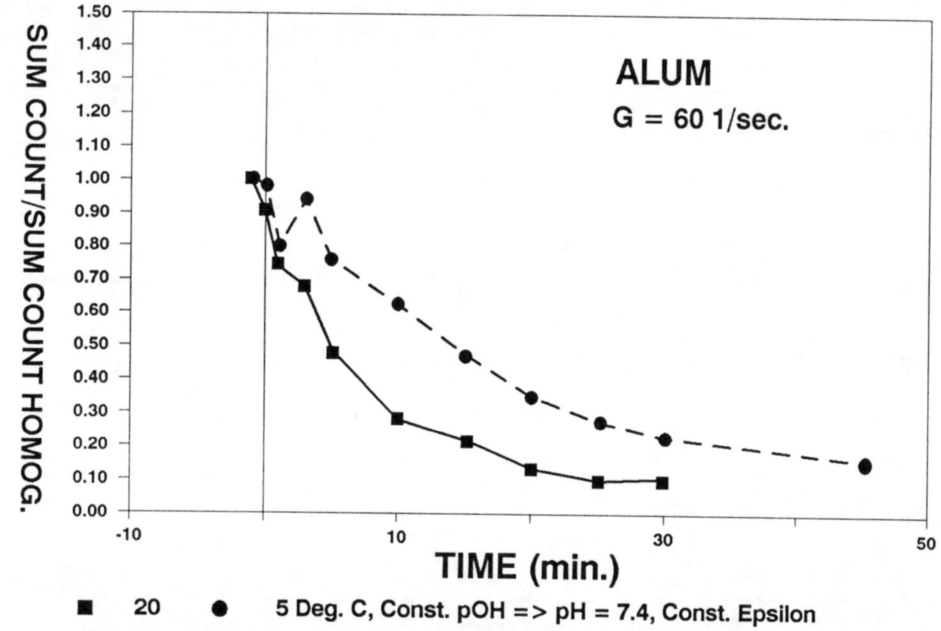

Figure 5. The effect of temperature on A/D flocculation in a batch reactor (Alum)

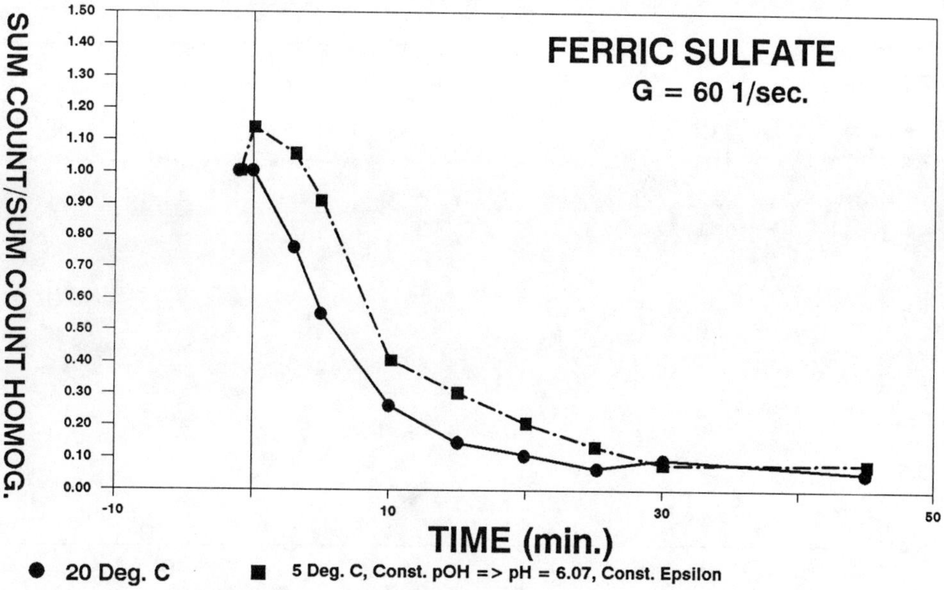

Figure 6. The effect of temperature on A/D flocculation in a batch reactor (Ferric Sulfate)

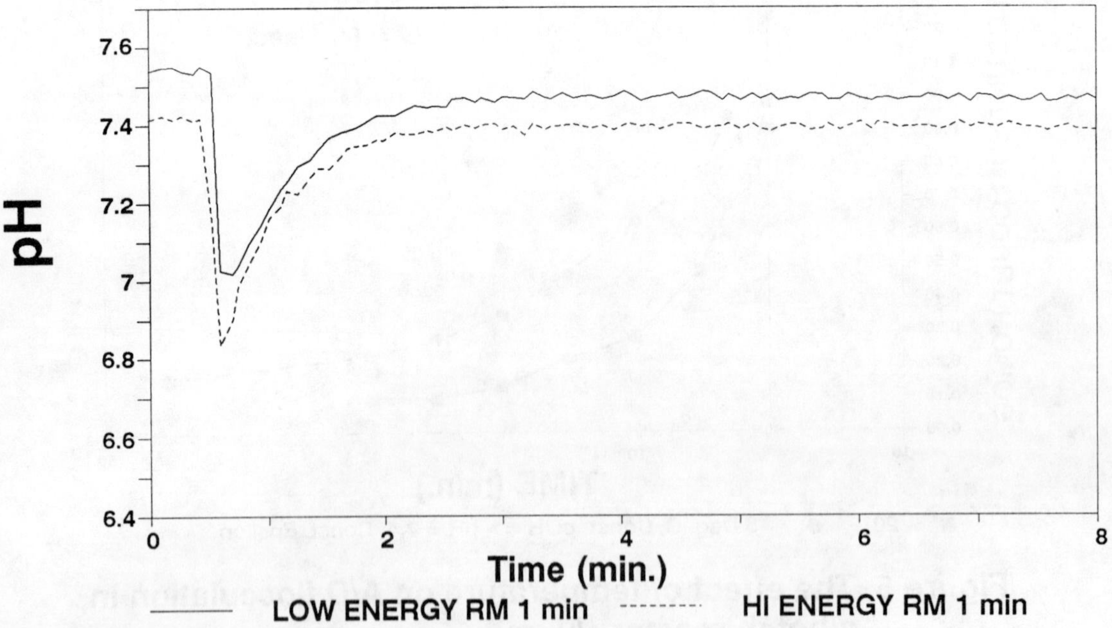

Figure 7. The effect of rapid mixing intensity on the pH recovery time after alum addition at 5 Deg. C in buffered tap water.

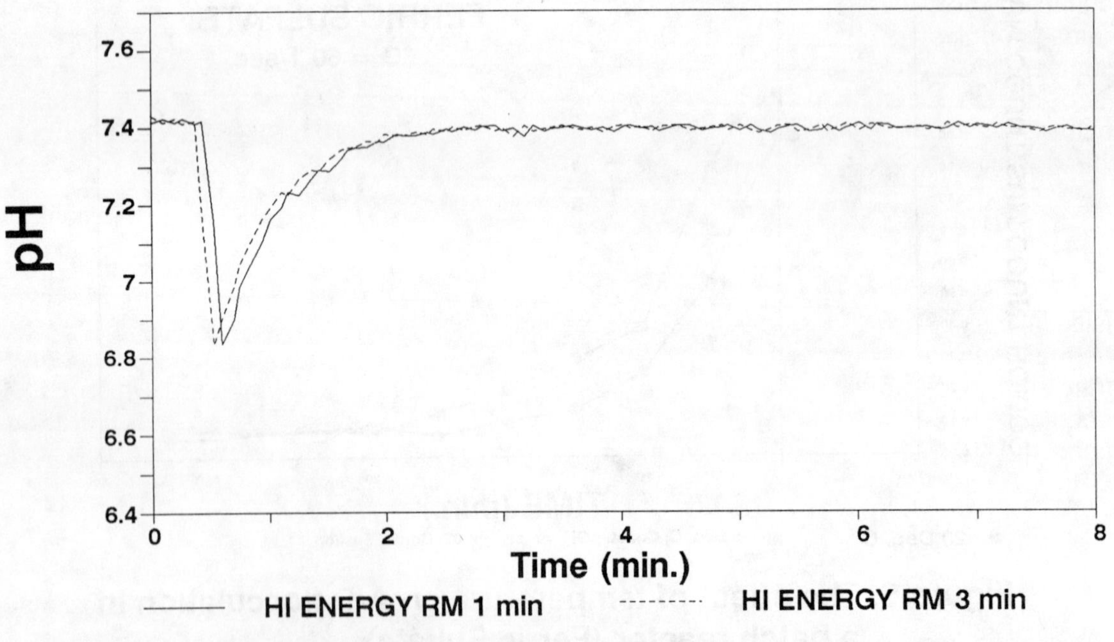

Figure 8. The effect of rapid mixing duration on the pH recovery time after alum addition at 5 Deg. C in buffered tap water.

NEW DEVELOPMENTS IN HIGH SHEAR CROSS FLOW MEMBRANE FILTRATION

Erik Dahlquist, ABB Corporate Research, S-721 78 Vasteras, Sweden, Telephone: +46 (0)21 323063
Jan Kastensson, ABB Water Filtration, S-721 63 Vasteras, Sweden, Telephone: +46 (0)21 107211

Background

Cross flow membrane filtration has been accepted as the best suitable technology for several applications during the 1980th. Generally tubular or plate and frame modules have been used. During the last two years ABB Water Filtration has commercialized a new technology for high shear cross flow filtration where a rotor is used to produce the shear forces. In this paper we will describe the principals for this technology, compare it with tubular membrane equipment and present results from some applications.

History

During the 1930th a patent (1) was presented showing cross flow filtration using a rotor. The rotor forced the water to flow parallel to the filter surface to prevent the build up of a filter cake. Since then several researchers have used the same principals in combination with membranes, e.g. Schock (2) och Murkes (3), but the technique has not been commercialized until 1987-88, when ABB Water Filtration made the first large scale installations in different process industries with the CR filter (CR = Cross-Rotational filter).

Operating principle

The principals for this technique is to use a rotor between two flat membranes. Several cells are stacked to a module, with a common motor drive, forcing the parallel rotors to rotate, and thus giving high shear cross flow conditions at the membrane surface. The volume reduction is determined by controlling the outlet valve and letting out a certain amount of concentrate.

Comparison between tubular- and CR-filters

Compared to tubular membrane systems the permeate flow is just slightly less than the feed-flow, as the shear forces are produced by rotors instead of a pump with the recirculation of the same feed several times. To produce a feed flow velocity of 6 m/s for example, you have to use a feed flow of appr. 6.8 m^3/m^2h running a system with tubes arranged in series with 2.5 cm tube diameter, 2.0 total length. For a permeate flux of 180 l/m^2h that makes 37 times higher feed flow than permeate flow. With the CR filter the corresponding feed flow is appr. 200 l/m^2h or just 1.1 times the permeate flow at 10 times volume reduction.

This makes the impact on components in the feed much less severe, as the effect of the rotor is very gentle compared to a pump. With the described tubular system the feed has to pass the pump 37 times instead of once with the CR filter to reach 10 times volume reduction. At our laboratory we have seen that the rotor gives an enhanced coalescence filtration efficiency in a cross flow coalescence filter. Very easily deformed oildrops were not split by the rotor, while the pump produced a very significant reduction of the drop size distribution, which shows that the pump actually gives a more severe impact on the media, than the rotor does.

By using rotors to produce shear forces you can use very low pressure drop over the membrane, compared to tubular modules, where you have to use a high feed pressure. To get the water through the tubular membrane at appr. 4-5 m/s feed flow velocity, you are normally up to at least 4-5 bars feed pressure. In the CR filter you can easily use 1-2 bar feed pressure and independent of the pressure choose a feed flow velocity up to 12 m/s or more.

With a lower feed pressure lower shear forces are required for the same "cleaning efficiency" on the membrane surface. Thus you can have twice as high average flux (l/m^2h) in the CR filter compared to a tubular filter, where the CR filter is using 1-3 bars, appr. 8 m/s and the tubular 4-5 bars feed pressure, 4 m/s. Although the energy consumption is higher per m^2 membrane area, it will still be lower per m^3 permeate, as the flux is higher. Another advantage is that the washing intervals will be longer, as the moderate pressure prevents the clogging of the membranes. The membrane replacement cost will also be lower, as the membrane area is smaller.

CR and tubular side-by-side experiment

The results above were found in a side-by-side comparison of the CR filter and a tubular ultra filtration system, where identical membranes were used for the tubular modules and for the CR filter. We concentrated cutting fluids, feeding both the tubular system, 25 mm and 7 mm diameter tubes, and the CR filter with the same spent mineral oil based cutting fluid. First we adjusted the rotor speed to get the same flux in both systems, appr 90 l/m^2h at appr 4.5 m/s flow velocity, at the same average feed pressure 2.5 bar.

For the tubular system this ment 4 bar feed pressure and 0.8 bar at the exit tube, which was the best possible performance for the tubular system due to the restriction in inlet pressure. The potential for the CR-filter was however much bigger. Operating the CR-filter under optimal conditions, in this case appr. 8-10 m/s average feed flow velocity and 3 bars pressure, the flux rate was twice as high at low concentrations, 2-3% oil, and three times as high at high concentrations, which was 30% oil in this case. This was still with identical membranes.

The next step was to try different membranes in the CR-filter wich is very flexible because it is possible to mount any membrane that is available on the market. With this effort we managed to improve the performance of the CR filter even more with maintained permeat quality.

Another difference in performance was observed after the concentration cycle. After diluting the concentrate with permeate to original concentration, the flux from the CR filter was relatively higher than the flux from the tubular system. This was probably an effect from more severe fouling in the tubular system.

Field tests

To get good information on the performance of a filter system we are using factor tests . According to a factor design like in Table 1.

<u>Table 1 - Example: Wite water, pulp plant</u>

rpm X_1	P X_2	VRF X_3	flux l/m^2h	deviation from model
0	0	-1.7	307	+24
1	- 1	- 1	307	-20
- 1	- 1	- 1	154	-11
- 1	1	- 1	166	- 5
1	1	- 1	346	-10
0	-1.7	0	218	14
1.7	0	0	384	12
0	0	0	230	12
-1.7	0	0	90	- 2
0	0	0	205	-13
0	1.7	0	218	- 5
1	- 1	1	267	- 1
- 1	- 1	1	128	2
- 1	1	1	128	9
- 1	1	1.7	115	3
1	1	1.7	269	- 3
1	1	1	294	8
0	0	1.7	174	-14

Fig. 1

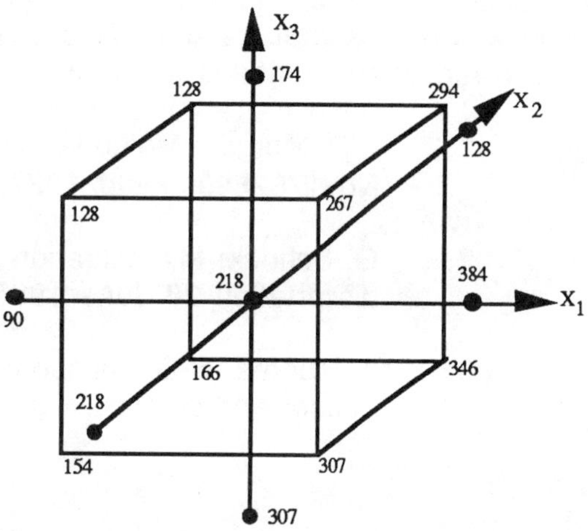

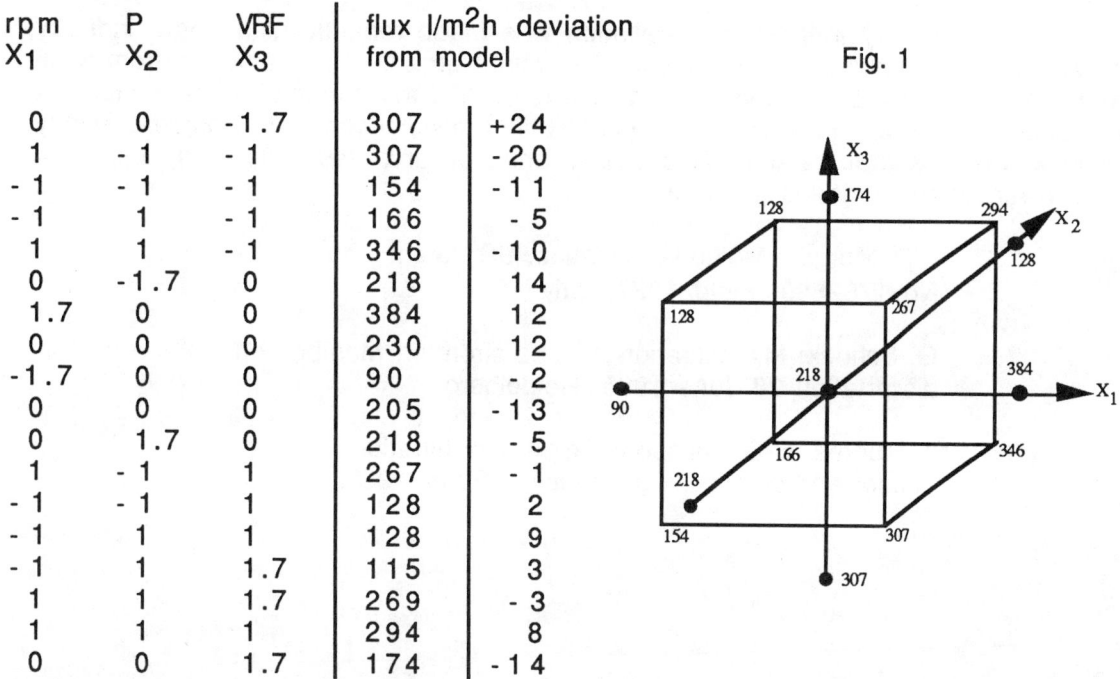

we get an equation describing the correlation between e.g. pressure, rotor speed and volume reduction factor:

$$\text{flux } l/m^2h = 218 + 82 \cdot X_1 + 6 \cdot X_2 - 27 \cdot X_3 + 5 \cdot X_1^2 - 2 \cdot X_2^2 + 6 \cdot X_3^2 + 6 \cdot X_1X_2 - 5 \cdot X_1X_3 - 3 \cdot X_2X_3$$

X_1	=	rotor speed	-1 = 5m/s;	0 = 7 m/s;	+1 = 9 m/s
X_2	=	pressure drop	-1 = 0.8 bar;	0 = 1.4 bar;	+1 = 2.0 bar
X_3	=	volume reduction factor	-1 = 2.8;	0 = 5.2;	+1 = 7.6

Normally the temperature is set by the process, but otherwise this can also be varied.

From such a factorized test a correlation can be determined by an equation, giving e.g. flux (l/m^2h) or separation efficiency (kg/m^2h). By plotting this eq. a good characterization of a certain process water can be found, as well as an indication of which parameter combination to use for further field-tests and preliminary system design. An example of such a factor test is shown in fig. 1.

Installations

During 1988 and 1989 ABB Water Filtration has built three major plants. Two of them for the Pulp & Paper Industry, and one for oily waste water treatment, which is shown in picture 1.

The Paper and Pulp industry has been very interested in this new type of large scale membrane technology and MoDo Husum (photo 2), one of the biggest European plants for bleached sulphate pulp, has installed a CR filter plant for removing organics (COD), and specially chlorinated organic compounds (AOX) from the alkaline bleach stage effluent. The plant, with 200 m^2 membrane area, has an average capacity of 50-60 m^3h and has now been running for 1.5 years. The separation efficiency depends very strongly on the total volume reduction factor (40-80% for volume reductions 5-25 times for the same system!). This shows the importance of considering the total system and not just uncritically accept some very good flux rates or separation efficiencies, given for laboratory tests at low volume reduction factors, sometimes presented.

Another very interesting installation was made recently at a board industry, where suspended solids and high molecular weight organics are removed from white water. The concentrate is evaporated and burned, while the permeate is recirculated to the process. By using this arrangement the CR filter system works as a process kidney, and makes an external water treatment plant unnecessary, also considering the tough Swedish regulatory demands (photo 3).

<u>Ref.</u>

1. C.D. Morgon: Method of filtration. US Patent No 1762560, Field 1927, July 15.

2. G. Schock: Microfiltration an überströmten Membranen, dissertation 28 June 1985, Heidelberg.

3. J. Murkes, C-G Carlsson: Cross flow filtration. Theory and practice. John Wiley & Sons, 1988.

AMERICAN FILTRATION SOCIETY

P.O. Box 6269
Kingwood, Texas 77325
Telephone: (713) 359-1894
FAX: (713) 358-3939

ABSTRACT & COPYRIGHT

Meeting Location: Arlington, VA **Meeting Date:** March 19-24, 1990

Session Chairman: Baumann, E.R. **Session Number:** IA

SPEAKER (AUTHOR):

Name: Delvin E. DeBoer
Company: South Dakota State University
Address: Civil Engr. Dept. CEH 104A
City, State, Zip: Brookings, SD 57007
Telephone: (605)688-4291 FAX: (605)688-5878

CO-AUTHOR:

Name:
Company:
Address:
City, State, Zip:
Telephone: FAX:

CO-AUTHOR:

Name:
Company:
Address:
City, State, Zip:
Telephone: FAX:

CO-AUTHOR:

Name:
Company:
Address:
City, State, Zip:
Telephone: FAX:

TITLE OF PAPER: Plating Wastewater Metals Removal Processes to Minimize Hazardous Sludge Production and Meet Municipal Wastewater Pretreatment Standards

TEXT OF ABSTRACT: (200 Words in Space Below)

Plating wastewater treatment typically precipitates metals from solution and produces a sludge cake which is classified as F006 hazardous waste. Liquid discharges from industrial pretreatment facilities are typically regulated by national and local pretreatment regulations. The combination of hazardous waste regulations and wastewater discharge regulations require the plating industry to evaluate the best course of action for wastewater treatment.

This paper presents results of investigations of plating wastewater treatment alternatives for a printed circuit board manufacturing facility. Metals precipitation using ferrous sulfate, sodium borohydride and dithiocarbamate as primary reducing agents are compared on the basis of sludge volume and metals content. Specific resistance tests are used to compare sludge dewaterability.

Results of ion exchange and electrowinning experiments are presented to provide information on metal recovery methods. Finally, the viability of ion exchange, electrowinning and precipitation methods is discussed.

****

PLEASE TYPE INFORMATION BECAUSE DATA WILL BE PUBLISHED IN THE ABSTRACT BOOKLET. RETURN FORM TO THE AMERICAN FILTRATION SOCIETY, P.O. Box 6269, Kingwood, Texas 77325, WITH A COPY TO YOUR SESSION CHAIRMAN.

NATIONAL EXPERIENCE IN HIGH RATE FILTRATION

Gregory L. Sindt
Bolton & Menk, Inc.
P.O. Box 1646
Ames, Iowa 50010

John L. Cleasby
496 Town Engineering
Iowa State University
Ames, Iowa 50011

E. Robert Baumann
496 Town Engineering
Iowa State University
Ames, Iowa 50011

A. Herman Dharmarajah
Bolton & Menk, Inc.
P.O. Box 1646
Ames, Iowa 50010

Many high rate, potable water, filtration plants are in operation in the U.S. and Canada. Over 150 such plants which were designed to operate at rates over 4.0 gpm per sq. ft. and to produce a filtered water turbidity less than 0.2 NTU were identified. Operating records were obtained from over 50 plants and 21 plants were visited to confirm operating methods and results.

This paper identifies design and operational practices at surface water treatment plants which have demonstrated the ability to produce low turbidity finished water at high filtration rates. These practices are particularly important as recent EPA regulations include mandatory filtration of all surface waters and a finished water turbidity requirement of 0.5 NTU for 95 percent of samples.

This study is sponsored by the American Water Works Association Research Foundation with a contract to Iowa State University and Bolton & Menk, Inc.

Introduction

Conventional surface water treatment, which consists of coagulation, sedimentation, and granular media filtration, has been for nearly a century. Even though conventional rapid sand filtration is very common today, in some instances its use has not been optimized. In recent years, there has been a growing use of direct filtration plants and direct, in-line, filtration plants on better quality raw waters. Direct filtration omits the sedimentation step of conventional treatment, and in-line filtration additionally omits the flocculation step of conventional treatment. These two process flow schemes are less forgiving of operator error or inattention because of short total detention time provided in the plant. The need for optimal design and operation is also essential in these plants, and has not always been attained.

The concern over organic contaminants and cysts of protozoan pathogenic organisms, such as Giardia lamblia and Cryptosporidium, has led to the development of the Federal Safe Drinking Water Act Amendments of 1986 and the EPA "Surface Water Treatment Rule". These rules will require, among other things, a filtered water turbidity of 0.5 NTU or less for 95% of the measurements each month.

At the same time that utilities face more stringent water quality goals, many of them are challenging their filters with higher flow rates per unit area out of economic necessity. Traditional filtration rates of 2 gpm/sq ft (9.8 m/h) are being replaced with higher rates of 4 gpm/sq ft (19.6 m/h) or even higher. Rates of 4 to 6 gpm/sq ft (9.8 to 14.6 m/h) are becoming common in new plants, and a few plants are designed for rates as high as 10 gpm/sq ft (24.4 m/h).

One new plant has a design filtration rate of 13.5 gpm/sq ft (32.9 m/h). Higher filtration rates mean lower capital investment to meet the water demands. But higher flow rates also mean more difficulty in achieving filtrate quality goals. In spite of this difficulty, the trend to higher filtration rates is growing. For the purposes of this study, high rate filtration has been defined as 4 gpm/sq ft (9.8 m/h) or higher.

Procedure

The objective of this work was to review design and operational practices at existing high rate filtration plants which are producing finished water with turbidity less than 0.2 NTU and to identify design and operational aspects which contribute to the success of these facilities in producing low turbidity water. Several of these aspects will be demonstrated by modification of design and operational practices at other full scale treatment facilities.

1. Existing treatment plants with successful operating experience at filtration rates at or above 4 gpm/sq ft producing filtered water turbidities less than 0.2 NTU were identified by telephone and mail inquiries. A total of 151 potential plants were identified in this step and initial plant information was collected by phone to help shorten the list of prospective plants.

2. Two to three years of operating data, a plant flow diagram, and engineering data were requested from the shorter list of prospective plants. The data thus obtained were evaluated to ascertain if the plants did, in fact, meet the desired criteria. Data were obtained by mail from 56 plants during this step.

3. Twenty-one plants were selected for visits by project staff to gather more detailed information. In selecting the 21 plants, consideration was given to geographic coverage as well as raw water and treatment process diversity.

4. The selected plants were visited to gather detailed plant and operational information. Twenty-one plants received full detailed visits and two additional plants received partial visits, primarily to study their chemical dosage control strategies. Samples of raw water, filter influent and effluent were collected during each full visit, shipped to Iowa State University on ice by overnight mail, and analyzed for particle size distribution and zeta potential.

5. Summaries of each full field visit were prepared including the important details about the physical facilities, flow diagram, operational strategies, and performance.

6. Key design and operational features were identified which contribute to the success of these plants in producing low filtered water turbidity at high filtration rates.

Results and Conclusions

The main conclusions of the plant surveys are presented below. These conclusions are based on detailed visits to 21 successful high rate filtration plants producing finished water turbidity less than 0.2 NTU. Most of the 21 plants treat relatively high quality raw water with peak turbidities less than 200 NTU, and turbidity generally less than 20 NTU. Three of the plants were treating challenging river sources with seasonally high turbidity and organic content. It was difficult to find more high rate filtration plants on difficult-to-treat raw water sources because conservative state regulations have generally discouraged high rate filtration on such sources. But there is much interest in uprating plants on such sources and there is adequate evidence to suggest that it can be done successfully.

The following conclusions relate to surface water treatment plants using rapid rate granular bed filters operating at or above 4 gpm/sq ft (9.8 m/h), at least during peak seasons, and that are successful in producing finished water turbidity less than 0.2 NTU. Factors contributing to successful high rate filtration are as follows:

1. Management must adopt a low turbidity goal, and convince the operators that this is a serious goal to be met. Inherent in such a decision is the willingness to budget adequate funds for whatever chemical dosages are required to achieve the goal.

2. Chemical costs for the visited surface water filtration plants, excluding one lime softening plant, ranged from $7 to $43 per million gallons treated, and were only a small part of the total operating cost. Therefore, improved chemical pretreatment should not be considered a costly method to improve plant performance.

3. Chemical pretreatment prior to filtration is more critical to success than the physical facilities at the plant. However, good physical facilities may make achievement of the goal easier and more economical.

4. Polymeric flocculation aids and/or filter aids are generally required at high rate filtration plants. Cationic polymers are most commonly used as flocculation aids added ahead of the flocculation tanks where they serve to augment the primary coagulant. Nonionic polymers are most commonly used as filter aids. High rate plants should be equipped to feed both polymers, as well as a primary coagulant.

5. The benefit of ozone over chlorine and ferric coagulants over alum in pretreatment has been demonstrated at a limited number of direct filtration plants treating high quality raw waters with low coagulant dosages. There is considerable current interest in these alternatives and a need to demonstrate whether such benefits will also be observed on a wide range of raw water sources.

6. The operating staff must use a well defined chemical control strategy that has been verified at that particular plant for the varying raw water qualities. The chemical control strategies are quite variable in different plants. Thirteen of the 23 visited plants use control devices such as streaming current detectors, zeta potential, pilot filters and particle counting in their control strategy. However, half of the visited plants use more traditional schemes such as jar testing, observing turbidity and pH of settled and filtered water, and filter head loss versus turbidity breakthrough performance.

7. Rapid mixing is an essential and important unit operation in chemical pretreatment. Various rapid mixing schemes were being used including back-mixed reactors, baffled tanks, in-line static or motor driven mixers, and hydraulic multi-jet injectors. Some plants reported significant chemical savings with improved rapid mixing, but this is not always observed and further observations on this issue would be desirable.

8. Flocculation appears less important than rapid mixing to successful high rate filtration, particularly at direct filtration plants. Success was being achieved with some or all of the flocculators out of service, or operated near their lowest available speed. Two of the visited plants were utilizing baffled flocculation tanks.

9. Paddle flocculators at the visited plants were considered superior to turbine or pitched blade turbine flocculators in producing large settleable floc in conventional plants.

Paddle flocculators are more prone to maintenance problems and have higher drive train power losses. Pitched blade turbines at the visited plants may not represent optimum blade configuration.

10. Most conventional plants consider settled water turbidity to be important in their control strategy, but the key issue should be the filterability of the solids reaching the filter and the resulting filtrate quality.

11. Flocculation effectiveness can be improved by increasing flocculator speed during cold water periods, use of proper baffling between stages to reduce short circuiting, and use of a ported baffle wall between flocculation chambers and sedimentation tanks to avoid floc breakup.

12. Dual or triple media filters were utilized at all of the visited high rate plants except for one plant which used a deep bed of coarse anthracite. There is considerable interest in the applicability of the latter option to other raw water sources.

13. Triple media is considered superior to dual media under stressful filtration conditions (e.g., high rates and sudden rate increases). However, the benefits to the filtered water quality are not well documented in published literature and additional work is needed on this issue. Triple media causes higher clean bed head losses, and this detriment may prevent its use in uprating existing plants with limited available filter head loss.

14. To produce the best filtrate quality, attention should be paid to minimizing periods when filtered water quality may be degraded, such as: (1) during the initial improvement period of the filter cycle (ripening period), and (2) during plant start up or increases in plant load. Filtering to waste, bringing filters slowly up to rate, and leaving some residual solids above the filter at the end of the backwash operation were some of the methods being used to reduce impact of the ripening period.

15. Individual continuous turbidity monitors on each filter are desirable to optimize filtrate quality. They are useful in detecting detrimental filtrate quality changes during the ripening period and at the end of the filter cycle, and in studying the effectiveness of corrective actions.

16. Post filtration chemical addition can have a detrimental impact on finished water turbidity and on filter performance. Lime and sodium silicofluoride addition can raise turbidity. Alternative chemicals or alternative feed points can correct this problem. Addition of polyphosphate or ammonia prior to the point where the backwash supply is withdrawn can worsen the water quality during the ripening period. An alternative location of chemical feed, or discontinuance of feed during backwashing can correct this problem.

17. Particle counting can be a useful and sensitive operational tool at a particular plant. However, different counting instruments and techniques yield different results. The particle counts are, therefore, not absolute values and cannot be directly compared without consideration of technique and instrumentation.

18. A fair correlation between turbidity and particle count, using a particular counting technique, for the raw and filtered water was developed for the 21 visited plants.

19. Good operator training and the building of operator pride in quality of the treated water are important steps in producing the best filtered water. Some plants utilize 12 hour operating shifts to give more continuity to plant operation, and a short period of shift overlap to provide for intershift communication related to the current treatment strategy.

The foregoing conclusions have been based on plant visits and detailed evaluations of a limited number of water treatment plants. Some of the conclusions could be strengthened by additional pilot or plant scale verification studies. Verification studies will be conducted in the next phase of this research project, and reported in a subsequent AWWARF report.

The complete summary of this work is published by and available from the American Water Works Association Research Foundation[1].

References:

1. J.L. Cleasby, A.H. Dharmarajah, G.L. Sindt, E.R. Baumann, Design and Operation Guidelines for the Optimization of the High-Rate Filtration Process: Plant Survey Results, American Water Works Association Research Foundation, 6666 West Quincy Ave., Denver, CO 80235, Sept 1989.

POLYMER PLANTS - LOCATION OF RAW MATERIAL AND PROCESS FILTERS

Harry W. Byrd
Southern Metal Processing Company
130 Allred Lane
Oxford, AL 36203
(205)831-8130

Approximately twenty-five years ago, major engineering firms for some of the largest USA chemical companies started construction of polymer plants throughout the country.

It was the start of the American and European joint ventures and, of course, the "BLACK BOX" of how to produce polymer.

Both large and small polyester plants were popping up all over the Southeast, Forecasting use of staple, filament and POY was increasing each month, and the major players started expansions before the first unit was brought on-stream.

To review and evaluate filtration's impact on polymer of all types today, we can look at Polyester, which was one, if not the first polymer to use fine filtration - better known today as "depth Media", produced with random laid fibers or sintered powder.

In 1966, start-ups were a monthly "happening" throughout the Southeast. Both batch and continuous polymer lines were off and running, and most plants were in a sold-out state. Customers would buy lower quality product if necessary to meet demands.

A continuous polymer plant was in fact a monster. Prime product was running as low as 80% in some plants, with 10% or more going to waste.

Plants were over-engineered with more vessels and systems than required for throughput, which made control almost impossible. Many of the plants that were designed in Europe had four transesterification systems, two monomer systems, two pre-polymer vessels, and two final polymer vessels. All this equipment with throughput being only 60,000 lbs/day for some plants.

A typical plant was a nightmare - vessel after vessel to control levels, temperature, etc. The demand was starting for higher grade polymer at lower prices. (See drawing I)

In the early seventies, better known as the "Debottlenecking years", plants started to increase throughput, larger vessels and less vessels would be used. In some cases, plants were so designed that half of the vessels were removed. These major changes in plant layouts and better understanding of polymer brought the first true control of the plants and new standards followed.

Around 1974, the market was changing again. Finer filaments were in demand, cost of raw materials was on the increase, and cost per pound was a major factor.

By 1976, many polymer plants had found that by adding filtration in the final stages, first quality and higher yields could be increased. Prime polymer was now approaching 90% or higher.

Then the "what if" came about. What effect would fine filtration have if introduced to all incoming raw materials? What would a monomer filter do in relation to the polymer filter system? Could you use only the monomer filter and make the same product?

In most cases, the monomer filter could not be used alone, but it did have an impact on the life of the polymer filter. Filtration of the raw materials also made an important difference in quality and life of the polymer filter. Soon to follow was the filtering of TiO_2 with metal elements as they would not unload, causing problems. (See drawings II, III, and IV)

Today, the location of the filtration units have not changed to a great degree from 1974. What has happened is the requirements for finer and finer depth media that will stay on stream and produce a product for the future. (See drawing V)

In 1970, a large polymer switch system was 30-50 sq. ft. Today, systems are approaching the 400 sq. ft. range, and from the old 40-60u abs. range to the 10-20u abs. range using new medias. From polyesters, some 24 odd polymers are filtered today. (See Drawing VI)

Recycled Polymer is the next step in filtration.

CHART I
TYPICAL 1966 CONTINUOUS UNIT POLYESTER PLANT
DMT

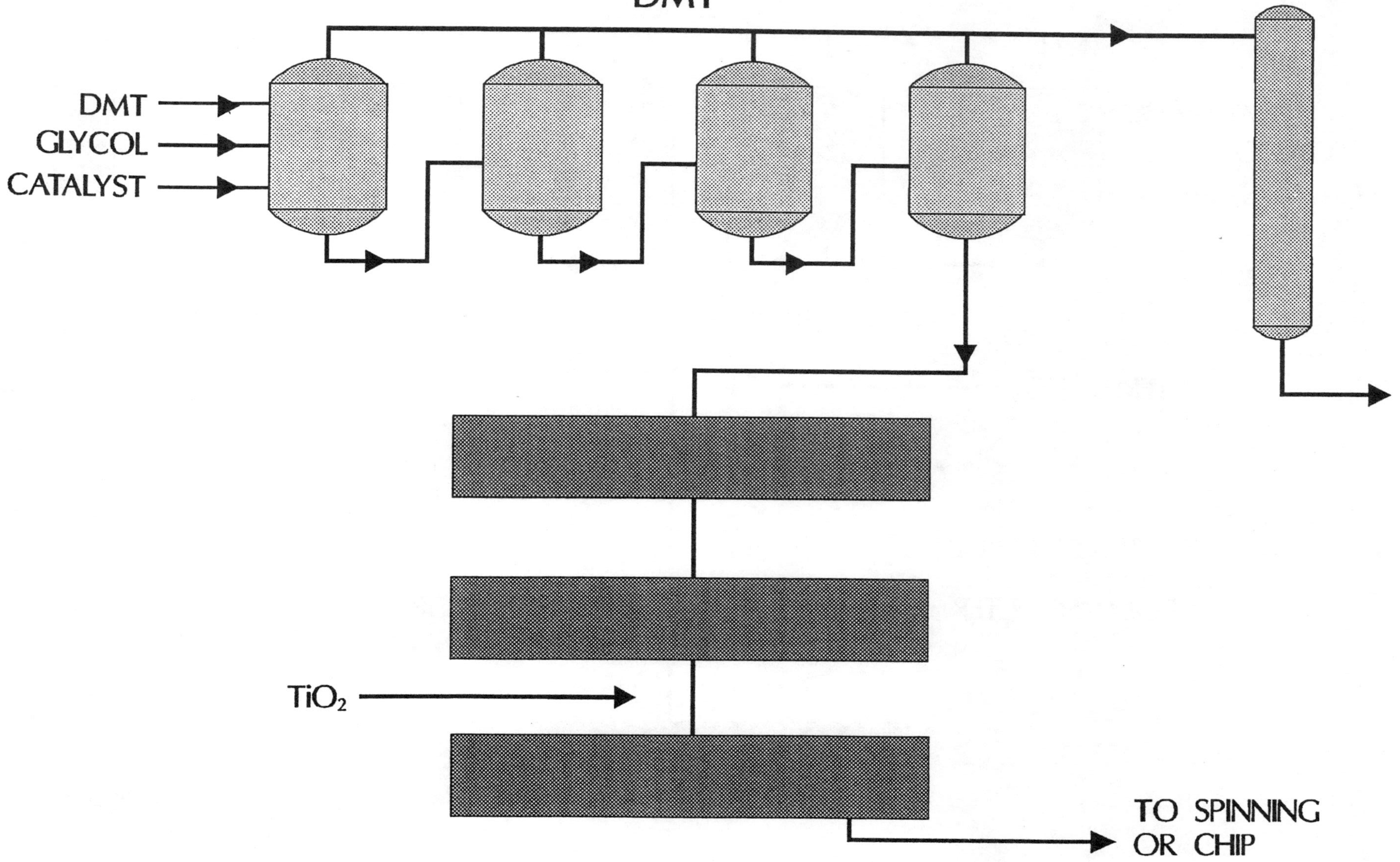

CHART II
TYPICAL 1976 CONTINUOUS UNIT
POLYESTER PLANT
DMT

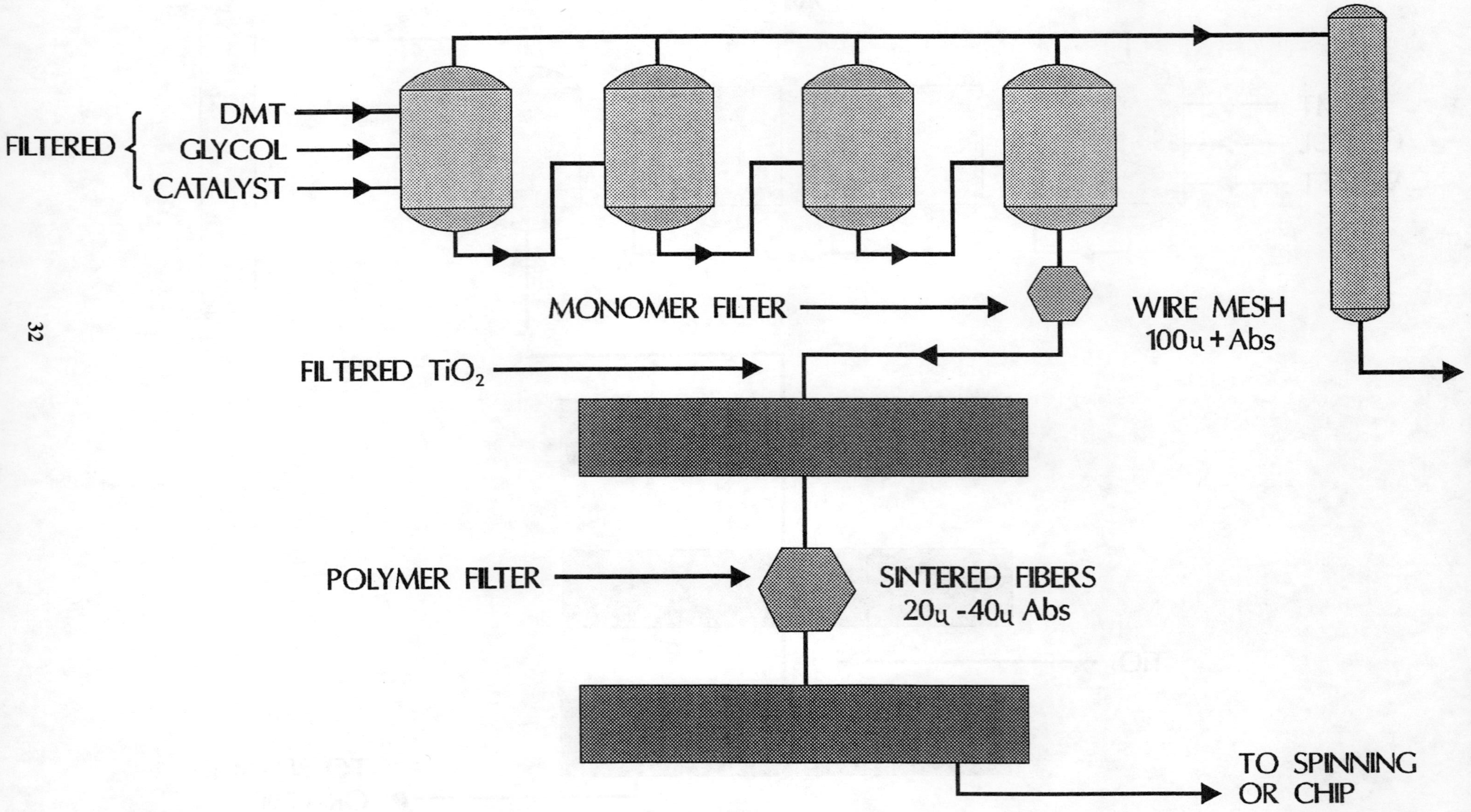

CHART III
TYPICAL GLYCOL FLOW
POLYESTER PLANT

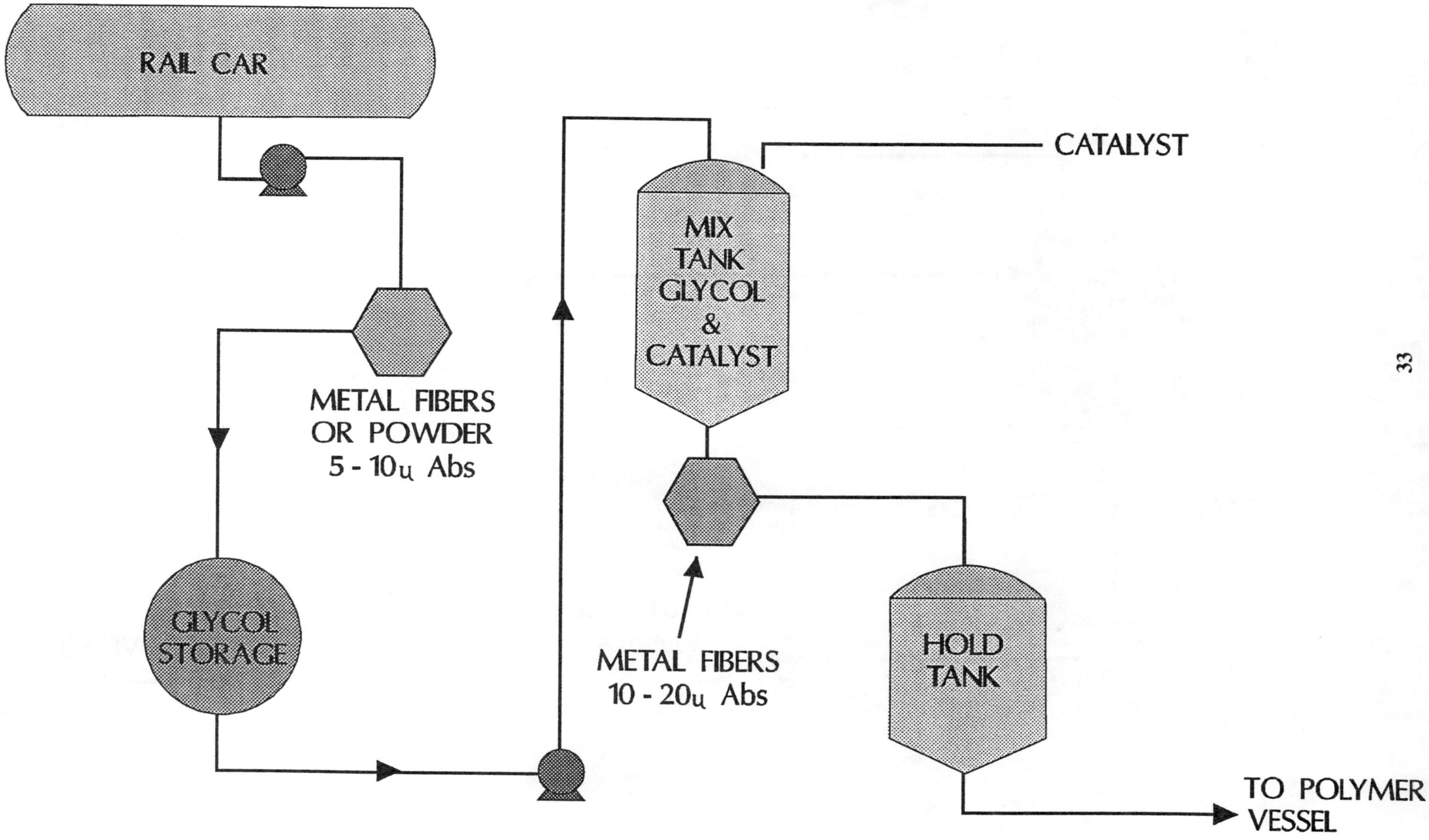

CHART IV
TYPICAL TiO_2 FILTRATION
POLYESTER PLANT

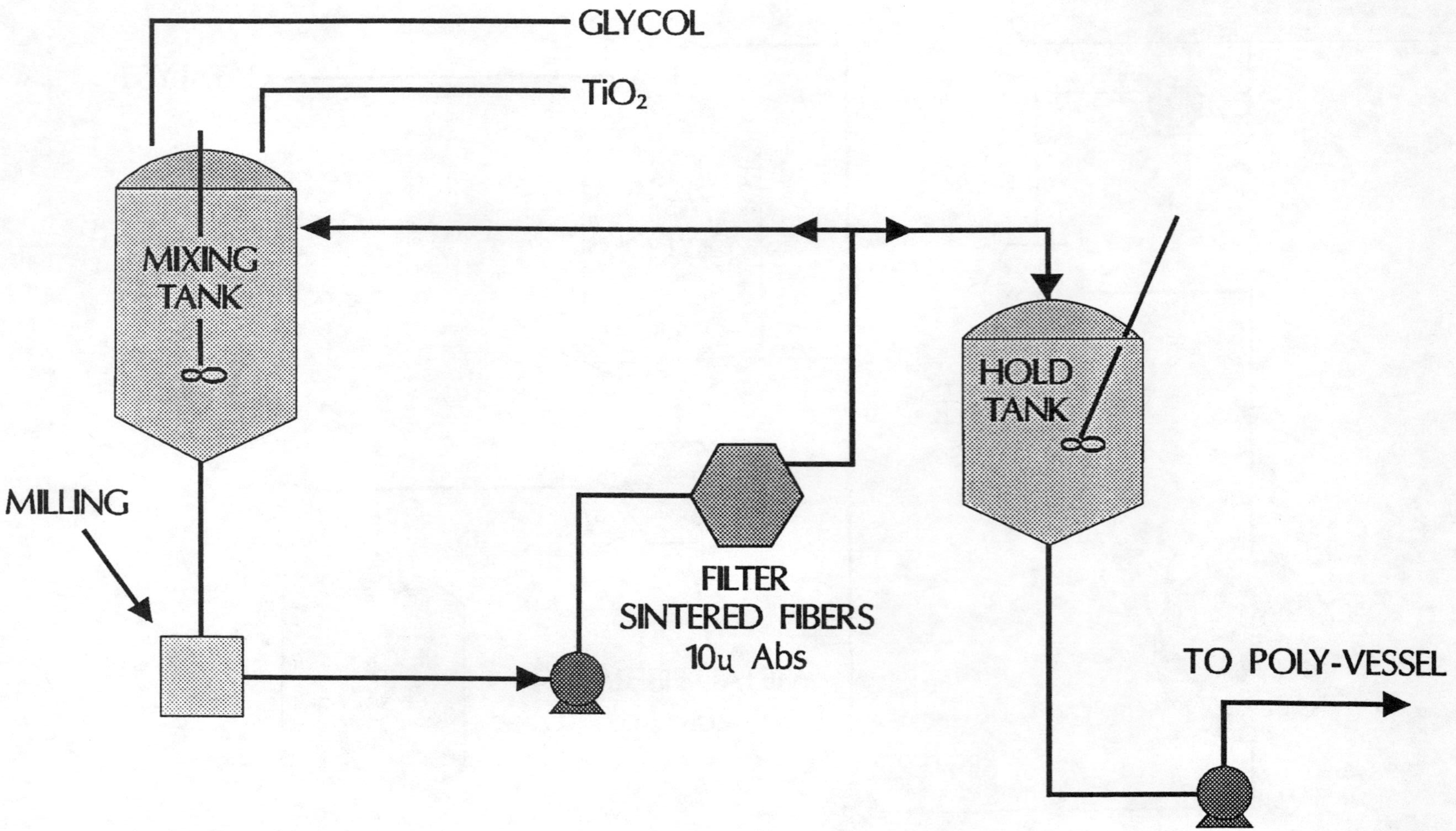

CHART V
TYPICAL 1989 CONTINUOUS UNIT POLYESTER PLANT DMT

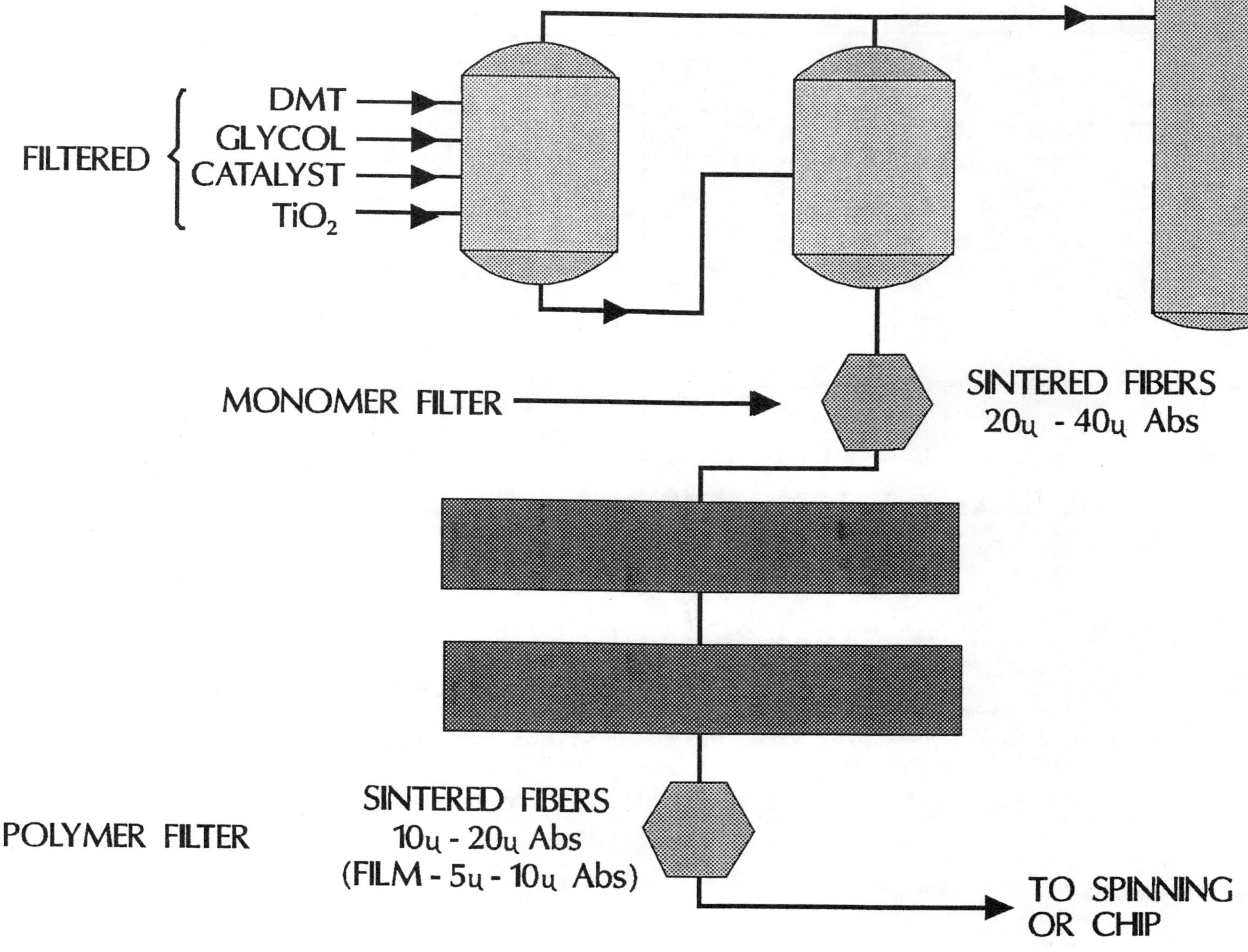

CHART VI

Polymer	Product	Micron Abs
Polyester	Stape	20-40
Polyester	Filament	10-20
Polyester	Bottle	20-40
Polyester	Film	5-10
Polycarbonate	Sheet	10-20
Polypropylene	Filament	10-100
Polyethylene	—	20-40
Polyurethane	—	20-40
Polyetherimide (PBI)	—	80-100
Polyvinyl Butryl (PVB)	Sheet	40
Polyetheretherketone (PEEK)	—	20
Polysulfone	—	20

Polymer	Product	Micron Abs
Polyimide	Film	20-60
Polyphenylenesulfide	—	20-60
Nylon 6 (66)	Staple	80-100
Nylon 6 (66)	Filament	20-40
Nylon 6	Monofilament	20
Acetate	Staple	40
Acrylic	Staple	20
Aramid	Filament	20
Fluorocarbon	—	20
Styreneacrylonitrile	—	100-200
Styrene Butadiene	Film	100-200
LLDPE	—	20

POLYMER FLOW MODELING

Fred W. Cole, P.E.
Steve Krasicky, P.E.
Purolator Products Company
Fluid Technology Division
8439 Triad Drive
Greensboro, NC 27409

A method for estimating clean pressure drop for molten polymer filters is given. Simple, low pressure, incompressible air flow tests measure filter or filter media permeability. This data is analyzed using Green's equation to separate viscous and inertial components of pressure drop. The viscous component is then extrapolated as a ratio of polymer/air viscosity to provide an engineering estimate of performance. Polymer viscosity estimates for thixotropic or shear-thinning fluids are discussed. Effective viscosity as a function of shear rate, using the simple "power law" model, is examined for woven wire cloth, fiber metal felt and sintered powder metal filter media. Flow distribution testing and analysis using the log-normal pore size distribution model is proposed and explicated.

Polymer Flow Modeling

Estimation of clean pressure drop for molten polymer flow through filters or filter media is an important engineering problem. Initial pressure drop is one indicator of probable comparative service life. Decisions to change filter media rating or type are partly based on expected clean pressure drop.

A variety of test methods and engineering interpretations of data have been used. Actual molten polymer flow tests, using the specific polymer in question, are ideal but expensive and inconvenient. Simulative room-temperature tests using high viscosity oils or silicone fluids are less arduous but introduce error because of different rheological properties compared to molten polymer. Simple, incompressible air flow test data can be analyzed and extrapolated to achieve an approximate and comparative measure of molten polymer flow performance.

Air flow is very different from molten polymer flow but not so different from oil flow which is commonly extrapolated to predict pressure drop. Air at room temperature has an absolute viscosity of 0.0186 centipoise compared to 300,000 cp (or less or more) for molten polymers and 11.7 cp for MIL-H-5606 hydraulic oil at 100F, for which there is much flow data. Air density is 0.0012 g/cm^3 and oil density is 0.83 g/cm^3. Kinematic viscosity, or ratio of absolute viscosity to density, is about 15.5 centistokes for air compared to 14.1 cs for hydraulic oil. Thus, Reynold's number and flow profile are similar for air and oil, and both are very different from polymer flow which is almost purely viscous. Oil spills, however, are harder to clean up than air spills.

Flow Analysis

Any flow test data can be analyzed to separate viscous and inertial components of pressure drop. Pressure drop attributable to viscous drag loss is directly proportional to flow velocity or flow rate. Pressure drop attributable to inertial loss, such as contraction-expansion or directional change, is proportional to the square of flow velocity. Total pressure drop is the sum of these components according to Osborne Reynold's superposition principle. This analysis has been applied to porous media by Forchheimer[1] in 1901 and to sintered powder metal by Green and Duwez[2] in 1951. Subsequent work by many others, including Armour and Cannon[3] in 1968, have used this analysis for wire cloth and other media.

The relationship of pressure drop to flow velocity and fluid viscosity and density can be expressed by the differential equation from Green[2]:

$$-dP/dx = \alpha\mu v + \beta\rho v^2 \qquad (1)$$

where P is pressure, x is flow path length, v is flow velocity, μ is absolute viscosity, and ρ is mass density of the fluid, and α and β are characteristic geometric constants of theporous medium with dimensions of L^{-2} and L^{-1}, respectively. This basic differential equation can be modified to model flow through a thick walled cylinder (-dP/dr), or isothermal compressible flow of a perfect gas (Green[2]). For the present simple case a constant geometry and incompressible flow is assumed, and since v = Q/Area, equation (1) is integrated:

$$\Delta P = A\mu Q + B\rho Q^2 \qquad (2)$$

where areas, flow path length and other constants are lumped in A and B and Q is volumetric flow rate. Equation (2) is useful for liquids and for incompressible gas flow where ΔP is less than about five percent of system pressure, P. When exhausting to atmosphere (14.7 PSID), maximum pressure drop is about 0.7 PSID or 20 in. W.C. for incompressible flow. For data analysis, equation (2) can be linearized:

$$\Delta P/Q = A\mu + B\rho Q \qquad (3)$$

which is an equation of a straight line. $\Delta P/Q$ is the y-axis (ordinate), Q is the x-axis (abscissa), $B\rho$ is the slope, and $A\mu$ is the intercept. This transform allows linear regression or "least squares" fitting of the data to the model. Thus, air flow test data can be analyzed to separate viscous ΔP ($A\mu Q$) from inertial ΔP ($B\rho Q^2$) as shown in equation (2).

Molten polymer flow is almost purely viscous with negligible inertial pressure drop. Absolute viscosities may be near 3000 poise or 300,000 cp. Therefore, the second inertial term of equation (2) can be ignored:

$$\Delta P = A\mu Q \qquad (4)$$

which is a model for polymer flow and A is substantially equivalent to the inverse of Darcy's permeability coefficient.

Flow Extrapolation

Equation (4) can be extrapolated for polymer flow using data derived from air flow tests. Since A is a lumped geometric constant, only absolute viscosity is different for the two fluids:

$$\Delta P = A(\mu_2/\mu_1)Q \qquad (5)$$

where μ_1 is air viscosity and μ_2 is polymer viscosity. The ratio μ_2/μ_1 may be about 300,000/0.0186 = 16,130,000.

This is a big extrapolation! But so is the μ_2/μ_1 ratio for hydraulic oil at about 25,640. The utility of this model in both cases is not so much in predicting actual molten polymer system pressure drop as in comparing two different filters or filter media. In any case, extrapolation is generally limited by the accuracy of the polymer viscosity term.

Polymer Viscosity

Polymers are generally thixotropic or shear-thinning. Unlike Newtonian fluids such as air or oil, the apparent absolute viscosity depends on the shear rate of the fluid flow and is not constant. Many models have been proposed to describe this rheology. One of the simple ones is described by Savins[4]:

$$\mu(\Gamma) = K(\Gamma)^{n-1} \quad (6)$$

where μ is apparent absolute viscosity, Γ is shear rate, K is a characteristic constant, and n describes Newtonian flow at n = 1 and increasing shear-thinning at n<1. This "power law" model applies to data where log-log plots of viscosity as a function of shear rate approximate a straight line from which K and n can be derived. This model applies, at least approximately, to many real polymers and may be used for flow extrapolation providing that shear rate through the filter medium is known or can be calculated.

Shear rate for flow through a circular capillary is defined by the equation:

$$\Gamma = 32Q_c/\pi\, d^3 \quad (7)$$

where Γ is shear rate (reciprocal seconds), Q_c is now flow rate through the singular capillary (not total flow rate), and d is capillary diameter. But no filter media pore in captivity remotely resembles a circular capillary! Still, the same concept can be used to model shear rate.

Filter Media

Filter media used for polymer filtration are mainly three types: woven wire cloth, fiber metal felt, and sintered powder metal. Each type is distinctly different in pore geometry and size distribution and there are many different specifications or "filter ratings" for each type. Yet all of these filter media can be measured and fitted to an appropriate pore size distribution model, generally a type of log-normal function.

Porosimetry

Pore size distributions for porous media can be measured by mercury intrusion or by two-phase flow testing. Mercury intrusion tests measure the absolute pressure required to inject successive incremental volumes of non-wetting mercury into the porous sample. Each progressively smaller pore size is inferred from the progressively larger pressure and the constant surface tension and contact angle of the mercury. Mercury intrusion also measures "vugs" or noncommunicating voids, especially in sintered powder metal.

Two-phase flow tests are similar in principle, except that the porous structure is first filled with a wetting liquid such as alcohol or, better, a very low vapor pressure fluorocarbon liquid. Air or inert gas pressure is applied to one side of the porous sample and the pressure is slowly increased until the first bubble breaks through. So far, this is the same as the "bubble point test" as described in SAE Aerospace Recommended Practice (ARP) 901 dated 1968. After the first bubble, corresponding to the largest pore or "absolute" rating, slowly increasing pressure produces slowly increasing flow until, finally, the wetting liquid is blown dry from the smallest pores.

Wet flow is then compared to dry flow for the same sample. The "flow pore size", actually the fourth moment of the "number pore size", is deduced from the pressure required to produce the cumulative flow fraction, using the bubble point equation:

$$d = K/P \qquad (8)$$

where d is circular capillary equivalent pore diameter, P is pressure and K is the characteristic bubble point constant. For the circular capillary model, $K = 4\sigma \cos \theta$, where σ is surface tension and θ is advancing contact angle, often assumed to be "zero" for perfect wetting.

This flow distribution test was first used in the 1950s by SAE to define a "mean flow pore" in a diesel fuel filter as the point where wet flow was 50 percent of dry flow at the same pressure drop. Later refinements were described by Cole[5,6] in 1968 and 1975, where the entire flow pore size distribution was measured and fitted to a log-normal model for analysis and engineering prediction of filter media performance. Coulter now manufactures and sells a "Porometer®" testing device which is fully automated and uses a microprocessor-type computer to record and reduce data. The work of hours is reduced to minutes with improved accuracy and repeatability. Thank you Mr. Coulter, wherever you are!

Log-Normal Distributions

The log-normal distribution function is simply a modified normal distribution in which the logarithm of the variate, rather than the variate itself, is normally distributed. The more familiar normal distribution function generates the well known "bell curve" showing frequency of attributes in Statistical Process Control or the height, weight, or IQ of a population. The log-normal distribution function is similar but skewed to the left and ranges from zero to infinity rather than minus to plus infinity. It provides a better fit for many natural populations such as particle sizes and pore sizes. The log-normal distribution was proposed by Hatch and Choate in 1933 and is explicated in many references, including Irani and Callis[7] who describe its application to particle size measurement and analysis.

Log-normal frequency (9) and cumulative (10) distribution functions are shown:

$$f(x) = \frac{1}{\sqrt{2\pi}\ln \sigma} \exp\{-[\ln(x/x)/\ln \sigma]^2\} \qquad (9)$$

where x is the variate (pore size, particle diameter, whatever), σ is the geometric standard deviation, and x is the geometric mean.

$$P = 50\text{-}100 \text{ erf } [\ln(x/x)/\ln \sigma] \qquad (10)$$

where P is percent greater than size x, and erf is the standard error integral. The cumulative distribution (or its conjugate, P percent smaller than size) plots as a straight line on log-probability graph paper such as Keuffel & Esser No. 468040 or 468080. This linearization allows least square fits for data and determination of both x and σ which define the function.

The 50 percentile line marks $\bar{x}$ and the ratio of $\bar{x}/x@{-\sigma} = x@{+\sigma}/\bar{x} = \sigma$, where $-\sigma$= 15.87 percent and $+\sigma$ = 84.13 percent.

This simple log-normal distribution model is generally adequate for most purposes. However, most pore size or particle size distributions have a smallest and largest size not at zero or infinity. This can result in a ''tailing off'' or curve of the straight line fit at the extreme ends of the probability scale. A better, but more laborious fit can be achieved by plotting a size parameter rather than the size itself:

$$z/\bar{z} = \frac{(x-x_o)(x_{oo}-x_o)}{\bar{x}(x_{oo}-x)} \tag{11}$$

where $z/\bar{z}$ is the new size parameter, x is size, x_o is smallest size, x_{oo} is largest size and $\bar{x}$ is geometric mean. Equation (11) reduces to the simple case, $x/\bar{x}$, when x_o is near zero and x_{oo} is large.

Log normal distributions have the useful mathematical property that all moments, or multiplications of the distribution by a power of the variate, are also log-normal distributions with a different geometric mean but with the same geometric standard deviation. Thus, a given number pore size distribution plot is parallel to its flow pore size distribution plot. These relationships for the first four moments are summarized from Irani and Callis[7]:

$$M_1 = \frac{n_i\ x_i}{n_i} = \exp\ (\ln \bar{x} + 1/2\ \ln^2\sigma) \tag{12}$$

$$M_2 = \frac{n_i\ x_i^{\ 2}}{n_i} = \exp\ (2\ \ln \bar{x} + 2\ \ln^2\sigma)$$

$$M_3 = \frac{n_i\ x_i^{\ 3}}{n_i} = \exp\ (3\ \ln \bar{x} + 9/2\ \ln^2\sigma)$$

$$M_4 = \frac{n_i\ x_i^{\ 4}}{n_i} = \exp\ (4\ \ln \bar{x} +8\ \ln^2\sigma)$$

where M_1 M_2, M_3, M_4,...,M_k are moments or averages of the indicated summations, n_i is the i th number and x_i is the i th variate.

Flow distribution or Porometer® data shows flow pore size which is proportional to the fourth power of number pore size for either viscous or inertial flow, and the geometric mean flow pore size is expressed:

$$\bar{x}_Q = [\sum_{i=1}^{i=n} x_i \ / \sum_{i=1}^{i=n} n_i]^{1/4} \tag{13}$$

where $\bar{x}_Q$ is geometric mean flow pore size and x_i is number pore size. Thus, geometric mean number pore size, using equations (12), is expressed:

$$\ln \overline{x}_n = \ln \overline{x}_Q - 2\ \ln^2\sigma \qquad (14)$$

Note that the mean by number, $\overline{x}_n$, may be much smaller than the mean by flow because of the much larger contribution to flow ($Q \propto d^4$) of the larger pores.

Mean Shear Rate

The shear rate equation (7) can now be applied to various porous media. Definitions have been based on the concept of the "circular capillary equivalent", recognizing that real filter media are not comprised of circular capillaries. However, it can be argued that this is a reasonable engineering approximation. Pores in wire cloth may be substantially square for square weave mesh or triangular for Dutch weave mesh. In both cases the inscribed circle defines the hydraulic radius for both flow and bubble point measurement and data analysis. Similarly, pore structures for fiber metal felt or sintered powder metal can be approximated by regular polygons for which the inscribed circle substantially defines the hydraulic radius.

For narrow, geometrically defined pore size distributions such as woven wire cloth, $\Gamma = 32Q_c/\pi\ d^3$ can be evaluated fairly easily. Capillary flow rate Q_c is estimated by dividing total flow rate Q by the total number of pores, n:

$$Q_c = \frac{Q}{n} = \frac{Q}{A} * \frac{A}{n} \qquad (15)$$

where Q/A is flow per unit area and n/A is pores per unit area. The latter can be calculated from the wire count. Volumetric flow rate Q_c must be expressed in units consistent with pore diameter d, say cubic micrometers per second. Pore diameter d can be taken as the simple mean, either measured or calculated, because the distribution is very narrow and other means are very close in numericalvalue.

Broader pore size distributions, such as fiber metal felt or sintered powder metal, make evaluation of mean shear rate more complicated. Flow distribution or Porometer® test data is used to define a log-normal flow pore size distribution as discussed in equations (9) through (14).

Just as number or flow pore sizes are log-normally distributed with the same geometric standard deviation and different geometric means, so are shear rate pore sizes. Shear rate is proportional to Q_c/d^3 and Q_c is proportional to d^4. Therefore, shear rate is proportional to $d^4/d^3 = d$ and the mean shear rate is:

$$\overline{x}_s = [\sum_{i=1}^{i=n} n_i\ x_i \ / \ \sum_{i=1}^{i=n} n_i] \qquad (16)$$

from equations (12), where $\overline{x}_s$ is mean shear rate pore size and x = d. This can be expressed in terms of the mean flow pore:

$$\ln \bar{x}_s = \ln \bar{x}_Q - 3/2 \ln^2\sigma \qquad (17)$$

Capillary volumetric flow rate is also harder to estimate because it requires a measure of pores per unit area, n/A, and cannot be directly calculated. However, pores can be counted using a microscope or estimated from the known or measured porosity of the medium and fiber or powder size. Flow rate Q_c through the mean shear rate pore size $\bar{x}_s$ is calculated by converting the flow pore size and number pore size distributions, equation (14), to frequency histograms with appropriate size ranges. Percent flow Q and number n are read from the mean shear rate pore size $\bar{x}_s$ and their ratio is multiplied by the ratio Q/n:

$$Q_c = \frac{Q\%}{n\%} * \frac{Q}{A} * \frac{A}{n} \qquad (18)$$

where single pore flow Q_c can then be used to calculate mean shear rate from the mean shear rate pore size $\bar{x}_s$ and equation (7).

Summary

Polymer flow modeling with porous filter media is a useful engineering tool for estimating clean pressure drop and, especially, comparing the relative performance of two or more different filter media or filters. Seven major topics are discussed:

1. Air flow testing is recommended for convenience and economy.
2. Flow test data is analyzed into viscous and inertial components of pressure drop.
3. Viscous pressure drop is extrapolated based on polymer viscosity.
4. Polymer viscosity is estimated as a function of shear rate using a power law model.
5. Filter media pore size distributions are measured using flow distribution or Porometer® testing.
6. Resulting test data is fitted to a log-normal distribution model.
7. The log-normal model, which is entirely defined by the geometric mean $\bar{x}$ and the geometric standard deviation σ, is used to estimate flow properties of the filter media, including shear rate.

These methods and tests have been used in filtration engineering for over two decades. Their application to polymer filtration improves filter design and reduces the total cost of filtration.

References:

1. Adrian E. Scheidegger, The Physics of Flow Through Porous Media, The MacMillan Co., New York. 1969, p. 162.

2. Leon Green, Jr. and Pol Duwez, "Flow through Porous Metals", Journal of Applied Mechanics, ASME, March 1951 pp. 39-45.

3. James C. Armour and Joseph N. Cannon (The Procter & Gamble Co.), "Fluid Flow Through Woven Screens", AIChE Journal, Vol. 14, No. 3, May 1968, pp. 415-420.

4. J. George Savins (Mobile Research and Development Corporation) "Non-Newtonian Flow through Porous Media" Flow Through Porous Media", American Chemical Society, Washington D.C., 1970, pp. 71-101.

5. Fred W. Cole, "Filter Media Rationalization" (Bendix Filter Division) Contamination and Filtration Panel, Society of Automotive Engineers, September 1968.

6. Fred W. Cole, "Filter Ratings - An Alternative to 'Black Art'", Filtration & Separation, January/February 1975, pp. 17-22.

7. Riyad R. Irani and Clayton F. Callis (Monsanto Chemical Company), Particle Size: Measurement, Interpretation, and Application, John Wiley & Sons, Inc., New York, 1963, pp. 39-55.

A SYSTEMATIC APPROACH FOR OPTIMAL FILTER SELECTION BY ANALYSIS OF CONSTITUENTS IN MELT POLYMER PROCESSES

Dr. Seyda Ayral
Pall Corporation
30 Sea Cliff Avenue
Glen Cove, New York 11542
516-671 4000

I. Introduction:

The dispersion quality of additives introduced into filled melt polymer processes often affect the quality of the final polymer product, as well as affecting the life of the filter employed for the filtration of this polymer. Poor dispersion quality contributes to defects in final polymer products[1] and shortens the life of final filters, leading to lowering of production yields and increasing operating costs.

A systematic approach is essential in order to obtain a high quality product[2] and best final filter life. This approach involves the evaluation of all the process streams and selection of high performance, efficient filters for each application. Achieving fine levels of filtration is easier for each component in a polymer process than attempting to accomplish the finest level filtration at the viscous melt polymer stage. This paper covers the protocol for classifying the dispersion levels and selecting optimum filtration systems for additive slurries, critical components of filled melt polymers, with the ultimate goal of improving product quality and manufacturing yields, as well as reducing operating costs.

The above mentioned protocol was applied to several different additive slurries of polyester processes in representative plants and the results were documented. A typical polyester process flow diagram is provided in Figure 1[3]. The results indicate that this approach allows the selection of the finest level of filtration without essentially altering the solids concentration of the dispersion; thus, providing high product quality and longer filter life. Typically, the average particle size distribution for these slurries is less than 1μm. Polymer manufacturers are concerned about larger size agglomerates (i.e., 3 or 5μm), which cause even larger size defects (ten times) in their finished products. The finest level of filtration within the above parameters was accomplished by absolute rated depth filters, i.e., depth filters which have controlled pore size distributions. Polymer manufacturers commonly resort to chemical (stabilizer addition) and/or physical (grinding and filtration) processes to control dispersion levels of additive slurries and avoid agglomeration[4,5].

Typically, polymer manufacturers employ nominally rated filters (depth or surface) for their additive slurries in their processes. Filters with absolute ratings provide accurate, measurable and reproducible performance in contrast to nominally rated filters of similar ratings. Filter removal ratings are specified in micrometers, or more commonly, microns (μm); that is 1/25,400 of an inch. There is no universal standard to which filter manufacturers adhere. Conventional depth filters, for example, are specified by a nominal rating, as defined by the National Fluid Power Association (NFPA) " an arbitrary micron value assigned by the filter manufacturer, based upon the weight removal of a percentage of all particles of a given size or larger. It is rarely well defined and not reproducible." The depth filters used in this study were absolute rated, based on the "Oklahoma State University F-2 Filter Performance Test" which was modified by Pall Corporation for use in water in the range from 0.5 to 25.0μm[6]. In this test an aqueous suspension of silicious contaminant, AC Fine Test Dust, is pumped through a single 10" module filter, while an automatic particle counter (upstream of the filter) records the influent particle levels and a second one (downstream) simultaneously records the effluent levels. Each counter can be preset at up to six particle diameters, and these counts are used to determine efficiencies at six or less diameters.

In the representative polyester plants selected for this study, the additives utilized were titanium dioxide, calcium carbonate, and china clay slurries in ethylene glycol in the manufacture of film or fiber products. Table I provides process details including the existing filtration systems employed for each additive slurry. Samples of different solutions were evaluated and the level of their dispersion determined. Filterability studies based on the above mentioned protocol were conducted to select the optimum absolute depth filter grade, which will provide good slurry quality and dispersion control. Also, a mathematical model was developed to compare the performance of different grades of absolute depth filters in controlling the level of a dispersion. The effect of system parameters on the dispersion quality was also examined.

Figure 1. Production of Polyester Film and Fiber

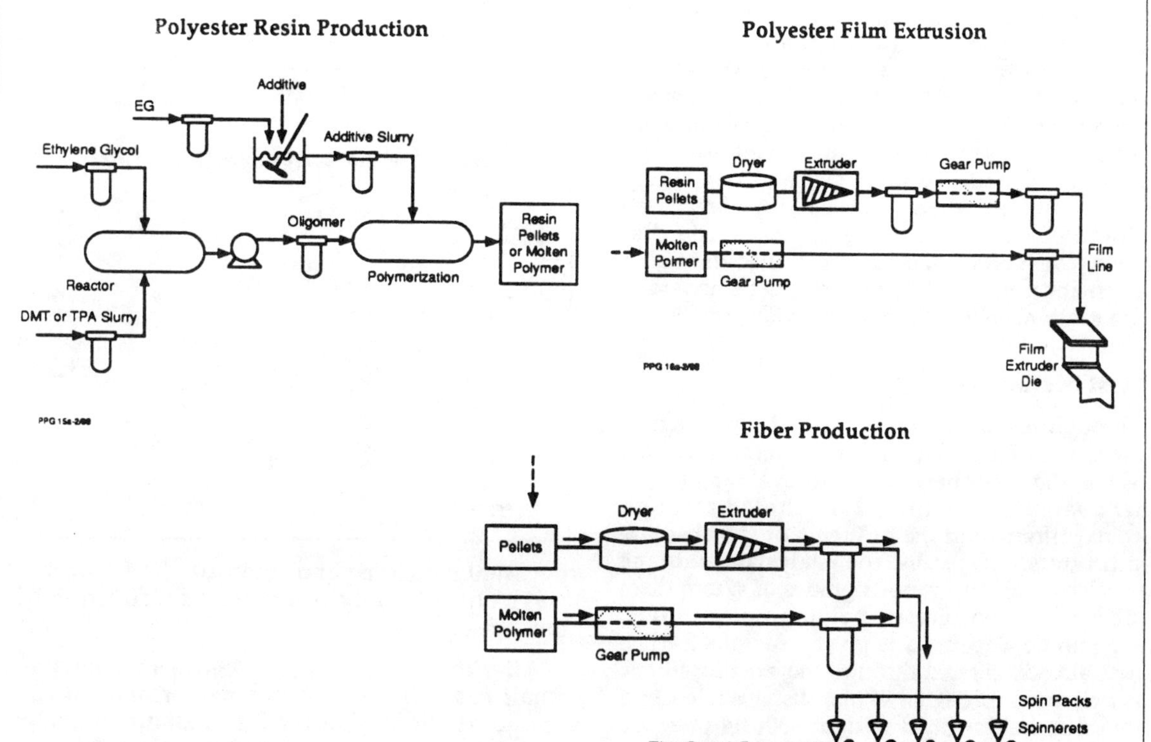

Table I:
Background Information on the Additive Slurries Evaluated

Additive Slurry	Solvent	Total Solids %	Utilized In Polyester Product	Present Filtration System and Size	Flow Rate gal/hr
Titanium Dioxide	Ethylene Glycol	15	Fiber	500 mesh, 10" (three stages)	66
China Clay	Ethylene Glycol	20	Film	40µm abs., 10" (recirculation)	9
Calcium Carbonate	Ethylene Glycol	20	Film	1µm nom., 2X10" (two stages)	1

II. Selection of Optimum Filtration Levels:

Filterability studies were conducted to select the optimum absolute depth filter grade or the finest filter grade which will provide good additive slurry quality. Figure 2 illustrates the test equipment involved, while the Flow Chart, Parts I and II, summarizes the test protocol employed in selecting an analysis membrane followed by the actual selection of the finest filter, which would not affect the solids concentration. With this method, the level of dispersion of the unfiltered slurry in addition to the level of filtration of the slurry filtered through existing nominal filters versus filtered through absolute rated depth filters were determined. Based on test results, the finest filter was incorporated in an optimal additive slurry filtration system where the dispersion level was maintained throughout the process. The dispersion level control was also quantified through a mathematical model developed for the slurry tank concentration, which is discussed in Section III.

A. Test Methods Used:

Through a trial and error method illustrated in Flow Chart, Part I, an analysis membrane was selected that was best suited to evaluate all the slurry samples (unfiltered or filtered through nominal filters, and the effluents of the absolute depth filters). Typically, the analysis membrane was selected among various grades of 47mm discs of absolute rated pleated polypropylene filters. The equipment required is given in Figure 2. The slurry, already filtered through the existing filters, was delivered through a 47mm disc installed in a 47mm disc holder fitted with a 0-30psig pressure gauge using a peristaltic pump (with Viton tubing) for the analysis membrane selection. The pressure drop across the membrane, flow rate and through-

Figure 2. Filtration Test Equipment

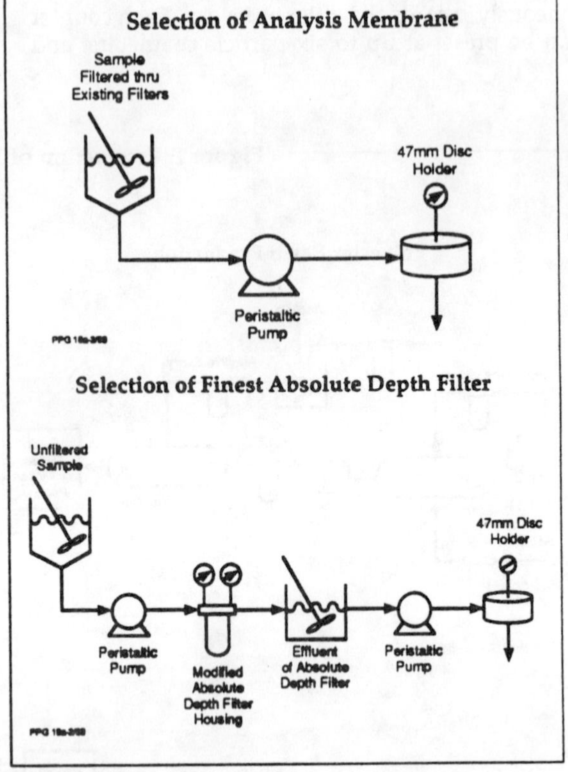

put were monitored and recorded. The flow rate was volumetrically measured as a function of pressure drop.

After the selection of the analysis membrane, small scale filtration was performed on the unfiltered additive slurry (i.e., titanium dioxide, calcium carbonate or china clay) using one inch sections of absolute rated polypropylene depth filters of various grades (Flow Chart Part II). The filter section was installed in a modified filter housing equipped with 0-30psig pressure gauges in the upstream and downstream sides and a peristaltic pump was used to filter the slurry (Figure 2). Approximately one liter effluent was generated through each filter grade. A system flow rate or a flow rate not to exceed an initial clean pressure drop of 10psid was selected. During the filterability study, the upstream and downstream pressures, flow rate and throughput were monitored. The total solids level of all the effluents of the depth filters was measured to determine if any essential solids were removed.

As shown in Flow Chart, Part II, all the slurry samples including the unfiltered and effluent of the absolute rated depth filters were filtered through the selected analysis membrane for a comparative study. The pressure drop and throughput data were compared to determine the level of unfiltered slurry and slurry filtered (through the present system) as well as to select the finest absolute rated depth filter providing effluent within total solids specification.

At a plant site a full scale study was conducted. One of the slurries, calcium carbonate dispersion, was filtered with the recommended finest absolute depth filters, after the completion of small scale testing mentioned above. This filtered slurry was utilized in the production of the plant's final product, polyester film. Data were collected on the film quality and the life of the final melt filter with the use of calcium carbonate filtered through absolute filters and compared to typical results obtained from utilizing slurry filtered with nominal filters.

Flow Chart for an Optimal Additive Slurry Filter Selection Part I

Selection of Analysis Membrane:

Objective: To select an analysis membrane for comparing slurry dispersion or filtration levels in terms of pressure drop and time at constant selected flow rate.

Criterion: When the slurry already filtered through the existing filters is delivered through the membrane, it should build up a pressure drop of 25 psid within 15-30 minutes at a constant flow rate. If a filtered sample is unavailable, the unfiltered sample should build up a pressure drop of at least 10psid within 15 to 30 minutes.

Approach: Trial and error method of varying the flow rate and/or the grade of the analysis membrane until the above criterion is met.

Test Parameters:

Fluid = Slurry effluent of existing system filters or unfiltered slurry
Filter = 47mm discs of pleated absolute filters as analysis membranes
Mode = Single pass

Terminology:

G = Disc grade in absolute
Q = Flow rate
ΔP = Differential pressure
t = time
abs. = absolute

Flow Chart Part I

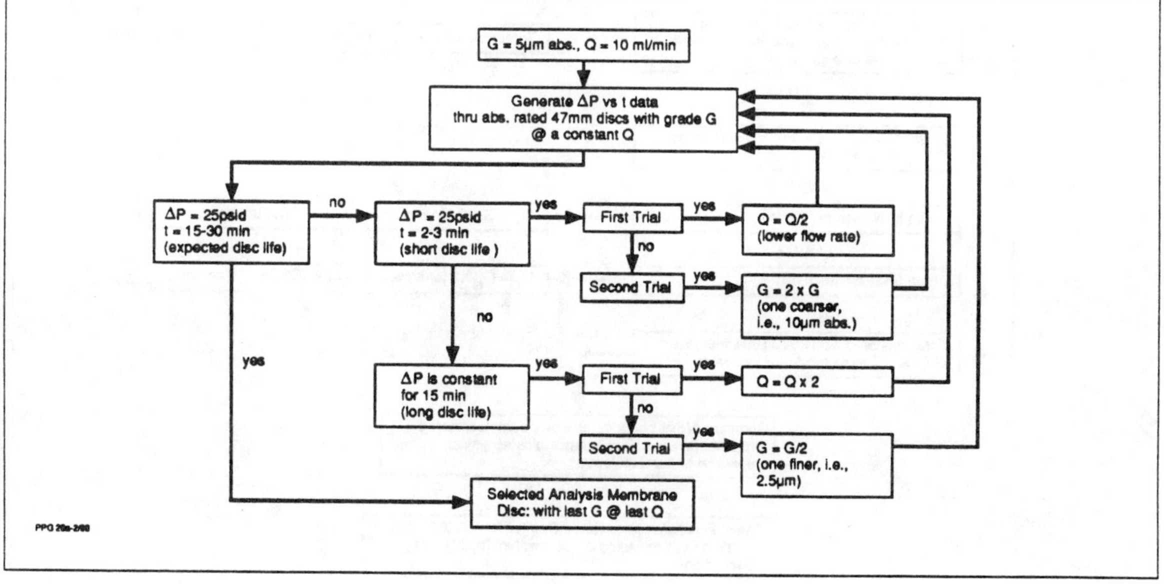

Flow Chart for an Optimal Additive Slurry Filter Selection Part II

Selection of Finest Absolute Depth Filter:

Objective: To select the finest grade absolute depth filter that will not change the solids concentration of the additive slurry, and characterizing the level of dispersion of the unfiltered slurry as well as the filtration levels of all filtered slurry samples.

Criteria: When the effluent of the finest filter is delivered through the selected analysis membrane, the membrane should build up a comparative pressure drop (of 25 psid or less) at a constant flow rate, providing a comparative throughput in the longest time period among all tested effluents and the total solids of this effluent should be within specifications.

Approach: Generation of effluents through different grades of absolute rated depth filters utilizing the unfiltered slurry and filtering all the slurry samples through the selected analysis membrane at the selected flow rate, comparing all the pressure and time data as well as the total solids level until the above criteria are met.

Test Parameters:

Fluid = Unfiltered Slurry
Filters = Different grades of absolute rated filters and 47mm discs of pleated absolute filters as analysis membranes
Mode = Single pass

Terminology:

G	=	Disc grade in absolute
G_d	=	Absolute depth filter grade
Q	=	Flow rate
ΔP	=	Differential pressure across the disc
ΔP_{in}	=	Initial differential pressure
t	=	Time
abs	=	Absolute
ARDF	=	Absolute rated depth filter
SAM	=	Selected analysis membrane

Flow Chart Part II

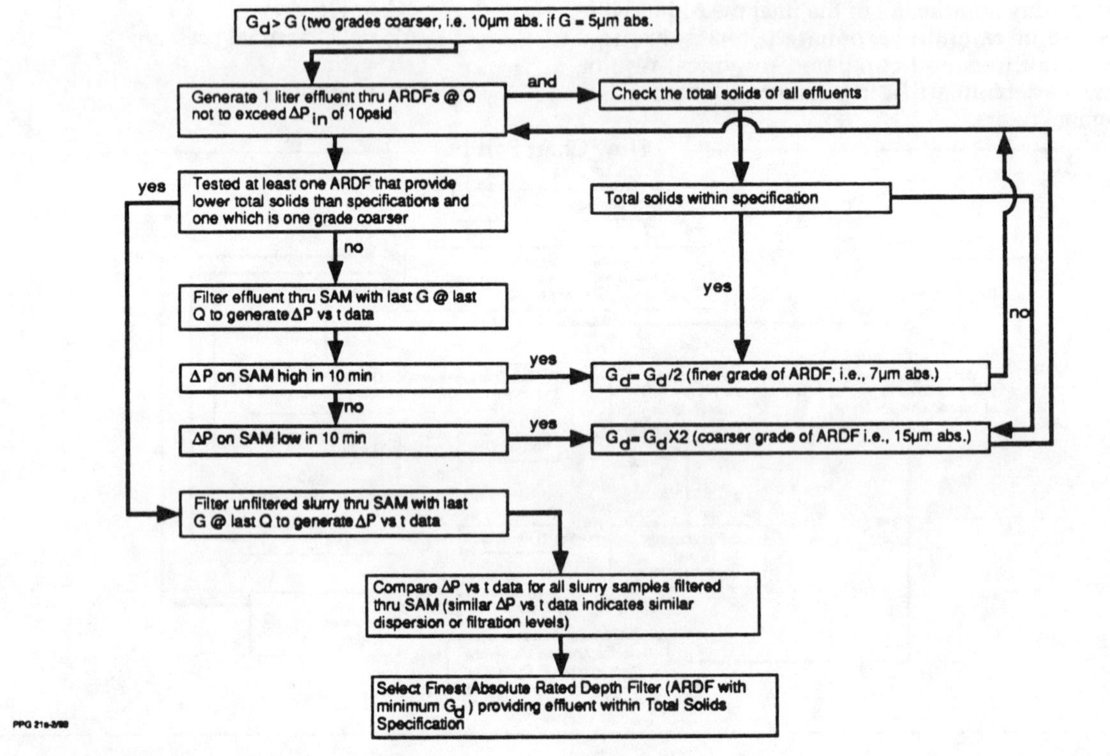

Gd> G (two grades coarser, i.e. 10µm abs. if G = 5µm abs.
Generate 1 liter effluent thru ARDFs @ Q not to exceed ΔPin of 10psid
and
Check the total solids of all effluents
yes
Tested at least one ARDF that provide lower total solids than specifications and one which is one grade coarser
no
Total solids within specification
yes
no
Filter effluent thru SAM with last G @ last Q to generate ΔP vs t data
ΔP on SAM high in 10 min
yes
Gd= Gd/2 (finer grade of ARDF, i.e., 7µm abs.)
no
ΔP on SAM low in 10 min
yes
Gd= GdX2 (coarser grade of ARDF i.e., 15µm abs.)
Filter unfiltered slurry thru SAM with last G @ last Q to generate ΔP vs t data
Compare ΔP vs t data for all slurry samples filtered thru SAM (similar ΔP vs t data indicates similar dispersion or filtration levels)
Select Finest Absolute Rated Depth Filter (ARDF with minimum Gd) providing effluent within Total Solids Specification

B. Results: Dispersion and Filtration Levels:

All the slurry effluent samples were within the total solids specifications. Tables IIA thru C and Figures 3A thru C provide the results on the filtration studies performed on titanium dioxide, china clay and calcium carbonate slurries. The tables summarize the filtration data; at constant flow rate, the pressure drop versus throughput data are given. In Figures 3A-C, the data of ΔP versus throughput illustrate the level of filtration or dispersion. The flatter each plot is, the better the level of filtration or dispersion is. Tables IIA-C include the filterability index values. The filterability index value indicates the level of filtration or dispersion, based on pressure and through-put. The higher the index value is, the worse the level of filtration or dispersion is.
The index value is defined as,

$$n = \Delta P/V$$

where,

ΔP = pressure drop
V = throughput

Two index values were selected at two different throughput values for each slurry. The first throughput value corresponds to data obtained from the most poorly dispersed sample, to which all the other samples were compared. The second throughput value corresponds to the final data points taken during the filtration cycle, and all samples providing equivalent throughputs were compared again.

Table IIA: Filtration Summary Data for Titanium Dioxide Slurry Using 4.5 μm absolute disc @ 16 ml/min

Sample	Time min	ΔP Initial psid	ΔP Final psid	Test Volume ml	Index n=ΔP/Vol.
Unfiltered	2.0	0	22.0	32.0	0.688[1]
500 mesh* Filtered	2.0	0	0.5	32.0	0.016[1]
"	18.8	0	0.5	300.0	0.002[2]
10 μm abs. Effluent	2.0	0	0.5	32.0	0.016[1]
"	7.5	0	20.0	120.0	0.167[3]
7 μm abs. Effluent	2.0	0	0.5	32.0	0.016[1]
"	18.8	0	1.5	300.0	0.005[2]
5 μm abs. Effluent	2.0	0	0.5	32.0	0.016[1]
"	18.8	0	0.5	300.0	0.002[2]

1 : Index at 32 ml throughput
2 : " "300 ml "
3 : " "120 ml "
* After 3 stages of filtration

The recommended finest filter was a 7μm absolute filter for the calcium carbonate slurry. With full scale operations, when the slurry filtered with these filters was used, the quality of the final film product (with 50% fewer drop outs) and the life performance of the final polymer melt filters (by more than two times) improved significantly in comparison to prior cases where nominal filters were employed.

C. Discussion of Results:

Typically, the average particle size distribution for these slurries is less than 1μm and larger size agglomerates (i.e., 3 or 5μm) cause defects in finished products. These fine levels of filtration can be achieved with absolute rated depth filters which provide reproducible effluent quality on a single pass basis. With nominal filters, fine dispersion levels may sometimes be achieved only with long cycles of recirculation and several filter stages. However, results are not always reproducible and the productivity of such systems is very low due to the time cycles involved.

It is easier and more economical to filter out the agglomerates and obtain fine dispersion levels in the beginning of the process than to filter the highly viscous melt polymers to the desired fine levels. Furthermore, the objective is to introduce clean streams into the polymer to minimize the burden on and improve the life of the final melt polymer filters. The melt polymer filters, which are typically costly to clean and especially to replace, should be removing gels and dirt intro-

Figure 3A: Filtration of TiO_2 Slurry using 4.5 μm absolute disc @ 16 ml/min

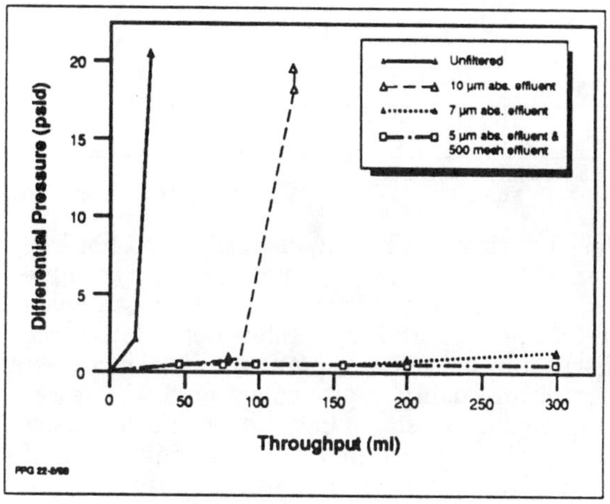

Table IIB: Filtration Summary Data for China Clay Slurry Using 25 μm absolute disc @ 10 ml/min

Sample	Time min	ΔP Initial psid	Final psid	Test Volume ml	Index n=ΔP/Vol.
Unfiltered	17.0	0	12.0	170.0	0.071[1]
"	30.0	0	25.0	300.0	0.083[2]
40 μm abs. Effluent	17.0	0	3.0	170.0	0.018[1]
"	30.0	0	6.0	300.0	0.020[2]
30 μm abs. Effluent	17.0	0	0.5	170.0	0.003[1]
"	30.0	0	4.3	300.0	0.014[2]
20 μm abs. Effluent	17.0	0	0.5	170.0	0.003[1]
"	30.0	0	0.5	300.0	0.002[2]

1 : Index at 170 ml throughput
2 : " " 300 ml "

Table IIC: Filtration Summary Data for Calcium Carbonate Slurry Using 5.0 μm absolute disc @ 10 ml/min

Sample	Time min	ΔP Initial psid	Final psid	Test Volume ml	Index n=ΔP/Vol.
Unfiltered	15.3	0	5.5	153.0	0.036[1]
"	20.0	0	11.0	200.0	0.053[2]
15 μm abs. Effluent	15.3	0	4.1	153.0	0.027[1]
"	20.0	0	8.4	200.0	0.036[2]
10 μm abs. Effluent	15.3	0	1.6	153.0	0.011[1]
"	20.0	0	3.0	200.0	0.013[2]
7 μm abs. Effluent	15.3	0	1.0	153.0	0.007[1]
"	20.0	0	1.4	200.0	0.005[2]

1 : Index at 153 ml throughput
2 : " " 200 ml "

duced during or after polymerization and not lose useful life in removing agglomerates of an additive slurry. In fact, the utilization of calcium carbonate slurry filtered with 7μm absolute depth filters (see discussion below) at a plant site demonstrated significant quality improvement with 50% fewer drop outs in the final film product in comparison to use of slurry filtered with existing nominal filters. Furthermore, during this production study, the final polymer melt filters provided more than twice the life cycle of melt filters exposed to polymer made from nominally filtered slurry.

Figure 3B: Filtration of China Clay Slurry using 25 μm absolute disc @ 10 ml/min

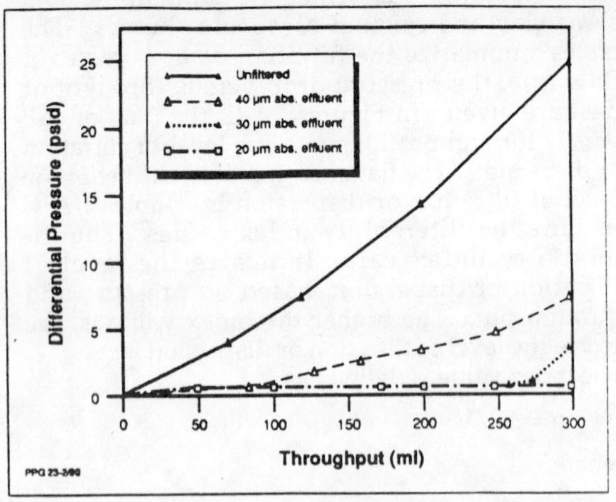

Figure 3C: Filtration of $CaCO_3$ Slurry using 5 μm absolute disc @ 10 ml/min

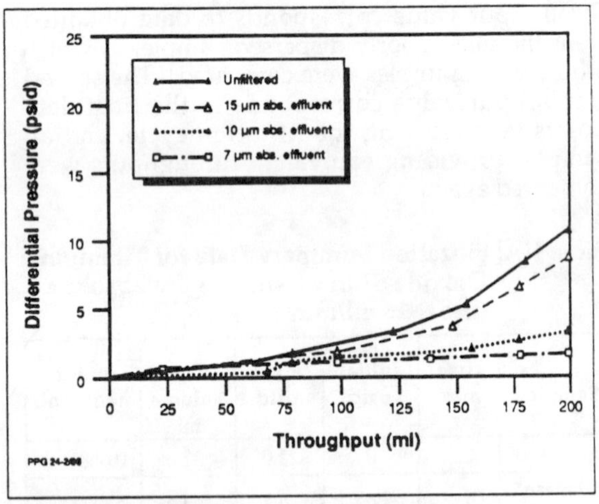

As seen in Figure 3A and Table IIA, the unfiltered titanium dioxide slurry has a low dispersion level with an index value (n) of 0.688 at a throughput of 32ml. All the absolute depth filters provide superior effluent quality with respect to the unfiltered slurry, demonstrating lower n values at 32ml. Superior effluent quality is obtained with the 5μm absolute depth filter on a single pass, single stage basis. Its index value of 0.002 at 300 ml is identical to that of the slurry filtered through the existing system (500 mesh final filter after two stages of 300 mesh filters and two grinding stages). Figure 3A illustrates the same results with the entire data sets. The finest filter not stripping any essential solids was the 5µm absolute rated depth filter.

Table IIB illustrates that the unfiltered slurry had a worse dispersion level than all the effluents of the absolute depth filters on a single pass basis with an index value of 0.083 at a throughput of 300ml. All the absolute depth filters provide superior effluent quality with respect to unfiltered with lower n values at 170 or 300ml. Best quality (lowest n of 0.002 at 300ml) is obtained with the 20µm absolute depth filter on a single pass basis. This absolute depth filter was the finest filter which did not alter the china clay slurry concentration. The same phenomena are observed in Figure 3B.

Table IIC indicates that all the absolute rated depth filters provided better slurry quality than the unfiltered calcium carbonate slurry at 153ml or 200ml. On a single pass basis, superior filtration level was obtained with a 7µm absolute rated depth filter which was the finest filter not affecting the calcium carbonate total solids level. Figure 3C provides similar detailed results.

For any kind of slurry sample it is important to filter through the finest absolute rated depth filter on a single pass basis initially as mentioned above. The filtration level can be easily controlled and essential solids removal avoided with fine grades of absolute depth filters on a single pass basis. It is more economical and easier to change filters frequently, if necessary, or make modifications when the slurry is far from the point of use. In most processes, holding the slurry for a certain period is required prior to introduction to the polymer process itself. It is significant to maintain the fine level of filtration or dispersion obtained previously in this holding stage as well. Perfect mixing and recirculating the batch through a coarser absolute depth filter than the finest filter ensures the maintenance of the dispersion without altering the solids concentration. This is further discussed through a mathematical model in the next section. If the unfiltered slurry is poorly dispersed, it should be recirculated through an absolute depth filter prior to filtration through the finest filter to extend its life. If there are long lines leading to the polymer process, it is also necessary to filter the slurry at the point of use through an absolute rated depth filter with an intermediate grade. This process scheme is outlined in Figure 4.

Figure 4: Optimal Filtration System for Additive Slurries

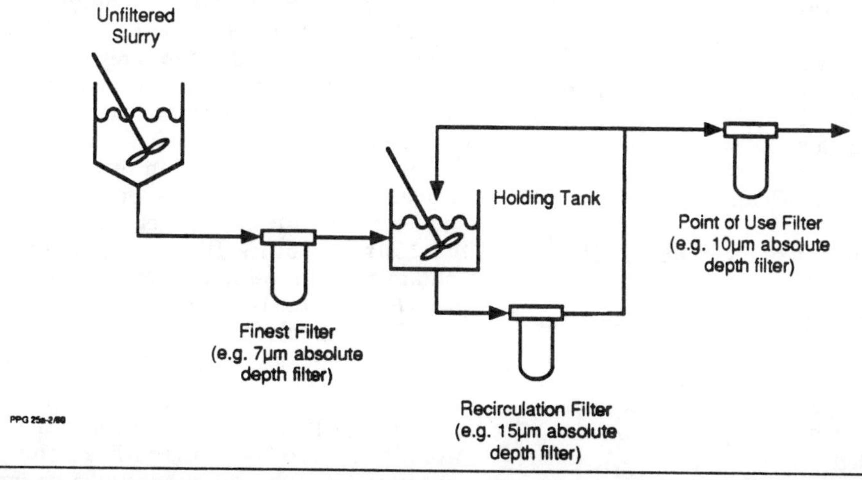

III. Dispersion Control:

A. Mathematical Model:

A mathematical model was developed for a slurry holding tank with continuous agglomeration of particles and recirculation through a filter (Figure 5)[7]. The agglomerate concentration distribution (or concentration as a function of time) was derived. The basis selected was particle size. This model could be utilized to study the effect of dispersion control for a slurry already filtered through the finest filter, being held for an extended period of time. Dispersion control is also significant for poorly dispersed unfiltered slurries. By recirculation through a coarse filter initially, the finest filter life may be improved.

The objective was to simulate concentration distributions for different grades of absolute rated filters on a recirculation basis and compare their performance in controlling the dispersion level in contrast to systems with no filtration. Also, the effects of the system parameters on the dispersion level were studied.

Figure 5: Holding Tank with a Recirculation Filtration System

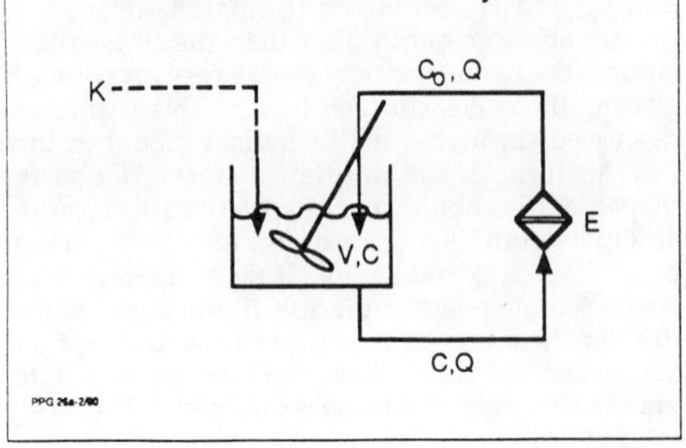

In the model development, the following terminology and units were used:

K = Rate of Particle Agglomeration (# of Agglomerates/Time)
V = Total Tank Volume (Volume)
C = Upstream and Tank Concentration of Agglomerates (# of Agglomerates/Volume)
C_o= Downstream Concentration of Agglomerates (# of Agglomerates/Volume)
Q = Flow Rate (Volume/Time)
E = Particle Reduction Factor

In deriving the modified agglomerate balance, the following assumptions were made:

1. Homogeneous, instant mixing of particles in the tank
2. Constant rate of agglomerate removal
3. Constant rate of agglomeration of particles
4. Constant tank volume, i.e., negligible volume decrease due to digression
5. Constant filtration flow rate

The total balance for the agglomerates for a given size range in the tank can be expressed as:

$$K + C_oQ - CQ = V\,dC/dt \qquad \text{——Eq.(1)}$$

where $$C_o = (1-E)C \qquad \text{——Eq.(2)}$$

The concentration distribution is obtained by integrating equation (1) as:

$$C=(1/QE)\left[K(1-e^{-QEt/V})+C_iQEe^{-QEt/V}\right] \qquad \text{—Eq.(3)}$$

With either initial boundary condition (C=0 or $C=C_i$), C_{max} or C_{equil}, maximum or equilibrium agglomeration concentration, at time $t \rightarrow \infty$, is

$$C_{max} \text{ (or } C_{equil}) = K/QE \qquad \text{——Eq.(4)}$$

B. Simulated Results:

Utilizing the mathematical model described above, agglomerate concentration distributions were simulated for 15 and 30μm absolute depth filters in a holding tank containing an additive slurry. The simulated results for absolute filters on a recirculation basis are included in Figure 6 (see page 12). The effect of agglomeration rate on the concentration is illustrated in Figure 6 as well. The details of the calculations can be found in Appendix A on page 13.

C. Discussion of Simulated Results:

If there is no agglomeration, then regardless of the filter grade used the concentration would be zero for particles greater than 10μm. However, this is an unrealistic situation. The mathematical model clearly illustrates the significance of employing a coarse filter (with respect to the finest filter) on a recirculation basis while holding a slurry. Without any filtration, based on the rate of agglomeration and the length of the holding stage, the original fine dispersion level could be destroyed. Based on simulated equilibrium for 15 and 30μm absolute depth filters, it is predicted that the 15μm absolute filters will maintain the contamination level at a minimum of 50% fewer particles (greater than 10μm) than that obtained with 30μm absolute filters, regardless of a finite rate of agglomeration (Figure 6 and Appendix A).

Assuming zero initial particle concentration greater than 10μm in the tank (since the slurry is already filtered by a finer filter), the particle concentration is directly proportional to the rate of agglomeration. Figure 6 shows particle concentrations for different assumed agglomeration rates. For both the absolute filters, the higher the agglomeration rate is, the worse the dispersion level is. Dispersion levels are inversely proportional to recirculation or tank turnover rates (Eq. 4).

IV. Conclusions:

1. Based on the application of the proposed test protocol, it was possible to determine the level of the dispersion of an unfiltered additive slurry or the level of filtration provided with an existing system. In most cases, the level of filtration with a single pass using an absolute depth filter was better than the existing nominal filter utilized on a recirculation basis for a long period of time. It was shown here that utilizing absolute rated depth filters on a single pass basis, with this test protocol the finest level of filtration (as fine as 5μm), could be selected for any additive slurry, without altering the solids content of the dispersion used in the production of polyester film or fiber.

2. The recommended system for common additive slurries is provided in Figure 4. After achieving the finest level of filtration, it is recommended to maintain the dispersion level through a recirculation system using coarse absolute rated depth filters. At the point of use, the slurry should be filtered through intermediate grade absolute depth filters on a single pass basis. The following modified filtration schemes are recommended for these exceptions:

a. If the unfiltered slurry is poorly dispersed (the finest filter builds up pressure very rapidly), it should be recirculated through a coarse absolute depth filter prior to filtering through the finest filter.

b. If the slurry is transferred to the polymer process through short piping, the point of use filter may be eliminated. The slurry can be directly delivered to the polymer process from the holding tank through the recirculation coarse absolute depth filter or through the finest filter, whichever the case may be.

3. The mathematical model developed for a holding tank with a filtration system on a recirculation basis demonstrated the significance of maintaining a fine dispersion level with coarse absolute rated depth filters. The example with 15μm absolute filter showed that the dispersion was maintained at a 50% lower particle concentration than a 30μm absolute filter. The model also illustrated that dispersion levels are directly proportional to agglomeration rates and inversely proportional to recirculation or turnover rates.

4. A systematic approach is essential in order to obtain a high quality product and best final filter life of a polymer process. This approach involves the evaluation of all the process streams and selection of high performance, efficient filters for each application as shown in Figure 1. This paper demonstrates the optimization of the filtration systems for additive slurries, critical components of filled melt polymers, with the ultimate goal of improving product quality and manufacturing yields, as well as reducing operating costs. This goal was achieved at a plant site; when calcium carbonate slurry filtered with 7μm absolute depth filters was used, the final polyester film product demonstrated 50% fewer drop outs and the final melt filters twice the life performance to reach terminal pressure drop in comparison to the prior cases where the slurry was nominally filtered.

Figure 6: Agglomerate Concentration versus Time for Slurry Tanks

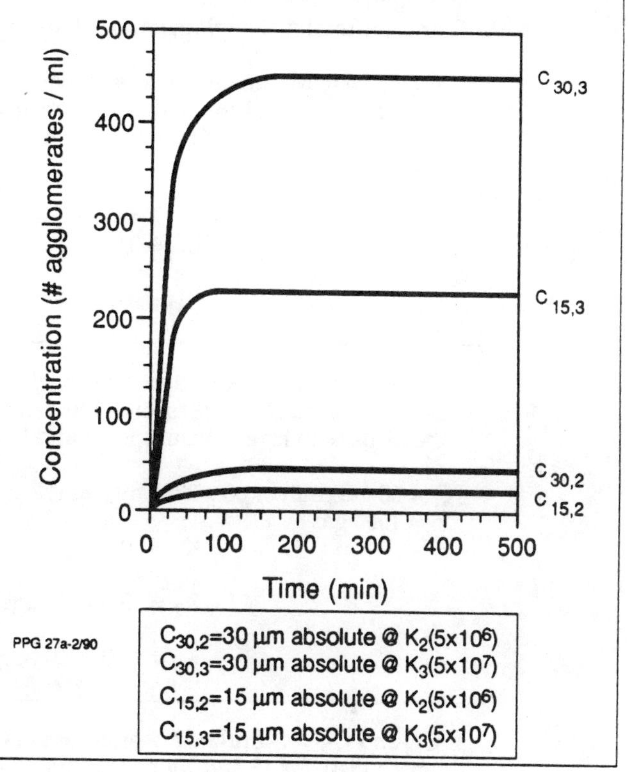

V. References:

1. Waters, A. G. and Loo, C. E., The Twelfth Australian Chemical Engineering Conference, Melbourne, 26-29 August, 1984, p. 831.

2. Kilham, L. B., 1988 Polymers, Laminations and Coating Conference Book 1, Tappi Press, Atlanta, December 1988, p.109.

3. "Pall Process Filtration Company Polymer Processing Group Application Guides PPG-1 and PPG-2," Pall Corporation Literature, East Hills, New York.

4. Powers, K. W. and Schatz, R. H., Exxon Research and Engineering Co., Patent # 4,358,560, February, 1981.

5. Bennett, D., Standridge Color Corporation, IFJ, August 1989, p. 12.

6. "Profile Field Service Report, #PR01a, 1986," Pall Corporation Literature, East Hills, New York.

7. Ayral, S., Proceedings of the Technical Program of NEPCON EAST' 89, Boston June 12-15,1989, p. 130.

APPENDIX A: Simulation of Results for Figure 6

The following approach was taken for the simulations and calculations:

Note: C or C_o = Total # of agglomerates ≥ 10μm /volume throughout the calculations

1. **E = Particle Reduction Factors** on particle size basis, E_{15} and E_{30} for 15 and 30 μm absolute depth filters respectively were obtained from literature data for removal efficiencies for particles greater than 10μm[6].

$$E_{15} = 0.99$$

$$E_{30} = 0.50$$

---

2. **K = Rate of Agglomeration** for particles ≥ 10μm in the holding tank: Different K values were assumed to determine the effect of agglomeration on the dispersion level. Values for V and Q were also assumed.

Q = 60 gal/min or 60 x 3785 ml/min
V = 1000 gal

$$K_1 = 0$$

$$K_2 = 5.0 \times 10^6 \text{ agglomerates/min}$$

$$K_3 = 5.0 \times 10^7 \text{ agglomerates/min}$$

---

3. **C_{15} and C_{30} = Simulated Agglomerate Concentration Distributions** were constructed for 15 and 30μm absolute depth filters respectively, in Figure 6. The effect of K on C is also illustrated in Figure 6.

With no filtration:

For $K_2 = 5.0 \times 10^6$ agglomerates/min,
in t = 10min, $C_2 = 13$ agglomerates/ml,
in t = 1000min, $C_2 = 1300$ agglomerates/ml

For $K_3 = 5.0 \times 10^7$ agglomerates/min,
in t = 10min, $C_3 = 130$ agglomerates/ml,
in t = 1000min, $C_3 = 13000$ agglomerates/ml, etc.

With absolute rated depth filters:

From Eq.(3) with $C_i = 0$, $C = (K/QE)\left[1 - e^{-QEt/V}\right]$

From K_1, K_2, K_3 and E above

$C_{15,1} = 0$, and $C_{30,1} = 0$

$$C_{15,2} = (5x10^6/60x3785x0.99)\left[1 - e^{-(60x0.99t)/1000}\right] \text{ agglomerates/ml}$$

$$= 22.2\left[1 - e^{-0.059t}\right] \text{ agglomerates/ml}$$

with $C_{equil,2} = 22.2$ at time $t \rightarrow \infty$ from equation (4).

$$C_{15,3} = 222\left[1 - e^{-0.059t}\right] \text{ agglomerates/ml}$$

with $C_{equil,3} = 222$ at time $t \rightarrow \infty$ from equation (4).

$$C_{30,2} = 44\left[1 - e^{-0.030t}\right] \text{ agglomerates/ml}$$

with $C_{equil,2} = 44$ at time $t \rightarrow \infty$ from equation (4).

$$C_{30,3} = 440\left[1 - e^{-0.030t}\right] \text{ agglomerates/ml}$$

with $C_{equil,3} = 440$ at time $t \rightarrow \infty$ from equation (4).

Paper Title : Residence Time Distribution in a Melt Polymer Filter.

Author : Michel J. Elghossain
Engineering Manager
Memtec America Corporation

Paper presented at the American Filtration Society Washington Conference, March 19-22, 1990.

ABSTRACT

A detailed analysis will be presented to examine the effects of internal surfaces and geometries on the residence time distribution in a melt polymer filtration system.

The paper will address the influence of streamlining flow passages and reduction of dead areas on the total residence time and its distribution within the system, as well as incremental pressure differential contribution. The associated costs and resultant benefits will be discussed.

INTRODUCTION:

It has been customary to calculate the residence time for a flowing fluid in a containment vessel by simply dividing the residence volume over the volumetric flow rate.

This method assumes plug flow characteristics of the fluid under consideration. While it is adequate to show the magnitude of residence time, this method greatly underestimates the effects of the infinitely possible geometrical shapes that lead to the same residence volume. Hence, the unlimited possibilities of residence time distribution functions.

First, we will analyze a Newtonian flow through a circular tube and show the effect of relatively simple geometry on the residence time distribution. Then, we will proceed to examine the effects of internal surfaces and geometries on the residence time and its distribution in a melt polymer filtration system.

THEORETICAL BACKGROUND:

The flow of a Newtonian fluid through a circular tube is governed by the following parabolic velocity equation:

$$v = Vmax\ (1-(r/R)^2) \qquad (1)$$

where $Vmax = DPR^2/4\mu L$. (See page 21 for nomenclature.)

The average residence time in the tube is obtained by dividing the residence volume ($\pi R^2 L$) over the volumetric flow rate obtained from the Hagen-Poiseuille law, and is as follows:

$$Tav = 8\mu L^2/DPR^2. \qquad (2)$$

Tav is based on the average velocity (Vav) which is equal to half (1/2) Vmax.

If we define Tm as a minimum residence time of the fluid portion at the maximum velocity (r = 0) then we obtain:

$$Tm = Tav/2 = 4\mu L^2/DPR^2 \qquad (3)$$

Making use of the velocity profile (equation 1), the cross sectional area and equation (3), we can easily obtain the following normalized residence time distribution function:

$$T/Tm = 1/(1-(r/R)^2) \qquad (4)$$

Figure 1 shows a plot of the residence time distribution as well as the velocity profile in function of normalized radial coordinate.

We can easily see (from Figure 1) that portions of the flow reside at orders of magnitude larger than the average residence time. Under these circumstances, it is obvious that if a molten polymer is here considered, it would be highly susceptible to degradation in the low flow regions.

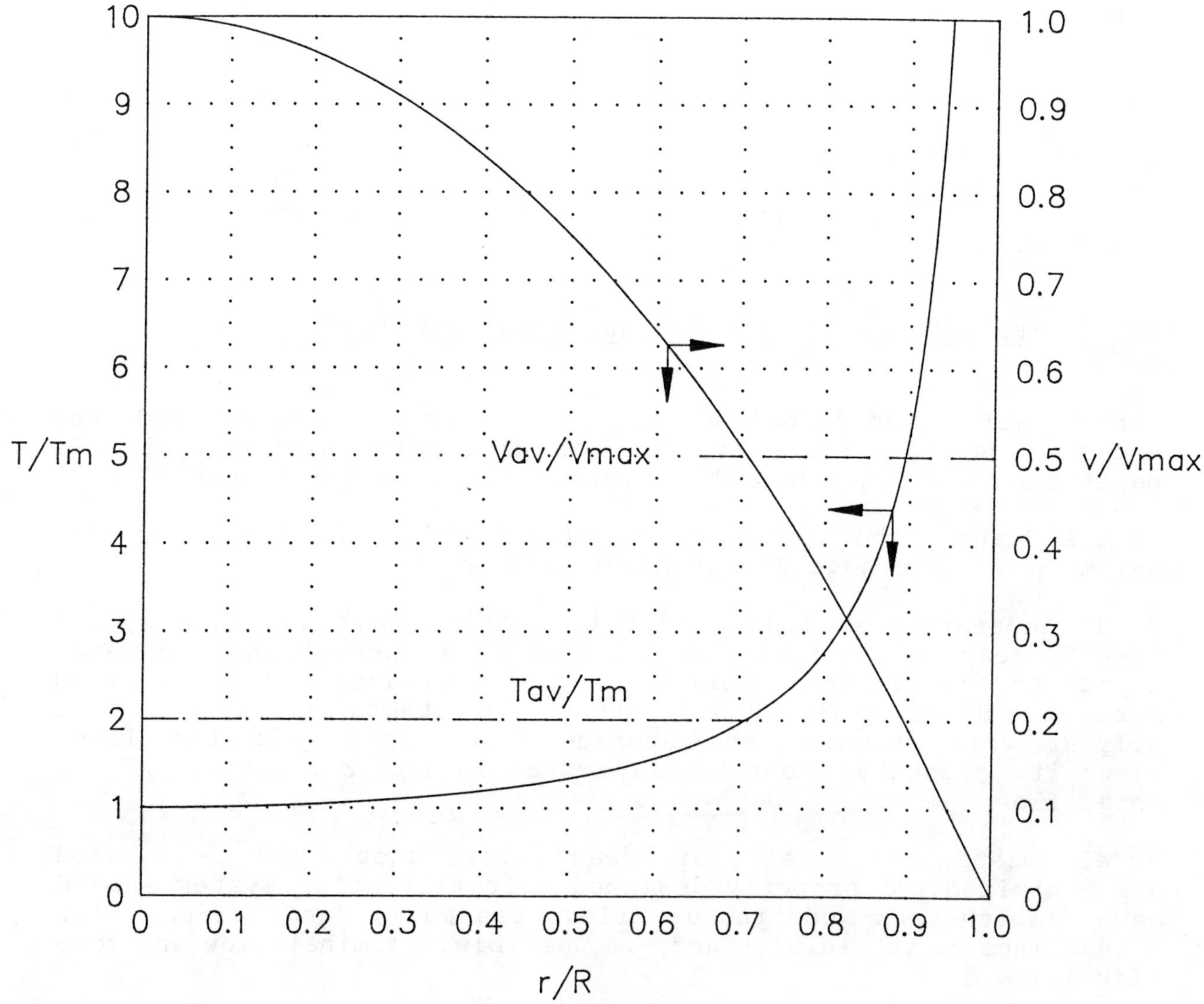
T/Tm
v/Vmax
r/R
Vav/Vmax
Tav/Tm
0
1
2
3
4
5
6
7
8
9
10
0.1
0.2
0.3
0.4
0.5
0.6
0.7
0.8
0.9
1.0

FIGURE 1

PROBLEM DEFINITION AND SOLUTION:

If we can determine precisely the velocity profiles in a system, such as a melt polymer filtration system, then we possibly can predict or calculate analytically the residence time distribution function. This task is certainly formidable and highly complex, which leads to an impractical approach to the solution.

Having a lesser goal and keeping in mind the simple residence time equation, we can now proceed to evaluate and obtain flow and thus time distribution in such a system by assuming that the velocity profile is flat in infinitesimally small sections. The availability of high speed low cost digital computers makes this task possible.

THE FILTRATION SYSTEM, ITS VARIABLES AND SOLUTION:

A flat leaf disc melt polymer filtration system is considered here and analyzed. The analysis in the proceeding sections of this paper can be adopted to other systems (i.e. candle elements).

Figure 2 shows the system under consideration and its associated variables. (See page 21 for nomenclature)

It is imperative at this moment to mention that the true actual "dead" areas in a system are not considered here to have a major impact on the residence time and age distribution of the flowing stream. Since these "dead" areas are stagnation points, the polymer will stagnate and degrade (i.e. form gels and lower viscosity products), but at an extremely small fraction of the total flow.

By all means, the effects of "dead" areas should not be ignored and minimized. A properly designed melt filtration system should have smooth and continuous flow passages, and offer flow streamlines so to minimize and, if possible, eliminate low and zero flow areas.

Having that in mind, let us proceed and analyze the age and time distribution for the important major bulk of the flow.

As mentioned earlier, the system under consideration is shown in Figure 2. It is basically a flat leaf disc filtration system.

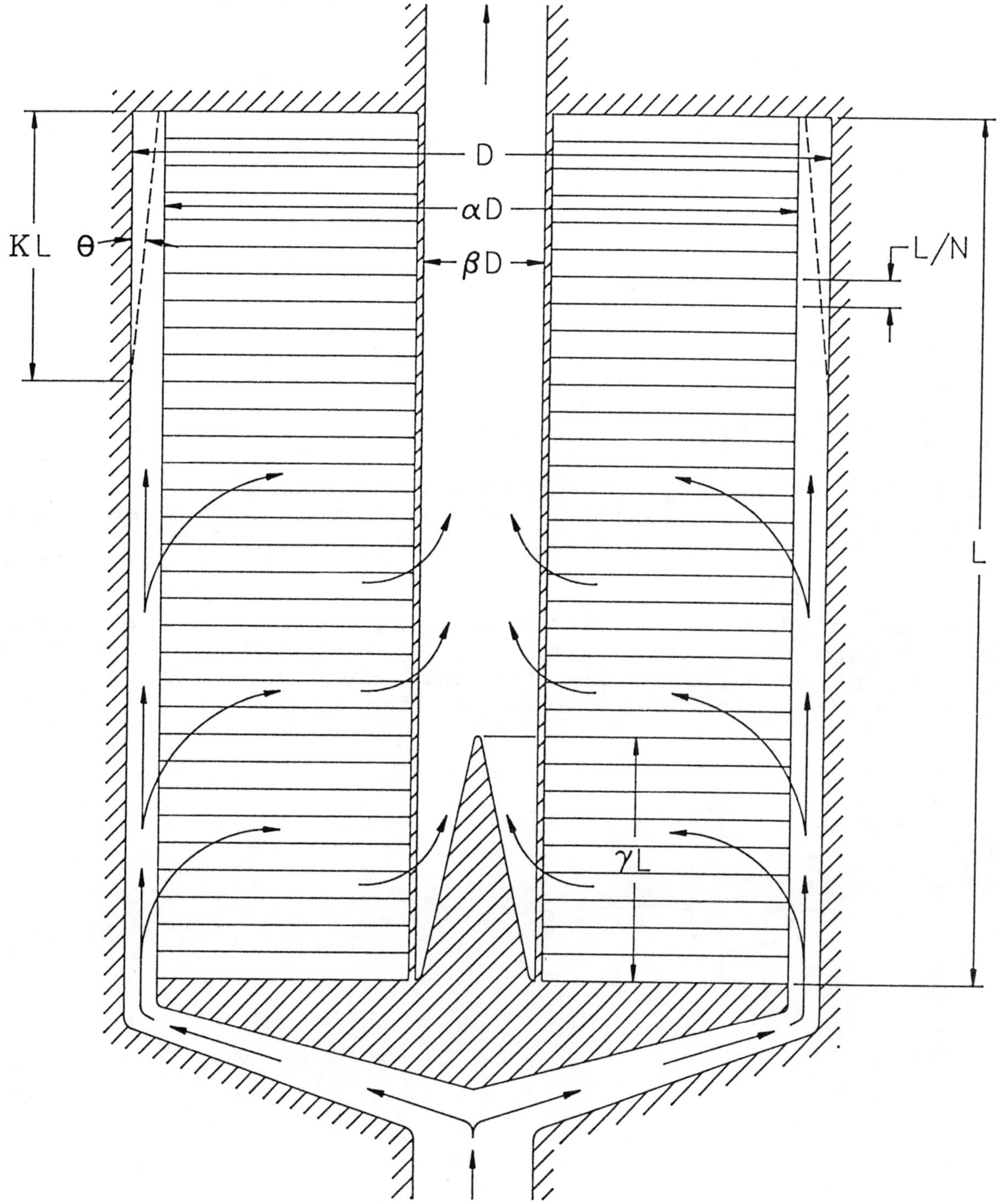

FIGURE 2

An individual leaf disc is constructed of filter media on the top and bottom surfaces with one or more drainage/support members sandwiched in between. Multiple leaf discs are usually stacked one on top of another over a centerpost structure to form a stack.

The leaf disc stack can be sectioned in multiple flow loops as shown in Figure 3 (only five leaf discs are shown). Resistances to flow are in the annular gap between the leaf disc stack and the inside wall of the housing, in the leaf discs themselves and in the centerpost.

The leaf disc assembly offers resistances to flow as well, which are in the gap between adjacent leaf discs, in the media, in the internal drainage support and in the hub section.

Since the flow is laminar, each of the flow loops leads to a linear equation of pressure loss in each of its branches (resistances). If the number of leaf discs is X and each one is divided into Y sections, then we end up with 4*X*Y + 3*X (3*X are for the annular gap, the hub and the centerpost) branches or resistances to flow.

Given a set of operating conditions and the characteristics of a given polymer, it is possible to calculate the flow in each branch of this network , its associated pressure drop and the residence time.

Having obtained the flow distribution, we can then proceed to sort the data and obtain the cumulative age distribution profile in function of normalized time (in this case being the ratio of residence time value over average residence time).

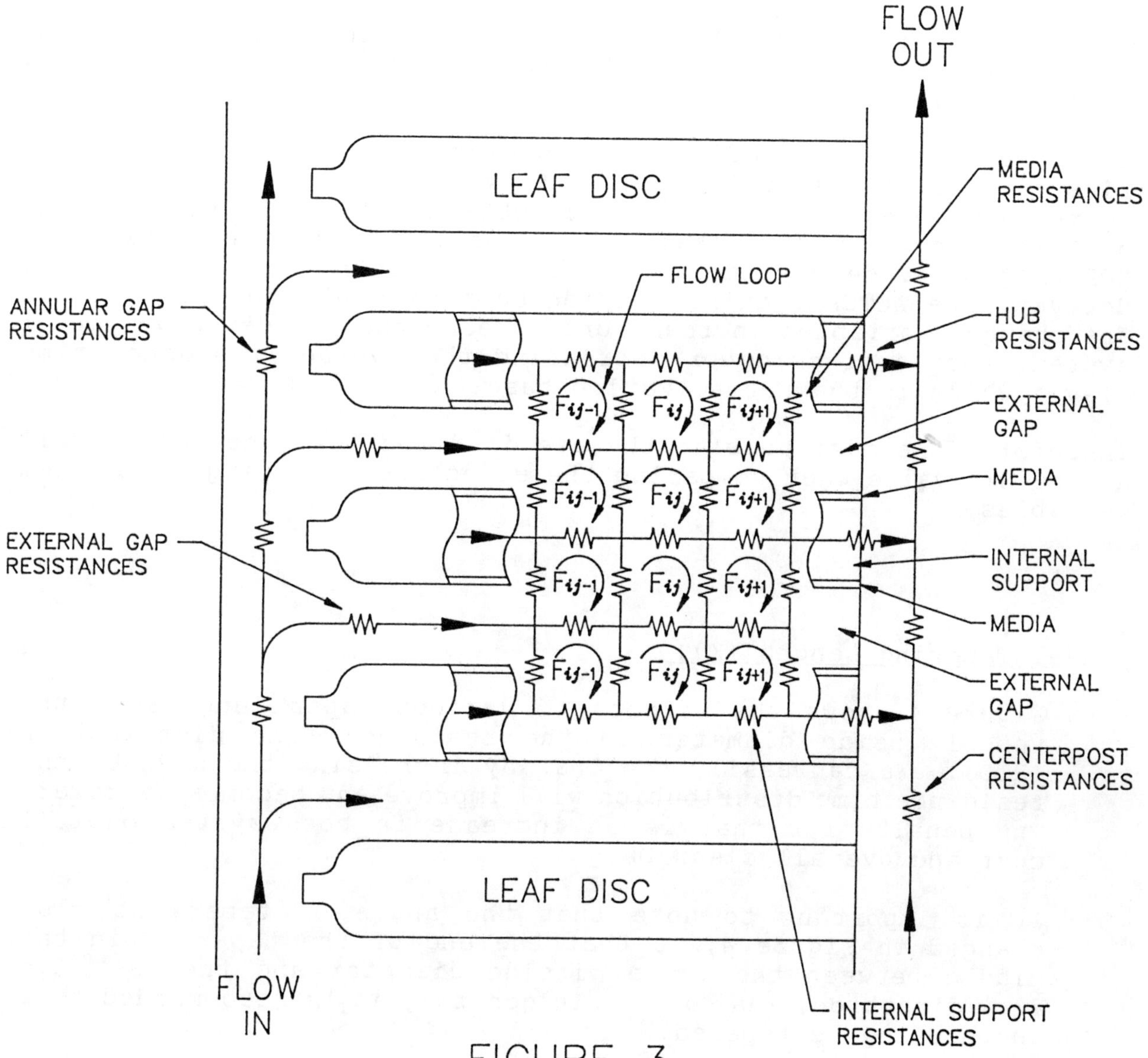
FLOW
OUT
LEAF DISC
MEDIA
RESISTANCES
FLOW LOOP
ANNULAR GAP
RESISTANCES
HUB
RESISTANCES
EXTERNAL
GAP
MEDIA
EXTERNAL GAP
RESISTANCES
INTERNAL
SUPPORT
MEDIA
EXTERNAL
GAP
CENTERPOST
RESISTANCES
LEAF DISC
FLOW
IN
INTERNAL SUPPORT
RESISTANCES

FIGURE 3

RESULTS AND DISCUSSION:

Before proceeding with the discussion, I would like to clarify a term, referred to as MTR, that appears on all of the remaining figures.

The Maximum residence Time Ratio(MTR) is the ratio of ACTUAL maximum residence time obtained with one set of filtration system variables over a set of another.

Let us assume, for example, that the maximum NORMALIZED residence time of a given system is known. If we change one or more of the system variables, it is possible to obtain relatively the same age distribution function. That is, the maximum (as well as minimum) NORMALIZED Residence Time is the same for both set of variables. However, the ACTUAL maximum residence time could be different if the changed variables increase or decrease the residence volume of system, thus increasing or decreasing the average residence time (Tav), which is the normalization base.

Therefore, one must combine the age distribution function as well as the MTR value when comparing the effects of changing the system variables.

Vessel Tapering Length (K):

Figure 4 shows the effects of various taper lengths of the vessel inside diameter on the residence time distribution (RTD). We can easily see that by increasing the K VALUE,the residence time distribution will improve and becomes narrower. The penalty for that is an increase in both system capital cost and overall clean DP.

It is important to note that the angle of taper was also changed in Figure 4, so that the end of the taper is in the middle between the vessel inside diameter and the last (in flow direction) leaf disc. In general, it is recommended that the housing be tapered.

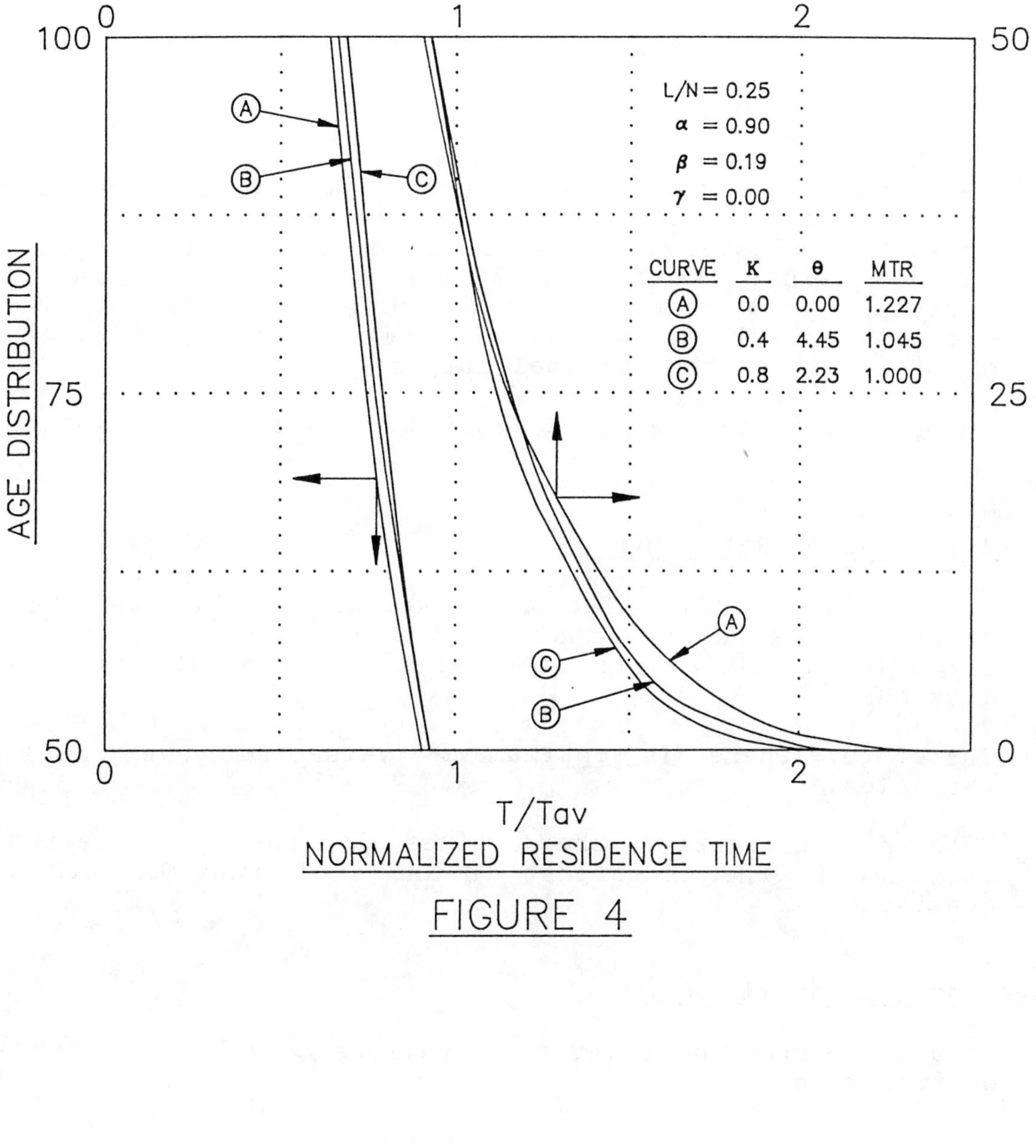

0
1
2
100
75
50
50
25
0
AGE DISTRIBUTION
L/N = 0.25
α = 0.90
β = 0.19
γ = 0.00
CURVE K θ MTR
A 0.0 0.00 1.227
B 0.4 4.45 1.045
C 0.8 2.23 1.000
A
B
C
T/Tav
NORMALIZED RESIDENCE TIME

FIGURE 4

Centerpost Cone Length (γ):

Figure 5 shows the effects of various centerpost cone lengths on the RTD. Although γ does not greatly effect the RTD, it does greatly increase the clean DP of the system if it is excessive.

A centerpost cone is always recommended with moderate to low values of γ. It eliminates a true dead area thus decreasing a degradation chance, at a small added cost.

Angle of Housing Taper (θ):

As indicated earlier, tapering the housing and using a centerpost cone will greatly narrow the RTD function as well as decrease the overall average residence time. Figure 6 shows the effect of various angles of the housing taper at a fixed K value. We can easily see how the MTR dramatically decreases as θ increases. However, the clean DP values increase as well and become excessive with higher values of α, thus negating the original purpose.

An optimum angle can be chosen for every set of variables.

Vessel ID/Disc OD Ratio (α):

Figure 7 shows the effects of various α values on the RTD function. We can see that, if the α values are either too large (high initial clean DP) or small, the MTR values will increase.

Therefore, there is optimum α value for every set of variables.

Note that in figure 7, the β values also changed. The reason for that is the centerpost ID and leaf disc OD were kept constant.

Centerpost ID/vessel ID (β):

Figure 8 shows the effects of various β values on the age distribution.

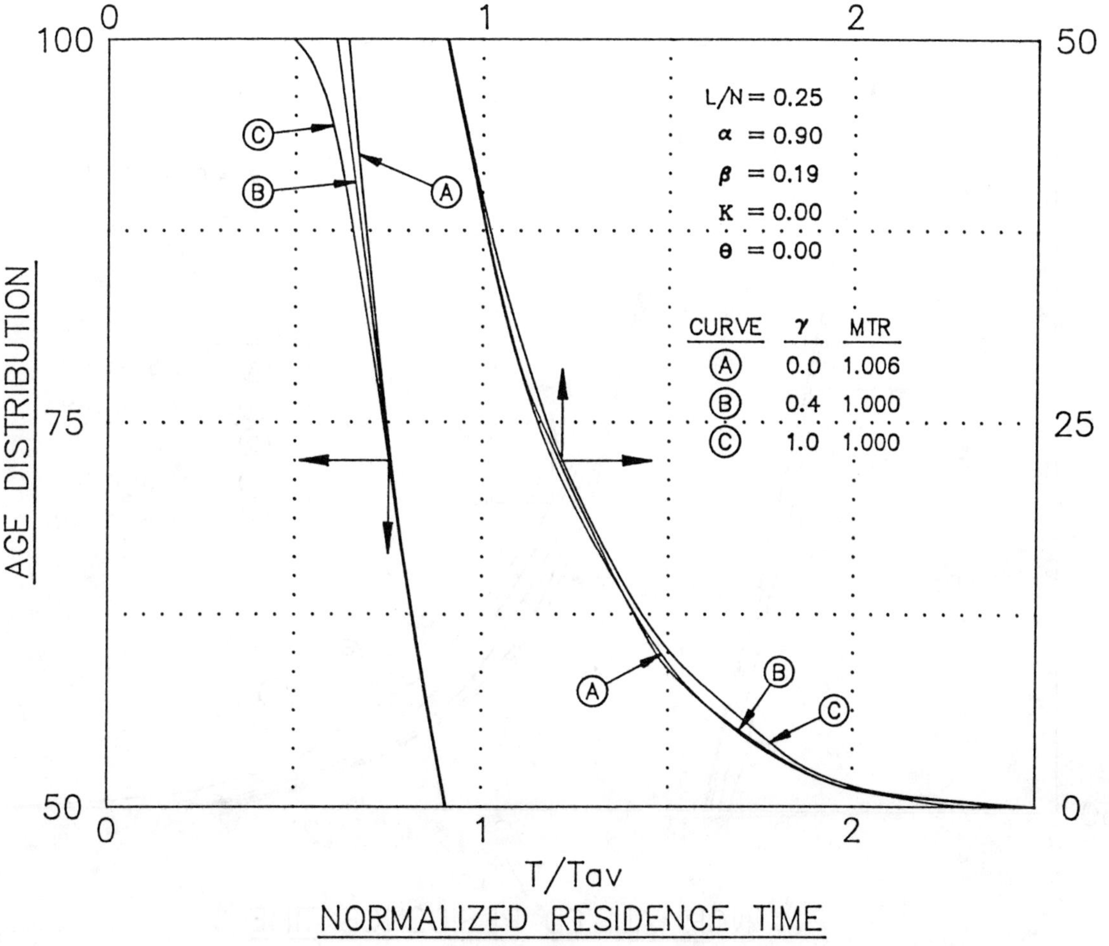

L/N = 0.25
α = 0.90
β = 0.19
K = 0.00
θ = 0.00
CURVE γ MTR
A 0.0 1.006
B 0.4 1.000
C 1.0 1.000
AGE DISTRIBUTION
T/Tav
NORMALIZED RESIDENCE TIME

FIGURE 5

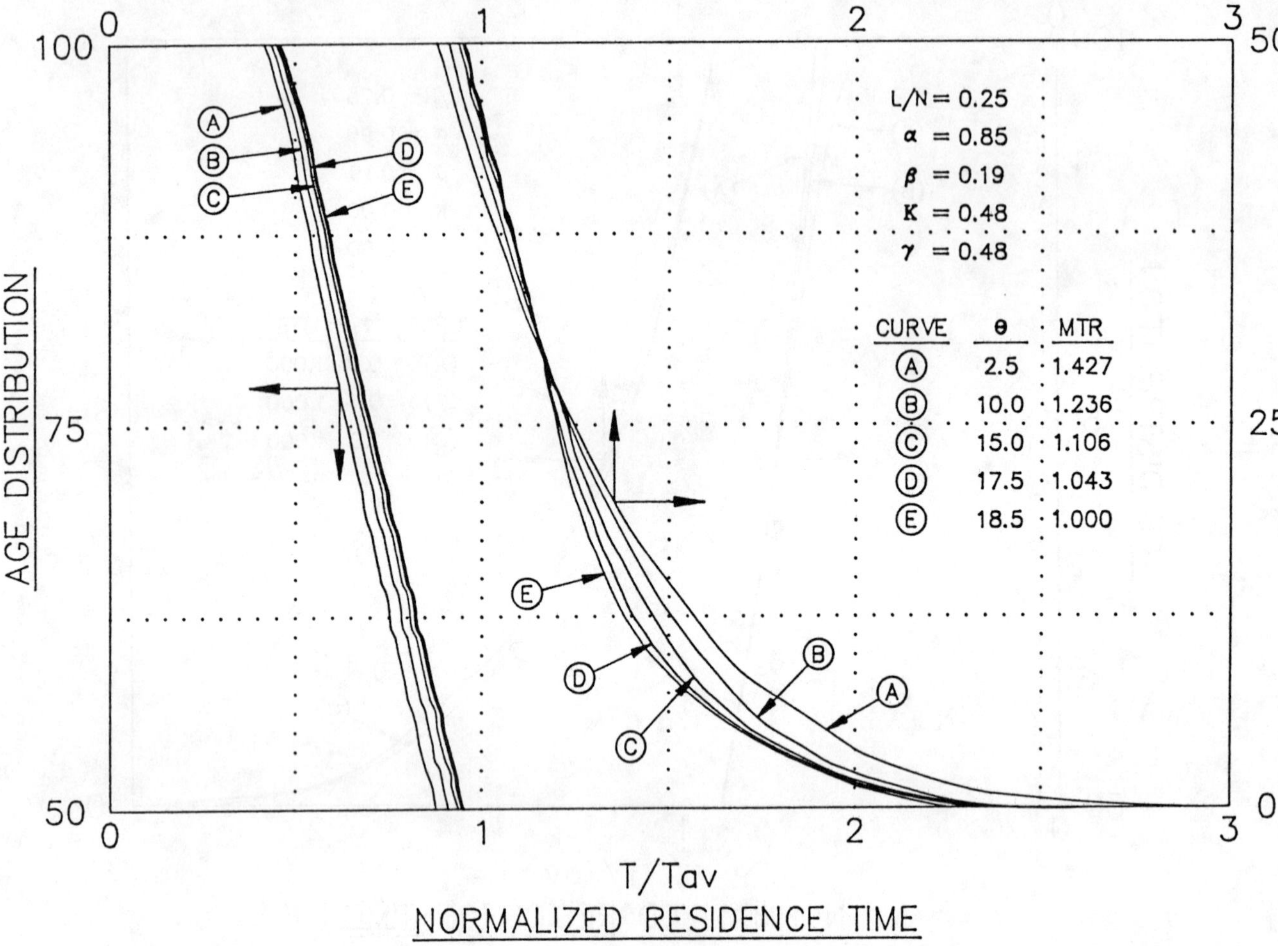

L/N = 0.25
α = 0.85
β = 0.19
K = 0.48
γ = 0.48
CURVE θ MTR
A 2.5 1.427
B 10.0 1.236
C 15.0 1.106
D 17.5 1.043
E 18.5 1.000
AGE DISTRIBUTION
100
75
50
0
1
2
3
25
T/Tav
NORMALIZED RESIDENCE TIME

FIGURE 6

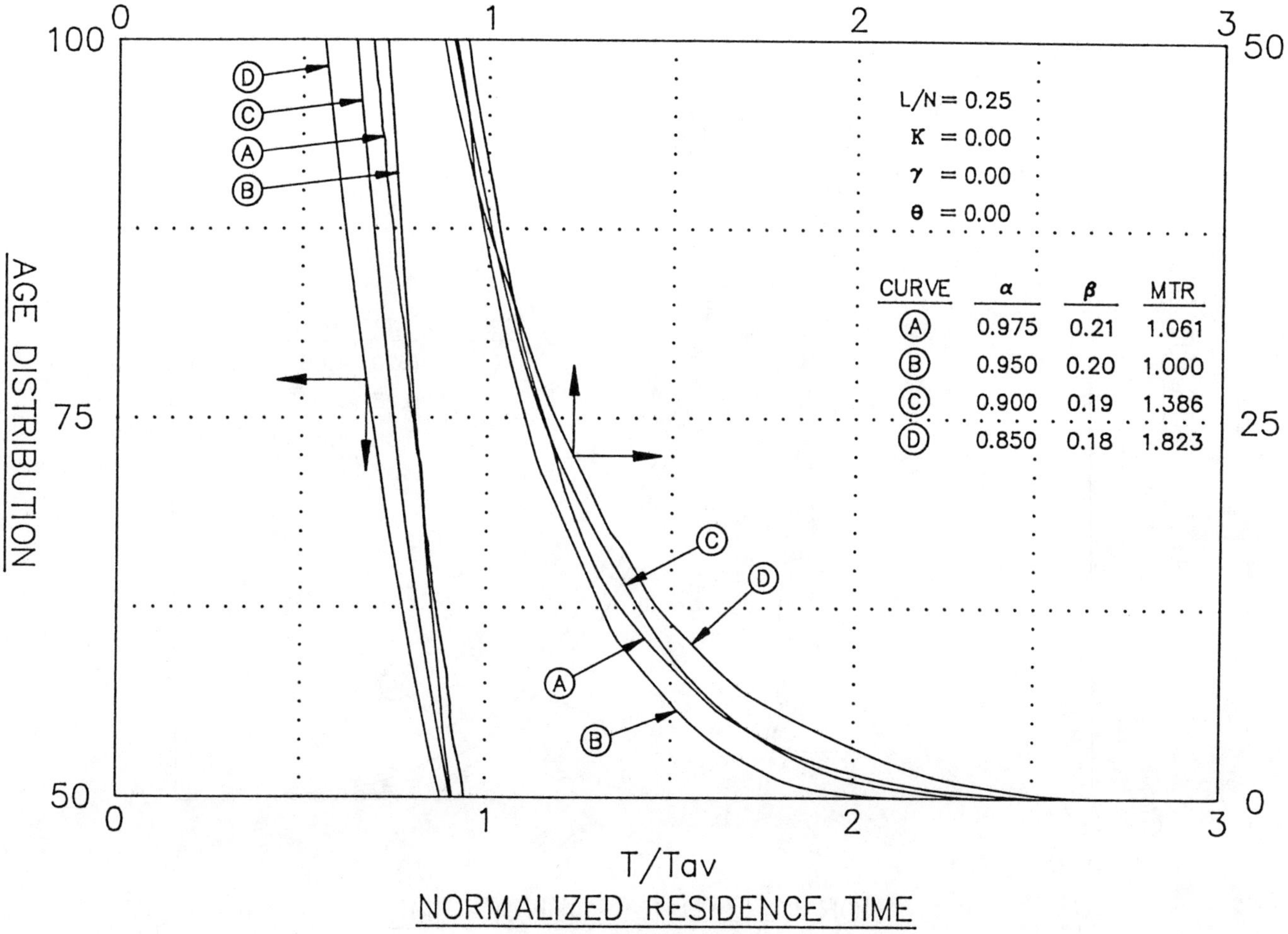

L/N = 0.25
K = 0.00
γ = 0.00
θ = 0.00
CURVE α β MTR
A 0.975 0.21 1.061
B 0.950 0.20 1.000
C 0.900 0.19 1.386
D 0.850 0.18 1.823
AGE DISTRIBUTION
T/Tav
NORMALIZED RESIDENCE TIME

FIGURE 7

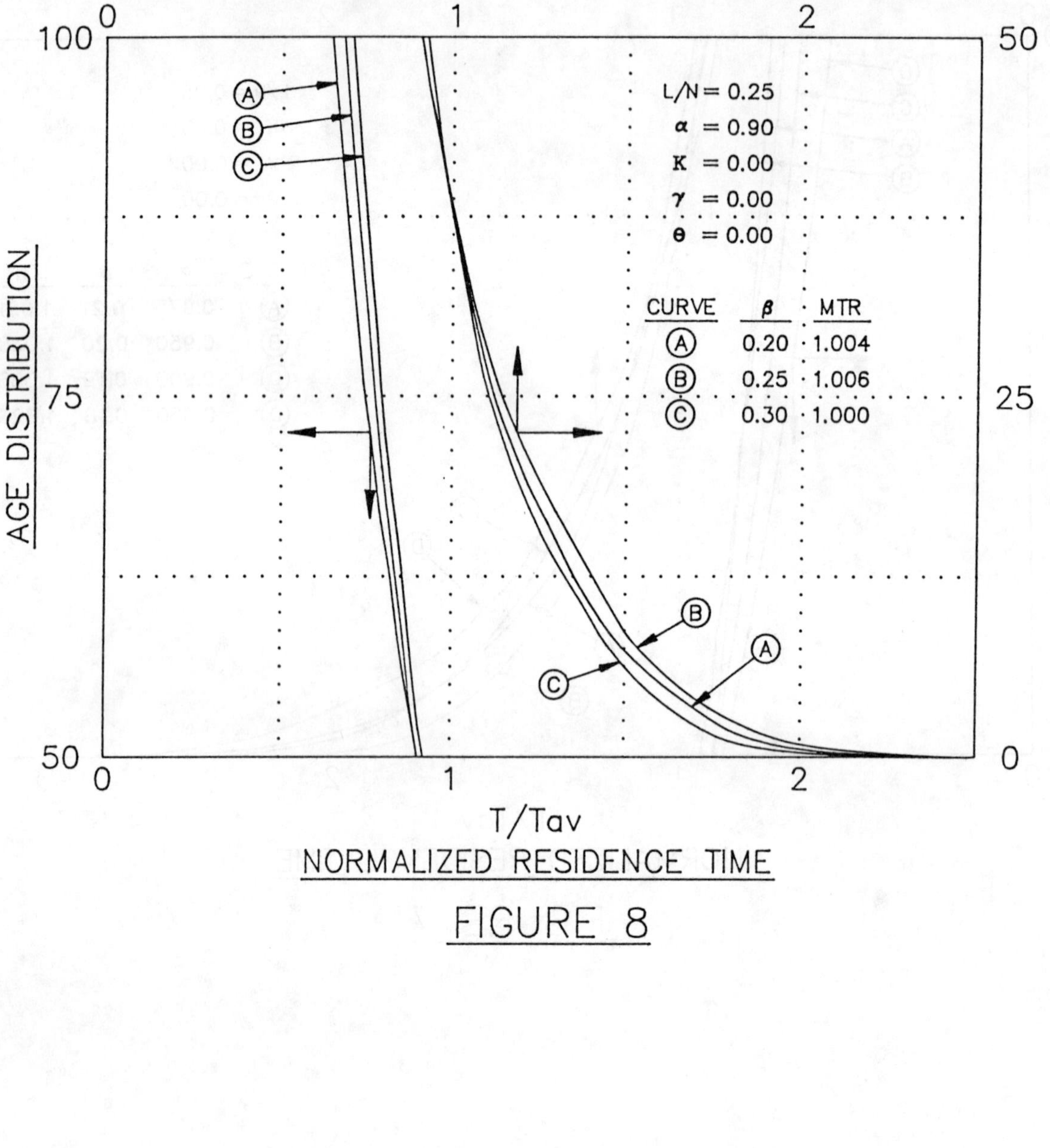

100
50
75
25
0
1
2
AGE DISTRIBUTION
L/N = 0.25
α = 0.90
K = 0.00
γ = 0.00
θ = 0.00
CURVE
β
MTR
A
0.20
1.004
B
0.25
1.006
C
0.30
1.000
T/Tav
NORMALIZED RESIDENCE TIME

FIGURE 8

In general, when β values are very small or very large they will affect the RTD function, but not in the range shown in figure 8. Those values were chosen to reflect some of the current industry standards of process piping and leaf discs inside diameters.

Resistance Ratio(RR):

The effects of drainage/support member(of leaf disc) over external slot (between two leaf discs) flow resistances ratio are shown in Figure 9.

In an earlier work[1], it was demonstrated that both a decrease in RR and a decrease in gap thickness (between two leaf discs) improve the uniformity of flux rate through the filter medium. Figure 9, explicitly demonstrates how a leaf disc is constructed can affect the age distribution as well as the MTR values.

Onstream Life:

Figure 10 shows how the RTD function changes with on stream life of filter medium. As filtration time increase, so does the DP of filter medium. This tends to improve the uniformity of flux rate through the medium.

In essence, figure 10 also shows the effects on the RTD when the filtration level is decreased (higher removal efficiency), by emulating an increase in medium resistance (Decrease in permeability).

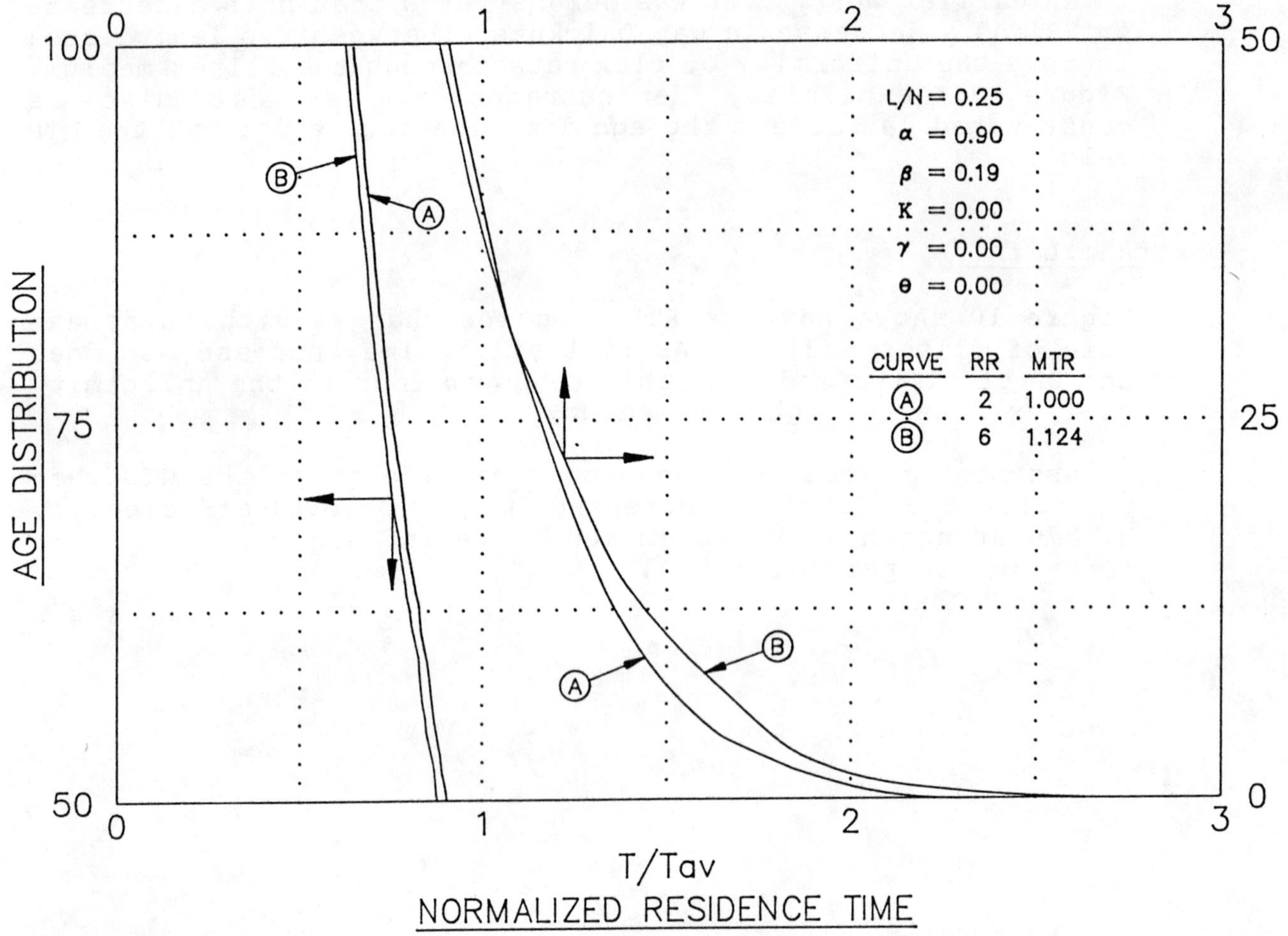
L/N = 0.25
α = 0.90
β = 0.19
K = 0.00
γ = 0.00
θ = 0.00
CURVE RR MTR
A 2 1.000
B 6 1.124
AGE DISTRIBUTION
100
75
50
0
1
2
3
50
25
0
T/Tav
NORMALIZED RESIDENCE TIME

FIGURE 9

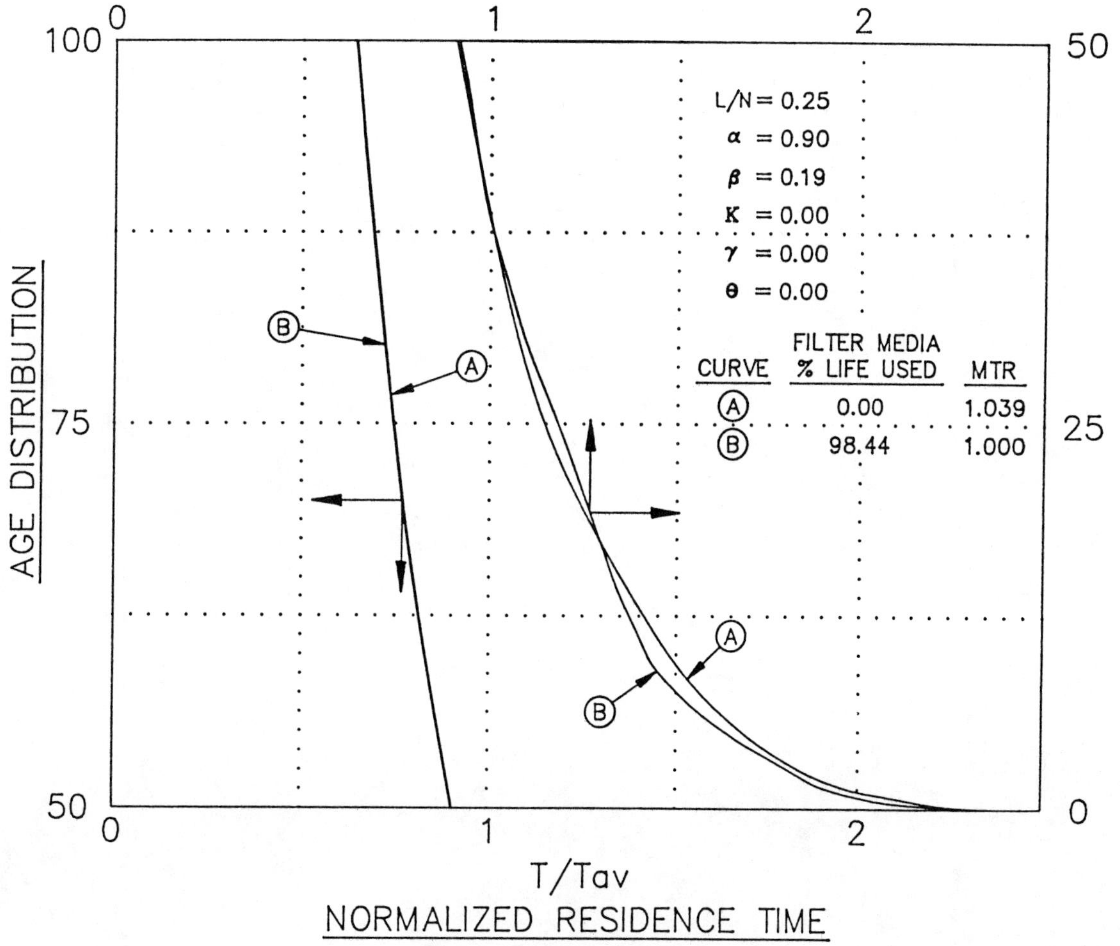

100
75
50
0
1
2
50
25
0
AGE DISTRIBUTION
L/N = 0.25
α = 0.90
β = 0.19
K = 0.00
γ = 0.00
θ = 0.00
CURVE
FILTER MEDIA % LIFE USED
MTR
A 0.00 1.039
B 98.44 1.000
A
B
T/Tav
NORMALIZED RESIDENCE TIME

FIGURE 10

SUMMARY:

It should be emphasised that this work is theoretical, and the results shown are as accurate as the assumptions can be. The predictions, while are not to be taken as absolutes, they however show us the trends and effects of the RTD when altering one or more of the system variables.

Knowledge of the Residence Time Distribution in a melt polymer filtration system offers us very useful information about the emerging filtrate and its quality.

It is recommended that a filtration system have smooth and continuous internal flow passages. It should also incorporate most of the influencing geometries and streamlines that will minimize dead areas and improve the overall RTD.

A system can and should be properly designed with an optimum combination of cost, clean differential pressure and narrow RTD function.

Ackowledgements:

The author wishes to acknowledge Mr. Walter Smith of Memtec America Corp. for his invaluable assistance in making the drawings on the CAD station.

References:

(1) Elghossain, Michel J. (1989) Filtration & Separation, Jan/Feb. "Design Optimization of a Melt Polymer Filtration System."(Page 43)

Also, "International Fiber Journal", April 1989, page 45.

NOMENCLATURE	UNITS l=length; m=mass; t=time
D = Vessel Inside Diameter	l
DP = Differential Pressure	m/lt^2
L = Length of Stack or Tube	l
MTR = Maximum Residence Time Ratio	
N = Number of Leaf Discs	
r = Radial Distance (coordinate)	l
RR = Resistance (to flow) Ratio of Support/Drainage Member over External gap between two Adjacent Leaf Discs.	
T = Residence Time	t
Tav = Average Residence Time	t
Tm = Minimum Residence Time	t
Vav = Average Velocity in Tube	l/t
Vmax = Maximum Velocity in Tube	l/t
v = Velocity	l/t
α = Ratio of Disc OD over Vessel ID	
β = Ratio of Centerpost ID to Vessel ID	
γ = Ratio of Centerpost Internal Cone length to stack length.	
θ = Angle of Vessel ID Taper	
K = Ratio of Taper Length over stack length	
μ = Viscocity	m/lt

Media Selection For Polymer Filtration

Terence Stott
Business Manager
Fairey Polymer Filtration
Standard Way
Fareham, England

Robert A. Smith
Applications Manager
Fairey Polymer Filtration
P.O. Box 27091
Greenville, South Carolina 29616

Increased quality and/or performance requirements of such thermoplastic materials as thin films and fine fibers has heightened the awareness of the importance of filtration in manufacturing.

In responding to the needs of the polymer industry, filter manufacturers have found that the characteristics of the various available filter medias combined with process conditions/ limitations present a very complicated picture. The focus of this paper will be on the characteristics of the available medias. In addition, the paper will specifically address the parameters affecting the performance of fiber medal media and experimental methods of predicting actual plant performance.

Filter Design and Media Selection for Polymer Filtration

Introduction

Within the last decade, the increasing demands on product quality and performance of various polymer based products like thin films and fine fibers has forced the industry to new levels of sophistication of both the polymeric materials as well as the processes involved. This in turn has been accompanied by an increasing recognition of the importance of the filtration operation in meeting these demands.

The fundamental structural, physical and chemical properties of a filtration medium which determine its performance are such that many characteristics are mutually exclusive (eg, permeability and filtration efficiency) and, as in so many situations, a compromise balance of properties must be found which is best suited to the particular process application. However, the demanding requirements of thermoplastics processing impose constraints additional to those which are commonly encountered. As will be discussed later, it is also vitally important to define filter performance under actual conditions, as these may be very different from those measured under experimental conditions.

The basic characteristics of a filter medium is that it must be porous to the fluid phase while being impervious to suspended contaminants. Media are often described as exhibiting either "depth" or "surface" characteristics. A depth medium is effective in three dimensions, and its voids/interstices are capable of holding contaminants throughout the structure in the direction of fluid flow. A surface medium is effective in two dimensions only, preventing contaminants from penetrating the media by a sieving action at the surface.

It is readily apparent that polymer processing has a number of features which require specific performance from the filtration media or imposes special constraints. This paper will discuss those features of existing media which make them suitable for polymer applications with specific attention given to the parameters affecting fiber metal media and their performance.

The Filter Medium

Wire Mesh

Because the gaps in a woven wire mesh are of well-defined dimensions, such media has very accurately defined filtration characteristics, whose performance can be predicted with a high level of confidence. Mesh will find application in situations where a very precise contaminant size cut-off is required. Mesh is essentially a "surface medium", which means that its contaminant holding capacity is very low. Improvements in filtration performance can be obtained by combining and sintering together several layers of mesh, arranged in order of progressively increasing fineness, so that succeeding layers trap increasingly smaller contaminant particles, hence giving the effect of a depth medium.

Stainless steel mesh filter elements are easy to clean, since they are very robust and most of the contaminant is on the outer surface. In contrast, the gel capture performance of mesh is poor which automatically eliminates it as a primary consideration for many applications.

Sintered Metal Powders

Bronze and stainless steel are the metals usually employed in sintered powder filters. Medium produced from spherical particles is close-packed and sinter-bonded at the points of contact forming a three dimensional network of interconnected voids. This pore structure is such that a high degree of shear is imparted to the polymer during processing. Some users consider that high shear is highly effective at breaking the bonds between those polymer molecules which constitute deformable gels; others consider that gels are effectively removed by entrapment within the labyrinth structure of a depth medium. It is probable that both of these mechanisms operate in practice.

Sintered bronze powders are relatively cheap to process and can be considered as a disposable filter medium. This gives the maximum assurance of filter cleanliness and eliminates the need to install cleaning and retesting facilities. However, despite its higher cost, stainless steel must be used in those situations where bronze might interact chemically with the polymer, promote degradation or, in the case of nylon, a discoloration. Stainless steel powders tend to be of a more irregular shape than bronze, but the media characteristics are generally very similar. Sintered powders are inherently very rigid, both bronze and stainless steel media being highly resistant to deformation.

Sintered Metal Fibers

Stainless steel fibers are available in a range of Zdiameters down to as low as 2 um. The fibers are randomly laid in sheet form, sintered, and then compressed to give a filtration medium with specific characteristics. When this medium is examined in cross-section, the void area is seen to be dramatically different from that in a sintered powder medium. The actual processing conditions can be varied to produce a range of fiber media having different properties. The principal process variable is the degree of bonding between adjacent fibers. A high degree of bonding produces strong sintered bonds between fibers which result in a so-called "hard-sintered" medium whose fiber network is relatively rigid. In contrast, a "soft-sintered" medium will have few such sinter bonds and will be relatively flexible. The claimed advantage of a soft-sintered medium is that it responds well to a final cleaning operation. However, it is always necessary to employ a fine wire mesh downstream of the fibers to provide integral support of the fibers and a final barrier against any contaminant that may force its way through the more flexible structure. The disadvantage of soft-sintered vs. hard-sintered media is one of absolute rating of the media under operating conditions and after cleaning. As will be discussed later, movement of the more flexible soft-sintered medium can lead to changes in filtration efficiency. While no specific documentation exist at the moment, it is further speculated that this movement can lead to premature failure of the media due to fatigue.

Sintered stainless steel fibers can readily be processed in the form of multi layer composites. Such an arrangement substantially improves the contaminant holding capacity of the overall medium, by enabling the fine fiber layer to be protected from blocking by coarser particles.

Filtration Performance

Fiber Materials

Fibers offer the greatest potential for the preparation of media having specific performance in polymer filtration applications. A bed of fibers which are cylindrical and of uniform circular cross-section constitutes a relatively simple geometrical structure. The principal parameters describing the medium are fiber diameter, relative density and overall thickness. There have been a number of attempts to derive theoretical expressions for the permeability and filtration efficiency of such a porous medium. Good correlation can be obtained with performance figures derived experimentally, usually using an apparatus in which a fluid such as a mineral oil is dosed with a contaminant powder and passed through the medium.

The dependance of permeability and filtration efficiency (represented by micron rating) upon fiber diameter and relative density is summarized in Fig 1 (for uniform fiber beds of fixed weight/unit area). This shows that the selection of a medium for a particular application must always be a compromise between decreasing permeability or increasing filtration efficiency.

Overall thickness affects permeability in a well defined way, permeability being inversely proportional to thickness. Hence, the medium can be described by a "permeability coefficient", which depends only upon fiber diameter and relative density. Although filtration efficiency increases with increasing thickness of a depth medium, this variation does not follow a simple relation and there is no evident way to define a "filtration efficiency coefficient" which would be strongly dependent on thickness alone. Nevertheless, it is to be expected that filtration efficiency is describable by the geometrical arrangement of the fibers (fiber diameter and relative density) together with a thickness factor representing the decreasing statistical probability of a contaminant particle penetrating further and further into a depth medium.

One limitation of such an analysis is that under practical conditions of polymer filtration, it is unlikely that the geometrical arrangement of fibers will remain fixed. For example, the medium is subjected to gross plastic deformation during a pleating operation so that the configuration of fibers in the region of a pleat may be different from that in undistorted sheet. Additionally during polymer filtration, there is a large and increasing pressure differential across the medium which will tend to compact the fibers.

Predicted Effects

In order to investigate the above affects, sections through a number of fibrous materials have been evaluated using an image analyzing computer. Such an analysis can provide a measure of "mean free distance", λ, which is defined as 'the average uninterrupted separation of particle/matrix interfaces measured linearly through the matrix (in all directions)'. An average value of λ was taken from five different fields within the cross-section of a sample. The results for four different media (different fiber diameter and densities) are shown in Fig 2, plotted against micron rating (as determined in a glass bead test). The values of λ are seen to be reproducible and to vary in a systematic way with the different measured micron ratings of the media. Using this type of calibration, it is then possible to obtain values on media subjected to the type of conditions likely to be encountered in practice.

For example, the effect of an increasing pressure difference across a fiber medium has been studied. Samples were impregnated with cold-curing resin and pressures up to 300 bar were applied across the sheet, the pressure being maintained until the resin had hardened. Fig 3 shows the compression (percentage change in thickness) of different media with increasing pressure. It is apparent that a "soft-sintered" medium compresses much more than a "hard-sintered" medium. The "mean free distance" of the soft-sintered samples was determined using the image analyzing computer. The corresponding values of micron rating are included in Fig 3, indicating how the filtration performance of this medium might vary in practice with applied pressure.

Similarly, the analysis could be applied to a section through a localized region of an actual filter element. A region of particular interest is the peak of the pleats where distortion of the fibers occurs.

Practical Performance

Although polymer filtration media can be evaluated using standard laboratory techniques (Permeability, filtration efficiency, contaminant holding capacity, etc.), and although the image analyzing computer can be used to predict localized effects which might occur in practice, there are nevertheless some features of polymer filtration which are not modelled by such tests. These features include:

- the presence of specific additives
- the non-Newtonian behavior of polymer melts
- the presence of deformable gel contaminants

In such circumstances, it is necessary to devise experimental procedures which are more representative of the practical situation. The effect of additives, which will themselves have a specific particle geometry and particle size distribution, can be studied by passing clear viscous liquid containing the dispersed additive through a cell holding the filter medium. Particle counters upstream and downstream of the cell determine the extent of interaction of the additive with the filter medium. Suitable liquids which would exhibit non-Newtonian behavior include silicone oils, polymer solutions, or modified polymers(PET could be modelled by an aliphatic co-polyester).

A still closer representation of the practical situation can be achieved with a small laboratory extruder, properly instrumented to record temperatures, pressures, flow rate etc. However, the important study of gel removal remains problematical and such studies must ultimately be conducted on actual production equipment in order to provide confirmation of the predicted on-stream life which is crucial to process economics.

The inescapable conclusion is that in order to improve the performance of filters used in the processing of high performance thermoplastics there must be a close collaboration between the filter manufacturer and the polymer processor, in order to ensure that the media and elements which are developed will truly meet the needs of the industry.

Acknowledgements. The authors acknowledge the assistance of their colleagues within the Fairey Filtration Division, particularly K.A. Bampfield and F.G. Wilson.

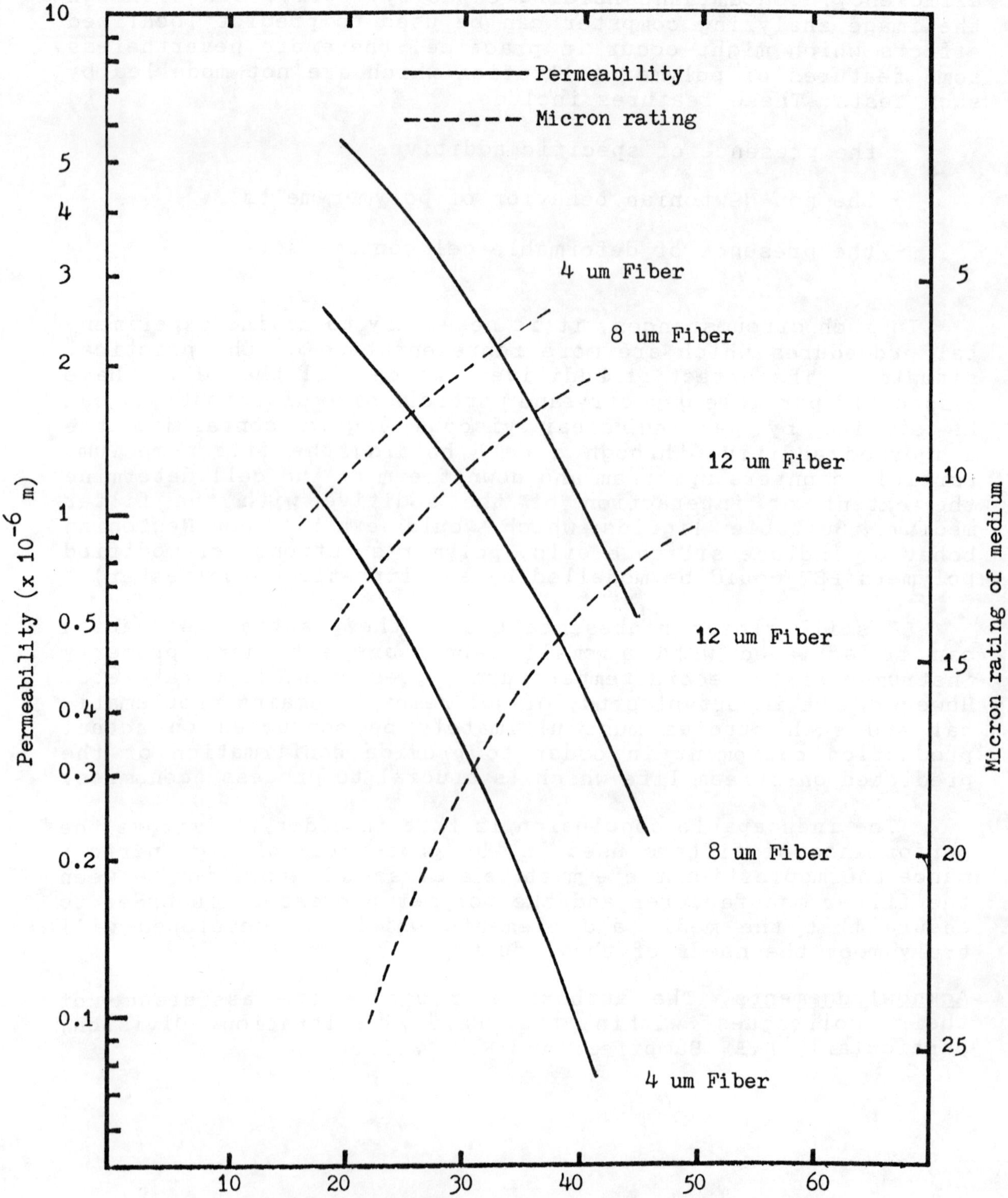

Figure 1 Variation of permeability and micron rating with relative density for media prepared from different diameter fibers.

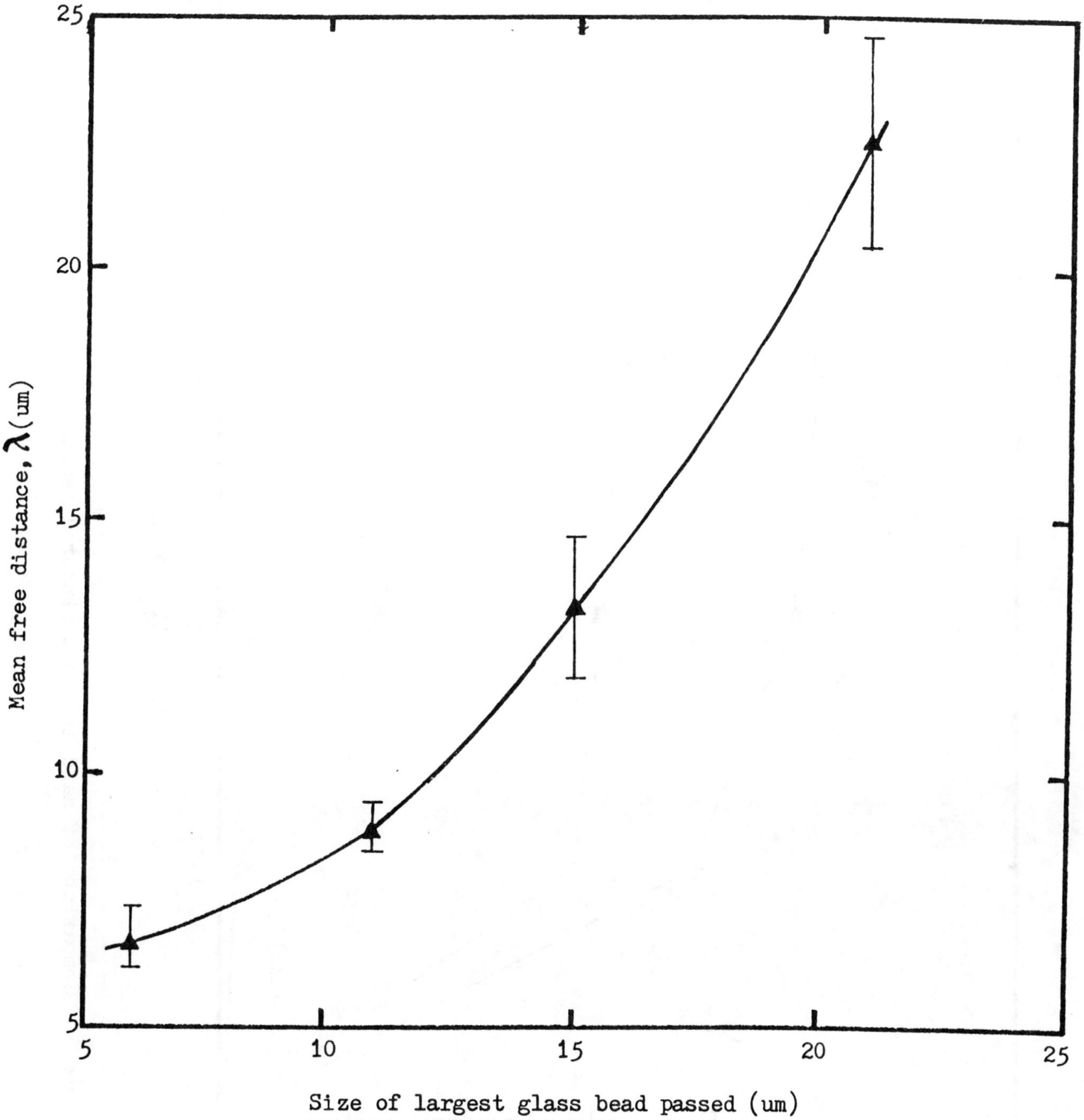

Figure 2 The relationship between mean free distance and the micron rating from measurements of four different media.

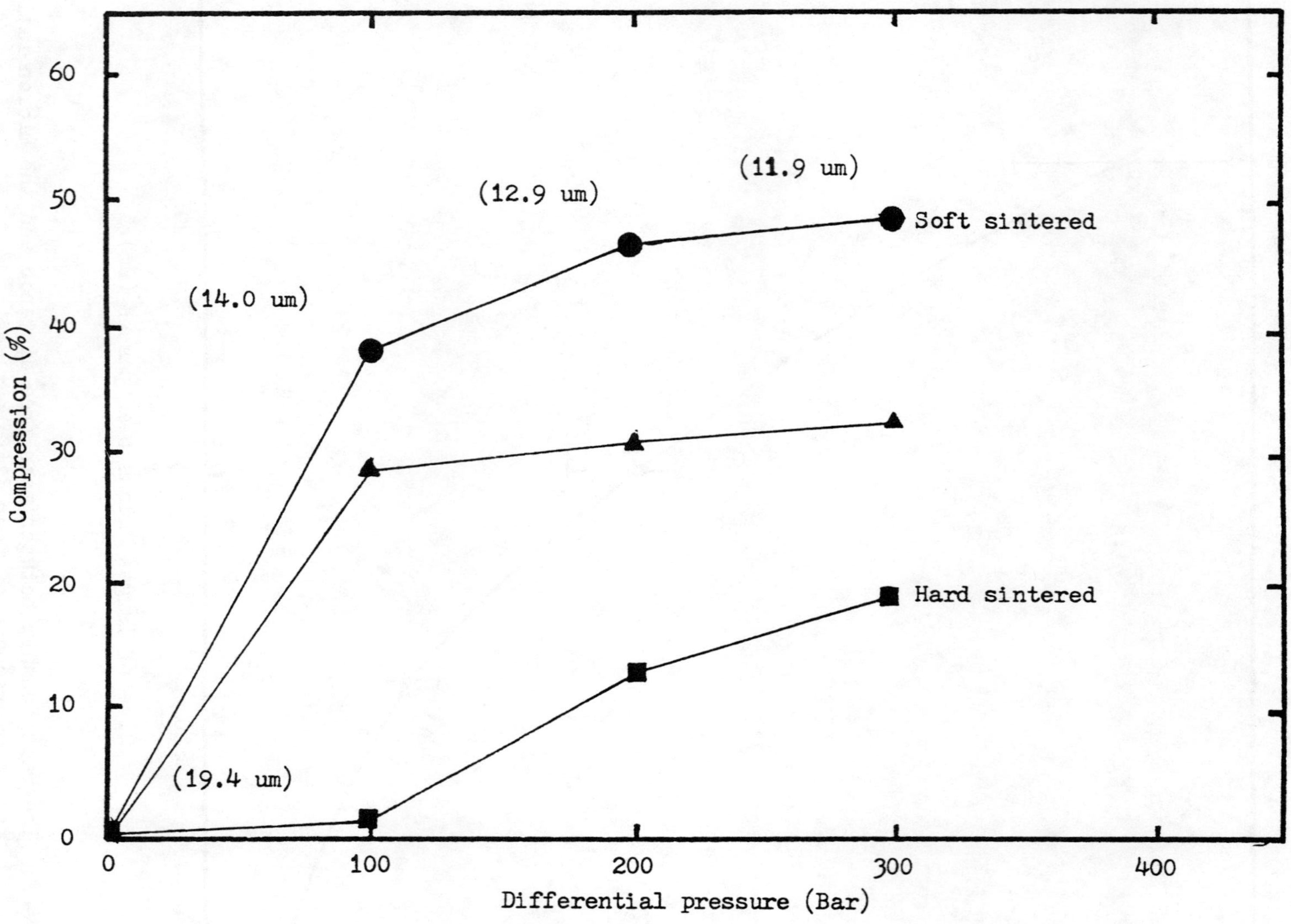

Figure 3 Comparison of hard and soft sintered media.

FILTRATION AND PUBLIC HEALTH

Wells Shoemaker
Filterex Inc
RD 1 Box 507
Shippensburg PA 17257
717 423 6218

The session in this conference relating to the title above comes at a critical time in search for new technologies for assist in preserving the quality of Planet Earth. The Conference in Ocean City in 1988 highlighted various of the needs for clean water, clean air, and improved medical technologies. Now, facing the 90's, concerns for "public health" are being addressed by the general public, governmental groups, and industry. Those of us involved with filtration and separation procedures must be aware of these pressures, and must be prepared to have available all the separation tools now in existance, and and those that must result from fufture development activities in academia and industry.

Terms cause difficulty... and to the filtration specialist, the words "public health" conjure up labels such as EPA, FDA. OSHA, ecologist, environmentalist, conservationist, preservationist, and many others. Some of the subject areas, such as water, indeed involve "filtration.." others, such as wildlife, are perhaps a little for removed. For this discussion and the speakers that follow, we will suggest that we concentrate on matters of air, water, foods...recognizing that pharmaceuticals and related FDA matters will be reviewed in other sessions.

Similarlarly, to the "public health" scientist, "filtration" must be viewed in a wide concept, involving all areas of fluid/particle separation. Thus we must encompass separations involving gas/gas, gas/liquid, gas/solid, liquid/liquid, solid/liquid... as well as single phase separations such as involved in membrane technology. Techniques including true barrier filtration, adsorption/absorption, coalescing, settling, ion exchange/molecular sieves, electrokinetically enhanced separations, electroacoustically enhanced separations (to be presented at this conference), and others must be considered. The public health scientist must have access to all such separation techniques.

Speakers in succeeding papers of this session will concentrate on outdoor air and water problems involving filtration and public health. The problem of indoor air will be considerd in this paper, but first comments are appropriate on related matters. Consider the entire problem of military protection against N-B-C- nuclear, chemical, and biological attack. Perhaps not a traditional public health concern, it ought to be of concern to the public in the now unexpected warfare. Certainly all three of these areas do require the utmost of separation technology.

Catastrophes are a further concern... what if a tank car full of poisonous pesticide were to tumble off a bridge into a river that sources a town's drinking water? What if a volcano put forth particles a la Mt. St. Helens? Or, toxic gases a la Bhopal? Increased smog a la Los Angeles? Or an oil spill a la Valdez in a populated area? Or a nuclear plant meltdown a la Chernobyl? Can't happen here? What can separation technology do for protection of the public in such instances?

Solid waste, as such, doesn't seem to be a candidate for filtration technology, but it happens that "solid" wastes often generate gaseous or liquid wastes. The recent Canadian tire fire generated air pollution and also created a gooey liquid that was a severe hazard for water.

Food and beverage problems, on the other hand, are a great concern and are handled carefully by FDA, Dept of Ag, state DER's, etc. Sanitation a major problem,such as in milk production, meat packing... virtually every area. Related to this is the tampering with foods and drugs. Even our day-to-day coffee made headlines recently in connection with the type of bleach used for the pulp in the coffee filter! In this regard, the attack on chrysotile asbestos following the discovery of a fiber in a wine bottle must be discussed later in this session.

For perspective purposes, the scope of the drinking water situation must be regarded as crucial for all parts of Planet Earth. Contaminants in water of sediment, short chain organics, inorganic salts, microorganisms, gasoline, agricultural and industrial wastes, and residual chlorine from disinfectation are a major problem for which the answer lies first, of course, in prevention, and second in treatment. One of the fastest growing segments of the filtration industry is the point-of-use systems now receiving such publicity and such merchandising hype.

Acid rain is the problem for outdoor air that vies with tail-pipe emissions as the target for chief concern of the public, the law-makers, and the users. But other sources add to the outdoor air problem, leading to particulate and gaseous contamination, such as combustion products from burning of trash.

Pollution of indoor air is another serious matter receiving increased attention of the public. The matter of "indoor" air arises in locations including homes, autos, offices, factories, hospitals, stores, schools, theaters, restaurants, hotels, airplanes, and judging from recent news stories, even the toll booths on New York City bridges! The type of contamination in these spaces can include: gases- CO, CO_2, NH_3, O_3. CH_2O; vapors- tobacco smells and chemicals, cooking byproducts, sweat, metal fumes, plasticizers, solvents, manure fumes, combustion byproducts, agricultural chemicals; mists- oil hazes; solids- dust, dust + radon, tobacco solids; and microorganisms. The two areas of greatest impact on the public have been smoking and radon.

Airline passengers now travel in smoke-free comfort, and more and more hotels, offices, restaurants, etc limit the areas for smoking. As an air pollutant, tobaco smoke has no equal in terms of numbers of situations. Radon has been reviewed at length in the media, and some of the hysteria seems to be waning. While it can be a problem to be sure, newer techniques have lessened its impact on family health and pocketbooks.

The family home has been a favorite source of concern. The formaldehyde from insulation... the plasticizer from plastics...the kitchen... the bathroom... all contribute to the indoor air pollution. Add in malfunctioning heating systems.. put all these contaminants in a home with tight weatherstripped doors and windows.. and the result is significantly poorer air quality inside than outdoors.

Two further problems involving the home have been reported and represent serious concerns. First is the inorganic powder stemming from the use of humidifiers. As the solids from the water are concentrated on the typical belt, the flow of air from the fan picks them off the belt and disperses them in the air for inhalation by the residents.

Second is the wide-reported problem with various carpets and the 4-PC coming from the latex backing, especially the situation at EPA's Waterside Mall complex where many workers became ill. Then there is the case of Mr Jack Benney of Florida who installed Stainmaster carpet, then used Raid in the rooms. Result: a thick white gas, severe illness for his family.

The entire question of "sick buildings" is now receiving attention by EPA and many others, and will be a significant problem in the 90's.

Microorganisms represent still further complications in all indoor air areas. These can include the following:

* protozoa
* bacteria...legionella pneumophila
* thermophyllic actinomycetes
* fungi
* virus

And they can occur in unexpected places, such as in a supermarket where there was a recent occurence of illness among the employees working in the fresh foods department presumably caused by bacteria carried in the water spray for the fresh veggies.

Another situation involves a schoolteacher who became disabled by a respiratory illness contracted from microorganisms carried in the school ventilation system. At first the Social Security system refused to consider the illness as one that would qualified for "disablement.." but the diagnosis of hypersensity pnemonitis has been accepted... and illustrates one more potential hazard in indoor air pollution.

Hospitals present a special case for indoor air problems,especially in regard to the dreaded nosocomial infections. Although airborne contaminants are serious, the real problem lies in physical transfer, not air transfer. Whenever people are concentrated, as in airplanes, subways, offices, schools, etc., the problem is magnified, and the challenge arises to the filtration problem to see what they can do.

The technology for air treating has been developing..and many suppliers have sophisticated cleaning systems combining electrokinetics, fiber filter media, and adsorption techniques with activated carbon, activated alumina, zeolite, molecular sieves, etc. There are savings to the user that are well presented by the industry: circulate the air, remove the unwanted materials, and reduce energy expense by preserving the heat, or lack of heat, in the air. Such technologies need to be encouraged, but as always, there will be resistance to expenditure. According to A J Palermo, many school systems need to replace the entire circulating air system to prevent continued growth of microorganisms, but school budgets can't afford it. If more teachers become ill, however, perhaps the outlook may change.

Mention must be made of the situation on asbestos...universally condemned as a carcinogen... without differentiating between the chrysotile asbestos used in filtration, and croccidolite used in insulation. The first type has never been shown to cause on health problems, while the second has been shown to be harmful to those who both inhale it and smoke. Once the latter is secure in a wall, or around a pipe, it is less harmful than when removal is attempted. This a public health matter that should be clarified.

The problems of indoor air, outdoor air, and drinking water are certainly not limited to our continent. Every country has them. Some efforts have been made to share new technology and should be encouraged. We in the US should be pleased that our efforts, through EPA and others, are ahead of the rest in many areas...but international cooperation is needed.

The bottom line on most of these matters is the cost of the program vs the benefits. There is the threeway relationships triangle:

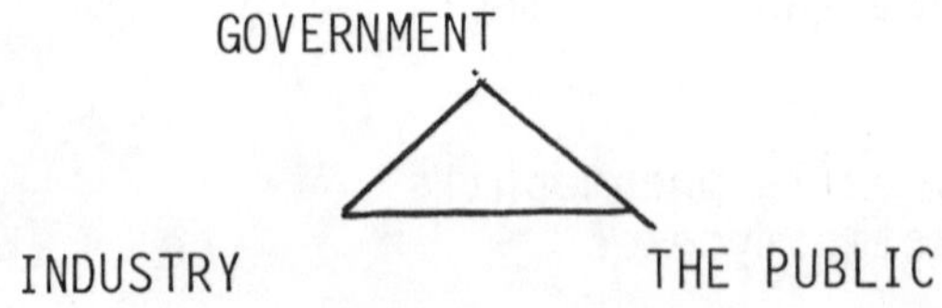

Each of the areas to be discussed can be viewed from these three positions. Who decides if the benefits are worth the cost? Certainly not the filter people... it's their job to reduce the cost.

The key people are those in the area of public health, EPA, any of the groups. They need money from the public and the government.. authority to make changes... confidence from the public that they know what they are doing... and from the filtration and separation industry, the tools to help them do their job.

Challenges exist: The filtration people need to develop products for use by the public health people. The public health people need to make their needs known. The general public must expect to pay more for a safer Planet Earth. And... technical organizations such as AFS must be prepared to assist in technology transfer.

AMERICAN FILTRATION SOCIETY

P.O. Box 6269
Kingwood, Texas 77325
Telephone: (713) 359-1894
FAX: (713) 358-3939

ABSTRACT & COPYRIGHT

Meeting Location: **Meeting Date:**

Session Chairman: **Session Number:**

SPEAKER (AUTHOR):

Name: THE HONORABLE JOHN BROUJOS
Company: Rep., COMMONWEALTH of PENNSYLVANIA
Address: 6 NORTH HANOVER STREET
City, State, Zip: CARLISLE, PA
Telephone: FAX:

CO-AUTHOR:

Name:
Company:
Address:
City, State, Zip:
Telephone: FAX:

CO-AUTHOR:

Name:
Company:
Address:
City, State, Zip:
Telephone: FAX:

CO-AUTHOR:

Name:
Company:
Address:
City, State, Zip:
Telephone: FAX:

TITLE OF PAPER:

TEXT OF ABSTRACT: (200 Words in Space Below)

ABSTRACT #2 JOHN BROUJOS ACID RAIN AND PUBLIC HEALTH

Acid rain has been recognized as a major ecological problem facing not only the US but the entire world. What is the magnitude of this situation? What will be the impact of acid rain on the public by 2000? What have legislators done to protect the public from acid rain? What should be done? Does the problem exist in other countries? What are they doing about it? How do sovereign nations exercise control over the environment when they have no extraterritorial jurisdiction?

****

PLEASE TYPE INFORMATION BECAUSE DATA WILL BE PUBLISHED IN THE ABSTRACT BOOKLET. RETURN FORM TO THE AMERICAN FILTRATION SOCIETY, P.O. Box 6269, Kingwood, Texas 77325, WITH A COPY TO YOUR SESSION CHAIRMAN.

Filtration Criteria for Public
Drinking Water Supplies

Joseph A. Cotruvo
U.S. EPA
Office of Drinking Water

The United States Environmental Protection Agency has recently issued new rules and criteria for filtration and disinfection of public drinking water supplies that use surface water as a source. Approximately 12,000 public water supplies are affected including virtually all of the largest communities.

These are designed to provide greater protection against microbiological contamination of drinking water and reduce the risk of waterborne disease.

All of these supplies must disinfect and those that do not achieve certain source criteria must also filter. All supplies, (-i.e. those that will only disinfect, those that currently filter and disinfect, and those that must introduce new treatment) must design and operate their plants to achieve at least 99.9% removal or inactivation of Giardia lamblia cysts, and 99.99% removal or inactivation of virus.

These rules will result in major new emphases being placed on filtration and disinfection processes in public water supplies.

SURFACE WATER TREATMENT REQUIREMENTS

- Proposal was published in Federal Register on November 3, 1987 (52 FR 42178)
- Notice of Availability, describing new regulatory options, was published in the Federal Register on May, 1988 (53 FR 16348).
- Final rule promulgated June, 1989.

Maximum Contaminant Level Goals

Giardia Lamblia	0
Viruses	0
Legionella	0
Turbidity	none
Heterotrophic Plate Count (HPC)	none

General Requirements

Coverage: All public water systems using any surface water or ground water under direct influence of surface water must disinfect, and may be required by the State to filter, unless certain water quality source requirements and site specific conditions are met.

Treatment technique requirements are established in lieu of MCLs for *Giardia*, viruses, heterotrophic plate count bacteria, *Legionella* and turbidity.

Treatment must achieve at least 99.9 percent removal and/or inactivation of *Giardia lamblia* cysts and 99.99 percent removal and/or inactivation of viruses.

All systems must be operated by qualified operators as determined by the State.

Criteria to be Met to Avoid Filtration

Source Water Criteria

Fecal coliform concentration must not exceed 20/100 ml or the total coliform concentration must not exceed 100/100 ml before disinfection in more than ten percent of the measurements for the previous six months, calculated each month.

Minimum sampling frequencies for fecal or total coliform determinations are:

SYSTEM SIZE (persons)	SAMPLES/WEEK
< 501	1
501-3,300	2
3,301-10,000	3
10,001-25,000	4
> 25,000	5

If not already conducted under the above requirements, a coliform test must be made each day that the turbidity exceeds 1 NTU.

Turbidity levels must be measured every four hours by grab sample or continuous monitoring. The turbidity level may not exceed 5 NTU. If the turbidity exceeds 5 NTU, the system must install filtration unless the State determines that the event is unusual or unpredictable, and the event does not occur more than two periods in any one year, or five times in any consecutive ten years. An "event" is one or more consecutive days when at least one turbidity measurement each day exceeds 5 NTU.

Site Specific Conditions

Disinfection

Disinfection must achieve at least a 99.9 and 99.99 percent inactivation of Giardia cysts and viruses, respectively. This must be demonstrated by the system meeting "CT" values in the rule ("CT" is the product of residual concentration (mg/l) and contact time (minutes) measured at peak hourly flow). "C" and "T" must be determined at or prior to the first customer. The total percent inactivation can be calculated based on unlimited disinfectant residual measurements in sequence prior to the first customer. Failure to meet this requirement on more than one day in a month is a violation. Filtration is required if a system has two or more violations in a year unless the State determines that the violation(s) were caused by unusual and unpredictable circumstances; regardless of such determinations by the State the system must filter if there are three or more violations in a year.

Disinfection systems must a) have redundant components including alternate power supply, automatic alarm and start-up to ensure continuous disinfection of the water during plant operation or b) have automatic shut-off of delivery of water to the distribution system whenever the disinfectant residual is less than 0.2 mg/L, provided that the State determines that a shut-off would not pose a potential health risk to the system.

For systems using chloramines if chlorine is added prior to ammonia, the CT values for achieving 99.9 percent inactivation of Giardia lamblia cysts can also be assumed to achieve 99.99 percent inactivation of viruses. Systems using chloramines and adding ammonia prior to chlorine must demonstrate with on-site studies that they achieve 99.99 percent inactivation of viruses. For systems using disinfectants other than chlorine, the system may demonstrate that other CT values than those specified in the rule, or other disinfection conditions are provided which achieve at least 99.9 and 99.99 percent inactivation of Giardia lamblia and viruses, respectively.

Disinfectant residuals in the distribution system cannot be undetectable in more than five percent of the samples, each month, for any two consecutive months. Samples must be taken at the same frequency as total coliforms under the revised coliform rule. A system may measure for HPC in lieu of disinfectant residual. If the HPC measurement is less than 500 colonies/ml, the site is considered to have a "detectable" residual for compliance purposes. Systems in

Criteria for Filtered Systems

Turbidity Monitoring

Turbidity must be measured every four hours by grab sample or continuous monitoring. For systems using slow sand filtration or filtration technologies other than conventional treatment, direct filtration or diatomaceous earth filtration, the State may reduce the sampling frequency to once per day. The State may reduce monitoring to one grab sample per day for all systems serving less than 500 people.

Turbidity Removal

Conventional filtration or direct filtration water must achieve a turbidity level in the filtered water at all times less than 5 NTU and not more than 0.5 NTU in more than five percent of the measurements taken each month. The State may increase the 0.5 NTU limit up to less than 1 NTU in greater than or equal to 95% of the measurements, without any demonstration by the system, if it determines that overalltreatment with disinfection achieves at least 99.9 percent and 99.99 percent removal/inactivation of *Giardia* cysts and viruses, respectively.

Slow sand filtration must achieve a turbidity level in the filtered water at all times less than 5 NTU and not more than 1 NTU in more than five percent of the samples taken each month. The turbidity limit of 1 NTU may be increased by the State (but at no time exceed 5 NTU) if it determines that there is no significant interference with disinfection.

Diatomaceous earth filtration must achieve a turbidity level in the filtered water at all times less than 5 NTU and of not more than 1 NTU in more than five percent of the samples taken each month.

Other filtration technologies may be used if the system demonstrates to the State that they achieve at least 99.9 and 99.99 percent removal/inactivation of *Giardia lamblia* cysts and viruses, respectively, and are approved by the State. Turbidity limits for these technologies are the same as those for slow sand filtration, including the allowance of increasing the turbidity limit of 1 NTU up to 5 NTU, but at no time exceeding 5 NTU upon approval by the State.

Disinfection Requirements

Disinfection with filtration must achieve at least 99.9 and 99.99 percent removal/inactivation of *Giardia* cysts and viruses, respectively. The States define the level of disinfection required, depending on technology and source water quality. Guidance on the use of CT values to make these determinations is available in the *Guidance Manual*. Recommended levels of inactivation are based on expected

Ground Water Systems Under Direct Influence of Surface Water

All systems using ground water under direct influence of surface water must meet the treatment requirements under the SWTR. States must determine which community and non-community ground water systems are under direct influence of surface water within 5 years and 10 years, respectively, following promulgation. Unfiltered systems under the direct influence of surface water must begin monitoring within 6 months following the determination of direct influence unless the State has determined that filtration is required. Systems under direct influence of surface water must begin meeting the criteria to avoid filtration 18 months after the determination of direct influence, unless the State has determined that filtration is required. Unfiltered systems under direct influence of surface water must install filtration within 18 months following the failure to meet any of the criteria to avoid filtration.

Variances

Variances are not applicable.

Exemptions

Exemptions are allowed for the requirement to filter. Systems using surface water must disinfect (i.e., no exemptions). Exemptions are allowed for the level of disinfection required.

AMERICAN FILTRATION SOCIETY

P.O. Box 6269
Kingwood, Texas 77325
Telephone: (713) 359-1894
FAX: (713) 358-3939

ABSTRACT & COPYRIGHT

Meeting Location: Alexandria, VA **Meeting Date:** March 18-22, 1990

Session Chairman: W. Shoemaker **Session Number:** 1D

SPEAKER (AUTHOR): Charles J. Drummond

Name: Charles J. Drummond
Company: U.S. Department of Energy
Address: PETC/P.O. Box 10940/M-S 922-H
City, State, Zip: Pgh., PA 15236
Telephone: (412) 892-4889 FAX: 892-4604

CO-AUTHOR:

Name:
Company:
Address:
City, State, Zip:
Telephone: FAX:

CO-AUTHOR:

Name:
Company:
Address:
City, State, Zip:
Telephone: FAX:

CO-AUTHOR:

Name:
Company:
Address:
City, State, Zip:
Telephone: FAX:

TITLE OF PAPER: Advanced Emissions Control Technology Being Developed for Acid Rain

TEXT OF ABSTRACT: (200 Words in Space Below)

Current debate centers around the requirements to be placed on the utility industry to achieve reductions in sulfur dioxide and nitrogen oxide emissions. How much, when, and by whom are still uncertain. It does seem, though, that revisions to the Clean Air Act, including attention to the Acid Rain issue, will become law this year.

Conventional technology, wet lime or limestone scrubbing, is available to meet the future requirements of this anticipated legislation. More interestingly, a number of advanced technologies are being developed to meet the unique demands of installing pollution control equipment at existing coal-fired electric power plants. These advanced technologies have the potential to reduce cost, decrease solid waste disposal requirements, and achieve nitrogen oxides removal in combination with sulfur dioxide removal.

This paper will describe these advanced technologies and the special demands that they will place on existing particulate control devices.

****

PLEASE TYPE INFORMATION BECAUSE DATA WILL BE PUBLISHED IN THE ABSTRACT BOOKLET. RETURN FORM TO THE AMERICAN FILTRATION SOCIETY, P.O. Box 6269, Kingwood, Texas 77325, WITH A COPY TO YOUR SESSION CHAIRMAN.

AMERICAN FILTRATION SOCIETY
P. O. Box 6269
Kingwood, Texas 77325
Telephone: (713) 359-1894
FAX: (713) 358-3939

ABSTRACT & COPYRIGHT

Meeting Location: Alexandria, Virginia **Meeting Date**: March 20-22, 1990

Session Chairman: Wells Shoemaker **Session Number**:

SPEAKER (AUTHOR):

Name: Benjamin W. Lykins, Jr.
Company: U.S. Environmental Protection Agency
Address: 26 W. Martin Luther King Drive
City, State, Zip: Cincinnati, OH 45268
Telephone: 513/569-7460 FAX: 513/569-7276

CO-AUTHOR:

Name: J. Keith Carswell
Company: U.S. Environmental Protection Agency
Address: 26 W. Martin Luther King Drive
City, State, Zip: Cincinnati, OH 45268
Telephone: 513/569-7460 FAX: 513/569-7276

CO-AUTHOR:

Name: Carol A. Fronk
Company: U.S. Environmental Protection Agency
Address: 26 W. Martin Luther King Drive
City, State, Zip: Cincinnati, OH 45268
Telephone: 513/569-7592 FAX: 513/569-7276

CO-AUTHOR:

Name:
Company:
Address:
City, State, Zip:
Telephone: FAX:

TITLE OF PAPER: Membranes For Removing Organics From Drinking Water

TEXT OF ABSTRACT: **(200 Words in Space Below)**

Until recently, membranes have mainly been used to remove salts and other inorganic compounds from water but both bench-scale and field studies have shown their effectiveness for removing organic compounds from drinking water. Two different membrane technologies have been evaluated: high pressure membranes and low pressure membranes. High pressure membranes are those using about 150 to 400 psig commonly called reverse osmosis. During bench-scale studies, reverse osmosis membranes tested included cellulose acetate, polyamide, and thin-film composite. These membranes were used to treat multisolute, aqueous solutions in the concentration range of 6 to 153 µg/L. Removal efficiencies for alkanes, alkenes, aromatics, and pesticides showed that the thin-film composite membranes were more effective than the polyamide or cellulose acetate membranes. At a research site in Suffolk County, New York, removal of agricultural contaminants by reverse osmosis was evaluated on the bench and in a pilot plant. Percent removals for aldicarb sulfone, aldicarb sulfoxide, 1,2-dichloropropane, carbofuran ranged from 53 to more than 95. Low pressure membranes are usually operated at or below 150 psig. These membranes (ultrafiltration) were evaluated at various sites in central Florida to investigate their efficiency for removing disinfection byproduct precursors. After membrane selection trails were completed, a mobile trailer was used to evaluate the performance of the selected membrane. With a system recovery of 75 percent at one groundwater site, the average reduction of trihalomethane formation potential and total organic halide was 95 percent and 96 percent from raw water averages of 456 µg/L and 977 µg/L respectively.

****

PLEASE TYPE INFORMATION BECAUSE DATA WILL BE PUBLISHED IN THE ABSTRACT BOOKLET. RETURN FORM TO THE AMERICAN FILTRATION SOCIETY, P. O. Box 6269, Kingwood, Texas 77325, WITH A COPY TO YOUR SESSION CHAIRMAN.

REAGENT ULTRAFILTRATION - NEW METHOD FOR SOLUTION OF PUBLIC PROBLEMS

Alex Svitsov
Mendeleev Institute of Chemical Technology,
Firm TECO
128850, Miusskaja sq.9, Moscow, USSR

Membrane separation of liguids and filtration of suspensions are often considered to be a single whole technological process though physico-chemical mechanism of substance penetration through a separating barrier is different. In the seguence: suspensions - emulsions - stable colloidal systems - solutions of high-molecular substances - solutions of low-molecular organic and mineral compounds the influence of usual sieving (filtering) continually decreases, but the influence of complicated interactions between membrane material, molecules of solution and particles of finely divided substance increases.

All separating processes: filtration, microfiltration, ultrafiltration and reverse osmosis - are very much alike due to the presence of a separating partition and the pressure gradient as a motive force. But in the sequence above the size of trapped particles decreases (from micrometers up to Angstroms), the pressure gradient uncreases, but the specific conductivity of the partition quickly drops. That is why the wish of technologists and engineers to unite the best properties of these processes is quite explicable. Besides it would have been good enough to raise a selectivity of the separating operation. Very often the objects, undergone treatment and particularly industrial sewage are irregular mixtures: there are mainly neutral safety salts and a bit (but much more than maximum permissible quantity) of contamination proper. The utmost case is liquid radioactive waste. There is no possibility to determine the quantity of radionuclides in mass concentrations and the total salt content reaches several hundred milligrams per liter.

We call the membrane separation process, which unites high productivity at low operation pressure, possibility to purify water from ionic components during selective separation of components the reagent ultrafiltration. Its main point lies in a shift of dissolved low-molecular components in a new associated molecular or colloidal state with the following separation of associated forms being formed on a large-pore membrane.

Reagent ultrafiltration has several variants.

1. Addition to the solution of water-soluble polyelectrolytes with such functional groups which can combine on themselves the components we are interested in through ion exchange, complexing or similar mechanisms. The solution remains homogeneous, its separation goes on according to usual laws of ultrafiltration;
2. Addition to the solution of the water-soluble compound which interacts with a necessary component through the extraction mechanism. The emulsion being formed is separated on a porous membrane with all corresponding requirements fulfilled.
3. Addition to the solution of a newly-made sol. The necessary component will be combined on sol particles due to physical adsorption.
4. The shift of dissolved ionic particles of the component into the colloidal phase by their hydrolysis while alkali is added to the solution. The variety of this case is the electro-membrane method of hydrolysis.
5. Addition to the solution of chemical reagents which interact with the necessary component and shift it into an insoluble form. The precipitate formation should be stopped at the stage fo the colloidal phase formation.

In the last three variants sols are undergone to membrane separation and it is important to keep their stability.

Some of the variants of the reagent ultrafiltration have serious scientific base and even an industrial use. The other ones are presently being developed.

In any case the process either by time or by place is devided into two stages: associated form formation and membrane separation. As a rule the first stage goes on in a usual capacity with a stirrer and reguires strict dosage of the substance being added. Stoichiometrical ratios, pH of the mixture, the time of reagent contact, stirring intensity, temperature influence the completeness of combination of the component being separated. Optical methods are the simplest ones to control the system condition.

Let's consider in detail the forth variant - salt hydrolysis. The hydrolysis proceeds with polynuclear complex formation when the concentration of the component being separated is more than 10^{-3} mol/l. The polynuclear hydrolysis goes through several stages including monomer and dimer formation their polymerization, amorphous colloidal particle formation and their transformation to a crystalline form. Each of the atoms being a polymer chain, can be twice or thrice hydrolyzed. It results in a chain branching. The identification of particles being formed is an extremely complicated process, but for practical use only one parameter - the size - is important. It is necessary to consider the fact that the size of particles can increase due to their coagulation even when alkali is added no more, it results in a loss of the agregative (block) stability of the colloidal system. There are several ways to prevent this phenomenon: addition of surface-active substances, change of temperature.

These ways plus complexing with the addition of low-molecular ligands and other "disguising" modes permit to use the reagent ultrafiltration not only for purification but for fractionation of multicomponent solutions.

The variety of the forth case is carring out the hydrolysis with the help of electrocorrection of pH for the solution, i.e. reagentless method of the reagent ultrafiltration, and it sounds as a pun. The question of the agregative (block) stability of the system is more critical here but the method at present is not studied.

The second stage of the process - membrane separation - becomes more complicated due to concentration polarization which is characteristic of all membrane processes. It practically always results in gel formation on the membrane surface. In the first variant of the process the gel layer forms from the dissolved molecules of the polyelectrolyte, and this point is well known by now, but in the rest of the cases the layer on the membrane consists of concentrated sols and this phenomenon has been studied much less.

The preliminary results of the investigation can be stated as follows:

1 - the particles of the size less than the pore size cause irreversible membrane fouling, they stick in the membrane pores;
2 - the charge of the colloidal particles is of great importance the electrostatic repulsion between the particles and the membrane becomes possible;
3 - it is possible to get the best conditions of ultrafiltration by controlling the process of colloidal system formation and combining the size and the charge of a particle;
4 - it is important to create the optimal hydrodynamic condition over the membrane excluding stagnant zones first of all.

The reagent ultrafiltration widens the field of application of membrane separation methods. It gives the possibility to use highly productive processes to treat liquid industrial sewage which earlier could be treated only by reverse osmosis. It lets the contaminating components be selectively removed from sewage without affecting salt ballast, this is particularly important while treating radioactive and galvanic sewage. A new fractionation method for liquid mixtures is being created which can become an analytical one after definite finishing development. And finally new approaches to reclamation and processing of toxic concentrates come into being.

LIQUID MEMBRANE PROCESS AS AN ALTERNATIVE TO FILTRATION FOR SOLID/LIQUID SEPARATION

S.Yu. Ivakhno, A.V. Afanasjev
and A.V. Rogatinsky
Mendeleev Institute of
Chemical Technology
9 Miusskaya sq., Moscow
USSR, 125190

Abstract

The recovery of 4-chlorobenzenesulphonamide (CBSA) from industrial aqueous solutions and suspensions by extraction across a bulk liquid membrane of chloroform or tetrachloroethylene into an alkaline receiving phase was studied. When applied to the treatment of waste water containing 1.0-1.5% of CBSA the liquid membrane process (LMP) ensured 99% extraction of CBSA for batch regime experiments and 95% for the semi-continuous regime. In the latter case a concentration coefficient as high as 8-12 for CBSA in the receiving phase with respect to the initial solution was obtained. LMP was then used instead of filtration for the treatment of an industrial suspension comprising: aqueous suspension of CBSA, pH 1.0-2.0/tetrachloroethylene/12-15% NaOH aqueous. This resulted in CBSA transfer into receiving solution with an intermediate process of dissolution into the organic liquid. Another important compound, dichlorodiphenylsulphone (DCDPS), present as an impurity in the initial pulp was accumulated in the tetrachloroethylene phase. When compared with filtration the LMP allowed not only the recovery of high purity CBSA but also the isolation of DCDPS as an additional product.

Introduction

LMP has recently demonstrated considerable potential as an effective tool for the recovery of different compounds from in dustrial solutions and waste waters [1]. Until now this technique has been only applied to the removal and concentration of solutes. However liquid membranes can also be used in some cases for treatment of suspensions in place of conventional filter materials. Unlike filtration LMP ensures chemical rather than mechanical separation resulting in higher selectivity. In adition LMP is faster and less energy intensive as it requires no pressure difference applied to the membrane.

The aim of this paper is to demonstrate the posibility of a solid/liquid separation by LMP using the process of CBSA extrac tion through a bulk liquid membrane as an example. CBSA is an important intermediate product in the technology of some chemi cals and drugs. In industry both diluted aqueous solutions and concentrated suspensions of CBSA need to be treated.

Experimental

Experiments on the LMP were carried out in a laboratory 4.5 litre diffusion cell with a bulk liquid membrane (Figure 1). Chloroform and tetrachlorethylene were tested as the membrane li quid phase sinse these solvents were most suitable for the CBSA technology. An aqueous solution of sodium hydroxide was used as the receiving phase. The concentration of CBSA in samples removed from the cell was analysed by UV-spectrophotometry. The distribution coefficients of CBSA under solvent extraction and stripping conditions were measured by standard methods.

Results and Discussion

CBSA behaves as a weak organic acid being extracted from neutral and acidic aqueous solutions into an organic solvent. The CBSA distribution coefficient depends strongly on the pH value of the aqueous phase (Figure 2) and is described by an equation which can be written in the logarithmic form [2]:

$$\log (d_1^0/d_1-1) = \log K_a + pH_1$$

where d_1 is the distribution coefficient at a given pH_1 value and d_1^0 is the real, i.e. maximum distribution coefficient.

By the length of a segment cut on the abscissa axis by the line $\log (d_1^0/d_1-1)$ as a function of pH (Figure 2) the CBSA ionization constant can be determined, $K_a = 3.16 \cdot 10^{-9}$ kmole m . At $pH_1 > 8.0$ a sharp decrease of d_1 was observed (Figure 2) because of salt formation:

$$ClC_6H_4SO_2NH_2 + NaOH \rightarrow ClC_6H_4SO_2NHNa + H_2O$$

This salt, unlike CBSA, is a strong electrolyte dissociating in aqueous solution so it can no longer exist in a non-polar organic liquid thus is almost completely stripped into the alcaline receiving phase.

Experiments on CBSA extraction from an industrial waste

water containing 1.0% of dissolved CBSA across a chloroform bulk liquid membrane into 10% NaOH aqueous solution proved LMP to be an effective tool for CBSA recovery (Figure 3). The degree of extraction for the batch regime exceeded 99% while for the semi-continuous regime, i.e. only the feed phase pumped through the contactor, it was over 95%. In the latter case CBSA was concentrated in the receiving phase up to 8-12% which corresponds to a concentration coefficient of 8-12 with respect to the initial solution. Substitution of chloroform by tetrachlorethylene affected neither the efficiency nor the rate of the process.

Taking into account the good results obtained with relatively dilute CBSA solutions the LMP was then used for CBSA recovery from concentrated industrial suspension which also contained several impurities including dichlordiphenylsulphone (DCDPS). According to the conventional flow-sheet this suspension would be filtered and washed to product a solid CBSA and a water filtrate containing the impurities. CBSA is then dissolved in alkaline solution and precipitated as a final product. DCDPS is discharged with waste water despite it being a valuable chemical.

LMP for the system: aqueous suspension of CBSA, pH 1.0-2.0/ tetrachloroethylene/12-15 NaOH aqueous , resulted in CBSA transfer into the receiving phase while DCDPS was accumulated in organic solvent (Table). CBSA concentration in the raffinate of 0.3-0.7% was definitely less when compared to conventional filtrates (1.0- 2.0%). CBSA with the melting point of 145-146°C finally obtained from the receiving phase by neutralization, precipitation, filtration and drying appeared to be a purer product than produced commercially by conventional technology. DCDPS (melting point 146- 148°C) isolated from the tetrachloroethylene phase corresponded to the industrial standard.

Conclusions

LMP ensures CBSA recovery from both industrial solutions and suspensions. When used for the treatment of suspensions. When used for the treatment of suspensions LMP produces CBSA of higher purity than obtained by filtration. An additional product of DCDPS could be isolated if LMP is applied rather than filtration.

References

1. S.Yu. Ivakhno, A.V. Afanasjev, G.A. Yagodin, Membrannaya extractiya neorganitchekych veschestv, VINITI, Moscow, 1985, pp. 120-125.

2. I.M. Korenman, Extractiya v analize organitcheskych vestchestv, Chimiya, Moscow, 1977, p. 22.

Table The final system composition resulting from the suspension treatment by LMP

N	Final concentration, %				
	Receiving alkaline solution		Tetrachloroethylene		Raffinate
	CBSA	NaOH	CBSA	DCDPS	CBSA
1	13.0	8.0	0.008	2.2	0.71
2	27.2	7.7	traces	1.2	0.30
3	18.7	5.9	0.01	1.2	0.34
4	21.1	6.8	traces	2.9	0.01

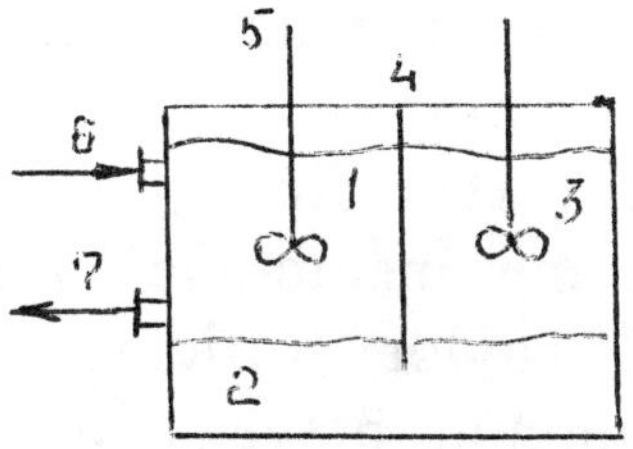

Figure 1

Diffusion cell with a bulk liquid membrane: 1- source phase, 2 - organic liquid, 3 - receiving phase, 4 - vertical bar, 5 - mixers, 6- initial solution inlet, 7 - raffinate outlet.

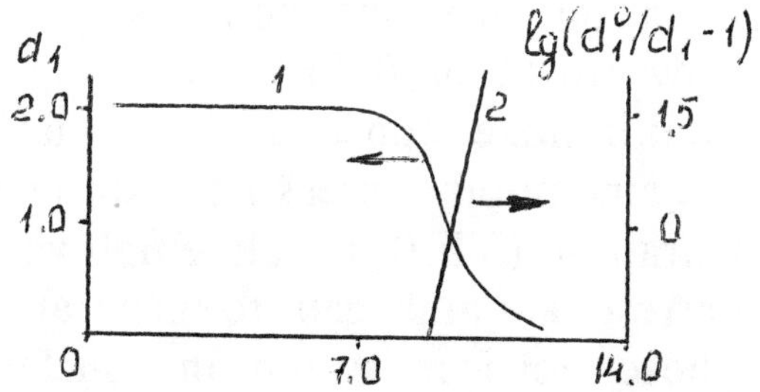

Figure 2

CBSA distribution coeffitient between chloroform and aqueous phases as a function of pH : 1 - in normal coordinates $d_1 = f(pH_1)$, 2 - in logarithmic coordinates $lg(d_1^0/d_1 - 1) = f(pH_1)$.

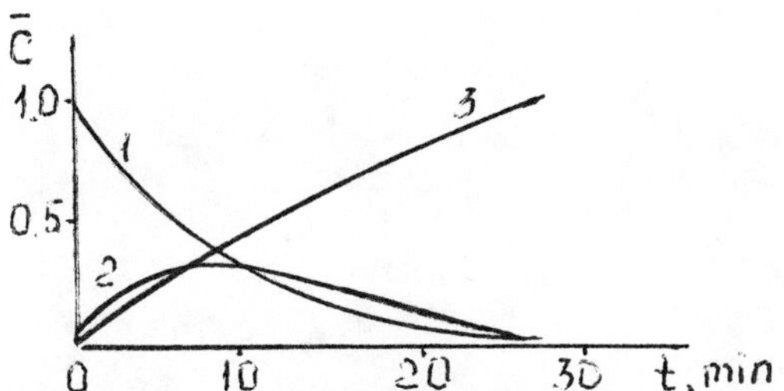

Figure 3

Dimensionless CBSA concentration in source (1), membrane (2) and receiving (3) phases as a function of time for the batch LMP in the system: 1.0% CBSA, pH 2.0/chloroform/10% NaOH

VORTEX FLOW FILTRATION: AN EXPERIMENTAL ANALYSIS

Charles L. Cooney, Ulrich Holechovsky and Gopal Agarwal
Department of Chemical Engineering and Biotechnology Process Engineering Center, Massachusetts Institute of Technology, Cambridge, MA 02139

Concentration polarization at a membrane surface restricts the filtration flux and controls molecular rejection. Advances in cross flow filtration (CFF) use a high velocity across the membrane surface to minimize concentration polarization but lead to a high transmembrane pressure. An alternative to CFF is vortex flow filtration (VFF) which employes a high speed rotating membrane in a narrow annular gap to generate a secondary flow, Taylor vortices. These vortices sweep the surface of the membrane and prevent or minimize concentration polarization. Results from studies on the concentration of bovine serum albumin will be described in which it is shown that one can achieve very high flux rates before encountering the pressure independent region which restricts CFF and further one can influence the molecular rejection by rotation rate as well as pressure. The result is that filtration performance is much more dependent on the membrane properties in VFF then in CFF.

MEMBRANE PLATE FILTER PRESS THEORY AND ASPECTS OF PRACTICAL APPLICATION

J. Stephen Slottee
Vaughn Weston
EIMCO Process Equipment Co.
P.O. Box 300
Salt Lake City, UT 84110
(801) 526-2315, 526-2329

ABSTRACT

For a broad range of applications, the membrane plate pressure filter has developed into an effective means of achieving higher cake solids for equal or shorter cycle times than conventional recessed plate pressure filters. Filter press sizing for membrane plates requires determination of the membrane squeeze time and pressure required to produce the desired cake solids. Total cycle time is determined by combining theoretical considerations, test results, and field experience with practical application.

Introduction

The typical filter press membrane plate utilizes a flexible membrane or bladder, made of a suitable chemically resistant elastomer, which is molded to the shape of the cavity of a recessed filter plate. After a cake is formed through filtration at a given feed pump pressure, the membrane is inflated which causes the cake to be compressed. Another type of plate utilizes a hollow core which can be pressurized to cause flexing and the desired compression effect. Filter media is installed as for any pressure filter.

Membrane, or variable volume plate filter presses are part of the arsenal of solid-liquid separation equipment used to achieve solutions to environmental, energy, and production problems where separation of suspended materials are required. Membrane plates are not new. There are descriptions of the use of membrane plates in 1914[1]. During the last 20 years, there have been steady improvements in membrane technology through improved membrane materials, plate materials, and plate sizes.

There have been a multitude of demonstrated advantages of using membrane plates including increased rates, increased cake solids, increased product recovery, reduced waste volume, reduced chemical costs, and improved product recovery. However, design engineers and end users who do not completely understand pressure filtration and membrane operations may incorrectly perceive the membrane plate filter as a panacea to all their problems. "Caveat Emptor" (buyers beware) applies to the acceptance of equipment performance claims. The objectives of this paper are to summarize the applicable theory and principles of operation, present examples of applications where the use of membrane plates do and do not offer advantages, and review some of the practical aspects of membrane filter press sizing and operation which may bear on the suitability for a specific application.

Principles of Operation

The membrane filter plate can effectively dewater compressible cakes which would present problems for a standard recessed plate filter press. The difference in principle of operation between these two plate types suggests the applications for which each plate type is best suited.

The following equation for pressure filtration describes the one dimensional flow of liquids through uniform, incompressible beds:

$$dV/dt = P/u(aW + R) \tag{1}$$

where:
- V = liquid volume per unit filtration area
- t = time
- P = applied pressure differential across the bed
- u = viscosity
- a = specific filtration resistance
- W = mass of dry cake per unit area
- R = filtration media resistance

With the exception of W and a, all variables are independent. The specific resistance, a, can be empirically expressed as a function of pressure:

$$a = C + bnP \tag{2}$$

where:
- C, b = empirical constants
- n = compressibility coefficient

As cake thickness increases with an increase in mass of dry cake per unit area, the resistance to flow through that cake also increases as predicted by equation (1). For noncompressible cakes (n=0) filtration flow rate can be maintained by increasing the applied pressure differential, P, with control of the feed pump discharge pressure. However, for compressible cakes (n>0) as pressure is increased to maintain flow rate, the specific filtration resistance increases as shown in equation (2), necessitating an additional pressure increase to maintain flow. With highly compressible cakes this cycle will continue until the maximum pressure of the pump is reached and filtration essentially stops, a condition called cake blinding.

The standard recessed chamber or plate and frame filter press is commonly used for relatively incompressible sludges. The use of the membrane plate addresses the problem of highly compressible sludges by terminating the slurry feed and replacing the pump pressure by the membrane squeeze pressure to express out filtrate. In addition to avoiding filtration shut down because of cake blinding, other benefits to filtration performance may be available, depending upon the application.

For compressible cakes, maximum cake solids are generated during the consolidation phase of the filtration cycle when typically 20% of the feed solids input can require 80% of the cycle. Employing the use of a membrane squeeze may shorten the cycle, while producing the same or higher cake solids than would eventually be generated for a full cycle using standard recessed chamber plates.

Squeezing a cake reduces void space and increases cake density throughout the cake. This consolidation can be useful when the cake is washed by ensuring a uniform capillary structure throughout the cake which improves contact of the wash water with all portions of the solids.

Membrane Plate Performance

Others have reported case studies demonstrating better dewatering performance of membrane plates as compared to standard recessed plates for cake solids, rate and operating costs for such varied sludges as pond acid waste dredgings, organic waste sludges, metal hydroxides, and latex waste[2,3]. However, membrane plate filtration is not applicable to all sludges. Comparison data for total cake solids from various applications of commercial scale and pilot scale membrane plate filter presses are shown in Table 1 for three applications to illustrate the range of applicability.

In the cases of clays and calcium carbonate, membrane plates provide a clear improvement in cake solids over filtration at 100 psig in standard recessed plates. The benefit in clays is usually a sufficient increase to achieve manufacturing requirements without further drying.

For mineral applications, such as copper, lead, iron, etc., membrane plates may not show an advantage. Recessed plate filters can produce the same cake solids provided the proper feed pressures are obtained before the chambers are filled.

For wastewater and water treatment applications involving biological solids, the advantage of using chambers varies with sludge type and must be demonstrated by testing.

Mechanical Considerations

Proper materials of construction for both membrane and plate material must be selected for the application. For example, in waste oil applications, rubber may not be acceptable and plastic would be necessary.

Some types of membranes may only be expanded 15-25% of the chamber volume, without immediate damage, while other membranes may be expanded up to 100%. Over-expanding will affect membrane life while under-expanding will affect solids content and/or production rates. The expansion limit and life expectancy of a membrane depend on the specific membrane type, chamber thickness, final cake thickness, and operating pressure.

Time required for the functioning of the membrane plate is often understated in sizing. Significant time may be required to fill and evacuate the membrane squeeze medium (air or water). For large press applications, the expanded membrane volume could be in the order of 375-550 gal. Time is also required for the cake to be maintained under membrane squeeze pressure from a few minutes to as many as 60-90 minutes. Most manufacturers suggest a minimum of 15 minutes to expand, reach and maintain pressure and deflate the membranes. Table II shows typical times that have been necessary in specific applications.

Expansion fluids may be liquid (water) or gas (air). Air is considered acceptable to about 7 bar. For safety, liquids are recommended above 7 bar. If the filter closing mechanism fails for some reason, the use of gas provides a large volume of compressed gas which may then be released with unexpected or unanticipated results. With liquid, the release results in a rapid drop of pressure with minimum loss of the fluid.

TABLE I
COMPARISON OF MEMBRANE AND RECESSED CHAMBER CAKE SOLIDS

	Recessed Chamber		Membrane
	Cake Solids wt%	Pressure psig	Cake Solids wt%
Coatings			
Calcium carbonate	48	100	62
Kaolin clay	56	100	60
Minerals			
Lead	89	100	91
	91	225	91
Zinc	91	100	92
Copper	89	100	92
	92	225	92
Talc	80	100	79
	82	225	79
Silica	47	100	61
Biological sludges			
100% WAS, muni (polymer)	27	100	27
50:50 pri:sec, muni (ferric & lime)	41	225	42
40:60 pri:sec, ind (lime)	34	160	40
100% WAS, ind (polymer)	26	200	46

TABLE II
TYPICAL MEMBRANE SQUEEZE TIMES

Application	Membrane Time, min
Municipal wastes (polymer treated)	30-90
Mineral concentrates	5-15
Clay (kaolin)	15
Water treatment (alum)	30-60

References

1. F.A. Bühler, Filters and Filter Presses, Norman Rodger, London. 1914, pp. 71-72.

2. Ernest Mayer, "Membrane Press Sludge Dewatering", AFS 2nd Annual Meeting. March 27-29, 1989.

3. C. Robert Steward, "Minimizing Waste, Utilizing High Pressure Membrane Filter Presses", Northwest Hazardous Waste Conference & Exhibitions, '88, Oct. 24-27, 1988.

A STRUCTURE BASED MODEL FOR CAKE FILTRATION

S. H. Chiang, Y. S. Cheng and J. W. Tierney

Chemical and Petroleum Engineering Department
University of Pittsburgh
Pittsburgh, PA 15261
Tel. 412-624-9636

Introduction

Filtration and dewatering are important unit operations commonly used in industry for separation of solid and liquid. The characteristics of filtration and dewatering processes, such as single phase permeabilities, filtration and dewatering rates, are controlled by the microscopic structures of filter cakes. The purpose of this study is to develop a network model for filtration and dewatering of fine coal slurries based on a quantitative description of the microstructures of filter cakes.

Filter Cake Characterization

To characterize the highly complicated nature of porous media, such as filter cakes, Debbas and Rumpf[1] introduced the idea of partitioning void space into "individual particles" and claimed that the logical position to sever the interconnected pore space should be at the constrictions. Based on this idea, a new pore structure analysis method was developed by Bayles[2] using a computerized Leitz TAS-Plus image analysis system. The interconnected pore space of a filter cake (Figure 1) was partitioned into discrete pores at the necks as shown in Figure 2. Based on the partitioned image, three major microscopic properties of the pore space were defined to quantify a cake structure-- the equivalent size distribution which characterizes the local porosity of the network, the hydraulic diameter distribution which measures the irregularity of the pores and the neck (constriction) size distribution which determines the capillary properties of the pores. These characterization parameters were used to establish a network model simulating the complex structure of the filter cakes.

The Network Model

A network model is a simplified geometric representation of the complicated structure of a filter cake and its flow paths. In a regular network, the fluid paths are connected in space based on some regular lattice. The fluid paths are referred to as bonds, and the junction points as nodes. We found[3] that a three dimensional simple cubic lattice with 15 nodes in each direction was suitable for modeling the dewatering process. This type of network was therefore chosen in this work.

In order to provide a generalized computational procedure, the number-based size, defined as the fraction of pores in the network which are larger than the given pore, was used. A number-based node size was assigned to each node using a set of random numbers uniformly distributed between zero and one. The number-based node sizes obtained were further modified by a bond-flow correlation.

$$d'_{ijk} = 1 - F\, d_{ijk} + F\, d_{1jk} \tag{1}$$

where d'_{ijk} is the modified size of a node, d_{ijk} is the size of the node before the modification, d_{1jk} is the size of the front face node directly in front of node ijk, and F, called the bond flow correlation factor, is a parameter having a value between 0.0 and 1.0. The value of the bond flow correlation factor for a filter cake was chosen so that the calculated single phase permeability agreed the best with the experimentally determined value. The bond flow correlation factor so determined was treated as a structural property of the cake and was used in the dewatering calculation. The number-based size of a bond was set equal to the larger of the number-based sizes of the adjacent nodes resulting in clusters of bonds with similar number-based sizes. The sizes were normalized and converted to actual sizes using the micrographically determined pore distribution. The pore volume associated with large pores was assigned to nodes and the rest to the bonds[4]. The irregular shapes of the bonds were taken into account by the form factors of pores measured in the cake structure analysis. The shape of the nodes was assumed to be spherical.

Application of Network Model to Filtration

To calculate the single phase permeability, the overall flow rate through the cake was determined by summing the flow rates through the bonds in a cross section perpendicular to the flow direction. The flow rate of filtrate q_{ij} through a bond is proportional to the pressure difference P_{ij} between the nodes at the ends of the bond,

$$q_{ij} = \pi\, d_{ij}^4\, P_{ij} \,/\, (128\, \mu\, l_{ij}) \tag{2}$$

where d_{ij} and l_{ij} are the hydraulic diameter and length of the bond, respectively. The use of the hydraulic diameter of the bond includes the effect of the bond shape irregularity on flow resistance. The pressure distribution at the nodes was determined by solving a system of linear equations based on a mass balance at each node. With the calculated flow rate, the side length of the network and known pressure drop across the network, the single phase permeability of the network was calculated using Darcy's equation.

The equivalent pore diameter and hydraulic diameter distributions of twelve filter cakes were used as inputs to the network model. The permeabilities of these cakes were calculated with the optimized bond flow correlation factors. It is seen from Table I that the calculated permeabilities are the same as the experimental values.

Application of Network Model to Dewatering

In dewatering, capillary forces play an important role. The effects of the capillary forces were considered by assigning a neck size to every bond, using number-based bond sizes without the bond flow correlation. The neck size of a bond was determined by comparing the number-based bond size with the neck size distribution at the same frequency.

A step-by-step desaturation of the network was carried out using a procedure described by Cheng[4]. At each step, pressures at the nodes were calculated by making a mass balance at each node and solving for pressures with consideration of air breakthrough in the network. The driving force across a bond was set equal to the pressure difference between the two nodes minus the capillary pressure determined with the neck size of the bond. However, it was set to zero if the bond was saturated with water and the two nodes were invaded by air. The water was, therefore, entrapped in the bond. The flow rate of liquid through the network was calculated in the same way as in the single phase flow calculation. In determining the saturation of the network at each step, the interface between liquid and air was located, and all the bonds which could open were checked. Two of the bonds were allowed to completely empty at a time, and all other bonds were partially dewatered. The volume of liquid displaced from these bonds was calculated, and the saturation of the network was determined. The calculations were repeated until no bond could be displaced. At this point, dewatering stopped, and the equilibrium saturation of the network at the given pressure drop was reached. The flow rate versus saturation data were integrated to obtain saturation versus time data. Several typical predicted dewatering curves are plotted together with the experimental data in Figures 3 to 6. Good agreement was found among the calculated and experimental values for all cases.

Conclusions

Structural characteristics of filter cakes can be quantitatively measured by equivalent diameter distributions, hydraulic diameter distributions, and neck size distributions using the pore isolation technique. With these structural properties of filter cakes, a three dimensional simple cubic lattice network model with a bond flow correlation can predict both filtration and dewatering properties.

Acknowledgment

The financial support provided by the United States Department of Energy under contract No. DE-AC22-85PC81582 is greatly acknowledged.

References

1. S. Debbas, H. Rumpf, "On the Randomness of Beds Packed with Spheres or Irregular Shaped Particles", Chem. Eng. Sci., 21, 583, (1966).

2. A. G. Bayles, New Pore Characterization Techniques for Flow Dynamics in Porous Media, PhD dissertation, School of Engineering, University of Pittsburgh, 1988.

3. I. Qamar, Application of a Three Dimensional Network Model to Coal Dewatering, PhD dissertation, School of Engineering, University of Pittsburgh, (1985).

4. Y. S. Cheng, An experimental Study and Theoretical Modeling of Fine Coal/Refuse Filtration and Dewatering, PhD dissertation, School of Engineering, University of Pittsburgh, (1988).

Table I. Calculated Single Phase Permeabilities.

Sample	Size (mesh)	Additive (Flocculant)	Permeability Cal. (mD)	Permeability Exp. (mD)	Bond-Flow Correlation Factor
Pittsburgh Coal	-32	none	214	214	0.65
Pittsburgh Coal	-32	Accoal-floc 204, 10 ppm	244	244	0.55
Pittsburgh Coal	-100	none	175	175	0.61
Pittsburgh Coal	-100	Accoal-floc 204, 40 ppm	288	288	0.81
Pittsburgh Coal	-200	none	142	142	0.57
Pittsburgh Coal	-200	Accoal-floc 204, 50 ppm	201	201	0.49
Taiyuan Coal	-32	none	54	54	0.72
Taiyuan Coal	-32	Accoal-floc 16, 7 ppm	128	128	0.72
Pingding Coal	-32	Accoal-floc 204, 4 ppm	93	93	0.23
Pingding Coal	-32	Accoal-floc 16, 10 ppm	45	45	0.51

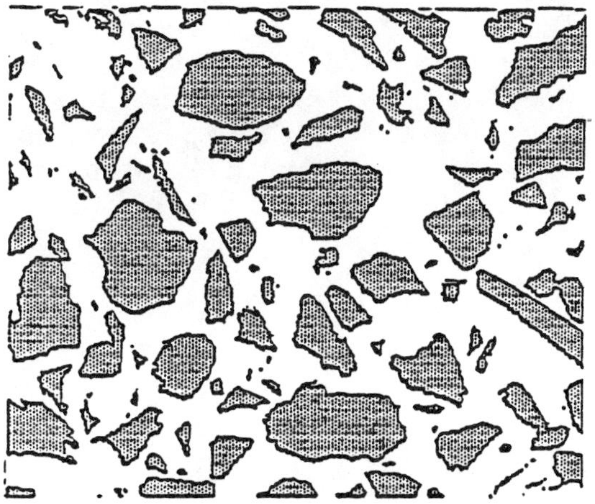

Figure 1. Detected Image of Filter Cake Structure. Shaded Areas are Particles.

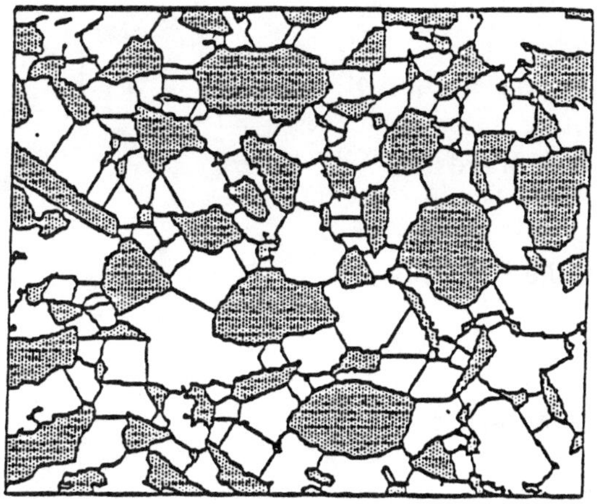

Figure 2. Partitioned Image of Pore Space at Necks.

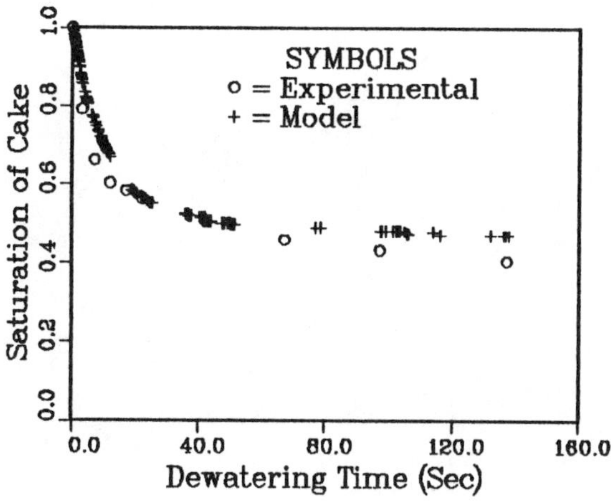

Figure 3. Dewatering Curve for -32 Mesh Pittsburgh Coal Cake.

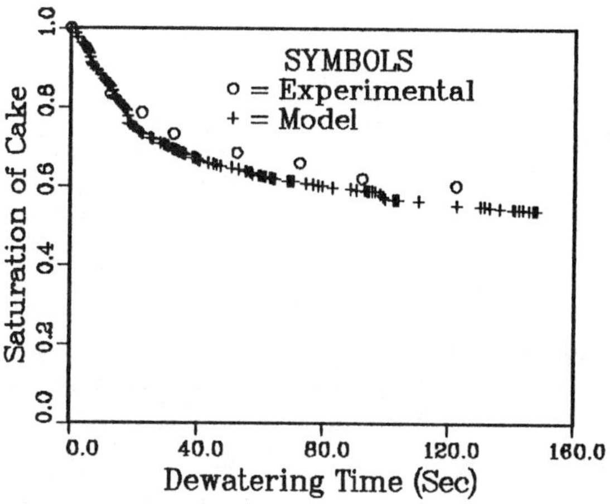

Figure 4. Dewatering Curve for -32 Mesh Taiyuan Coal Cake.

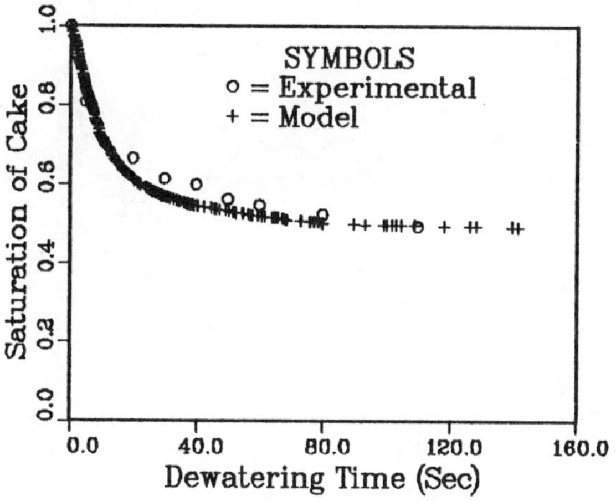

Figure 5. Dewatering Curve for -200 Mesh Pittsburgh Coal Cake.

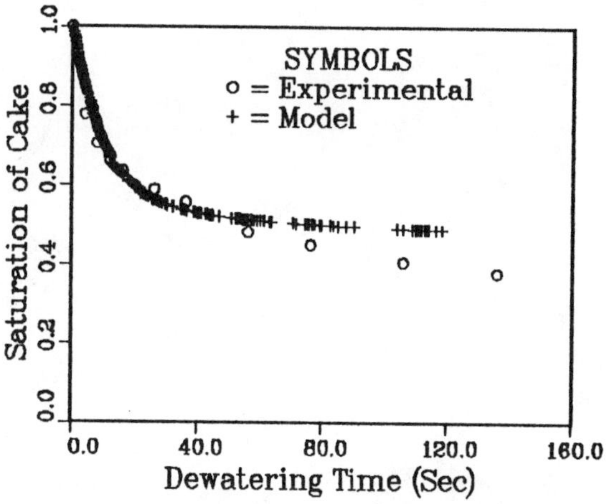

Figure 6. Dewatering Curve for -200 Mesh Pittsburgh Coal Cake Formed with 50 ppm of Accoal-floc 204.

Key Words

Cake Filtration

Capillary Pressure

Dewatering

Filter Cake Structure

Micrographic Analysis

Neck Size

Network Model

Permeability

Pore Size

Porous Media

Single Phase Flow

Two Phase Flow

EFFECTS OF INTERNAL REORDERING ON SEDIMENTATION WAVES IN CONCENTRATED INCOMPRESSIBLE SUSPENSIONS

Harri K. Kytömaa
Department of Mechanical Engineering, bldg 3-258
Massachusetts Institute of Technology
77 Massachusetts avenue
Cambridge, Massachusetts 02144

1. INTRODUCTION

The analysis of one-dimensional particle sedimentation was first discussed in depth in the classical paper by Kynch (1952) in which he used the method of characteristics to describe batch sedimentation of monodisperse incompressible particles. His approach has since been extended to include additional effects in the context of continuous thickening, filtration, soil mechanics and numerous other applications, some of which are discussed below.

The effect elasticity, although neglected by Kynch, was studied by Terzaghi (1943) in the context of soil consolidation, and it has since been combined with Kynch's analysis by Shirato *et al* (1970), Tiller (1981) and Buscall and White (1987) to include the effects of solid matrix elasticity due to either particle networks caused by solid-solid contacts or thermally driven osmotic pressure. The recent works of Auzerais *et al* (1988) and Davis & Russel (1989) generalize the Kynch formulation to include the effects of Brownian diffusion and an osmotic solid phase stress on the profile of concentration shocks. The analysis by Davis and Russel presents an asymptotically matched solution between an inner region (the shock) where diffusion and convection are of equal importance, and a convection dominated outer region which is modeled as a perturbation in Péclet number about the classical Kynch solution. The present study addresses the behavior of assemblies of large monodisperse particles for which the Péclet number is effectively infinite, and thermal effects are negligible. Stokes-Einstein diffusion is therefore neglected, and wave propagation reduces to the Kynch solution in the far field. However, while thermally driven osmotic pressure is negligible in this case, the solids phase pressure due to solid-solid contact between particles is not, and must be included. It is this effect of reordering in the absence of diffusion that is presently analysed to describe the structure of concentration shocks in globally incompressible mono-disperse suspensions.

2. RESULTS

The sedimentation wave propagation in non-Brownian, monodisperse solid-liquid mixtures was studied, taking into account the elasticity of the settled mixture due the reordering that occurs immediately after the particles settle. This is modeled in terms of a stress-solids fraction relationship that allows for an increase in solids fraction under increasing load and that asymptotes to infinite stiffness at a maximum solids fraction. Experimentally derived normal stress - volume fraction curves are shown in Figure (1) along with the idealized curve used in the analysis. The resulting sedimentation wave speed is equal to the Kynch result. For sedimentation of concentrated suspensions, the wave is shown to achieve magnitudes that are markedly larger than the unhindered terminal velocity of a single grain by an order of magnitude. The wave structure presented in terms of the distribution of the flow variables differs significantly from the classical incompressible Kynch result due to reordering below the sedimented interface.

The sedimentation characteristics of the particulate material are quantified in terms of ν_{crit}, the concentration at which it begins to transmit normal stresses, and ν_{abs}, the maximum concentration that it can reach under load. The analysis of the shock structure was formulated with the use of momentum and mass conservation principles and yielded the following results:

- A material of low ν_{crit} has thicker sedimentation waves, and
- consequently, for such materials, the solid normal stress remains low in a larger vertical region. Since the normal stress has a strong influence on the shear strength of a settled mixture, a large region of low normal stress may thus affect its resistance to shear failure.

The results of this study show that an effective way of measuring the local degree of suspension, that is the ratio of the weight of overlying matter supported by the solids to that supported by the pore fluid. This is done by measuring the local pore pressure gradient with a transducer pair. The pore pressure diff erence between two closely spaced transducers assumes different values for suspended and settled mixtures.

This pressure difference was simulated using the present sedimentation wave analysis. The results exhibited a high sensitivity to the load bearing characteristics of the particulate material as can be seen in Figure (2), and indicate that by making dynamic measurements of pore pressure differences, these properties could me indirectly determined more easily than by stress-strain experiments (of the kind used to construct Figure 1), which require accurate means of monitoring the solids volume fraction.

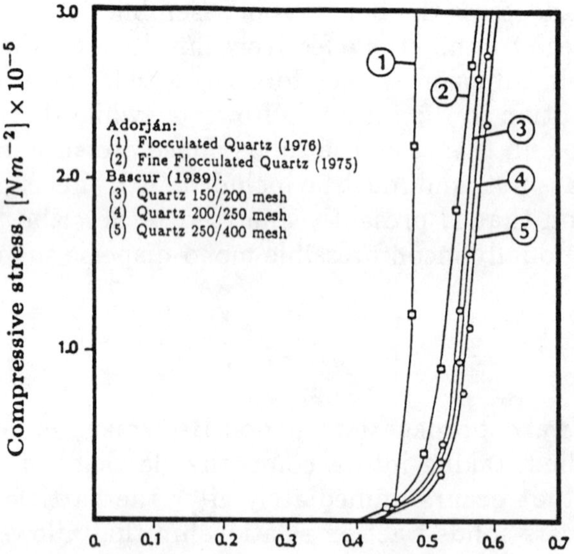

Figure 1a Uniaxial compression data for quartz taken from Bascur (1989), showing his and Adorján results.

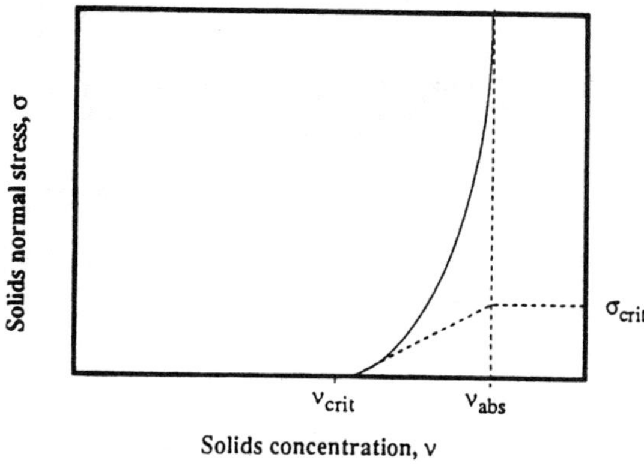

Figure 1b **Exponential solids stress - volume fraction relation for incompressible solid-liquid mixture in which reordering occurs at low stresses.**

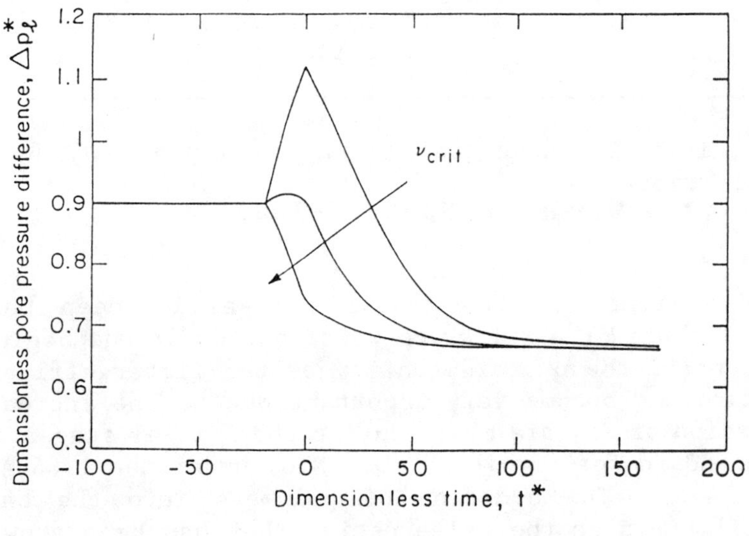

Figure 2 **Pressure difference between two pressure taps separated by a dimensionless distance of 1.0. Note the marked difference in nature between the low and high ν_{crit} curves. The sudden rise in Δp is predicted to occur with materials that "solidify" at low values of ν_{crit}. Fibreous materials are speculated to fall into this category. (ν_0 = .614, ν_{abs} = .620, ν_{crit} = .615, .617, and .619).**

AMERICAN FILTRATION SOCIETY

P.O. Box 6269
Kingwood, Texas 77325
Telephone: (713) 359-1894
FAX: (713) 358-3939

ABSTRACT & COPYRIGHT

Meeting Location: Washington, D.C. **Meeting Date:** March 19-22, 1990

Session Chairman: Dr. Frank Tiller and Dr. Wallace Leung **Session Number:** III or VI

SPEAKER (AUTHOR):

Name: Appiah Amirtharajah
Company: Georgia Institute of Tech.
Address: School of Civil Engineering
City, State, Zip: Atlanta, GA 30332
Telephone: (404)894-2265 **FAX:** (404)894-2278

CO-AUTHOR:

Name: Kuo-Wei Chiu
Company: Georgia Institute of Tech.
Address: School of Civil Engineering
City, State, Zip: Atlanta, GA 30332
Telephone: (404)894-2265 **FAX:**

CO-AUTHOR:

Name: G. Ricky Bennett
Company: Cobb County-Marietta Water Authority
Address: P.O. Box 540
City, State, Zip: Acworth, GA 30101
Telephone: (404)974-4286 **FAX:**

CO-AUTHOR:

Name:
Company:
Address:
City, State, Zip:
Telephone: **FAX:**

TITLE OF PAPER: Initial Effluent Quality and Its Improvement By Coagulants in Backwash

TEXT OF ABSTRACT: (200 Words in Space Below)

The quality of filtrate at the initial stages of deep bed filtration is comparatively poor. The use of a filter to waste period at the beginning of a filter run may be used to control the overall quality of the filter effluent from a plant. Recently, the procedure has become very important due to the increasing concern for the possible transmission of *Giardia* cysts during the initial stages of filtration and the proposed lower standard for turbidity (0.5 NTU) under the 1986 Amendments to the Safe Drinking Water Act. The addition of polymers into the backwash water to "precondition" the filter is another alternative that has been shown to improve the initial effluent quality. One pilot scale study has shown that addition of alum to the backwash water is significantly more efficient than addition of polymer to improve the initial effluent quality. The current research reports on an extensive plant scale demonstration, the first of its kind, to show the efficacy of adding alum into the backwash water to improve overall quality.

The paper will present a conceptual two peak model for the causes of poor quality water at the initial stages of filtration and demonstrate the optimum duration and dosages of alum addition to the backwash water to improve the overall quality of filtrate from tests conducted at a large water treatment plant in the greater Atlanta area (50 MGD).

****

PLEASE TYPE INFORMATION BECAUSE DATA WILL BE PUBLISHED IN THE ABSTRACT BOOKLET. <u>RETURN FORM TO THE AMERICAN FILTRATION SOCIETY</u>, P.O. Box 6269, Kingwood, Texas 77325, WITH A COPY TO YOUR SESSION CHAIRMAN.

AMERICAN FILTRATION SOCIETY

P.O. Box 6269
Kingwood, Texas 77325
Telephone: (713) 359-1894
FAX: (713) 358-3939

ABSTRACT & COPYRIGHT

Meeting Location: **Meeting Date:**

Session Chairman: **Session Number:**

SPEAKER (AUTHOR):

Name: Richard V. Levy, Manager
Company: Millipore Corporation
Address: 80 Ashby Road
City, State, Zip: Bedford, MA 01730
Telephone: 617-275-9200 **FAX:** 275-5550

CO-AUTHOR:

Name: Kathleen Souza, Manager
Company: Millipore Corporation
Address: 80 Ashby Road
City, State, Zip: Bedford, MA 01730
Telephone: 617-275-9200 **FAX:** 275-5550

CO-AUTHOR:

Name: Carol Neville, Validation Mgr.
Company: Millipore Corporation
Address: 80 Ashby Road
City, State, Zip: Bedford, MA 01730
Telephone: 275-9200 **FAX:** 275-5550

CO-AUTHOR:

Name:
Company:
Address:
City, State, Zip:
Telephone: **FAX:**

TITLE OF PAPER: Validation of Microbial Retention in Sterilizing Grade Filters

TEXT OF ABSTRACT: (200 Words in Space Below)

The "FDA Guideline on Sterile Drug Products Produced by Aseptic Processing" (June 1978) recommends that the performance of filters used in the aseptic processing be verified. Validation of sterilizing grade filters should include microbial retention testing of the filter product under simulated process filtration conditions with actual drug products. Other supportive data generated should include integrity test specifications established in the drug products and filter/drug product compatibility claims. Bacterial challenge test data demonstrate that with at least one sterilizing grade filter, retention is independent of liquid vehicle attributes and process conditions. A matrix of process conditions and solution attributes tested will be presented.

****

PLEASE TYPE INFORMATION BECAUSE DATA WILL BE PUBLISHED IN THE ABSTRACT BOOKLET. RETURN FORM TO THE AMERICAN FILTRATION SOCIETY, P.O. Box 6269, Kingwood, Texas 77325, WITH A COPY TO YOUR SESSION CHAIRMAN.

VALIDATION OF 0.1 um RETENTIVE MEMBRANES

by Peter J. Waibel
Peter Wolber

While filtration has been very successfully used for the sterilization of pharmaceutical and biological solutions, evidence is accumulating that current 0.2 um rated membranes have been penetrated by newly isolated bacteria originating from purified water supplies. Pharmaceutical product recalls involving nasal and oral liquid dosage forms have been occasioned by organisms originating in the water used in formulating the drugs. One such highly publicized recall involved providone-iodine solutions contaminated with Pseudomonas cepacia from deionization resin beds(1,2). Since the recall incident involving P. cepacia, use has increased of tighter filters for this purpose. There has even been an advocacy of the use of 0.1 um rated membranes for this purpose, prompted by the findings that certain well waters naturally contain Pseudomonads small enough to negotiate 0.2 um rated membranes(3). Additionally, with the advent of more sophisticated biotechnological techniques has arisen the need for even finer filtration for the removal of Mycoplasmas that may be found as a contaminant in microbiological and cell culture media.

The qualification of sterilizing grade membrane filters is currently based on the bacterial challenge test as outlined by the HIMA Guidelines(4) using Ps. diminuta. This bacterial strain originally isolated as a contaminant in a pharmaceutical solution can be cultivated under minimal growth conditions to produce a 0.3 um coccibaccili capable of penetrating membranes with retention ratings greater than 0.2 um. Other genera of Pseudomonas, in particular Ps. cepacia or Ps. mesophilia, can be found with smaller morphological sizes capable of penetrating validated 0.2 um filter membranes. The source of these organisms has been traced to the purified water systems where the lack of nutritional resources favor the growth of small and starved organisms.

In order to validate the use of membranes finer than 0.2 um filters, for example 0.1 or 0.07 um membrane systems, the following Guidelines and recommendations are available for the validation of these sub-0.2 um Membranes:

a. Guideline on Sterile Drug Products Produced by Aseptic Processing, FDA, July 1987.(8)
b. HIMA Document "Microbiological Evaluation of Filters for Sterilizing Liquids" - for the validation of 0.2 and 0.1 um membrane filters.(4)
c. Filter Manufacture's Recommendations - ie. The Sartorius 0.07 um Cellulose Acetate Membrane Cartridge with Respect to Mycoplasma Retention, October 1987.(5)

These guidelines, coupled with filter manufacturer's recommendations and other validation experience lead to the following standard test organisms for the validation of 0.2, 0.1 and 0.07 um sterilizing grade membranes:

a. 0.2 um Membrane Systems - Pseudomonas diminuta, ATCC 19146,
b. 0.1 um Membrane Systems - Pseudomonas cepacia, ATCC 25416,
c. 0.07 um Membrane Systems - Mycoplasma (Acholeplasma laidlawii).

The selection criteria for these organisms will be discussed later.

Certain criteria for the development of validation protocols for these given membrane systems are based on the recommendations from the FDA Guidelines. Considerations for validation testing are based on both product and process parameters. The product parameter are pH, viscosity, concentration of components, osmotic pressure, ionic strength, additives, preservatives and specific gravity. These parameters can be used to asses the testing requirements for given products to be sterile filtered. The process parameters are related to flow rates, operating and differential pressures, batch sizes, fluid/filter contact time, temperature and filter configuration.

These parameter form the basis of the validation testing. Other things to consider under the production and process conditions are to use the "Worst Case" conditions and using membranes/cartridges for the actual testing. When analyzing product parameters, one must consider the chemical compatibility of the drug product with the given membranes, toxicity of the drug product to the test organism, effects of the product on membrane integrity and pharmaceutical classifications.

0.1 um Filter Validation

For the validation of 0.1 um membrane systems, Pseudomonas cepacia is the organism of choice for validation of these filters. The organism that is used for this testing is Ps. cepacia, ATCC 25416. Selection criteria for the use of this organism are based on the following characteristics - Size of the organism, shape of the organism, ease and safety of handling, ease of cultivation and relationship to the pharmaceutical environment.

The performance of the actual filtration runs, for the validation of 0.1 um membranes being challenged with Ps. cepacia, after the equipment is set-up and the challenge solutions are prepared, the tests are conducted under the "worst case" conditions specified in the test protocols. Analysis of the effluent can be done either by a direct filtration assay during the filtration trial, or by analysis of the total effluent being collected for a later determination for microbial content. Before and after the filtration trial, the filters should be integrity tested.

Actual Test Results:

Table 1. Bacterial Challenge Results for Cellulose Acetate 0.1 um Membrane Material, Sartorius reorder number 12358.

Lot No.	BP Range(1)	Challenge Conc(2)	Result	LRV
2030/8	55 - 60	> 1 x 10(7)	Sterile	> 7
2042/8	56 - 57	> 1 x 10(7)	Sterile	> 7
2049/8	56 - 61	> 1 x 10(7)	Sterile	> 7
2063/8	53 - 59	> 1 x 10(7)	Sterile	> 7
2064/8	55 - 60	> 1 x 10(7)	Sterile	> 7
2085/8	55 - 56	> 1 x 10(7)	Sterile	> 7
2090/8	56 - 59	> 1 x 10(7)	Sterile	> 7
2093/8	53 - 59	> 1 x 10(7)	Sterile	> 7
88-103	53 - 56	> 1 x 10(7)	Sterile	> 7
89-025	57 - 59	> 1 x 10(8)	Sterile	> 8

[1]Bubble Point Values in psi

[2]Concentration in cells/cm^2 of filtration area

The control of these organisms has to be seriously considered to avoid spontaneous sterility failures due to the presence of these organisms. Additionally, seasonal variations have been demonstrated to effect the quality of the water resulting in changes in the level and compositions of the naturally occurring bacterial flora(5). Filtration of all products to the level of 0.1 um may not be feasible, but filtration of the water prior to utilization in the formulation of pharmaceutical or biological solutions or general plant use will ensure the isolation of these microorganisms from the water system.

0.07 um Filter Validation

With the use of finer filtration in the Biotechnology Industry, the need to validate the use of 0.07 um prior to use in production, is essential. Most of the process and product considerations are the same as that for the 0.2 and 0.1 um membrane filters. For the Mycoplasma retention validation, the test organism of choice is Acholeplasma laidlawii. The criteria for selection of this organism is due to the organisms lack of a rigid cell wall, small size, deformability and these mycoplasma can be cultured quantitatively. Important validation considerations to keep in mind in developing these validation protocols includes microbial concentration, differential pressure, pulsation residence time and serial filtration.

The validation of the use of 0.07 um membranes for the retention of mycoplasma has been divided into two phases. Phase One is a simulation of actual filtration runs with filtration at various increasing pressures. In Phase Two, the total mycoplasma challenge is filtered during the first 10 liters to establish a high initial concentration on the membrane. This is followed by saline being filtered through the challenged filters at various pressure profiles using a positive displacement pump to add the effects of pulsations.

Actual Test Results:

Table 2. Mycoplasma Challenge results for Cellulose Acetate 0.07 um Filter Membrane Material

Mycoplasma Analysis for Phase One Samples

Sample No.	Inoculum	Effluent Cell/ml	LRV
1	1.6 X 10(9) cfu/70l	0.0	>9.2
2	1.6 X 10(9) cfu/70l	0.0	>9.2
3	1.6 X 10(9) cfu/70l	0.0	>9.2
4	1.6 X 10(9) cfu/70l	0.0	>9.2
5	1.6 X 10(9) cfu/70l	0.0	>9.2
6	1.6 X 10(9) cfu/70l	0.0	>9.2

Mycoplasma Analysis for Phase Two Samples

Sample No.	Inoculum	Effluent Cell/ml	LRV
1	3.7 X 10(8) cfu/10l	0.0	>8.57
2	3.7 X 10(8) cfu/10l	0.0	>8.57
3	3.7 X 10(8) cfu/10l	0.0	>8.57
4	3.7 X 10(8) cfu/10l	0.0	>8.57
5	3.7 X 10(8) cfu/10l	0.0	>8.57
6	3.7 X 10(8) cfu/10l	0.0	>8.57

Mycoplasma, because of their small size and plasticity, are difficult to remove from liquids by filtration. This is particularly true when high pressures are used in processing serum or cell culture media. Attempts to remove the mycoplasma have include triple membrane filtration through individual 0.1 um filters. A simpler and more efficient way to remove mycoplasma is to use the above mentioned 0.07 um membrane filters. As the data suggests, even at high pressures or systems prone to pulsation pressures, mycoplasma can be effectively removed.

Conclusion:

The results demonstrate that the absolute removal of organisms, such as Ps. cepacia and Mycoplasma, can be achieved in absolute terms under the test conditions using the 0.1 or 0.07 um cellulose acetate membrane. These retentive qualities were also confirmed in field trials in purified water systems, where previously the passage of microorganisms through standard 0.2 um rated membrane was observed.

The use of membrane filters of at least 0.2 um rating downstream from DI beds for the removal of most microorganisms seems warranted. Such employment can be effective provided adequate sanitization, to control grow through, and/or replacement, to minimize pyrogens, is assured by a disciplined practice(6). When, however, the naturally occurring bacterial flora of any of the components in the formulation are known to contain organisms that can penetrate 0.2 um rated sterilizing grade filters, the use of newly developed 0.1 um membranes can greatly reduce the risk of spontaneous sterility failure.

References:

(1) Micheals, D. (1981) "Letter to the Pharmaceutical Industry: Validation and Control of Deionized Water Systems," FDA, Rockville, MD. Bureau of Drugs, Assoc. Dir. of Compliance.

(2) Motise, P. (1982) "FDA Viewpoint of Water and Its Use," Proceedings of PMA Water Seminar Program, Atlanta, February, 1982.

(3) Duberstein, R. and Howard, G. (1980) " a Case of Penetration of 0.2 um rated Membrane Filters by Bacteria." J. Parenteral Drug Assoc., 34(2), 95 - 102.

(4) HIMA Document Number 3, Volume 4 (1982). "Microbiological Evaluation of Filters for Sterilizing Liquids." Health Industry Manufacturing Association, Washington, D.C.

(5) Product Evaluation Study, "The Sartorius 0.070 um Cellulose Acetate Membrane Cartridge with Respect to Mycoplasma Retention." October 1987. Available from Sartorius Corporation.

(6) Meltzer, T. Filtration in the Pharmaceutical Industry, Marcel Dekker, Inc. New York, pp 832 - 833, 1986.

(7) Wallhausser, K.H. (1983) "Grow Through and Blow Through Effects in Long Term Sterilization Processes," Die Pharmazeutische Industrie 45(5), 527 - 531.

(8) Guideline on Sterile Drug Products Produced by Aseptic Processing, Rockville, MD, Food and Drug Administration, July, 1987.

Crossflow Microfiltration

A Possible Tool for Large Volume Sterilization

Dr. Amos Korin, Tech. Dir., Separations Venture
Exxon Chemical Company

ABSRACT

Micoporous membrane filtration, mostly in cartridge form is commonly used as a method for sterilization of pharmaceutical solutions. this technique is associated with relatively high operation expenses due to high cost of cartridge disposal. Crossflow filtration mode has been known for more than ten years has a possible solution to filter clogging. The possibility of reducing sterilizing filtration expenses by application of cross flow operation mode has been raised several times. Various aspects of this possibility are discussed in this presentation.

* * *

Microporous membrane filtration is a proven technique for achieving fluid sterilization in many fields. However, this technique is restricted to relatively costly applications such as the pharmaceutical or food and beverage industries. Sterilization filtration is mostly carried out in dead end filtration mode which restricts the membrane sterilization process to relatively clean fluids.

For many purposes sterility of fluids is best obtained by a filtration process. Filtration is used for sterilization when heat or one of the other more robust methods cannot be used. Filter sterilization is frequently used in the pharmaceutical industry and the food and beverage industries. In the electronic, aerospace, and chemical industries, filtration of fluids to remove bacteria is commonly practiced.

The prime requirement of a sterilizing filter is that it remove all viable bacteria. In most situations, this requires the use of a filter with rated pore size of 0.2 micron. There are a few bacterial types, such as certain mycoplasmas, that are small enough or flexible enough to penetrate 0.2 micron filters, and if these tiny organisms are to be removed then 0.1 micron filters must be used. However, since the flow rate of 0.1 micron filter is so much slower than that of a 0.2 micron filter, the 0.1 micron filter would be used only when specifically required. (Brock 1983).

The U.S. Pharmacopeia, which provides guidance for pharmaceutical manufacturers, prescribes sterility tests employing several different culture media. If a viable organism present in a liquid is not able to grow in one of these media, then its presence will not be suspected and liquid will be falsely reported to be sterile. It cannot be too strongly emphasized that sterility is not an absolute property, but is operationally defined. (Lukaszewicz 1979)

According to Johnston and Meltzer there is no absolute sterilization by membrane filters. The sterilization effect can be expressed in a logarithm form and is called Log Reduction Value (LRV). LRV of 6 is equivalent to reduction of 6 orders of magnitude in the microbe population. By using microporous membranes, one will be able to maintain LRV large enough to control the microbe population in the solution

In - biotechnological industry and research, membrane filters find considerable use for separating cells from liquid media. Membrane separation has been used to obtain the filtrate or to obtain the cells in an unaltered way by elusion from the membrane surface.

The simplest way of separating organism with membrane filters is by use of the dead-end filtration mode, in which the liquid is passed directly through the filter and the organisms obtained by scraping or back-washing (Thomson and Foster 1970) Dead-end filtration was, until recently, the only alternative available in microfiltration, but dead-end filtration has several drawbacks. It is handicapped by the rapid decay in flux that occurs due to a continuous buildup of a particle layer on the membrane surface. The particle layer buildup leads to low overall flux rates and requires repeated cleaning or cartridge replacement. These factors result in increased downtime and high operation and cartridge replacement costs.

Crossflow Microfiltration (MF) has been known for more than 10 years. A lot of experimental and theoretical work has been done on it. This work has been prompted by the attractive feature of a crossflow, presumably eliminating all deposits on membrane surface. However, the success of such processes is very often compromised by the complexity of the interactions between the membrane surface and the suspension to be filtered.

This process was mostly on a laboratory scale or a small scale system. Recently there has been penetration of this technology to large scale separation applications.

In cross-flow filtration mode, the influent is passed under pressure laterally cross the membrane, resulting in two streams - the retentate (nonfiltered material), and the filtrate. As the fluid is passed tangentially across the filter membrane surface, the washing action keeps most of the particulate material in suspensions. and recirculation returns the concentrated particles back to the feed reservoir. Concentration of cells in this manner is much more rapid than with conventional filtration. Tangential-flow filtration was first developed in the ultrafiltration and reverse osmosis fields, and was only later applied to micro-porous membrane filtration of particles. (11 Tanny and Hauk 1980)

In the crossflow process, one would use microporous membrane of pore size in the range between 0.01 micron to 3 micron in a similar tangential flow configuration to that which is normally used in processes such as reverse osmosis and ultra-filtration.

Tangential flow mode had been applied successfully for inhibiting either concentration polarization in reverse osmosis or gel polarization in ultra filtration. About a decade ago people assumed that combining this concept with microporous filtration would result in clog-free type filtration.

Fig. 1 shows the conceptual difference between dead-end and crossflow filtration. The new technology showed great promise as a means to solve difficult filtration problems but too often it failed to live up to expectations. Many people have failed to make the technique work consistently or at all and have tended to become disenchanted. It was found that crossflow can't simply be defined as clog free microporous filtration but the situations involved with this process are more complex.

Fig. 2 shows the generic pure water fluxes of microporous membranes of various pore size. The fluxes are given in GFD [gallons per square foot per day] for trans membrane pressure of 1 psi and 20 psi. (In this presentation flux units will be given either in GFD units or $L/M^2/HR$ [liter per square meter per hour]. One has to remember that 1 GFD = 1.7 $L/M^2/HR$.) The pure water fluxes shown in the plot are in the order of magnitude of thousands and tens of thousands, depending on the pore size.

The actual fluxes obtained in process crossflow microfiltration are about two order of magnitudes lower than these fluxes. Table No. 1 shows typical process fluxes observed in crossflow microfiltration of various feedstreams.

These fluxes are of the same order of magnitude commonly observed in ultra filtration processes. There were early attempts to correlate the flux to cake formation and shear effects (Rushton et al 1979). However, the phenomena involved are more complex than simple cake formation.

Fig. 3 shows the general flux pressure diagram observed in crossflow filtration under laminar flow conditions. This diagram is similar to what has been observed in ultra filtration (Blat et al 1970). As shown in the diagram compared to pure water flux which is pressure dependent, the observed process fluxes are in the plateau range where the flux is pressure independent and is velocity dependent. In this range the flux is controlled by mass transfer phenomena.

As shown in Fig. 4, suspension containing fine particles or cells is passed under moderate pressure through a tube or channel having microporous membrane walls. Purified liquid passes through the membranes as permeate whereas the retained particles form a thin cake or fouling layer adjacent to the membrane surface. This fouling layer, which is analogous to gel layer for ultra filtration of micro molecules, reduces the permeate flux. However, the action of the bulk flow tangent to the membrane limits the accumulation of the immobile cake layer (Davis and Sherwood 1989). Several attempts have been made to predict the steady state permeate flux as a function of the operating condition. Early works were based on the classical concentration-polarization film theory (Porter 1972: Henry, 1972). Recently, (Zydny and Colton, 1986) also used the concentration-polarization model but they replaced the diffusivities of the particles with a constant shear induced hydrodynamic diffusivity. This yielded improved agreement with experiments. Further work has been done by Romero & Davis 1988.

Fig. 5 shows the Zydney and Colton equation. This equation is used in this work as the basis for the process calculation. The reason for choosing an analytical approach to the process was the need to obtain a generic base for process comparison which would not be restricted or limited to any proprietary design or data and will enable one to obtain information of unit operation type. This information could be used as objective criteria to evaluate the comparative economics of the crossflow membrane process.

However, calculation of the process variant based on this equation is restricting the process to a range of laminar flow conditions, namely a condition where Reynolds number is lower than 2000.

The shear rate in the equation could be calculated as follows:

$$= 8\ V/d \text{ for circular tubes}$$
$$= 6\ V/b \text{ for rectangular channel}$$

Where V is the average velocity, d is the pipe diameter and b the channel depth. All the calculations in this work were done for tubular configuration.

Fig. 6 shows calculated concentration lines for suspensions of various particle sizes. The initial partial volume ratio for this plot is 80 for all the suspensions. One may see that the effect of particle size on the flux is substantial. Another point that should be noted is that all the concentration lines are extrapolated to zero flux at the concentration ratio (factor) which is equal to the initial partial volume ratio. Where initial volume ratio is the ratio between the volumetric fraction of the particles near the membrane surface to volumetric fraction of particles in the bulk for the feed suspension. This fact enables one to find the initial partial volume ratio of any feed stream by running a small scale concentration experiment and extrapolating the flux concentration line to zero.

Fig. 7 shows the effect of the tube diameter on flux for various particle sizes. One may see that the diameter has a dramatic effect on the flux at any fluid velocity under Laminar flow condition. If all other parameters are held constant, reducing the tube diameter as small as possible will benefit the flux. This would occur due to the increase of the shear rate which contributes to the fouling layer removal .

The effect of fluid velocity on the concentration process is shown in Fig. 8. All the concentration lines were calculated under laminar flow conditions. This plot conforms to what has been shown generally in Fig. 3. Each data point in this plot could conform to a plateau on the pressure flux plot. One may see that increasing the velocity will move a system to a lower concentration curve. We will return to this graph when we discuss the concept of the multistage system.

When the concept of crossflow filtration is to be applied to a process, the options are either batch or continuous system.

For the base of the equipment comparison in this work we choose the concept of a continuous multistage system which is shown in Fig. 9. The system consists of 5 stages. A process simulation computer program was written. The program emulates the process by using the mass transfer equations and calculates the process variants as different operation parameters are entered as input.

The number of tubes in each stage is adjusted and reduced so the average velocity in this stage will be 2m/sec and the process is returned to the upper concentration line.

In order to obtain economical variants that will be applicable for comparison with dead end cartridge filtration, a common process denominator had to be selected. It was decided to base the economical comparison on process volume unit. The system size variants were calculated as cost per thousand liters. This volume base is common in the pharmaceutical industry and its use would provide two benefits:

- it will provide an objective criterion for unit optimization
- it will enable an apples to apples comparison with different types of processes

Fig. 10 would provide an example of process optimization in relation to fluid velocity for a fixed concentrate factor.

As shown in the plot, there are two counter trends: the cost of power will increase with velocity and the specific membrane and system cost will decrease.

Fig 11 shows the filtration cost of dead end filtration process based on cartridge thruput. The base for cost is similar to that used in the previous figure thus one may compare the two types of process and decide at what performance range one is better than another. One should remember that the numbers in this work are based on theoretical calculation, however similar rational may be used for comparison of actual performance data

REFERENCES

BLAT, W.R.; DAVID, A.; MICHAELS, A.S. and NELSON, L., 1970. Solute polarization and cake formation in membrane ultrafiltration causes consequence and control techniques. In Membrane Science and Technology, (Edited by J.E. Flinn, pp 47-97, Plenum Press, NY, 1970.)

BROCK, T.D., Membrane Filtration, A User Guide and a Reference Manual, Science Tech, Inc. Madison WI 1983

DAVIS, R.H.; SHERWOOD, J.D.. A similarity solution for steady-state crossflow microfilter (to be published).

HENRY, J.D. Crossflow filtration in recent development in Separation Science, pp 205-225, CRC Press, Cleveland, (1972)

JOHNSON, J.N. Crossflow ultra filtration using polypropylene hollow fibers. ASTM Symposium, Filtration Society, Philadelphia, PA, (1986)

JOHNSTON P.R. and T. H MELTZER,1979, Comments on organism-challenge levels in sterilizing-filters efficiency testing, Pharmaceutical Technology 3: 66-70, 110.

LUKASZEWICZ, R.C. Microporous filter sanitization of large volume product and feed water systems, Pharmaceutical Technology volume 3 August 1979

PORTER, M.C. Concentration polarization with membrane ultrafiltration, Ind. Chemical Engineering Product Research Development 11,236, pp 234-248, (1972).

ROMERO, C.A.; DAVIS, R.H. Global model of crossflow microfiltration. based on hydrodynamic diffusion. Journal Membrane Science 39, pp 157-185, (1988)

RUSHTON, A.; HOSSEINI, M.; RUSHTON, A. Shear effect in cake membrane formation mechanisms. Filtration & Separation, Sept./Oct., 1979.

REFERENCES (Cont'd)

SHEEHAN, J.J.; HAMILTON, B.K.; LEVY, P.F. Pilot scale membrane filtration process for the recovery of an extra cellular bacterial protease. A.C.S. Conference, 1988.

THOMSON,R.O and W.H FOSTER. 1970. Harvesting and clarification of culture-storage, Methods In Microbiology Volume 2 ,Academic Press, London.

TANNY G.B and D. HAUK, 1980, Filteration of particulates with a pleated thin channel cross-flow module, Separation Science and Technology, 15: 317-337

ZYDNEY, A.L; COLTON, C.K. A concentration polarization model for the filtrate flux in crossflow microfiltration of particulate suspensions. Chemical Engineering Communication 47, pp 1-21, (1986)

Fig. 1

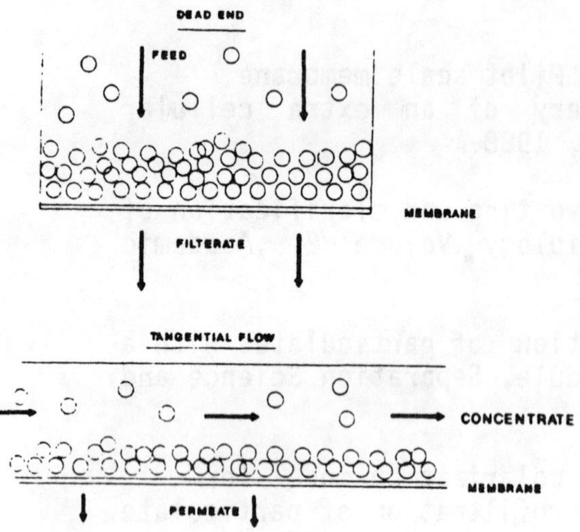

Fig. 2

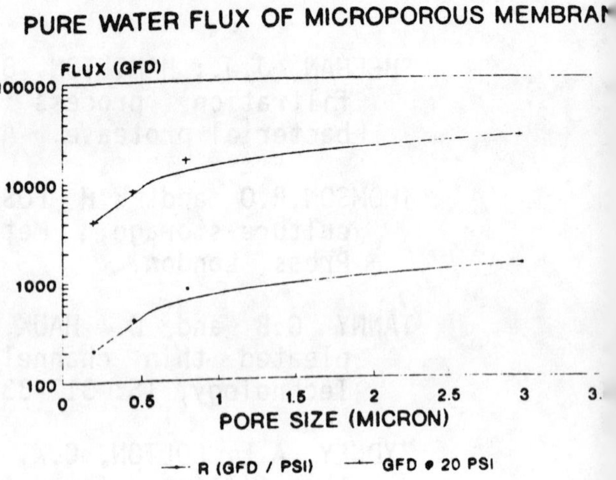

SOURCE: GELMAN

Fig. 3

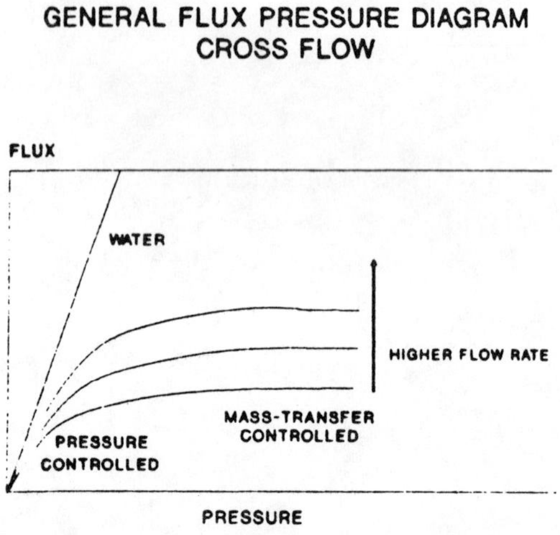

Fig. 4

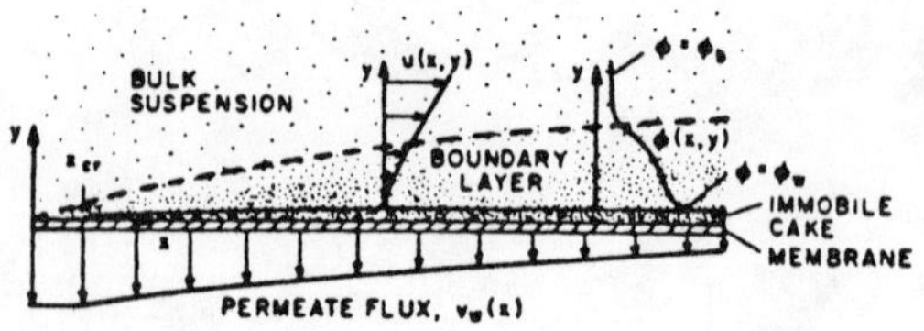

From:

Dr. Robert H. Davis
University of Colorado

Fig. 5

Zydney & Colton Equation

$$\bar{V}_w = 0.078 \left(\frac{a^4}{L}\right)^{1/3} \dot{\gamma}_o \ln\left(\frac{\phi_w}{\phi_o}\right)$$

$\bar{V}_w$	length averaged permeate flux, cm/sec
a	particle radius, cm
L	length of filter, cm
$\dot{\gamma}_o$	shear rate, sec^{-1}
ϕ_w	particle volume fraction at wall
ϕ_o	particle volume fraction at bulk

Fig. 6

CONCENTRATION LINES
EFFECT OF PARTICLE SIZE

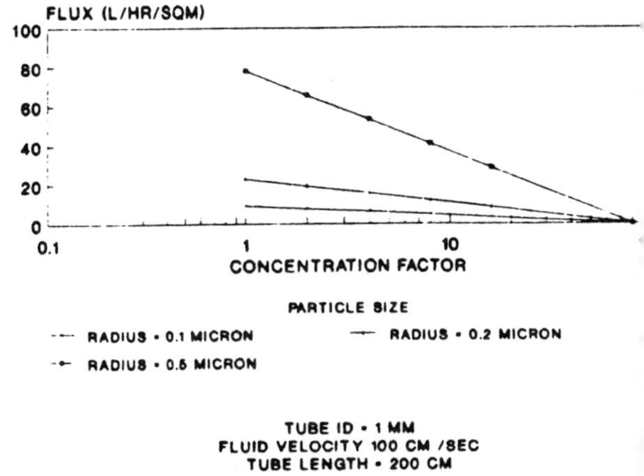

Fig. 7

AVERAGE FLUX VS. TUBE DIAMETER

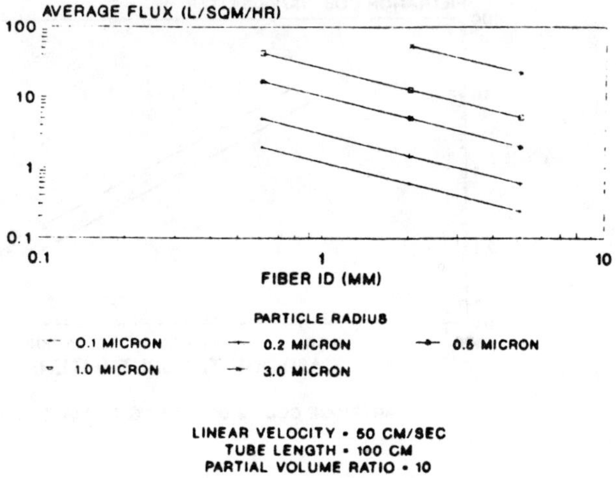

Fig. 8

CONCENTRATION LINES
EFFECT OF VELOCITY

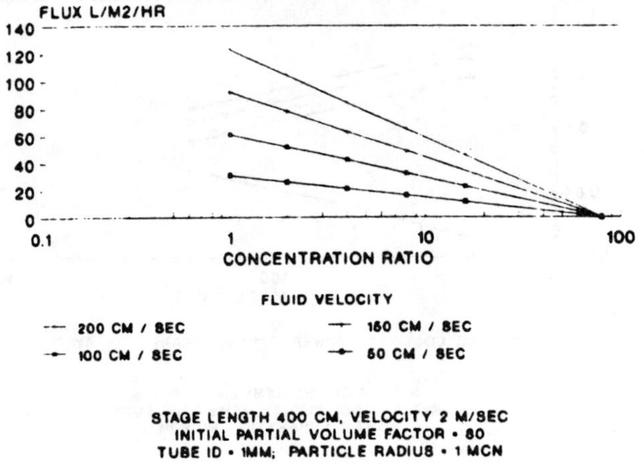

Fig. 9

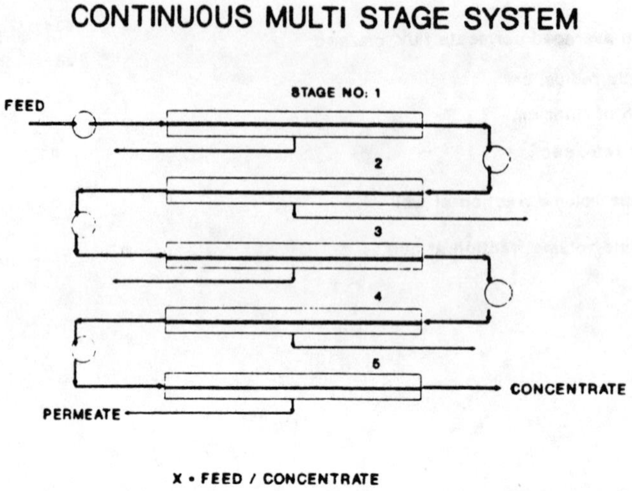

Fig. 10

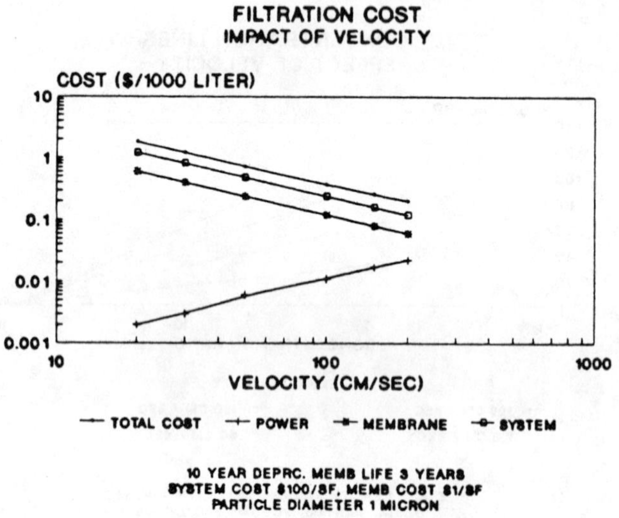

Fig. 11

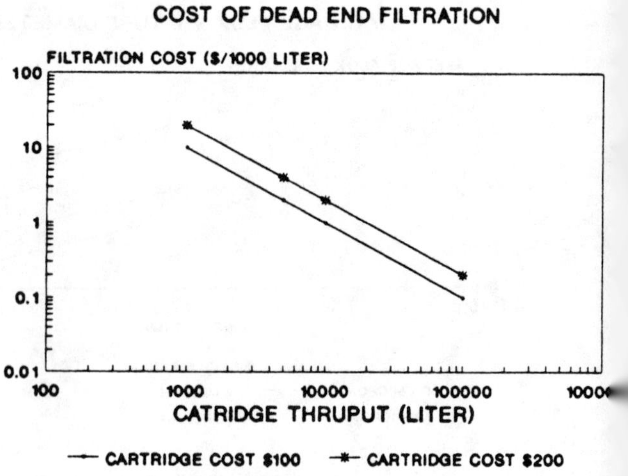

Table 1

PROCESS FLUX OF FEEDSTREAMS
CROSS FLOW MICROPOROUS FILTRATION

FEADSTREAM	FLUX RANGE (GFD)
EMULSIFIED CUTTING OIL	70 TO 105
SUSPENSION OF ALUMINIUM HYDROX	60 TO 100
GRINDING WATER	58 TO 98
DYE WASTE WATER	40 TO 70
VINEGAR FERMENTATION BROTH	17 TO 35
E. COLI BROTH	7 TO 29
MARINE BACTERIUM BROTH	5 TO 23

SOURCES: JOHNSON; SHEEHAN

American Filtration Society
Annual Meeting - March 19-21, 1990

Abstract

"Cartridge Filtration in Downstream Processing"

Presented by:
Jerry Martin, M.S.
Senior Marketing Manager
Pall Ultrafine Filtration Company

Membrane and depth filter cartridges find many useful applications in process-scale bioprocessing and downstream purification. This presentation will describe efficient through-flow filtration schemes for high flow/low sheer cell and cell debris removal, alone and in conjunction with centrifugation and tangential flow filtration. Also discussed will be the use of membrane filters for virus removal. Additional topics are clarification and sterilization of intermediate feeds, buffers, and solvents to protect ultrafilters and chromatography columns, sterile gas/air/vent filtration, and final sterilization of purified protein products.

EMPIRICAL RELATIONSHIPS BETWEEN PORE SIZE AND AUTOMOTIVE SINGLE PASS EFFICIENCY IN NONWOVEN LUBE MEDIA

ANTHONY CARONIA, DIRECTOR TECHNICAL SERVICES

JAMES W MARTIN, SENIOR ENGINEER

ALLIED SIGNAL AFTERMARKET FILTER GROUP
55 PAWTUCKET AVENUE
EAST RPOVIDENCE, RI 02916

INTRODUCTION

The ability of an automotive filter to remove 10 to 20 micron diameter contaminants in a single pass has been shown to reduce abrasive engine wear and to increase the effective life of a vehicle's engine [6]. The current study presents the results of an investigation into the development of empirical relationships between porosity parameters such as weighted average (WAPS) and maximum pore to single pass efficiency (SPE) of nonwoven medias. These results are applicable under strictly specified; but, industry accepted test conditions delineated under SAE HS J806 [5] and using a standard filter design. Figures 1 and 2 show the filter cartridge and assembly designs. Figure 3 shows the SAE test stand schematics.

The pore distributions of the nonwoven medias were determined using mercury intrusion porisimetry. An estimate of the relative mean pore size was calculated and termed the weighted average pore size (WAPS). This parameter is a measure of the central location of the pore size distribution. A second parameter, called the maximum pore size, was determined using a different technique which is described under the experimental method section of this paper. These two parameters were sufficient to develop a high correlation to the SPE performance of our standard automotive lube oil filter.

The derivation of empirical relationships between flat sheet porosity and efficiency are not new. Kinsley [4] recently postulated, using a polygon model developed by Corte, that the effective size of the polygon (pore) decreases as the fiber diameter decreases and as the density of the media increases relative to a constant composition in the nonwoven. Kinsley found high linear correlations between fiber diameter and flat sheet efficiency for various types of fibers including cellulose, polyester and glass. The smaller the diameter of the fiber the higher the associated overall efficiency given a constant packing density. This work investigated the effects of 2 and 30 micron sized contaminants. High linear correlations between efficiency and fiber diameter were also found when using AC fine test dust as the contaminant. Johnson found similar results [3].

In related work Thomas and Kebea [7] found that the " maximum efficiency at any particle size would indicate the need for a finer denier , or higher basis weight in order to increase efficiency". This study found that decreasing the fiber diameter increased efficiency at every contaminant size analyzed (10 to 30 microns).

Thus empirical relationships between pore structures and nonwoven media particulate removel efficiency can be postulated based on the work described above [3,4,7] and form the basis of the present paper.

EXPERIMENTAL METHOD

Nine nonwoven medias consisting of combinations of polyester, cellulose and glass fibers with phenolic resin binders were analyzed. Each media was cured in a convection oven at 300 F for 10 minutes to cure and condition the samples prior to analysis.

Table 1 lists these medias associated with some key physical characteristics including maximum and weighted average pore values. Table 2 shows how a typical WAPS parameter is determined from the pore size distribution. The WAPS is calculated by multiplying the cell midpoint of the summarized distribution by the intruded fraction of the media associated with that pore size range. The result is a weighted average estimate of the pore distribution which is comparable to the average permeability of the media. The WAPS is apparently a more sensitive measure (high F-ratio) of the central location of the pore size distribution in a nonwoven media.

The mercury intrusion scans were conducted using the Autoscan-500 Porosimeter manufactured by Quantachrome Corporation located in Syosset, NY (516-935-2240). Each scan represented approximately 2 square inches in area. Two to three scans were run for each media.

Since mercury is non-wetting, the amount of mercury entering a porous material versus the applied pressure allows calculation of the pore size distribution assuming a cylindrical pore shape model. The smaller the pore diameter, the higher the pressure necessary for intrusion of the media. The Washburn equation provides the essential relationship [1]:

$$P = \frac{-4 \gamma \cos \theta}{d} \qquad (1)$$

where: θ is the contact angle.
γ is the surface energy.
d is the pore diameter.
P is the intrusion pressure.

Mercury has a contact angle of 130 degrees and a surface energy approximately 0.485 J/ sq meters. The pore size distribution is calculated from the intrusion sequence. The extrusion sequence provides information on the pore shape and open pore surface area through the degree of hysteresis of the extrusion curve. This information is related to the retention force experienced on depressurization [1]. Efficiency relationships are impacted by pore size and shape distributions. Capacity has been postulated to be related by the degree of tortuousity revealed by the hysteresis of the extrusion vs intrusion curves [1].

Maximum pore is determined (in our facility) by using a highly refined Kerosene called Deobase which has a surface tension of 25 +/- .5 dynes per centimeter and a specific gravity of .82 +/- .04. At the first bubble the pressure differential is recorded in inches of water.

The single pass efficiencies were accumulated for each media as part of differing design projects over a one year time frame. Each media was developed for a specific automotive lube oil application. Some applications required low SPE targets (50%) while others very high SPE results (95%). The media samples were from the same experimental slit pad (lot) as the original SPE performance tests.

The basis weight measurements were determined using a gram scale. The valley caliper measurements were determined using a Randall and Stickney Model 580 dial caliper gage with a 4 psi load applied to an area of .25 square inches. The percent glass was determined by burning off all other constituents of the nonwoven sheet at elevated temperature.

RESULTS

Table 3 shows the average SPE value with its associated WAPS value. This data was fitted to a linear regression model (r^2 = .97115) and is plotted in figure 4. The stepwise regression results are shown in table 4. The final model is shown in table 5. The regression model shows that 97.1% of the variability in the SPE results in the range of 53.7% to 96% can be explained as a linear combination of WAPS and maximum pore. The model is significant because two flat sheet porosity measurements provide empirical correlation to SPE performance for this key filter.

Working back towards the raw material constituents, we see in table 6 that the apparent density and the percent glass content of the sheet are highly correlated to SPE. This is logical since the average pore size tends to be smaller in sheets with higher percentages of smaller diameter fibers. Glass is typically such a fiber. The higher the density and glass content of a given media the smaller the expected WAPS and the higher the maximum pore size (pressure differential at the first bubble).

CONCLUSION

In lube oil filtration media, the WAPS and maximum pore appear to be good predictors of single pass efficiency in the SPE range of 54% to 96% when the media is challenged with 10 to 20 micron contaminant. This contaminant range has been shown to be responsible for increased abrasive engine wear [6].

The ability to quickly screen filtration media for their suitability in single pass efficiency functions prior to building a prototype filter will decrease the prototype development process. Alternatively, knowledge of the pore structure of a media in the context of standard test conditions and contaminant distribution will provide information useful in the design of lube oil filtration media and in the development of incoming quality control techniques.

REFERENCES

1. R.M. German, Powder Metallurgy Science, Metal Powder Industries Federation Princeton, NJ. 1984, pp.228-47.

2. H. Batchu, J.G. Harfield, and R.A. Wenman, " A Comparison of Filtration Performance Using Particle Counting and Pore Sizing Techniques", Fluid/ Particle Separation Journal. Vol. 2, No. 1, March, 1989, pp. 5-10.

3. Peter R. Johnston, "The Viscous Permeability of a Mat of Randomly Arrayed Fibers as a Function of Fiber Diameter and Packing Density", Fluid/ Particle Separation Journal. Vol. 2, No. 1, March, 1989, pp. 15-6.

4. Horman B. Kinsley, "The Relationship Between Fiber Diameter and Filtration", Tappi Nonwovens Conference Proceedings 1989, pp. 209-12.

5. Society of Automotive Engineers Oil Filter Test Procedure HS J806, June 1989, pp. 19-22.

6. " Correlating Lube Oil Filtration Performance Efficiencies With Engine Wear", Society of Automotive Engineers Publication No. 881825, 1988.

7. Thomas Bryan and Robert Keban, " Filtration Properties of Wetlaid Polyester Nonwovens", Tappi Nonwovens Conference Proceedings, 1988, pp. 229-36.

Figure 1   Standard Automotive Lube Filter

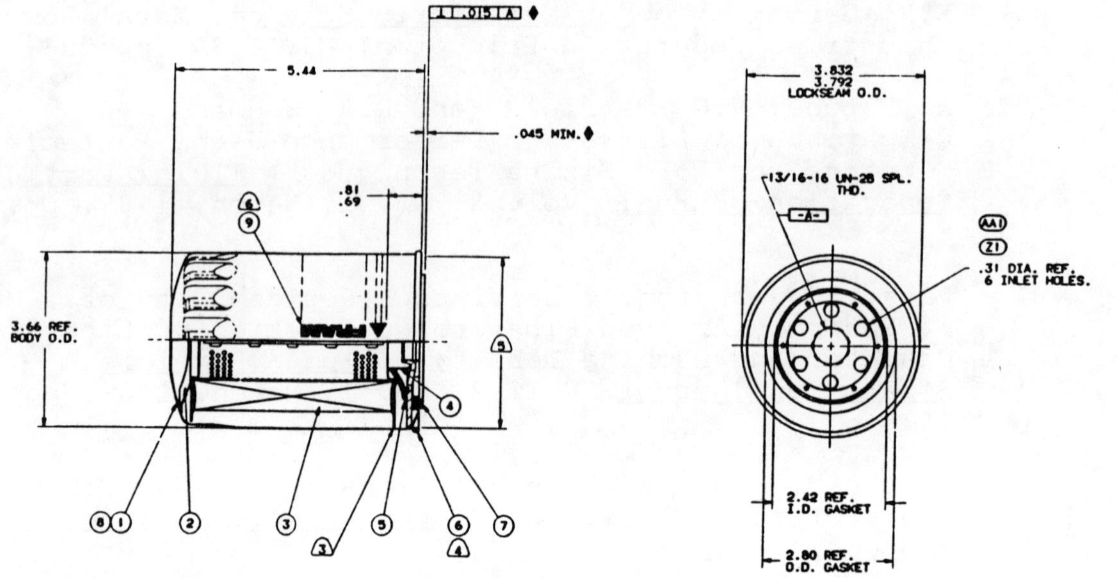

Figure 2   Associated Filtration Cartridge

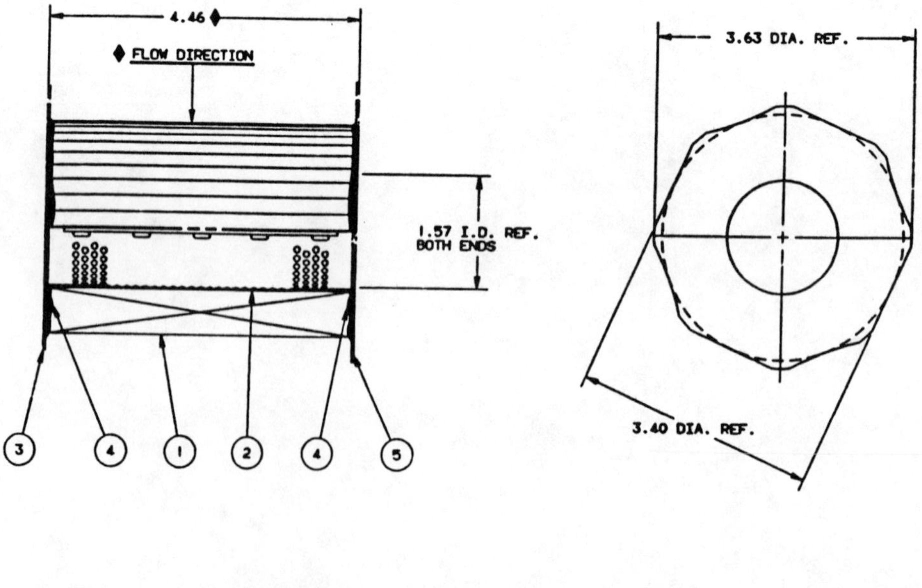

Figure 3 Single Pass Efficiency Test Stand

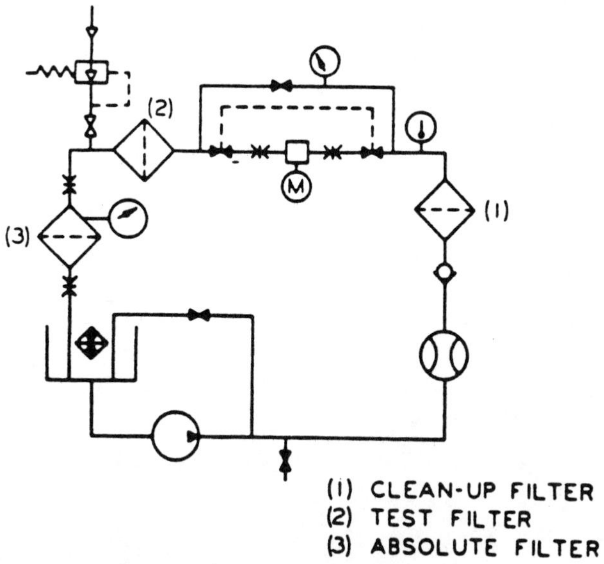

Fig. 7 - Single Pass Test Stand

$$\text{Test filter single pass efficiency (\%)} = \frac{W_1 - W_2 - W_4}{W_1 + W_3 - W_4} \times 100$$

where:

W_1 = weight of test contaminant presented to filter (paragraph 6.5)

W_2 = weight of test contaminant collected on absolute filter (paragraph 6.10) (that is, total weight on absolute filter - W_3)

W_3 = tare value from test system cleanliness (paragraph 6.3)

W_4 = weight of test contaminant left in mixer (paragraph 6.11)

Figure 4 Plot of Actual Versus Predicted Single Pass Efficiency

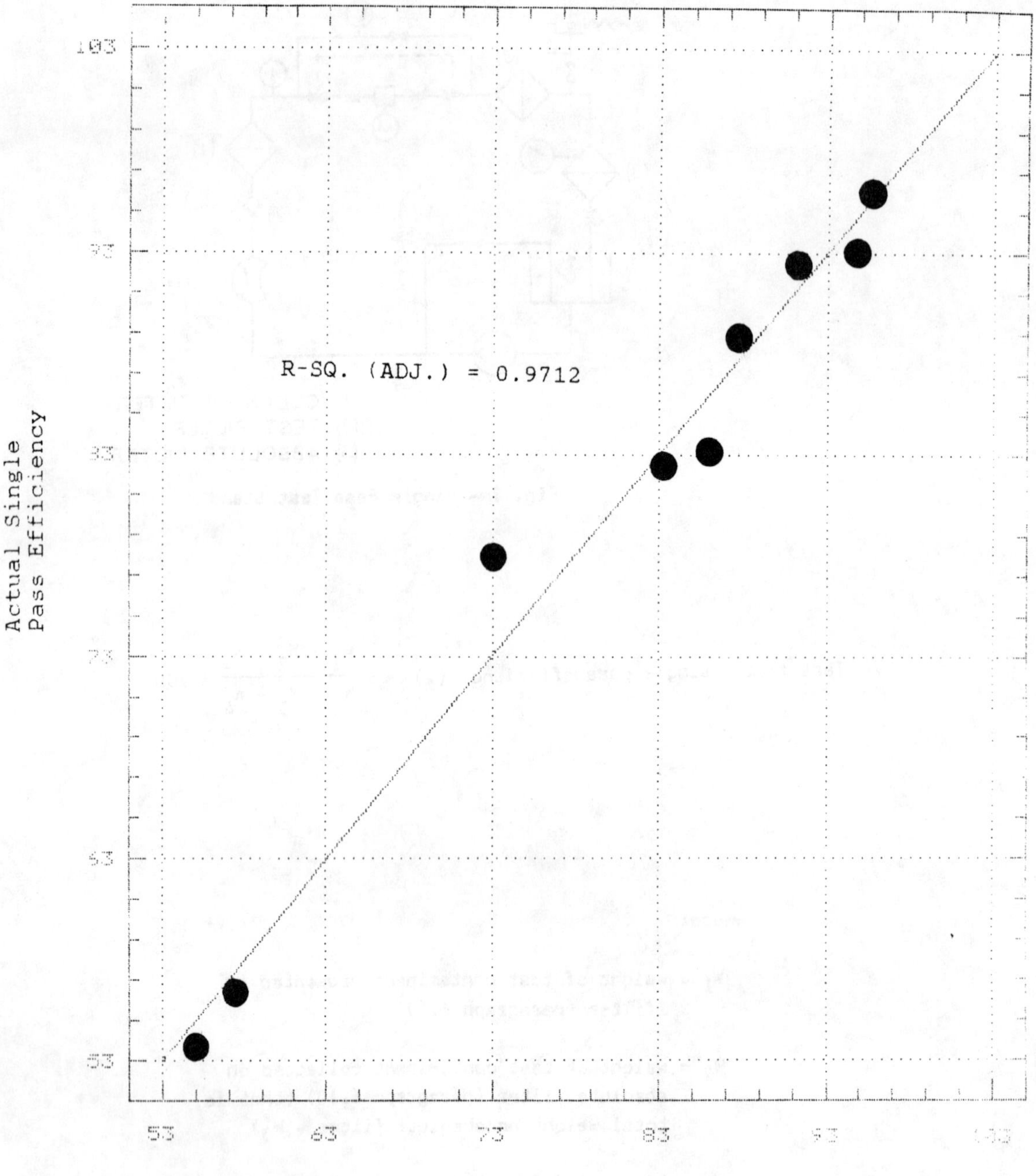

Predicted Single Pass Efficiency = -.989182 (WAPS) + 5.672209 (MAX) + 101.154

Table 1 Physical Properties of the Medias

Media Type	Basis Weight[1]	Caliper[2]	Density[3]	(WAPS)[4]	Max Pore[5]	% Glass	Mean Pore[6]
1	120.0	.035	3.591	49.0	5.4	6.1	44.5
2	131.1	.048	2.859	72.5	4.5	3.3	112.0
3	131.5	.037	3.719	58.3	5.2	0.4	45.6
4	153.5	.040	4.019	47.9	7.4	4.6	20.0
5	154.1	.041	3.936	45.4	6.8	10.8	32.7
6	143.7	.055	2.733	68.9	4.3	0.6	79.6
7	108.4	.032	3.549	51.4	6.3	6.7	42.0
8	119.1	.043	2.895	50.6	7.1	16.5	42.2
9	115.6	.037	3.260	50.9	6.5	14.8	44.0

Units:

[1]Basis Weight equals pounds per 3,000 square feet.

[2]Caliper equals inches.

[3]Density equals grams per square inch.

[4]Weighted Average Pore Size equals microns.

[5]Max Pore equals inches of water.

[6]Mean Pore (permeability) equals cubic feet per minute.

Table 2 Typical Calculation of the Weighted Average Pore Size For Media 3

Pore Size (microns)	Midpoint (2)	Intrusion Volume (3)	Fraction Intruded (4)	WAPS (2x4)
0 - 10	5	.017	.021	.105
11 - 20	15	.044	.054	.813
21 - 30	25	.052	.064	1.601
31 - 40	35	.063	.077	2.716
41 - 50	45	.119	.146	6.595
51 - 60	55	.120	.148	8.128
61 - 70	65	.143	.176	11.447
71 - 80	75	.078	.096	7.204
81 - 90	85	.077	.095	8.060
90 - 100	95	.016	.019	1.872
101 - 150	125	.059	.073	9.083
151 - 200	175	.024	.030	5.172
		.812	.999	63.336

Table 3 Single Pass Efficiency With Associated Weighted Average Pore Size

Media	Average SPE (%)	Average WAPS (microns)
1	82.5	49.0
2	53.7	72.5
3	78.0	58.3
4	96.0	47.9
5	93.2	45.4
6	56.4	68.9
7	83.2	51.4
8	92.4	50.6
9	88.0	50.9

Table 4 Results of Stepwise Regression Relating Single Pass Efficiency to the Maximum and Weighted Average Pore Sizes of the Medias

Selection: Forward	Maximum steps: 500	F-to-enter: 4.00	
Control: Automatic	Step: 2	F-to-remove: 4.00	
R-squared: .97836	Adjusted: .97115	MSE: 6.96041	d.f.: 6

Variables in Model	Coeff.	F-Remove	Variables Not in Model	P.Corr.	F-Enter
1. Maximum Pore Size	5.67221	12.6604	3. Mean Pore (Permeability)	.5403	2.0614
2. WAPS	-0.98918	27.6910			

Table 5 Results of The Model Fit For Predicting Single Pass Efficiency As a Linear Function of WAPS

Independent variable	coefficient	std. error	t-value	sig.level
CONSTANT	101.153649	19.109165	5.2935	0.0018
Maximum Pore Size	5.672209	1.594146	3.5581	0.0120
WAPS	-0.989182	0.187978	-5.2622	0.0019

R-SQ. (ADJ.) =	0.9712	SE=	2.638260	MAE=	1.672571	DurbWat=	2.029
Previously:	0.0000		0.000000		0.000000		0.000

9 observations fitted, forecast(s) computed for 0 missing val. of dep. var.

Table 6 Results of Stepwise Regression Relating Single Pass Efficiency to Percent Glass and Density of the Various Medias

Selection: Forward	Maximum steps: 500	F-to-enter: 4.00	
Control: Automatic	Step: 2	F-to-remove: 4.00	
R-squared: .90097	Adjusted: .86796	MSE: 31.8591	d.f.: 6

Variables in Model	Coeff.	F-Remove	Variables Not in Model	P.Corr.	F-Enter
3. Density of Media	22.9761	30.4077	1. Basis Weight	.0237	.00[illegible]
4. Percent Glass	1.81016	27.6396	2. Caliper of Media	.0063	.00[illegible]

THEORETICAL MODEL OF PRESSURE DROP LOSS IN DUST LOADED FIBROUS FILTERS

Tadeusz Ptak, Department of Physics,
University of Virginia, Charlottesville,
VA 22901
Tadeusz Jaroszczyk, Corporate Research Dept.,
Nelson Industries, Inc., P. O. Box 600,
Stoughton, WI 53589-0600 608-873-2423

INTRODUCTION

An external dust cake, typical of surface type filters, is unlikely to occur in high porosity depth type nonwoven filter media. Due to great porosity, dust particles are accumulated inside the filter. This type of filter loading has a "volume" character.[1] Generally, two types of dust distribution on fiber surfaces are distinguished to represent the filter clogging process: dendrite-type distribution and uniform distribution.

For very high porosity filter media operating under high flow velocities, the dust mass distribution on the fiber surface is more likely to be uniform.[2] An example of this type of media is prefilters for ventilation systems and some engine air filters.

In the "volume" model of dust loading, particles are deposited inside the filter media.[1] Dust particles are also uniformly distributed on a fiber surface. Filter solidity in this case is a sum of the pure filter solidity and volume of dust particles. Since particles are deposited on the fiber, the corresponding fiber aerodynamic diameter changes with time. By introducing the hydraulic radius of a clogging filter into the Fuchs-Stechkina equation,[3] Juda and Chrosciel[1] developed a mathematical model to describe filter restriction in the non-stationary region of filtration. However, this model does not include dust mass distribution into the filter. Therefore, it may be used only to determine pressure drop of thin, high porosity filter media.

EXPONENTIAL MODEL OF DUST MASS DISTRIBUTION INTO THE FILTER

Assuming uniform dust particle distribution on the filter surface and a non-linear dust mass distribution across the filter, the aerodynamic drag of unit length of a fiber perpendicular to the flow direction may be expressed by an equation

$$dF(x) = C(x)\cdot\rho_o\cdot v^2(x)\cdot d_f(x)dL \qquad (1)$$

Since the fiber length in a filter with a unit surface area, solidity β, and thickness, b, is equal to

$$L = 4\beta b/\pi d_f^2 \qquad (2)$$

and

$$dL = \frac{4\beta}{\pi d_f^2}dx$$

Therefore, the increase of filter restriction, $d\Delta P$, may be expressed as:

$$d\Delta P(x) = C(x)\cdot\rho_o\cdot v^2(x)\cdot d_f(x)\cdot\frac{4\beta}{\pi d_f^2}dx \qquad (3)$$

The influence of neighboring fibers is included by introducing a coefficient $C\beta(x)$. $C\beta(x)$ is a function of solidity $\beta(x)$. $C\beta(x) = f[\beta(x)]$. As a result, the equation describing pressure drop increase will have the form:

$$d\Delta P = C(x)\cdot f[\beta(x)]\cdot\rho_o\cdot v^2(x)\cdot d_f(x)\cdot\frac{4\beta}{\pi d_f^2}dx \qquad (4)$$

The exponential dust mass distribution across a filter, $M(x)$, is given by:

$$M(x) = \frac{Mp}{1-e^{-z}}\cdot(1-e^{-zx/b}) \qquad (5)$$

and

$$dM(x) = \frac{Mp}{b}\cdot\frac{z}{1-e^{-z}}\cdot e^{-zx/b}dx = \rho(x)dx.$$

It is obvious that $\int_0^b \rho(x)dx = Mp$ (6)

if $z=0$, $\rho(x) = Mp/b$. This is a case of a uniform dust mass distribution across a filter.

The exponential distribution is typical for dry aerosol filtration through depth-type nonwoven synthetics. Previous results have shown that this phenomenon depends on filter porosity[4]. The first layer of the filter accumulates 50-90% of the dust, the second 5-20%, while the last 0.5-5%. This specific experiment was performed on five-layer synthetic nonwovens with the following parameters: fiber diameter = 23-27 μm, layer thickness = 7-10 mm, filter solidity = 0.006 to 0.012, and aerosol velocity = 1 m/s. Silica fine test dust was used at terminal loadings of 1000-1400 g/m^2. It was learned that as the filter porosity becomes higher, the dust mass distribution across the filter becomes closer to a linear function.

FILTER RESTRICTION FOR EXPONENTIAL DUST MASS DISTRIBUTION

For uniform distribution of dust particles on the surface of a fiber with exponential decreasing dust mass distribution into the media, changes in filter solidity and fiber geometric diameter during dust loading may be described as follows:

$$\beta(x) = \beta\cdot\left(\frac{mp\cdot z\cdot e^{-zx/b}}{(1-\varepsilon)\cdot\rho_p\cdot\beta\cdot b\cdot(1-e^{-z})}+1\right) = \alpha\cdot e^{-zx/b}+\beta \tag{7}$$

and

$$d_f(x) = A_o(x)\cdot d_f\left(\frac{\beta(x)}{\beta}\right)^{0.5}$$

where $\alpha = \dfrac{mp\cdot z}{(1-\varepsilon)\cdot\rho_p\cdot b\cdot(1-e^{-z})}$ and $mp = \dfrac{M_p}{S}$.

In general, $f[\beta(x)] = [\beta(x)]^n$ (8)

The value of index n is in the range of 0.35-0.55. For this work it was assumed to be n=0.5.

A coefficient $A_o(x)$ can be counted based on the aerosol and filter parameters responsible for dust particle deposition on the fiber surface. The most important parameters are dust particle size, particle density and aerosol velocity. This coefficent may be described by the equation

$$A_o(x) = 1-K_p\cdot\alpha\cdot e^{-zx/b}\cdot Rep \tag{9}$$

The value of $A_o(x)$ decreases with increasing particle momentum. With high aerosol velocity the particles are mainly collected on the fiber face exposed to normal flow.

By including all assumptions, the increase of pressure drop across a dust loaded filter is given by

$$d\Delta P(x) = \rho_o\cdot C(x)\cdot[\beta(x)]^{0.5}\cdot\left[\frac{Vo}{1-\beta(x)}\right]^2\cdot d_f(x)\cdot\frac{4\beta}{\pi d_f^2}dx \tag{10}$$

$\beta(x)$ and $d_f(x)$ have been expressed by equation 7 whereas the C(x) coefficient depends on the type of flow. Generally, it can be given as

$$C(x) = \frac{Kd}{Re(x)} + B \tag{11}$$

By knowing the increase of pressure drop $d\Delta P(x)$, the total pressure drop across a dust-loaded fibrous filter is given by

$$\Delta P = \int_o^b k\cdot d\Delta P(x) \tag{12}$$

When $m_p = 0$, $\Delta P=\Delta P_o$ = initial pressure drop of a pure filter.

SUMMARY

A mathematical model was developed to describe pressure loss for typical high porosity volume type filters. This model predicts filter restriction performance for filter media with uniform dust particle distribution on fiber surfaces and exponential dust mass distribution into the filter. With uniform particle dust distributions, the typical dust cake formed on fiber surfaces has an eccentric offset shape with the majority of particles collected on the leading half of the fiber surface exposed to aerosol flow.[1] Also, with exponential dust mass distribution across the filter, the mass of deposited dust decreases exponentially with the thickness of the filter. Future work is ongoing to quantify and explain the behavior of the dust mass distribution index z and other constants. Determination of z index and constants Kp and k for assumed type of aerosol flow will be the subject of a future publication.

NOMENCLATURE

$A_o(x)$	coefficient of non-uniformity in dust deposition
B,Kd,Kp,k	constants
b	filter thickness
C(x)	drag coefficient of an isolated fiber
$C\beta(x)$	coefficient of the influence of neighboring fibers
d_f, $d_f(x)$	pure fiber diameter, diameter of dust loaded fiber
L	fiber length in the filter
Mp	mass of dust accumulated in the filter
M(x)	dust mass distribution across the filter
mp	dust mass loaded on a unit of filter surface area
n	index of the function describing influence of neighboring fibers
Re, Rep	Reynolds number, particle Reynolds number
S	filter surface area
V_o	filter media face flow velocity
v(x)	flow velocity in the filter
z	mass distribution parameter
β, $\beta(x)$	initial filter solidity, solidity of dust loaded filter
ΔP_o	initial pressure drop
$\Delta P(x)$	pressure loss in a dust loaded filter
ε	dust porosity
ρ_o, ρ_p	aerosol density, dust particle density
$\rho(x)$	density of dust mass distribution

LITERATURE

1. J. Juda and S. Chrosciel, "Ein theoretisches Modell der Druckverlusterhöhung beim Filtrationsvorgang," Staub-Reinhalt, Luft 30(5): 196 (1970).

2. T. Jaroszczyk, "Experimental Study of Nonwoven Filter Performance Using Second Order Orthogonal Design," Part. Sci. and Technol. 5: 271 (1987).

3. N. A. Fuchs and I. B. Stechkina, "A Note on the Theory of Fibrous Aerosol Filters," Ann. Occup. Hyg. 6: 27 (1963).

4. T. Jaroszczyk, J. Lis, and J. Nowicki, "Filter Performance of Lentex-type Nonwovens," Ochrona Powietrza (Air Protection) 4: 94 (1980).

The Care and Filtration of a
Phosphate Ester Gas Turbine Lube System

Leslie R. White
Senior Research and Development Engineer
The Hilliard Corporation
100 West 4th Street
Elmira, NY 14901
607-733-7121

ABSTRACT

The care and filtration, otherwise known as contamination control, in triaryl phosphate ester fluids for gas turbine lubrication has a dual purpose. Maintaining the correct fluid cleanliness will both protect the vital clearances of the mechanical components to provide a trouble free and long service life and also extend the life and stability of the vital fluids that lubricate and/or power control systems. Solid particulate filtration , water removal, and acid control methods are discussed as the contamination control measures to accomplish this. Also covered are the recommended physical layout of the filter systems. The history and application of adsorbants such as fullers earth, activated alumina, and Selexsorb[R] GT are also covered.

INTRODUCTION

Turbine lube oil is the life-blood of a turbine system. The turbine lube oil must not only provide a lubricating film between bearing surfaces; it must also transfer heat, and transmit power.

Phosphate ester fluids are often selected as gas turbine lubricants based on both their excellent lubricity and fire resistance. When subjected to flame, they will not continue to burn once the source of ignition is removed.

The prime disadvantage in using phosphate ester as a turbine lube is the high cost of up to five times that of petroleum based lubrication fluids plus the disposal costs once contaminated.

The solution to this dilemma is to extend the life of the phosphate ester through proper care and filtration so that it is an economically viable option. Triaryl phosphate esters have demonstrated the potential for virtually unlimited life when acidity, moisture and solid contaminants are controlled with side loop fluid conditioning systems.

The selection of the correct type of oil filters or conditioner on a turbine lube system will both extend the life of the turbine and extend the life of its phosphate ester lube oil.

TURBINE LIFE

The key to extended turbine bearing life is in controlling solid particulate contamination that can wear or score main and thrust bearing journal surfaces. As particles abrade surfaces within a bearing clearance, more particles are generated to cause further wear in an accelerating process. These solid particles not only cause bearing problems, they can also jam electro-hydraulic control (EHC) valves. Centrifugal force in high speed, lubricated couplings forces solid particulates to concentrate in the coupling, inducing accelerated wear in this area.

Filtration is the process of removing suspended contaminants, usually by passing the oil through a porous medium to break this cycle of tribological wear. Each portion of a system has its unique filtration requirements. One has to consider the maximum particle size that can be tolerated without causing undue wear, and select a filter capable of maintaining the required cleanliness level. Table 1 shows the cleanliness levels as SAE Class numbers and the equivalent ISO Code numbers. Westinghouse requires a SAE Class 2 cleanliness and most other manufacturers require a Class 3 or better. All new oil should be a Class 3 or better but could go as high as a Class 6 in an "as received" condition. For this reason, one should always practice the golden rule of fluid handling. "Always filter the fluid anytime it is pumped" even if it is only to transfer it from a drum or one tank to another.

One may consider a turbine as a relatively clean running machine. Solid contaminants, however, may enter the system with the introduction of makeup oil if the golden rule of filtering the oil anytime it is pumped has been broken. Particulates may also enter through reservoir breathers or be generated from within the system. Wear generates particles that create more particles that create more wear and so on. Fluid oxidation from hot spots can also create deposits of solid matter. This continuous influx of particles into, or generated by, a system is called ingression. The amount of particles per unit of time is the ingression rate. Under very clean conditions, an ingression rate of 5×10^6 particles per minute might be expected. Under operating conditions that may include blowing sand or industrial and construction dirt, an ingression rate of 5×10^8 or more is possible.

Given the ingression rate (R) and the filter efficiency, the system cleanliness can be predicted mathematically, or given the desired cleanliness, the filtration efficiency requirements can be calculated. Filtration performance is expressed as the ratio of the number of particles per unit of measure of a given size and larger, upstream of a filter (Nu) to the number of the same sized particles per unit of measure downstream of the filter (Nd). This ratio is called the beta ratio. The beta ratio for a hypothetical filter with 1000 particles per ml greater than ten microns upstream and 100 particles per ml downstream would be expressed as follows:

$$B = Nu/Nd$$
$$B_{10} = 1000/100 = 10$$

Where:

B = Beta ratio
R = Ingression rate
Q = Flow rate

The particle population immediately downstream of the filter can be expressed as follows:

$$Nd = R/([B-1][Q])$$

A lube or hydraulic system will stabilize at a particle population that is in equilibrium with the ingression rate and removal rate. Therefore, Nd when plotted against time becomes a steady state value. In an actual system with an off-stream filter loop or fluid conditioning console, the reservoir contamination level would be expressed as:

$$Nu = (B)(Nd)$$
$$Nu = (B)(R)/([B-1][Q])$$

Figure 1 shows reservoir cleanliness as a function of flow rate for three filtration efficiencies that might be typically used for turbine applications. Nu is plotted as the number of particles per ml greater than 15 microns in size using the formula above with an ingression rate of 5×10^7. The filter element designated $B_{15} = 1.5$ will reduce this contamination ingression load to 10,000 particles greater than 15 microns per 100 ml at 400 GPM. By increasing the flow to 600 GPM, the particle population is reduced even further to 6600 particles per 100 ml, making it a good choice for an SAE Class 3 requirement. The element labeled $B_{15} = 2.5$ will maintain a reservoir population of 3700 particles per 100 ml at 600 GPM qualifying for Class 2 service at this ingression rate. The $B_{15} = 500$ will keep it down to 2200 particles per 100 ml which is probably cleaner than most requirements dictate for lube oil. For EHC oil and lubricated couplings, however, this would be a very good choice.

By carefully tailoring flow rates and filter efficiency to system cleanliness requirements and particle ingression rates, turbine bearing life will be maximized. Corrosion, however, may also affect bearing life as much or more than solid particle wear. Corrosion control is a function of acid and water control which will be addressed in the following section on oil life.

OIL LIFE

To achieve long fluid life, the focus of contamination control will be on acid and water removal or control. Acid phosphates and water can corrosively attack bearing materials and destabilize the oil. Phosphate ester oil stability is defined as the ability of the synthetic oil to endure extended operating conditions without the loss or change of vital fluid properties. Table 2 lists the physical properties of the fluid that are typically evaluated in an active monitoring program. Most of these properties relate to how the fluid will function. The acid number, however, is the key to long life for a synthetic phosphate ester turbine lube oil. When contamination is monitored and controlled, it is possible to operate for over 50,000 hours with no significant problems. Ideally, a phosphate ester turbine lube oil should be maintained at total acid number (TAN) of .05. A TAN of .2 to .3 is considered the maximum acidity that is feasible to recondition. As total acid numbers rise above the .3 to .5 range, the acid number increase will begin to accelerate rapidly, making further control difficult if not impossible. In an advanced state of hydrolysis, or chemical breakdown as indicated by the higher TAN, the fluid becomes very unstable. Acidity promotes more acidity; that is the breakdown reaction feeds upon itself auto-catalytically at a very rapid rate. High acid number oil is corrosive and can erode servo valves and bearings. Symptoms associated with a high acid oil include excessive foaming, varnish build-up on valves, and polymeric gel formation. Acid phosphates, pyro phosphates, and higher polymers are generated and phenols and alkylphenols may also be generated.

Users of phosphate ester fluids should work closely with their fluid representative on monitoring, conditioning, replacing or even replenishing their oil. For example, if an excessive foaming condition exists, one might be tempted to simply add more anti-foam additive. An overbalance of additive, however, could in itself create more foaming and only add to the problem. The only successful way to treat this problem is to bring the acid number under control first, then anti-foam additive can be carefully added under the direction of your fluid representative if still necessary.

To reduce acidity, fresh fluid is often added to dilute it to acceptable levels. Unfortunately, this method will only have a temporary effect. Because the degradation reaction is auto-catalytic, the new oil simply feeds the reaction and is quickly degraded.

RECLAMATION

The first line of defense for maintaining phosphate ester fluid stability is to keep the total acid numbers less than .2 and preferably between .05 and .1. There are two schools of thought on the best method to reduce the acid level. One is to allow the TAN to reach an upper limit; then condition the oil to bring it back into specification. The other way is to continuously condition the oil so as to maintain it at a constant low TAN. The most effective method is to use a constant fluid circulation through an adsorptive media. Continuous treatment prevents the fluid from reaching a dangerous TAN level in the first place. By maintaining a minimum acid level, the adsorptive media need only be changed when the TAN begins to rise signaling that the adsorptive media is saturated. Figure 2 shows that at an initial TAN of .1, a dosage of one half pound of activated alumina per gallon of fluid would bring the TAN down to .05. At a starting point of .5, however, it would require over three times as much activated alumina to bring the TAN down to .05. Constant treatment is more effective because it doesn't allow the acid number to ever get high enough to trigger the self-destructive autocatalytic reaction. This, in turn, prevents the formation of problem causing polymers and varnishes.

ADSORBENTS

The adsorbents used for phosphate ester conditioning are fullers earth, activated alumina, and Selexsorb^R GT. Fullers earth was the original adsorbent to be used for phosphate ester conditioning, coming into use in the early 1950's. Activated alumina became the adsorbent of choice in the 1970's and now most recently, a new material called Selexsorb^R GT developed by Ackzo Chemical Inc. specifically for treatment of phosphate ester turbine lubes is beginning to make inroads upon activated alumina usage.

Fullers earth is still widely used although it is not as effective as activated alumina in removing acid phosphates. Problems with metallic salts leaching out and reacting with the base oil to form varnishes and gels led to the joint development of activated alumina by General Electric, Stauffer Chemicals (now Ackzo Chemicals Inc.), and The Hilliard Corporation, for phosphate ester turbine lube conditioning. Activated alumina reduces acid phosphates through neutralization of the media surface as does fullers earth. Activated alumina, however, was found to be 2-3 times as effective as fullers earth in its acid reduction ability. Table 3 is composed of data from acid reduction comparisons of fullers earth vs. activated alumina on phosphate ester lube oils taken from three different turbine sites.

Although activated alumina was a big improvement over fullers earth, there were still some problems. Because fullers earth and activated alumina are both good dessicants, they would attract moisture from the oil which would occupy active sites in the media which in turn reduces the available space for acid phosphates. Also, in some systems, polymer formation from the reaction with metallic salts in the activated alumina still presented a problem. This led Ackzo Chemical to work with Alcoa in the development of an enhanced form of activated alumina called Selexsorb^R GT, that solved both of these deficiencies. Selexsorb^R GT is similar to activated alumina in its acid removal rate but instead of neutralizing the acid phosphates, it chemically bonds them to the media surface. This prevents the liberation of metallic salts thus eliminating the gel formation associated with them. Selexsorb^R GT also will remove water but will preferentially bond acid phosphates first. This means that water will not cause a reduction in the ability to hold acid phosphates which can greatly extend cartridge life.

When using either activated alumina or Selexsorb[R] GT, a sideloop filtration system using packed cartridges as shown in Figure 3 is recommended. Activated alumina typically would be used at a dosage of .5 lb per gallon of oil in the system to be conditioned at a flow rate of .5 to 2 GPM per 718 sized cartridge. The lower the flow rate the better. Selexsorb[R] GT is used at .1 per gallon of fluid to be treated at a flow rate not to exceed .5 GPM. With either material a .5 micron nominal particulate trap filter downstream with a beta ratio of 75 for particle sizes 5 microns and larger is recommended. The trap filter is placed in series with, and downstream of, the adsorbent filter. Because Selexsorb[R] GT preferentially adsorbs acid to water and even generates water as it chemically bonds acids to its surface, a water absorbent particulate filter should be used in series downstream of it.

WATER REMOVAL

Water in phosphate ester fluids accelerates hydrolysis, causing an increase in the total acid number. The higher the water content, the higher the rate of hydrolysis. Therefore, to successfully maintain control of the acidity, one also has to control the moisture content. It is recommended that the moisture content not exceed .1% to .3%. A water content higher than .15% could indicate the possibility of water entering from an outside source such as a leaking heat exchanger or defective breather.

The use of adsorbents such as fullers earth and activated alumina has been the traditional approach to moisture control. As pointed out previously, however, they both will preferentially adsorb water. To maximize their acid reduction ability, it is best to use a water absorbent or dryer type filter cartridge to handle the water removal load. Another reason for using a water absorbent cartridge is that when activated alumina or fullers earth becomes saturated, they can desorb water back into the system. Also there is no way to tell when they are saturated with water until the monitoring program picks it up or water related problems begin to occur.

Modern water absorbent particulate filters combine pleated filter media in various filtration efficiencies with a super-absorbent polymer that chemically bonds the water in the media so that it can not desorb when saturated. The super-absorbent polymers also show a marked pressure increase at their saturation point to signal that a changeout is required. The advantage in using these super-absorbent particulate filters is that they enhance the effectiveness of the acid control adsorbents by providing a dry oil that cannot sustain a hydrolytic breakdown reaction.

Dessicant breathers will provide dry air to the reservoir as fluid levels fluctuate in the reservoir. Phosphate ester fluids are hygroscopic and readily absorb moisture from the atmosphere.

Vacuum distillation will continuously remove water without requiring changing cartridges and avoids the cost of replacing cartridges. Where water is a chronic problem, vacuum distillation will cost less to operate although it will be a higher initial expense. An absorbent cartridge in a 718 size might typically hold about 2 liters of water. That means it costs the price of the cartridge plus labor for every 2 liters of water removed. Another related application of a vacuum reclaimer is to degas compressor seal oil.

CONCLUSION

Water, acids, and solid particulates are continuously entering or being generated within a turbine lube oil system. The most effective and economical method of control is a continuous filtration loop that can clean and condition even when the turbine is shut down or on standby. New filtration materials such as Selexsorb[R] GT and super-absorbent particulate filters can insure long life for both the turbine's lubricated parts and the phosphate ester oil that gets the job done.

REFERENCES

1. Anzenberger Sr., J.F., "Evaluation of Phosphate Ester Fluids to Determine Stability and Suitability for Continued Service in Gas Turbines," Lubrication Engineering, Vol 43, 7, 528-532, July 1987.

2. Young, W. C., Roberton, R.S., "Turbine Oil Monitoring," STP 1021, ASTM, 1916 Race St. Philadelphia, PA 19103.

3. Fitch, E. C., "An Encyclopedia of Fluid Contamination Control for Hydraulic Systems," Fluid Power Research Center, Oklahoma State University, Stillwater, OK (1979).

4. Shade, W. N., "Recent Field Experience With Degraded Synthetic Lubricants," Proceedings Compressor Workshop "Engineering Solutions to Field Problems," held in conjunction with the Annual Fall Meeting of the Pacific Energy Association at Anaheim, CA, 24-26 October, 1984.

5. Steele, F. M., "Contamination Control in Gas Turbine Systems," Paper No. 84-GT-155, ASME Gas Turbine Division Annual Meeting, Amsterdam, Holland, June 4-7, 1984.

6. Steele, F. M., "Filtration and Reclamation of Turbine Oils," Lubrication Engineering, Vol 34, 5, 252-257, May 1978.

TABLE 2 Standard Phosphate Ester Turbine Lube Monitoring Tests

Property	Test Method
Viscosity, m^2/s, at 37.8°C	ASTM D 445
Total Acid Number, mg KOH/g	ASTM D 974
Water Content, %	ASTM D 1744
Mineral Oil Content, %	ASTM D 1399
Particle Count per 100 ml	SAE-A-6D

Fig. 1, Reservoir cleanliness vs. flow

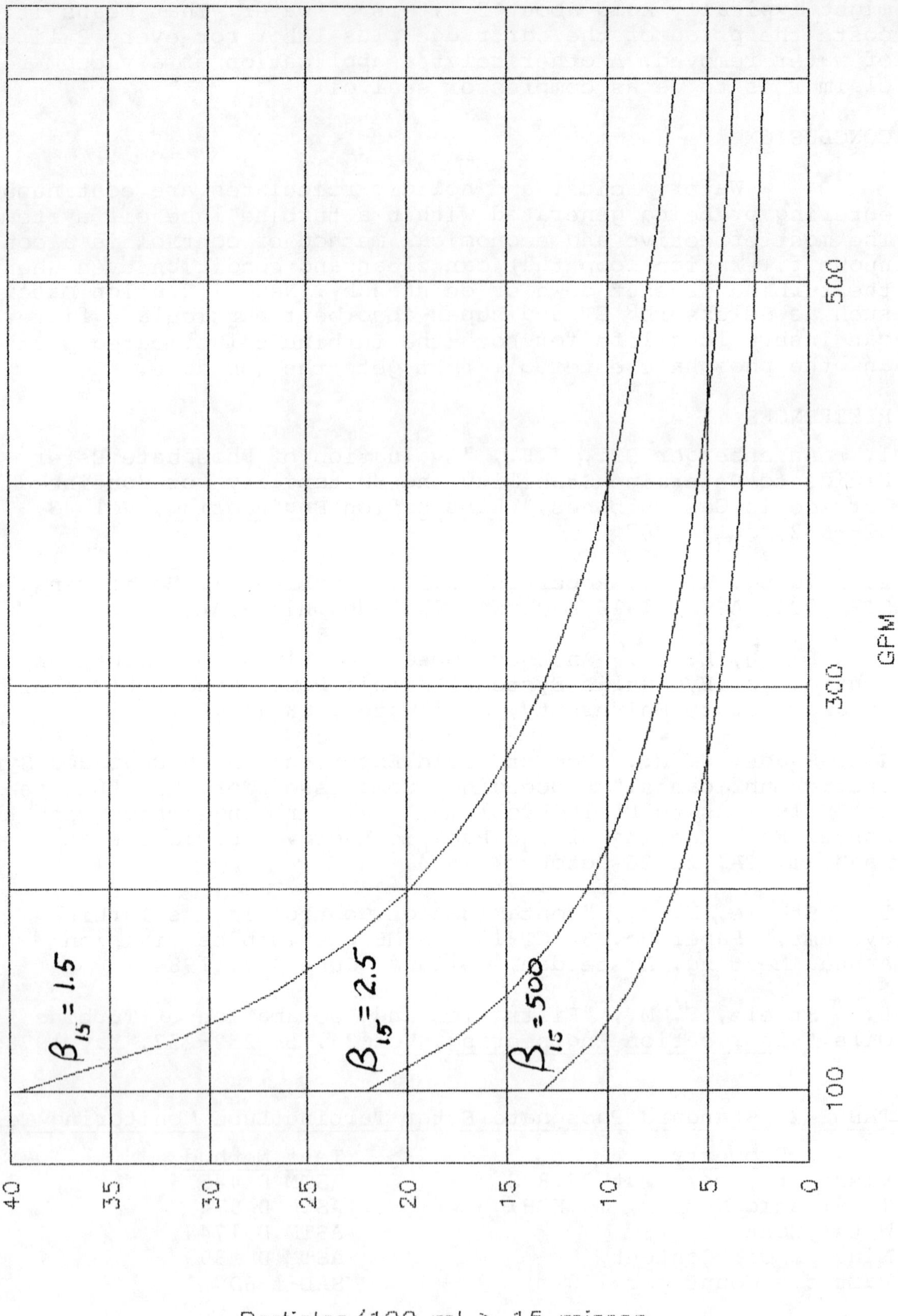

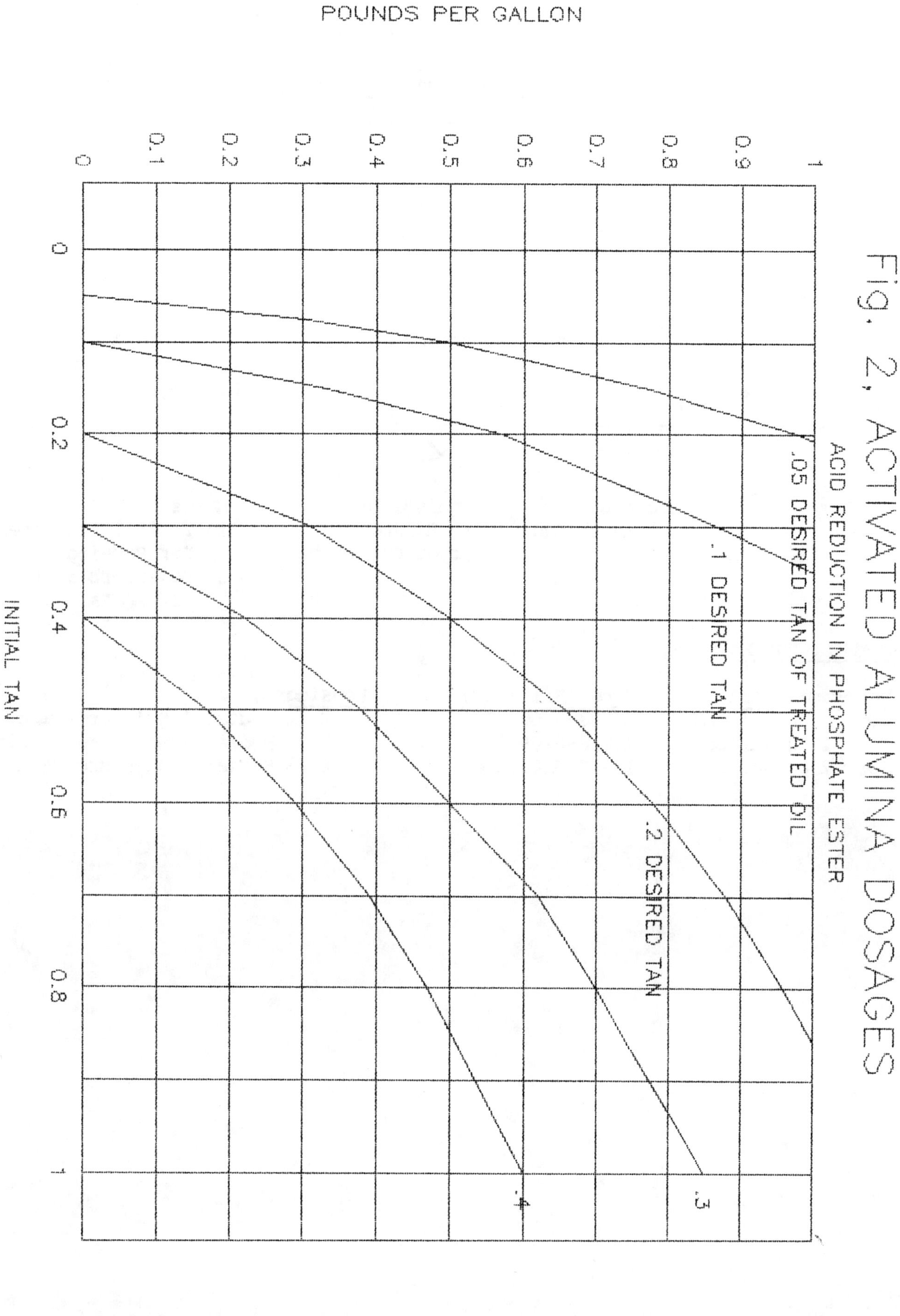
Fig. 2, ACTIVATED ALUMINA DOSAGES
ACID REDUCTION IN PHOSPHATE ESTER
POUNDS PER GALLON
INITIAL TAN
.05 DESIRED TAN OF TREATED OIL
.1 DESIRED TAN
.2 DESIRED TAN
.3
.4
0
0.1
0.2
0.3
0.4
0.5
0.6
0.7
0.8
0.9
1
0
0.2
0.4
0.6
0.8
1

FIGURE 3

FILTER SYSTEM SCHEMATICS/SELEXSORB GT)
HILLIARD DESIGN

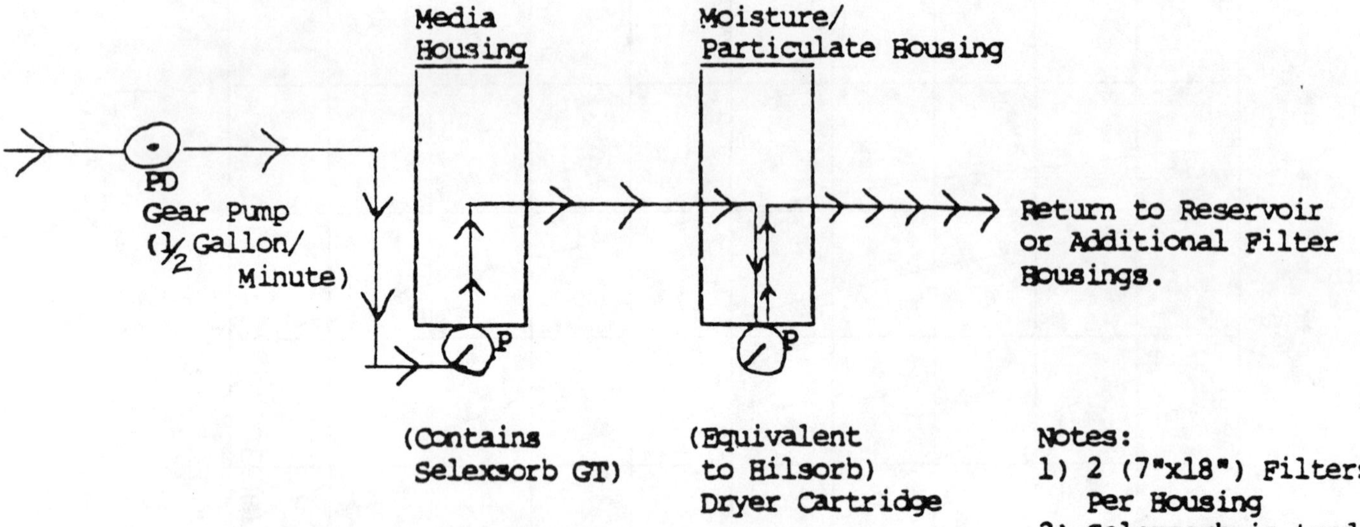

Notes:
1) 2 (7"x18") Filters Per Housing
2) Selexsorb is trademark of ALCOA.

RECOMMENDATIONS:

Reservoir Size	Type (Hilliard)	Housings
To 500 gallons	13718-0100-2A04	2 (1 Media; 1 Moisture/Particulates
To 2500 gallons	13718-0100-2A06	3 (2 Media; 1 Moisture/Particulates
Over 2500 gallons	13718-0100-2A08	4 (3 Media; 1 Moisture/Particulates

Table 1	SAE Contamination Criteria					
		Particle Counts per 100 ml.				
ISO Code	SAE-A-6D Class	Particle Size Range (Microns)				
		5-10	10-25	25-50	50-100	>100
12/10	0	2700	670	93	16	1
13/11	1	4600	1340	210	28	3
14/12	2	9700	2680	380	56	5
15/13	3	24000	5360	780	110	11
16/14	4	32000	10700	1510	225	21
17/15	5	87000	21400	3130	430	41
18/16	6	128000	42000	6500	1000	92

Table 3, Samples of used Fyrquel GT from different Frame 7001C Gas Turbine Engines were treated with varying amounts of fullers earth and activated alumina in a closed loop filter system.

Fluid	Original		Final	Metals Found, PPM					
Source	TAN	Treatment	TAN	Cu	Al	Fe	Zn	Mg	Na
Gallipolis Ferry	2.74	Untreated- circulated	2.4	236		1	110	10	
		0.53 lb. FE / gal.	1.84	189	25	11	140	50	
		1.1 lb. FE / gal.	1.44	145	45	25	150	80	
		1.67 lb. FE / gal.	1.19	111	38	19	130	70	
		Untreated- circulated	2.2	180	0		140	10	1
		.6 lb. Alumina / gal.	0.97	165	15		160	10	65
		1.3 lb. Alumina / gal.	0.35	63	13		150	0	36
Houston Texas	0.39	Untreated- circulated	0.34	19					
		0.54 lb. FE / gal.	0.2	10	3	2			
		1.11 lb. FE / gal.	0.1	9	2	2	20		
		1.72 lb. FE / gal.	0.04	5		1			
		Untreated- circulated	0.31	13	0		20	0	7
		0.6 lb. alumina / gal.	0.05	8	5		20	0	7
		1.73 lb. alumina / gal.	0.04	6	5		20	0	1
Waterloo Iowa	0.75	Untreated- circulated	0.67	25			30		
		0.47 lb. FE / gal.	0.39	17	11	6	20	20	
		0.98 lb. FE / gal.	0.24	9	11	7	10	20	
		1.5 lb. FE / gal.	0.13	3	7	3			
		Untreated- circulated	0.74	25	0		30	0	0
		0.27 lb. alumina / gal.	0.39	22	2		30	0	15
		0.64 lb. alumina / gal.	0.14	15	6		0	0	15

PREDICTION OF PRESSURE DROP PERFORMANCE IN AUTOMOTIVE AIR INDUCTION SYSTEMS

V. GURUMOORTHY, A. J. BAWABE, G. A. BROWN, R. C. LESSMANN
Fram/Allied Filtration Research Laboratory
Department of Mechanical Engineering
University of Rhode Island, Kingston Rhode Island, 02881

Abstract

A typical automotive air induction system contains several different components. Since it is relatively straight forward to predict the pressure drop for most of these; this paper examines approaches to analyzing the more complicated filter panel-plenum assembly. The use of a detailed computational fluid dynamics (CFD) simulation is discussed, and an example calculation of this type is presented for a geometrically simple plenum. This approach was found to be time consuming and requires a sophisticated understanding of the numerical methods employed in the chosen computer code if the results are to be meaningful. Alternately, a generic numerical method was developed which automatically generates a simplified body fitted computational grid for any filter plenum. Using this method is quite simple, however the computational time is still significant. Lastly, utilizing aerodynamic pressure loss coefficients, available in the literature, a hand computation is shown to provide reasonable results. When compared to experiments all of the aforementioned techniques were with in 10% to 15% of the data. Parametric studies are now under way to develop a customized loss coefficient model for use in the design process.

Introduction

The most critical aspect in the design of automotive air induction systems is overall pressure drop. Most of the system components, such as the ones shown in figure 1, are easily characterized by aerodynamic loss coefficients. These can be obtained from several fluid mechanics handbooks[1,2]. Thus, an induction system may be modeled as the summation of individual losses. Unfortunately, the filter panel and plenum cannot be handled so easily. Here we present three approaches to modeling an automotive filter plenum. First a detailed computational model of a simple geometry is studied as an example. Second, a generic, user friendly computational method for use in modeling any filter plenum is presented. This makes use of several geometrical approximations. Finally, a very simple approach which employs Bernoulli loss coefficients in a hand calculation is evaluated.

CFD Analysis

Typical automotive air induction systems work within a flow rate range of 100-350 cfm, consequently the flow is fully turbulent. The complicated geometry of filter plenums gives rise to sudden flow contractions and expansions, flow separations, and recirculation zones. The above difficulties, combined with the task of properly simulating the filter panel, makes CFD analysis difficult. We have used the commercial software code PHOENICS [3] to do our simulations, all of which were steady and three-dimensional, and used the k-ε turbulence model [4].

The governing equations are the continuity equation, the momentum equations and the turbulence equations. PHOENICS solves these for the primitive variables of velocity and pressure on a user input computational grid using the method of Spalding and Patankar [5,6]. Strictly speaking, the momentum equations do not apply inside the filter panel, where the flow is laminar, and a modified form of Darcy's equation [7] is applicable. An exact simulation of the filter panel-plenum assembly would involve dividing the problem into three zones: a plenum upstream of the filter, the filter itself, and a plenum down stream of the filter. This means solving three different problems and then patching the solutions together in an iterative fashion. Such an approach is very difficult and has not been attempted. Instead we used an approach by Vafai and Tien[7] which bridges the gap between the modified Darcy's equation and the Navier-Stokes equation. Accordingly, inside the region of the filter the momentum equation takes the form

$$\frac{\nu}{K}\vec{V} + \frac{b}{2}|\vec{V}|\vec{V} = -\frac{D\vec{V}}{Dt} - \frac{1}{\rho}\nabla P + \nu\nabla^2\vec{V} \qquad (1)$$

where K is an effective permeability, and b is an inertial correction, both determined experimentally. Taken together these completely characterize the filter element in our simulations. As the permeability K increases, and b tends towards zero, the left hand side of (1) vanishes, and the conventional form of the Navier-Stokes equation is recovered. The extra pressure drop terms on the left hand side of (1) may be included, for those computational cells comprising the filter, as an extra source term in the momentum equation. Thus the computer will solve the usual momentum equation for all cells outside the filter panel, and it will solve equation (1) inside the filter, making the simulation a one domain problem.

Detailed Analysis Of Simplified Geometry

This CFD technique was applied to a simplified rectangular filter-plenum geometry as shown in Figure 2. The unit consists of a long rectangular inlet duct which leads into a rectangular plenum just below the panel. The exit duct above the filter panel is also rectangular. The inlet duct is asymmetrically located, and runs for some distance into the lower plenum rather than exhausting flush with the walls. This geometry was chosen because it lends itself nicely to being described in Cartesian coordinates, hence simple rectangular griding is appropriate, and as an example it exhibits many of the flow pathologies common to real units. Dimensions were chosen to be characteristic of a certain real system, and an experimental model was constructed to allow for retrieval of data to verify computer predictions

Wall pressures and internal pressure profiles were determined experimentally and were compared with the CFD predictions. Simulations were performed for various flow rates and grid configurations. The average agreement with the measured wall pressures was within 10 - 15% and for the internal pressures agreement was within 5 - 10% . In this paper we have shown the results of the internal pressure comparisons for the 200 cfm case only. Readers are referred to reference 8 for further details.

Figures 3 and 4 show the comparisons of internal pressures. Figure 3 shows the pressure profile at traverse 1, figure 4 shows traverse 2 which was just below the filter panel. The solid lines on these figures show the lower and upper uncertainty limits of the computational results. These are due to the uncertainty associated with measuring the flow rate. The experimental data are shown as rectangles with their uncertainty bars. We see that except for one location in each figure the experiment and computation overlap.Traverse 3 was made above the filter panel and the agreement between the experiment and the prediction was within 5%. Thus from these results we conclude that CFD techniques do a good job of predicting the pressure drop across the plenum-panel assembly, at least for this example.

The disadvantage of this approach is that detailed knowledge of CFD techniques is required for such an analysis. Once the problem has been set up the solution must be converged. Convergence is usually a major problem, and the user has to spend time varying the relaxation factors before a physically meaningful result is finally obtained. The next problem is the time involved in the actual execution of the program. We ran the simple geometry for three grid configurations: 12000, 23000 and 65000 nodes. The convergence times were 8,10 and 20 hours respectively on a SUN 4/110 work station. The results improved by 3-4% from case 1 to 2 and the improvement was 1-2% from case 2 to 3. This clearly indicates the grid sensitive nature of the CFD technique. Thus the user also has to decide how finely he wants to grid the assembly. Lastly, griding is a problem in itself. Most real systems have geometries which are substantially more complicated then that in our simple example. The use of Body Fitted Coordinates would be a must for such panel-plenum assemblies. BFC is very involved, and the griding can be a project in itself.

Given the above, one must ask if it is worth the effort to perform a complete CFD analysis. The answer is probably no if all that is required is the overall pressure drop, however CFD can provide a great deal of other detailed information which may be of interest to the designer. As an example, the complete flow field is solved, and this may be used to analyze flow distribution through the filter panel, and to identify separated flow regions which lead to excessive pressure loss and noise generation. These latter considerations can have a major impact in the analysis of the acoustics of air induction systems. Figure 5 shows the flow distribution through the filter panel for our simple geometry. This figure shows the % Flow, and the corresponding % Area of flow for different velocity ranges R. As an example, R = 0.85 is the velocity range from 0.8 - 0.9 of the mean velocity for the plane of the filter.

The above arguments led us to believe that one could still use CFD techniques for pressure drop prediction, but not simulate each system in detail. Instead we identify a few critical geometric parameters and input them into a general purpose code, which then automatically generates the grid, performs the analysis, and predicts the pressure drop. This is the Generic CFD concept.

The Generic Model

Having seen the difficulties associated with developing complete, individual, numerical models, we endeavored to develop a computational model to suite any filter plenum assembly. Though this requires several geometric inaccuracies for each plenum modeled, the major parameters governing pressure drop performance are preserved. The majority of filter plenum assemblies utilize a box-like shape. Consequently, the generic model assumes a box geometry. Two basic configurations were required; one in which the inlet and outlet ducts are located in the end walls, figure 6, and another with the ducts in the side walls, figure 7. In the end/end model the calculation domain is a box whose cross sectional dimensions are taken to be those of the filter panel. The inlet and outlet ducts are located in the walls opposite the filter. Thus, by prescribing the location of the duct in the wall, as well as the distance to the filter, one has defined the entire domain. With the side/side configuration the box cross sectional area is still dictated by the filter panel, however the ducts are located in the side walls, adjacent to the panel. Here we prescribe the location of the duct in the side wall, and take the end walls to be located flush with the duct, see figure 7.

In the early stages of the generic model, three vector components were used to duplicate the inlet flow. This led to a multitude of problems. The computation would rarely converge and one could do nothing to set the outlet flow direction without over specifying the problem. The solution was to attach inlet and outlet ducts to the flow domain in order to allow the flow sufficient development length. To preserve the actual flow directions of the real plenum, a new model was created that utilized body fitted coordinate grids.

To model any given plenum assembly, one needs only to select the appropriate version and input several geometric parameters. The data entry phase is handled by an auxiliary program which prompts the user for filter dimensions and characteristics, as well as the inlet/outlet areas, locations, and flow directions. The computational grid is then automatically generated by PHOENICS and one is ready to run.

All of the filter plenums modeled were made by the Ford Motor Company . The first plenum selected was of the end/end configuration. As can be seen from figure 8a, we found good agreement with experimental data. The average error was 8% while the maximum was only 12%. The next plenum simulated was of the side/side configuration. Here we had excellent results with a maximum error of only 4% as shown in figure 8b. The third Ford unit modeled was also a side/side type with the computational predictions, on figure 8c, falling within 15% of experiment. This is still a good prediction for a generic model.

Thus we have an excellent model which takes only minutes to implement, yields accurate results, but takes 37 cpu hours to run on a VAX 8600. Obviously this is not a good interactive design tool. Consequently, we have engaged in a parametric study to extend the series loss coefficient concept to the filter plenum problem.

Bernoulli Coefficients

Here the total loss of the filter plenum would be taken as a summation of pressure loss coefficients; a sudden expansion from the inlet, a sudden contraction at the outlet, and a "geometry coefficient". This would be added to the pressure loss for the prescribed filter panel to obtain the overall pressure drop. In order to reduce design time the sudden expansion and sudden contraction terms are calculated using the same geometric approximations as the generic program. Thus we have the equation

$$\Delta P_{TOTAL} = \frac{1}{2} \rho V^2 (K_{SE} + K_{SC} + K_{GEOMETRY}) + \Delta P_{FILTER} \tag{2}$$

and from reference 9

$$K_{SE} = \left(1 - \left(\frac{d}{D}\right)^2_{inlet}\right)^2 \tag{3}$$

$$K_{SC} \cong 0.42 \left(1 - \left(\frac{d}{D}\right)^2_{outlet}\right) \tag{4}$$

In equations (3) and (4) d/D is the appropriate ratio of the inlet or outlet duct diameter to the hydraulic diameter of the face of the generic plenum containing the inlet or outlet duct . The filter pressure drop term in equation (2) is evaluated from the same empirical curve fit used to obtain the the constants in equation (1).

A hand calculation was used to predict the pressure drop across the three units modeled above. Neglecting the geometry coefficient, the calculation still predicts pressure drop within 15% of experiment as shown in figures 9a thru 9c. In figure 9a there are two sets of predictions. This is because the actual plenum had a small internal bellmouth leading to the outlet duct. This made a substantial difference and consequently an additional coefficient was added to the calculation. The presence of this bellmouth was not accounted for in the generic numerical simulation of the same plenum show in figure 8a. The loss coefficient approach appears quite promising for the design process. A small computer program will yield these predictions in less than a second, with accuracies comparable to numerical predictions.

All of these results may still be improved upon with the addition of the geometry coefficient. Determining the nature of this coefficient is our next task. The generic model is an excellent tool for determining the effect of each geometrical parameter on the overall pressure drop. Thus by analyzing each of these parameters individually, we hope to develop empirical relations for a geometry coefficient.

Acknowledgement

The authors wish to gratefully recognize the Allied Aftermarket Division of the Allied Signal Corporation and the Ford Motor Company for the financial support which made this work possible. Also, this paper is dedicated to the late Professor George A. Brown. George did not live to see this work published, but his guidance and inspiration was very much on the minds of the authors during the preparation of this manuscript.

References

1. Idelchik, I. E., "Handbook of Hydraulic Resistance," Hemisphere, Washington D. C., 1986.
2. Blevins, R. D., "Applied Fluid Dynamics Handbook," Van Nostrand Reinhold Company, New York, 1984.
3. Rosten, H. I., and D. B. Spalding, "The Phoenics Reference Manual," CHAM Report TR/200, October, 1987.
4. Rodi, W., "Turbulence Models and Their Applications in Hydraulics -- A State of the Art Review," International Association for Hydraulic Research, The Netherlands, June, 1980.
5. Patankar, S. V., "Numerical Heat Transfer and Fluid Flow," Hemisphere, Washington D. C., 1980.
6. Spalding, D. B., " A Novel Finite Difference Formulation for Differential Expressions involving both First and Second Derivatives, " *Int. J. Numer. Methods Eng*, vol. 4, pp. 551--559, 1972.
7. Bejan, A., " Convective Heat Transfer in Porous Media," Handbook of Single-Phase Convective Heat Transfer, John Wiley & Sons, New York, 1987, pp 16.3--16.6.
8. Gurumoorthy, V., " Computational Fluid Dynamics Modeling of An Air Induction System," Master of Science Thesis, Dept. of Mech. Engr., University of Rhode Island, Kingston, R.I., February, 1990.
9. White, F. M., "Fluid Mechanics," McGraw-Hill book Company, 1986.

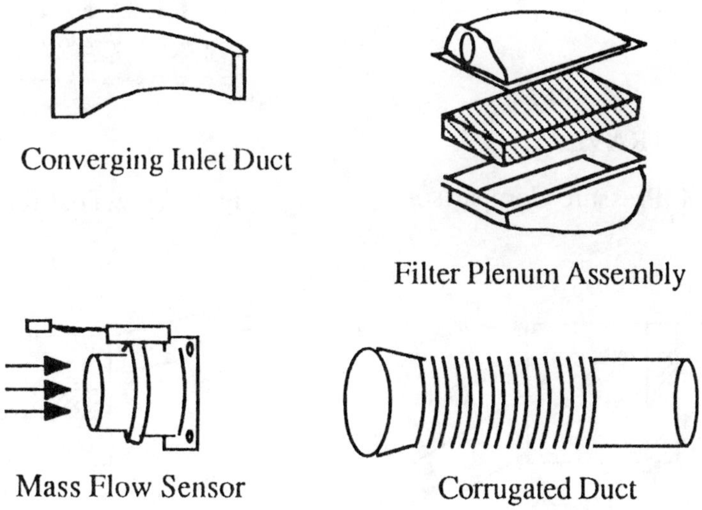

Fig. 1 Typical Air Induction Systems Components

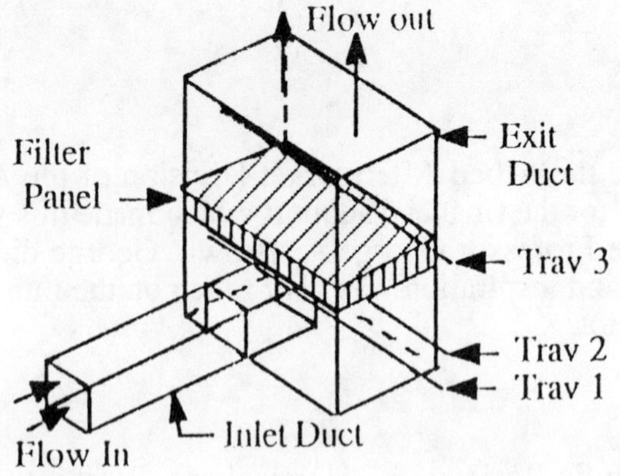

Fig. 2 The Simplified Geometry

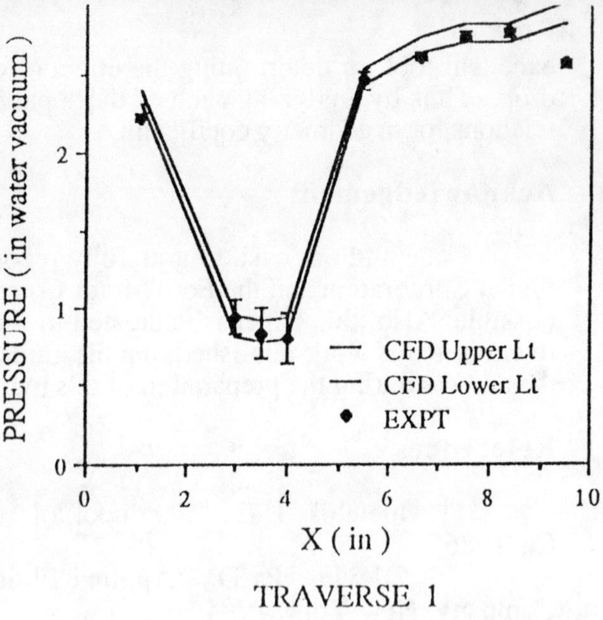

Fig. 3 Pressure Comparison

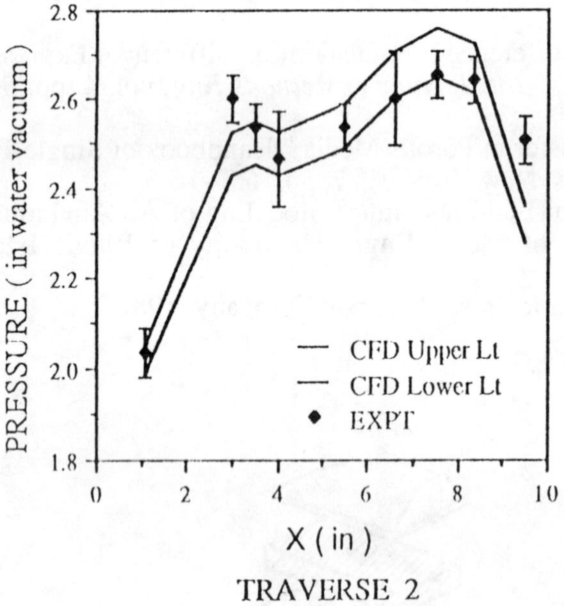

Fig. 4 Pressure Comparison

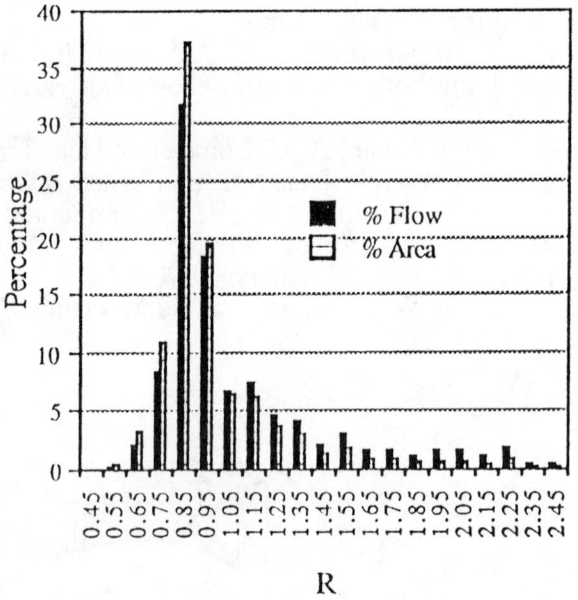

Fig. 5 Flow Distribution In Filter Panel

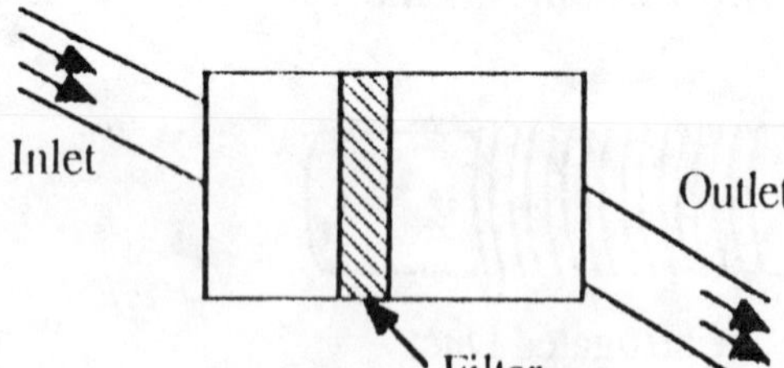

Fig. 6: Generic End/End Configuration

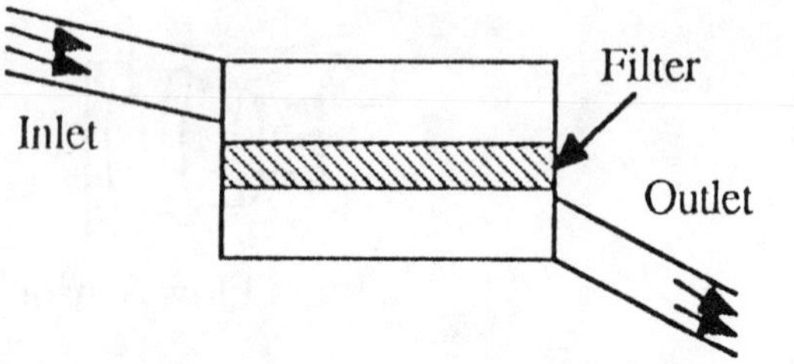

Fig.7: Generic Side/Side Configuration

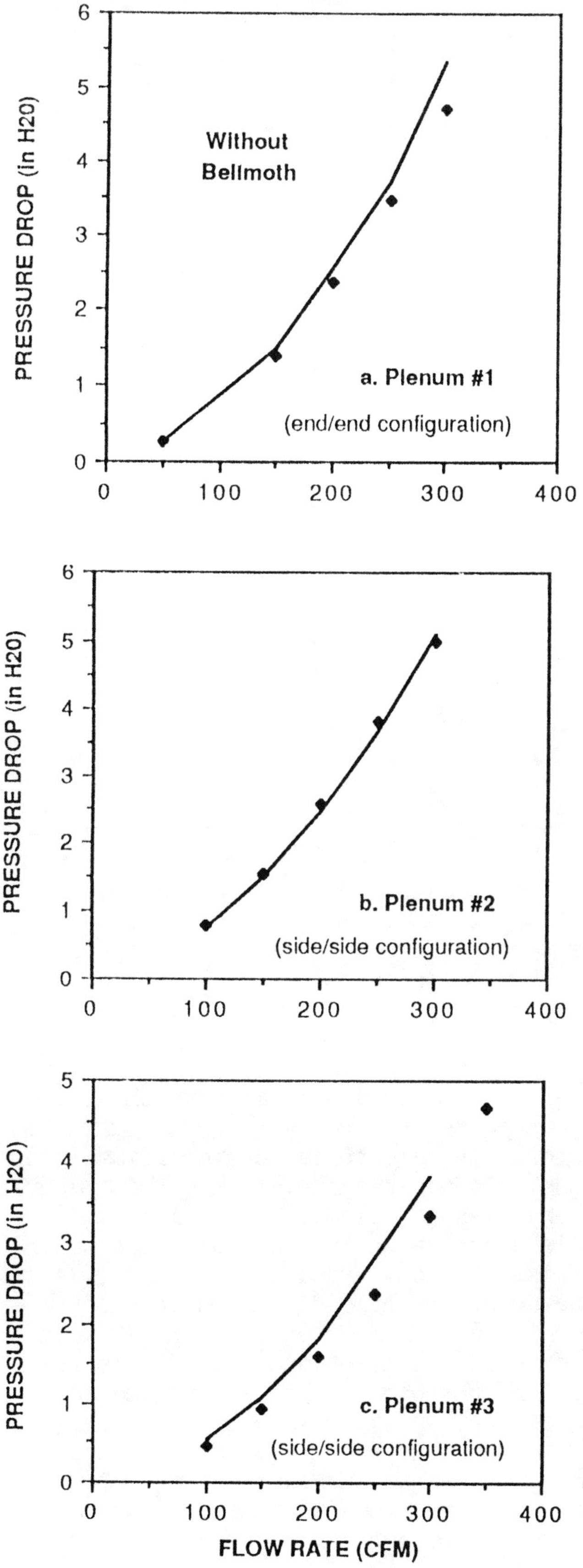

Fig. 8 Generic Model Predictions
Compared to Data

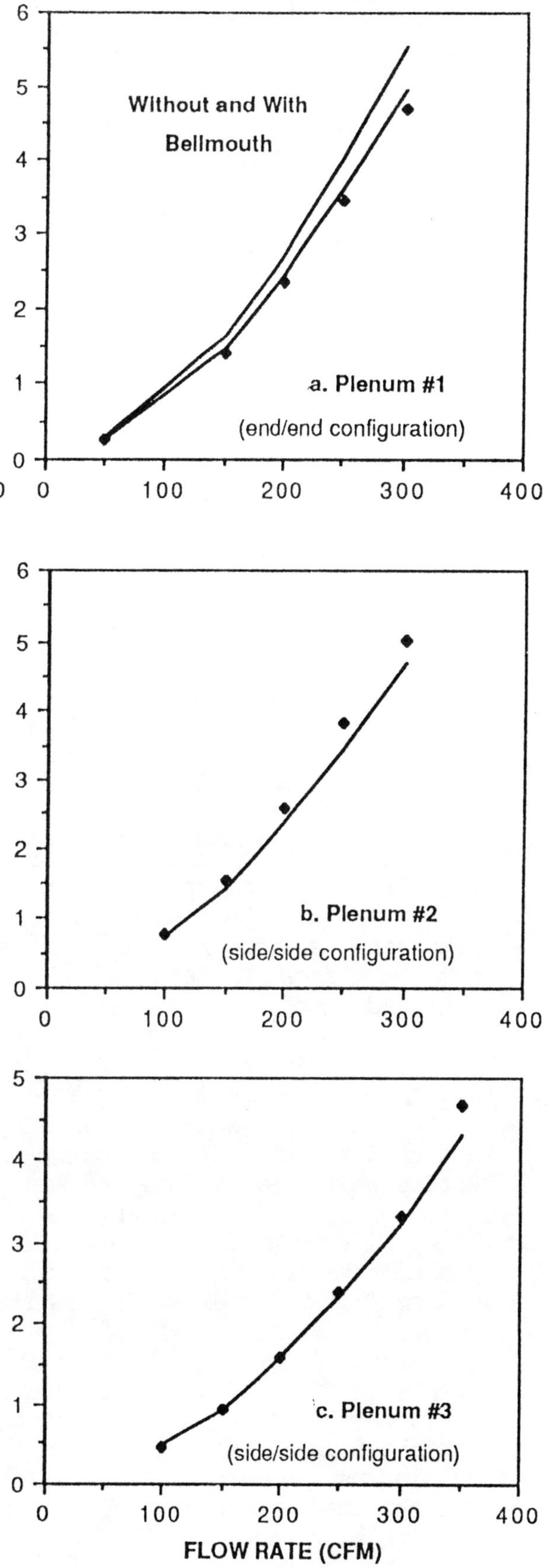

Fig. 9 Loss Coefficient Prediction
Compared to Data

THE USE OF POLYMERIC RESINS IN PAPERMAKING

Albany International
Press Fabrics Division
Post Office Box 1109
Albany, NY 12201

Abstract: The use of synthetic organic resins to impart improved paper properties to paper and also improve the retention of paper fines, fillers, and pigments to the sheet of paper being made is now widespread in the paper industry. This presentation is a review of the basic properties and nomenclature of the types of products used for this purpose, and some of the physical and chemical processes and phenomenon observed in this application.

Papermaking in Brief: The paper industry within the last two decades has been using polymeric resinous materials to improve the properties of paper and to improve the papermaking process. These synthetic and natural materials have many interesting electrical and colloidal properties which have become indispensable in papermaking. Since this papermaking process is in water, the properties of polymeric materials and other fillers used are present while dissolved or dispersed in the water phase. The solid phase which also is the paper product being made, is wet and exhibits similar surface characteristics while dispersed in this watery environment. Briefly, paper is made from wood fibers which when dispersed in water is known as paper pulp. Conventionally the pulp is dewatered at the fourdrinier section of the paper machine, and the sheet is forwarded to the press section for further water removal. Thereafter the sheet is dried through a dryer section and wound. This process is continuous and much time and effort is spent to increase the efficiency and the speed of this process. Presently machines that make tissue are running at speeds of 5000 fpm and the most modern tissue machines are approaching 7000 fpm in speed.

In practice, the pulp can vary significantly, depending on the delignification process which varies from chemical cooking to grinding the wood mechanically. A few other variables include the type, species and geography of the trees, and the use of recycling of waste paper.

Pulp contains long wood fibers, as well as a large population of microscopic paper fines and resins such as wood pitch. To this is added, depending on the characteristics sought, many polymeric natural and synthetic materials such as resin size, starch, and inorganic pigments, including synthetic polymeric materials.

It is the intent of this paper to describe the nature of some of these additives and how a papermaker manipulates these additive properties to obtain the retention of small particles to the sheet of paper, thereby giving paper some special characteristics.

Most of the additives used including paper fibers possess an electrical charge at the surface which may be characterized in terms of zeta potential. This zeta potential is measured using an electrophoresis apparatus. A microscope is used to observe colloidal particles inside a chamber called an electrophoresis cell. Electrodes placed in each end of the chamber are connected to a power supply which creates an electric field causing the particles of colloid to move. The movement and direction is related to zeta potential.

The zeta potential is obtained by measuring the slipping plane between the particle and the flowing liquid via a streaming potential measurement or vice versa, as the particle is migrating toward an anode or cathode in a polarized cell. The slipping plane is located somewhere in the diffuse layer and not exactly at the interface. This is shown in Figure 1.

Paper and paper fines, as well as most inorganic fillers such as clays, titanium and calcium carbonate are all negatively charged.

The polymeric synthetic resins used in papermaking are often made cationic or possess a positive zeta potential. By carefully titrating the amount used, it is possible to cause flocculation and retention of these fines to paper.

When a cationic alum or a low molecular weight cationic retention aid is used, it is also possible to interact these substances with an anionic polymer and cause the exhaustion of the anionic polymer to the sheet, by virtue of its cationic outer surface. The following are but some examples of the types of attraction, repulsion and floc formations that are possible by using the characteristics of paper making resins or lower molecular weight retention aids and flocculants.

Figure 2 shows the forces at work when contaminants such as polyvinyl acetate are present with intentionally added wet strength resins, as are used in tissue and towelling grades.

Figure 3 enlarges on the possible combination of reaction products and possible entanglements and adsorption of these products to fibers and fines. Figure 3 also shows that there is no limit to the size and weight of some agglomerations that may feasibly form and fall out of water dispersions in voluminous flocs.

Similar reactions are also responsible to the plugging up of pressing fabrics which have to be kept clean continuously while in operation on the paper machine. These products if left on the fabric will eventually cure and become permanent in paper machine clothing thus causing their removal at great expense.

Wet strength resins and retention aids change in surface charge characteristics depending on pH, salt content of the water and the use of other surface active agents such as a dispersing agent. Figure 4 is a pictorial description of changes in surface charge that would occur from anionic polyvinyl acetate above and a cationic wet strength resin as a result of pH or the addition of PVA a dispersing agent.

Figure 5 shows how the water drainage of press fabrics is effected with two types of wet strength resins used in the industry depending on the pH of the paper stock. This difference is brought about because of the magnitude of the charge in each of the resins shown in this figure. At low pH, kymene 557 is extremely cationic and if used at pH 5, it tends to cause fabric filling. On the other hand, when used at pH's above 7.0, its cationic nature is modified, resulting in its widespread acceptance as a wet strength resin under neutral and alkaline ranges. At these pH's the damage to the fabrics is not hampered with this resin.

Parez, another wet strength resin on the other hand is a weaker cationic resin than kymene. Its cationic nature is improved in acid stock. It does not interfere with fabric drainage at the lower pH's. However, as the charge dissipates on the alkaline end, it loses its surface charge and the product forms a gel.

Zeta Potential vs. pH - The explanation of the presence and intensity of charges may be explainable from a molecular level based on some concepts of charges resulting from the donation or sharing of electrons.

For instance, Figure 6 shows how the magnitude and sign of the surface charge is varied depending on the pH for protein in wool. Similarly, wet and dry strength resins used in paper as well as retention and flocculants show varying activity at different pH's. The table below shows how the zeta potential varies with pH for a polyamine-type wet strength resin and for corn starch.

Charge vs. pH

pH	Polyamine Resin	Starch
2	+ 45.0	- 5
4	+ 35.0	- 10
6	+ 27.0	- 18
8	+ 14.0	- 26
9	+ 6.9	- 37
11	± 0.0	- 45

It is important to understand factors that are involved in controlled flocculation of papermaking additives and paper fines. The first pass retention of these fines with the sheet of paper being manufactured is becoming more important in closed water systems in paper mills. Many sizes and additives used, hydrolyze rapidly. For this reason it is important that the processes used for flocculation are operating with good controls set for monitoring all aspects of floc formation and the retention of these flocs to the sheet. The papermaker relies on empirical information based on runnability, which is a subjective measure in the use of these active products.

The papermaker needs to determine the type, amounts and points where he needs to introduce his active ingredients to the pulp. The idea is not only to obtain improvements in sheet properties and total retention but to also insure that premature flocculation will not build excessive residue in the system.

Poor control will result in costly down time, in clean up of machine equipment and replacement of the costly paper machine clothing.

Conclusion - The use of polymers to improve paper properties and improve the retention of small solids to the sheet has been increasing during the last two decades.

These products have added additional complications to the existing multi-variable process of papermaking.

Even though theory does not always coincide with practice in complex systems, some knowledge of how particles react and grow in size and by doing so deposit in the sheet of paper, provides ideas for attacking real problems that occur on the paper machine and on pressing fabrics.

STERN DOUBLE LAYER

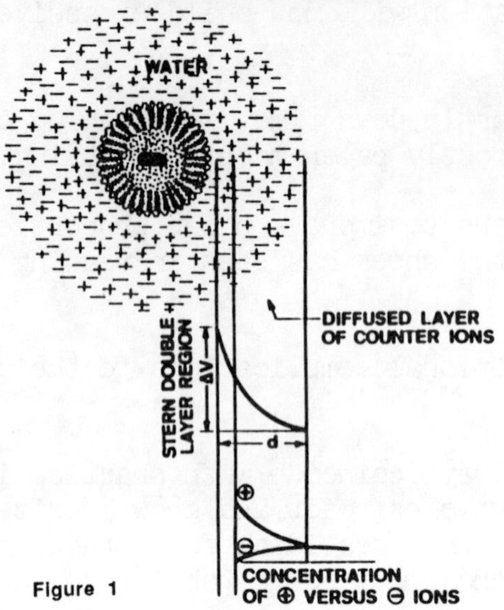

Figure 1

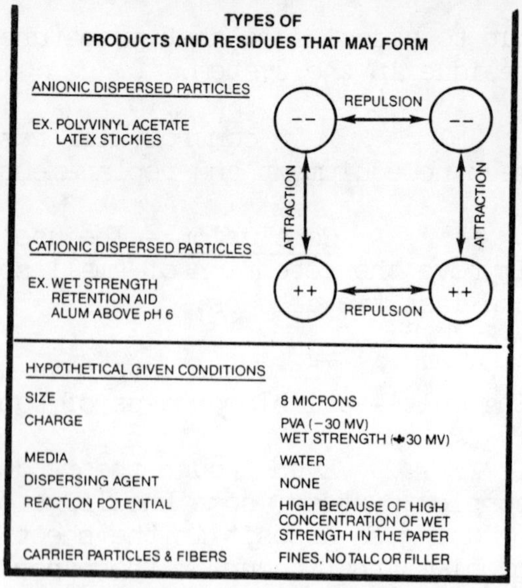

TYPES OF PRODUCTS AND RESIDUES THAT MAY FORM
ANIONIC DISPERSED PARTICLES
EX. POLYVINYL ACETATE LATEX STICKIES
REPULSION
ATTRACTION
ATTRACTION
CATIONIC DISPERSED PARTICLES
EX. WET STRENGTH RETENTION AID ALUM ABOVE pH 6
REPULSION
HYPOTHETICAL GIVEN CONDITIONS
SIZE
8 MICRONS
CHARGE
PVA (−30 MV) WET STRENGTH (+30 MV)
MEDIA
WATER
DISPERSING AGENT
NONE
REACTION POTENTIAL
HIGH BECAUSE OF HIGH CONCENTRATION OF WET STRENGTH IN THE PAPER
CARRIER PARTICLES & FIBERS
FINES, NO TALC OR FILLER

Figure 2

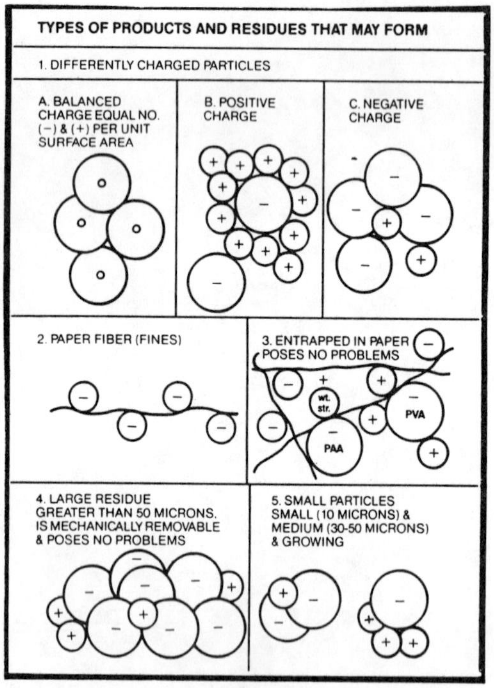

Figure 3

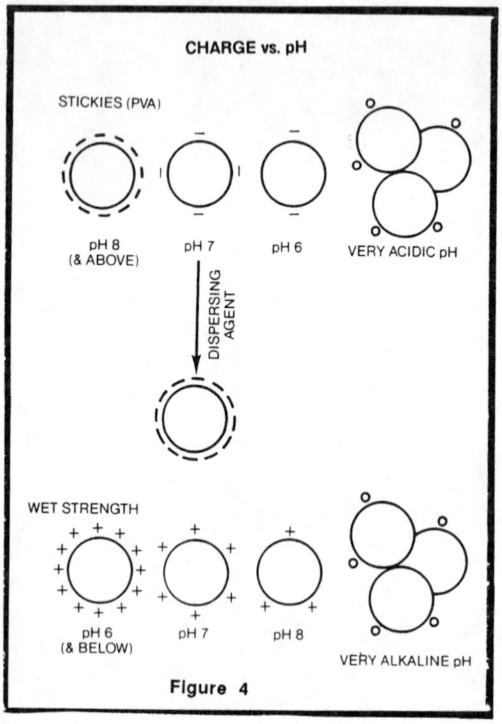

Figure 4

RELATIONSHIP OF FELT DRAINAGE TO KYMENE & PAREZ

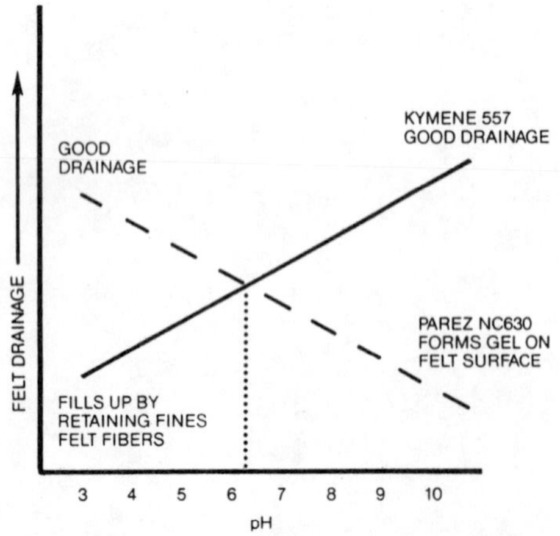

FOR GOOD FELT DRAINAGE, USE KYMENE 557 AT HIGH pH
PAREZ NC630 AT LOW pH

Figure 5

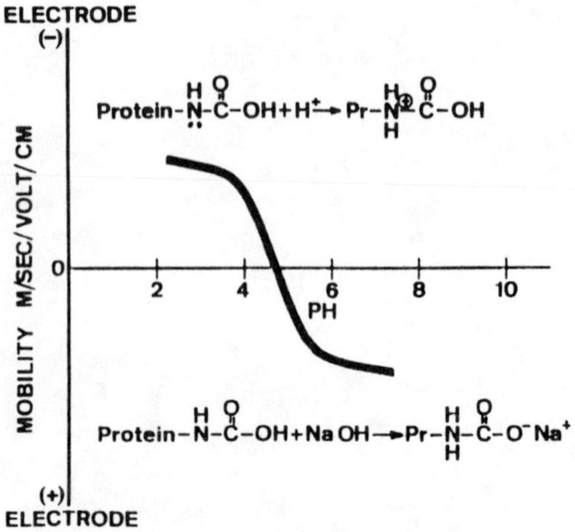

Figure 6

THE USE OF ELECTRIC FIELDS FOR FILTER CLEANING

Professor Stuart A. Hoenig
Dept. of Electrical and Computer Engineering
University of Arizona
Tucson, AZ 85721

The problem of filter cleaning has been with us since the invention of the filter itself. If cleaning is done on the "dirty side" with an air hose we can expect fair to poor results. The factors are not hard to identify:

1) The flow of air against the dust will be hampered by the rapid build up of a gaseous boundary layer. Once the boundary layer builds up the forces on the dust particles "inside the boundary layer" will be very small.

2) Another problem is that the flow from the air hose is actually driving the major part of the dust into the fabric rather than of into the air. This is the reason that filters are changed regularly, they simply cannot be effectively cleaned.

One very obvious solution here is to blow the filter from the inside. Now all the dust will driven away from the filter and cleaning will be much better. Unfortunately most large filter systems are not easily blown from the inside. This system while very good in theory is seldom used in practice.

Another solution is the application of electric fields to the filter during deposition. This will be discussed in more detail in another paper at this conference, here we shall only comment that a much lighter and more easily removed dust cake is generated. Naturally a dust cake of this type is more easily removed, and the problems aerodynamic boundary layer and dust being driven into the fabric are somewhat alleviated. We might note that some filters are designed for a reverse flow during blowoff. This filters are designed for a reverse flow during blowoff. This certainly helps, but observations have indicated that the flow of air is not uniform, so that some parts of the filter are not cleaned by this technique.

The problem of designing the filter for this type of application is not a simple one as we shall see in the other paper to be given at this session. Using this technique on large filters that are already built is almost impossible. For new filters it is relatively easy to add the system and this is seen more and more.

The University has done a short program (with National Science Foundation support) on the concept of simply applying a metal grid to the filter itself. When the filter is operating the grid is held at a DC voltage (typically -12 KV) to form the softer and thicker dust cake. When this field is present blowoff occurs naturally when the airflow is stopped. No externally applied is needed. If externally applied air is used the process is accelerated and dust removal is further enhanced.

The voltage is maintained during blowoff to assist in the process. Some experimental photographs are shown in Figures 1 and 2. It is clear that the system operates as described, but there are still many questions to be answered. The voltage required, the sign of the voltage, whether varying the voltage, or pulsing it would be an advantage.

The application of this technique to a particular problem is dependent upon several safety variables. Since the voltage is applied directly to the filter bag, there is some hazard to personnel who might be working near to the bag. The current per bag is actually likely to be quite small and it should be possible to limit the hazard to the point that there is no danger of injury. We suggest that initially the system be applied to nonflammable dust. When more experience in the laboratory and the field has been gained it may be possible to use the system on some "weakly" flammable dusts.

In summary we have shown that it is possible to enhance the blowoff of dust from a filter by a grid applied directly to the filter. The work is still limited to the laboratory because of the many unknowns. Never the less it has been shown that electrical effects can be important in filter blowoff.

Figure 1 Filter which is loaded (with no voltage on it)

Figure 2 Filter with 500 volts DC and 0.25 ampere on it

EFFECTS OF PRECHARGING AND ELECTRIC FIELD ON COLLECTION

Professor Stuart A. Hoenig
Dept. of Electrical and Computer Engineering
University of Arizona
Tucson, AZ 85721

The action of conventional filters is reasonably well understood. Large particles (all these terms are defined in terms of the filter being used) will be caught in the meshes of the filter. Very small particle are caught by Brownian motion as they move through the filter. Naturally the deeper the filter - the more of these particles will be caught. (Depth in a filter is limited by the pressure drop that can be tolerated. All the discussion here will pertain to depth filters rather than surface filters.)

The particles in between these limits (neither large nor small) are the ones most likely to get through the filter. In HEPA filters used by the semiconductor industry this size ranges from 0.3 to 1 micron. The filters "get" a very large fraction of the dust so it is not a large problem, at least for "new" filters.

For small particles we used the term "sticking" - it is strongly influenced by the kind of material used for the filter and by the properties of the dust. At present there is very little known about the whole question of "sticking". Essentially manufactures go by experience.

At present there is little agreement among manufacturers as to the type of material HEPA best for HEPA filters. In fact the construction details vary from one manufacturer to another. This is no small thing since a small distortion can lead to the leakage of particles. Testing of HEPA filters is a question of great interest that I cannot discuss here, suffice to say that most users do not do a very good job in this area.

In fact HEPA filters are generally very good when they are "new" and most of the data in the literature refers the "new" filters. The problems that have arisen refer to:

1) "old" filters for whatever that may mean.

2) filters used in areas that are hot (around furnaces.)

3) filters that are exposed to chemical fumes.

4) the response of filters to bacteria, fungus or other material that might be regarded as "alive". The present experience is that these living materials reproduce in the filter and "come through" very fast. They then take root on wet areas, it may be that much of the surface contamination we see is bacteria, fungus, etc. In Table I we show data taken in an operating clean room. The HEPA filters are seriously contaminated and "wet" surfaces in the room have picked up the bacteria, etc. At present the Japanese are trying to meet the problem by sterilizing the air before it gets to the HEPA filters. Another technique is to stabilize the air pressure so that the HEPA filters are not exposed to varying air pressures. These are only "partial" solutions and this is a topic that will have to be taken up at length in the future.

Every filter user would like to work with the minimum pressure and have the maximum life between filter changes. This has led to many studies of electrically charged filters where the dust would have a polarity of the proper sign thereby leading to more efficient collection. With a system like this the filter thickness might be less thereby reducing the pressure drop.

The problem with this has been finding a filter cloth that would hold a charge for the useful life of the filter. This has now been done by several companies around the world and the material is for sale through various outlets. The problem of dust charging is a matter of finding a simple system that does not raise the danger of igniting flammable dusts. One system is now for sale and seems to be operating very well.

In Figure 1 we show the effect of simply "charging" the dust negatively before it reaches a very low quality filter. The change is dramatic and other experiments have indicated that the dust on the filter is not lost if the charging system is disconnected temporarily. The filter simply drops back to its "normal" performance.

In Figure 2 we show a commercial system developed in Japan. The dust is first filtered to remove the larger particles since they tend to charge to a polarity opposite to that of the smaller particles. The smaller particles (below about 10 microns) are then charged and allowed to flow into the HEPA filter. Note that no charge is applied to the HEPA filter itself and that the biased elements are separated by a grounded grid.

The extension in life with charging is shown in the Figure 3. This effect appears due to a change the way the dust builds in the filter when a field is applied. The improvement in collection efficiency is shown in Figure 4 which clearly shows that the electric filter is better. The operation of an electrical filter results in a reduction in the pressure drop.

In conclusion we might say the following:

1) HEPA filters deserve more attention from semiconductor users than they get at the moment. Questions are the construction, lifetime, exposure to gases, and heat.

2) The question of bacteria, fungus, etc, getting rapidly through the filter and then plating out in the clean room should be settled.

3) Electrical technique may be an important way to reduce pressure drop and extend filter life. More studies of this should be done in the USA.

ACKNOWLEDGMENTS:

This work was supported by a grant from the National Science Foundation.

Table I   Bacteria found in the cleanroom

Location	Colony Counts	Morphology Gram Rx
#1 Wet Hood 22D Horizontal Surfaces at front	1 Bacteria 1 fungus grew at room temp. 48 hours	Bacillus G+
#2 Wet Hood 22D Horizontal Surface at Middle	3	1 Staphylococcus hemolytic(aureus) 2 Bacillus G+
#3 Wet Hood 22D DI Rinse Tank Cover	25	15 Staphylococcus hemolytic(aureus) 2 Bacilli hemolytic 6 Bacilli G+ 2 Sarcina
#4 Plastic Grill below HEPA filter	TMTC* ← spreader colonies	Bacilli G+ Staph hemolytic (aureus)
#5 21D Microscope Inspection Table	0	Still 0 after incubation at room temp. 6 days
#6 21D Microscope Stage Surface	1 1 Fungi grew at room temp. 48 hours	Bacillus G+
#7 1 BA existing HEPA filter with AC inlet air	4	1 Bacillus G+ 3 Staph aureus (hemolytic)
#8 Laminar Flow Hood pre-filter	17	2 Sarcina 7 Staph epi 2 Staph hemolytic (aureus) 3 Bacillus G+ 1 Bacillus G-
#9 Laminar Flow Hood polycarbonate air deflector	0	Still 0 after 6 days at room temp.
#10 Spin Rinse Dryer Loading Door	1 after 48 hrs. 37° (TMTC* 150) room temp. ← 48 hours	Staph. Fungi

TMTC* = Too Many Too Count

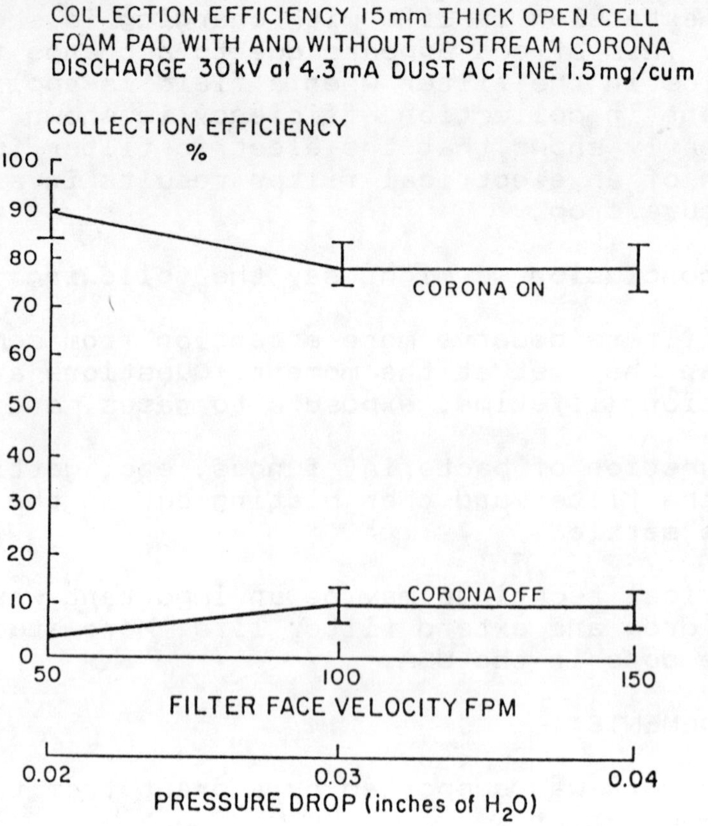

Figure 1 Collection efficiency of filter

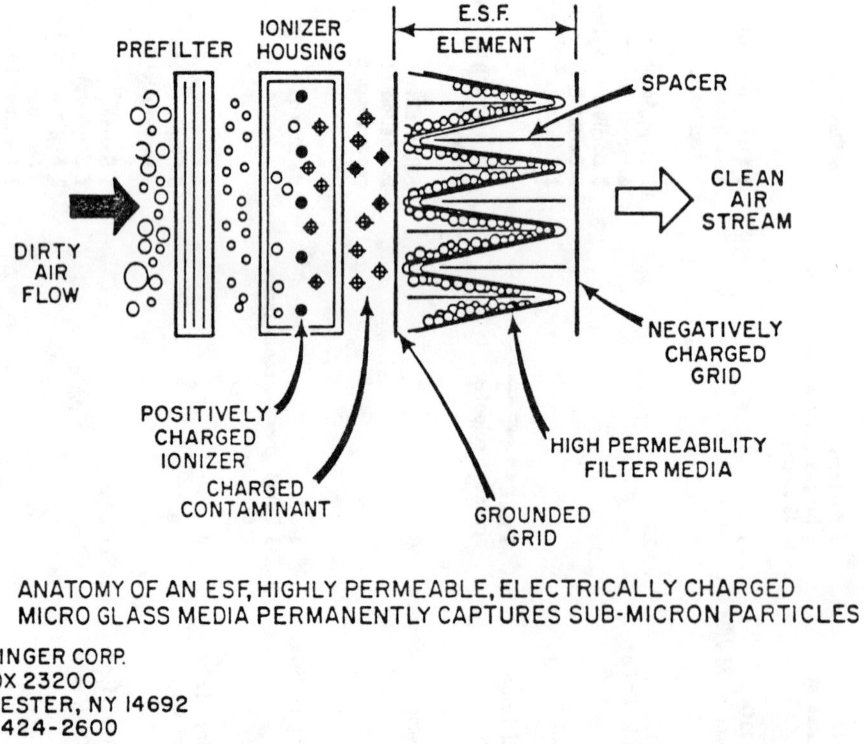

Figure 2 Anatomy of an ESF filter

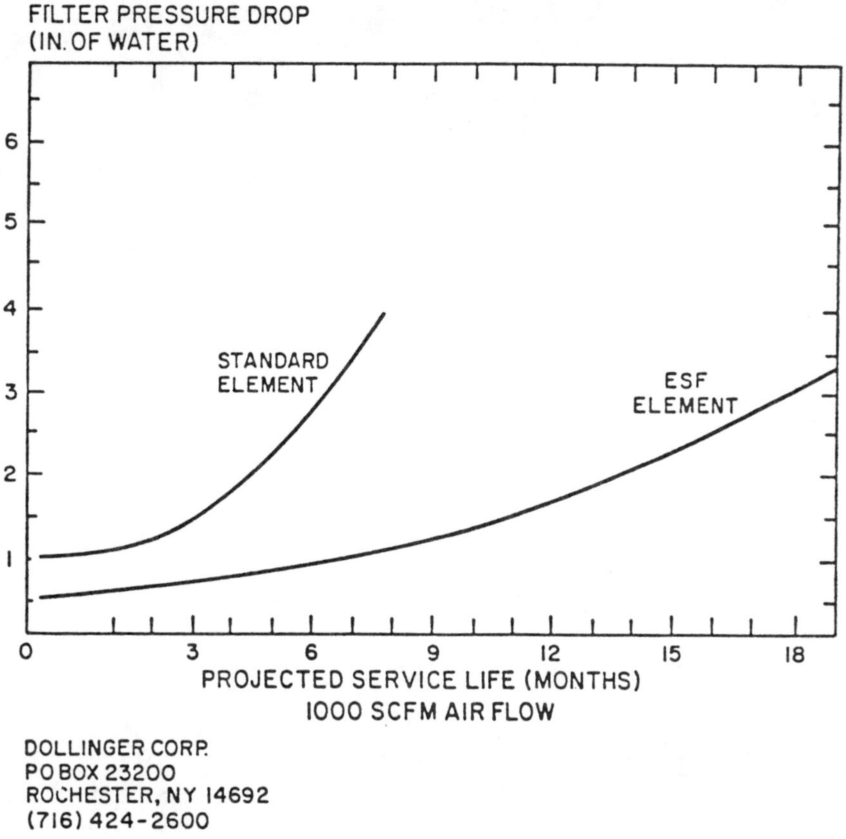

Figure 3 Comparison of standard and ESF filter service life

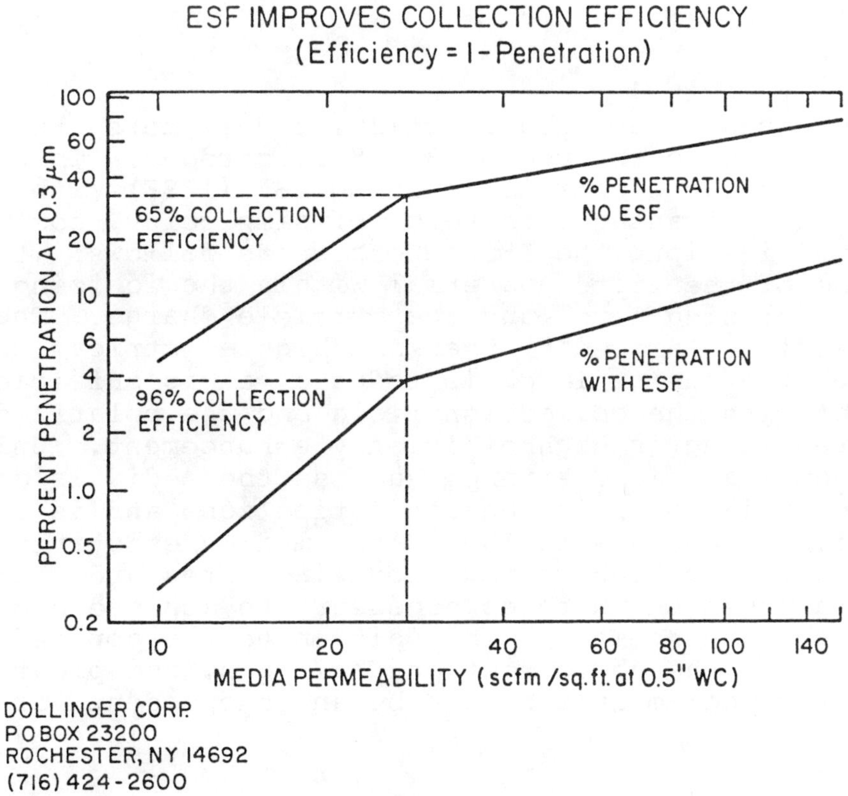

Figure 4 Comparison of efficiency of standard and ESF filter

A TWO ELECTRODE IONIZING ELECTRICALLY STIMULATED FILTER

Rajan A. Jaisinghani & Neville J. Bugli
American Filtrona Corporation
7527 Whitepine Rd.
Richmond, VA 23237

Abstract

Conventional electrically stimulated filters (ESFs) with ionizers require the use of 3 - 5 electrodes. This paper describes a two electrode ionizing ESF (IESF) with significant advantages in terms of simplicity, cost, and space utilization. The IESF operates as follows: the presence of the filter material within the ionizing field reduces ionizing level and the particle charge. The filter material is polarized by the same high electrical field. Although the charge level is reduced, it is sufficient to interact with the collection field and the polarized filter media resulting in high efficiency enhancement. Unlike ESFs, the electrical field strength across the media is dependent upon potential drop in the ionization zone and is also sensitive to changes in the filter media resistivity. An air gap between the high voltage ionizing wires and the filter media has been shown to be necessary to achieve high efficiency enhancement. In addition to the applied field strength and the above mentioned air gap, the pleat depth of the filter medium is shown to be an independent variable.

1.0 Introduction

One of the pioneering efforts in electrically stimulated filtration was by Zebel[1] and an excellent, in-depth coverage of the subject matter is presented by Bergman et al.[2] Contributions regarding design of such filters have also been made by Masuda et al.[3] and Jaisinghani and Hamade[4], among others. These ESF designs utilize between 3 to 5 electrodes, as illustrated in Figure 1, if the incoming particulate matter is to be charged. Typically these ESFs require two high voltage electrodes. All of this contributes to increased complexity, cost, and poor space utilization. Consequently these devices often are not attractive enough to warrant their use over conventional filtration devices.

These limitations are overcome to a large extent by our development of a two electrode ionizing ESF (IESF) (patent applied for by the authors; patent rights assigned to American Filtrona Corporation). The filter medium is positioned within the ionizing field (as shown in Figure 1(d)) such that particle charging and collection is accomplished by one electrical field. The filter medium is typically in an accordion like pleated form. This results in design simplicity, better space utilization, and lower cost with one high voltage electrode (the ionizing wires). Yet this device achieves extremely high efficiency enhancement.

Although the basic mechanisms of collection for an IESF are similar to conventional ESFs, there are significant differences between the two in terms of operational characteristics. For example, the field strength across the IESF filter medium is not known or cannot be calculated. Further, unlike an ESF where the applied field strength is constant within limits of power supply, the field strength across the IESF filter medium depends on the drop in potential due to ionization and is sensitive to the conductivity of the filter medium.

The purpose of this paper is to experimentally study the operational characteristics of the IESF. It should be noted that the performance advantages of the IESF over conventional filtration have been reported by Jaisinghani and Bugli[5]. The performance advantages are clearly summarized in Figure 6, in terms of dust loading at different velocities, for equal sized small scale filter elements.

2.0 Experimental Set Up

A simple schematic drawing of the test set up is shown in Figure 2. All experiments were performed in a temperature and humidity controlled aerosol filtration laboratory, capable of maintaining temperature at 70°F ± 2°F and relative humidity at 50% ± 5%.

2.1 Efficiency Measurement

All filter elements were challenged using cold polydisperse DOP aerosol. The polydisperse aerosol was generated using an

2

atomizer. Both upstream and downstream particle counts were measured using a light scattering optical particle counter with 0.3 micrometer particle size sensitivity. The optical particle counter was coupled with a suitable dilution system capable of approximately 350:1 dilution ratio.

2.2 Charge Measurements

The charge levels were measured using a Faraday cell. The Faraday cell contained a dry, pre-weighed 0.2 μm fiberglass filter (greater than 99.98% efficiency at 0.3 μm) sandwiched (with contact) between two perforated electrodes. A sample of the aerosol charged by the ionizer was made to pass through the Faraday cell for approximately 15 minutes at a constant flow rate. The current drained from both the perforated electrodes was measured via a high impedance electrometer. The electrometer was coupled to a real time, computer data acquisition system. The time average value of the drained current was then calculated for the whole sampling period. The filter was then dried and reweighed to obtain the differential weight gain on the filter due to aerosol. The average charge level per weight of aerosol was then calculated by dividing the charge drained by the differential weight gain of the filter. The mass mean size of the aerosol was approximately 0.5 μm.

3.0 Results and Discussion

3.1 Effect of Filter Media on Ionization and Efficiency

<u>3.1.1 Ionization/Charge Level</u>. In order to evaluate how the presence of the filter media affects the particle charge, two sets of experiments were conducted. In the first case the particle charge was measured at various distances within the two electrodes and just beyond the perforated ground electrode, without the presence of the filter material. The aerosol sample was withdrawn within the electrical field by means of an isolated, moveable sampling probe. For the second set of measurements, a widely spaced pleated (2.54 cm pleat separation) fiberglass paper medium (50% DOP efficiency) was used. Further, a 2 cm gap was maintained between the peaks of the pleats and the ionizing wires. The moveable, insulated probe was then positioned at various positions within a pleat, until the probe inlet almost touched the valley of a pleat. In this case, due to the high efficiency of the enhanced filter medium, it was not possible to measure the particle charge downstream of the ground electrode. Since the intent was to relatively compare the charge levels, the possible errors due to the local distortion of the electrical field by the sampling probe, were ignored.

With an applied potential of 28 KV the charge level increases with distance (Figure 3) within the ionizing field, achieving a maximum charge level of 140 μC/g when no filter medium is used. According to White[6] the residence time to achieve 90% of the saturation charge level should be

3

approximately 0.024s based on field charging theory. In this case, the residence time is 0.027s.

With the introduction of the filter medium, however, the charge level dropped to 43 - 52 μC/g. In this range, the charge levels seem to be relatively independent of the distance from the ionizing wires.

Unlike conventional ESFs, where separate ionizing and collection fields are used, the current consumption, and particle charge are simultaneously affected by the presence of the dielectric, non-conductive fiber glass medium in the IESF. The presence of the pleated filter medium decreases the current consumption by about 60% as shown in Table I. Table I also shows the current consumption for different filter media with different rated efficiency. Within experimental error, there is no charge or current dependence on the mechanical efficiency of the media. This is contrary to our prior expectation since, as discussed in section 3.1.3, the efficiency enhancement is greater for lower mechanical efficiency filter media; this led us to believe that for lower mechanical efficiency media the charge levels should be higher. Without the filter media, the charge level is about 140-165 μC/g. This is reduced to an average level of about 59 μC/g in the presence of the filter media.

Although these materials have different average fiber sizes, their bulk porosities are identical (92.3%). Measured values of media surface and volume resistivities fall in the range of 1.6×10^{13} to 1.2×10^{14} ohm and 1.69×10^{13} to 5.9×10^{13} ohm cm respectively. Given the equal porosity, the narrow range of resistivity should be the determining factor in establishing the surface potential of the media near the fixed voltage ionizer wires. Higher media potential values reduce the effective ionizing field strength, thus reducing the ionizing current. This narrow resistivity range explains the current and charge independence with respect to these filter media.

3.1.2 Air Gap Between Ionizing Wires and Filter Media. Experiments were designed to illustrate the importance of having an air gap or space between the ionizing wires and filter medium. In the previous section we have shown that the ionization is partially suppressed by presence of the highly porous filter media, with an air gap between the ionizing wires and filter medium. Without this air gap we expected the ionization level to be further suppressed by charge transfer to the dielectric filter medium, resulting in a reduction in efficiency enhancement. In order to confirm this, four filter configurations or cases, shown in Figure 4, were evaluated. Case 1 is simply a mechanical filter, designed to establish a reference point for the electrical enhancement of efficiency. In Case 2 both the ground electrode and high voltage ionizing wires are in close proximity (i.e. approximately 1-2 mm space) to the filter medium. At some points there were localized contacts between the wires and the fiber ends of the filter medium. In Case 3 the ionizing wires are in close proximity to the filter medium (as above), however the ground electrode is clearly separated from the filter medium by a 2 cm gap. Case 4 is the IESF configuration with the ground electrode in close proximity to the filter medium and the ionizing wires clearly separated from the filter medium by an air space of 2 cm. This work was done on a small scale, using 15.24 x 15.24 cm cross section, flat (non-pleated) 50% fiber glass paper at an air flow velocity of 0.28 m/s.

The performance results for these configurations are shown in Table II. Both Cases 2 and 3 result in only marginal enhancement in efficiency, while the IESF configuration (Case 4) results in a high level of efficiency enhancement. This suggests that at similar applied electrical field strengths the close proximity of the dielectric filter medium, in both Cases 2 and 3, almost totally suppresses ionization. This is supported by the negligible current consumption at the above field strength (see Table II). Case 2 is thus reduced to the case of an ESF with a polarized filter medium and no pre-charger. Further, due to the low surface area of the high voltage wires, the filter is not effectively polarized and hence only about a 5% increase in 0.3 μm DOP efficiency is observed. Our prior experience has shown that an ESF with two perforated (with approximately 55% electrode area) electrodes achieves a slightly higher (about a 8-10%) increase in efficiency without a pre-charger. ESFs (as in Case 2) exhibit only a marginal increase in small particle removal efficiency due to the negligible polarization of small (0.3 μm) particles and consequently low dielectrophoretic mobility. Our prior experience has also shown that by using only a pre-charger and the same filter medium (without an electric field across it) only about 80-85%, 0.3 μm DOP efficiency can be achieved with this filter medium. The IESF, on the other hand, achieves over 99% efficiency at the same conditions. This suggests that, although ionizing is partially suppressed in the IESF, the IESF does in fact achieve sufficient particle charging. Both electrophoretic and dielectrophoretic interactions between the particles and the polarized filter medium play an important role in particle collection in an IESF. The air gap is necessary for achieving particle charging; the close proximity of the filter medium to the ionizer wire almost completely suppresses ionization probably due to the small but finite charge transfer by the filter medium.

3.1.3 Efficiency Enhancement for Different Media. The efficiency enhancement is also dependent upon the mechanical efficiency of the filter material. Table III shows the 0.3 μm efficiencies (mechanical and enhanced) of various grades of fiber glass filter paper. Due to the excessively high pressure drop, the HEPA type filter media (mechanical efficiencies greater than 98%) were tested at lower flow rates as shown in Table III. The percent increase in efficiency, fractional reduction in penetration ($\overline{E}$) and penetration ratios are also calculated in Table III. Examination of the percent increase in efficiency shows that this device (IESF) is most effective for lower efficiency filter media. In high efficiency applications, the penetration reduction is a more meaningful quantity than the efficiency. On this basis the IESF is most effective for media with mechanical efficiencies in the 20%-70% range. In this range, the percent reduction in 0.3 μm penetration is almost constant (Table III).

The above results may be explained as follows. Due to almost constant porosity and resistivity of these various filter media and due to the charge and current levels being independent of filter media (Table I), it is reasonable to assume that the electrical collection forces are constant for these media. Hence, it is reasonable to expect a higher ratio of electrical collection force to mechanical collection forces for lower efficiency media. Consequently, the efficiency enhancement would be greater for lower mechanical efficiency media.

3.1.4 Effect of Applied Field Strength, and Pleat Depth on Efficiency. Since one electric field is utilized for combined charging and collection of particles, it is clear that the field strength across the filter medium is lower than the applied field strength across the two electrodes. The particle collection field should be affected by the applied field strength and volume resistivity. The collection field depends on the applied potential, the drop in potential in the ionization gap and on the pleat depth. The drop in potential in the ionization region depends not only on the applied potential and distance between the two electrodes, but also on the resistivity of the filter material. This is so because the ionization level depends on the potential gradient between the wires and the filter media. The potential at the media surface (for a fixed resistivity), in turn, depends on the degree of ionization. Due to this complex interrelationship, it was unclear whether the filter media pleat depth is an independent design variable. The following experiments were conducted to study this effect.

We have shown that an air gap between the ionizing wires and filter medium is a variable affecting IESF performance. Hence in the following set of experiments a constant air gap of 2 cm was utilized. Filters of 61 cm x 61 cm cross section with two pleat depths, 4.45 cm and 6.98 cm, were used and the applied potential was varied from 0 - 30 KV to result in a range of applied field strength from 0 - 4.5 KV/cm. Both 15% and 50% (at 0.3 μm) media were evaluated at 4.45 cm pleat depth, while only the 15% filter medium was used at 6.98 cm pleat depths. These filters were evaluated at two flow rates, 2039 and 1444 m^3/hr. The efficiency results are shown in Figure 5. In order to simplify interpretation of these figures and to separate the variables the following convention is applied:

solid lines = 2039 m^3/hr.; dashed lines = 1444 m^3/hr.
square symbol = 15% media IESF; circle = 50% media IESF
all open symbols = 4.45cm pleat depth; all dark symbols = 6.98cm pleat depth

6

In order to evaluate the 15% and 50% media on an equivalent basis, a dimensionless efficiency was used,

$$\bar{E} = \frac{E - E_o}{100 - E_o} = \frac{Po - P}{Po}$$

where E and P are the 0.3 μm, percent efficiency and penetration respectively, with the use of the electric field and E_o and Po are the corresponding mechanical efficiency and penetration values. Note that $\bar{E}$ is simply the fractional reduction in penetration. Figure 5 is a plot of $\bar{E}$ versus the applied field strength. For all situations (Figure 5) an S-shaped dependence on field strength is evident. In addition the flow rate, pleat depth, and the mechanical efficiency of the filters also affect $\bar{E}$. At the same applied field strength, $\bar{E}$ is higher for the larger pleat depth. Further, $\bar{E}$ is lower at higher flow rates. As the applied field strength exceeds about 3.5 KV/cm - 4 KV/cm, the dependence of $\bar{E}$ on pleat depth and flow rate diminishes, as evident in Figure 5. These experiments suggest that the pleat depth is an independent design variable only for medium applied field strengths.

4.0 Conclusions

Although the basic mechanisms of particle collection in conventional ESF and IESF are similar, the IESF and ESF differ significantly in terms of operational characteristics. Unlike the ESF, the collection field in the IESF is dependent on many variables including applied voltage, electrode distance, ionization suppression, pleat depth and resistivity of the filter medium. This new cost effective device results in a certain amount of ionization suppression and thus particle charge reduction. However, sufficient particle charge is achieved to result in a highly efficient mode of filtration. Further, due to the use of highly resistive filter medium, the possibility of back carona discharge is virtually eliminated.

In order to compare theory to experimental penetration measurements, an expression based on Zebel's[1] theory has been used by Bergman et al.[2] among others. This expression relates the logarithm of the penetration ratio, P/Po, to the particle charge, size and dielectric properties, applied field strength, face velocity, fiber dielectric constant and porosity. It is important to note that this expression (and other similarly derived expressions) cannot be readily used in the case of the IESF. This is because the applied potential difference across the IESF filter material cannot be determined at this point. Further, except by direct measurement involving some uncertainty (as discussed previously), it is not possible to estimate the particle charge from theory since in the case of the IESF, the electrical field strength in the ionizing section (air space) is also not known.

References

1. G. Zebel, "Deposition of aerosol flow past a cylindrical fiber in a uniform electric field", J. Colloid Sci. 20: 552 (1965).

2. W. Bergman, A. Beirmann, W. Kuhl, B. Lum, A. Bogdanoff, H. Hebard, M. Hall, D. Banks, M. Mazumder, J. Johnson, Electric Air Filtration: Theory, Laboratory Studies, Hardware Development, and Field Evaluations, Lawrence Livermore National Laboratory, Rep # UCID-19952 (1984).

3. Masuda and Sugita, United States Patent No. 4,357,150, November 2, 1982, and United States Patent No. 4,509,958 (April 9, 1985).

4. R.A. Jaisinghani and T.A. Hamade, "Effect of relative humidity on electrically stimulated filter performance", JAPCA 37: 823-828 (1987).

5. R.A. Jaisinghani and N.J. Bugli, Advantages of Electrically Stimulated Filtration Over Conventional Filtration, presented at International Technical Conference, Am. Filt. Soc., Ocean City, MD, (March 21-24, 1988).

6. H. J. White, Industrial Electrostatic Precipitation, 134, Addison-Wesley, Reading, MA, (1963).

TABLE I. EFFECT OF FILTER MEDIA ON IONIZATION

RATED DOP EFF, %	AREA m^2	CURRENT mA	CHARGE μCOUL/g	ENHANCED 0.3 μm EFF, %
15	0.1	0.020	62.8	94.0
50	0.1	0.026	52.3	94.0
65	0.1	0.023	66.9	95.0
96	0.1	0.024	54.8	99.9
NO MEDIA	N/A	0.04-0.05	140.0-165.0	N/A

15.24 cm X 15.24 cm CROSS SECTION IESF WITH 4.45 CM DEEP PLEATS

8

TABLE II. THE EFFECT OF IONIZATION GAP ON EFFICIENCY ENHANCEMENT

CASES	APPLIED VOLT-TAGE KV	ELECT-TRODE GAP cm	APPLIED FIELD STRENGTH KV/cm	CURRENT mA	ENHANCED 0.3 μm EFF, %
1	0	N/A	N/A	N/A	67.0
2	2.5	0.5	5.0	0.00	72.0
2	11,0	0.5	22.0	0.22	71.2
3	11.0	2.0	5.5	0.01	73.0
4	11.0	2.0	5.5	0.18	99.3

TABLE III. EFFICIENCY ENHANCEMENT FOR DIFFERENT FIBERGLASS PAPERS

FLOW RATE m^3/HR	MEAS. MECH. 0.3 μm EFF, %	ENHANCED 0.3 μm EFF, %	% EFF. INCREASE	% PEN. REDUCTION $\bar{E}$ X 100
135.9	18.0	98.00	80.0	97.6
135.9	55.0	99.65	44.6	99.2
135.9	71.0	99.87	28.9	99.6
71.3	98.0	99.86	1.9	93.0
47.6	99.4	99.95	0.6	91.7
47.6	99.9	99.99	0.1	90.0

CONDITIONS: APPLIED FIELD STRENGTH = 4.3 KV/cm, 15.24 cm X 1524 cm CROSS SECTION, 4.45 CM DEEP PLEATS.

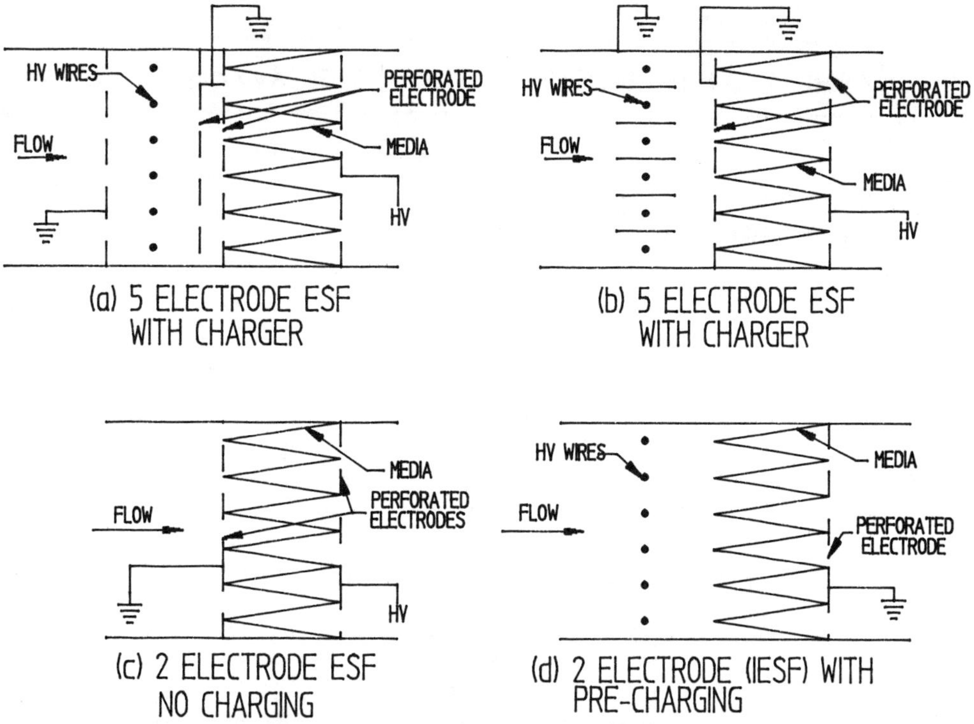

"Figure 1"   Electrode configurations.

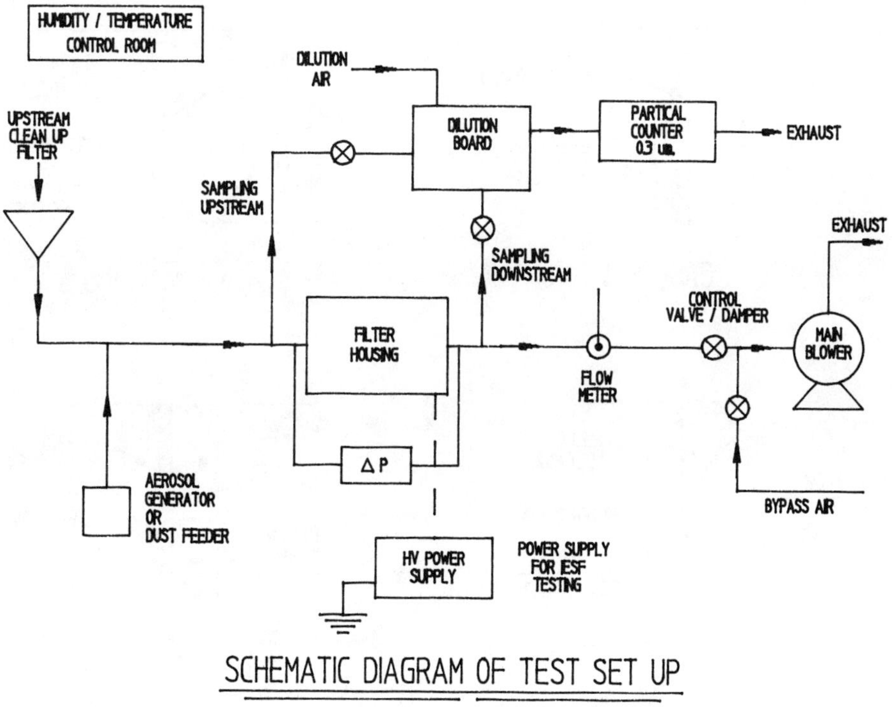

"Figure 2"   Schematic of Experimental set-up.

10

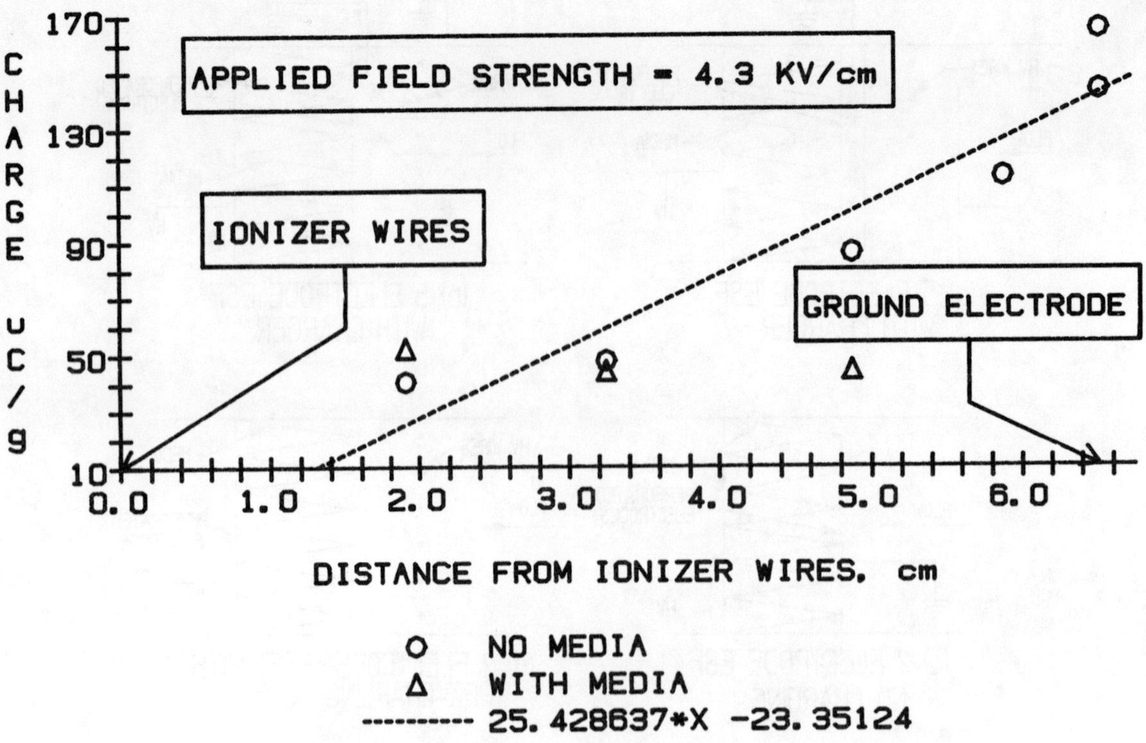

"Figure 3"   Charge levels within the IESF.

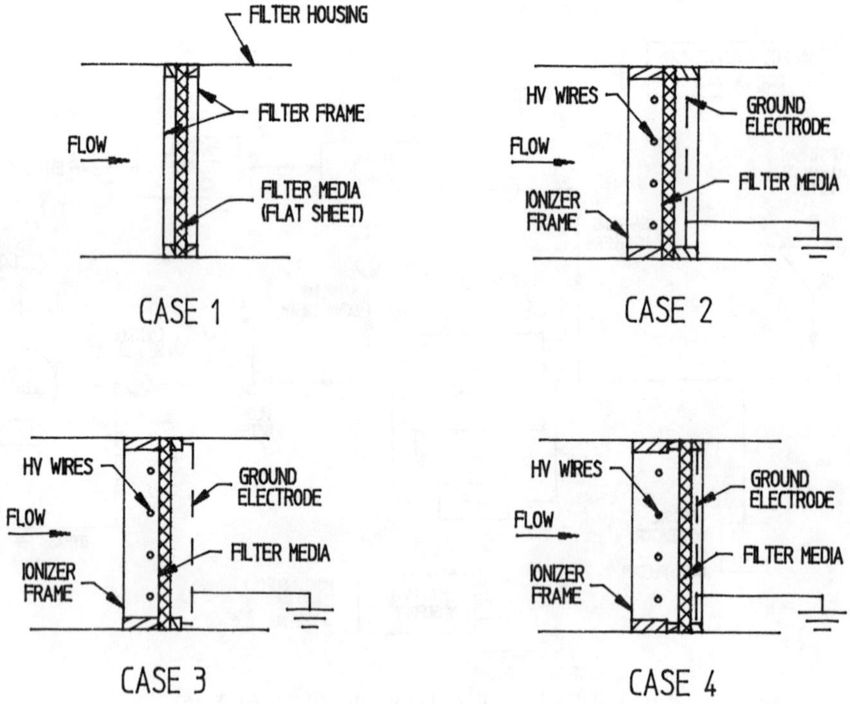

"Figure 4"   Experimental design to illustrate importance of the air gap between ionizer wires and media.

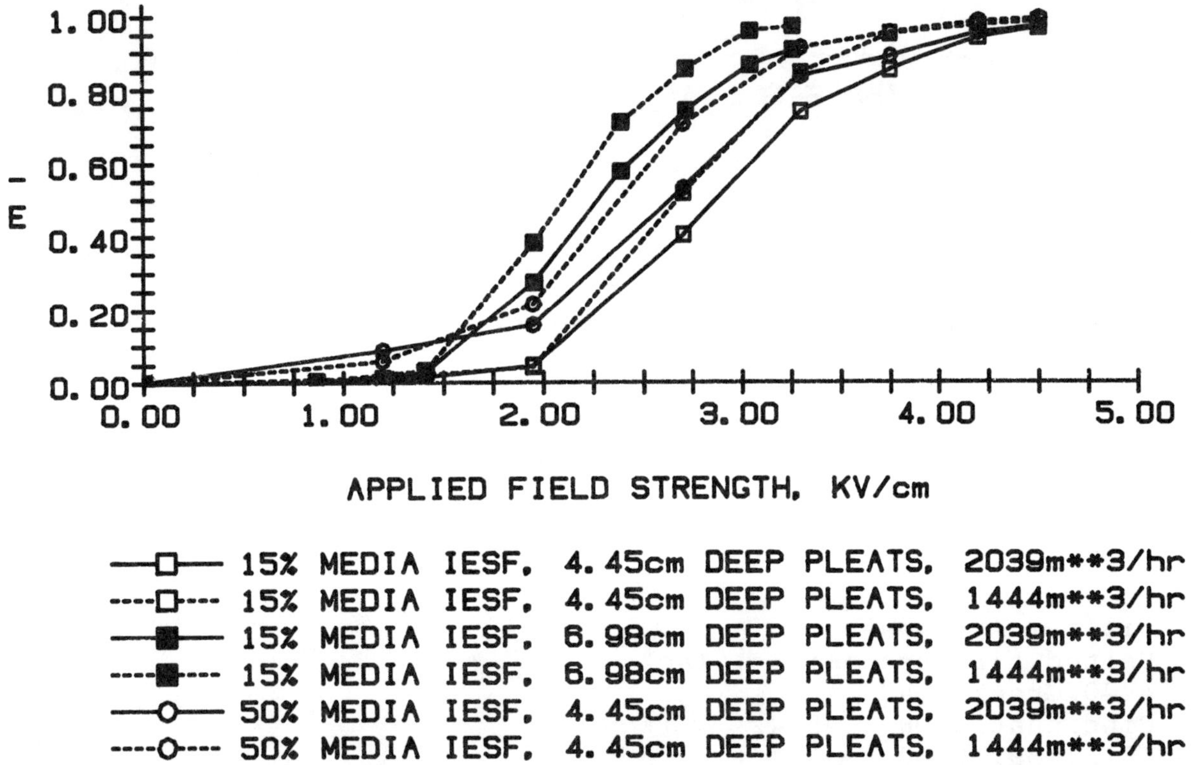

"Figure 5" *Dependence of fractional reduction of penetration, $\overline{E}$, on applied field strength.*

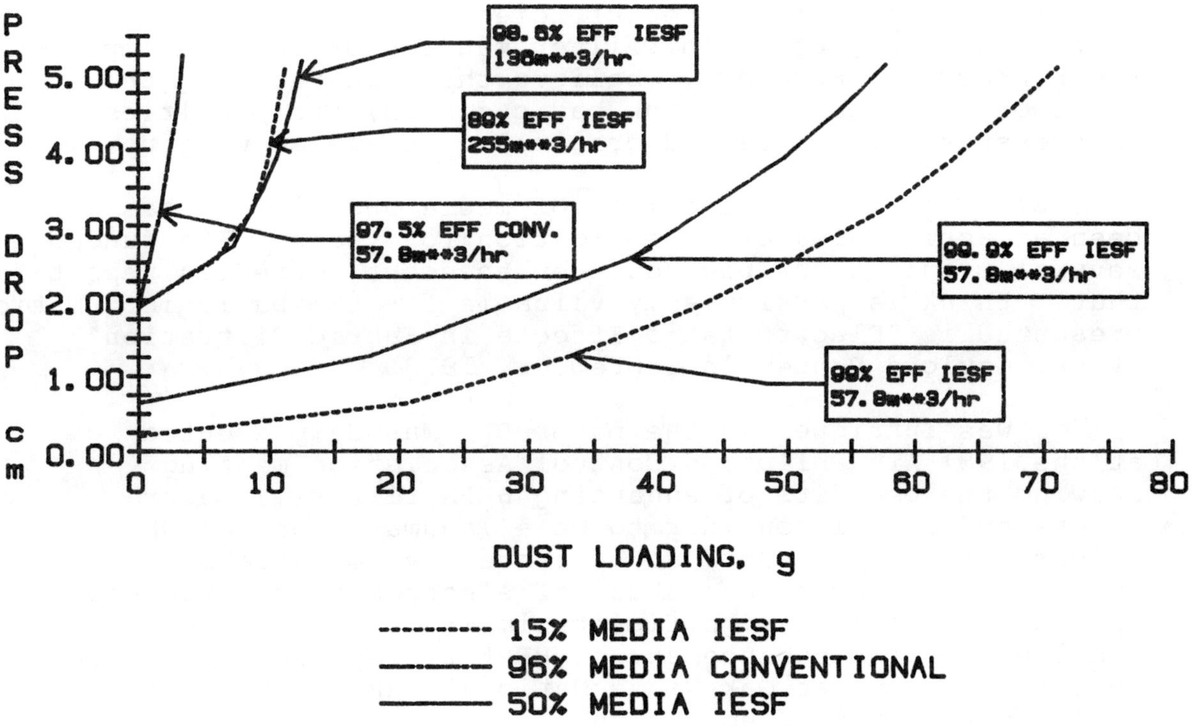

"Figure 6" *Equivalent basis (small scale) comparison of IESF and conventional filtration.*

A NEW METHOD OF ELECTROSTATIC ENHANCEMENT OF HEPA FILTERS

William A. Cheney
United Air Specialists, Inc.
4440 Creek Road
Cincinnati, OH 45242
(513) 891-0400

As the need for higher efficiency removal of submicron particulate has increased, filter manufacturers have improved their product. However, I believe they are reaching the limit of practicality in the basic design, and penalties in pressure drop, life and original cost are showing up.

Electrostatic enhancement of media and filters has been an active area of investigation for years. EPA funded several projects for bag house enhancement. One EPA report that I think is particularly valuable for the basic information presented is "Electrostatic Effects in Fabric Filtration" by Dr. Gaylord Penney in September, 1978.

I was intrigued by the paper Dr. Masuda presented at the 1981 Air Pollution Control Association meeting showing the benefits of enhancing HEPA filters. Later I obtained what I consider to be a landmark document by Werner Bergman and staff at Lawrence Livermore Labs summarizing a seven year study of electrostatic enhancement. My paper explains a significant simplification in the construction for enhancement of HEPA filters and documents test results run at Oak Ridge National Laboratories to evaluate the simplified design.

In Figure 1, you see the enhancement construction as proposed by Masuda in his 1981 APCA paper. By putting a precharger (ionizer) ahead of the HEPA and attaching alternate aluminum dividers together electrically, he created a field between the dividers that reduced penetration nearly two orders of magnitude.

The original enhancement design work done by Masuda and Bergman did not gain hoped-for acceptance. In addition to the ever present resistance to change, there was legitimate concern that arcing between divider plates in the HEPA could punch holes in the media and in the atomic energy field, and there was concern about a fire hazard. By tying alternate dividers together, arc energy draws power not only from the power supply, but also from the capacitance of all the dividers as they release energy in an arc.

In Figure 2, you see a modification of the original concept where dividers no longer have direct electrical contact with a high voltage power source. The upstream exposed plates maintain a voltage of about 1.5 kv from the air ions produced by the ionizer. The alternate adjoining plates are grounded to create a nominal voltage gradient through the media in excess of 4 kv per cm. This creates and electrical field condition that is the equivalent of the first Masuda design. The upstream dividers are now at positive potential instead of ground.

The modification to the HEPA for this design is relatively simple. It requires only that a conductive strip or spring be applied to the dividers that serve as ground in the enhanced HEPA. This conductive strip can easily be extended to a ground connection. The electrical energy in an arc in this HEPA with individually electrified plates will be reduced more than 98% compared to a unit that is directly connected for electrification. About half of this reduction comes from isolation of the high voltage transformer source from the charged divider plates. In my simplified design, the energy in any short circuit between dividers is limited to the capacitive discharge between two adjoining aluminum dividers.

Another advantage is that any momentary arc de-energizes the enhancement of only two air passages instead of the entire HEPA unit.

Evaluation of the effectiveness of the enhancement took me into new areas of technology. As a manufacturer of two stage precipitators, my company has done experimental work on efficiency by particle size, but because of inherent by-pass in a precipitator, we were not concerned about measuring efficiencies above 99%.

I found that Oak Ridge National Laboratories had a penetration measuring system that was developed over a 6-year period at the Los Alamos National Laboratory to qualify all nuclear grade HEPAs. It produces a particle size distribution with a geometric mean diameter of just under 0.2 micrometers. Sufficient particles on down to less than 0.1 micrometers are generated to permit penetration evaluation by particle size, in 14 classifications, from about .090 to .370 micrometers (as shown in Figure 3).

The HEPA used was a 24" x 24" x 12" fiberboard frame filter furnished by HEPA Corporation of Anaheim, California. This filter had been tested at 99.993% efficiency by HEPA Corporation. All tests reported in the attached Figures were run at 1,000 cfm. Sampling was conducted for 2 minutes on the upstream side of the filter and 12 minutes on the downstream side.

Tests were run with no enhancement and then repeated with enhancement. It was observed that the inlet test sample concentration always lowered when the enhancement ionizer was turned on. This indicated that the ionizer charging field must be having some impact upstream. A second set of tests confirmed this phenomenon. The challenge particle count dropped from 1527 to 946 when the ionizer was turned on.

At 0.12 micrometer, the penetration of an unenhanced filter compared to an enhanced is indicated to increase almost two orders of magnitude. This is shown in Figure 4.

The cost of adding the enhancement to a "HEPA construction" filter would add less than 5% to the filter cost. Since a conductive strip and ground terminal is the only change in the filter, it is possible to field retrofit most filters without removing them and having to reseal them into position.

The upstream ionization function would involve a one-time installation that would range from about $0.15 to $0.25 per cfm. There is an attractive payback on this investment in reduced power cost to move air to get the same efficiency as a higher air resistance media.

Runs were made at the 20% velocity rating also. At this flow rate, both showed relative insensitivity to particle size for both the enhanced and unenhanced penetration, but the enhancement showed an improvement of about one order of magnitude throughout the size range.

No field tests have yet been run with this design. Based on the reports of Masuda and Bergman, it is reasonable to anticipate that the lower pressure drop using the enhanced HEPA grade to obtain ULPA performance may be one of the most attractive economic aspects of this design. The longer life that was also documented by Bergman does not seem to be a factor in typical recirculation air clean room applications. However, in cleaning makeup air, it could be very beneficial.

Since I started with a 99.993% HEPA, the test results presented here do not offer proof that 99.97% HEPA filters can be upgraded to ULPAs, but it is a strong possibility.

Use in clean room benches to produce islands of clean air looks attractive because the forward curved blowers commonly used would react unfavorably to the increased pressure drop in the ULPA and the faster increase with loading would put an additional strain on maintenance of rated air flow.

No tests have yet been run on ULPA filters, but there is good reason to feel that their performance could be upgraded also by enhancement. Enhancement holds good promise as a solution to the problem of escalating pressure drop and media cost to obtain the super efficiencies now being considered.

The enhancement benefit carries over into lower efficiency media also. When we tested a 60% rated "HEPA" construction filter in an AHAM chamber filled with tobacco smoke, we found that it performed like a 95% rated HEPA in clearing tobacco smoke from the air. There was a 445% improvement in the equivalent tobacco smoke removal rate for the 60% unit with 57% less pressure drop than the equivalent 95% HEPA.

In summary, the simple electrostatic enhancement of aluminum plate divider HEPA type construction offers real benefits in improved efficiency, reduced cost to move air and extended life of a filter. This would extend from a 60% filter on to the highest efficiency that can be attained with media now available or that could be devised in the future.

References.

1. G. W. Penney, Electrostatic Effects in Fabric Filtration: Vol I, EPA - 600/7-142a, Sept., 1978.

2. S. Masuda, Electrostatically Enhanced HEPA Filter, Paper #81-45.3, Air Pollution Control Association, June, 1981.

3. W. Bergman, Electric Air Filtration: Theory, Laboratory Studies, Hardware Development and Field Evaluation, Lawrence Livermore National Laboratory, DOE Contract W-7405-ENG-48.

"Figure 1" Conventional enhancement.

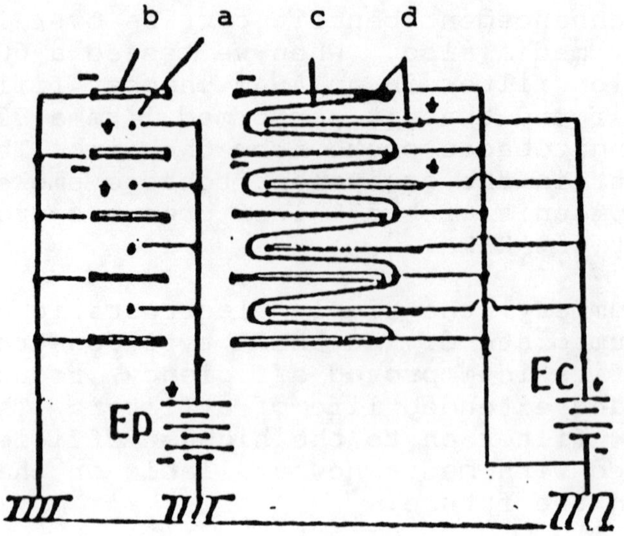

a - Corona wires

b - Ground electrodes for precharger

c - Filter media

d - Aluminum separators

E_p - High voltage source for precharger

E_c - High voltage source for collector

-5

"Figure 2" Simplified enhancement.

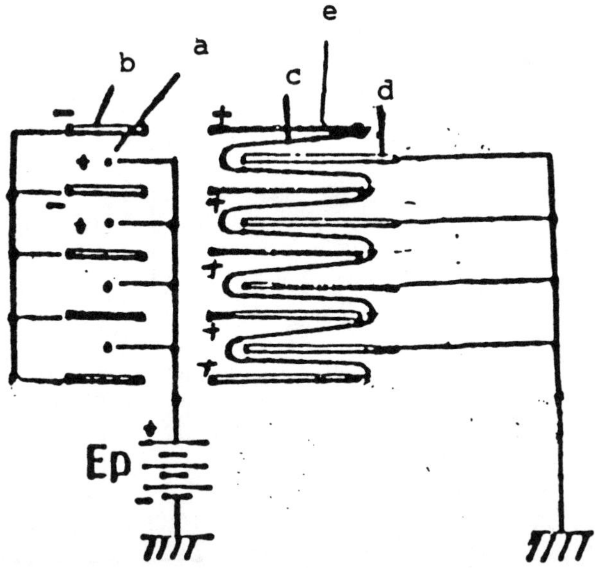

a - Corona wires

b - Ground plate for precharger

c - Filter paper

d - Grounded aluminum separators

e - Indirectly charged aluminum separators

E_p - High voltage source for precharger

"Figure 3" Particle size distribution in Oak Ridge test.

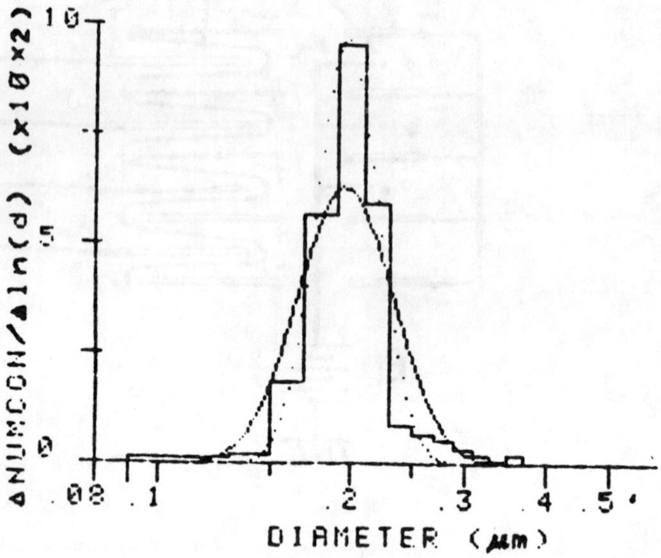

"Figure 4"   Enhanced HEPA efficiency comparison.

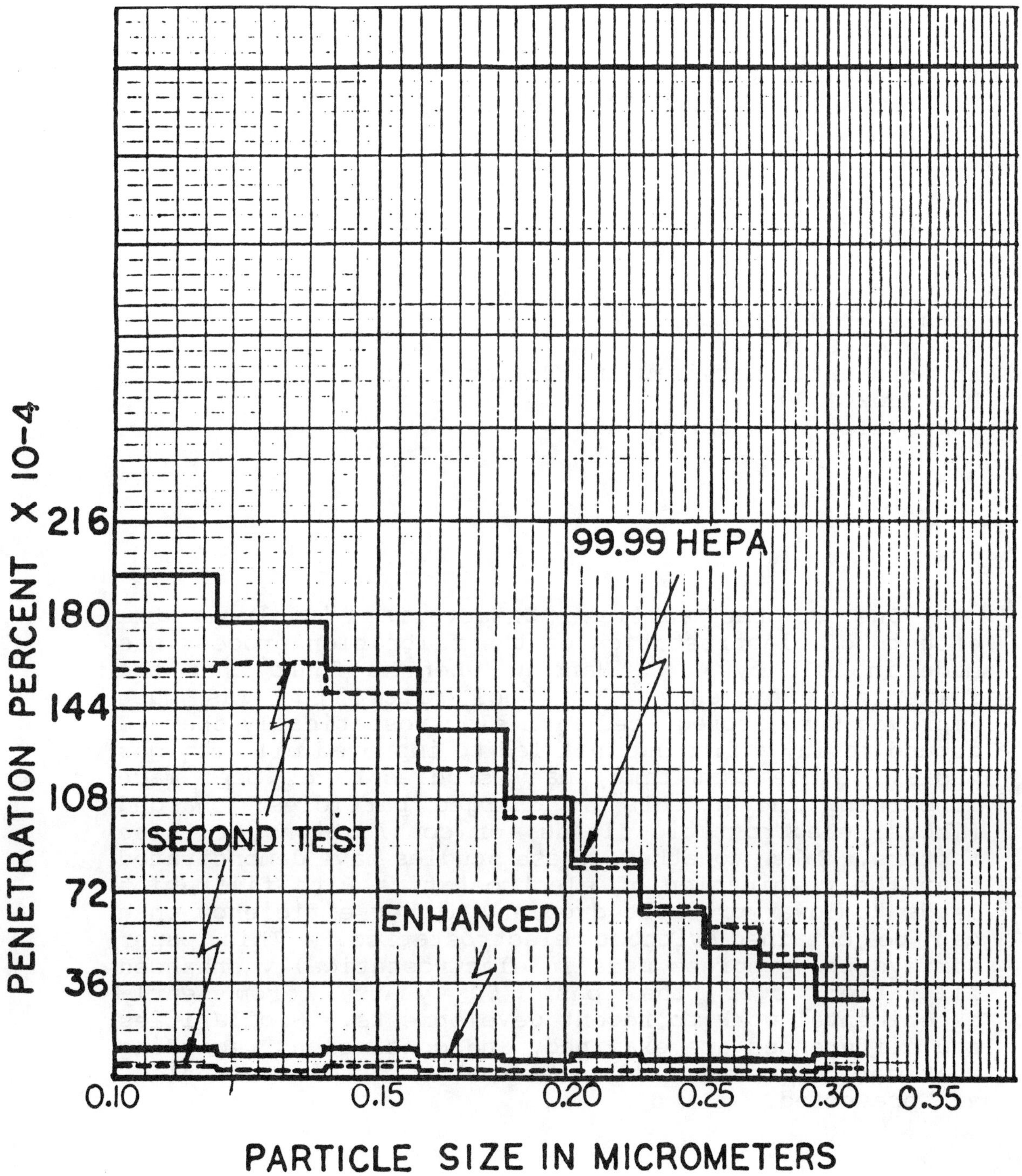

THE DEVELOPMENT OF
ELECTROSTATIC FILTRATION TECHNOLOGY

Dennis J. Helfritch
Research-Cottrell
Environmental Services and Technologies
P.O. Box 1500
Somerville, New Jersey 08876

One of the earliest discoveries involving the effects of electrostatics on the filtration process was made in 1930 by N. J. Hansen, when he found that the filtering efficiency of a wool filter pad was much improved when impregnated with electrostatically charged ground resin, a finding that led to improved military gas masks. Since that time, many studies have been made concerning the effects of charged particles, charged fibers, electrostatic fields, or combinations of these factors. Taken together, these studies have demonstrated that the combination of electrostatics with filtration leads to lower pressure drop and higher efficiency than can be obtained without electrostatics. This paper surveys the progress made in electrostatically enhanced filtration over the past 50 years, from early experimental work to recent developments. Similarities and differences are examined, and conclusions regarding what can be commercially expected from this technology are presented.

INTRODUCTION

Since the early findings of Hansen, a great many modeling and experimental studies have been undertaken. Lundgren and Whitby[1] investigated the effect of particle charge on the filtration efficiency of uncharged fibers. Inculet and Castle[2] describe a device in which particles are charged and a field is applied across the filter. Zebel[3] gives equations for the collision efficiency between a charged particle and a dielectric fiber in the presence of an external electrostatic field. Ariman and Tang[4] extend Zebel's results by including the effects of neighboring fibers.

Virtually all of these early investigations indicate that a substantial increase in collection efficiency can be realized when particles are charged or fields are applied. More recent work has concentrated on the effects of electrostatics on filter pressure drop, and it will be seen that significant reductions in pressure drop are realized when electrostatics are employed.

CHARGED PARTICLES

Much of the work done to date on electrostatically enhanced filters has been concerned with the filtration of charged particles. Lembach and Penney[5] performed laboratory studies of particles charged by impingement on a tungsten carbide surface and subsequently filtered. They found that nodular deposits were formed due to the establishment of localized electric fields. The development of a coarse, porous deposit has been observed by most investigators of charged particulate filtration, and this phenomenon is generally cited as the reason for low pressure drop.

The APITRON fabric filter utilized conventional filter bag construction with corona precharging. It was pilot scale tested during three very dissimilar applications - foundry cleaning room dust, redispersed silica dust, and welding fumes. It was found that particle charging prior to filtration gave rise to approximately a threefold decrease in penetration and a twofold decrease in filter drag for each case.

ELECTRIC FIELDS

Electric fields can be applied across filters perpendicular to the filter plane by means of electrode grids on opposite sides of the filter, or parallel to the filter plane by means of alternately charged wire electrodes on or within the filter. Chambers et al.[6] conducted a coal-fired boiler pilot scale test utilizing wire electrodes woven into full scale fiberglass filter bags. They found that for an applied field strength of 2.5 kv/cm, the filter pressure drop was about one half of that for no applied field. Chudleigh[7] tested various fabrics in a bench scale apparatus in which an electric

2

field could be applied normally to the filter plane. He found that the pressure drop across the filter decreased proportionately with an increase in the magnitude of the applied field and that the filter drag with maximum electrostatic enhancement was approximately one-half that for no electrostatics. Bergman[8] applied fields across fibrous filters and found that the application of a 10kv/cm field resulted in an approximate threefold decrease in penetration for particles between .1 and 1 micron.

CHARGED FILTERS

Utilization of natural or imposed charges on filtration fibers (electret fibers) is by far the simpliest technique for electrostatic enhancement since it requires no external electrical circuitry; however, the fiber charge must have a lifetime on the order of years to have practical use. Ackley[9] has found that high temperature, high humidity conditions greatly accelerate the charge depletion of fibers, and so electret fibers should not be considered for such service.

Van Turnhout[10], et. al., studied electret filters composed of dipolar fibers. They compared the efficiency of commercial face mask filters with electret filters in filtering sodium chloride aerosol. They found an order of magnitude decrease in penetration for electrostatic filtration. Brown and Blackford[11] found a five-fold decrease in penetration for charged filter elements lying parallel to the airflow as compared to the same, uncharged, filter.

SUMMARY

One can estimate the relative effectiveness of each type of electrostatic enhancement by comparing the radial forces between fibers and particles. These forces are:

Enhancement	Force
Charged Particles	$k_1\ Q_p^2/r^2$
Applied Field	$k_2\ R_p^3 R_f^2 E^2/r^3$
Applied Field + Charged Particles	$k_3\ EQ_p$
Charged Fibers	$k_4\ R_p^3\ Q_f^2/r^3$
Charged Fibers + Charged Particles	$k_5\ Q_p Q_f/r$

Where k_1, k_2, k_3, k_4, k_5 are all order of magnitude 1.
r = radial distance between fiber and particle
E = applied electric field
R_p, R_f = particle, fiber radii
Q_p, Q_f = particle, fiber charge

It is clear that the forces represented above can have vastly different magnitudes. It is therefore somewhat surprising that the cited experiments achieved remarkably similar electrostatic enhancements for efficiency and pressure drop.

The commercial utilization of electrostatically enhanced filters will certainly take advantage of the potential order of magnitude decrease in particle penetration. The recently promulgated PM_{10} regulations will promote the application of charged particle filters for industrial emissions. It is likely that electret filters will experience increasing usage as HEPA and as respiration filters. The future of this technology appears to be bright.

REFERENCES

1. D. Lundgren and K. Whitby, I & EC Process Design, Vol. 4, October 1965, pp. 345-349.

2. I. Inculet and G. Castle, G.S.P., ASHRAE Journal, March 1971, pp. 47-52.

3. G. Zebel, J. Colloid Sci, Vol. 20, 1965, pp. 522-543.

4. T. Ariman and L. Tang, Atmos, Environ., Vol. 10, 1976, pp. 205-210.

5. R. Lembach and G. Penney, JAPCA, Vol. 29, 1979 pp. 823-826.

6. R. Chambers, J. Spivey, and D. Harmon, EPRI CS-4404, Feb. 1986.

7. P. Chudleigh, Filtration & Separation, May/June 1983, pp. 213-216.

8. W. Bergman, Proceedings-Electrostatics in Filtration, November 1979.

9. M. Ackley, Filtration & Separation, July/Aug 1985, pp. 239-242.

10. J. Van Turnhout, W. Hoeneveld, J. Adamse and L. Van Rossen, IEEE Trans Ind. Appl, Vol 1A-17, 1981, pp. 240-248.

11. R. Brown and D. Blackford, Filtration & Separation, Sept/Oct 1983, pp 349-351.

PARTICLE DEPOSITION IN GRANULAR MEDIA UNDER UNFAVORABLE SURFACE CONDITIONS

Rajasekar Vaidyanathan*and Chi Tien
Dept. of Chem. Eng & Mat. Sci.
Syracuse University, Syracuse, NY 13244.
*Bird Machine Company, 100 Neponset St.
So. Walpole, MA 02071.
Tel: (508) 668 0400. Fax: (508)-668-6855

Particle deposition onto granular media is examined under the assumption the deposition surface is not homogeneous. Expressions are derived for the double layer interaction of particles with charged, patch-like regions on the granular surface and for the extent of deposition onto such a surface. Results of these calculations are compared with experimental data for hydrosol filtration under conditions usually considered unfavorable for deposition based on the well-known Deryaguin-Landau-Verwey-Overbeek (DLVO) theory. Under some conditions, the significant amounts of deposition observed experimentally could be explained by the present calculations.

Introduction

Particle deposition in granular porous media occurs and affects important processes such as deep bed filtration, oil-recovery operations and the movement of groundwater through soils. A proper understanding of these processes therefore requires a knowledge of the mechanisms of particle deposition involved. Earlier theoretical investigations of deep-bed filtration[1-3] have successfully identified these mechanisms and estimated the amounts of deposition or collection. However, their results agree with experiments only when filtration conditions are 'favorable', i.e., when the surface interaction forces, namely, the London and double layer forces between hydrosol and filter grain, are attractive based on the classical Deryaguin- Landau- Verwey- Overbeek (DLVO) theory. Under unfavorable conditions (repulsive surface forces), reduced but significant amounts of deposition still occur[4], whereas theoretically there should be zero deposition. The present work defines a new theoretical approach that examines the influence of heterogeneities in the double layer forces on particle deposition.

Model

The granular materials commonly used as filter media such as sand and glass beads are composed chiefly of silica and silicates. Silica surfaces have chemical heterogeneities as determined by several investigators[5-7]. These investigators found that at least two types of hydroxyl groups exist on silica surfaces, and their dissociative tendencies differ. Consequently, when immersed in an aqueous phase, different parts of the surface of a filter grain may be expected to acquire different electrostatic charge and exert different double layer forces on an approaching hydrosol particle.

In this work, the filter grain surface is assumed to contain two types of surface groups -- the first type dissociates completely and contributes entirely to the surface charge/ potential while the second type remains inert. The charged regions are assumed to occur in circular patches of radius r on the grain surface which interact electrostatically with an approaching particle at a distance s from the surface (Figure 1). The double layer interaction energy, V_{DL}, between the hydrosol particle and the granular surface may be determined using Deryaguin's method[8] together with the expression for the energy of two parallel plates interacting at constant surface potential, Vo[9] :

$$Vo(g_1, g_2, z) = [(g_1^2 + g_2^2)(1-\coth(Kz)) + 2g_1 g_2 \operatorname{cosech}(Kz)] \qquad (1)$$

In the above expression, g_1 and g_2 are the surface potentials of the two plates, z is the separation distance between them and K is the reciprocal Debye length. Deryaguin's method considers the sum total interaction of parallel rings on the particle and grain surface (Figure 1). For this situation, g_1 may be substituted by g_p, the surface potential of the particle and g_2

by the surface potential on the granular surface given by:

$$g_2 = \begin{cases} 0 & \text{outside the patch} \\ g_r & \text{within the patch} \end{cases} \qquad (2)$$

If the charged patches are well separated from each other, a particle will interact with at most one patch at a time. Application of Deryaguin's method leads to the following expression for V_{DL}:

$$V_{DL} = \int_0^{d_p/2} \int_{-\pi}^{\pi} Vo(g_p, g_2, z)\, d\theta R dR \qquad (3)$$

where (R, θ) are cylindrical coordinates on the grain surface as shown in Figure 1 and z is the separation distance between two rings. We expect that the surface potential of the patch, g_r lies in the interval $[g_g, g_g/f]$, where f is the fraction of granular surface covered by charged patches and g_g the streaming/zeta potential of the filter grain. In this study we restrict ourselves to the lower limit in the above interval.

The London interaction potential, V_{LO}, is not influenced by the surface charge heterogenieties and we may use the expressions for V_{LO} already available in the literature[10-11].

The procedure for determining the extent of hydrosol deposition onto a filter grain is as follows. For sufficiently small particles, the sum of London and double layer forces, F_{TOT}, is the dominant factor that determines whether the particles can or cannot deposit on the granular surface. Hydrodynamic forces which are related to the bed superficial velocity have little influence in making the above determination. Let w be the distance of the projection of the particle center onto the granular surface from the center of the patch (Figure 1). The possibility of deposition is dictated by the sign of F_{TOT}. For fixed w, particle deposition will take place if:

$$F_{TOT} = -[\nabla (V_{DL} + V_{LO})] < 0 \text{ for all s} \qquad (4)$$

where ∇ is the gradient operator. By testing equation (4) for various values of w, we can determine a circular region of radius w_c centered on the patch such that:

$R < w_c$ equation (4) not valid, no particle deposition
$R > w_c$ equation (4) valid, particle deposition (5)

Under favorable conditions, $w_c \equiv 0$ and equation (4) is applicable for all w. As the surface conditions become progressively unfavorable, the value of w_c will increase as a circular portion of the granular surface concentric with the patch becomes unavailable for deposition. Under highly unfavorable conditions we could have $w_c > r$. The fraction of the grain surface <u>available</u> for deposition is

$$\lambda/\lambda_{Fav} = 1 - f.(w_c/r)^2 \qquad (6)$$

where λ/λ_{Fav} is the ratio of the actual filter coefficient to its value under favorable surface conditions, (i.e., when the entire surface is available for deposition).

Experimental

Filtration experiments were performed by passing suspensions of latex spheres through granular beds composed of soda-lime glass beads. The latices were obtained in narrow size ranges with mean diameter dp = 4 & 11.9 µm and were surfactant-free styrene-divinylbenzene copolymer spheres. The glass beads were sieved into a narrow size range with mean diameter dg= 525 µm. The porosity of the granular beds was 0.4. Prefiltered distilled water was used to suspend the latex particles and sodium chloride (NaCl) was used as the background electrolyte. Conditions for particle deposition could be made favorable or unfavorable by varying the background NaCl concentration between 0.171 and 0.0001 M. Surface potentials on the latices and glass beads were estimated using a Rank Bros. Mark II unit, while latex deposition was determined by measuring particle concentrations entering and leaving the filter bed using a Coulter Counter model ZB. The superficial velocity, u_S, was maintained at a constant value during each experiment.

In the initial stages of deposition, the filter coefficient may be determined by the expression:

$$\lambda = (1/L).\log_e[\text{inlet particle concentration/outlet concn}] \quad (7)$$

where L is the height of the granular bed.

Results

The value of f, the fraction of glass bead surface that is charged, was chosen to be 0.2, which is in the range of values for silica surfaces[5-7]. The patch radius, r, is a free parameter whose value is difficult to estimate independently. In this work r was taken = 0.5 µm, which is in the size range of the geometric imperfections on the glass bead surface. V_{DL} and F_{TOT} (equations 3 & 4) were evaluated numerically and the value of w_c determined. Subsequently, λ/λ_{Fav} was determined from equation (6).

Figures 2-3 show results from the experiments and their comparison with the theoretical model. The dimensionless filter coefficient, λ/λ_{Fav}, is plotted vs. the concentration of NaCl. As the NaCl concentration decreases, λ/λ_{Fav} decreases because conditions become progressively unfavorable for deposition as the repulsive double layer force between particles and granular surface increases. One can see that the present theory is remarkably better at predicting the qualitative and quantitative trend of the experimental observations, as compared to the earlier theory[3] (dotted line), which does not account for surface heterogeneities. Further, as anticipated, the experimental results show little or no dependence on the superficial velocity of the suspension through the filter bed.

Discussion

The patch model used in this study appears capable of explaining and possibly predicting the significant amounts of deposition observed in porous media under unfavorable conditions. More work is needed in clearly establishing the model parameters such as the patch size for a given granular surface and this in itself is a very difficult problem. On the theoretical side suitable techniques need to be devised to calculate the double layer interaction energy when the patches are clustered close together rather than far apart. Finally, it must be noted that the present model ignores all forms of surface interactions other than the London and van der Waals type. One may encounter other interaction forces such as steric forces in many real systems.

References

1. J.A. FitzPatrick, "Mechanisms of particle capture in water filtration", Ph.D. dissertation, Harvard University, Cambridge, MA 1972.

2. A.C. Payatakes, "A new model for granular porous media - application to filtration through packed beds", Ph.D. dissertation, Syracuse University, Syracuse, NY 1973.

3. R. Rajagopalan, C. Tien, "Trajectory analysis of deep bed filtration with the sphere-in-cell porous media model," AIChEJ, 22: 523 (1976).

4. R. Vaidyanathan, C. Tien, "Hydrosol filtration in granular beds - an experimental study," Chem. Eng. Comm., 81: 123 (1989).

5. C.G. Armistead, A.J. Tyler, F.H. Hambleton, S.A. Mitchell, J.A. Hockey, "The surface hydroxylation of silicas", J. Phys. Chem., 73: 3947 (1969).

6. J.R. Goates, C. Anderson, "Acidic properties of Quartz," Soil Science, 81: 277 (1956).

7. M.L. Hair, W. Hertl, "Reactions of chlorosilanes with Silica surfaces," J. Phys. Chem., 73: 2372 (1969).

8. B.V. Deryaguin, "On the repulsive forces between charged colloidal particles..", Trans. Far. Soc., 36: 203 (1940).

9. R. Hogg, T.W. Healy, D.W. Fuerstenau, "Mutual coagulation of colloidal dispersions", Trans. Far. Soc., 62: 1638 (1966).

10. J. Czarnecki, "van der waals attraction energy between sphere and half-space", J. Colloid Interface Sci., 72: 361 (1979).

11. N.F.H. Ho, W.I. Higuchi, "Preferential aggregation and coalescence in heterogeneous systems", J. Pharm. Sci., 57: 436 (1968).

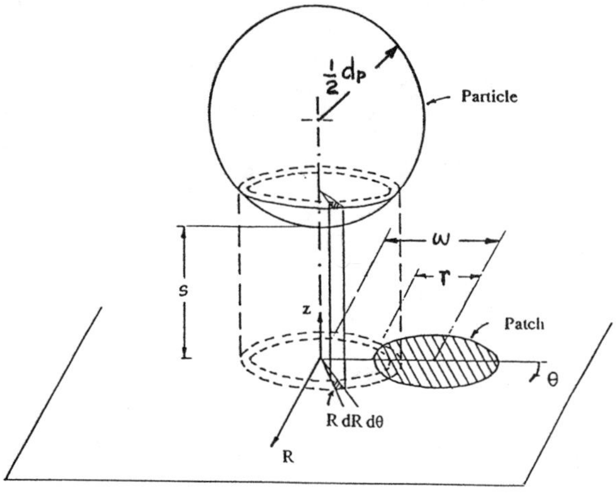

Fig. 1 Interaction of a Sphere With a Plane Containing a Solitary Charged Patch.

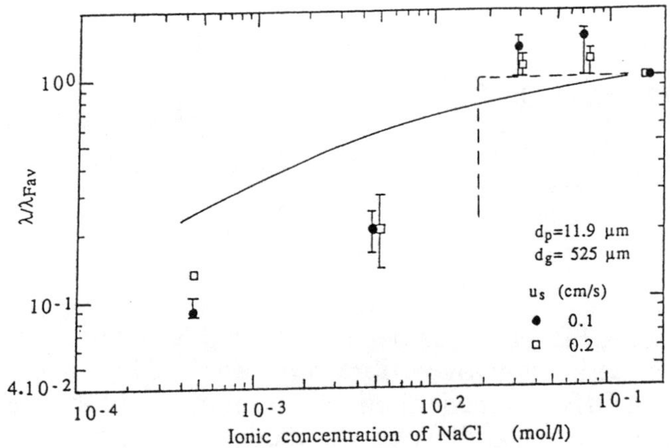

Fig. 3 Results of the Patch Model (Solid line, f=0.2, τ =0.5μm), Compared with Experiments.

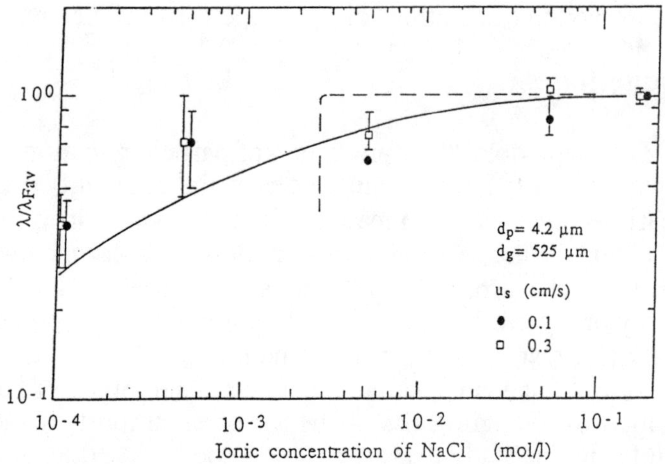

Fig. 2 Results of the Patch Model (Solid line, f=0.2, τ =0.5μm), Compared with Experiments.

THE EFFECT OF DIFFERENT FLOW FIELD MODELS ON AEROSOL PARTICLE CAPTURE IN NONWOVEN FIBROUS FILTERS

Majid Zia, Mohammad Faghri and Richard C. Lessmann
Fram/Allied Filtration Research Laboratory
Department of Mechanical Engineering
University of Rhode Island,Kingston Rhode Island, 02881

Abstract

Recently a user friendly computer code, FILTER, has been developed to predict particle capture and pressure drop for aerosol filtration by nonwoven fibrous materials. This code uses results from a complicated three-dimensional Offset Screen micro-structural model of the media to calculate pressure drop; but it uses a much simpler parallel cylinder model to calculate particle capture in single fiber size materials for small fiber Reynolds numbers. The original version of FILTER employed the Kuwabara flow field to model the flow around individual fiber collectors. This study examines the effect of other collector flow field models on the prediction of particle capture efficiency, including those due to Happel and Sangani and Acrivos. Capture efficiency is calculated by a unified particle trajectory mapping approach which includes the three mechanisms of interception, inertial impaction, and Brownian diffusion. Variations on the order of ten percent were observed in calculations of the single fiber efficiency.

Introduction

The methodology for prediction of particle retention within a simulated filter can be divided into two parts. First, a suitable model of the undisturbed flow field within the filter is needed. The earliest such models idealized the filter as an array of isolated cylinders (Langmuir[1] and Friedlander[2]), whereas more recent models consider the hydrodynamic interactions between the elements of the array (Kuwabara[3] and Happel[4]). Because the forces which determine the trajectory of a particle depend upon the flow field, it is preferable to choose a model which yields an analytical result. This is true of the models developed by Kuwabara[3], Happel[4], and Sangani and Acrivos[5] . Second, a suitable mathematical model of particle motion is needed. This may be developed by including the influence of all important mechanisms which effect a particle's trajectory. In this study these are taken to be Stokes drag and Brownian diffusion.

The purpose of the present study is to compare the approximate flow fields of Kuwabara[3] and Happel[4] to the more exact result of Sangani and Acrivos[5] and to examine the effects on these different models on particle collection efficiency. The results to be discussed have been generated by the new version of the computer code FILTER. The original code, which was configured for execution on a PC, used the simple model of Kuwabara[3]. This has been expanded to incorporate the flow models of Happel[4] and Sangani and Acrivos[5], but in its present form requires a larger computer. FILTER incorporates the effect of Brownian diffusion directly in the trajectory problem through the use of an effective particle radius[6]. This makes it unnecessary to separately solve a convection diffusion equations for small particles, and a trajectory equation for particles with significant inertia. The trajectory equation is solved using a fourth order Runge-Kutta integration algorithm, and a Perturbation scheme applicable for small stokes number is employed.

Formulation of the problem

In this study, a non-woven single fiber size filter is modeled as a hexagonal array of cylinders placed perpendicular to the main flow direction as shown on figure 1. Such an idealization enables the analysis to be simplified by computing the flow only in a representative unit cell. Examples are shown on figure 2. Sangani and Acrivos[5] obtained an approximate truncated series solution for the flow stream function, in the exact geometry of the unit cell of figure 2a, by an unconventional numerical method. For low solidities, Kuwabara[3] and Happel[4] approximated the unit cell as an imaginary fluid filled cylinder coaxial with a fiber, whose volume fraction was equal to the filter solidity as in figure 2b. This leads to very simple closed form solutions.

The flow fields

The steady motion of an incompressible fluid, at very small fiber Reynolds number, through an array of fibers is governed by the stokes equations.

$$\nabla.\mathbf{u} = 0 \tag{1}$$

$$\nabla p = \mu \nabla^2 \mathbf{u} \tag{2}$$

For two-dimensional flow these creeping flow equations can be written in terms of the stream function ψ and vorticity ω as:

$$\nabla^2 \Psi = \omega \tag{3}$$

$$\nabla^2 \omega = 0 \tag{4}$$

Under the assumption of a fully developed periodic flow the associated boundary conditions for the unit cell of figure 2a are:

$$\begin{gathered} \Psi = \omega \text{ on BC}, \quad \frac{\partial \Psi}{\partial x} = \frac{\partial \omega}{\partial x} = 0 \text{ on CD and GH}, \quad \Psi = \frac{\partial \Psi}{\partial n} = 0 \text{ on BH} \\ \Psi = \sqrt{3}/2, \quad \omega = 0 \text{ on FG}, \quad \Psi = \sqrt{3}/2, \quad \frac{\partial \Psi}{\partial n} = 0 \text{ on DF} \end{gathered} \tag{5}$$

Sangani and Acrivos[5] applied the boundary conditions along GH, HB, and BC directly to the solution of equations (3) and (4) in polar coordinates to arrive at the stream function given as:

$$\begin{aligned} \Psi_s = \Big[a_1 r^3 \Big\{ \frac{4\ln(r)}{2\ln(r)+1} \left(\frac{R}{r}\right)^2 + \frac{2\ln(R)-1}{2\ln(R)+1} \left(\frac{R}{r}\right)^4 \Big\} + b_1 R^2 r \Big\{ 1 - \frac{2\ln(r)}{2\ln(R)+1} - \\ \frac{1}{2\ln(R)+1} \left(\frac{R}{r}\right)^2 \Big\} \Big] \sin(\theta) + \sum_{n=2}^{N} \Big[a_n r^{2n+1} \Big\{ 1 - 2n\left(\frac{R}{r}\right)^{4n-2} + (2n-1)\left(\frac{R}{r}\right)^{4n} \Big\} + \\ b_n R^2 r^{2n-1} \Big\{ 1 - (2n-1)\left(\frac{R}{r}\right)^{4n-4} + 2(n-1)\left(\frac{R}{r}\right)^{4n-2} \Big\} \Big] \sin((2n-1)\theta) \end{aligned} \tag{6}$$

The remaining boundary conditions along DF, FG, and CD are forced to be satisfied at M>N arbitrarily chosen points. This gives rise to 2M linear algebraic equations in 2N unknowns, which can not be satisfied simultaneously. Hence the unknowns a_n and b_n were determined such that these equations were satisfied in a least square sense by using a Single Value Decomposition Algorithm. This is a difficult numerical procedure requiring double precision arithmetic for accuracy and is not suited for execution on a personal computer.

Sangani and Acrivos[5] used their solution to evaluate the drag per unit length of a fiber which determines the pressure drop of the array. Based on their study minimum values of M and N required to give an accurate drag result, for the full range of solidities were, 90 and 40 respectively. A similar sensitivity analysis for this study, led to the choice of M and N equal to 150 and 60.

Kuwabara[3] and Happel[4] approximated the unit cell boundary as shown on figure 2b. Both these investigators assumed that the fluid beyond the virtual surface was is in uniform flow. Therefore at the cell boundary

$$u_r = - u_0 \cos(\theta) \tag{7}$$

To uncouple the inner and outer cell flows Kuwabara[3] proposed a zero vorticity boundary condition on the virtual surface and Happel[4] suggested a zero shear stress condition on this boundary. Thus, the following expressions result from Kuwabara[3] and Happel's[4] approximate solutions to the stream function:

$$\Psi_k = \frac{u_0 r}{2k} \left\{ 2\ln\left(\frac{r}{R}\right) - 1 + c + \left(1 - \frac{c}{2}\right)\left(\frac{R}{r}\right)^2 - \frac{c}{2}\left(\frac{r}{R}\right)^2 \right\} \sin(\theta) \tag{8}$$

$$\Psi_h = \frac{u_0 r}{2h} \left\{ 2(c^2+1)\ln\left(\frac{r}{R}\right) + \frac{c^2-1}{c^2+1} + \frac{1}{c^2+1}\left(\frac{R}{r}\right)^2 - \frac{c^2}{c^2+1}\left(\frac{r}{R}\right)^2 \right\} \sin(\theta) \tag{9}$$

where the Kuwabara[3] and Happel[4] coefficients, k and h, are given as:

$$k = -\frac{1}{2}\ln(c) - 0.75 + c - \frac{c^2}{4} \tag{10}$$

$$h = -\frac{1}{2}\left\{ \ln(c) + \frac{1 - c^2}{1 + c^2} \right\} \tag{11}$$

The trajectory equation

In order to predict particle capture it is necessary to track a particle from its upstream launch position until it is either captured by a fiber, or misses and is past the fiber by the flow. The critical or grazing trajectory is the path of a particle which all particles traveling below this path are captured by the fiber, whereas those above this path escape collection. If the number of particles are evenly distributed in the flow then the single fiber efficiency, which is defined as the ratio of the number of particles striking the fiber to the number of particles which would strike if the flow was not diverted by the fiber, can be expressed as:

$$\eta = \frac{\Psi_{critical}}{u_0 R} \tag{12}$$

The effective particle radius by Lessmann[6] ,which accounts for the influence of Brownian diffusion, will be used for the capture criteria in this analysis. Capture is assumed when the particle and fiber center to center distance is less than the value given by the following equation:

$$r_{eff} = \sqrt{4\beta^2 Dt + R_p^2} \quad (13)$$

where the Eienstein diffusion coefficient is given as:

$$D = \frac{CKT}{6\pi\mu R_p} \quad (14)$$

The governing equation of motion for a small rigid spherical particle in a viscous flow is given as:

$$m_p \frac{dv}{dt} = 6\pi\mu R_p \ (\mathbf{u} - \mathbf{v}) \quad (15)$$

This equation may be non-dimensionalized using the fiber radius and the mean flow speed as length and velocity scales, respectively. The result of this is:

$$\sigma \frac{d\mathbf{v}^*}{dt^*} = \mathbf{u}^* - \mathbf{v}^* \quad (16)$$

The trajectory of a particle is determined by integrating Equation (16) by the use of a fourth order Runge Kutta Algorithm. For low values of the stokes number, $\sigma<1$, a perturbation method described by Banks and Kurowski[7] is used in this computation. The grazing trajectory is determined by a conventional "hit or miss" criteria. The particles are injected from the upstream boundary with a velocity equal to the local fluid velocity in the main flow direction. Once two trajectories are determine in which one misses and the other hits the fiber a particle is injected with a mean vertical position of the initial vertical coordinates of the hit and missed trajectories. This process is repeated until the percentage difference in efficiency between the hit and missed trajectories is less than 1%.

Results and Discussion

Comparison of the flow fields

To examine differences in the three flow field models their stream lines, and their radial, tangential, and absolute velocity components were examined for two packing densities: c=0.0086 and c= 0.299.

Figure 3a shows the stream lines for the three flow models for a solidity of 0.0086. The solid, dashed, and dotted lines represent the stream lines associated with the Sangani and Acrivos[5], Kuwabara[3], and Happel[4] flow models, respectively. The solid-dashed arc circle represents the virtual surface for the Kuwabara[3] and Happel[4] models. It is seen from this figure that the stream lines of Kuwabara[3] better approximate the flow obtained by Sangani and Acrivos. Happel's model yields stream lines which are always farther from the fiber surface. It is also apparent that the stream lines obtain by the Kuwabara[3] and Happel[4] model converge at the virtual boundary.

In figure 3b the non-dimensional radial velocity components of the three flow field models have been plotted against non-dimensional radial position for various angles. As in figure 3a, the solid, dashed, and dotted lines represent the results associated with Sangani and Acrivos, Kuwabara[3], and Happel[4]. The radial velocity components have been non-dimensionalized with respect to the mean flow velocity u_0. The non-dimensional radial positions of 0 and 1 represent positions on the fiber and virtual surfaces, respectively. It is seen from these figures that the Kuwabara[3] flow model agrees well with Sangani and Acrivos except near the virtual boundary. Their discrepancies become very pronounced at larger angles. Happel's[4] model results in radial velocities which everywhere higher than those predicted by Kuwabara[3]. This might lead one to believe that, at least for low packing densities, the Happel flow field should give higher predicted efficiencies than either of the other two models. This, however, is not the case.

In figure 3c the non-dimensional tangential velocity component of the three flow fields have been plotted against non-dimensional radial distance. The trend in these figures is similar to those observed in figure 3b. The velocities predicted by Kuwabara[3] and Happel[4] deviate at distances close to the virtual boundary. These deviations are more pronounced at smaller angles.

Figures 4a through 4c represent the same types of plots as figures 3a through 3c but for a solidity of 0.299. The trends are the same as in the low solidity case except that the discrepancies between Sangani and Acrivos and the other two models have become much larger, and moved closer to the fiber surface. This might have been expected since the model of Sangani and Acrivos was developed to better represent the flow field at higher packing densities.

Comparison single fiber efficiency computations

In figure 5 the percentage difference between single fiber efficiency calculations using the flow fields of Kuwabara, or Happel, and Sangani and Acrivos have been plotted against the interception parameter, $I=R_p/R$, assuming a packing density of 0.0086. The solid line represents predictions from either the Kuwabara or Happel models for high particle inertia (large stokes number). This shows a negative percentage difference of about 12% over the entire range.

The dashed and the dashed-dot lines represent results from the Kuwabara and Happel flow model predictions for low particle inertia (small stokes number). The single fiber efficiency resulting from Kuwabara's model agrees with that predicted by Sangani and Acrivos with essentially no discrepancy. Happel's model under predicted by 12%.

In figure 6 the same analysis has been preformed for a solidity of 0.299. The single fiber efficiency calculations using the Kuwabara and Happel flow models, for particles with large Stokes number, showed only less than a 1% negative difference up to an interception parameter of 0.01. Beyond this point the difference increases dramatically reaching thirty percent near I=0.5. Such large differences should not be taken as two significant as the trajectory model employed in this study does not include near fiber surface modifications to the hydrodynamic drag which would be important for correctly predicting the motion of particles with large interception parameters.

For low Stokes number using the flow field of Kuwabara generally over predicts by 12 to 13%, while using the Happel field under predicts by around 5%. Again the larger discrepancies apparent at higher interception parameters are most likely not modeled correctly.

Comparison with experimental data

Figures 7 and 8 show single fiber efficiency comparisons with data taken from the measurements of Lee[8]. These cases represent extremes from the reported data base with solidities of 0.0086 and 0.299. The experiments were done using DOP as the aerosol and filter mats composed of 11 micron diameter dacron fibers. In these figures the solid, dashed, and dotted lines represent results obtained by the use of Sangani and Acrivos, Kuwabara, and Happel flow field models, respectively. The face velocity for both cases is 30 cm/sec.

As shown on figure 7, for the low packing density, the predictions obtained by the use of the Kuwabara and Sangani and Acrivos models are identical, and Happel's predictions are slightly lower everywhere. At higher packing densities, as on figure 8, results using Kuwabara and Happel are lower then those using Sangani and Acrivos for small particle size where diffusion dominates as a capture mechanism; however all models give reasonable predictions in this region. For larger particle sizes, where particle inertia becomes important, the curves cross and the model by Sangani and Acrivos gives lower predictions which are more characteristic of the data.

Acknowledgement

The authors wish to gratefully recognize the Allied Aftermarket Division of the Allied Signal Corporation, the Hollingsworth and Vose Corporation, and the Donaldson Company for the financial support which made this work possible.

Nomenclature

a_n	stream function coefficient	t	time
b_n	stream function coefficient	T	temperature
c	packing density or solidity	**u**	fluid velocity vector
C	Cunningham slip factor	u_r	fluid radial velocity
h	Happel factor	u_q	fluid tangential velocity
k	Kuwabara factor	u_0	mean fluid velocity
K	Boltzmann's constant	v	particle velocity vector
m_p	particle mass	β	diffusion coefficient
n	outward unit normal	θ	angular position
p	pressure	θ_c	initial critical angular position
r	radial position	μ	fluid viscosity
R	fiber radius	η	single fiber efficiency
R_p	particle radius	ψ	stream function
R_v	virtual radius	ω	vorticity function

References

1. Langmuir, I., "Report on smokes and Filters", Section I U.S. Office of Scientific Research and Development, No 865, part IV (1942).
2. Friedlander, S. K., "Theory of aerosol filtration," Ind. Eng. Chem. 50, 1161-1164 (1958).
3. Kuwabara, S., "The forces experienced by randomly distributed parallel cylinders or spheres in viscous flow at small Reynolds number," J. Phs. Soc. Japan, 14, 257 (1959).
4. Happel, J., "Viscous flow relative to arrays of cylinders," AIChE Journal, 5, 174 (1959).
5. Sangani, A. S., and Acrivos, A., "Slow flow past arrays of cylinders with application to heat transfer," Int. J. Multiphase Flow, 8, 193 (1982).
6. Lessmann, R. C., Faghri, M., Khan A., and Zia, M., "A unified trajectory mapping approach to predicting particle capture in fibrous filters," Proceedings of the Annual Technical Conference on Filtration and Separation, American Filtration Society, 171 (1989)
7. Banks, D. O., and Kurowski, G. J., "A perturbation method for approximation of the inertial collection efficiency for fibrous filters with electrical enhancement," J. Aerosol Sci., 4, 463 (1983).
8. Lee, K. W., "Filtration of submicron aerosols by fibrous filters," PhD Dissertation, Department of Mechanical engineering, University of Minnesota (1977).

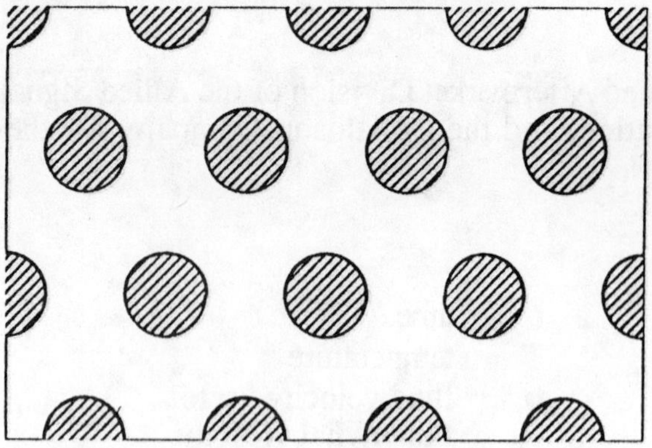

Fig. 1 - Hexagonal Array

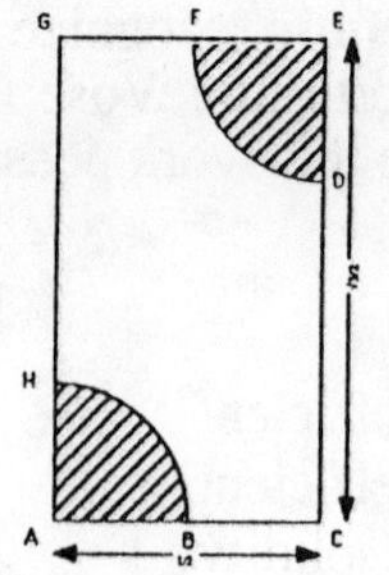

Fig. 2a - Sangani & Acrivos Unit Cell

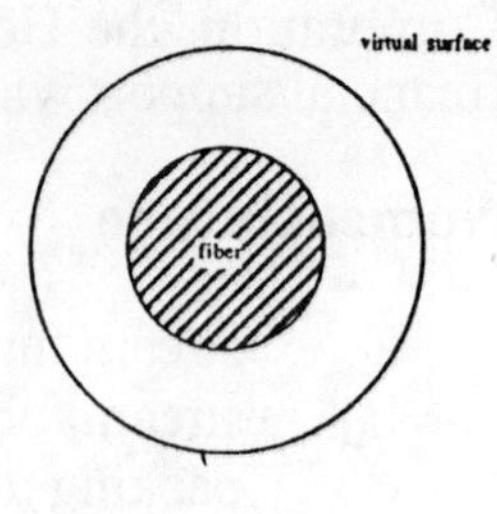

Fig. 2b - Kuwabara & Happel Unit Cell

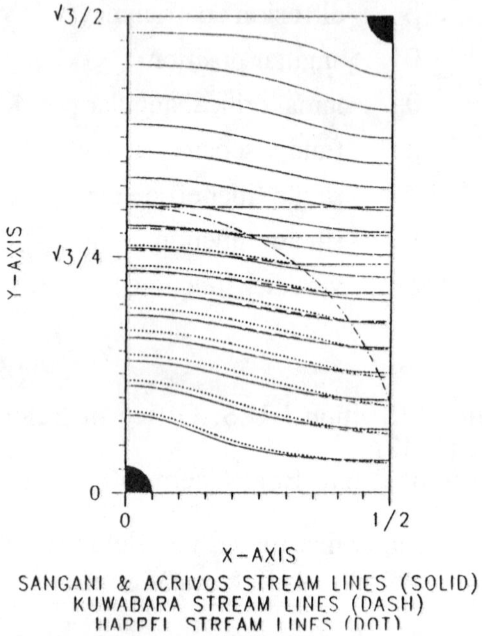

Fig. 3a - Streamlines, c = 0.0086

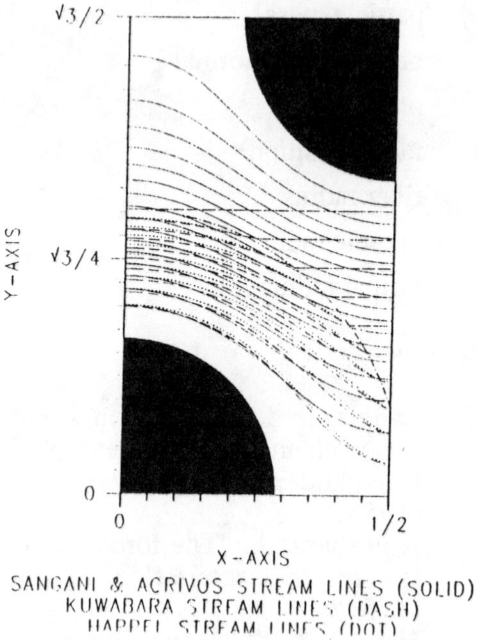

Fig. 4a - Streamlines, c = 0.299

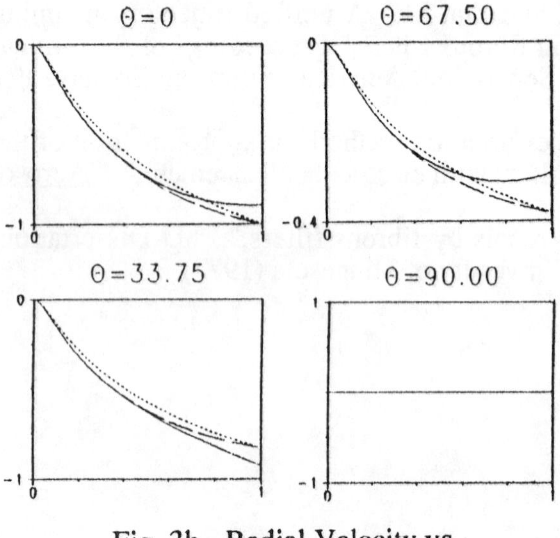

Fig. 3b - Radial Velocity vs. Radial Position, c = 0.0086

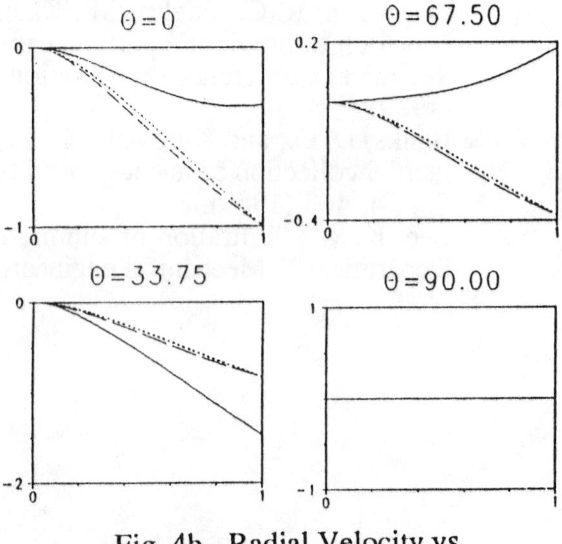

Fig. 4b - Radial Velocity vs. Radial Position, c = 0.299

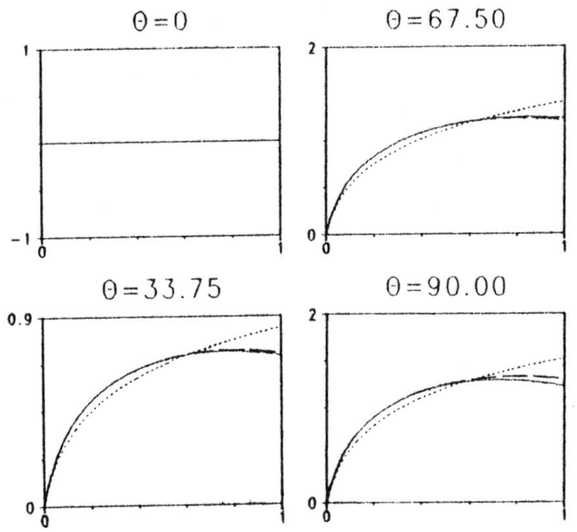

Fig. 3c - Tangential Velocity vs. Radial Position, c = 0.0086

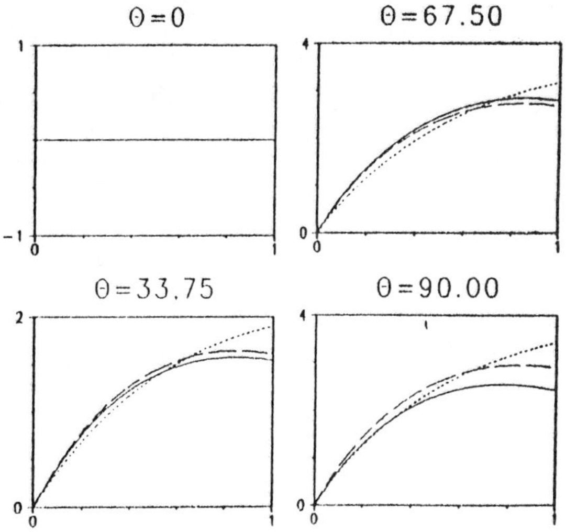

Fig. 4c - Tangential Velocity vs. Radial Position, c = 0.299

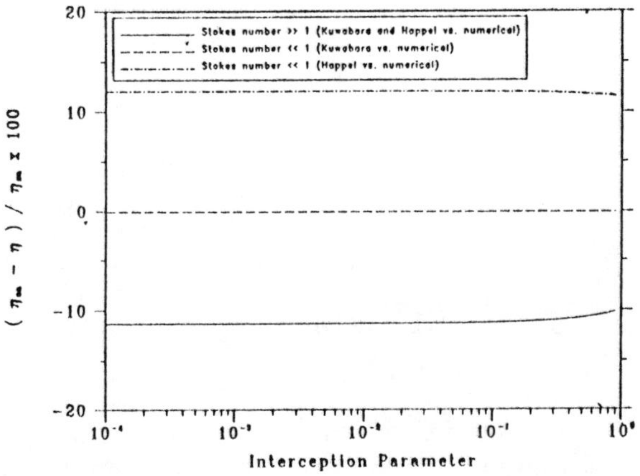

Fig. 5 - Predicted Percentage Difference of Single Fiber Efficency, c = 0.0086

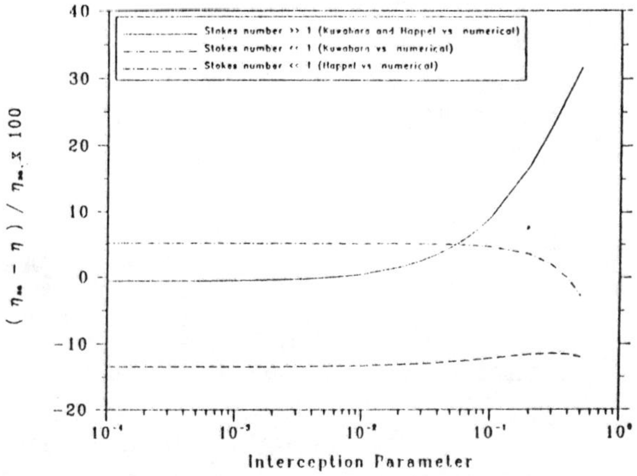

Fig. 6 - Predicted Percentage Difference of Single Fiber Efficency, c = 0.299

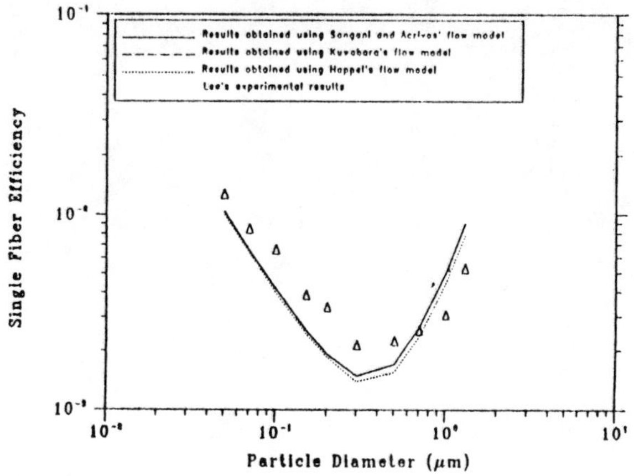

Fig. 7 - Single Fiber Efficency Compared to Data, c = 0.0086

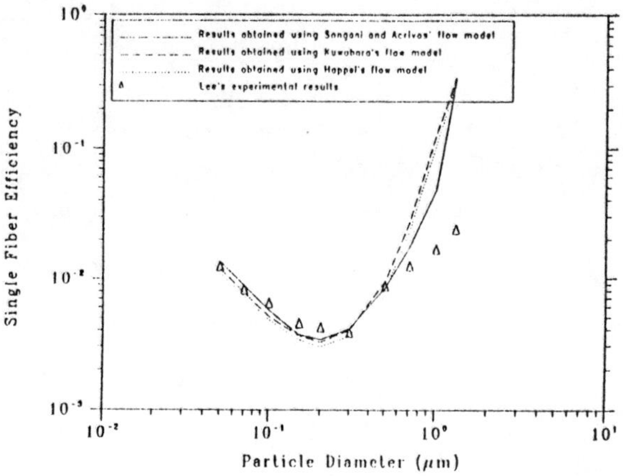

Fig. 8 - Single Fiber Efficency Compared to Data, c = 0.299

THEORY OF VACUUM FILTRATION AND PROCESS APPLICATIONS

Mary L. Robison
EIMCO Process Equipment Company
P. O. Box 300
Salt Lake City, UT 84110
(801) 526-2391

ABSTRACT

This paper will address the practical selection and process applications of vacuum filters.

The theory of data analysis for vacuum filtration equipment sizing will be presented along with scale-up factors, vacuum filter types, and operating parameters that can effect filter performance. Process applications for drum filters, disc filters, and horizontal filters will be discussed.

Introduction

The information involved in the selection and sizing of continuous vacuum filtration equipment has been well documented. It should be recognized at the outset that in some situations more than one option may be available to achieve the desired target. In some cases, purely technical objectives may be sacrificed because of cost considerations, design factors or other elements in order to satisfy optimization within the flowsheet and/or to provide economical results.

Before considering available options, and prior to the next step of sizing the selected filter, the design and operating criteria should be established. The test work to develop selection and sizing data will be based on collection of these criteria and the final product characteristics desired.

Process Information

The results which are obtained in any bench-scale testing program can only be as good as the sample which is tested. It is absolutely essential that the sample used, be representative of that which is or will be obtained in the full-scale plant, and that it be tested under the conditions that prevail in the process. Basic process information that is generally required, should encompass a complete understanding of the upstream variables and desired results of the filtration application.

Filter Types

Continuous filters may be characterized by the relative positions of the slurry and filter surface (bottom feed or top feed), the general shape of the filter (drum, disc, horizontal belt), the means of providing the driving force (vacuum, pressure or gravity) and the primary function of the filtration step (high rate filtration, dewatering, washing, thermal drying, steam drying). The choice of filter type is dictated first by the characteristics of the slurry, secondly by the operations which must be performed on the filter cake solids and finally by overall economics, if the desired results can be obtained on two or more types.

Filters may operate either with or against the force of gravity depending upon whether the filtering surface is rotating in a slurry tank or the slurry is fed onto the filter surface. Units falling into the first category are normally referred to as bottom feed units while the others fall into the general category of top feed units.

When considering the types of units which are available, one must also bear in mind the different unit operations which may be carried out on a filter. Any filter cycle will consist of cake formation and cake discharge, and will usually involve deliquoring or dewatering, washing, thermal drying and steam drying.

The number of operations which are required are dictated by the process flowsheet. The number of operations which may be performed on a given filter type is, however, dictated by the configuration of the filtration surface. Tables I and II present the various types and configuration limits.

Sizing Determination

Laboratory sizing of vacuum filters is a universally applied practice. The simulation involves using an apparatus to measure: 1) the cake formation times for various cake thicknesses, 2) if needed, the time for various quantities of wash water to go through the cakes, 3) the wash effectiveness by analyzing the cakes to determine the quantity of solute remaining in the cake, and 4) the rate of moisture removal during the drying zone of the filter. The apparatus consists of a circular filter leaf with the appropriate filter cloth attached, a vacuum pump with pressure control, filtrate receiver, and gas meter. The techniques used to determine all of the measured values are extremely important in generating accurate information.

Figure 1 illustrates the typical relationship of cake weight vs cake thickness. Figure 2 illustrates the typical relationship of cake weight as a function of cake formation time. Figure 3 illustrates the typical relationship of wash time as a function of wash volume and cake weight. Figure 4 illustrates the relationship of remaining soluble salt vs wash ratio or volume. Figure 5 illustrates the relationship of cake moisture vs the parameters that control cake dewatering (cake weight, dry time, pressure differential, air rate, and liquid viscosity). Figure 6 illustrates the typical relationship of air rate as a function of cake drying time at constant cake thickness.

As mentioned previously, the nature of the solids (i.e., size, shape) is the principal limiting factor for the rate of filtration. The concentration of solids in the feed to all vacuum filters is also very influential on rate. The rate of cake formation is increased substantially by decreasing the quantity of liquid that must pass through the cake. Thus, many vacuum filters are preceded by thickeners to increase the solids concentration, since this will decrease the size of the filter installation. Often, the savings in the cost of vacuum filtration equipment offsets the cost of the thickener.

During the testing for vacuum filtration, the requirements for the vacuum pump capacity are also measured on a lab scale by measuring the vacuum level and quantity of air drawn through the cake. Thus, vacuum pumps can be sized.

The data from measuring all the times and rates associated with the given application are then combined to form an overall filtration cycle time for the filter in question. A scale-up factor that is based on the relationship of laboratory testing data to full-scale operation is applied. The scale-up factor permits some variation in filtration performance without leading to system failure.

Process Applications

Table III illustrates several ranges of actual vacuum filtration performance. The variability is caused by different process flowsheets used upstream of the filters and by different types of slurries normally encountered. Generally, higher rates are a result of solids that have a larger average particle size. Accompanying higher filtration rates are higher air flow rates required for vacuum pump sizing. These data clearly illustrate the need to define each application through some form of testing.

Conclusions

The use of vacuum filters is still an economical and often desirable method of filtration in many process applications. The advantage of a continuous operation still holds precendence in many plants today.

TABLE 1

TYPICAL MINIMUM CAKE THICKNESS FOR VACUUM FILTER DISCHARGE

Filter Type	Minimum Design Thickness mm
Disc	10-13
Drum:	
Belt	3-5
Roll Discharge	1
Std. Scraper	6
Precoat	0-3 max
Horizontal Belt	3-5
Tilting Pan	20-25

TABLE II

TYPICAL VACUUM FILTER FUNCTIONAL LIMITS

Percent of Total Cycle Available For:

Vacuum Filter Type	Cake Formation		Cake Washing	Cake Drying	Discharge
	Apparent	Active			
Disc	35	28	N/A	40	25
Drum:					
Scraper Discharge	35	30	0-30	20-50	20
Belt Discharge	35	30	0-30	20-45	25
Roll Discharge	35	30	0-30	20-50	20
Precoat Discharge	--	35-55	0-30	10-60	5
Horizontal Belt	No Limit		No Limit	No Limit	0
Tilting Pan	75		75	75	25

TABLE III

TYPICAL VACUUM FILTER PERFORMANCES

Application	Filter Type	Rate kg/hr/m²	Feed Solids wt%	Cake Moisture wt%	Flocculent
Coal	Disc, Drum or Horizontal Belt	200-400	15-30	16-30	Anionic
Coal Refuse	Disc	40-160	25-30	25-35	Anionic
Copper					
Chalcocite	Disc	170-250	50-60	12-20	None
Chalcopyrite	Disc or Drum	250-425	55-70	9-14	None
Cement	Disc or Horizontal Belt	200-800	30-65	12-35	None
Fluorspar (Fluorite)	Disc or Drum	200-400	50	12-20	Anionic
Glauber Salt	Drum or Horizontal Belt	400	---	---	None
Gold (Cyanide Leach)	Drum or Horizontal Belt	150-350	50-60	20-30	Anionic
Iron Ore					
Magnetite	Disc	250-1,500	40-65	8-11	None
Hematite	Disc	200-2,500	60-75	6-11	Cat and/or anionic
Lead Concentrate	Disc	325-650	70-85	6-8	None
Magnesium Hydroxide	Disc or Drum	25-125	15-30	50	Anionic
Phosphoric Acid	Horizontal Belt	200-350 (as P_2O_5)	25-40	20	None
Potash	Drum or Horizontal Belt	3,300-6,400	35-50	4-7	None
Silver (Cyanide Leach)	Drum or Horizontal Belt	150-350	50-60	20-30	Anionic
Zinc Concentrate	Disc	250-400	55-65	10-15	None
Zinc Leach Residue	Drum	40-120	20-40	50-55	Anionic

Figure 1
Dry Cake Density Correlation

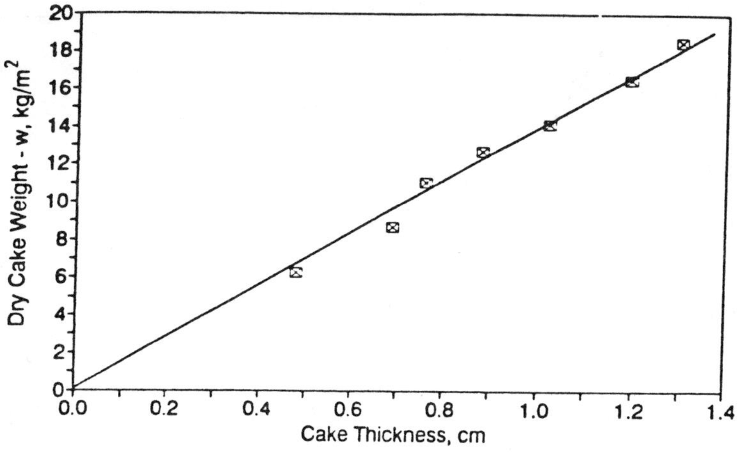

Figure 2
Cake Formation Correlation

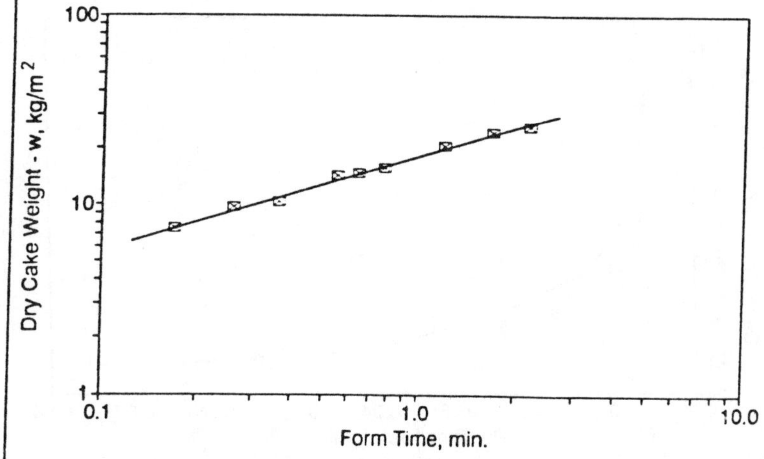

Figure 3
Cake Wash Time Correlation

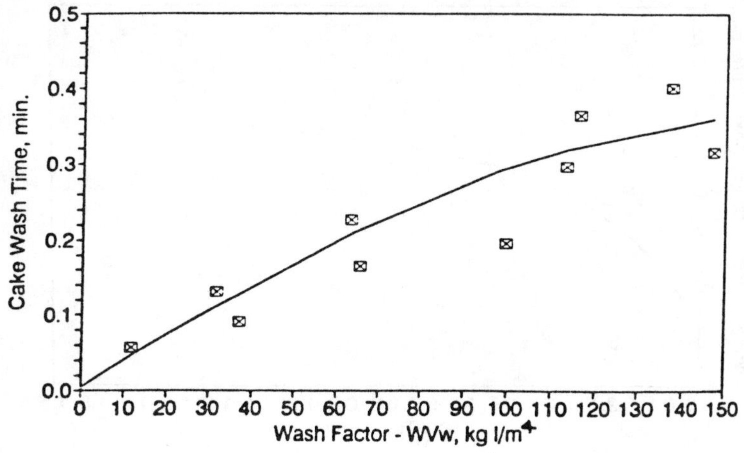

Figure 4
Wash Effectiveness Correlation

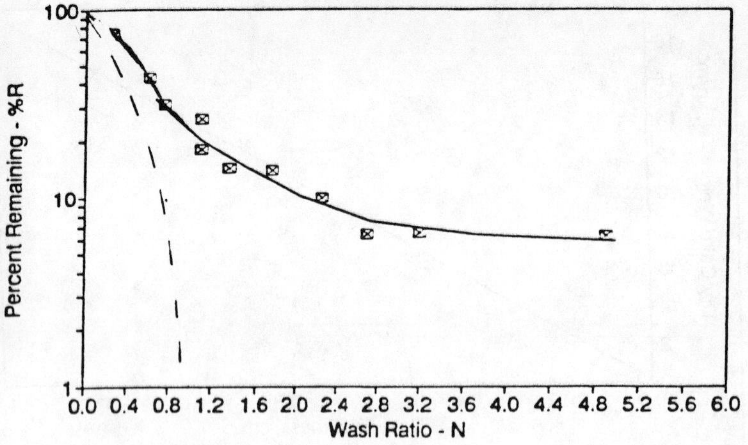

Figure 5
Cake Moisture Correlation

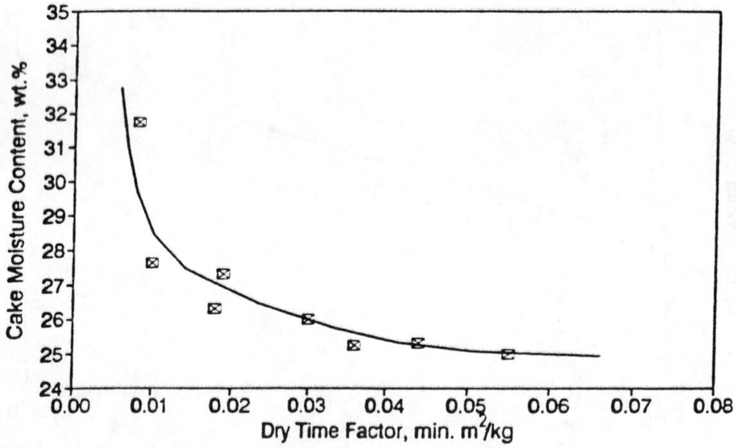

Figure 6
Air Flow Through Cake Correlation

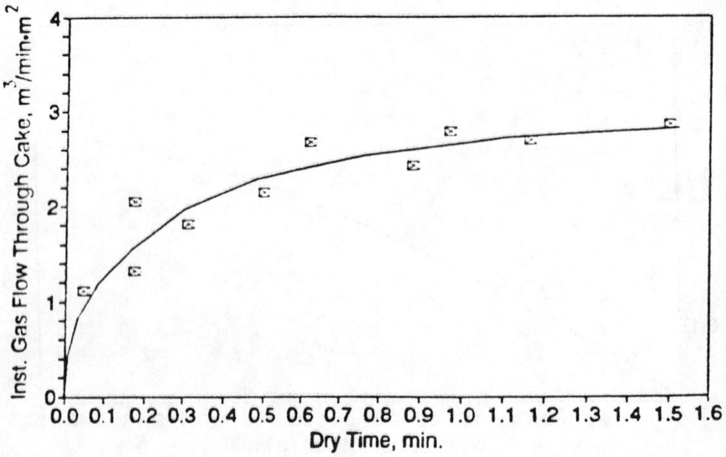

DETERMINATION OF A FRICTION FACTOR AND PERMEABILITY OF NONWOVEN MATERIAL

D. M. Cecala, K. J. Choi*, D. W. Yarbrough

Department of Chemical Engineering
Tennessee Technological University
Cookeville, Tennessee 38505

*Fleetguard, Inc.
Cookeville, Tennessee 38501
Fax: (615) 528-9583, Tel: (615) 528-9409, Attn: Choi

ABSTRACT

One-dimensional laminar flow through nonwoven webs was studied. Pressure drops across the media for various flow rates were experimentally determined and used to calculate a friction factor as a function of Reynolds number. The friction factor and permeability were calculated by means of a model analogous to viscous flow through a conduit, which incorporates mean flow pore diameter, porosity, and tortuosity. Results were verified using two different test fluids with nonwovens of various porosities and mean flow pore diameters. Permeabilities are calculated by Darcy's law directly from experimental data and are compared to values calculated by the model.

INTRODUCTION

Nonwoven webs are a common material used for filtration in a wide range of processes. They are composed of a network of overlapping fibers which create pores through which the process fluid flows. Particles are generally trapped in the pores or adhere upon the fibers. This paper describes the relationship between pressure drop and flow rate for nonwoven materials for one-phase laminar flow.

Darcy's Law is the most commonly accepted model for one-phase laminar flow through porous media at low phase velocity. For horizontal flow it is commonly expressed as:

$$Q = - kA(P_1 - P_2)/\mu L \qquad (1)$$

where "Q" is the volumetric flow rate, "A" is the cross-sectional area, "P_1 and P_2" are the upstream and downstream pressures, "μ" is the fluid viscosity, and "L" is the thickness of the medium. The proportionality constant "k" defined by equation (1) is the permeability of the medium.

Another common method of studying single phase laminar flow through porous media is to make an analogy of viscous flow through the media to viscous flow through a conduit. The approach taken here is similar to that of Collins[1] and van der Sluys[2] except that a tortuous path for fluid flow is considered by using the approximation[3]

$$T = 1 / \epsilon \qquad (2)$$

where "T" is the tortuosity and "ϵ" is the porosity. The characteristic length used for the Reynolds number is the mean flow pore diameter (D_M). Substitution of these values into the equations for viscous flow through a conduit leads to the following expressions for friction factor (f) and Reynolds number (Re).

$$f = (DP) \cdot D_M \cdot A^2 \cdot \epsilon^3 \cdot g_c \ / \ 2 \cdot L \cdot Q^2 \cdot \tau \qquad (3)$$

$$Re = D_M \cdot Q \cdot \tau \ / \ A \cdot \epsilon \cdot \mu \qquad (4)$$

where "τ" is the fluid density and "DP" is the pressure drop. A plot of ln f versus ln Re will be linear and the data can be described by equation (5).

$$f = c / Re \qquad . \qquad (5)$$

Substitution of equations (3), (4), and (5) into (1) and rearranging for 'k' gives

$$k = D_M^2 \cdot \epsilon^2 g_c \ / \ 2 \cdot c \qquad . \qquad (6)$$

Thus, an expression for permeability based upon physical properties of the media can be obtained. The agreement of k values calculated with equation (1) provides a check on the adequacy of the assumptions used to obtain equation (6).

EXPERIMENTAL

Values for f were determined experimentally by measuring DP and Q for steady-state flow. The remaining terms in equation (1) were evaluated by additional measurements on the nonwoven materials or the fluids. Data were obtained for water and an aqueous solution containing 50 wt % of ethylene-glycol. Fluid properties at 20 C, the test temperature are given in Table I. Figure 1 contains a schematic diagram of the experimental apparatus used to measure this relationship.

D_M was determined by ASTM F 316 by means of a model 0647 Coulter Porometer. The porosity in this research is calculated by ASTM method D 4197. Twelve different polyester (PT) nonwovens with D_M's ranging from 17.0 to 26.4 μm, four different polypropylene (PP) nonwovens with D_M's ranging from 16 to 20 μm, and one fiberglass (HV) nonwoven with a D_M of 11.3 were investigated. The porosities for all the nonwovens ranged from 0.86 to 0.90 .

RESULTS AND DISCUSSION

The results for one-phase flow gave a linear relationship between DP and Q for each type of filter as predicted by Darcy's Law. The type of material from which the filter was made affected the amount of DP. For each filter, the permeability was calculated by Darcy's Law (1) by using the Method of Least Squares to derive a relationship between pressure drop and volumetric flow rate which passes through the origin. The pressure drop versus flow rate data was then entered into equations (3) and (4) and ln f was plotted versus ln Re as shown in Figure 2. The "c" values shown in the figure were independent of the test fluid. Table II shows the permeabilities calculated by (1) to those calculated by (6). It can be seen that the average difference between the model and Darcy's law is 3.6 %. The maximum difference observed was 6.1 %.

CONCLUSIONS

The flow model proposed in this paper has been checked experimentally against Darcy's Law for three filter types. The results for ten sets of data show an average deviation of 3.6 % between k obtained by the two methods. The proportionality constant in the f - Re equation has been determined for three filter types. These constants depend upon the type of nonwoven being used which indicates that nonwovens composed of different materials with similar mean flow pore diameters do not exhibit the same permeability.

REFERENCES

1. Collins, R.E., Flow of Fluids Through Porous Materials, Reinhold, New York, 1961, 51-52.

2. van der Sluys, L., Geotextiles and Geomembranes, Vol 5., 1987, 283-295.

3. Johnston, P. R., and Hatch G. L., Journal of Testing and Evaluation, Vol. 10, 188, 1982.

Table I. Test Fluid Properties @ 20 C

Fluid	τ	μ
Glycol-Water*	1.08 +/- .02	4.4 +/- .05
Water**	0.998	1.002

* measured
** CRC Handbook of Chemistry and Physics 65th Ed.

Table II. Comparison of Permeabilties Calculated by Darcy's Law to Permeabilities Calculated from the Friction Model

Sample	D_M (μm)	k (Darcy) (μm^2)	k (Model) (μm^2)	Percent Difference
PT1	17.56	25.65	25.99	1.3
PT2	18.76	29.82	29.27	-1.8
PT3	19.96	31.47	33.05	5.0
PT4	22.82	41.10	42.74	4.0
PT6	22.08	40.76	39.21	-3.8
PT11	23.33	43.65	45.98	-5.3
PT12	26.25	60.10	58.00	-3.5
PP1	16.54	40.33	37.85	-6.1
PP2	17.57	46.58	45.21	-2.9
HV1	11.32	48.34	49.62	2.6

% diff = ((k/model)-(k(Darcy)) * 100 / k(Darcy)

KEYWORDS

Permeability
Friction
Darcy's Law
Nonwoven
Mean Flow Pore Diameter

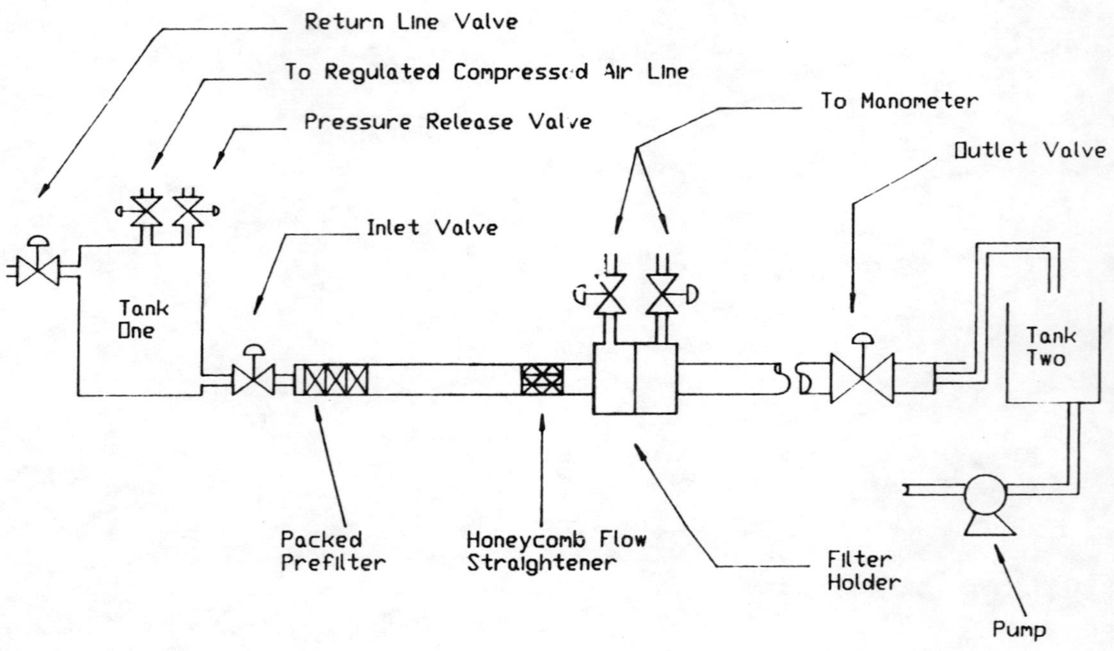

Figure 1. Experimental Apparatus for One-Phase Flow Analysis.

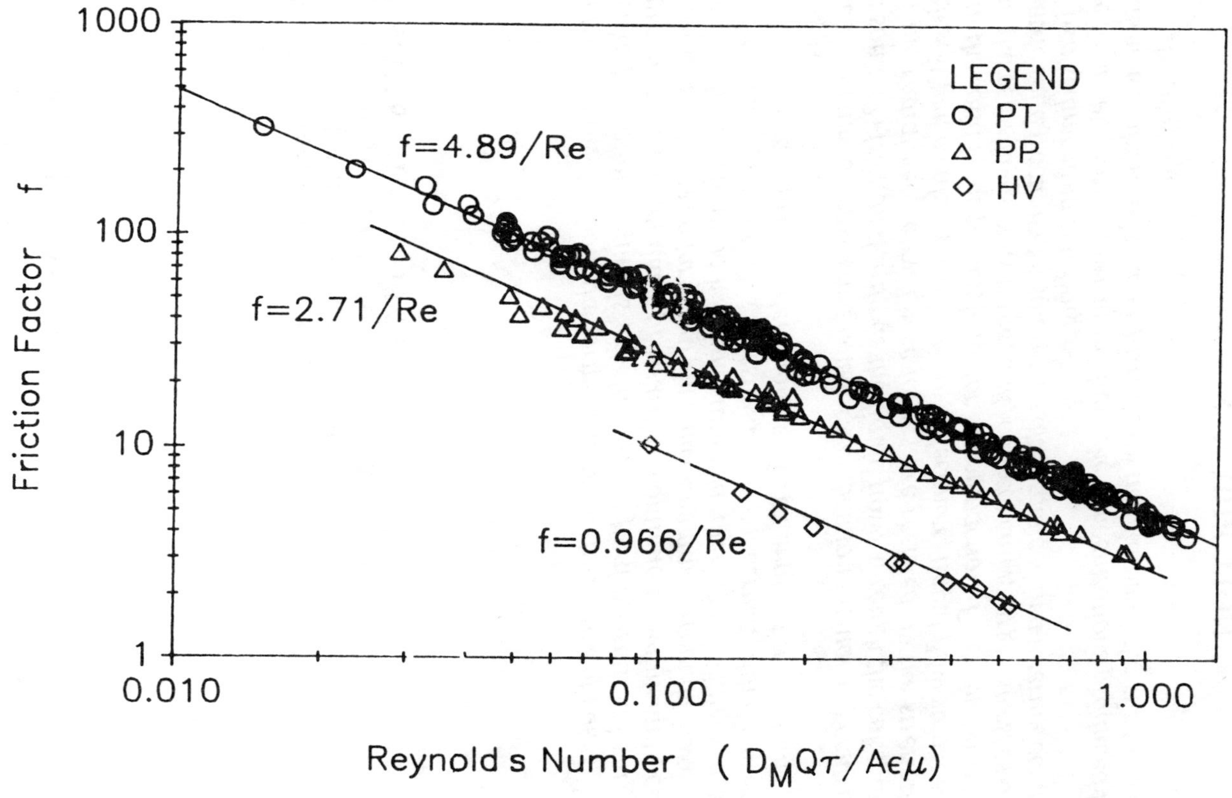

Figure 2. Friction Factor for Three Media Types as a Function of Reynolds Number.

COMPARISON OF COMPACTED CAKES IN SEDIMENTING AND FILTERING CENTRIFUGES

Frank M. Tiller and N.B. Hysung
Department of Chemical Engineering
University of Houston, Houston, TX 77204-4792
713-749-4316 FAX 713-747-6323

Centrifuges involve both the formation and deliquoring of cakes. In a filtering centrifuge, after cake formation, liquid is expelled in the final operational stage when the centrifugal pressure exceeds the capillary pressure. In continuous sedimenting centrifuges, action of the conveyor (scroll) plays a large part in squeezing liquid from the cake as it is transported from the pond to the beach. In both sedimenting and filtering centrifuges, a saturated cake is formed before deliquoring begins. This investigation is concerned with the average volumetric fraction of solids in the cake prior to deliquoring. An example of predictions for a highly flocculated material is provided.

Stresses acting on cakes and causing compaction include (1) centrifugal body force acting on the solid, (2) frictional drag as the liquid flows through the cake, and (3) an arching effect due to the radial geometry. These forces are resisted by the strength of the matrix of solid particles. When a cake is produced in a decanter, liquid flow stops; and the centrifugal body force in the solid is the principle factor in compaction. Frictional drag in a filtering centrifuge adds to the centrifugal body force; and, consequently, larger stresses are present. The average volumetric solid fraction of a saturated cake produced in a filtering centrifuge is larger than that found in a sedimenting centrifuge.

The final deliquoring operation alters the concentrations produced in the first stage of cake formation. Results of this paper provide an indication of the state of cakes at an intermediate stage of the process.

INTRODUCTION

Centrifuges are employed in filtering and sedimenting modes. The terms *perforated bowl* and *solid bowl* differentiate the two types of operations. The steps involved in batch-type filtering centrifuges include (1) filling, (2) sedimenting of the solid while filtering (assuming that the solids are denser than the liquid), (3) passage of liquid through the cake or syphoning the clear liquid out a side-flow centrifuge (Zeitsch, 1981), (4) washing, and (5) drainage of the resulting cake (Flynn and Rutter, 1989). In a solid bowl centrifuge, the solids are (1) sedimented and (2) transported and discharged continuously by means of a conveyor (scroll) which operates at a lower speed than the bowl. The conveyance process compacts the cake and results in a drier product (Reif and Stahl, 1989). Recently Albertson (190) reported on the increased solid content of cake resulting from very small differential speeds between the conveyor and bowl.

TYPES OF STRESS

In both types of centrifuges, a wet, saturated cake is produced prior to the final deliquoring produced by (1) centrifugal expulsion of liquid in the perforated bowl and by (2) scroll transport in the solid bowl unit. Our investigation concerns the porosity distribution in the cake prior to the final deliquoring step. Four types of forces involved in the centrifugation process are (1) centrifugal body forces on liquid and solid, (2) frictional drag due to the relative motion of the liquid with respect to the solids, (3) an arching effect arising from the radial geometry, and (4) inter-particle stress resisting compaction. Deposition of a cake is a transient process which may be lengthy when sub-micron particles are present or very short where large, dense solids are involved. In this treatment, it will be assumed that the cakes have reached a steady-state condition and that the sedimentation process ins short compared to the total detention time.

DARCY EQUATION

Darcy's equation in spatial coordinates takes the form

$$\frac{dp_L}{dr} - \rho_L r \Omega^2 = -\frac{\mu q}{K} = -\frac{\mu}{K}\frac{Q}{2\pi r} \tag{1}$$

where p_L = liquid pressure, r = radius, ρ_L = liquid density, μ = viscosity, k = permeability, q = flow rate/unit area, and Q = flow rate per unit height. Eq. 1 applies to flow through a centrifugally deposited cake. For a sedimented cake that has reached an equilibrium state, there is not flow and $q = 0$. During the transient deposition and compaction of sediment in decanter centrifuges, flow of liquid is in a negative direction (away from the bowl); and the resulting drag on the particles delays sedimentation and compaction.

When the sedimentation process is terminated in a solid bowl centrifuge and $q = 0$, Eq. 1 can be integrated to give

$$p_L = 0.5\, \rho_L\, (r^2 - r_s^2)\Omega^2 \tag{2}$$

where r_s represents the inner radius of the liquid (initial position of the slurry) or the radius of the surface of the pond. When compressible cakes are formed in a filtering centrifuge, Eq. 1 cannot be integrated because the variable K is not a function of p_L. It depends upon the stress p_s on the particles.

STRESS BALANCE

A momentum balance reduces to a simple force balance when the inertial terms are neglected. A force balance over a differential element is required to relate the effective or compressive pressure p_s to the hydraulic pressure p_L. Cakes are similar to soils, and a lateral stress is developed which we assume to be proportional to the normal stress on the particles. If F_s is the stress/unit height on the particles between the fake radius r_c and an arbitrary radius r, the effective pressure developed by the accumulative frictional drag is defined by

$$p_s = F_s / 2\pi r \tag{3}$$

The lateral stress orthogonal to a radius vector is given by $k_o F_s$. Assuming momentum changes to be negligible and utilizing Fig.1, a force balance yields

$$d\,(rp_L)\, d\,\theta + d\,(rp_s) d\,\theta - (p_L + k_o p_s) dr \sin d\,\theta = [p_L \varepsilon + \rho_s \varepsilon_s] r^2 \Omega^2 dr d\,\theta \tag{4}$$

where point contact among particles is assumed so that the liquid pressure is effective over the entire cross-sectional area. Assuming $d\theta = \sin d\theta$, Eq. 4 becomes (Shirato and Aragaki, 1969)

$$\frac{dp_L}{dr} + \frac{dp_s}{dr} + (1 - k_o)\frac{p_s}{r} - (\rho_L\varepsilon_L + \rho_s\varepsilon_s)r\Omega^2 = 0 \tag{5}$$

The value of k_o generally ranges from 0.4 to 0.7.

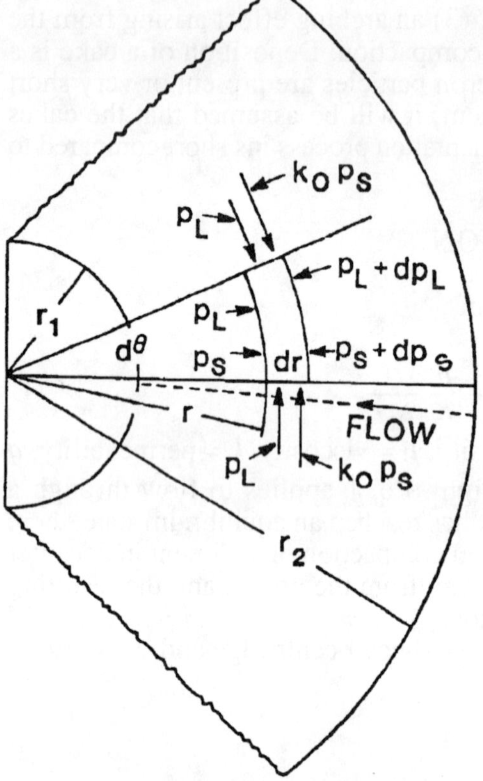

Figure 1. Liquid and solid compressive pressures in radial flow.

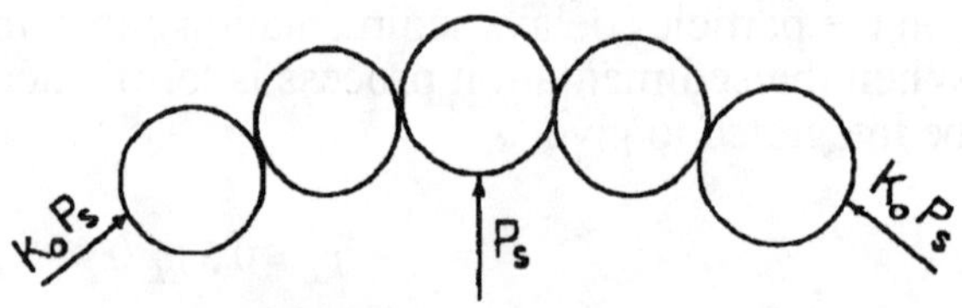

Figure 2. Illustrating the arch effect which tends to increase the compaction effect as compared with external deposition in a candle filter with cylindrical elements.

THE PARTICULATE STRUCTURE EQUATION

Neither Eq. 1 or 5 can be solved by itself for a compactible cake because of the presence of p_L. Consequently dp_L/dr must be eliminated between Eqs. 1 and 4 to give the following equation involving ε_S, K, p_S, and r

$$\frac{dp_s}{dr} + (1 - k_o)\frac{p_s}{r} - (\rho_s - \rho_L)\varepsilon_s r\Omega^2 = \frac{\mu}{K}\frac{Q}{2\pi r} \tag{6}$$

As the solidosity ε_S and permeability K are assumed to be unique functions of the effective pressure p_S, Eq. 6 can be solved numerically or in some cases analytically. Solution of Eq. 6 leads to p_S as a function of r. In turn K and ε_S can be obtained as a functions of r, and the cake structure is thus established. The variation of the liquid pressure can then be found by integrating Eq. 1 as follows

$$p_L = 0.5\, \rho_L (r^2 - r_s^2)\Omega^2 - \frac{\mu Q}{2\pi} \int_{rc}^{r} \frac{dr}{rK} \qquad (7)$$

where the value of K is obtained from the p_s vs r relation. Comparing Eqs. 2 and 7, it can be seen that p_L in a filtering centrifuge is less than the centrifugally developed pressure by the value of the integral in Eq. 7.

For interpretative purposed, Eq. 6 is best written in the form

$$\frac{d}{dr}(rp_s) = (\rho_s - \rho_L)\varepsilon_s r^2\Omega^2 + k_o p_s \; + \mu Q/2\pi K \qquad (8)$$

The previously discussed types of stress are represented by the terms in Eq. 8.

Term	Description
$(\rho_s - \rho_L)\varepsilon_s r^2\Omega^2$	Body force on solids. This effect is small when there is little difference in the densities.
$k_o p_s$	As shown in Fig. 2, particles arranged in radial fashion act like an arch. In a centrifuge the components arising from $k_o p_s$ add to compaction. With a cake deposited externally in a candle filter, the arching effect opposes compaction.
$\mu Q/2\pi K$	Darcian friction. In filtering centrifuges, frictional flow of the liquid adds to the compaction. In the transient stage in decanter centrifuges, liquid flow ($Q < 0$) opposes sedimentation. When equilibrium is reached, $Q = 0$; and this term disappears.
$\frac{d}{dr}(rp_s)$	Summing of the gradients of all the stresses leads to the gradient of the effective pressure causing cake compaction.

SEDIMENTING VS. FILTERING CENTRIFUGE

The effective stress in a filtering centrifuge will be larger than it is in a sedimenting centrifuge because of the effect of frictional drag represented by $\mu Q/2\pi rK$. Because of the higher stresses, a cake produced in a perforated bowl centrifuge will have a higher average solidosity than the same cake in a solid bowl centrifuge. We shall explore the effective pressure distribution in incompressible and compressible beds. Although the p_s distribution in an incompressible cake does not change the porosity, it provides a good starting place to analyze compactibility.

ANALYTICAL FORMULAS FOR p_s

If ε_s and K are constant, both Eqs. 1 (or Eq. 6) and 5 can be integrated. assuming a medium resistance R_m is related to the flow rate by

$$p_L(r_B) = \frac{\mu Q}{2\pi r_b} R_m \qquad (9)$$

Integrating Eq. 1 from r_c to r_B yields

$$\frac{\mu Q}{2\pi r_b} R_m = 0.5\, \rho_L (r_b^2 - r_s^2)\Omega^2 - \frac{\mu Q}{2\mu K} \ln r_b/r_c \tag{10}$$

Solving for Q yields

$$\frac{\mu Q}{2\pi r_b} = \frac{0.5\rho_L (r_b^2 - r_s^2)\Omega^2}{\frac{r_b}{K}\ln\frac{r_b}{r_c} + R_m} \tag{11}$$

Eq. 10 or its equivalent is frequently encountered in the literature. The numerator represents the driving force or centrifugally developed pressure. The denominator consists of the sum of the cake and medium resistances.

Integration of Eq. 6 yields

$$p_s = \Delta\rho\Omega^2 r_c^2 \varepsilon_s \frac{(r/r_c)^2 - (r_c/r)^{1/2}}{3 - k_o} + \frac{\mu Q}{2\pi K}\frac{1 - (r_c/r)^{1/2}}{1 - k_o} \tag{12}$$

At the cake surface, $p_s = 0$. Its maximum value occurs at the bowl where $r = r_b$. The term $\mu Q/2\pi$ can be eliminated by means of Eq. 11 to give

$$p_s = \Delta\rho\Omega^2 r_c^2 \varepsilon_s \frac{(r/r_c)^2 - (r_c/r)^{1/2}}{3 - k_o} + 0.5\, \rho_L \Omega^2 r_c^2 \frac{(r_b/r_c)^2 - (r_s/r_c)^2}{\ln\frac{r_b}{r_c} + \frac{R_m K}{r_b}}\left[\frac{1 - (r_c/r)^{1/2}}{1 - k_o}\right] \tag{13}$$

When $r = r_b$, p_s reaches its maximum value.

The first term on the *RHS* of Eqs. 12 and 13 with $r = r_b$ represents the value of the effective pressure at the bowl in a sedimenting centrifuge. The second term give the additional stress due to the Darcian drag in a filtering centrifuge. As an unexpected result, the p_s distribution of an incomressible material is virtually independent of the rate and permeability. The appearance of the ratio Q/K in Eq. 12 provides the first clue to lack of importance of either rate or permeability. In the absence of a medium resistance, the rate is proportional to the permeability; and the ratio is constant. In Eq. 13, the permeability appears in the dimensionless group KR_m/r_b. The product KR_m represents the equivalent cake thickness having the same resistance as the medium. Although there are few published data on operating medium resistance, the value probably has a maximum in the range of 0.5 - 1.0 cm, leading to a range of 0.005 - 0.01 for KR_m. With r_B ranging from 0.25 - 1.0 m, the group KR_m/r_b would seldom be expected to reach a value of 0.04. Except for very thin cakes, the logarithmic term $\ln(r_b/r_c)$ in Eq. 13 would be expected to dominate the group KR_m/r_b.

The first term in Eq. 12 represents the stress developed in a sedimented cake. The ratio of the maximum stresses at the bowl produced in filtering and sedimenting centrifuges is given by

$$\frac{p_s\,(\text{Filt.})}{p_s(\text{Sed.})} = 1 + \frac{0.5}{(\sigma - 1)\varepsilon_s}\left(\frac{3 - k_o}{1 - k_o}\right)\left(\frac{1 - R_s^2}{G - \ln R_c}\right)\left(\frac{1 - R_c^{0.5}}{1 - R_c^{2.5}}\right) \tag{14}$$

where $\sigma = \rho_s/\rho_L$, $R_s = r_s/r_b$, $R_c = r_c/r_b$, and $G = KR_m/r_b$. In Table 1, values of the ratio are shown as a function of R_c for $\sigma = 2$, $k_o = 0.5$, $G = 0.005$, and $R_s = 0.75$.

TABLE 1

Rc	0.75	0.80	0.85	0.90	0.95	1.00
Ratio	2.95	3.36	4.05	5.39	9.17	87.5

For thin cakes where R_c lies between 0.95 and 1.0, the ratio of the effective pressures at the bowl is quite high. Even for thick cakes, the effective pressure for filtering centrifuges is several times its value in sedimenting centrifuges. Although, the results for incompressible materials cannot be directly used for compressible cakes, the large ratios are indicative of the important compaction effect of the drag forces. Just as Zeitsch (1981) noted that a dimensionless liquid pressure vs. dimensionless radius was independent of permeability and porosity, we note that the effective pressure ratio is independent of K but not ε_s. The magnitude of the solidosity has an important effect on the centrifugal body forces.

COMPACTIBLE CAKES

With the exception of a few cases, analytical solutions are not available where compactibility is present. In order to integrate the Darcy and particulate structure equations, it is necessary to have relations involving ε_s, K, and p_s. Although almost never available, experimental data can be used directly in numerical integrations. There are no theoretical relations involving the cake structure and effective pressure p_s, and empiricism rules supreme. The authors have employed power functions, polynomials, and exponentials in the general forms

$$\alpha = \alpha_o f_1(p_s) \text{ and } \varepsilon_s = \varepsilon_{so} f_2(p_s) \tag{15}$$

where $f_i(0) = 1$ and $df_i/dp_s > 0$. One form of Eq. 15 adapted from work of Carman and Shirato takes the form

$$(\varepsilon_s/\varepsilon_{so})^{1/\beta} = (K_o/K)^{1/\delta} = 1 + p_s/p_a \tag{16}$$

where p_a is an empirical parameter. When $p_s = 0$, $\varepsilon_s = \varepsilon_{so}$ and $K = K_o$. Another set of equations is

$$\left(\frac{K}{K_i}\right)^{-1/\delta} = \left(\frac{\varepsilon_s}{\varepsilon_{si}}\right)^{1/\beta} = \frac{p_s}{p_{si}} \tag{17}$$

where K_i and ε_{si} respectively are values of the permeability and solidosity at the relatively low pressure p_{si}. When $p_s \le p_{si}$, K and ε_s are assumed to be constant and related to the empirical constants J and B by

$$J = K_i p_{si}^{\delta} \qquad B = \varepsilon_{si} p_{si}^{-\beta} \tag{18}$$

Eq. 17 can also be written as

$$K = J p_s^{-\delta} \qquad \varepsilon_s = B p_s^{\beta} \tag{19}$$

The exponent δ represents the negative of the slope of a logarithmic plot of K vs p_s. It is related to the degree of compactibility and markedly affects cake behavior in filters and centrifuges. High values of δ are associated with large flocs. When $\delta > 1$, various adverse effects in centrifugal filtration occur primarily due to formation of cakes with wide porosity variations. The cake surface tends to be wet and soupy while a tight skin forms at the medium.

Most of the resistance to flow resides in the skin, and the system behaves as through the medium were being clogged.

Although permeability and its variation with effective pressure is important in the transient stages of centrifugal sedimentation, the final equilibrium is independent of K.

The two terms in Eq. 12 illustrate the difference between sedimenting and filtering centrifuges. When $Q = 0$, the equation applies to a decanter; and the solidosity and density difference are the cake parameters affecting the effective pressure. The second term involving Q and K represents the frictional affect of flow on effective pressure. With a buoyant solid for which $\Delta\rho = 0$, the first term disappears, and no sedimentation effects are present. Parameters affecting centrifugation can be summarized as follows:

Parameters	**Sedimenting**	**Filtering**
$\varepsilon_s, \varepsilon_{so}, B, \beta$	Determine final cake structure	Partially affect final cake structure
K, K_o, J, δ	Affect time of transient cake formation	Affect flow rate and have major effect on final cake structure

Operations following deposition are affected by the structure of the cakes initially formed.

NUMERICAL PROCEDURES

It is necessary to solve Eqs. 1 and 6 simultaneously. Assuming a negligible medium resistance, p_L equals zero at the bowl where $r = r_b$. At the cake surface, p_L is given by Eq. 2 with $r = r_c$. the effective pressure $p_s = 0$ at the cake surface and assumes an unknown value at the bowl. To start the process, a value of Q is chosen; and p_s is determined as a function of r. Values of K and ε_s can then be substituted into Eq. 1 and p_L found as a function of r. In general p_L will not be zero at r_b, and the process must be repeated with new values of q until ρ_L reaches zero at the bowl.

Rather than use the differential equations, we shall employ the integrated forms for a non-compactible material over small changes in radius. Values of ε_s and K are modified for each interval in accord with the progressive values of p_s. The integrated equations take the form

$$p_{s\,2} = F^{\,1-k_o} p_{s\,1} + \Delta\rho\Omega^2 \frac{1-F^{\,3-k_o}}{3-k_o}(\varepsilon_s r_2^2) + \frac{\mu Q}{2\pi}\,\frac{1-F^{\,1-k_o}}{1-k_o}\left(\frac{1}{K}\right) \tag{20}$$

$$p_{L\,2} = p_{L\,1} + 0.5\rho_L\Omega^2(1-F^2)\,r_2^2 + \frac{\mu Q}{2\pi}\ln F\left(\frac{1}{K}\right) \tag{21}$$

where $F = r_1/r_2$. By maintaining a constant ratio r_1/r_2 rather than a constant difference (r_2 - r_1), the numerical solution is simplified. For the case in which $r_s = 0.3$m, $r_c = 0.4$m, $r_b =$ 0.5m, $\rho_s = 2000$ kgm/m^3, $\rho_L = 1000$ kgm/m^3, $\mu = 0.001$ Pa·s, $\Omega = 35$ rad./s (334.2 RPM), and $F = 0.99$

$$p_{s\,2} = 0.99487\, p_{s\,1} + 1.2158\text{E}(4)\ \varepsilon_s r_2^2 + 1.59555\text{E}\ (-6)\ Q/K \tag{22}$$

$$p_{L\,2} = p_{L\,1} + 1.2189\text{E}(4)\ r_2^2 + 1.59996\text{E}(-6)\ Q/K \tag{23}$$

Eqs. 22 and 23 are relatively easy to solve. They must be combined with one set of constitutive equations (Eq. 16 or 17) or experimental data. Only the first two terms on the RHS of Eqs. 22 and 23 are needed for the sedimenting centrifuge.

In Fig. 3, the results for p_L and p_s are shown for very low speeds of 20 rad/s (191 RPM) and 35 rad/s (334 RPM) for the filtering case. A highly flocculated material with the following characteristics was used for model calculations

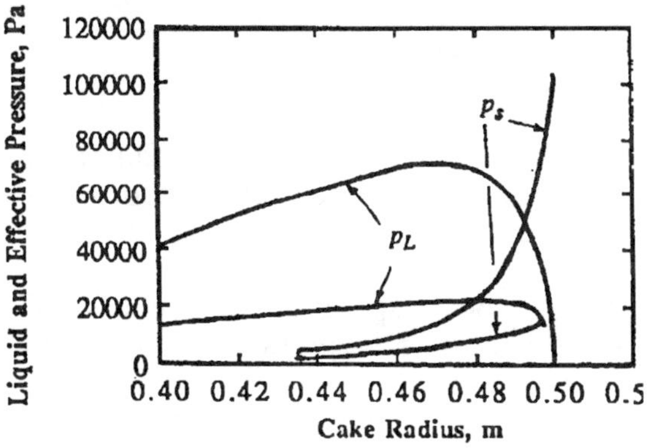

Figure 3. Comparison of p_L and p_s for bowl speeds of 20 and 35 rad/s.

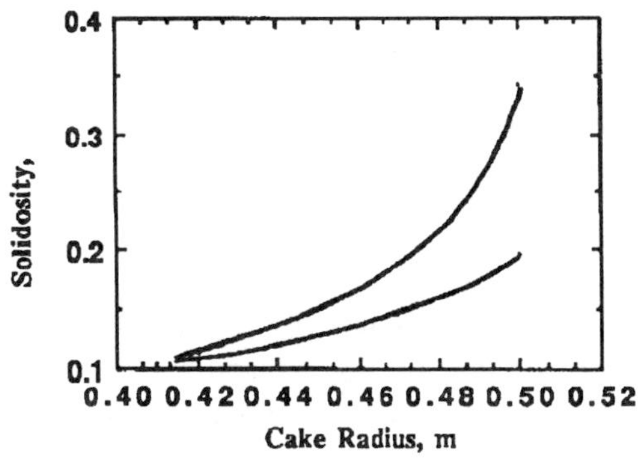

Figure 4. The volume fraction of solids in the cake is shown for speeds of 20 and 35 rad/s.

However, when the speed is increased to 334 RPM, there is a sudden increase in p_s at the bowl to over 100 kPa. The volume fractions of solids plotted against r shows the effects of the increased pressure in Fig. 4 where ε_s is rising rapidly near $r_b = 0.5$. A resistant skin is being formed.

In Fig. 5, the permeability is shown as function of cake radius for the calculations made at 334 RPM. There permeability drop to a very low value in the last 2.0 cm. of the cake emphasizes the adverse effect of the tight skin formed at the bowl surface. Raising the bowl speed would further aggravate the skin effect. Raising the RPM would cause very little increase in either centrate rate or the average volume fraction of solids in the cake because the average permeability drops rapidly with increasing speeds.

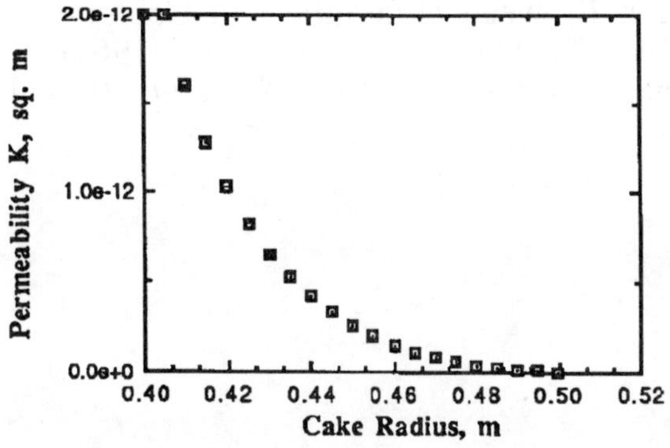

Figure 5. At 35 rad/s, the permeability of the highly flocculated cake approaches a small value at the bowl radius.

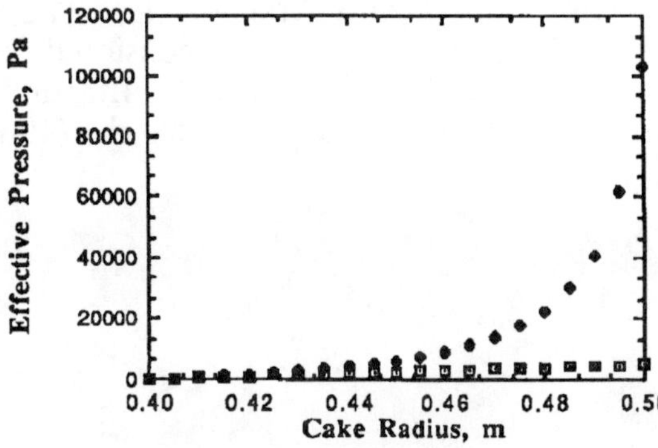

Figure 6. Comparison of effective pressure variation for filtering and sedimenting centrifuges. At the same bowl speed, Darcian friction adds substantially to the stress in a filtering centrifuge.

A comparison of the effective pressure developed for the sedimenting and filtering centrifuges at 35 rad/s is illustrated in Fig. 6. Very little compaction occurs in the solid bowl centrifuge because of the absence of Darcian friction. In contrast to the effect of increased bowl speed on the highly flocculated cake in the perforated bowl centrifuge, increasing RPM would be highly beneficial in improving the average solidosity in the solid bowl. There is approximately a ratio of 20 between the maximum values of p_s developed in the two types of centrifuges in Fig. 6. An increase of the RPM by 4 or 5 fold to 1200-1500 RPM in the solid bowl would produce a cake having a solid content comparable to that predicted for the filtering centrifuge.

We again emphasize that the condition of the cakes will be markedly affected by subsequent processing in the centrifuges.

CONCLUSIONS

A method has been developed for predicting the volume fraction of solids of compactible materials as a function of radius for both solid and screen bowl centrifuges. In addition, the centrate and cake rates can be predicted for the filtering centrifuge as a function of bowl speed, permeability, and degree of compactibility.

For a highly flocculated material, high perforated bowl speds are ineffective. The appearance of a resistant skin at the bowl surface absorbs most of the pressure drop and tends to nullify increases in Darcian friction due to high bowl speeds. However, with the solid bowl, Darcian friction does not affect the final cake; and high bowl speeds are effective in producing drier cakes.

ACKNOWLEDGMENT

The authors express their appreciation to the Division of Basic Chemical Sciences, Office of Basic Energy Sciences, Office of Energy Research, U.S. Department of Energy for support which has enabled them to carry on fundamental research in the theory of solid-liquid separation.

LITERATURE CITED

1. Albertson, Orris, "Improved Centrifuge Design for High Cake Solids," *Fluid Particle Sep. J.*, **3**, (1990).
2. Flynn, S.A. and Rutter, S.A., "Factors Affecting the Dewatering of Small Particles in Perforate Bowl Centrifuge," *Fil. and Sep.*, **25**, 348 (1989).
3. Reif, F. and Stahl, W.F., "Transportation of Moist Solids in Decanter Centrifuges," *Chem Eng. Prog.*, **57** (Nov. 1989).
4. Shirato, M. and Aragaki, T., "The Relation Between Hydraulic and Compressive Pressure in Non-Unidimensional Filter Cakes," *Kagaku Kogaku* **33**, 205 (1969).
5. Zeitsch, K., "Centrifugal Filtration," Chap. 14 in *Solid-Liquid Separation*, L. Savarovsky, ed., Butterworths (1981).

NONFILTRATION USES OF MICROPOROUS MEMBRANES IN BIOTECHNOLOGY

Borek Janik
Gelman Sciences Inc.
Corporate Technical Service
600 S. Wagner Road
Ann Arbor, MI 48106

The impressive advances in molecular genetics and biotechnology of the past decade have created a host of opportunities for the use of membranes. The particular applications of microporous membranes in bioengineering processes primarily entail separation as a part of upstream processing, bioreactor system, and downstream processing. In all of these applications, the microporous membranes act as a filter, i.e., an effective barrier for particulate matter in the micrometer range (0.1-10μm) suspended in liquid or gaseous fluids. The retention mechanisms primarily include sieving effects frequently working in concert with adsorptive and other interactive phenomena. In bioengineering's nonfiltration applications, the microporous membranes can be used to immobilize whole cells or individual enzymes to effect biochemical reactions. The subject of this presentation is another nonfiltration application which concerns recombinant DNA technology, an important part of biotechnology.

Biotechnology in its broadest definition refers to practical applications of biology in technology to make or modify products. A major segment of modern biotechnology is genetic engineering, which is the artificial manipulation of genetic material to design cells or organisms meeting certain preset, practical goals. The backbone of genetic engineering is recombinant DNA technology. The insertion of foreign fragments of DNA into the genetic material (the genome) of microbial, plant, or animal cells is the essential step common to all recombinant DNA experiments.

Recombinant DNA technology can be used in a wide range of industrial applications to develop organisms that produce new products, or existing products more efficiently or in larger quantities. Also, organisms can be developed that themselves are useful, such as microorganisms that degrade toxic wastes or new strains of agriculturally important plants and animals. Recombinant DNA technology together with a set of other techniques can lead to protein engineering: the stability, bioavailability, intrinsic bioactivity, pharmacokinetics, and other properties of proteins can be affected to achieve desirable functional characteristics. Still distant, yet realistic, applications will undoubtedly include the area of human gene therapy. Insertion of normal genes into the somatic cells of patients could be used to correct an inherited or acquired disorder through a synthesis in vivo of a missing or defective gene products. Finally, as a spin-off of recombinant DNA technology, tests are being developed for diagnostic or forensic applications which surpass the traditional counterparts both in sensitivity and specificity.

It is the recombinant DNA technology, as well as immunology, where biotechnology researchers found nonfiltration applications for microporous binding membranes. Some semblance with microfiltration remains. However, the retained "particles" are macromolecules and the capture mechanisms are strong, relatively nonspecific adsorptive forces. The latter primarily include hydrophobic forces often augmented by electrostatic (Coulombic) interactions. The microporous membranes act as a solid phase matrix attracting and binding nucleic acids or proteins so that these can be detected and identified. Why microporous membranes? Their internal surface area is much larger than that of conventional binding matrices (e.g., solid plastics and beads), and strong adsorptive forces allow the bound macromolecules to reach high concentration per surface area unit. At the same time, the detection reagents -- specific macromolecular probes -- can relatively freely diffuse in the highly porous membrane and react with target molecules, and the unreacted excess can be easily washed out. The molecules (nucleic acids or proteins) to be bound can be applied to membranes by three basic techniques (Figure 1):

1. Transfer (Blotting) Techniques. These are techniques used to transfer electrophoretically separated biomolecules to binding membranes. Depending on the class of molecule transferred, DNA, RNA, or protein, the technique is traditionally, albeit improperly, referred to in jargon as Southern, Northern, or Western blotting, respectively.

   The transfer starts with the electrophoretic fractionation of (a) DNA that has been cut into fragments by specific enzymes (restriction endonucleases), (b) a mixture of purified mRNA, or (c) mixture of proteins. Agarose gels are typically used to electrophorese DNA, RNA, and polyacrylamide gels for proteins. The fractionated material is transferred (often with variable efficiency) from the gel onto a binding membrane by using various driving forces such as capillary action, diffusion, suction, or electromotive force. Once bound to the porous membrane, the material can be analyzed using specific molecular probes. The detection assay is called *hybridization* or *immunoblotting* when complementary nucleic acid sequences or specific antibodies, respectively, are used to probe. The probes are labeled with reporters to allow visualization. Prior to reacting the blot with the probe the unoccupied yet fully reactive sites on the membrane must be blocked to prevent nonspecific binding of the probe. This is done by saturating the membrane with an agent inert to the probe. It should be noted that the pore structure of an electrophoretic gel is too tight to allow the probe to migrate freely either to react with the target molecule or when unreacted, to be washed out.

Blotting is a valuable qualitative tool which helps researchers study gene structure and expression, reveal differences between genes from various sources, detect gene abnormalities, select DNA fragments for insertion into cell genome, analyze RNA, or to identify proteins.

2. <u>Dot Blotting (Dotting) Techniques</u>. These are the techniques of applying biomolecules directly to the membrane either by pipetting or filtration. Dotting is used when quantitative information is required. It is also the technique of choice for diagnostic kit makers who desire assays that are simple, rapid, and do not require accessory equipment (e.g., electrophoresis apparatus).

   Typically, the samples are applied in serial dilutions to the membrane using minifiltration devices. The detection of target macromolecules in the sample is done using procedures similar to those described in transfer techniques. Application of the sample in serial manner assures that some of the developed spots are within the linear range of detection (e.g., autoradiographic, colorimetric, etc.). The unknown values are usually related directly to a serially diluted sample with a known concentration value.

   The dot blotting technique is primarily used for screening a large number of samples for the presence, absence, or amount of certain molecules.

3. <u>Lifting Techniques</u>. The last method of applying biomolecules to membranes is known as "lifting." Certain membranes, by virtue of inherent adsorptive properties, are able to lift an imprint of various substances much in the same way that the children's toy, Silly Putty, is able to lift newsprint. The technique is used in applications known as colony and plaque lift hybridizations.

   These two applications have similar procedures and share the common goal, which is to identify bacteria or bacteriophages (bacterial viruses) harboring DNA of interest. In genetic engineering experiments, this DNA is usually a functional fragment of foreigner DNA which has been integrated in a host bacterial cell. Whether the DNA represents the gene for a disease or the gene for a therapeutic protein, the investigator is faced with the task of identifying a few organisms out of millions which carry the target DNA. A description of colony lift hybridization is outlined below. Note that plaque lifts concern bacteriophages.

   Briefly, a population of bacteria carrying DNA is grown on an agar plate at densities low enough that neighboring bacteria are distinctly separated. Depending upon the DNA being studied, as little as 0.0001% of the total bacteria population may carry the DNA of interest. Following a period of incubation, each bacterium will have multiplied several times giving rise to thousands of bacteria in the form of visible colonies. Each colony, therefore, originated from a single bacterium and will, therefore, contain the same DNA as the original bacterium. A piece of binding membrane is placed over the colonies thereby transferring or "lifting" bacteria from the agar plate to the membrane. The configuration of colonies on the membrane is a replica of the colony configuration on the agar plate. The bacteria on the membrane are disrupted by exposure to detergent, which causes the release of DNA. The released DNA is bound to the membrane in a concentrated, localized spot. Through the use of a radiolabeled DNA probe, the spot containing the DNA of interest can be detected in a hybridization assay and visualized by autoradiography. Since the membrane is a replica of the original agar plate, the investigator now has a means of identifying the bacterial colonies on the plate which carry the target DNA. The specific colony can be isolated and grown in large quantities to produce the desired product.

Biotechnology together with its subdisciplines is a fertile field, verdant with new developments and ideas. The orientation and attention of researchers is molded by economical needs and social trends. These will undoubtedly expand the conceptual use of binding membranes in genetic engineering into new areas of diagnostic, therapeutic, and other applications.

Figure 1 Techniques for the Use of Binding Membranes in Genetic Engineering Experiments

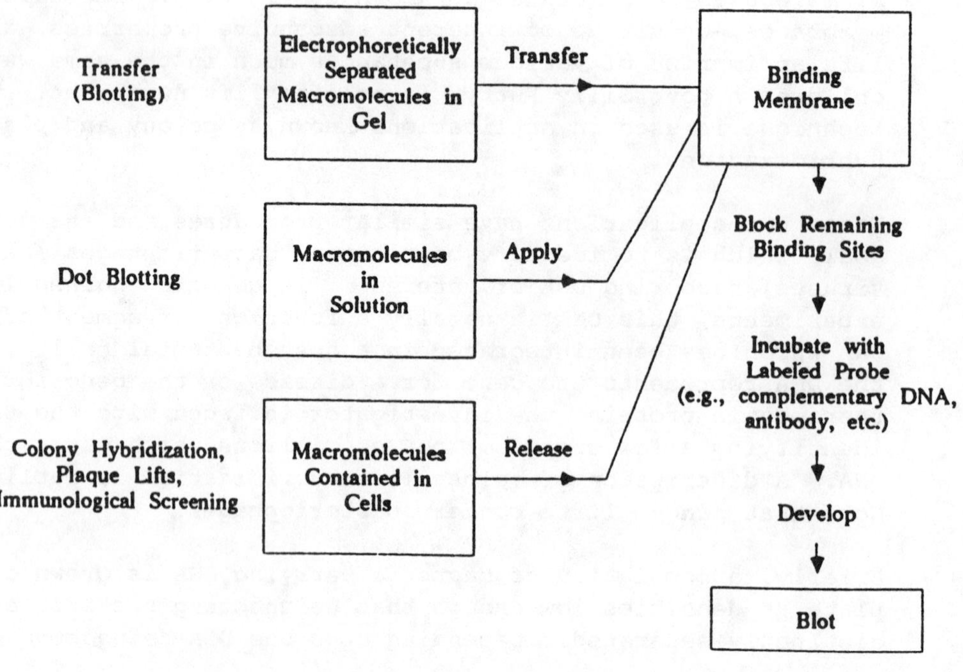

Figure 1 Techniques for the Use of Binding Membranes in Genetic Engineering Experiments

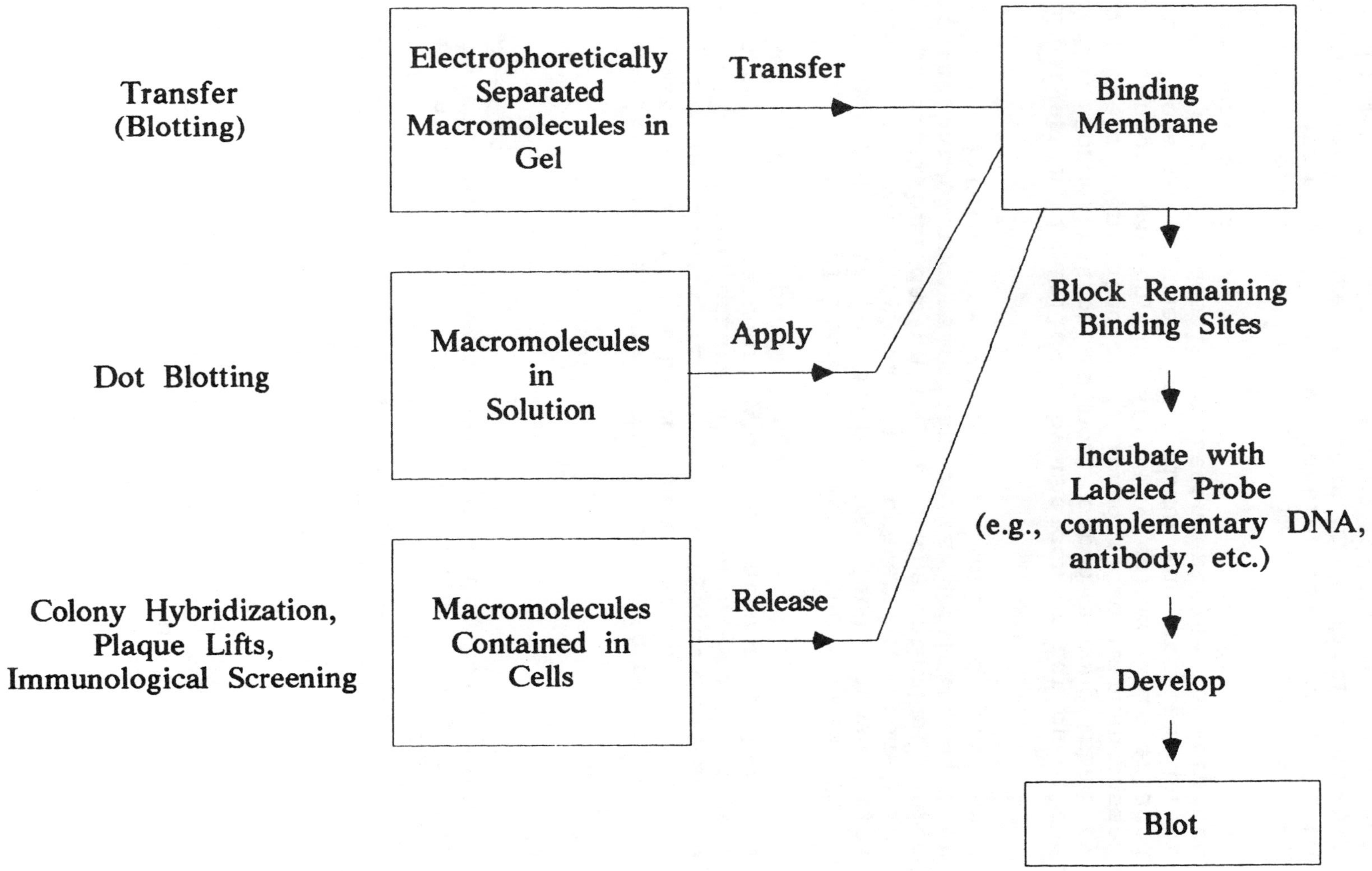

Design Considerations for the Integration of Cross Flow Processes in Biopharmaceutical Applications.

Michael Dosmar, Peter Wolber and James Banks

Abstract

Cross flow filtration has become more prevalent in the pharmaceutical marketplace with the commercial development of products derived from cell culture techniques. The applications, for which cross flow is commonly used, includes the separation of cell laden harvest fluids and the concentration and diafiltration of cell free conditioned media. The application of these systems in the pharmaceutical process require special design considerations in order to conform with cGMP's. Furthermore, integration of these processes offers the production facility the ability to minimize the handling of the products, shorten the process time, reduce the tankage used in the process, and to miniaturize the equipment. Further integration of these processes can lead linking cell culture system to the down stream process in order to operate in a continuous closed mode.

Introduction

Biotech processing involves various clarification and purification steps for the separation of active biomolecules. After the desired levels of the active peptides of interest are synthesized in the bioreactor, the cells are then separated from the product. The cell free conditioned media is subsequently concentrated to reduce the volume for further downstream processing.

Cell separation may be accomplished via several techniques. These include centrifugation, static filtration and cross flow filtration. Since low sheer systems are preferred, cross flow devices using a parallel leaf design are often the method of choice especially for the harvesting mammalian cells. Typically 0.2 or 0.45 um membrane filters are primarily employed for this application.

Typically cell harvesting is followed by the concentration of the conditioned media. Concentration is most often accomplished by ultrafiltration. The concentrated conditioned media is then passed over a series of anion, cation, or affinity columns. Interlaced between each of these steps one often finds the utilization of ultrafiltration in order to desalt or reconcentrate the product prior to further purification.

As bioengineered products entered the manufacturing stage, regulatory requirements and critical validation issues specific to this industry had to be addressed. Cross flow filtration has undergone a development process over the last few years as a direct result of the increasing requirements of the biotech industry.

Implementation of these systems into the biopharmaceutical industry has raised issues regarding sanitization, cleaning and integrity testing (1). In response, users have looked to the vendors of this equipment for additional support in the form of more stringent controls, improved documentation and increased capabilities such as steam tolerance of tangential flow membranes, clean-in-place installations, and automated process control. These efforts began with the introduction of ultrafilters that offered a retention correlated, and validated integrity test. In 1987 steam/autoclavable microfilter and ultrafilter cross flow modules were introduced. As the need to supply complete systems increased, a new relationship developed between vendor and user. Engineering staffs from both sides were required to sit side by side and define the specifications to which these systems should be designed. These efforts have led to a system design where the microfiltration and ultrafiltration processes are integrated into a single package. Since each production step adds to the overall processing time and requires multiple systems, simultaneous operation of two sequential stages offers a substantial advantages which specifically result in:

the reduction of process time
allows for the down sizing of equipment
decreases the equipment foot print
decreases tankage requirements
decreases product loss and equipment hold up
reduces production costs

In order to achieve these design goals there were many criteria that would have to be met. The integrated system design contains a single inlet and outlet and is designed to be self regulating. Feed from the bioreactor is directed into the cell harvesting loop where the cells are retained by microporous membranes. Cell free conditioned media then permeates through the microfilter and flows into the ultrafiltration loop where it is concentrated. Exiting the system from the outlet are the low molecular weight components that may either be reused as wash buffer, sent to waste, or supplemented and used as growth media. Self regulation is achieved by having a closed system similar to those used for diafiltration. The rate limiting ultrafiltration step controls both the microfiltration flux and flow into the system from the bioreactor. The levels in the two tanks remain unchanged because the tanks operate under a pressure vapor lock. Optimal system operation is predicated on the proper sizing of each of the cross flow systems and establishing the optimal cross flow velocities, and trans membrane pressures (2).

The basic components of an integrated system include 2 cross flow loops (microfilter, ultrafilter), a cart, 2 pumps, 2 pressure vessels and an assortment of valves and tubing. Specific component selection varies in response to the operational and plant specifications.

3

Cross Flow System Design Criteria

When selecting the materials of construction for a cross flow system one must consider the design specifications. Selection of components become a direct result of the specification.
These considerations include the support structure on to which the hardware will be mounted, the tank design requirements, the manner in which the temperature will be controlled, the type of pumps to be employed, the valving requirements, and the degree of instrumentation and process control.

In order to logically approach the materials selection process one must outline and fix the specifications for the system. Only after these are in place can the process begin. Furthermore, understanding the user's needs will help the system designers direct their energies appropriately. The design specifications must be kept within the confines of the operational specification of each of the component parts. Since designs are generally a compromise between an ideal and the realities of a budget. Each of the above mentioned component areas must be addressed individually, but not considered exclusive of the companion components. Ultimately, one must try to make the proper decision with regards to the overall scale of the project.

The following rules apply:

- -Process tanks should be sized to accommodate the largest anticipated batch with the lowest concentration requirement.
- -Pumps should be capable of driving the largest anticipated filtration area.
- -Filtration hardware should be sized based on the desired flow rates and throughput.
- -Filter hardware configurations should be capable of expansion and contraction to meet the changing needs of the user.

Framework

The first item to be evaluated is the frame work onto which the system will be built. The design engineers must select skid components. These may range anywhere from angle steel to 316 SS square tube. Generally 304 SS is acceptable for the system frame work. However, those users intending to autoclave the system may select 316 SS to insure that the skid does not rust.
The selection of the appropriate framework components is not only a utilitarian decision but one of aesthetics and budget. Angle steel is the least costly and the most difficult to clean, it is also the least appealing aesthetically. Circular tube stock as compared to square tube stock is a trade off between cost and strength where the square stock provides the greater strength for the greater price. Even when budgetary constraints are tight our preference has been to use 304 SS square tubing. The difference to the total system costs generally does not justify the less

expensive alternatives. Furthermore, tube stock is immanently more cleanable than the angle steel. Using a lower grade metal would not be in keeping with the other components on the system.

Assembly

Weld versus bracket assembly is primarily an aesthetic decision. The spacial relationships between pumps, tanks and filter holders are fixed in order to maintain a small and known hold up volume. However, if post fabrication modifications are required, they are easily made only when the system utilizes bracket mounts.

System Configuration

Special care must be taken with regards to the location of components so as to assure that there will be sufficient clearance to reach all portions of the system regardless the surface area. Furthermore, the design must include provisions for adequate drainage, and suitably placed product recovery ports. A variety of system geometries have been explored. Tower configurations offer reduced floor space requirements but create accessibility and maintenance problems. Furthermore, flow distribution becomes difficult as a result of the gravitational effects of this design. Horizontal configurations while requiring slightly greater floor space provide configurations that are more efficient and easier to use.

Operational Specifications

Critical to the design approach is the intended method of cleaning and sanitization. At issue is whether the system will include CIP capability, and is to be sanitized using chemicals or steam (SIP, autoclaving). Current membrane technology makes it possible for the end users to select among these options. Depending on the methods of choice component selection will be affected.

A CIP system must include spray balls for the tanks and the system tanks, tubing and pumps must be consistent with CIP designs. Tank spray balls must be of a material and finish consistent with the rest of the system. Spray ball location must be such as to promote good CIP solution contact along drop tube lines and tank walls without casting shadows. These requirements are greatly simplified when small tanks are used (24 inch diameter and less). Low pressure spray balls (20 psi, or less) can be used to supply CIP solution to these small tanks. A flow rate between 0.1 to 0.5 GPM/ (sq. ft. tank surface area) creates a free film Reynolds number of 200 which insures proper cleaning. High spray devices (greater than 30 psi) are unnecessary for small tanks. Furthermore, they may reduce cleaning effectiveness by fluid atomization that can occur (3).

Autoclavability, requires that the system have detachable pump motors and instrumentation. Consideration must also be given to the dimensional constraints imposed by the autoclave chamber. Also, the valving must be compatible with the process. Diaphragm valves designated for use in an autoclave must be configured with 316 SS bonnets. SIP systems, though, more rigorous, and stressful to the membranes, allows for the construction of larger systems. These systems do not have the added costs associated with stainless steel valve bonnets but do require that all the vent filters be encased with in stainless steel filter holders and that all vessels be ASME coded.

Piping

Piping decisions are a function of the associated equipment and design specifications. A system that is totally hard piped limits the users with regards to the degrees of freedom associated with the assembly of the system. A combination of flexible tubing and off the shelf sanitary fittings provides the desired flexibility with out compromising the systems design. Of major concern, though, is the assurance that all piping diameters provide sufficient flow velocity (>5 ft/sec) so as to assure that the system is cleanable. Care must be taken not to include any dead legs in the system. The standard is 7 pipe diameters, though we attempt to design to <5 pipe diameters (4). Designs utilizing a single set of piping, for the full range of filtration areas that may be required, will employ a variety of hoses, standard sanitary clamps and fittings, between the filter holder and each process tank. This provides the flexibility to accommodate 5 to 50 sq. ft. of filtration area without changing piping or the hold up volume appreciably. Reinforced silicone is the hose of choice. These hoses are available with polished stainless steel clamp ends allowing them to meet "SIP" specifications. Flexible stainless steel conduit is also available as armor over the silicone hose which makes the line cut proof and virtually indestructible under normal operating conditions. Sanitary clamp fittings can be rotated to provide additional degrees of freedom that facilitate a flexible design.

Valves

A variety of valves may be selected for use on the system. The valves of choice are diaphragm but in certain instances one might consider a butterfly design where adherence to 3A rating is not required. Valve selection will be depending on the expected service to be provided by the system and particular valve. In the case of autoclaving a stainless steel or nickel clad bonnet must be used. If the system is to be maintained in a sanitary mode of operation, then all valves interfacing with off skid components should be bossed so that sterile connections can be made. As with the piping layout, the valve design and size should be such that the valve does not represent a dead leg.

Tanks

Tank selection is based on a number of issues. The size of the tank should be large enough to meets the requirements of the process. The size and the shape of the tank must take in to consideration the intended concentration factors, so that the appropriate sloped or conical bottoms are used. The use of steam (SIP) requires that the tanks be ASME coded, and be rated for full vacuum. The location of the tanks in the system must also be such that the tank discharge be a low point in the system so as to prevent product hold up. Tanks may be affixed to the skid so that they are immovable or supported in such a way that they are somewhat mobile. Since each system includes a recirculation loop, anti-vortex approaches should be included. Furthermore, since these systems are generally handling proteinaceous solutions, drop tubes should be included so that foaming is kept at a minimum. Depending on the temperature requirements the environment, and the nature of the solution a jacket or heat exchanger may be included for temperature control. We find that jacketed tanks are the most economical approach to temperature control. Generally, the preferred cross flow designs do not require external heat exchanges, however, if they are included the selection process must consider if the system is self draining, provides the appropriate heat exchange, and is compatible with the process overall and compiles with the required codes. Primarily, tube and shell heat exchangers are favored for their cooling efficiency and sanitary design.

Tanks may be configured with a variety of controls which include high and low level, and pressure sensors. In addition sight glasses of various designs are available. Special consideration must be taken with regard to insuring that all the internal wetted surfaces drain freely.

Pumps

Pump selection is a process dependent decision. Generally, because of low shear requirements the selection is restricted to peristaltic and rotary lobe pumps. For mammalian cell handling applications peristaltic pumps are usually the pump of choice. These pumps allow for the complete containment of the solutions, are gentle and thereby pose little risk of cell rupture. Depending on the chemical and thermal stresses placed on the system care must be taken to select an appropriate tubing material. Marprene, for example, is a material that is steam tolerant when supported by an outer braid, resistant to leaching, and possesses wide chemical compatibility. Each pump manufacturer offers different features.

Considering the applications, and cost of materials, one needs to select a pump head that assures good tube durability. Mostly the tubing will be changed after each production run. Through efficient space utilization the tubing lengths are kept at a minimum so that this does not become a cost consideration. When peristaltic pump are specified for the cell harvesting step, pumping capacity may be the rate limiting step. Therefore, users should closely evaluate cell harvesting equipment with respect to hydraulic efficiency (ratio of cross flow to filtration rates). These potential limitations may be overcome through the use of multiple pump heads. Peristaltic pumps, though, are limited to maximum discharge pressures of 30 psi and thus have limited applications in ultrafiltration.

Because they are capable of generating moderately high pressures and flow at low operating speeds, rotary lobe pumps are preferred for most ultrafiltration applications. Like peristaltic pumps, rotary lobe pumps are capable of being steamed when proper design considerations are met. Specifically, steam sterilizable rotary lobe pumps have larger clearance between rotors and the head plate to allow axial thermal expansion and may have a drain port for condensate. These pumps also require speed control. Variable frequency speed controllers are a convenient and reliable means to control speed, pumping direction and emergency shut down.

Centrifugal pumps offer a low cost alternative to both the peristaltic and rotary lobe pumps. These pumps though, operate with higher sheers and require flooded inlets since they are unable draw a vacuum. For Biotech applications we have opted not to utilize these unless specifically requested by the client.

Instrumentation

Regardless of the sophistication of the system, temperature, pressure, and flow rates should be monitored. Pressure may be sensed via either a standard analog pressure gauge or by a remote pressure sensor. The same is true for temperature sensing. Remote sensors allow use of digital out put devices coupled to strip chart or circular recorders. This provides hard copy on-line documentation of the process, as well as the ability to better define and manipulate the operational parameters.

Flow meters utilizing different technologies are currently available. The major categories, by increasing cost and improving sanitary design are rotometers, magnetic, and ultrasonic meters. When product contact flow meters are acceptable, rotometers are generally the choice because of cost. For applications that demand a non-product contacting meter, ultrasonic meters are preferred over magnetic flow meters because of accuracy and flexibility. A less expensive alternative would be to use a peristaltic pump that has an output signal proportional to pump speed. The pump can be attached to a meter calibrated to transform pump speed to flow rate. Each of these non product contacting units eliminates any concerns regarding cleaning, and sanitizing of the flow sensor.

An additional level of process control can be achieved through the inclusion of simple on/off switches. In case of either the loss of temperature or pressure control, the use of set point controls and on and off switches may power down the system automatically and prevent catastrophic product loss. This is best illustrated by pump shut down as a result of over pressurization due to membrane fouling. In addition operator notification of a system malfunction may be relayed through visual or audible alarms. This type of instrumentation is relatively inexpensive and provides a significant level of assurance.

Subsequent levels of automation utilize process controllers that can detect variations in the process conditions and make the required adjustments. This is best exemplified by the control of pump acceleration and valves thereby maintaining the optimum pressure, and flux profiles. At this level of process control systems may now be configured with automatic feed reversal and filtrate back flush capabilities in order to maximize performance.

The highest level of process control employs computerization of the processes. Computers makes it possible to completely automate the process. This level of control allows for constant on stream adjustments assuring peak system performance at all times. Furthermore, the computer can collect, tabulate, and transmit the operational parameters to a printer or another computer. Large scale system can usually justify this degree of automation due to the relative cost.

Bibliography

1. Wolber, P. and M. Dosmar, "Depyrogenation of Pharmaceutical Solutions by Ultrafiltration: Aspects of Validation". Pharm Tech 11, (9) 38-43 (1987)

2. Wolber, P., M. Dosmar, and J. Banks "Cell Harvesting Scale-up: Parallel-Leaf Cross Flow Microfiltration Methods" Biopharm 1, (6) 38-45 (1988)

3. Adams, D.G., and D. Agarwal "Clean in Place System Design" Biopharm 2, (6) 48-57 (1989)

4. International Association of Milk, Food, and Environmental Sanitarians, "3-A Accepted Practices for Permanently Installed Sanitary Product-Pipelines and Cleaning Systems," no. 605-02 (Ames, Iowa)

The Validation of Membrane Filtration Devices for Pharmaceutical Uses

C. Thomas Badenhop, Ph. D.
CUNO Inc.
400 Research Parkway
Meriden, CT 06450

One great advantage to the use of membrane filters in the final filtration of critical products, such as in the case of pharmaceuticals, is the ability to pretest the integrity of the filters prior to use, and at the end of the filtration step, with an integrity test. Recently, there has been some discussion as to what degree of reliability is guaranteed with these tests. Is it necessary to test the filter's performance with the actual material being filtered? Do these tests really prove the infallibility of the devices?

The theories behind the validation testing and just exactly what is being measured is discussed for a better understanding of the risks involved with the use of these products.

The Validation of Membrane Filtration Devices for Pharmaceutical Uses

Abstract

One great advantage to the use of membrane filters in the final filtration of critical products is the ability to test the integrity of the filters prior to use and at the end of the filtration step. This is standard practice in the pharmaceutical industry. Recently there has been some discussion as to what degree of reliability is guaranteed with these tests. Is it necessary to test the filter's performance with the actual material being filtered? Do these tests really prove the infallibility of the devices?

The theories behind the validation testing and just exactly what is being measured is discussed for a better understanding of the risks involved with the use of these products.

Diffusion and the Bubble Point

There have been a number of articles written to either justify the use of diffusion testing to validate the use of membrane filters in pharmaceutical applications, or show that even with diffusion testing there exist potential problems with the media not shown by this testing. It is evident, that with large areas of membrane, bubble point testing is not possible. The diffusion of air through the wet membranes makes the determination of the bubble point impossible to distinguish from air diffusion.

The amount of air diffusing through a wetted membrane can be predicted with the Fick equation (Equation 1 below) which relates the rate of diffusion to the membrane thickness. The diffusion rate is virtually independent of the pore structure of a microporous membrane, but it is proportional to the thickness of the water film being supported by the membrane structure. In order to utilize a diffusion test to measure not only the integrity, but the pore size of the media, it is necessary to test close to the bubble point of the membrane. Testing the filter at a value of air pressure of 80% of the bubble point will make certain that the media is at the pore size characterized by the bubble point. This is now common practice. Testing at pressures so near the bubble point can, however, result in overlooking small defects in media. This will be discussed and examined in detail.

Unless the full diffusion curve of the membrane is measured, which is time consuming and usually not necessary, the true bubble point will be known to be greater than the diffusion test pressure. Thus, errors in the labeling of the filter will not be discovered if the media is actually finer than that being tested. This offers no problem with retention: only with flow rates and possibly total throughput.

The aim of the diffusion test is to ascertain if the cartridges are indeed free of defects, By-pass, that can be caused by flaws in the media, or in the construction and sealing of the cartridge should be identified in the testing. We may consider two types of defects which can result in filters being difficult to test using standard integrity test methods, be it diffusion, or pressure hold type of testing.

Equation 1

$$V = \frac{k(P_1 - P_2)}{L}$$

Where:

V	is the volume flow of gas diffusing through the membrane.
L	is the membrane thickness.
P_1	is the upstream air pressure on the membrane.
P_2	is the downstream air pressure, usually atmospheric.
k	is the combined constant that establishes proportionality. and is related to the solubility of the test gas in water or the wetting fluid and may be defined as $k = R_a(C_1\text{-}C_2)$
Ra	is the diffusion constant

The fact that membranes, that are wettable, will exhibit a rather high bubble point, have shown an advantage in integrity testability over media previously used such as depth filter sheets, which served to filter microorganisms out of pharmaceutical and beverage fluids. The real significance of the bubble point is not that it measures the largest pore size, and that this pore size is too small for bacterial passage, but that it is related to the mean pore size of the membrane.

The mean pore size of the membrane (which in present usage is the mean flow pore) is the parameter that controls the retention characteristics of the membrane. There is a nominal ratio between mean flow pore and the maximum pore size which differs somewhat depending upon the polymer used to make the membrane. It is the maximum pore size that is evaluated in the measurement of the bubble point. Equation 2 shows this relationship which is called the Cantor equation:

Equation 2

$$D = \frac{4 ß \cos Ø}{P_1 - P_2}$$

where:

	D	is the maximum pore size
	P_1-P_2	is the air pressure difference across the membrane
	ß	is the surface tension of water or test fluid
and,	Ø	is the wetting angle.

The bubble point of the media is related only to the surface tension of the wetting fluid, the geometry of the pore and the wetting angle between the membrane surface and the wetting fluid.

When only one fluid is to be used in testing, the wetting angle relationship can be neglected as the cosine will be very close to one for any fluid seeming to wet the membrane. As soon as a different fluid is used, as is the case when filtrate is used to establish integrity, or if the surface of the membrane is affected by the test fluid, this factor must be considered to avoid the rejection of perfectly sound membrane. This can be a significant problem if the findings of high diffusion is at the end of the filtration run. The entire lot of filtrate will than be suspect.

The effective surface tension of pure water is about 72 to 74 dynes per cm. This is a very high surface tension for a wetting liquid. Most organics show much lower values. Contamination of pure water with small levels of organics can materially reduce this surface tension. Membranes have also become less wetting due to the adsorption of organics from the materials being filtered. The effect of adsorbing organics can be the lowering of the bubble point, through the change in the wetting angle, measured without a change in the pore size. In some cases the membrane can become hydrophobic and the bubble point can be lost. Only by the use of wetting agents to re-wet the membrane can one determine if the membrane still has integrity.

Indeed, difficulty in obtaining good wetting is a far greater problem in the acceptance of integral membranes than problems in not identifying and accepting bad membrane as good. More accurate results are always obtained with good wetting fluids than with water. The difficulty is that a wetting fluid, other than water, will contaminate the filtrate.

The first of the possible defects that might pass inspection during the integrity testing of the filtration system is the interface between the membrane and the encapsulating plastic, which is also an area where wetting problems can occur. Low surface tension liquids will show that a joint can be sound whereas water, in not wetting the interface, will show the cartridge to be defective. If, in the construction of the cartridge, a fissure in the encapsulating plastic is formed and the plastic shows good wetting, the defect will not be observed during integrity testing. The fissure can be several millimeters long and have an effective diameter of 10 µm. Water will accumulate above the end encapsulation and drain through the fissure. The time for only a small amount of water to drain from the cartridge and release air during testing will be so long as not to be detected. I have seen epoxy end-caps that were completely detached from the membrane after many steam sterilizations and still test integral.

HAGEN-POISEUILLE LAW- EQUATION 3

$$V = \frac{3.14\, D^4 (P_1 - P_2)}{128 L n}$$

Where:

V	is the volume flow through the capillary.
P1-P2	pressure difference across the capillary.
n	is the viscosity of the fluid.
D	is the capillary diameter.
and	
L	the capillary length.

The relationship which predicts the rate of flow through end-cap and media defects is the Hagen-Poiseuille law. This is shown in Equation 3 above. From the equation we see that the amount of fluid passing through the defect is proportional to the diameter of the defect to the fourth power; directly proportional to the pressure difference and inversely proportional to the capillary length. Calculations made of the time it would require to drain water through an end-cap defect, and the effect on retention of the cartridge, are shown in the tables to follow. The time required to drain the cartridge and the amount of air that will actually bleed through the end-cap defect during the diffusion testing, assuming sufficient patience is exercised, will be excessive and the defect will not be noticed in the testing. A single capillary of 10 µm. in diameter and with a capillary length of 1 mm will only flow 0.18 ml/hour with a two bar (29 psi) pressure across the capillary. It would be very unlikely that this cartridge would be considered defective in any test due to the amount of water that would have to flow through the capillary before air could pass. The consequences of the defect not being detected and the effect on the filtration, has also been calculated.

The same 10 µm hole in the cartridge media itself could go easily undetected. The following tables and charts show the effect of a media defect in the diffusion testing of cartridges. In testing the media itself, a 10 µm hole would be found in a small filter as the flow of air is sizable with respect to the diffusion flow from the small membrane area.

In a large filter with several cartridges in the housing a single large pore will not be detected with any test of integrity, either pressure hold or diffusion, For this reason, a battery of tests are needed to give a degree of security to the integrity test. The integrity test is still a test of probability. The probability is high that if the cartridges in a large membrane filter assembly pass the integrity test, the system is sound and free of defects that will pass bacteria. There is no absolute guarantee that the system is without defects.

Cartridges are tested by the producer by challenging the filters with live bacteria. A test pressure somewhat below the bubble point is used to establish the diffusion rate for varying degrees of membrane porosities. Retention capacity versus the diffusion rate is empirically determined to indicate the point where the likelihood that there will be bacterial passage is very small. Thus the acceptable range of diffusion values determined in the statistical evaluations has shown that there is almost no chance for a failure of the cartridge. The usage of these cartridges in pharmaceutical and beverage applications has shown very few failures. With this history and test protocol, good security in the application of membrane filters can be accepted.

The achievement of absolute assurance in the integrity of a filtration system is possible if the level of contamination to be removed by the system approaches zero. There can be complete assurance of a sterile filtrate if the feed is also sterile. In many cases this is the condition of the final filtration. It is true in the filtration of beverages which have been prefiltered very sharply to obtain economical life of the final membrane filters. In this case the membranes are present to prevent passage of bacteria or other contaminants if the upstream filtration becomes compromised.

Statistical Considerations of a Defect.

If we evaluate the effect on the filtrate of a 10 µm defect in the end cap and the media as a function of an incoming bacterial load we can see just how this defect will compromise the final product. Let us assume that the defect is in a single element cartridge filter with a flow rate of 2 gallons per minute at a one bar pressure difference. In addition, let us assume that the contamination is uniformly distributed in the feed and that the bacteria is one hundred percent removed by the membrane with the exception of the defect.

TABLE 1

THE FLOW THROUGH A 10 μM CAPILLARY DEFECT

Capillary Diameter cm	Pressure Diff. Bar	Capillary length cm	Water Flow cc/min	% of 2 gal/min
0.001	1.00	0.1	.0015	.0003
0.001	1.00	0.2	.0007	.0001
0.001	1.00	0.3	.0005	.0001

The above results showing the percentage leakage through the defects has the following effect:

At very low bacterial contamination in the feed, only some of the filtrate will be contaminated. If the filtrate is being filled into one liter containers, the number of containers having at least one bacteria as compared to the number of containers without bacteria is illustrated in the next two tables.

In Table 2, the first row is the number of bacteria per ml of feed to the filtration unit. The number below indicates the number of bacteria per liter of filtrate. As can be seen at the 100 level all the filtrate has at least on bacteria in the filtrate per liter.

TABLE 2

BACTERIA PER LITER FILTRATE AS A FUNCTION OF THE FEED LEVEL

Cap Length cm	Feed Level 1/cc: 0.01	0.1	1	10	100
0.1	0	0	0	0	3
0.2	0	0	0	0	1
0.3	0	0	0	0	1

In Table 3, the first line is the number of bacteria per ml of feed and the columns of numbers under the bacterial level are the number of liters free of contaminant per liter of filtrate containing at least one bacterium. It is evident form the data and the charts graphically illustrating the data that the length of the capillary likely to be encountered does not improve the statistics in avoiding contamination. In Tables 4 and 5 the potential for contamination with small amounts of bacteria in the feed is even more severe. With the capillary defect in the end-caps there is virtually no chance of detection with any diffusion type of test due to the time required to remove casual water from around the capillary.

TABLE 3

THE NUMBER OF BACTERIA FREE LITERS OF FILTRATE PER CONTAMINATED LITER.

Cap Length cm	Feed Level 1/cc 0.01	0.1	1	10	100
0.1	3593	359	36	4	0
0.2	7186	719	72	7	1
0.3	10779	1078	108	11	1

CHART 1

BACTERIA PER LITER FILTRATE AS A FUNCTION OF THE FEED LEVEL

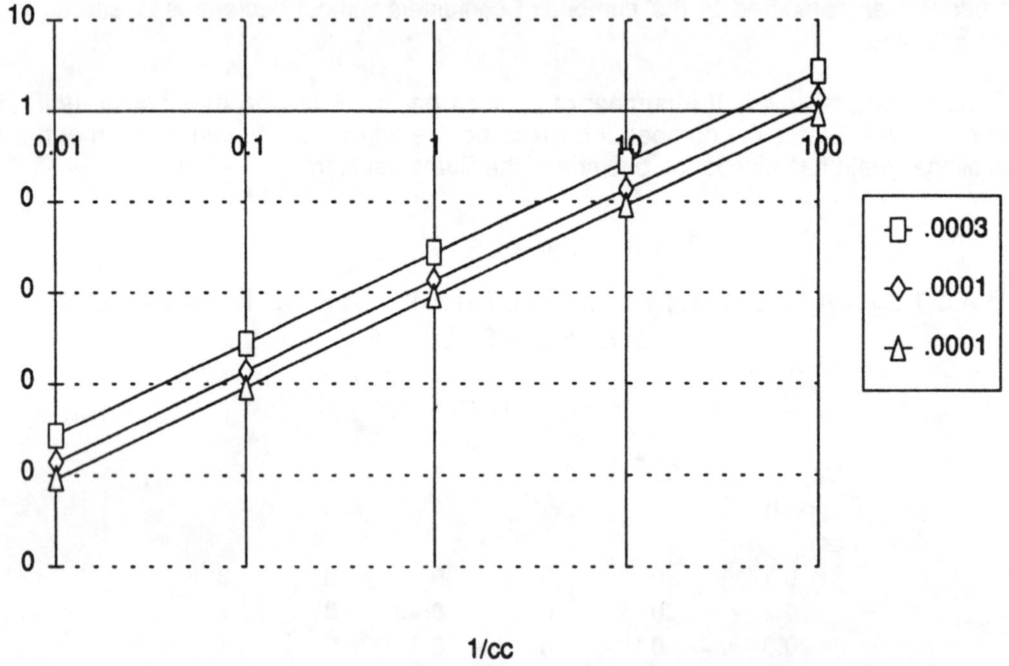

Legend:

.0003 >.0003 % of the 2 gal per min cartridge flow as leakage through defect.
.0001 .0003 - .0001 % of the 2 gal per min cartridge flow as leakage through defect.
.0001 <.0001 % of the 2 gal per min cartridge flow as leakage through defect.

Table 1 shows the probable level of contamination of filtrate being packaged in one liter containers as a function of the incoming feed contamination. The defect in the cartridge is a 10 µm capillary (with capillary lengths of 1, 2, and 3 mm) in the end-cap. As is noticed, the contribution to flow from the defect is of the same magnitude as the diffusion flow rate.

CHART 2

CAPILLARY AIR FLOW THROUGH A MEDIA DEFECT OF 10 µM IN A SINGLE AND DOUBLE LAYER CARTRIDGE

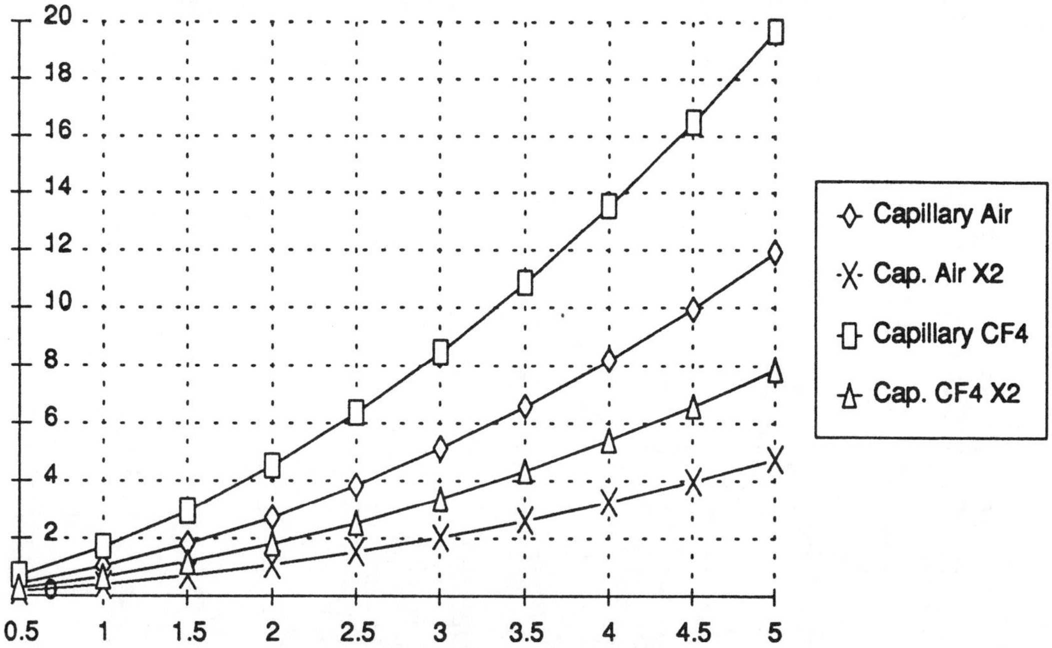

Pressure in Bar— Flow in Ml/Min

Legend:

Capillary Air	-is the air flow through a single layer media with a 10 µm capillary defect below the bubble point
Cap. Air X2	-Same as above with double layer membrane.
Capillary CF_4	-Same as above in single layer using Freon 12 as test gas.
Cap. CF_4	-Same as above using Freon 12 in a Double layer cartridge.

Defect Determination

From the equations of Fick's law it is possible to construct a method for the evaluation of defects in the media of a large filtration array. Assuming that we are examining only one filter cartridge in a filtration configuration, the volume of air due to the diffusion of air or test gas through the filter can be compared with the air leaking through the capillary defect of 10 µm. The difference in air flow rate between the calculated rate of diffusion and the sum of the diffusion rate and the rate of leakage is not great with air.

Chart 3 shows the relative calculated diffusion rates through a 7 ft^2 cartridge filter at varying pressures. The values for the CF_4 and SF_6 are compared with that of air. Both the Freon and the sulphur hexafloride are non-toxic gases that are very slightly soluble in water. Their size makes them much slower in diffusion than air, with diffusion constants lower than that of air. The viscosity of the CF_4 is also lower than that of air as calculated from the kinetic theory. SF_6 on the other hand, has a higher viscosity than air with the result that the flow differential between the diffusional flow and the leakage rate will not be as high as the Freon 12.

CHART 3

RELATIVE DIFFUSION RATES BETWEEN AIR, CF_4, AND SF_6 CALCULATED AT 7 FT^2 MEDIA AREA IN CARTRIDGE

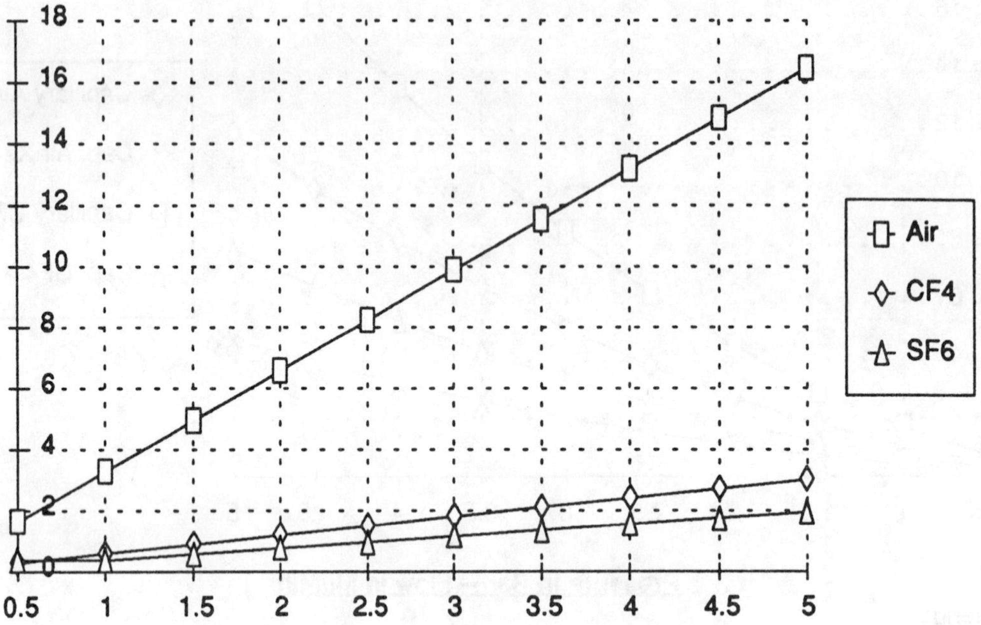

Pressure- Bar -- Flow- Ml/Min

Generally, the calculated diffusion rate is usually higher than the rate measured in practice. This is most likely due to the thickness of the water layer being indeterminate in the cartridge. The diffusion rate also requires low time at constant differential pressure to equilibrate. This will result in even a greater difference in the flow rates with and without defects.

Air Flow Rates Through Capillary Defects in Media and End-caps.

The flow equation for this flow is shown in Equation 4 below. The values for the viscosity of the CF_4 was calculated from the viscosity of air and consideration of the kinetic theory of gases.

EQUATION 4

GAS FLOW THROUGH SMALL CAPILLARIES

$$V = \frac{3.14\, D^4 \frac{(P_1+P_2)}{2} (P_1-P_2)}{128\, n\, L} \left[1 + \frac{8Kn}{D(P_1+P_2)}\right]$$

Kn is the Knudsen Factor

From this equation the following table was prepared to calculate the leakage through the 10 μm defect.

Air Flow Rate - Media Defect

D	P_1	P_2	avgP	diffP	L	n x 10000	flow / min. cc/min
0.001	1.00E+6	6.00E+6	3.50E+6	5.00E+6	0.012	1.83	11.91
0.001	1.00E+6	5.00E+6	3.00E+6	4.00E+6	0.012	1.83	8.19
0.001	1.00E+6	4.00E+6	2.50E+6	3.00E+6	0.012	1.83	5.14
0.001	1.00E+6	3.00E+6	2.00E+6	2.00E+6	0.012	1.83	2.75
0.001	1.00E+6	2.00E+6	1.50E+6	1.00E+6	0.012	1.83	1.04
0.001	1.00E+6	6.00E+6	3.50E+6	5.00E+6	0.03	1.83	5.00
0.001	1.00E+6	5.00E+6	3.00E+6	4.00E+6	0.03	1.83	3.28
0.001	1.00E+6	4.00E+6	2.50E+6	3.00E+6	0.03	1.83	2.05
0.001	1.00E+6	3.00E+6	2.00E+6	2.00E+6	0.03	1.83	1.10
0.001	1.00E+6	2.00E+6	1.50E+6	1.00E+6	0.03	1.83	0.42

CF_4 - Flow Through the Same Defect

0.001	1.0E+6	6.0E+6	3.5E+6	5.0E+6	0.012	1.09	19.88
0.001	1.0E+6	5.0E+6	3.0E+6	4.0E+6	0.012	1.09	13.65
0.001	1.0E+6	4.0E+6	2.5E+6	3.0E+6	0.012	1.09	8.55
0.001	1.0E+6	3.0E+6	2.0E+6	2.0E+6	0.012	1.09	4.57
0.001	1.0E+6	2.0E+6	1.5E+6	1.0E+6	0.012	1.09	1.72
0.001	1.0E+6	6.0E+6	3.5E+6	5.0E+6	0.03	1.09	7.95
0.001	1.0E+6	5.0E+6	3.0E+6	4.0E+6	0.03	1.09	5.46
0.001	1.0E+6	4.0E+6	2.5E+6	3.0E+6	0.03	1.09	3.42
0.001	1.0E+6	3.0E+6	2.0E+6	2.0E+6	0.03	1.09	1.83
0.001	1.0E+6	2.0E+6	1.5E+6	1.0E+6	0.03	1.09	0.69

From the calculations made at the 12 mil (300 μm) thickness the next charts show the relative flow between diffusional flow and the combination of the diffusional flow and the leakage.

Chart 4

Air Diffusion and Air Flow Through Capillary Defect of 10 μm

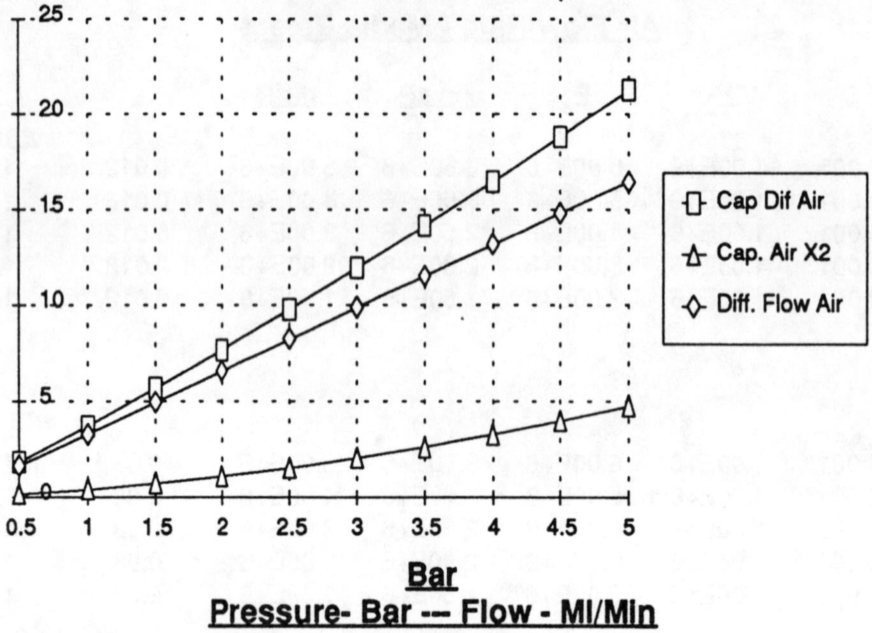

Chart 5

CF_4 Diffusion and CF_4 Flow Through Capillary Defect of 10 μm

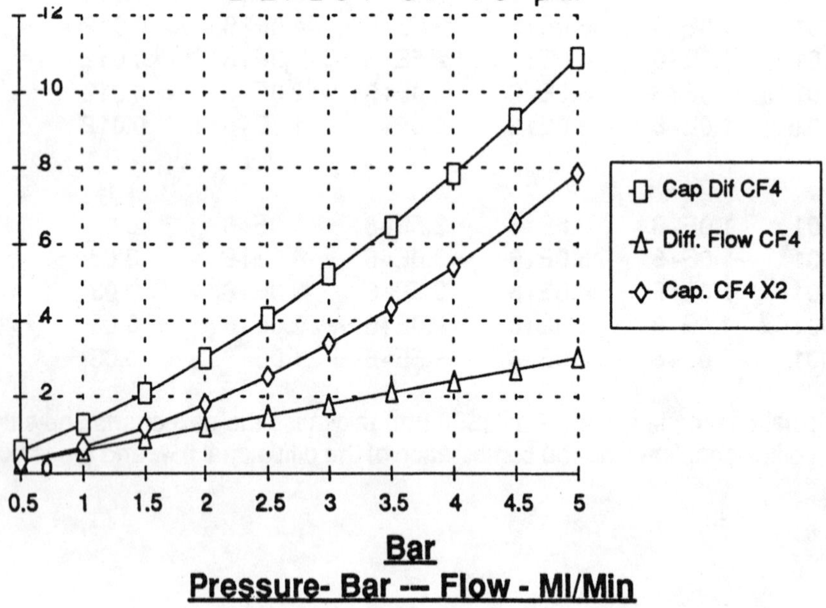

Conclusions:

Testing the integrity of filters in large filter arrays is possible with the diffusion type of test procedures. There is a degree of risk associated with the testing. This is due to the relatively low level of flow associated with single defects. Defects in the end-cap construction, when very small, are virtually undetectable. For this reason it is not possible to achieve a 100% guarantee that all filters that test good in diffusional testing are without flaw.

The manufacturers of filters have test programs which significantly reduce the risk of defective material being shipped to the end user. The system requires extensive testing to obtain an acceptable level of acceptability.

Further methods of testing using gases such as CF_4 or SF_6 are possible due to the ease of detection of these halogen containing gases. Both are safe to use and can take into account the time necessary to establish diffusion.

Optimal Prefiltration of
Plasma-Derived Products

Patricia Waters Schwartz, PH.D.
Pall Corporation
30 Sea Cliff Ave.
Glen Cove, N.Y. 11542
516 671 4000

The leaders in the filtration industry have consistently pioneered new developments in the construction of finer and increasingly complex filters. The current filter elements available to the pharmaceutical industry for microfiltration span the range of pore sizes from 0.04 to 120 microns. Depth and membrane filters are composed of a variety of materials such as glass fiber, nylon, polypropylene, cellulose esters and polyvinyldifluoride. Such a variety of choices in media alone demonstrates the need for a discussion of the parameters involved in designing a filtration system. In any filtration process the key to success is the optimalization of prefiltration.

In a filtration system, the final filter performs the prime objective of the filtration - the removal of contaminants of a particular size or character. The final filter selection specifies the final characteristics of the eluate.

The function of a prefilter is to remove contaminants that would quickly block the fine pore structure of the sterilizing grade filter and in doing so, extend its life.

In performing this function, especially in plasma or serum-derived products, the prefiltration process functions in an additional mode, it "organizes" the proteins in the the solution to allow passage through the fine pores of the sterilizing filter. By using a succession of finer and finer grade filters, large conglomerations of protein complexes are subtly restructured into smaller units. The initial random aggregates are shaped into conformations that can pass through the very fine pores of the final filter.

Consideration of low protein binding media and also holdup volume is just as important in the selection of prefilters as it is in final filter selection. The process fluid is in contact with the prefilter as much as the final filter. Frequently prefilters have larger dimensions than the final filters which translates into a greater surface area for potential adsorption. The overall adsorption of protein from the product by the complete filtration scheme should be investigated when designing a filtration system.

SELECTION OF PORE SIZE

The selection of prefilter pore size can only be made on an individual basis with respect to the size and nature of the contaminants. The functions of prefilters in plasma-derived as well as other products are to increase throughput and reduce cost. A general rule of thumb is to initially select a prefilter about three sizes larger than the final filter such as a 0.65 to 0.7 um to protect a 0.2 um final filter.

If this prefilter effectively protects the final filter, the differential pressure across the final filter will have decreased. The prefilter should extend the life of the final filter at least three times to be cost effective. Since in general, the cost of a prefilter is inversely proportional to pore size, the next step would be to determine the coarsest pore size that effectively functions.

If on experimentation, the initial prefilter selected does not meet the criteria for protecting the final filter, one would select a finer grade. The contaminants might be of a smaller size such that the tighter prefilter functions better. One might also consider a prefilter of a different media. A 1.0 um depth filter will protect a 0.2 um final filter by a different mechanism than a nylon membrane.

In filtering serum, the age of the animal corresponds to an increase in initial prefilter size. Fetal calf serum can use in many cases a 1 um initial prefilter. As the age of the donor increases to grown cow, the animal builds antibodies and the circulating lipid content increases. This makes the serum much more difficult to filter and necessitates a more complex filtration scheme with a coarser initial prefilter such as a 6-10 um pore size.

FILTERABILITY TESTING

Since plasma derived products are extremely expensive it is prudent that large scale filtration schemes be tested for feasibility on small amounts of product. The most common method is filterability testing with 47 mm discs of the same material used in the pleated membrane cartridges. A 47 mm disc with effective filtration area of _0.015 ft^2 uses almost 1/330th of the material a ten inch element with 5^2 of area would use. The flow rate is decreased proportionately. This allows several prefilter-final filter combinations to be tested with as little as 1 L of material. Samples of effluent can be tested for clarity, protein activity or any other required criteria.

to give a consequential increase in flow rate to process parameters. This will minimize blinding of the filter by gel polarization and gel compression.

POSITIVE ZETA POTENTIAL FILTERS

Asbestos filters have in the past been used extensively in filtering parenterals and biologicals specifically in the Cohn procedure for the fractionation of blood plasma. Chrysotile asbestos exhibits two characteristics which are responsible for its excellent properties in filtration. As essentially a bundle of 0.2 micron diameter fibrils, chrysotile has an extremely large surface area which enhances it ability to remove particles by sieving. In addition, it exhibits a high positive zeta potential with a very large number of positively charged sites. This allows chrysotile to capture particles by electrostatic attraction. In 1975, the USFDA banned the use of asbestos-containing filters in the processing of parenterals intended for injection in humans.

Many filter manufacturers are marketing filters with positive zeta potentials. Since most particles and organisms are negatively charged in aqueous solutions, the positively charged media can bind more particles of a smaller size than its absolute pore size rating. By mimicking the charge behavior of asbestos, a filter element increases clarification of a filtrate and binds endotoxin much more efficiently. In a system designed with positively charged prefilters, the load on the filters downstream is decreased and the net effect is longer life for the final filters.

PROTEIN ADSORTION

Protein adsorption to membranes during processing is of critical concern to manufacturers because of the extremely high cost of blood-derived products. Solution conditions such as pH, temperature, buffers , and ionic strength are very important considerations in protein binding but the critical determinants in the extent of adsorption are the individual structure of the protein and the surface chemistry of the membrane. The membrane chemistry, specifically what kind of molecules are exposed to the exterior of the protein and available for interaction and potential binding, determines to a great extent how much adsorption will occur.

By and large proteins in aqueous solution adopt a three dimensional conformation in which most of the hydrophobic amino acid functional groups are buried in the interior of the folded molecule. Hydrophillic groups are concentrated on the surface of the protein exposed to the aqueous solution. A membrane that is inherently hydrophillic can be said to be passive with respect to protein binding or considered to be biologically inert.

Protein adsortion to filter media is of special and most important concern when the fluid to be processed is very dilute with a low concentration of protein. Binding of 100 ug/cm^2 in a 100 L batch of 0.05 % protein approaches a loss of 1 % for a 5 ft^2 cartridge. The use of a biologically inert membrane with as low an adsorption as 1 ug/cm^2 significantly reduces cost by increasing the net yield of the product.

TEMPERATURE

In practice serum is stored and/or filtered at low temperatures. Since viscosity is inversely proportional to temperature, this should increase the already elevated viscosity of the fluid making it more difficult to filter. A decrease of temperature below the T_m (melting temperature) of the lipid components of the serum causes these molecules to become more rigid in conformation. The lipoidal components become more particulate in nature. As less fluid materials they are less likely to gel or smear the filtration surface.

Chilling or freezing sera samples can result in the formation of a cryoprecipitate. This decreases the viscosity of the supernatant solution which may allow more throughput volume to pass through the filtration system. Removal of the cryoprecipitate ,however in some systems can cause undesirable changes in the composition of the supernatant. Lyophilization and resuspension of protein solutions can also increase filterability by structurally disrupting lipoproteins and causing the molecules to aggregate.

Attempts have been made to increase filterability by increasing the filtration temperature. This is based on the principle that the increased temperature will lower the viscosity of the product and consequently improve flow rates. In actual practice, few pharmaceutical firms employ this technique in all but special instances. Many proteins are heat sensitive and have reduced activity at elevated temperatures. They may also undergo undesired conformational changes. Equally important is the realization that increased temperature facilitates microbial growth.

FLOW RATES AND DIFFERENTIAL PRESSURES

As a biologically unstable material susceptible to small changes in pH, etc. and as an excellent material for microbial growth, rapid filtration of sera- type products minimizes deterioration. With decreased filtration time, additional processing steps such as lyophilization and filling may be performed on the product in one working day. High flow rates decrease processing time and labor costs and therefore capital expenses.

Soluble macromolecules such as lipoproteins are deformable and have a tendency to compress into the fine pore structure of filter media. They form gels at the filter/solution interface and essentially plug the filter, a phenomena known as gel polarization. As the differential pressure across a filter increases, the likelihood of this occurring increases. High flow rates through membranes can lead to gel blinding of the media surface and shortened filter life.

Ideally any filtration scheme for serum or plasma products should be sized for the lowest differential pressure feasible for the reasons already discussed. In general pharmaceutical firms recognize the difficulty of serum filtrations. Although flow rate is an important factor, the primary operational concern is the throughput volume. Ideally the initial differential pressure on the filters should be minimized such that it is just enough to initiate flow of the product through the filters. As the filtration progresses the differential pressure may be increased

In the pharmaceutical and biomedical industries, whole human and/or animal blood is the basic source for the three major subgroups of blood-derived products.

1. therapeutic agents - such as antihemophilic factor, immune serum globulins, albumin.
2. diagnostic reagents - blood typing antisera, controls for blood chemistries.
3. tissue culture media - fetal bovine serum.

Human plasma is isolated on a large scale through fractionation of anticoagulated whole blood by centrifugation. Plasma is basically a protein solution - sixty percent albumin, thirty-five percent globulins and five percent fibrinogen. Currently more than sixty plasma proteins have been isolated, characterized, and purified to some degree.

Human plasma is generally utilized in specific subfractions. The major therapeutic components produced are summarized below:

<u>Protein</u>	<u>Physiological Function</u>	<u>Clinical Indications and Use</u>
Albumin	Maintenance of Osmotic Pressure	Plasma Volume Expander, Drug Transport
Immunoglobulins	Antibodies Removal of Foreign Bacteria	Immunization Inherited Deficiencies
Factor VIII:C	Procoagulant activity	Classic Hemophilia A
Factor VIII:RC	Platelet Aggregation	von Willebrand's Disease Bleeding Time
Factor IX	Blood Clotting	Hemophilia B
Fibrinogen	Precursor of Fibrin which forms clot	Rare Inherited Deficiency Topical Application after trauma
Fibronectin opsonin	Removal of Particulates	Trauma, burns
Anti-thrombin (Factor III)	Inhibition of Blood Clotting	Inherited Deficiency Anti-coagulant

Although pharmaceutical firms may process plasma, serum, or any of the other subfractions, some general factors can be considered when developing a filtration scheme. Generally serum from most sources is a pale, straw yellow colored liquid of viscosity 10-20 CP. The various complex particulates to be considered in any purification process include: residual cell fragments, bacteria, fibrin, proteins, lipoidal organic molecules, organic molecules (amino acids, saccharides, & urea), salts, and chemical agents used in the purification process.

Sera and plasma products, even when considering only one individual fraction such as an specific immune complex, contain quite different particulate loading. Their composition varies enormously not only when comparing the species of origin, such as IGG from horse versus human, but even from day to day with the same donor animal. The blood components are affected by breed, age, sex, diet, and method of exsanguination as examples. In addition to these factors the chemical nature of the major serum components, proteins and lipids, plays a more critical role. Time, temperature, pH, cation and anion concentrations, and exposure to air can cause alterations in their physical properties.

SUMMARY

To design the optimal prefiltration system for plasma-derived products, the primary factors for consideration can be summarized as:

1. media selection - compatibility and/or enhancement of product recovery
2. flux rate and throughput
3. economics

By enhancement of the prefiltration process with respect to the first two factors, the end result is an increased recovery of a purer, cleaner product in a minimum of process time. The overall effect is economics - the best product at the minimum cost.

AMERICAN FILTRATION SOCIETY

P.O. Box 6269
Kingwood, Texas 77325
Telephone: (713) 359-1894
FAX: (713) 358-3939

ABSTRACT & COPYRIGHT

Meeting Location: **Meeting Date:**

Session Chairman: **Session Number:**

SPEAKER (AUTHOR):

Name: Amy Penticoff
Company: Dow Corning Corporation
Address: 2200 W. Salzburg Road
City, State, Zip: Midland, MI 48686
Telephone: 517-496-5853 **FAX:** 496-4654

CO-AUTHOR:

Name: Chris Combs
Company: Dow Corning Corporation
Address: 2200 W. Salzburg Road
City, State, Zip: Midland, MI 48686
Telephone: 517-496-5316 **FAX:** 496-4654

CO-AUTHOR:

Name:
Company:
Address:
City, State, Zip:
Telephone: **FAX:**

CO-AUTHOR:

Name:
Company:
Address:
City, State, Zip:
Telephone: **FAX:**

TITLE OF PAPER:

TEXT OF ABSTRACT: (200 Words in Space Below)

TITLE: FOAM CONTROL FOR ULTRAFILTRATION SYSTEMS

ABSTRACT:

Advances in biotechnology, fermentation, and food processing in recent years have resulted in the widespread use of ultrafiltration as a means of separation and purification. With this new technology, it has been reported that commonly recommended antifoams used in these types of systems can cause severe and even irreversible fouling of the ultrafiltration membranes. This paper summarizes research work which was conducted to determine the various factors contributing to the antifoam fouling problem. Variables tested include the components of silicone and organic antifoams and their effects on the reduction in permeate flux of various ultrafiltration membranes. The paper reveals recent work conducted which has resulted in the development of a new antifoam product, designed to alleviate the problems associated with antifoam-membrane fouling in ultrafiltration processes. Various ultrafiltration applications which encounter foaming during processing will also be discussed.

****

PLEASE TYPE INFORMATION BECAUSE DATA WILL BE PUBLISHED IN THE ABSTRACT BOOKLET. RETURN FORM TO THE AMERICAN FILTRATION SOCIETY, P.O. Box 6269, Kingwood, Texas 77325 WITH A COPY TO YOUR SESSION CHAIRMAN

OVERVIEW OF HIGH TEMPERATURE PARTICULATE CONTROL TECHNOLOGY

Robert C. Bedick
U. S. Department of Energy
Morgantown Energy Technology Center
P. O. Box 880
Morgantown, West Virginia 26505

INTRODUCTION

The Morgantown Energy Technology Center (METC) of the U.S. Department of Energy is actively sponsoring research to develop coal-fueled power generation systems that use coal more efficiently and economically and with lower emissions than current pulverized coal power plants. Specific systems include: pressurized fluidized-bed combustion (PFBC), integrated gasification combined cycle (IGCC), and direct coal-fueled turbines (DCFT). Each of these systems rely on gas turbines to produce all or a portion of the electrical power. Particulate control is necessary in advanced power generation systems to meet environmental regulations and to protect the gas turbine and other major system components.

A research and development (R&D) program was established at METC in 1979 to develop high temperature and high pressure (HTHP) particulate control devices for use in advanced energy conversion systems. The basic premise behind this research program is that removal of particles, as well as gaseous contaminants, at system temperature and pressure would result in more efficient and less complex energy production systems. General process conditions and particulate control goals for each of the advanced power generation systems are shown in Table 1.[1] Conditions include those for both reducing gases, such as in IGCC systems, and oxidizing gases in PFBC systems. The specific system has important implications for numerous reasons including selection of construction materials and particle characteristics such as mass loading, size distribution, composition, shape, density, and morphology.

DISCUSSION

A number of HTHP particulate control concepts have been evaluated for use in advanced power generation systems and are summarized in Table 2. This test summary is not intended to represent all of the testing that has been done, but only some of the more representative tests. All values in this table should be considered nominal to give an indication of the scale of testing at the most representative gas conditions, as well as typical filter operating conditions. Testing of control concepts has been accomplished at relatively small scale and for short periods. In addition many of the tests have not been done under actual PFBC or gasification conditions. For example, while over 17,000 hours of testing were attained with a single set of woven ceramic bag filters, these tests were conducted at atmospheric pressure and less than 1,000°F at a pulverized coal fired boiler. A major emphasis of continued R&D is to prove commercial readiness by conducting larger scale tests under actual operating conditions for extended periods.

While some of the concepts listed are commercially available, such as the electrostatic precipitator (ESP) and electrostatically enhanced granular bed filter (GBF), the commercial application is for much lower temperatures than indicated in Table 1. For all practical purposes, R&D for HTHP conditions began in the early 1960's for the ESP and the mid- to late-1970's for the other concepts.[2] Evaluations to date indicate that particulate control at HTHP is feasible and there are no fundamental, theoretical reasons why high particle collection efficiencies cannot be attained.[3] Major issues yet to be resolved include component durability, suitability of materials and optimum equipment design.

Conventional high efficiency cyclones are the only particulate control devices that currently are considered reliable enough to protect the gas turbine, but their use in most systems requires a post-turbine baghouse or ESP to meet environmental limitations. For some IGCC systems, using a fixed-bed gasifier, there are indications that cyclones can meet the cleanup requirements in Table 1. There have been numerous efforts to improve the efficiency of cyclones, but these efforts were largely unsuccessful. The use of high intensity acoustic energy to agglomerate submicron particles and shift the total mass distribution of particles toward particles above 10 microns diameter, where cyclones are most effective, continues to show significant potential. Acoustic agglomeration is being further studied with an emphasis on developing a reliable sound source based on pulsed combustion.

Several types of GBFs have been investigated, but only a moving bed with the flow of granules downward, countercurrent to the upward flow of particulate laden gas has demonstrated the required high collection efficiencies. The moving GBF has shown excellent particle collection efficiency for PFBC exhaust gases. The high collection efficiency results from the formation of a filter cake where the dirty gas enters the granular bed, thus collection is not totally dependent on impaction, diffusion and direct interception as with other GBFs. Electrostatic enhancements to panel bed filters have not proven practical nor effective, although there is interest in Japan to further develop this concept for lower temperature gasification environments. Efforts to use a strong magnetic field to preferentially orient a bed of cobalt granules to achieve higher gas throughput were too costly and showed no positive effect on particle collection. Static and jetted GBFs consist of a shallow bed of granules supported on a distributor plate. In the static design the dirty gas flows downward through the granules and distributor plate and is cleaned by a reverse gas flow. A specially designed distributor plate in the jetted GBF directed high velocity gas up through the shallow bed. With both shallow GBFs significant design issues were never resolved.

Several attempts to demonstrate HTHP ESPs resulted in inconclusive tests. The preferred operating conditions were never attained due to process upsets and failure of the high voltage ceramic bushings. While the true potential of ESPs has not been demonstrated, METC is not supporting further development due to concerns over materials and the unfavorable projected cost of the system. Current work in electrostatics is limited to more fundamental effects on particle agglomeration and repulsion of particles from surfaces.

Both ceramic bag and rigid barrier filters are analogous to pulse jet baghouses except filtration media have been adapted for high temperature. The HTHP filters are suspended from a tube sheet with the gas flow from the outside to the inside and are cleaned by a reverse pulse of high pressure gas. All of these filters have demonstrated high collection efficiency, but there are concerns about material durability in the hot, corrosive environment, long term filtration and cleaning performance, and the reliability of components, such as solenoid valves for pulse cleaning.

Two of the more promising concepts are the ceramic candle and ceramic cross flow filters. These two devices are receiving the greatest support in the U. S., while European R&D is focused on the candle filter. Candle filters are typically 5 ft long and 2.4 in. outside diameter, with one end closed and the other end attached to the tube sheet. Candles have been made of two layers of silicon carbide; a porous, large grain support structure is covered by a thin small pore outer layer. Cross flow filters are constructed of porous ceramic plates with ribs to form alternating gas channels at 90 degrees to each other. The cross flow filter is a 12 in. by 12 in. by 4 in. block that is sealed on one 12 in. by 4 in. end and attached to a clean air duct on the other. Both filters have demonstrated near absolute filtration when operating properly.

CONCLUSIONS

From the initial HTHP particle control concepts evaluated, R&D is continuing on the screenless GBF, acoustic agglomeration, and ceramic cross flow and candle filter systems.[1,4] Tests of these devices have demonstrated the ability to meet NSPS for particulate emissions from coal-fired power plants and particulate tolerance limits for gas turbines. Major issues yet to be resolved include component durability and reliability, optimum engineering design, system integration, and confirmation of performance for long duration at larger scale. When these technical issues are resolved, one or more of these technologies can realize its commercial potential for incorporation into advanced coal-based power generation systems.

REFERENCES

1. S. J. Bossart, "Advanced particle control for coal based power generation systems," Proceedings of the Joint ASME/IEEE Power Generation Conference, Dallas, Texas, October 22 -26, 1989.

2. Rubow, L. N. et al, "Technical and Economic Evaluation of Ten High Temperature, High Pressure Particulate Cleanup Systems for Pressurized Fluidized Bed Combustion," NTIS Report Number DE84012004, 1984.

3. M. W. First, "High Temperature Gas Filtration Research Needs," NTIS Report Number DE85003819, 1984.

4. R. C. Bedick, M. C. Williams, "Gas Stream Cleanup, Technology Status Report," NTIS Report Number DE89000925, 1988.

Table 1. Particle control conditions and cleanup goals

	PFBC	IGCC	DCFT
Temperature (degrees F)	1,500-1,700	1,000-1,800	1,800-2,250
Pressure (psig)	100-240	120-1,500	120-500
Inlet Particle Loading (ppmw)*	1,000-20,000	200-3,000	Undefined
Outlet Particle Loading (ppmw)**	Lower than NSPS *** 15-30	Lower than NSPS *** 120-300	Lower than NSPS *** 15-30
Outlet Size	No particles > 5-12 micron	No particles > 5-12 micron	No particles > 5-12 micron

* Possible ranges at inlet of advanced particle control filter

** At outlet of the advanced particle control filter

*** New Source Performance Standards for particulate emissions from coal-fired power plants is 0.03 lbs./MM BTU

Table 2. Summary of high temperature particulate control testing

DEVICE	NOMINAL GAS FLOW (ACFM)	HOURS OF OPERATION	POTENTIALLY MEETS GOALS FOR CLEANUP	NOMINAL DELTA P (IN H_2O)/ VELOCITY (fpm)	MAJOR ISSUES
CYCLONES					
High efficiency two in series	Commercial	NA	No	305/145 fps	Does not meet emission limits for all systems
High efficiency w/acoustic agglomeration	20*	<10	Yes	60/145 fps	Reliable/efficient sound source, operability
Electrostatically enhanced	3,400*	<20	No	40/140 fps	Electrostatics ineffective
GRANULAR BED FILTERS					
Static beds	500*	50	No	30/45	Cleaning, complexity
Moving bed w/ electrostatics	500*	75	No	1/200	Electrostatics ineffective, plugging
Moving bed w/magnetic stabilization	45**	<10	No	12/290	Complexity, cost
Moving bed w/counter-current flow	1,200**	160	Yes	24/60	Complexity, cost, Component durability
Staged, jetted bed	22*	<10	No	11/65	Complexity, cost
ELECTROSTATIC PRECIP.	1,100**	362	Undetermined	5/200	Component durability
CERAMIC BAG FILTERS					
Woven	900*	17,000	Yes	10/6	Component durability, dust cake stability
Felted	820**	140	Yes	25/15	Component durability
RIGID BARRIER FILTERS					
Sintered Metal	200**	571	Yes	20/4	Blinding, corrosion
Ceramic cross flow	1,000**	85	Yes	15/10	Component durability, long term operation
Ceramic candles	2,800**	790	Yes	55/15	Component durability, long term operation

* Denotes simulated or not actual HTHP conditions ** Denotes actual HTHP conditions

LONG-TERM HIGH TEMPERATURE DEGRADATION MECHANISMS IN POROUS CERAMIC FILTERS

Mary Anne Alvin
Westinghouse Science & Technology Center
1310 Beulah Road
Pittsburgh, PA 15235

John Sawyer
Acurex Corporation
485 Clyde Avenue, P.O.Box 7044
Mountain View, CA 94039

Introduction

Advanced coal utilization processes as Pressurized Fluidized-Bed Combustion (PFBC), Integrated Gasification Combined-Cycle (IGCC), and Direct Coal Fueled Turbines (DCFT) systems will require hot gas cleaning in order to comply with Federal particulate emissions regulations, and to protect downstream equipment (i.e., turbine). Based on their demonstrated particulate removal collection efficiencies of >99.9% at filter process conditions, both the ceramic candle and cross flow filters are viewed as technically viable for use, in particulate removal systems.

Ceramic candle are rigid tubular filters formed by bonding ceramic fibers and/or grains with an aluminosilicate binder. Candles are typically 1 to 1.5 meters long, and have a 60 mm outer diameter, and 10-15 mm wall thickness. They are flanged at the top and closed at the bottom. During operation, the flange portion rests in a tubesheet, providing support for the candle and sealing the clean gas plenum from the dirty gas plenum.

Ceramic cross flow filters are rigid elements constructed from multiple layers of thin, porous ceramic plates that contain ribs to form gas flow channels. Consecutive layers of the ceramic plate are oriented such that the channels of alternating plates are perpendicular to each other. Current cross flow filter dimensions are 12x12x4 inches. This configuration contains approximately 7-8 square feet of surface area, producing a high surface area per unit volume ratio. One end of the long 12 inch filter is sealed, while the opposite end is mounted to the clean gas plenum.

Although both filters have demonstrated excellent particulate removal efficiencies at high temperature, their current effective use has been hampered by crack formations. Candle filter typically crack due to thermal or mechanical shock, as well as lose up to 50% of their material strength during process operation. Cross flow filters have typically failed by delamination between gas channel plates. Although crystallization within the cross flow filter substrate material has been observed, loss of cross flow material strength has not been confirmed.

Degradation of the ceramic filter materials may result from both thermal/mechanical and chemical effects. The filter systems are subject to temperature variations caused by cold pulse gas cleaning. In addition, the filters are subject to vibrations due to the operating environment, and large scale temperature variations during start-up and shut-down cycles. The ceramic filters which consist of oxides, mixed-oxides, and non-oxide materials are also exposed to process gases containing volatile alkali, chloride, and sulfur species, water vapor, and fines that may be enriched with calcium, magnesium, and/or iron. This paper reviews current efforts being conducted at Westinghouse and Acurex to identify potential long-term, high temperature degradation mechanisms which may occur in the porous candle and cross flow materials at process operating conditions.

Approach

A comprehensive literature review was performed, identifying process gas and particulate chemistry, expected filter operating conditions, and ceramic filter material (i.e., alumina, mullite, cordierite, silicon carbide, and silicon nitride) property data. In addition candle or cross flow filters that have been exposed in various process systems were subjected to chemical and physical characterization, and compared with as-fabricated filter material. Experimental efforts are currently focused on identifying the effect of alkali on ceramic filter materials, and investigating the effect of pulse gas cleaning on material strength. The possible synergistic influence of alkali and thermal cycling is also being explored.

Results

Literature Review

Literature indicates that the effects of thermal stress and chemical reactions are considered to be potentially detrimental to the service life and performance of the ceramic filter materials. Chemical reaction of alkali with the aluminosilicate binder phase in silicon carbide candle filters, as well as with glass phases present in the alumina/mullite or cordierite cross flow filter materials could occur, forming viscous phase melts. This is expected to reduce material strength as a result of increased creep in the ceramic substrate, and subsequent loss of structural integrity. The amorphous or glassy phases which could form in either the oxide, mixed oxide or non-oxide filter materials are considered to have increased thermal expansion coefficients, which could promote crack formations under thermally induced stress conditions. Non-oxide ceramics (silicon carbide and silicon nitride) are considered to undergo rapid surface or grain boundary oxidation at elevated process temperatures, particularly in the presence of steam and volatile alkali species. Thermodynamic equilibrium calculations have identified the potential high temperature reaction of gas phase sulfur species with the non-oxide ceramic materials.

Field-Testing

Chemical and physical characterization of silicon carbide candle filters which experienced PFBC or IGCC conditions indicate the occurrence of several degradation mechanisms, the most serious of which is depletion of the bond material between adjoining grains. Bond depletion is considered to be caused by the formation of an alkali-rich viscous liquid binder phase at temperatures as low as 650°C.

Chemical and physical characterization of cross flow filter materials that experienced PFBC conditions indicate that crystallization of anorthite results in the amorphous phase of the alumina/mullite filter matrix. The amorphous alumina/mullite filter material phase has also been shown to be the preferred site for adherence of ash or char fines.

Bench-Scale Testing

Kaolinite is considered to be a binder-like phase in candle filters. Initial experimental efforts have demonstrated reaction of kaolinite and sodium at temperatures as low as 820°C. At temperatures greater than 925°C, a reaction layer forms on the surface of the material, which can spall, exposing unreacted material. In comparison to sodium, reaction with potassium is less severe at temperatures below 925°C, but is equally severe at 1225°C.

When mullite, another binder-like phase in candle filters, is exposed to sodium at 1225°C, extensive crack formations result. XRD analyses indicate that mullite undergoes a phase transformation to carnegieite and sodium aluminosilicate.

Experimental efforts have also demonstrated high temperature absorption of volatile alkali with the alumina/mullite cross flow filter material. Preliminary bench-scale data indicate that the principal reaction site for alkali absorption is temperature specific. For cross flow filter materials exposed to temperatures of 815°C, 92% of the absorbed alkali appears to result within a water soluble alkali phase (i.e., NaCl); at temperatures of 925°C, 98% of the absorbed alkali is driven into either an acid soluble or insoluble phase, possibly representing reactions with the amorphous or "glassy" grain boundary phase of the alumina/mullite structure.

Based on statistical and finite element stress analyses and bench-scale testing, both filters are considered to experience thermal stress, particularly during pulse gas cleaning. The thermal stress levels may not cause the filter elements to break, but may introduce microcracks which weaken the material.

Conclusions

Loss of material strength, phase transformations, and microcracking all could occur during operation of porous ceramic filters at advanced coal process conditions. These could result from exposure of the filters to high operating temperatures; thermal gradients induced during pulse gas cleaning or system transients; chemical reactions with the oxidizing or reducing process gases; reactions with trace or minor constituents as volatile alkali or sulfur species; or interactions with condensed phases or particulates which impact the filter surface. Potential degradation mechanisms in ceramic filter materials are considered to include both short term (<1000 hours) thermal/mechanical effects, and long-term (>1000 hours) chemical reactions, particularly with alkali species. Synergistic thermal/chemical effects are expected to affect long-term ceramic filter material stability.

Ceramic candle filters have been shown to experience a loss of 50% of their original strength after exposure in advanced coal utilization processes. Significant binder loss has been observed in candles after short-term exposure at temperatures as low as 650°C. Sodium has been shown to react with binder-like phases of the candle filters at 820°C. Chemical reaction(s) of silicon carbide and silicon nitride with steam and alkali in the process gases have been identified. Efforts are needed to demonstrate whether these reactions are detrimental to candle filter material strength and long-term durability.

Crystallization within the amorphous phase of the alumina/ mullite cross flow filter material occurs after exposure to PFBC conditions. The alumina/mullite cross flow filter material is known to absorb volatile alkali. Absorption is considered at the grain boundary phase. Similar to candle filters, efforts are needed to demonstrate whether alkali absorption is detrimental to cross flow filter material strength and long-term durability.

Acknowledgements

The authors wish to acknowledge the technical contributions of Jesse and Nancy Brown at VPI, Richard Tressler at Penn State University, and Tom Lippert and Jay Lane at Westinghouse to this work.

ATTRIBUTES OF PARTICLES AND DUST CAKES RESULTING FROM HOT GAS CLEANUP IN ADVANCED PROCESSES FOR COAL UTILIZATION

Duane H. Pontius
Southern Research Institute
2000 Ninth Avenue South
P.O. Box 55305
Birmingham, Alabama 35255-5305
(205)581-2000

Advanced methods for deriving energy from coal include gasifiers and combined-cycle approaches. Obtaining maximum efficiency from such systems requires particulate removal from gases at high temperatures and pressures. In addition to the manifest difficulties in developing materials and approaches for filtering under these extreme conditions, the characteristics of the particles may lead to problems in the long-term reliability of filters for these processes. A number of techniques have been applied to the characterization of these particles, with varying degrees of success. Results of several types of measurements are described in this paper, and comparisons are drawn between particles emitted from conventional coal-fired boilers and those derived from advanced coal-based technologies. Sharp contrasts are evident in terms of size distribution, dust cake permeability, cohesivity, and particle shape. The effects of these differences on filtration characteristics are discussed.

INTRODUCTION

Fundamental studies of filtration properties often begin with an investigation into the basic mechanisms by which individual particles become attached to separate fibers, granules, or the walls of pores in a filter material. In large industrial filters, however, the interactions among particles are normally far more important than those between particles and filter. In such devices, the filter serves chiefly as a substrate on which a dust cake is formed. Once a thin layer of dust covers the filter surface, very few particles will penetrate the dust cake to the filter. The properties of the dust cake are therefore very important to the overall performance of a filter system. These properties include porosity, bulk density, pore size distribution, and cohesivity, which depend upon such particulate characteristics as particle size distribution, shape, and chemical composition. Among the more promising methods for particulate removal from gas streams at high temperatures and pressures is the use of porous ceramic filters. These filters differ significantly from conventional fabric filters, particularly as regards the cleaning mechanisms. Whether a fabric filter is cleaned by means of reverse gas, shaking, or reverse pulse jet, an important part of the process is the flexure that occurs during cleaning. As the fabric flexes, the dust cake breaks up, facilitating the removal of large agglomerates. No such action occurs in ceramic filters because of the rigidity of the material.

Most ceramic filters are designed to be cleaned by an energetic pulse of gas in the reverse direction. Since these filters are rigid, the force of the pulse must work directly on the accumulated dust layer. Since there is an abrupt change in the physical nature of the medium as the pressure pulse passes from the filter into the dust cake, the sharpest change in the gradient of the pressure occurs at that boundary. The tensile strength of the dust cake is extremely weak in comparison with that of the ceramic filter material, so the dust cake tends to break away at a layer very near the surface of the filter. Those few particles that have actually penetrated into the pores of the ceramic during filtration may not be so effectively removed, however. These particles are very small, and attached firmly to the walls of the pores. Even if broken loose by the aerodynamic forces of the reverse gas pulse, they are subject to immediate reattachment inside the porous structure of the ceramic.

Evidently, in view of the above remarks, there is reason to be concerned about the long-term vulnerability of rigid filters to a gradual accumulation of very fine particles in the pores, resulting in an irreversible increase in pressure drop as the filter ages.

A more direct concern is the rate of increase in pressure drop during an ordinary filtration cycle. Again, the particles at the smaller end of the scale are the most important factor. As a general rule, the cohesive effects become more important as particle size is reduced, since the ratio of surface area to mass is inversely proportional to particle size. Similarly, particles having rough, irregular shapes tend to be more cohesive. The effects of increased cohesiveness in a dust cake are generally beneficial because of the concomitant higher porosity. A reduction in pore size, however, may offset any benefits attributable to increased porosity.

PARTICLE CHARACTERIZATION METHODS

Laboratory tests have been established for characterizing fly ash in terms of its filtration properties, and a data base has been developed. Although most of the data are from pulverized-coal boilers, recent work has extended the files to include characteristics of particles from advanced coal-fueled systems. The tests include methods for measuring particle size distribution, bulk porosity as a function of compressive stress, effective angle of internal friction (a measure of cohesiveness), and specific surface area. Chemical analysis and observations by scanning electron microscope (SEM) are also useful.

In many advanced systems that generate power from the combustion of coal, the suspended particulate matter in the exhaust gases are substantially different from those issuing from conventional boilers. In fluidized bed combustors (FBC), both atmospheric and pressurized, the temperature is kept relatively low to control the production of nitrogen oxides. Consequently, the ash particles are not heated above their melting point, and their shapes are irregular, rather than nearly spherical like the particles found in the emissions from conventional boilers where melting does take place. In the exhaust from gasifiers the suspended material is even more complex, because it contains both ash and char.

Since filtration effects depend principally on resistance to gas flow, measurements of particle size distribution should be done with a procedure that separates particles on the basis of aerodynamic resistance, such as the Bahco classifier. Sedigraphic measurements are also applicable. Measurements by Coulter counter are not well correlated with filtration properties in general, because the instrument is based on the attenuation of an electrical signal in an electrolyte by the passage of a particle through a fine orifice. For spherical particles, a reasonable correlation can be made with aerodynamic measurements, but for particles of extremely irregular shape, the results are less useable. Carbonaceous particles, as from the exhaust of a gasifier, are especially troublesome, since their low electrical resistivity makes them appear to be substantially smaller than they are.

Measurements of the cohesivity of particulate samples are difficult to carry out accurately. One of the more reliable techniques is the ring-shear method, which uses a special apparatus to measure the shear stress and strain across an annular layer of particulate material. The result of the ring-shear analysis is the effective angle of internal friction. For conventional fly ashes, this quantity normally lies in the range between about 38 and 46 degrees. Results obtained for samples taken from fluidized bed combustors and gasifiers run somewhat higher, but cover a narrower range, typically from 46 to 48 degrees.

Most fluidized bed combustors and gasifiers have mechanical particulate collectors (usually cyclones) on the exhaust stream to provide for recycling usable material and for first-stage particulate control in the outlet gas. Consequently, the particle size distribution measured at the outlet depends nearly as strongly on the design of the cyclone as on the nature of the process. Since cyclones are efficient collectors of large particles, but very poor for particles less than a few micrometers in diameter, the size distribution in these exit streams tends toward the smaller end of the spectrum. Mass median diameters (mmd) are typically 3 to 6 μm.

One of the more interesting aspects of the effluents from the systems under discussion is the shape of the particles. The specific surface area of the particles is the ratio of the total surface, as measured by the Brunnauer-Emmett-Teller (BET) method or equivalent, to the mass of a sample. Since the mass is directly proportional to the volume, it follows that for a given particle shape, the specific surface area is inversely proportional to the diameter or equivalent linear dimension. The specific surface area also gives an indication as to the regularity of shape; smaller numbers are associated with regular, smooth shapes. For example, a monodisperse aerosol with mmd of 5.6 μm and density of 2.7 g/cm^3 would have a specific surface area of 0.2 m^2/g. The polydisperse fly ash from conventional, pulverized-coal-fired boilers typically has a specific surface area ranging from about 1 m^2/g for smooth, nearly spherical particles to about 4 m^2/g for rough-looking particles with many surface asperities. Particulate samples from FBC and gasifier systems have much higher values of specific surface area, usually above 10 m^2/g. SEM observations have confirmed that these particles are extremely rough and jagged in appearance.

RESULTS AND CONCLUSIONS

The relative gas flow resistance of dust cake samples taken from operating systems has been found to depend on the porosity of the sample and on the size and shape of the pores. Thus, the size distribution, the cohesivity, and the shape of the particles all affect the relationship between flow rate and pressure drop in a filter.

It can be concluded from correlations between particulate properties and effects on filtration that the importance of specific characteristics can be ranked approximately as follows:

- Particle size distribution--This quantity clearly dominates the filtration process.
- Particle shape--There is a strong influence on both cohesivity and aerodynamic drag that results from the shape of the particles.
- Chemical composition of the particles--This factor may play some part in determining the cohesivity of the particles, but, compared with the above two items, it results in only second-order effects.
- Mass loading--While not strictly a particulate characteristic, the mass loading may be considered another descriptive quantity relevant to the particulate source. If all other characteristics are the same, the mass loading affects only the rate of dust cake buildup, and hence, the tolerable length of the filtering cycle.

EVALUATION OF CERAMIC CANDLE-FILTER PERFORMANCE IN A HOT PARTICULATE-LADEN STREAM

Charles M. Zeh, Ta-Kuan Chiang, and Larry D. Strickland
U.S. Department of Energy
Morgantown Energy Technology Center
P.O. Box 880, Morgantown, WV 26505
304/291-4265 or FTS 923-4265

Introduction

Ceramic candle barrier filters can effectively remove fine particulate matter from the hot pressurized gas streams that are typical of pressurized fluidized-bed combustors. Operating parameters such as face velocity and cleaning technique affect dust cake formation and filter performance. The effect of these parameters on filter performance must be understood to effectively design particulate cleanup devices.

There has been considerable research on barrier filtration. A series of tests was performed to better understand barrier filtration and its application to advanced coal-fueled combustion systems. Four tests were conducted to evaluate the performance of silicon-carbide ceramic candle-filters in a high-temperature, high-pressure environment. Pressure drop across the filter, dust cake permeability, cleaning efficiency, and specific resistance coefficients of the dust cake were analyzed to characterize the performance of candle filters.

Experimental Methods

A limited test matrix was conducted in a bench-scale pressurized candle-filter test apparatus, Figure 1. A simulated, hot particulate-laden stream was generated by combusting natural gas. Then, fly ash (nominal 4 micrometers [μm] in size) from a pressurized, fluidized-bed combustor was injected. The hot, pressurized ash-laden stream was then filtered through two ceramic candle-filter elements (nominal 1 m length, Industrial Filter and Pump, Laycer™ 50/10).

Operating conditions for the four tests are shown in Table I. Pressure drop across the filter elements was monitored continuously throughout each test period (e.g., Figure 2). The dust cake collected on the filters was periodically removed by back flushing each filter with a high-pressure one-second pulse of nitrogen. During each test, the dust cake was initially removed every 20 minutes, and then subsequently when the pressure drop reached a predetermined maximum value. The nitrogen pulse was generated by opening an electric solenoid valve that was connected to a nitrogen reservoir maintained at 360 psig. These pulse conditions were found to effectively remove the filter dust cake in each of the four tests.

Results

During each of the four tests, the pressure drop measurements following each cleaning cycle were observed to increase rapidly; within 10 to 24 hours, these measurements exhibited a much slower rate of increase. Conversely, the permeability was observed to decrease over the same period of time at a similar inverse rate. Neither pressure drop nor permeability were found to reach a stable value over any of the four tests.

The average pressure drop during time-based cleaning, and the time period between cleaning cycles during pressure-based cleaning, were dependent upon the face velocity, cleaning technique, and the specific resistance coefficient of the dust cake. Table I shows that increasing the face velocity increased the average pressure drop during time-based cleaning and decreased the time period between pressure-based cleaning cycles. Off-line cleaning was observed to significantly increase the time required between pressure-based cleaning cycles over on-line cleaning.

Two independent, dust-cake resistance coefficients were deduced to characterize the dust cake. Resistance coefficients were deduced for the dust cake prior to removal from the filter surface (K_2) and for the residual dust cake remaining on the filter subsequent to cleaning (K_2'). Estimated values for K_2' were slightly less than or comparable to K_2 values. This observation tends to confirm previous analytical assumptions that K_2' is equivalent to K_2, but contradicts some previous research by Dennis and Klemm.[1] Dennis and Klemm claim that K_2', composed of redeposited dust, may be five to six times less than K_2. Dust cake K_2 was observed to be dependent upon the operating conditions of the candle filter. K_2 was observed to increase with increasing face velocity for tests performed at similar temperature and pressure conditions. Table I shows that the average K_2 during pressure-based cleaning increased from 12.8 to 26.4 inches $H_2O\cdot ft\cdot min/lbm$ when the face velocity increased from 12 to 19 ft/min. Likewise, K_2 was observed to decrease from 26.4 to 19.5 inches $H_2O\cdot ft\cdot min/lbm$ for on-line versus off-line cleaning at 19 ft/min face velocity. Results from test No. 4 appear to contradict the previous theory that higher face velocity produces more resistive dust cakes. The observed higher K_2 value for test No. 4, performed at a face veloctiy of 7.5 ft/min, is believed to be attributable to the lower operating temperature and higher pressure, although the increased gas density and lower viscosity cannot fully explain the observed variation. Further analysis is required to determine the effect of pressure and temperature on K_2.

Permeability was observed to decrease in all four tests and to reach a slow rate of decrease after roughly 10 to 24 hours. Trends similar to those observed for K_2 between face velocity and permeability were also found. Permeability was found to decrease with increasing face velocity for tests performed at similar temperature and pressure conditions.

Cleaning efficiency, defined as the amount of dust cake removed versus that remaining on the candle filter, was calculated for each cleaning cycle. Analysis of this parameter provided little insight into candle filter performance other than a relatively constant amount of dust remained on the filters subsequent to cleaning after 10 to 24 hours of testing.

Conclusions

Dust cake resistance is dependent on the face velocity, temperature, and pressure at which the dust cake is formed. A higher face velocity yields a higher K_2 for tests performed at the same temperature and pressure.

Off-line cleaning produces dust cakes with a lower K_2 than on-line cleaning.

Dust cakes formed from pressurized, fluidized-bed combustor ash exhibit significantly lower specific resistance coefficients than dust cakes formed from pulverized coal-fired boiler ash (when compared at standard conditions of temperature and pressure).

Acknowledgments

The perseverance and hard work of Mr. Richard Griffith, Mr. Mark Tucker, Mr. Charles Carter, Mr. David Turner, and other members of the Project Support Staff at the Morgantown Energy Technology Center made this work possible.

References

1. R. Dennis and H.A. Klemm, "Modeling concepts for pulse jet filtration," J. Air Poll. Control Assoc. 30: 38 (1980).

TABLE I. Summary of High-Temperature/High-Pressure Test Conditions and Results

PARAMETER	TEST NO. 1	TEST NO. 2	TEST NO. 3	TEST NO. 4
Combustor Pressure (atm)	8.5	8.5	8.5	9.9
Filter Vessel Temp (°F)	1550	1550	1550	1350
Face Velocity (ft/min)	12	19	19	7.5
Particulate Loading (ppmw)	775	840	840	838
Cleaning Technique	On-line	On-line	Off-line	On-line
Test Period (hr)	196	93	160	192
TIME CLEANING				
Duration - hours	48	75	60	30
Duration - cleaning cycles	144	225	180	90
Dust Cake Properties				
- avg. K_2 (in H_2O·ft·min/lbm)	17.2	21.7	19.1	22.4
- avg. K_2' (in H_2O·ft·min/lbm)	18.1	19.8	17.4	25.3
- avg. areal density (lbm/ft^2)	0.03	0.05	0.06	0.03
- avg. residual areal density (lbm/ft^2)	0.05	0.12	0.10	0.08
PRESSURE CLEANING				
- set point (psi)	1.9	3.5/4.5	3.0	2.0
- duration (hr)	148	8/10	100	162
- cleaning cycles	81	21/33	166	58
Dust Cake Properties				
- avg. K_2 (in H_2O·ft·min/lbm)	12.8	26.4	19.5	17.8
- avg. K_2' (in H_2O·ft·min/lbm)	13.3	22.5	18.6	20.7
- avg. areal density (lbm/ft^2)	0.15	0.05	0.10	0.25
- avg. residual areal density (lbm/ft^2)	0.10	0.15	0.10	0.11

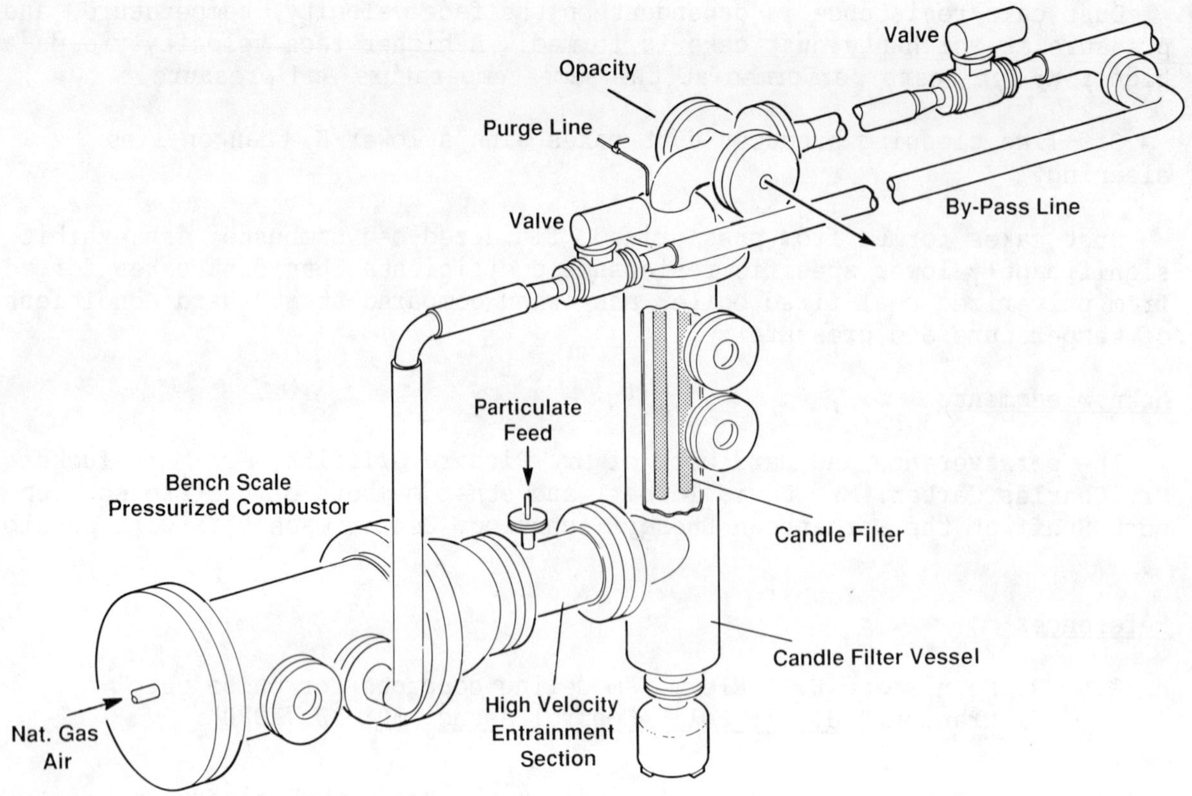

FIGURE 1 Pressurized, entrained combustor candle-filter test apparatus.

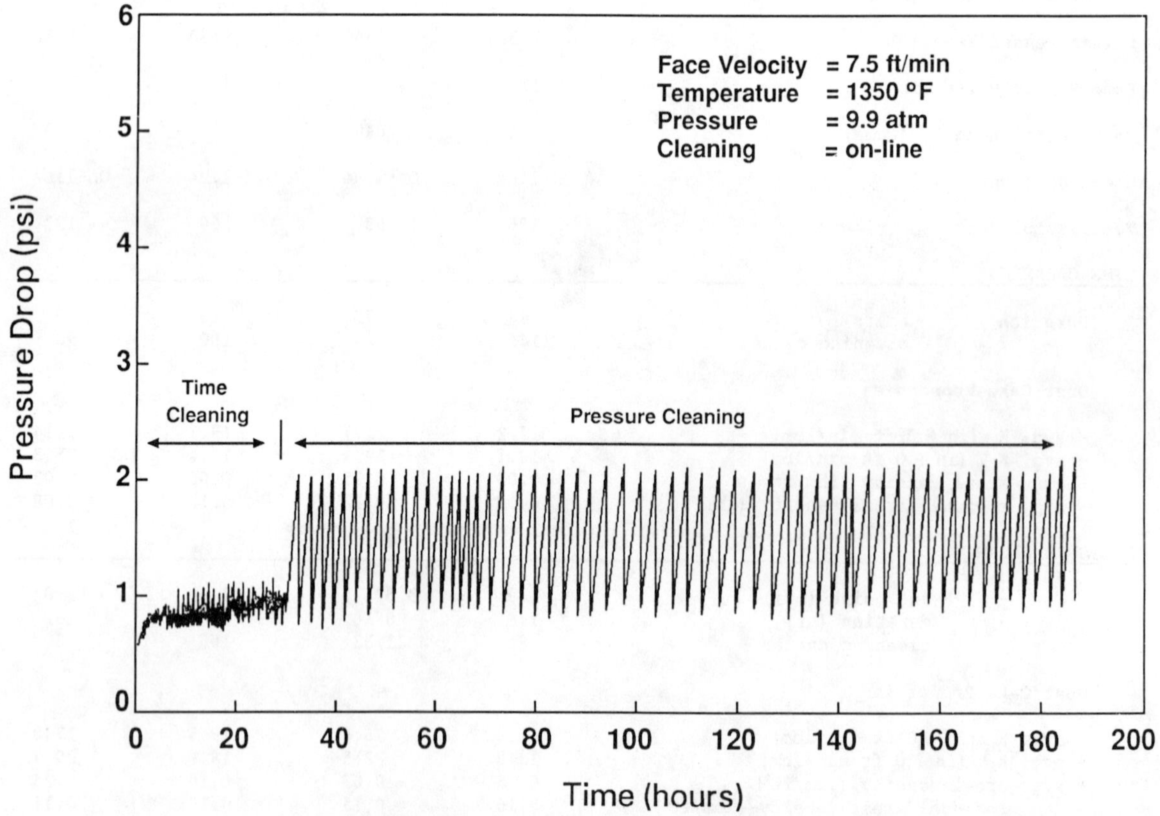

FIGURE 2 Candle-filter pressure drop, test no. 4.

CHARACTERIZATION OF DUST CAKE FILTRATION USING LABORATORY MEASURED QUANTITIES

Ta-Kuan Chiang, Richard A. Dennis, Larry D. Strickland, Charles M. Zeh
U.S. Department of Energy
Morgantown Energy Technology Center
P.O. Box 880
Morgantown, WV 26505

The characteristics of dust cakes formed on porous ceramic filters were investigated to examine high-pressure and high-temperature gas filtration in pressurized fluidized-bed combustion (PFBC). Dimensionless particulate and flow characteristics were used to develop empirical correlations for filter cake porosity and the filter cake specific resistance coefficient. Operating data from full-scale utility fabric filter installations were introduced to validate and to extend the applicable range of laboratory test data.

Introduction

Two test apparatuses were configured to evaluate the performance of ceramic candle filters in support of a clean-coal PFBC demonstration project: (1) a laboratory-scale, coal-fired, atmospheric fluidized-bed combustion (AFBC) system[1] comprising a ceramic disc filter to provide fundamental information on filter cake porosity under actual coal combustion and sorbent injection, and (2) a laboratory-scale pressurized, natural-gas combustor[2] comprising two full-size ceramic candle filters with re-injected particulate to provide application data under elevated temperature and pressure. The re-injected particulate was primary cyclone dust from the New York University (NYU) sub-pilot-scale PFBC test facility[3], ground to a mass median diameter (MMD) of 4.68-μm with a geometric standard deviation of 1.62. The purpose was to compare these two basically different dust cakes to see if laboratory-simulated particulate testing is representative of actual operating conditions. These data were then compared with operating data available from full-scale and pilot-scale fabric filter installations for pulverized-coal (PC) utility boilers to see if laboratory data can be extended to full-scale applications.

Experimental Methods

Ten tests were conducted in an AFBC disc-filter apparatus to provide filter cake porosities and specific resistance coefficients at various face velocities and particle size distributions. Particle size distribution was varied by using different size cyclones (or no cyclone) upstream of the disc filter. A virgin disc filter was used for each test. Prior to each test, the clean disc weight, and the clean disc pressure drop versus face velocity were established. Filter cakes were formed under various steady flow conditions but no pulse-jet cleaning cycles.

After a filter cake formed with the desired thickness at a specific face velocity, the disc filter was by-passed from the AFBC flow. A hot nitrogen flow was then directed through the established filter cake and the disc filter to examine the specific resistance coefficient of the filter cake at various face velocities. At the completion of this test, the disc filter and filter cake were carefully removed from the test apparatus for final weighing and cake thickness measurements. The filter cake porosity was then calculated from the particulate bulk density, which was derived from its cake weight and cake volume.

Four tests were conducted in the pressurized natural-gas combustor to examine the effects of face velocity and cleaning technique on the specific resistance coefficient of the ceramic candle filter. Virgin candles were used for each test. Prior to each test, the virgin candle weight, and the virgin candle pressure drop versus face velocity were established. Filter cakes were formed and removed repeatedly. Continuous testing was aimed for each test. A total of 641 h of testing were conducted, including two tests lasting for 192 and 196 h.

Laboratory measurement on particle parameters included particle size distribution by Coulter counter assuming log-normal distribution, specific surface area of the particles by Brunaeur-Emmett-Teller (BET) analyzer, and particle specific gravity by helium pycnometer.

Results

Differences in testing conditions between the two test apparatuses precluded direct comparison of the two sets of dust cake filtration data. Dimensionless groups that characterize dust cake filtration and gas flow properties were thus used to establish correlations. Using the general analysis of flow through a packed bed, dimensionless groups of interest included the flow Reynolds number Re_f, the particle Reynolds number Re_p, the particle shape factor λ, the dust cake porosity ε, and the skin-friction coefficient c_f. If ρ_g is assumed to be the gas density, μ the gas viscosity, v_f the face velocity, ρ_p the particle density, S the particle specific surface area, d_{vsm} the particle volume-surface mean diameter, τ_p the particle relaxation time constant, K_2 the specific resistance coefficient of the dust cake, ΔP_c the dust cake pressure drop, and σ_c the dust cake areal density, by analogy with flow through a long pipe and the defining Fanning equation, it can be shown that

$$Re_f = (2/3)Re_p/[(1 - \varepsilon)\lambda] \text{ , and} \quad (1)$$

$$c_f = (6\,\varepsilon^3/\lambda)(K_2\,\tau_p/Re_p) \text{ ,} \quad (2)$$

where $Re_p = \rho_g\, v_f\, d_{vsm}/\mu$, $\lambda = (1/6)d_{vsm}\,\rho_p\, S$, $\tau_p = (1/18)\,\rho_p\, d_{vsm}^2/\mu$, and $K_2 = \Delta P_c/(\sigma_c\, v_f)$. Temperature and pressure information are contained in the gas density and gas viscosity terms.

Using the empirical Blasius law, a logarithmic plot of Equations 1 and 2 leads to a linear relationship in the viscous flow region. Figure 1 is a plot using the experimental data obtained from the laboratory AFBC test series. Figure 1 demonstrates this linear relationship, given by

$$c_f = 8.766\ Re_f^{-0.615}\ . \tag{3}$$

To develop an empirical relationship for dust cake porosity, it is physically plausible that the modified dust cake solidity $(1 - \varepsilon)\lambda$ is a function of the normalized particle residence time $2\varepsilon\lambda^2/STK$ where STK is the modified Stokes number, based on the particle's volume-surface mean diameter. The normalized particle residence time is defined as the ratio of time τ_r taken for a particle entrained in a gas stream to go through a dusk cake channel of an effective particle diameter, and the particle relaxation time constant τ_p. If d_e is assumed to be an effective particle diameter in the local flow direction, and v_e the effective flow velocity $(= v_f/\varepsilon)$, then the normalized particle residence time is

$$\tau_r/\tau_p \equiv [d_e/(v_f/\varepsilon)]/[\rho_p\ d_{vsm}^{\ 2}/(18\ \mu)]\ . \tag{4}$$

Assuming that d_e is proportional to d_{vsm}, λ and S account for various particle shapes and surface areas: $d_e \propto \lambda^2 d_{vsm}$, since $S \propto \lambda$, the normalized particle residence time defined in Equation 4 reduces to

$$\tau_r/\tau_p \propto 2\ \varepsilon\ \lambda^2/STK\ ,\ \text{with}\ \ STK \equiv \rho_p\ v_f\ d_{vsm}^{\ 2}/(9\ \mu\ d_{vsm})\ .$$

Figure 2 is a logarithmic plot of the modified dust cake solidity $(1 - \varepsilon)\lambda$ versus the normalized particle residence time $2\varepsilon\lambda^2/STK$, using data from the laboratory AFBC test series and data from operating full-scale and research pilot-scale utility, fabric-filter installations.[4] The linear relationship established in Figure 2 is given by

$$(1 - \varepsilon)\varepsilon^{-0.293} = (0.558/\lambda)\ (\lambda^2/STK)^{0.293}. \tag{5}$$

Data on full-size ceramic candle filters obtained from the laboratory PFBC test series were used in Equation 5 to establish dust cake porosities. Values on c_f and Re_f for the laboratory PFBC test series were then plotted along with the data from laboratory AFBC test series as shown in Figure 3. Figure 3 also includes data from operating full-scale and research pilot-scale utility, fabric-filter installations. Figure 3 shows a general trend in accordance with the characteristics of Equation 3 for all data points, regardless of the origin or scale. A noticeable departure exists only for the fabric filter installation at Monticello.

Equation 3 can also be written explicitly in terms of particle and flow charcteristics:

$$K_2 = 1.869\ \varepsilon^{-3}[(1 - \varepsilon)\mu\ \rho_p\ S^2]^{0.615}(\rho_g\ v_f)^{0.385}. \tag{6}$$

Equation 6 indicates that the specific resistance coefficient is gas density or pressure dependent, although the dust cake porosity is not according to Equation 5.

Conclusions

Dimensionless grouping and normalization of particle and gas flow characteristics indicate that dust cake filtration data can be compared, regardless of the operating condition (PC, AFBC, PFBC), scale (full, pilot, laboratory), particulate source (coal-burning, coal-burning with sorbent-injection, injected flyash), and type of filter used (woven glass-fiber, porous ceramic candle). Dust cake porosity and the specific resistance coefficient can now be predicted under actual operating conditions, provided that a representative particle sample can be analyzed in the laboratory, or a design specification is given.

Dust cake porosity is found to be independent of gas pressure or density; the specific resistance coefficient of the dust cake is, however, gas pressure dependent. Particle shape factor and its Stokes number affect the dust cake porosity in a complicated way. In general, a smaller parti-cle size constitutes a smaller porosity; a larger particle shape factor, however, leads to a increased porosity, thus counter-balancing the diminishing porosity caused by smaller particles. A higher face velocity leads to a larger dust cake porosity, but also leads to a larger specific resistance coefficient. The net result is a higher pressure drop.

References

1. Dennis, R.A. Evaluation of Dust Cake Filtration at High Temperature With Particulate From an Atmospheric Fluidized-Bed Combustor, Technical Note. U.S. DOE, Morgantown Energy Technology Center (in press).

2. Zeh, C.M., T-K. Chiang, and W.J. Ayers. Evaluation of Ceramic Candle Filter Performance in a Hot Particulate Laden Stream, Technical Note. U.S. DOE, Morgantown Energy Technology Center, (in press).

3. V. Zakkay, et al. Construction and Utilisation of a Large-Scale Coal Fired Pressurized Fluidized Bed Facility. New York University, for the U.S. DOE, Morgantown Energy Technology Center, DOE/MC/14322-1800, NTIS/DE85013685, August 1985, 245 p.

4. Bush, P.V., T.R. Snyder, and R.L. Chang, "Determination of Baghouse Performance From Coal and Ash Properties: Part I." JAPCA 39(228) 228-237 (1989).

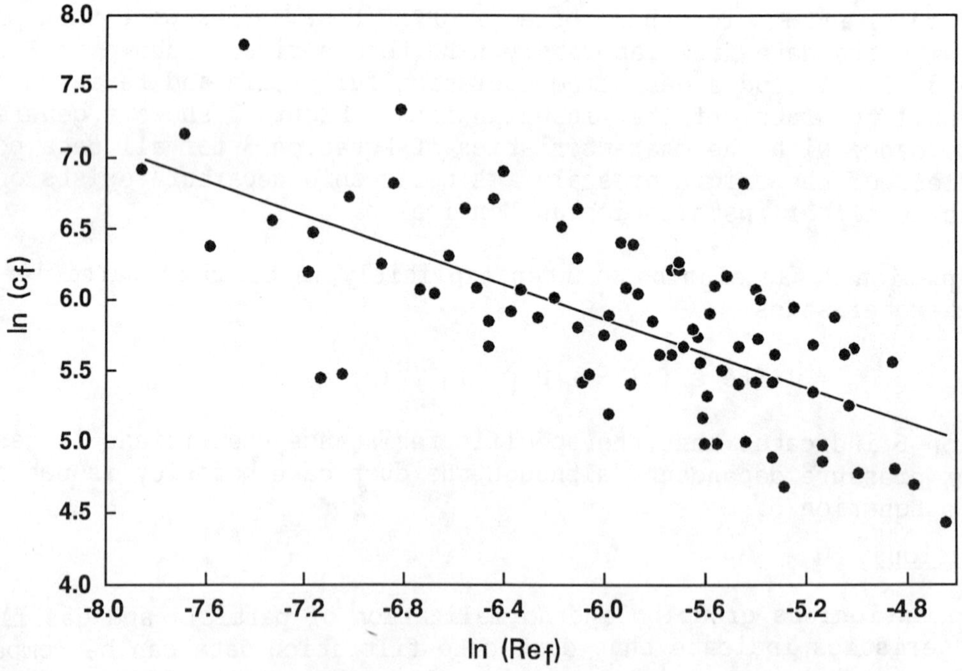

Figure 1. Laboratory AFBC test series, skin resistance coefficient versus flow Reynolds number.

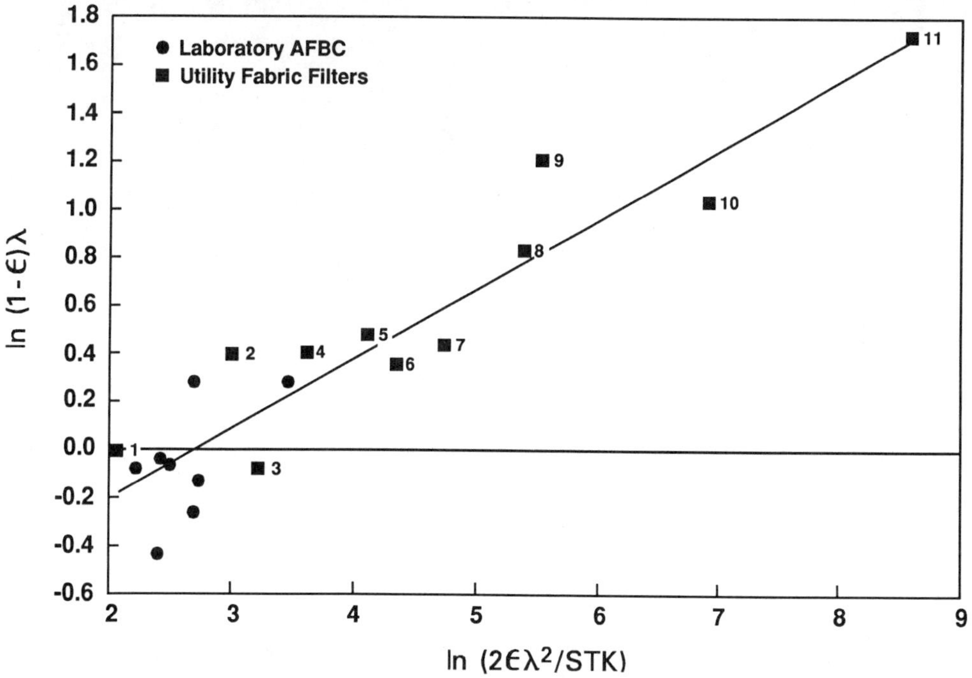

Legend

1-Monticello; 2-Intermountain Power Project Unit 1; 3-Brunner Island; 4-Arapahoe Pilot; 5-Arapahoe Unit 4; 6-Arapahoe Ecolaire; 7-Arapahoe Pilot Downstream of Cyclone; 8-Scholz High-Sulfur Pilot; 9-Nixon; 10-EPRI High-Sulfur Test Center; 11-TVA AFBC Pilot

Figure 2. Modified filter cake solidity versus normalized residence time for laboratory AFBC and utility fabric filters.

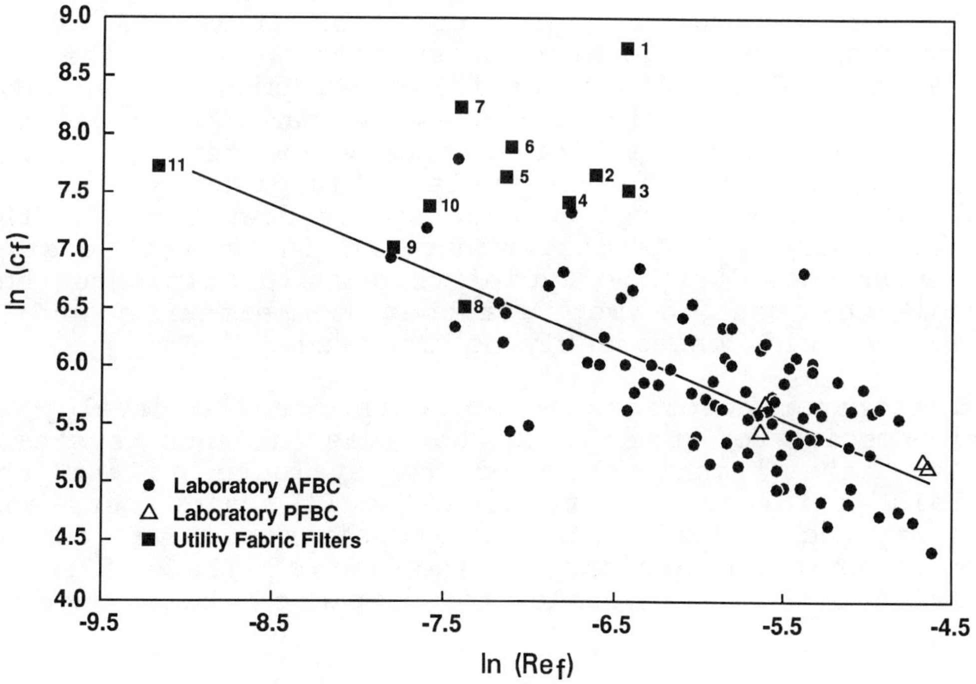

Legend

1-Monticello; 2-Intermountain Power Project Unit 1; 3-Brunner Island; 4-Arapahoe Pilot; 5-Arapahoe Unit 4; 6-Arapahoe Ecolaire; 7-Arapahoe Pilot Downstream of Cyclone; 8-Scholz High-Sulfur Pilot; 9-Nixon; 10-EPRI High-Sulfur Test Center; 11-TVA AFBC Pilot

Figure 3. Skin resistance coefficient versus flow Reynolds number for laboratory AFBC, laboratory PFBC, and utility fabric filters.

DEVELOPMENT OF CERAMIC MEDIA FOR THE FILTRATION OF HOT GASES

A N Twigg, C J Bower, G J Kelsall and
D M Hudson
British Coal Corporation
Coal Research Establishment
Stoke Orchard
Gloucestershire UK GL52 4RZ

The filtration of coal derived gases at high temperature and pressure affords the potential for increased efficiency from advanced power generation systems such as the British Coal 'Topping Cycle'. Successful development of an advanced hot gas filtration device for use upstream of the gas turbine would eliminate the need for secondary and tertiary cyclones, and in the case of the Pressurised Fluidised Bed Combustion (PFBC) cycle, for a large cleanup system downstream of the gas turbine, giving a potential reduction in capital costs. For the British Coal 'Topping Cycle' this would eliminate the need to cool the gas to wet scrubber temperatures and cycle efficiency would consequently be improved.

This paper describes an approach for the development of ceramic media for use in a hot gas cleanup system. It discusses the techniques used to evaluate filtration and materials performance of media in small scale tests and how these may be related to the requirements of commercial systems. Unproven aspects of the use of these filters are identified and a rationale to complete their development presented.

INTRODUCTION

Worldwide interest in technology for cleaning gases at high temperature and pressure has been stimulated mainly by the demands of advanced systems for generation of electrical power from coal. The lack of a fully developed system for hot particulate removal is a major obstacle in development of such advanced systems. Cyclones are currently used for the removal of particulates from gases at 10-20 bar and 500-1000°C. However, even with two or three stages of cyclonic cleaning particulate concentrations would exceed those required for acceptable turbine blade life in a high temperature (1260°C) gas turbine.

Development of an advanced hot gas filtration device for use upstream of the gas turbine in the British Coal 'Topping Cycle' (Figure 1) would eliminate the need for secondary and tertiary cyclones, giving a potential reduction in capital costs. This would also eliminate the need to cool the gas to wet scrubber entry temperatures (circa 120°C) and thus improve cycle efficiency.

Similarly, an advanced hot gas filtration device upstream of the gas turbine in a PFBC cycle would eliminate the need for secondary and tertiary cyclones and for a cleanup system downstream of the turbine, giving a potential reduction in capital cost.

Systems for high temperature, high pressure duty are currently being developed by workers in Europe, Japan and the United States[1,2,3,4]. Of these the silicon carbide based porous ceramics are probably nearest to commercial exploitation. However whilst they have been shown to be effective filters for this duty[5], their durability over commercial timescales remains unproven.

DEVELOPMENT OF CERAMIC FILTER MEDIA

This paper describes the approach taken by British Coal in the development of ceramic filter systems and outlines a programme which commenced in July 1989 to resolve the outstanding issues required for their commercialisation.

Filter Criteria

Successful filtration of hot coal derived gas in advanced power generation systems requires a filter medium with the following features:

High filtration efficiency. The medium must be capable of reducing particulate concentrations to below, say, 10 ppm (all <2 μm) to restrict erosion by, and deposition of, particulates on gas turbine blades which would reduce their life to uneconomic timescales.

Exhibit surface filtration. Filtration should ideally occur on the surface of the filter medium to ensure that filtration efficiency and high permeability is maintained for commercial operating periods. The stress required for effective filter cake release and hence the energy requirement for the pulse cleaning gas has also shown to be lower for surface filtration in recent work by Cheung[6].

High permeability. A sustained high permeability is important as it reduces the number of elements and the size of filter vessel and thus the capital cost of a hot gas cleanup system. It also reduces the overall pressure drop across the filter vessel, reducing the energy requirement of the booster compressor to the gasifier or combustor, and hence operating costs in a combined cycle plant.

Adequate strength. Filters should be sufficiently strong to resist robust handling and mechanical fixing. This is particularly important in the neck region where the highest stresses are experienced.

Chemically inert and resistant to thermal fatigue. Filters must maintain adequate strength over extended periods of operation. Loss in strength may occur when volatile species such as alkali metal salts react with the ceramic medium to form new phases which are weaker or have different thermal expansion properties than the parent material. Thermal fatigue is also thought to be responsible for strength loss. Fatigue can result from the cumulative effects of thermal shock caused by relatively cold pulse gas passing through a filter medium during cleaning. In addition, fatigue may result from maintaining filter elements at elevated temperatures for extended duration. This may become increasingly important for high temperature duties such as the gasifier stream of the British Coal 'Topping Cycle' (where filtration at 950°C is the preferred option) and for PFBC based systems where gas temperature (850°C) must be maintained to sustain high cycle efficiency.

Assessment of new filter media

The first step in the development of a new medium is to assess its ability to filter dust laden gas. This would normally be carried out by the manufacturer using a test dust in an ambient or high temperature test facility. Whilst this gives an indication of the filtration efficiency of the filter, it gives little information on how the medium would perform under the severe duty associated with combined cycle power plants. Therefore it is recommended that in order to prevent the waste of resources by the filter manufacturer, a promising medium is evaluated at or near the conditions for which it has been designed early on in its development. Failures or problems resulting from use under these severe conditions can then be highlighted and used to direct the next stage of filter development.

British Coal have established a range of test facilities to assist in filter media development. These include hot coupon, single candle and multi candle test rigs. The type and duration of the test would obviously depend on the development status of the filter medium. Initially a medium or range of media would be screened in coupon tests where their filtration performance can be compared with that of the best existing media. Promising media would be used to fabricate candles for longer duration tests where their suitability as filters would be rigorously examined. Media which performed well in the candle tests would then be included in a durability programme where their resistance to thermal fatigue and chemical attack would be determined and their service life predicted.

A medium which performed well as a filter and demonstrated adequate durability would be a strong candidate for a larger scale demonstration test, for example in the 130 candle unit at the Grimethorpe PFBC Establishment. Successful demonstration on such a scale would lend considerable commercial credibility to a new filter medium. A schematic of the British Coal approach to filter medium development is given in Figure 2.

Hot Coupon Tests

Coupons of ceramic filter media (60 mm diameter by 10 mm thick) are tested on a rig supplied with a sidestream of dust laden fuel gas from a 0.15 m fluid bed gasifier (Figure 3). This facility allows the pressure drop and cake release properties of a filter medium to be determined under conditions representative of those encountered in the gasifier stream of the British Coal Topping Cycle. The rig is mobile and could be used on the sidestream of any atmospheric pressure gasifier or combustor.

In the test, the medium is conditioned by repeatedly building up a layer of dust cake on the underside of the coupon and removing this cake by the application of a pulse from above. After 50-100 of such cleaning cycles the residual (cleaned) pressure drop of the medium would normally reach a constant value indicating that the medium had been conditioned. The stress required for complete cake removal is determined by building up a filter cake and removing it by sequentially increasing the pulse pressure until the pressure drop of the medium returns to its cleaned value. Filtration efficiency, although not currently measured could be determined by means of an absolute filter on the clean side of the rig, or by the addition of an instrumental on-line dust analyser.

A good filter medium will reach a stable pressure drop after some 30-50 cleaning cycles. The magnitude of this 'conditioned' pressure drop will depend on the porosity of the medium and on the extent of dust penetration. If no dust penetration occurs (surface filtration) and the bulk of the medium is highly porous, the increase in pressure drop from the virgin to conditioned state will be minimal. For example, the pressure drop of the Schumacher Dia Schumalith medium with its surface coating (pore size 10 μm) increases from 3.3 kPa to 4.2 kPa during conditioning, an increase of 27%, whereas the Schumacher HTHP medium, a material with a uniform pore size of 30 μm, increases by 230% to give a conditioned pressure drop of 13.3 kPa.

Measurement of the stresses required for cake removal from a conditioned filter coupon is also important as they are directly related to the degree of dust penetration. Surface ('ideal') filters will require lower stresses for cake release. For example, complete cake removal requires 9 kPa for the Schumacher Dia Schumalith medium compared to 50 kPa for HTHP (Figure 4); a further indication of the effectiveness of the microporous surface coating on the former medium. The cleanability of a filter medium is an important criterion as it relates to pulse cleaning gas usage and consequently to the energy consumption of a combined cycle power plant.

The coupon test facility has been used extensively in the development of the Foseco fibre based filter where it was found to be an invaluable tool for screening new filter variants. However the test does have its limitations. The pressure drops determined for conditioned media are considerably lower than those measured under similar conditions in candle tests. This is presumably due to the relative ease of cake removal from a small flat coupon compared to a cylindrical candle of up to 1.5 metres in length. Optimisation of candle cleaning may reduce this discrepancy but it is unlikely that candles can be cleaned as effectively as coupons without excessive pulse gas usage. There is also a considerable spread in the stress measurement data for the detachment of cakes under similar conditions, due presumably to differences in dust size distribution or particle orientation between filter cakes. However the difference between a good and a poor medium can be easily resolved. The effects of gross changes to the gas temperature, size distribution or composition of the dust cake (e.g. addition of in-bed limestone to the gasifier) can also be determined by comparing the stress curves from several cake detachments under similar conditions.

In conclusion the coupon test facility is a rapid, inexpensive screening test for new ceramic filter media.

<u>Filter Tests</u>

A medium which performed well in the coupon tests would be used to manufacture test candles for extended filtration tests and material evaluation. Tests would normally be undertaken in a four candle unit. The unit takes the full flow of fuel gas from a dedicated 0.15 m fluid bed gasifier. The unlined ASI310 stainless steel vessel is maintained at up to 1000°C using ceramic covered heaters mounted externally. The dust laden fuel gas passes through one stage of cyclonic cleaning where its dust concentration is typically reduced to 5000 ppm and its average size to typically 7 μm. As for the coupon tests, limestone can be fed with the coal into the fluidised bed to simulate power plant conditions.

Dust concentrations are measured at the inlet and outlet of the filter system using removable isokinetic probes. The outlet probe contains a silica glass liner designed to avoid contamination from the probe, which corrodes rapidly. Sampled dust is collected in chloroform filled bubblers for subsequent coulter size analysis (inlet samples only) or on glass fibre filter papers for mass loading measurements. Pressure drop across the tubesheet is measured using a purged differential pressure cell. The filters are cleaned on-line using reverse pulses of nitrogen. The duration and frequency of the pulse are kept constant whilst its pressure is increased until a stable differential pressure across the tubesheet is attained. Test duration varies but is typically 100-200 hours (500-2000 cleaning cycles). The three main objectives of these tests are to study the filtration performance, permeability and materials performance of the filters under realistic process conditions.

Filtration performance

In determining filtration efficiency it is vital that the seals between the candles and tubesheet do not permit the passage of dust. As an expedient for these tests, effective sealing is achieved by holding the filters in their seats using piston driven counterplates. This ensures that the concentration of dust measured on the outlet of the filter vessel is only due to dust passing through the filters. A correction is made for corrosion products from ductwork upstream of the sampling probe which can significantly increase the dust loading. A good filter should reduce dust concentrations to below the level required to minimise erosion of gas turbine blades, that is 10 ppm (all <2 μm).

Permeability

The permeability of the virgin medium (although useful as an indication of filter porosity) is not directly related to the permeability of a conditioned medium. It is the magnitude of the conditioned permeability which is of key concern in power plant applications, since minimising the filter system pressure drop will reduce the energy requirements of the booster compressor, and thus increase the cycle efficiency. In order to establish the necessary confidence in the stability of filter permability it is necessary to conduct tests over extended durations (100 hour).

Differential pressure (inversely proportional to permeability) is continually measured throughout the test to monitor the conditioning behaviour of the filter. Good filters will condition rapidly to a residual (cleaned) permeability of between 30-50% of its virgin value. The gradient of the differential pressure trace following a pulse clean is also examined to investigate cleaning behaviour of the filters. Ineffective (patchy) cleaning or extensive re-entrainment of detached filter cake will significantly increase the initial gradient of the differential pressure trace following a pulse clean.

On completion of a test, the dust is removed from the filters by applying gentle suction to the outer surface. The permeability of the filters at ambient conditions are then compared to measurements made on the virgin filters. Filters exhibiting 'ideal' (i.e. surface filters) behaviour will have similar used and unused permeabilities indicating little pore plugging or depth penetration.

Materials Performance

'O' and 'C' ring tests (which measure the change in failure strain of circular element walls in tension and compression) are traditionally used to indicate changes in material strength with use. An additional test to determine the strength required to resist the forces exerted during pulse cleaning has also been developed for use with fibrous media. Where possible, these destructive strength tests are related to non-destructive tests (NDT) such as ultrasonic time of flight and resonant frequency measurements. It is useful to develop these NDT tests as they may ultimately find widespread use for quality control and in service monitoring by filter manufacturers and users.

X-ray diffraction, scanning electron microscopy and energy dispersive analysis of X-rays are also used to detect phase changes in the materials occurring during the filter tests.

It is possible for a medium to give acceptable filtration performance but to suffer serious degradation in 100-200 hour filter test. For example, the original fibrous medium Cerafil 12H10 manufactured by Foseco International showed comparable performance to Schumacher Dia Schumalith (a high quality rigid medium) in coupon tests, but suffered extensive delamination after only 50 hours exposure in a filter test. This highlighted the need for a substantial increase in inter-fibre bond strength, particularly at the filter surface. This was achieved by increasing the bulk density and applying a surface treatment to the medium. The improved medium, designated 2001i has since been successfully tested at 950°C.

A filter which performs well in the filtration tests, strength tests and material examination would be sufficiently promising to warrant inclusion in the durability programme recently initiated at the Coal Research Establishment.

Durability Programme

It is recognised that ceramic filters can be very effective filters capable of reducing dust concentrations to less than 10 ppmw. However their long term durability remains unproven. Thermal fatigue induced during pulse cleaning and chemical degradation (principally from alkali metal and calcium salts) through long term exposure to coal derived gas, may cause embrittlement and thus significantly reduce filter life.

A durability programme is currently examining the factors responsible for loss in strength of filters with use. Thermal fatigue will be examined in a six candle durability rig capable of operation at up to 950°C and 18 bar. Filters will typically be subjected to 120,000 pulse cleaning cycles (corresponding to about three years of normal operation) over a period of three weeks. The pulse cleaning system will be well characterised during commissioning to ensure that the thermal degradation experienced by the filters can be quantified and correlated to commercial scale systems. Strength measurements will be made before, during and after these tests to establish the relationship between strength loss and cleaning history.

Chemical degradation will be investigated in a suite of laboratory tests where components of promising filter media will be exposed to filter cake materials and to known gaseous corrodants. Solid-solid and gas-solid reactions will be studied using fine particles of filter material and high corrodant concentrations to accelerate the degradation processes. The relative rates of corrosion of the binder, grit and fibres will be determined and related to strength loss of filter elements. The effects of thermal fatigue and chemical degradation will then be combined in an overall strength loss model to predict filter life for a given duty.

CONCLUSION

British Coal has identified the need for a high performance, highly durable filter medium for use in combined cycle power plant. Extensive test facilities have been established, and a programme aimed at improving their long term durability has been undertaken. Collaboration with filter media manufacturers is being actively pursued, and the use of British Coal test facilities on a commercial basis for filter development is encouraged.

ACKNOWLEDGEMENTS

The authors wish to thank the European Coal and Steel Community (ECSC) for financial support of this work. Any views expressed are those of the authors and not necessarily those of the supporting organisations.

REFERENCES

1. Schiffer, H.P., Rens, U. and Tassicher, O.J. "Hot gas filtration research at the RWTH Aachen PFBC Facilities", 10th International Conference on Fluidised Bed Combustion, San Francisco, USA 30 April - 3 May 1989.

2. Fukatsu, Y., Oda, N., Watanabe, H. and Takehara, T. "Ceramic Tube Filter for High Temperature Gas", Gas Cleaning at High Temperatures, I Chem E Symposium Series No. 99, Surrey, September 1986.

3. Lippert, T.E., Smeltzer, E.E. and Meyer, J.H. "Performance evaluation of Ceramic Cross Flow Filter at New York University PFBC", 5th Annual Coal-Fuelled Heat Engines and Gas Stream Cleanup Systems Contractors Review Meeting, Morgantown WV, March 1989.

4. Tassicker, O.J., Bernard, G.K., Leitch, A.J. and Reed, G.P. "Performance of a large filter module utilising porous ceramics and pressurised fluidised bed combustor", 10th International Conference on Fluidised

Bed Combustion, San Francisco, USA, 30 April - 3 May 1989.

5. Bower, C.J., Arnold, MStJ., Oakey, J.E. and Cross, P.J.I. Hot particulate removal for coal gasification systems, Filtech Conference '89 Karlsruhe, W. Germany, September 1989.

6. Cheung, W. "Filtration and cleaning characteristics of ceramic media", PhD Thesis, University of Surrey, January 1989.

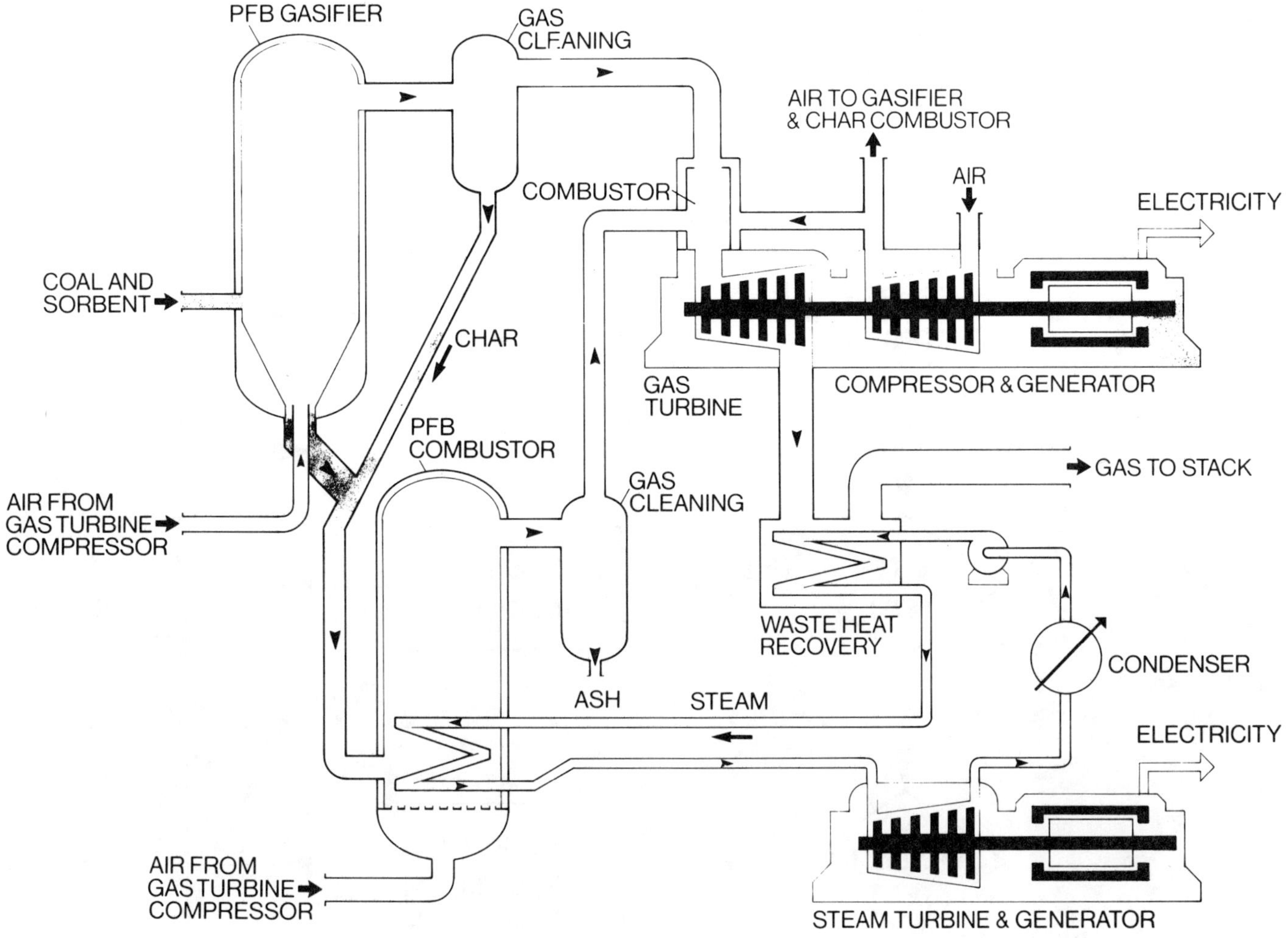

Figure 1 PFBC Topping Cycle.

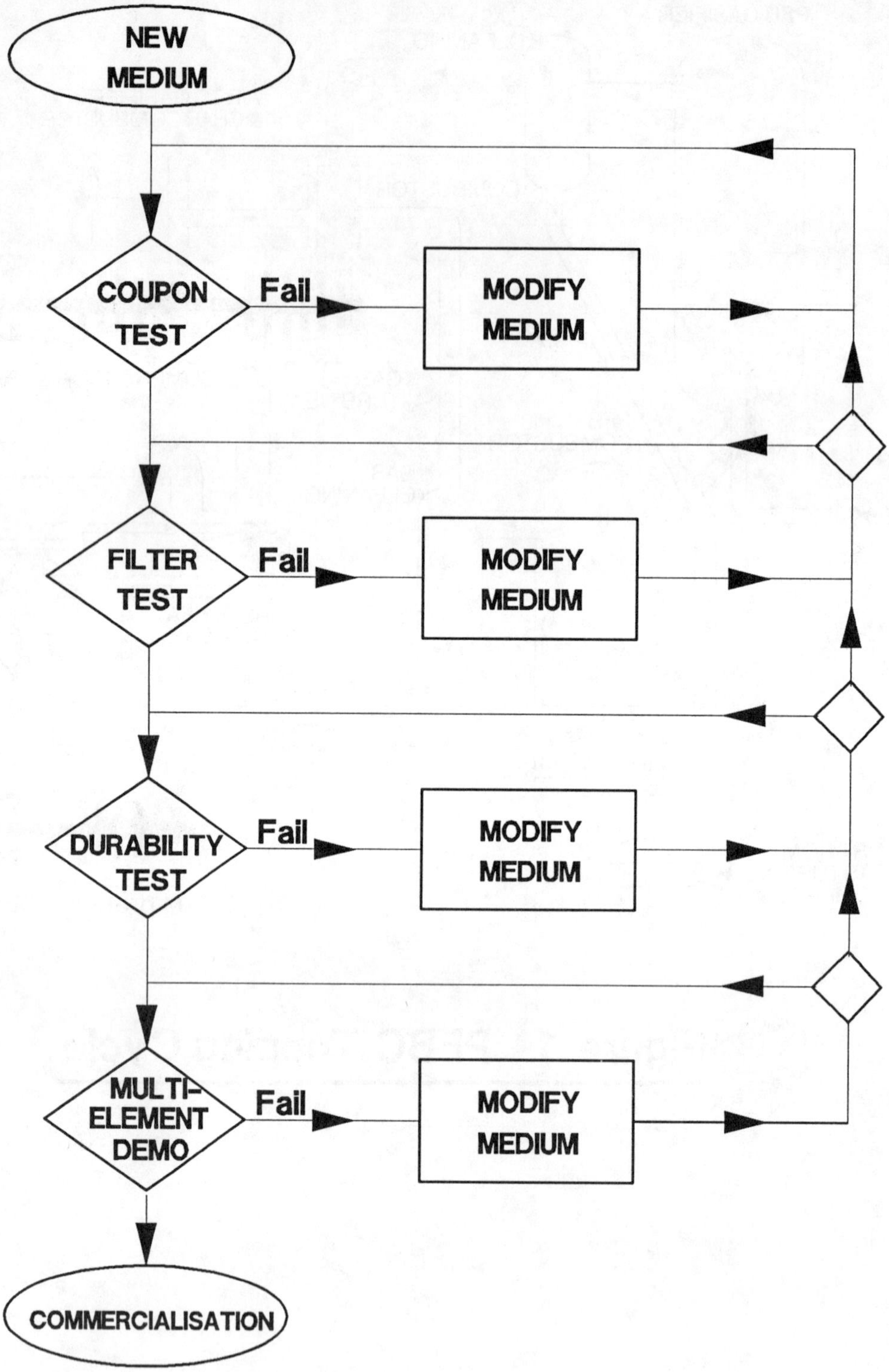

Figure 2 British Coal approach to filter medium development.

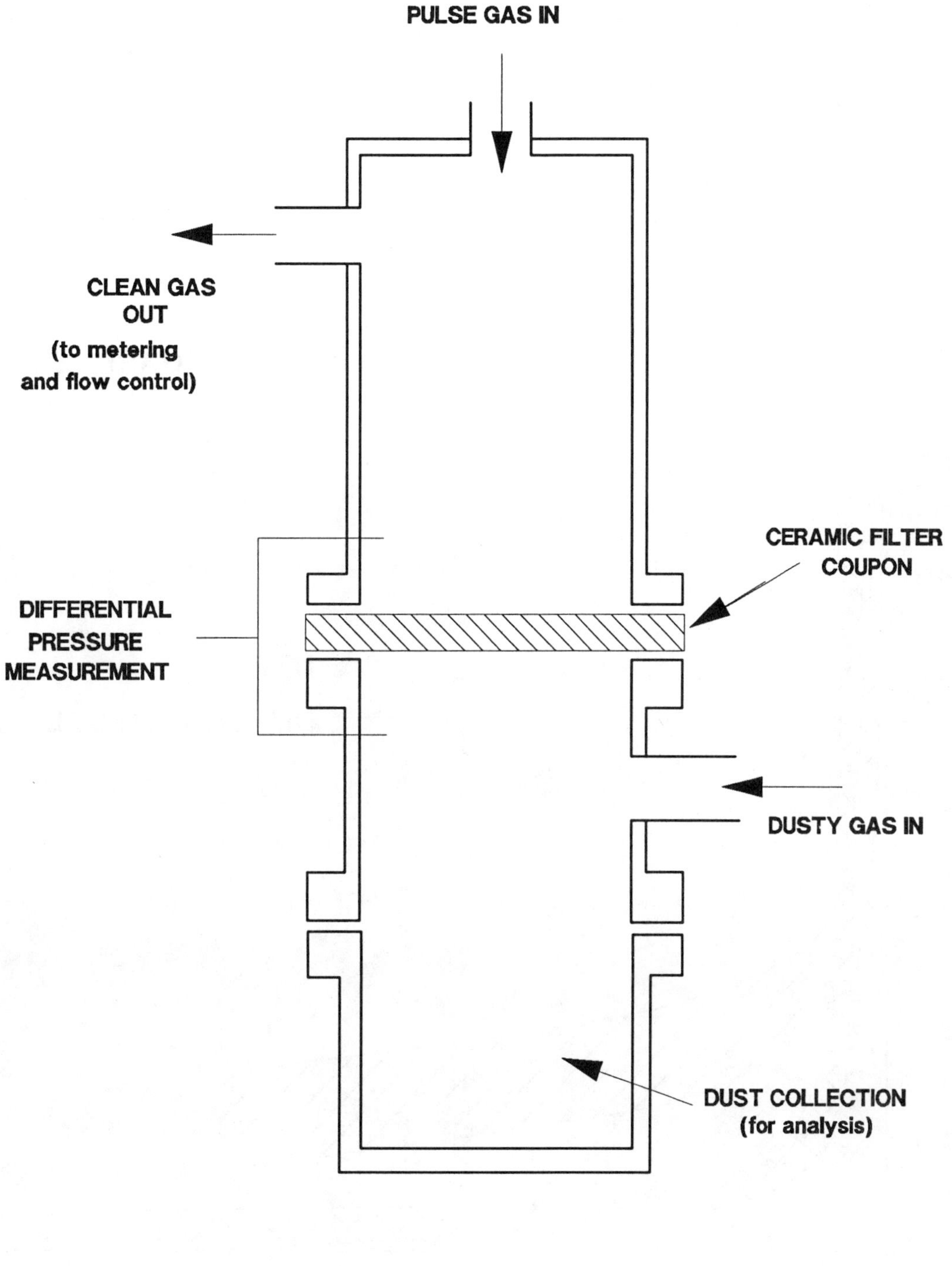

Figure 3 Ceramic filter coupon test unit.

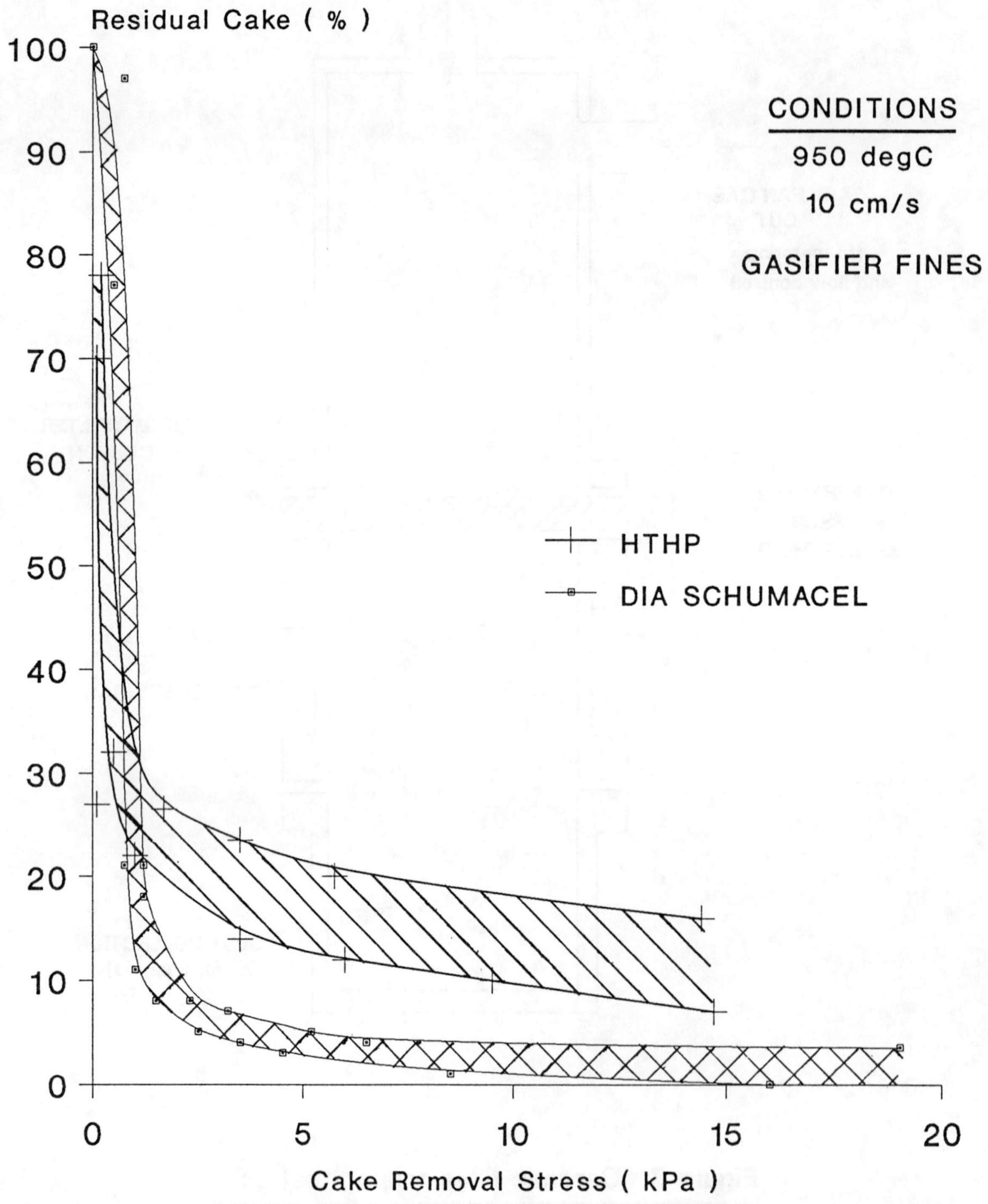

Figure 4 Cake removal stress for Dia Schumacel & HTHP.

A STRATEGY FOR QUALITY - Move Filter Testing Back to the Filter Vendor

Donald M. Forster
Eastman Kodak Company
Kodak Park
Rochester, NY 14652-3702

First, I'd like to give you a bit of background on my experience and responsibilities at Kodak which may help in understanding my point of view.

I am responsible for a Liquid Filtration Consulting Group charged with solving filtration problems of fluids that are used in the production of photographic products. In that capacity, we design filter media, filters, and filter hardware. We also design filter tests and equipment. We correlate these test results with filter performance and ultimately, the effect of a filter on product quality, which is the bottom line.

Now at this point, I'd like to stop and ask you a question. What is this? (Hold up a flanged cartridge filter)

This is the absence of physical imperfections on coated product. It is water, acceptable for manufacturing micro chips. It is a vat of wine, clear enough to bottle.

The point is that users of filters and manufacturers of filters often think of the filter, and not of its function. They measure the filter performance, but not the ultimate product quality. If you are a user, I'd like to ask you what is the value of your product compared to the cost of your filtration.

What is the consequence of waste to you or your customer? I suspect that this could be a ratio of 3-4 orders of magnitude. With this at stake, it has been my feeling that the filter user must do the following:

1. Understand your product, the potential imperfections and how filters can prevent these imperfections.

2. Understand how your filters work. This involves the classic filter tests of particle removal, dirt holding capacity, fiber erosion, and compatibility with the fluid.

3. Establish a dialogue with your filter supplier so that ultimately, the filters perform to meet your needs.

I'd like to digress for a few minutes. If you look at filtration sales literature, you will often find micron ratings, AC dust tests, beta ratios, and flow vs. pressure drop comparisons with water. In my view, this is a very simplistic picture of filter performance.

I think I understand why this happens. I think it is because the filter sales representative very often interacts with a filter buyer who doesn't understand the chemical processes or how filters work. The filter is all negatives: It costs money; it causes production delays; operators hate to change it, clean it, and assemble its holder. Have you ever heard, "The filter plugged! What's wrong with that filter? It's the filters fault!"

I won't imply that all users or sales people take this rather simplistic, negative view, but my 30 years of experience in dealing with many filter users is that the view is common.

So what can we do about this and how does it apply to filter testing? I would suggest that filter testing applies in two forms:

1. Evaluation of the physical parameters of filter media:
   Basis weight, fiber diameter distribution, tensile strength.

2. Evaluation of filtration parameters of the filter:
   Particle removal, permeability, dirt holding capacity.

There are also a variety of reasons for testing:

1. The media/filter manufacturer must design a product.

2. The manufacturer must control his process.

3. The user must test to find a filter to fit a product need.

Each of these different reasons may require changes in fluids, contaminants, and process parameters. The test conditions must be close to the process or there is a likelihood that what you want to filter will not be accurately represented. If we are dealing with room air or process air, there is a strong similarity from process to process. In the chemical process industry, the variety of fluids, contaminants, process conditions, and filtrate needs is infinite.

As complicated as this may sound, I believe there is an answer.

1. As a user, you need to understand your process and how the filter removes the unwanted contaminant.

2. You need to devise test methods for measuring the differences between unfiltered and filtered product, and between good and bad product.

3. You need to correlate your product test with your filter tests.

4. You need to work with your media and filter suppliers to incorporate tests that they perform which represent your needs.

5. Last, but most important, you need to develop tests with your supplier that can be used to control the filter and media manufacturing processes.

My bottom line is that I want media and filters that meet my ultimate product needs and are invariant Day to day, lot to lot, the filter must avoid imperfections. This is not as easy to do as it is to talk about. It requires an understanding that may have to be developed. It requires close cooperation between you and your suppliers. However, the pay-off is usually a significant reduction in cost, waste, and an improved quality to your customer.

Later on the agenda, I will present a suggested test method.

Validation of integrity test values for cleanable porous stainless steel polymer filters.

Tore H. Lindstrom, Ph.D.
Staff Scientist
Scientific and Laboratory Services Department
Pall Corporation
Glen Cove, New York

Most filter applications, in the polymer industry, have a critical impact on product quality and as a user it has become more and more important to be able to predict in-service performance of a new or cleaned filter assembly prior to actual installation. One important characteristic of such a filter assembly is integrity. This paper addresses the commonly used method of 1st Bubble Point determination, for integrity evaluation.

In order to draw conclusions regarding the information obtained by 1st bubble point determination, three bubble pointed 7" diameter powder metal segments were subjected to a F-2 filter performance test. To obtain a sub-specification value, for one segment, a 0.0145" diameter hole was drilled (through the medium on one side of the segment and into the drainage layer in the center of the segment).

The value of the 1st Bubble Point measurement for a filter assembly provides important integrity information since the maximum attainable filter efficiency will be finite, if the 1st Bubble Point falls below a threshold value (between 0.5 and 9.1 "wc, for the segments evaluated in this study).

A minimum acceptable 1st bubble point value (ensuring a desired filter removal at a particle size) can be determined. This threshold value will be dependent on several parameters (i.e. viscosity and density of fluid filtered, terminal pressure differential, flowrate, and accuracy of bypass path model).

Background

Most filter applications, in the polymer industry, have a critical impact on product quality. A number of these applications involve the use of cleanable porous metal filter assemblies (pleated candles or segments), to high pressure differentials (1500 psid is not unusual). As a user it has become more and more important to be able to predict in-service performance of a new or cleaned filter assembly prior to actual installation. Evaluations of these filter assemblies should be simple, quick and inexpensive. One important characteristic of the filter assembly is integrity.

This paper addresses a commonly used method of 1st Bubble Point determination (see Appendix I). It should be noted that this method will not be able to detect certain failures, such as loss of sinter bonds, since the associated change in pore sizes (fiber deformation etc.) is undetectable at low pressure differentials.

Test procedures and results

In order to draw conclusions regarding the information obtained by 1st bubble point determination, three (S/N's: A1, A2, and A3) 7" diameter powder metal segments were subjected to 1st bubble point evaluation followed by a F-2 filter performance test (see Appendix II).

F-2 testing does not destroy (irreversibly alter) the filter assembly under evaluation, but the evaluated assembly will require cleaning after the test, due to retained test contaminant.

The three segments to be tested all passed quality assurance requirements for 1st and open bubble point values. In order to obtain a sub-specification value, for segment A1, a 0.0145" diameter (368 μm) hole was drilled (through the medium on one side of the segment and into the drainage layer in the center of the segment).

The following bubble point values were obtained for the segments in question:

S/N	1st Bubble point (" water column)	Remarks
A0	9.1 "wc	Without hole
A1	0.5 "wc	A0 with 368 μm hole
A2	13.3 "wc	
A3	9.1 "wc	

The measured bubble point values are also shown in Figure I. A 1st bubble point of 0.5 "wc (segment A1) is equivalent to a straight cylindrical pore with a diameter lager than 238 μm, based on the bubble point formula (Appendix I) with a displacement coefficient (δ) of 238 (straight cylindrical pore), and an accuracy of ± .5 "wc (at the low end of the pressure scale).

F-2 filter efficiencies (see Appendix II) obtained for each of the three segments (time averaged to 40 and 100 psid) are shown in Figures II through IV.

The following observations can be made in these figures:

a. The sub-specification segment (A1) show a leveling off of the removal curve at a finite β-ratio (140-200).

b. The removal curves of the other segments (A2 and A3) becomes infinite (>5000) for a particle size below 20 μm.

c. When comparing the efficiency curves to 40 and 100 psid, one can see that only the sub-specification segment (A1) shows lower efficiency values to a higher pressure.

These observations are consistent with the presence of a bypass (a path for the fluid around but not through the media), in segment A1.

The efficiency (β-ratio) of a filter without a bypass can be calculated as follows:

$$\beta_x = \frac{N_u}{N_d}$$

Where: N_u = # of upstream particles (larger than x μm)
N_d = # of downstream particles (larger than x μm)

If a bypass is present this relationship changes to:

$$\beta_x^* = \frac{N_u}{\alpha N_u + (1-\alpha) N_d} = \frac{1}{\alpha + \frac{(1-\alpha)}{\beta_x}}$$

Where: α = ratio between bypass and total flow.

If the limiting particle size, x, is set high enough (20 μm for the evaluated segments) the filter efficiency, β_x, become infinite and the filter efficiency measured in the presence of a bypass, β_x^*, becomes $1/\alpha$.

The measured β (where x is 20 μm) for segment A1 as a function of test duration can be found in Figure V. If this data is combined with the differential pressure (ΔP) across the segment during the same test (shown in Figure VI) β can be presented as a function of ΔP (shown in Figure VII).

Based on the geometry of the hole (a simple cylinder) the following derivation can be made of the bypass ratio:

$$\alpha = \frac{Q_{bp}}{Q_{tot}}$$

Where: Q_{tot} = total flow (8.2 l/min for segment A1)
Q_{bp} = bypass flow

$$= \frac{\pi * d^4 * \Delta P}{128 * l * \mu}$$

Where: d = diameter of hole
ΔP = diff. pressure
l = length of hole (1/16", for A1)
μ = viscosity of fluid (13.95 cP, for A1)

This can be reduced to a linear function:

$$\alpha = C * \Delta P = 1/\beta$$

$$\text{Where: } C = \frac{\pi * d^4}{128 * l * \mu * Q_{tot}}$$

In Figure VIII $1/\beta$ (>20 μm) for segment A1 is shown as a function of ΔP. It should be noted that the spread in the data points are due to the statistical nature of the particle counting. A linear fit provided a value of 0.00024 psid^{-1} for C. Based on this value the size of the hole is estimated at 256 μm.

The deviation between this value and the known diameter of the drill used (368 μm) can be explained by the hole being obtained by drilling into the drainage layer and the clearance between this drainage layer and the medium being unknown. The flow through the hole could be restricted by the opening between the drainage layer and the medium, if this clearance is significantly small.

Figure IX show the β-ratio (>20 μm) for segment A1 as a function of ΔP. Superimposed on the data points is a curve showing infinite filtration efficiency based on the assumption that there is a cylindrical bypass with a diameter of 256 μm.

Figures X and XI show the β-ratio (>20 μm) for segments A2 and A3 respectively. As can be seen the β-ratio (> 20 μm) for neither of these segments show a decline with increasing pressure differential, even to 1500 psid. The occasional low values (<5000) are again due to the statistical nature of the particle counting process.

A curve of threshold values versus differential pressure can be obtained, based on a specific flow rate, a specific fluid (viscosity), a desired filter efficiency (for particles larger than the absolute rating of the filter assembly), and a model of the bypass path. An example of such a curve is shown in Figure XII. This particular curve is based on the following criteria:

Fluid: Polymer (at 575°F, μ=2000 P and ρ=1.2 kg/liter)
Flow rate: 0.053 liter/min (20 lbs/hr/ft^2, and 0.42 ft^2 per 7" segment)
Minimum filter efficiency: 99.98% (β=5000)
Bypass path model: Straight cylinder (1/16 inch long)

Conclusions

The value of the 1st Bubble Point measurement for a filter assembly provides important integrity information since the maximum attainable efficiency will be finite, if the 1st Bubble Point falls below a threshold value (between 0.5 and 9.1 "wc, for the segments evaluated in this study).

A minimum acceptable 1st bubble point value (ensuring a desired filter removal at a particle size) can be determined. This threshold value will be dependent on several parameters (i.e. viscosity and density of fluid filtered, terminal pressure differential, flowrate, and accuracy of bypass path model).

Figure I.

1st Bubble Point values for

7" diameter powder metal segments

S/N: A1, A2, and A3

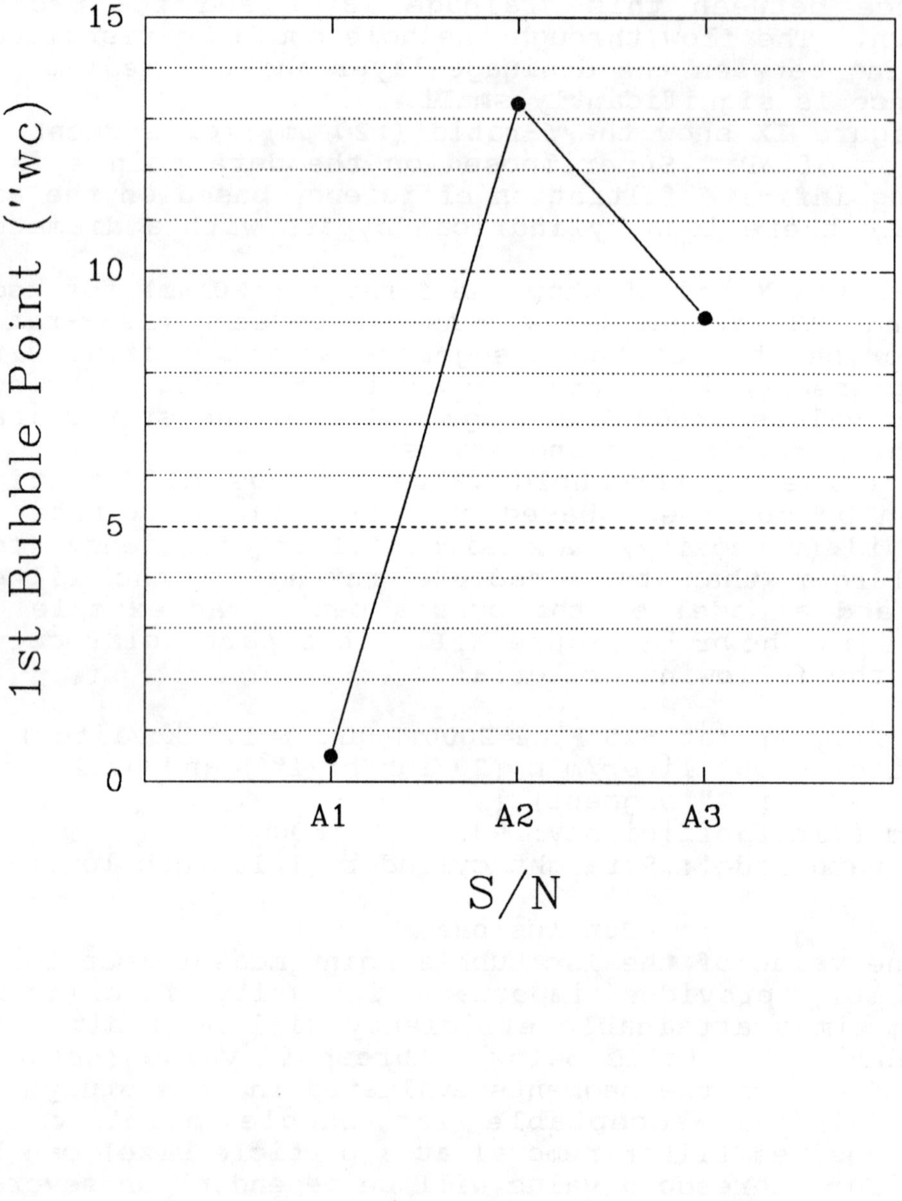

Figure II.

Time average β−ratio versus particle size for 7" diameter powder metal segment (S/N:A1)

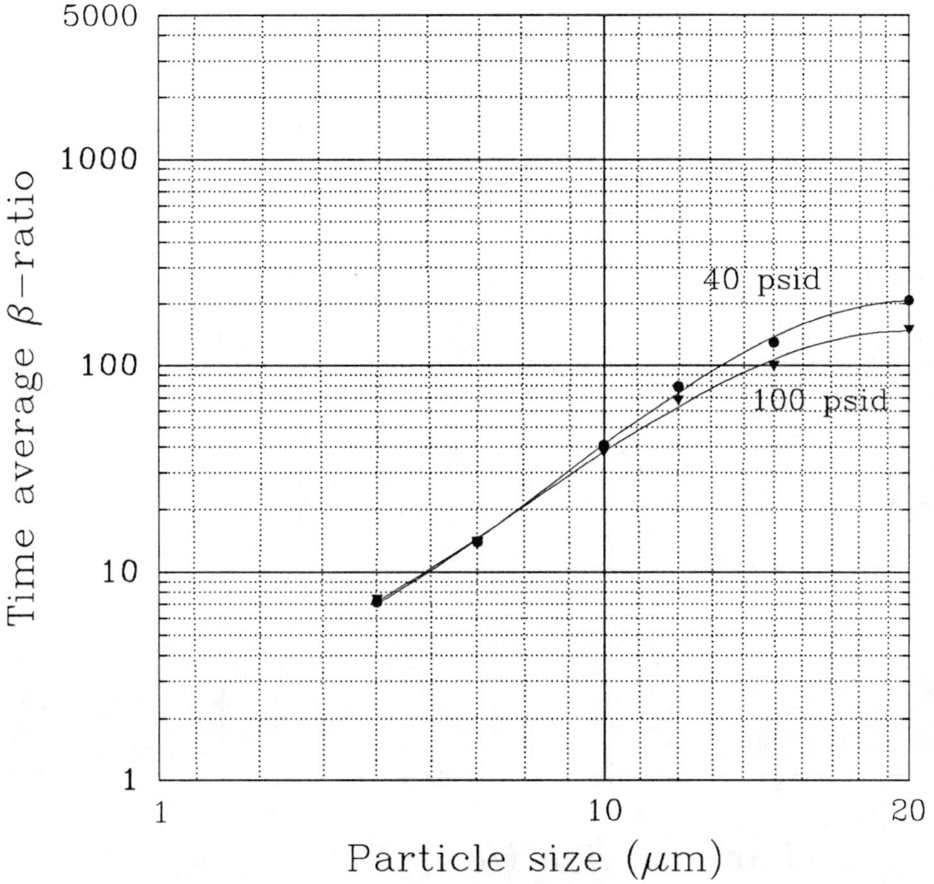

Figure III.

Time average β−ratio
versus particle size for
7″ diameter powder metal segment (S/N:A2)

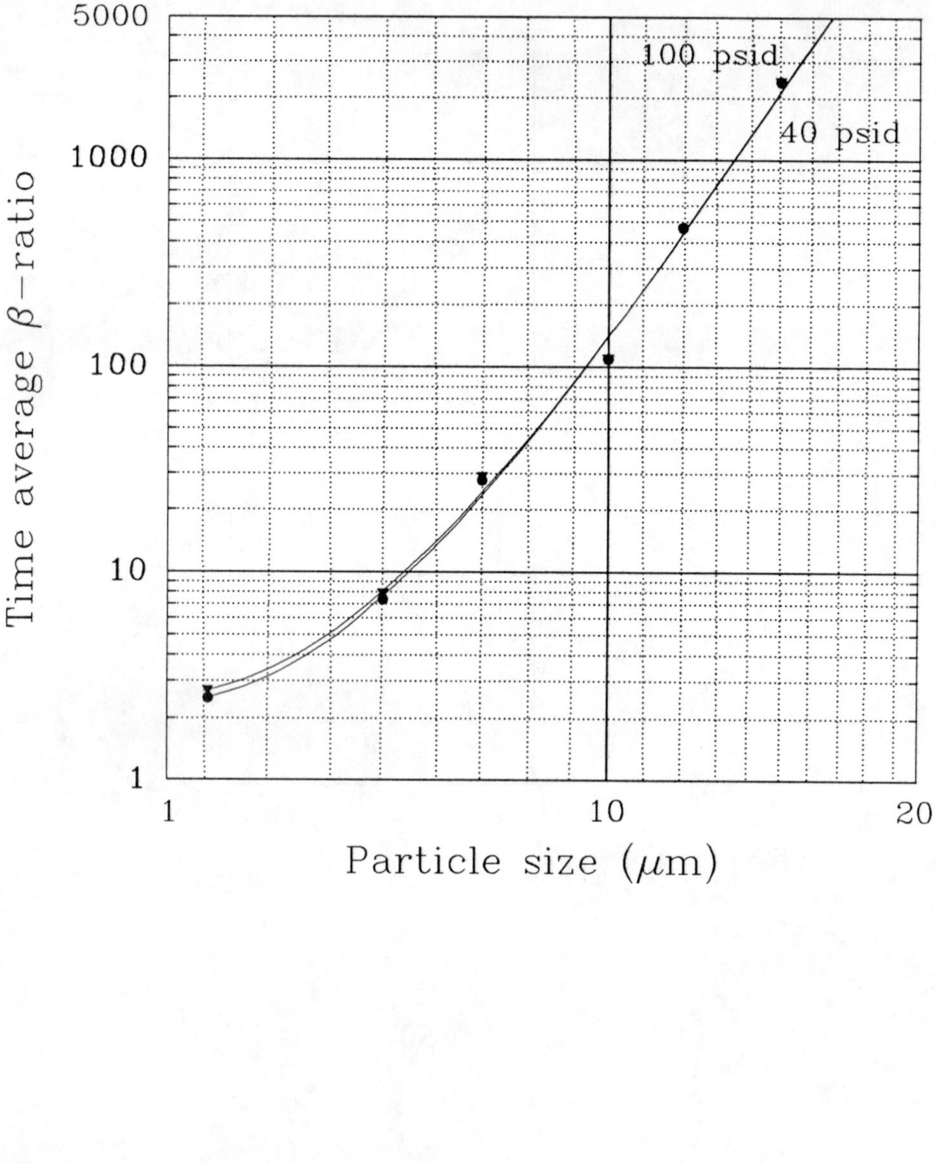

Figure IV.

Time average β-ratio versus particle size for 7" diameter powder metal segment (S/N:A3)

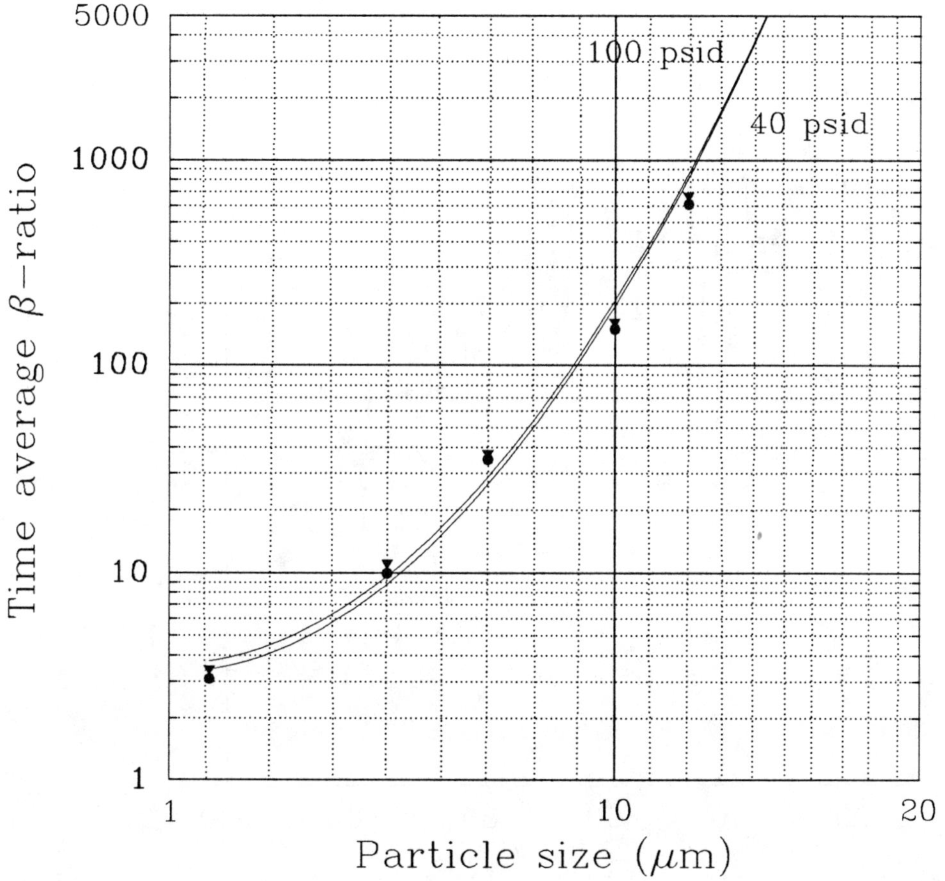

Figure V.

β-ratio (>20μm) versus
challenge duration for
7″ diameter powder metal segment (S/N:A1)

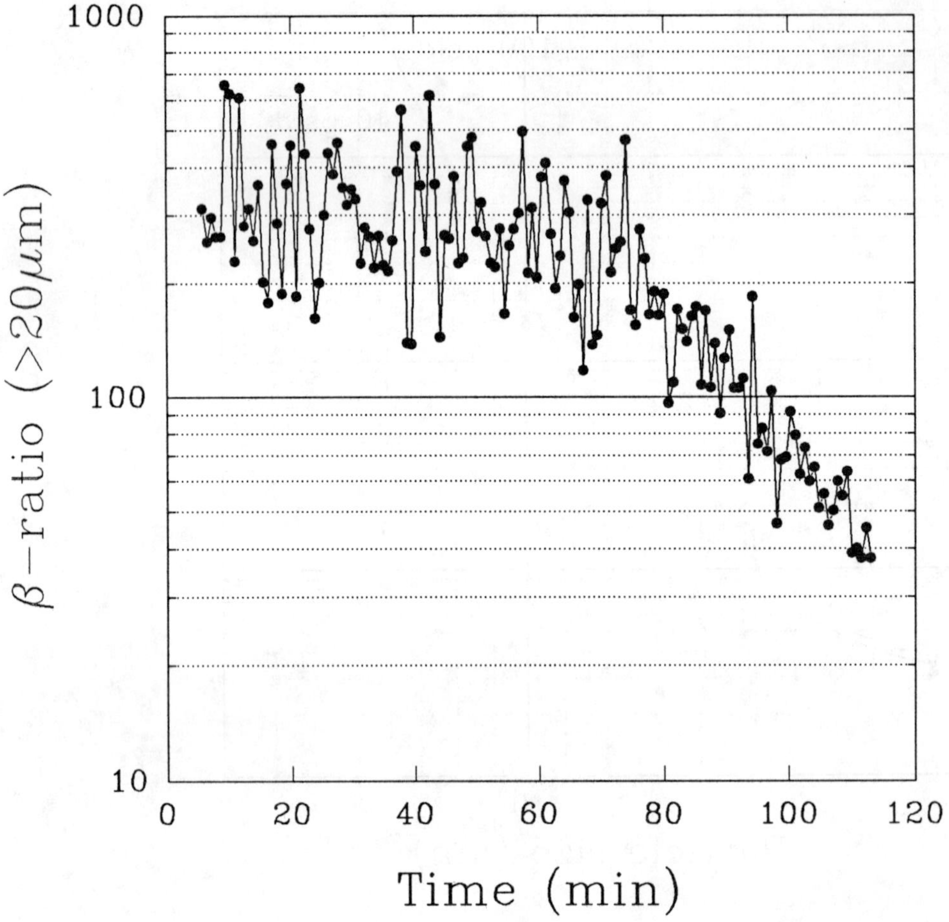

Figure VI.

ΔP versus challenge duration for
7" diameter powder metal segment (S/N:A1)

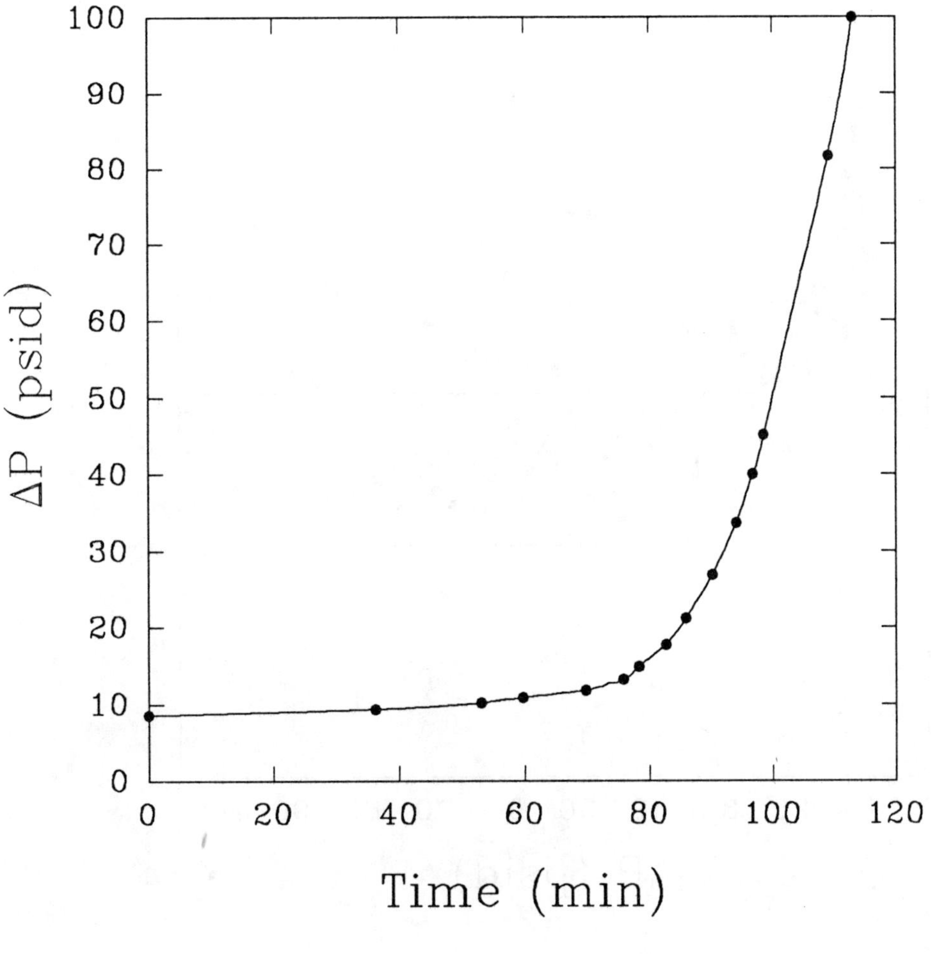

Figure VII.

β-ratio (>20μm) versus ΔP for
7″ diameter powder metal segment (S/N:A1)

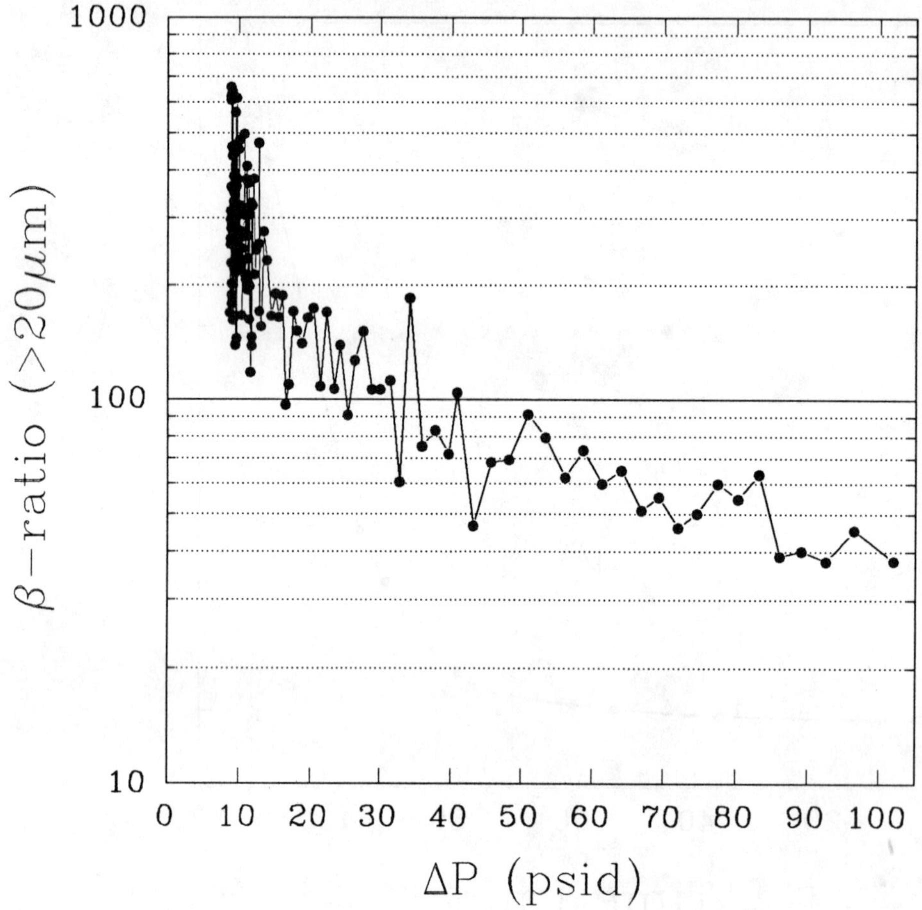

Figure VIII.

1/β (⟩20μm) versus ΔP for
7" diameter powder metal segment (S/N:A1)

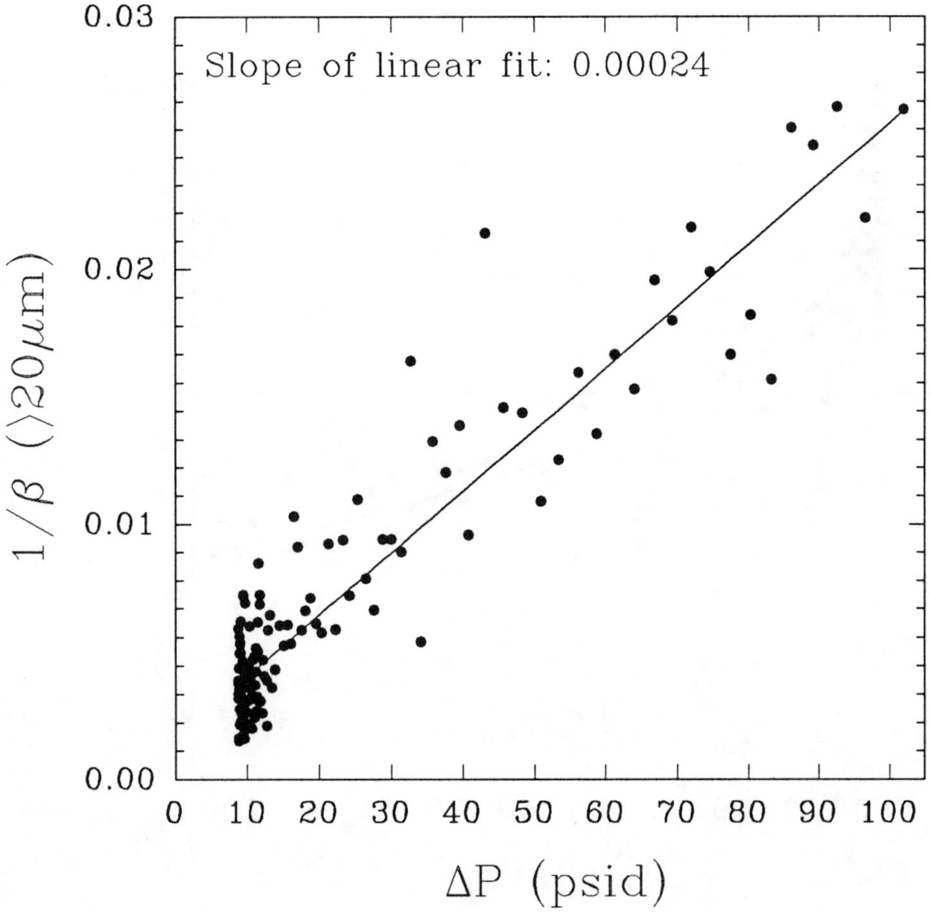

Figure IX.

β–ratio (>20μm) versus ΔP for 7" diameter powder metal segment (S/N:A1)

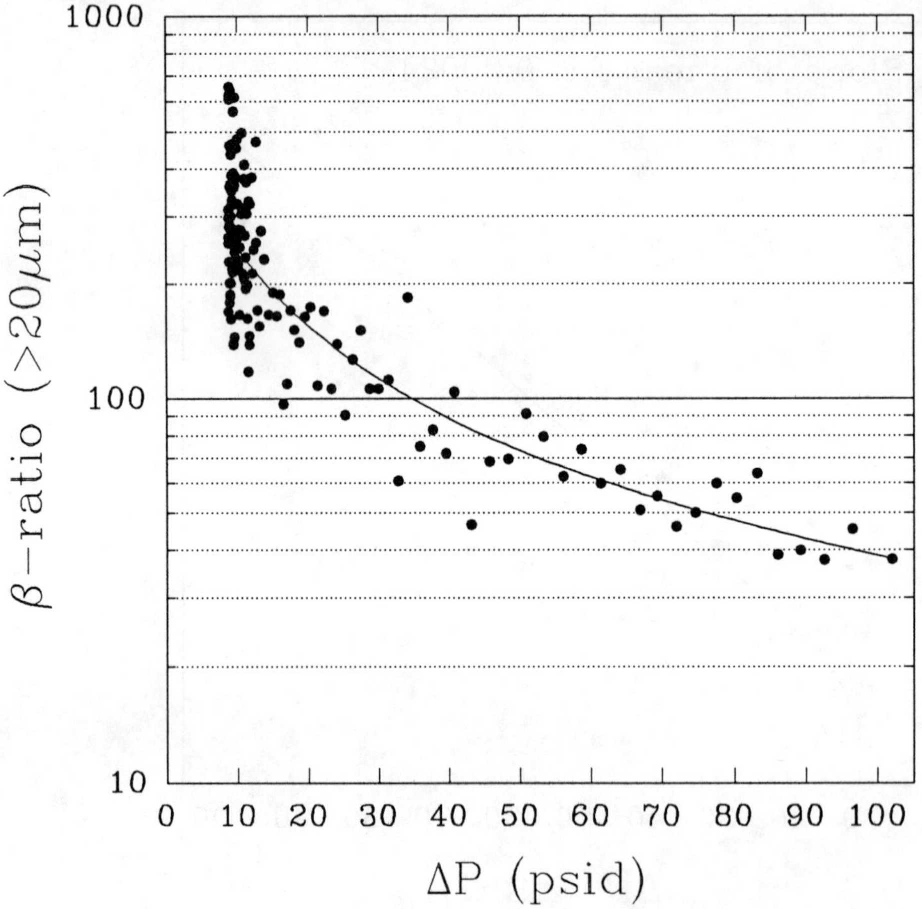

Figure X.

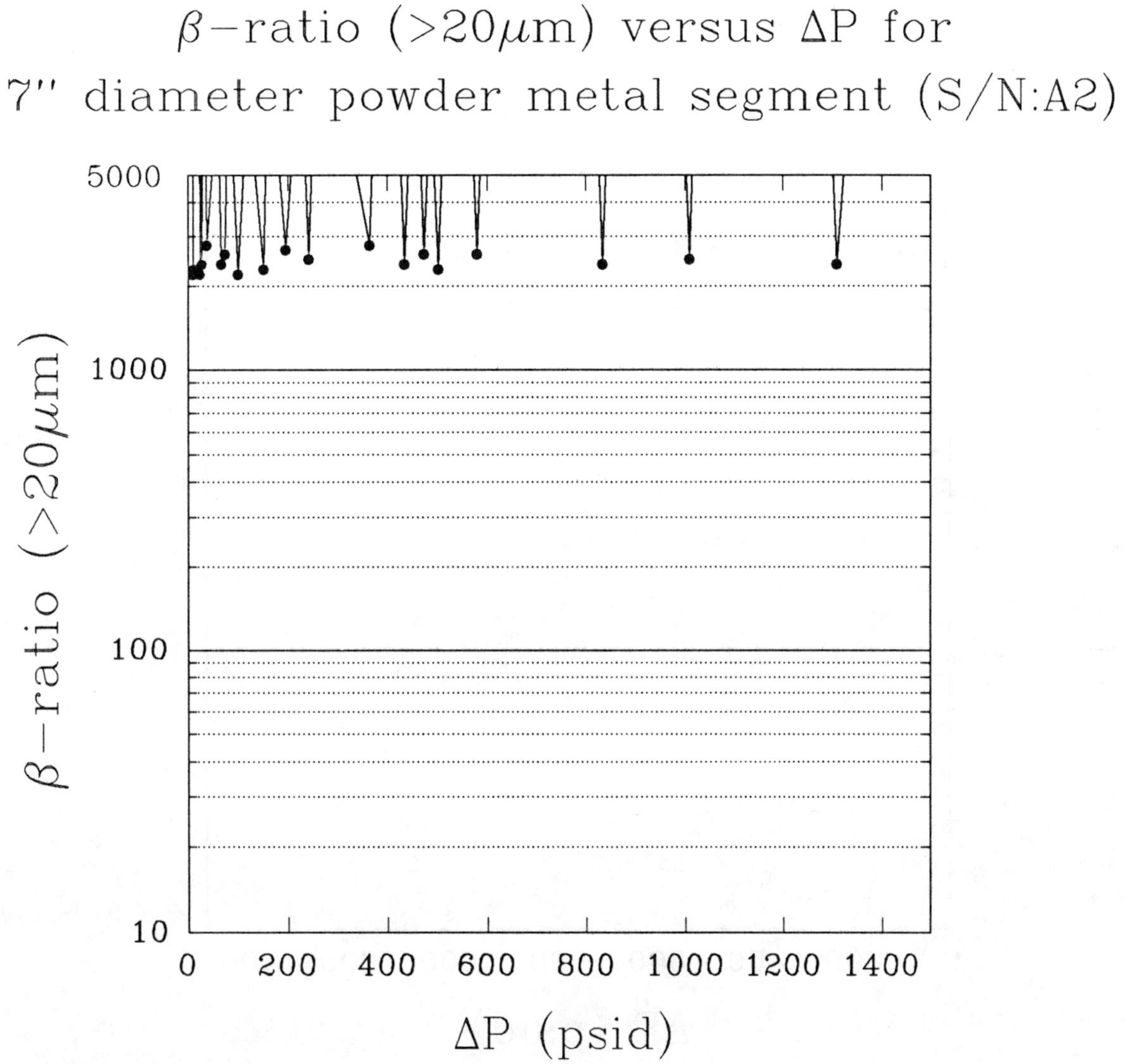
β−ratio (>20μm) versus ΔP for
7″ diameter powder metal segment (S/N:A2)
5000
1000
100
10
β−ratio (>20μm)
0
200
400
600
800
1000
1200
1400
ΔP (psid)

Figure XI.

β–ratio (>20μm) versus ΔP for
7" diameter powder metal segment (S/N:A3)

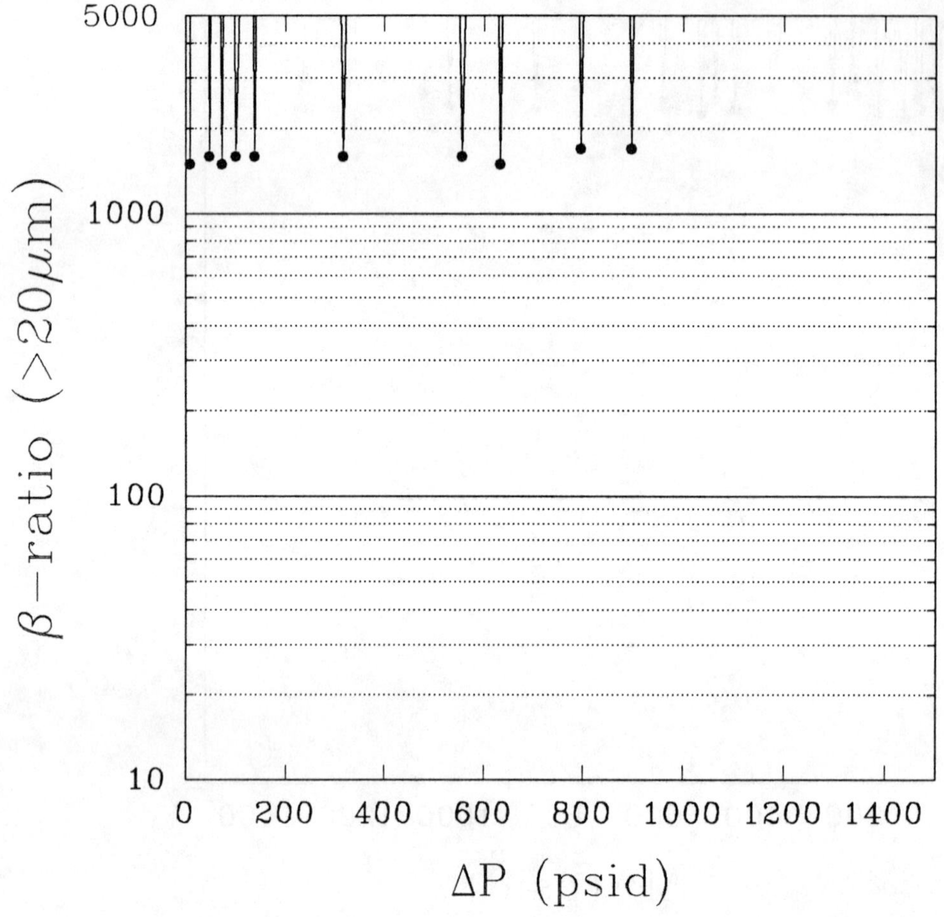

Figure XII.

Theoretical minimum 1st bubble points and maximum largest cylindrical pore diameter for filtration of a polymer at 575 °F, 2000 Poise, 1.2 kg/liter and 0.053 liter/minute while requiring a minimum β−ratio of 5000 for particles larger than the absolute rating of the filter.

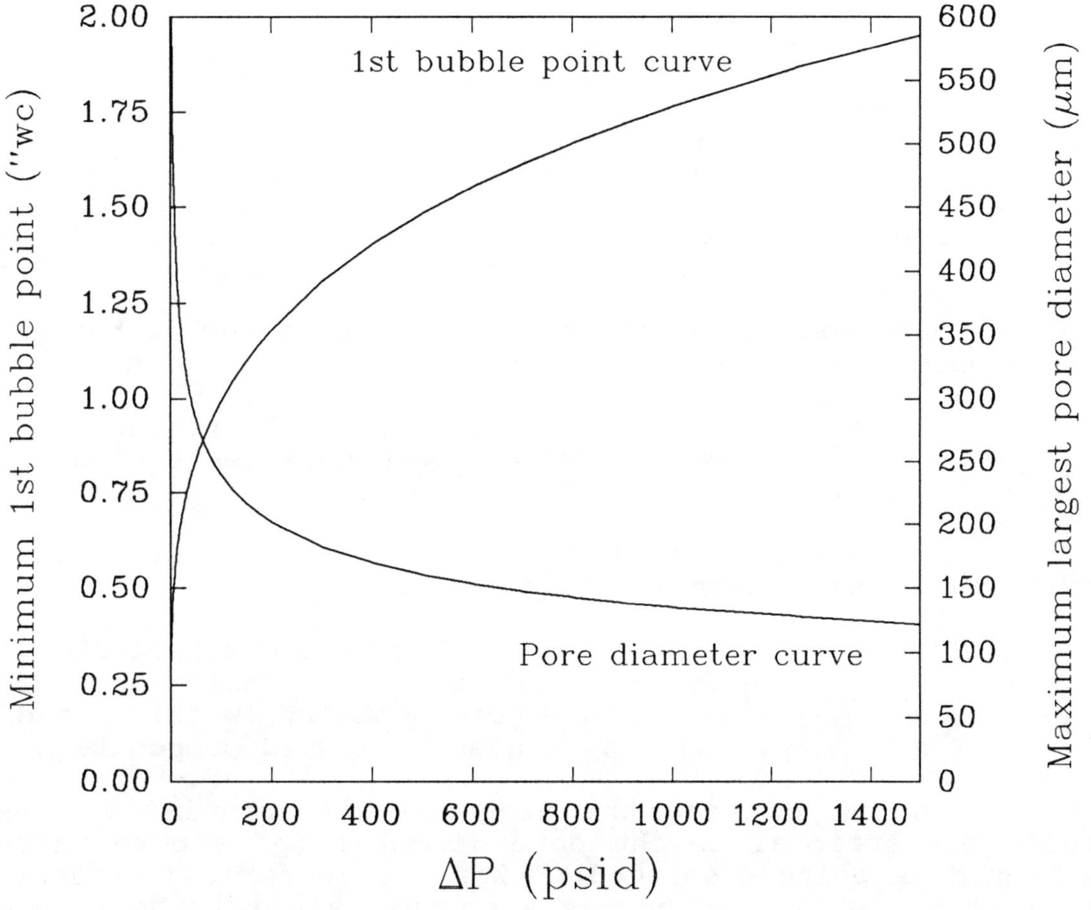

APPENDIX I

1st Bubble Point Evaluation

Theoretical derivation

When there is a difference in gas pressure between two sides of a fully wetted (i.e. all pores completely filled a liquid) porous material, a balance exists between the surface tension of the liquid and the exerted gas pressure. As the pressure differential is increased the gas will strive to displace the wetting liquid in some of the pores.

To simplify the derivation and form of a formula for this displacement the following assumptions are necessary:

1. The pores are simple cylinders.
2. Complete wetting occurs between porous material and wetting liquid (i.e. all pores are completely filled with the wetting liquid).
3. The increase of pressure differential from none (fully wetted material) to the point where a measurement is to be taken, is made so slowly that surface tension and pressure forces are always in balance.
4. The flow through the porous material and as a result the friction losses in the medium are negligible.

Based on these assumptions the relationship between pore diameter and pressure differential can be written as:

$$\Delta p = \frac{4 \cdot S}{d}$$

Where: Δp = Differential pressure
S = Liquid surface tension
d = Pore diameter

Deviations from the above mentioned assumptions results in a more commonly used relationship:

$$\Delta P = \frac{\delta}{d}$$

Where: δ = Displacement coefficient dependent on medium make-up, pore geometry, wetting, and gas and liquid composition.

As can be seen the pressure differential is inversely proportional to the pore diameter (of a pore where the liquid is being displaced). When an increasing pressure differential is exerted across a porous material the liquid in the largest pores will be displaced first, and the liquid in the smallest pores last.

The displacement coefficient, δ, is a constant, as long as the pressure differential associated with flow, through the porous material, is negligible when compared to that of liquid displacement.

Test procedure

1. Assemble a measuring system equivalent to that in Figure I-A.
2. Fully wet the filter assembly (element or segment) with the wetting liquid.
3. Submerge the filter assembly in the wetting liquid.

4. Connect test adapter to the filter fitting and seal any other filter openings. This should be done such that the gas is allowed to enter inside the filter assembly and displace the liquid out through the medium.

5. Slowly increase the pressure differential until the first stream of bubbles can be detected, while continuously rotating or turning the filter assembly (so that the entire filter surface has been subjected to the minimum submersion depth at each pressure differential setting), and record the pressure differential when this occurs (see interpretation of results).

6. Calculate the corrected pressure differential by subtracting the immersion depth in the fluid, as a static pressure. The corrected pressure differential is dependent on the temperature and the type of liquid used. Known surface tensions for different liquids at different temperatures can be used to convert the measurement to a value for a different liquid and/or a different temperature.

$$\Delta p_2 = \frac{S_2}{S_1} \cdot \Delta p_1$$

Where: S_x = Surface tension liquid #x,@ temp. t_x
Δp_x= Diff. pressure for liquid #x,@ temp. t_x

Interpretation of results

A Bubble Point measurement is dependent on medium make-up (material), pore geometry, wetting, gas and liquid composition, and temperature. As a result care should be taken when using Bubble Point values for comparisons.

Unless careful correction can be made (extremely complicated) for all of the above mentioned parameters, <u>comparison of Bubble Point values should be restricted to the same filter assembly and medium type</u>.

Based on the theoretical relationship presented above, the 1st Bubble Point value is related to the largest pore in the filter assembly, and, as a result, <u>can be used as quality control measurement of medium and filter assembly integrity</u>. Small flaws in medium and seals in the filter assembly can be detected by a low 1st Bubble Point value.

Since most porous media have some form of bell-shaped pore size distribution and the 1st Bubble Point measurement is related to the size of the largest pore, <u>the validity of use for anything but integrity verification, is questionable</u>.

References

1. Aerospace Recommended Practice, ARP 901 (1968), <u>Bubble-point test method</u>, Society of Automotive Engineers, Inc., Warrendale. Theoretical derivation of and test instructions for bubble-point determination.

FIGURE I-A

Schematic of measuring system for Bubble-points

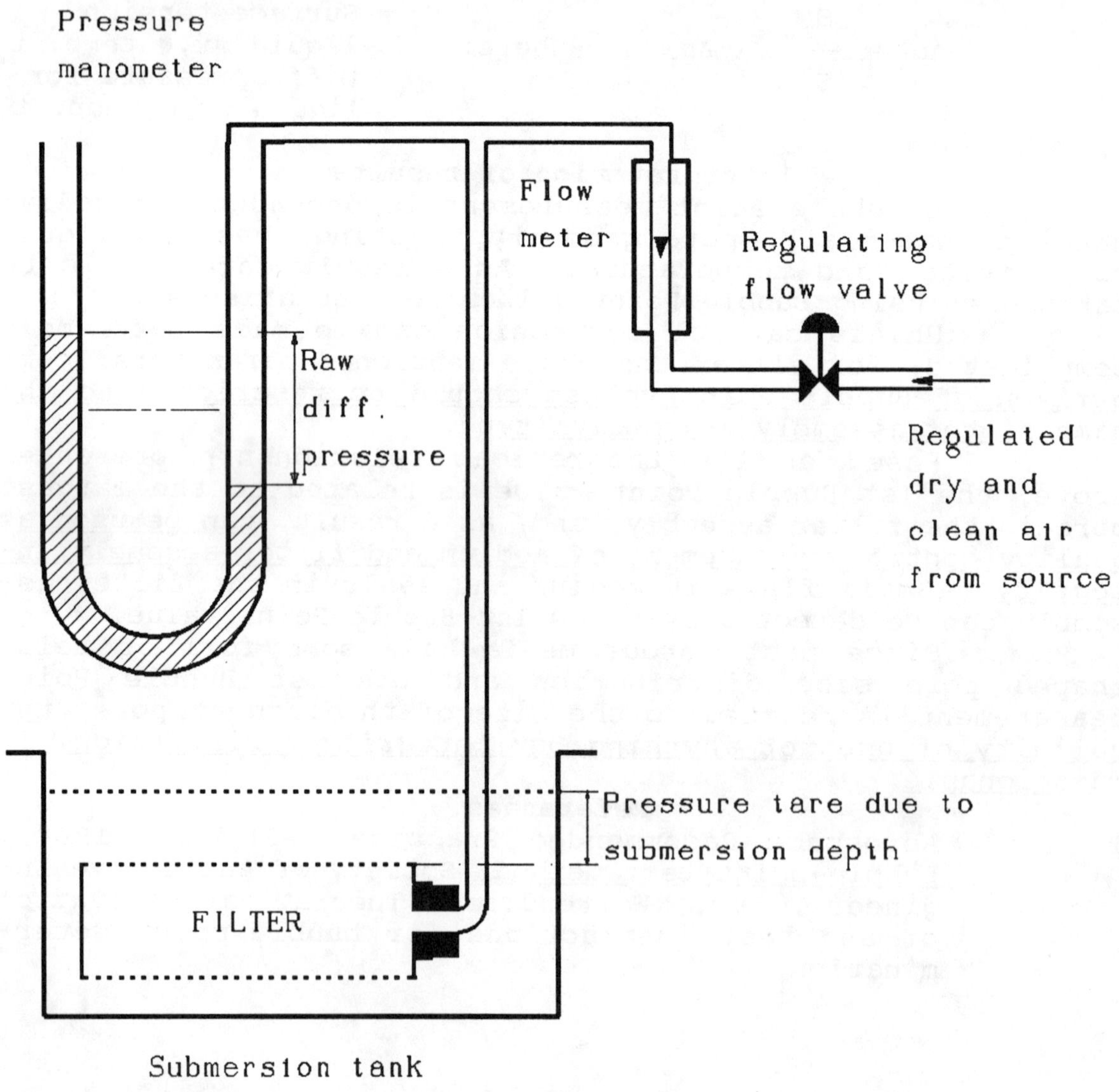

APPENDIX II

F-2 Test Procedure

Background

The F-2 test establishes the degree of efficiency of a filter and its contaminant capacity. This is achieved by providing a continuous supply of ingressed contaminant and allowing constant monitoring of the performance characteristics of the filter.

The test procedure employed is an adaptation of the "Oklahoma State University F-2 Filter Performance Test", adopted by the American National Standards Institute as their approved procedure ANSI B93.31-73. The original procedure was developed for the evaluation of hydraulic filters, but has been modified for the rapid, semi-automated testing of filters with aqueous liquids in single pass mode.

The scope of the test includes a single pass filtration performance test with continuous contaminant injection in aqueous media and determination of the filter contaminant capacity and particulate removal characteristic at various pressures.

A typical F-2 test apparatus is schematically represented in Figure II-A.

The aqueous test procedure employs a test contaminant prepared from AC Fine Test Dust (ACFTD), an irregularly shaped, naturally occurring siliceous dust used as a standard in many industries and specified in the original procedure. ACFTD is dispersed in water by a Cowless mixer and then agitated with an air driven "Lightning" mixer for 4-6 weeks. This procedure overcomes the difficulty of achieving a reproducible dispersion via less vigorous methods of mixing.

Two automatic particle systems are installed to monitor, in-line, the contaminant level of particles of interest upstream and downstream of the filter under evaluation.

Procedure

1. The test is started by setting a required flow rate through the test filter housing. In general a test flow of 10 LPM per 10" filter element is recommended.
2. While the clean water is recirculated, particle counting equipment monitors the number of particles present. As the water passes through the system clean-up filter (usually rated at 0.2 μm), the number of particles present in the recirculation stream decreases and finally reaches a predetermined low level of particles in each size range. This is known as blanking out the system.
3. After the required cleanliness of the system is achieved, the test filter is installed in the test housing, followed by bleeding the system of air.
4. A slurry contaminant tank is charged from a stock suspension, with a known concentration and constant size distribution. The slurry concentration is based upon the expected upstream gravimetric level, desired test duration and injection flow rate.

5. The differential pressure across the filter is established while running clean water through the filter at the specified flow rate.
6. The test filter is challenged with the contaminant, by injection into the recirculation loop.
7. Filtration efficiency is determined with two automatic particle counters installed in-line, each equipped with a particle sensor. Particle counts are obtained simultaneously at six different particle diameters. Upstream and downstream counts are recorded automatically (upstream and downstream of the test filter) at each of the designated particle diameters.
8. All test parameters are held constant while the pressure drop across the test filter and test time is being monitored.
9. When the pressure across the filter reaches a specified terminal differential pressure the test is terminated.

The automatically recorded particle counts upstream and downstream of the filter permits the calculation of beta efficiency, or filtration ratio, over the course of the test, as follows:

$$\beta_x = \frac{N_u}{N_d}$$

Where:

β_x = Filtration ration @ x
N_u = # of part. upstream > x
N_d = # of part. downstream > x
x = Particle diameter (μm)

A reciprocal time average of the beta efficiencies throughout the test is calculated, for each particle size evaluated, and a contaminant holding capacity is determined.

FIGURE II-A

Schematic of F-2 test apparatus

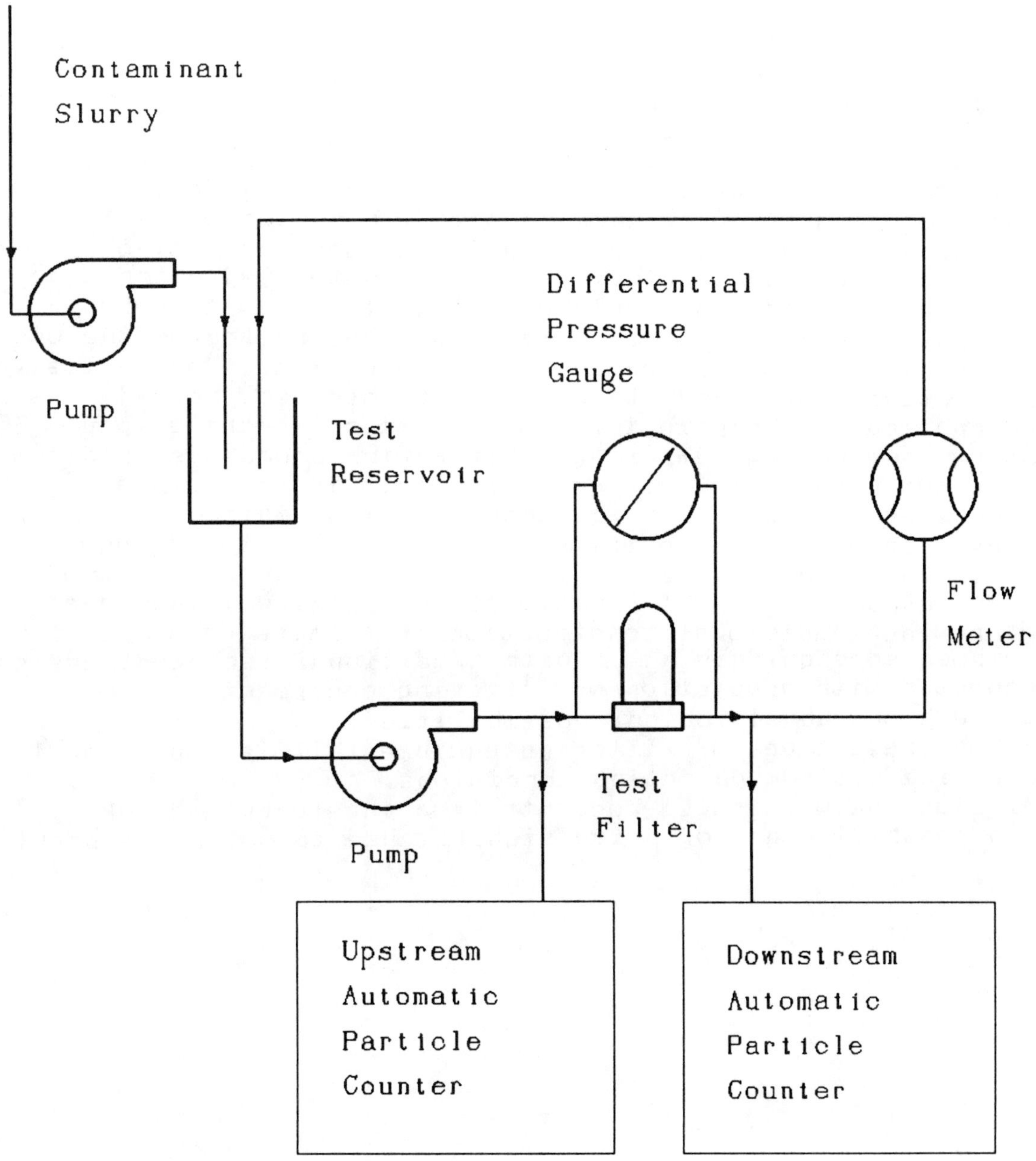

Summaries

THE ANALYSIS OF THE DEVELOPMENT OF FILTERING EQUIPMENT AS FLEXIBLE TECHNOLOGY

A.I.Yelshin, Polytechnic Institute of Novopolotsk, USSR

The analysis showed that today a new direction in filter disigning was formed besides the development of traditional types of a filtering equipment was formed. Here the filters are considered as a unite of a flexible technology. The basic kinds of the flexible technology and the role of filtration in it are given.

Outlines of a apparatus which combine a processes of heat, and mass transfer and filtration are considered. Complete analysis of two perspective tendencies in disigning of the filtering equipment is done. First, the application of robot principles in the automated operation of the batch filtering equipment. It will allow to make qualitative change in the decision of the problem of the use of filters in conditions dangerous for people, such as explosive, radioactive, toxic and other hazard substances. Second, the application of principles of rotational machines and rotary-conveyor systems in the filtration processes is discussed. These two branches of disign will permit in the future to make use of filtering equipment in : automatic shops, mining at the bottom of the sea, space technology and in lunar developing.

The expansion of the functional possibility is achieved by: new nontraditional construction of filters and equipment design, giving up with traditional technical design connected with adaptation of filtering equipment to the human physiological possibilities.

For these types of filtering equipment it is importand to determine optimum operating conditions.
Relations between solid concentration in slurry and cake compressibility are of prime significance to solve the problem.

THE COST CORRELATION OF SOME TYPES OF FILTERING EQUIPMENT

A. Yelshin

The tendency which should be considered in this paper is typical for the manufacture of filtering equipment in general.

Analysis showed quite definite correlation between filters cost (C), mass of filter (M) and total filtering area (S) within the same type of equipment.

The simple dependences between (C), (M) and (S) for the most representative types of filtering equipment can be presented.

At the Fig.1 and 2 the data for C = f(S) are presented. On the Fig.3 the M vs. S is given.

Note should we taken that the data given here are approximate. The cost of filtering equipment is depending to a considerable extent on the constructive materials used for filter manufacturing (carbon steel, stainless steel etc.) and on filter constraction peculiarity.

For filters presented in the Tables the following regressive equations were used:

$$C = a*S^{b}, \qquad (1)$$

$$M = a_1*S^{b1}, \qquad (2)$$

where a, a_1 and b, b_1 are coefficients.

The values a, a_1 and b, b_1 are presented in Tables 1 and 2.

Table 1

THE VELUES OF COEFFICIENTS a AND b FOR DIFFERENT TYPES OF FILTERING EQUIPMENT

Type of filter	a, roubles	b	The correlation coefficient
1. Automatic filter-presses FPAKM - type	16185	0.292	0.999
2. Manytier chember filter-presses with paper filtering ribbon	11779	0.146	0.998
3. Plates and frames mechanized filter-presses	3136	0.37	0.7
4. Nonmechanized filter-presses	554	0.61	0.76
5. Special types of filter-presses	423	0.857	0.888
6. Belt vacuum filters	4211	0.866	0.992
7. Disc vacuum filters	C=18488*exp(S*6.5E(-2))		0.931
	or		
	3023	0.568	0.928
8. Drum vacuum filters	5060	0.523	0.969
9. Cartridge filters	2625	0.739	0.972
10. Vertical leaf filters	877	0.843	0.922
11. Horizontal leaf filters	1236	0.583	0.455

Relation of cost to filter mass C/M (Fig.4,Table 2) is more comfortable for comparing the different types of filters:

$$C/M = (a/a_1)*S^{(b-b_1)} \qquad (3)$$

Table 2

THE VELUES OF COEFFICIENTS a_1 AND b_1 IN EQUATION (2) AND a/a_1, $b-b_1$ IN EQUATION (3)

Type of filter	a_1, kg	b_1	Coefficient of correlation	a/a_1 rub/kg	$b-b_1$
1.FPAKM	3745	0.42	0.989	4.32	-0.128
2.Chamber filter-presses with paper filtering ribbon	2488	0.347	0.987	4.73	-0.201
3.Plates and frames mechanized filter-presses	1182	0.535	0.973	2.65	-0.165
4.Special types of filter-presses	377	0.7	0.921	1.12	+0.157
5.Belt vacuum filters	2600	0.672	0.963	1.16	+0.194
6.Disc vacuum filters	329	0.839	0.842	9.19	-0.271
7.Drum vacuum filters	909	0.88	0.994	5.57	-0.357
8.Catrtdge filters	352	0.785	0.981	7.46	-0.046
9.Vertical leaf filters	87.6	0.95	0.968	10	-0.107
10.Horizontal leaf filters	60.5	1.06	0.725	20	-0.477

As can seen, for special types of filter-presses and belt vacuum filters the growth of cost per unit of filter mass take place with an increase of total filtering area. It can be explained by the constructive filter particularities. The catridge filters as well as vertical leaf filters have a high level of specific cost.

The data presented here do not pretend to analyses of problem fully. So, we do not take into account here the working conditions of filters (continuous or periodical), the type of filtering media, slurry property, the capital investments for assembling the filters at enterprtse and level of operating costs.

However on the basis of this data we can be decid on the advisability and the way of modernization of the appointed types of filtering equipment.

One of many ways to improve the filtering equipment characteristics on the stage of manufacturing is:

- increasing of total filtering area with fixed basic filter sizes. It can be done with the help of pleated filter units (filter elements), optimum filter units (filter elemens) forms and with the help of optimizing thier placing in the filter chamber (for leaf and catridge filters);
- passage to more light and cheap constructive materials (for all types of filters) and especially for belt, disc and drum filters);
- increasing the total filter area at the expense of filter size (filters FPAKM, filter-presses).

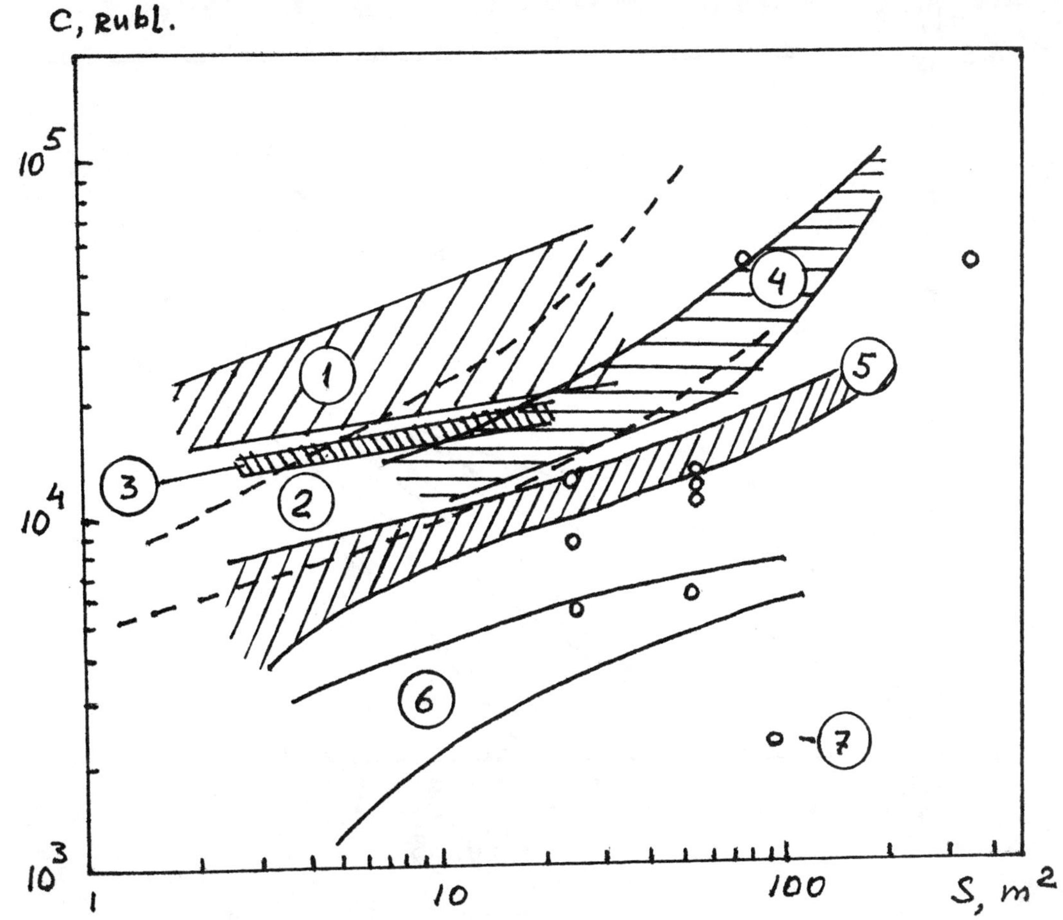
C, Rubl.
10^5
10^4
10^3
1
10
100
S, m^2
1
2
3
4
5
6
7

Fig. 1

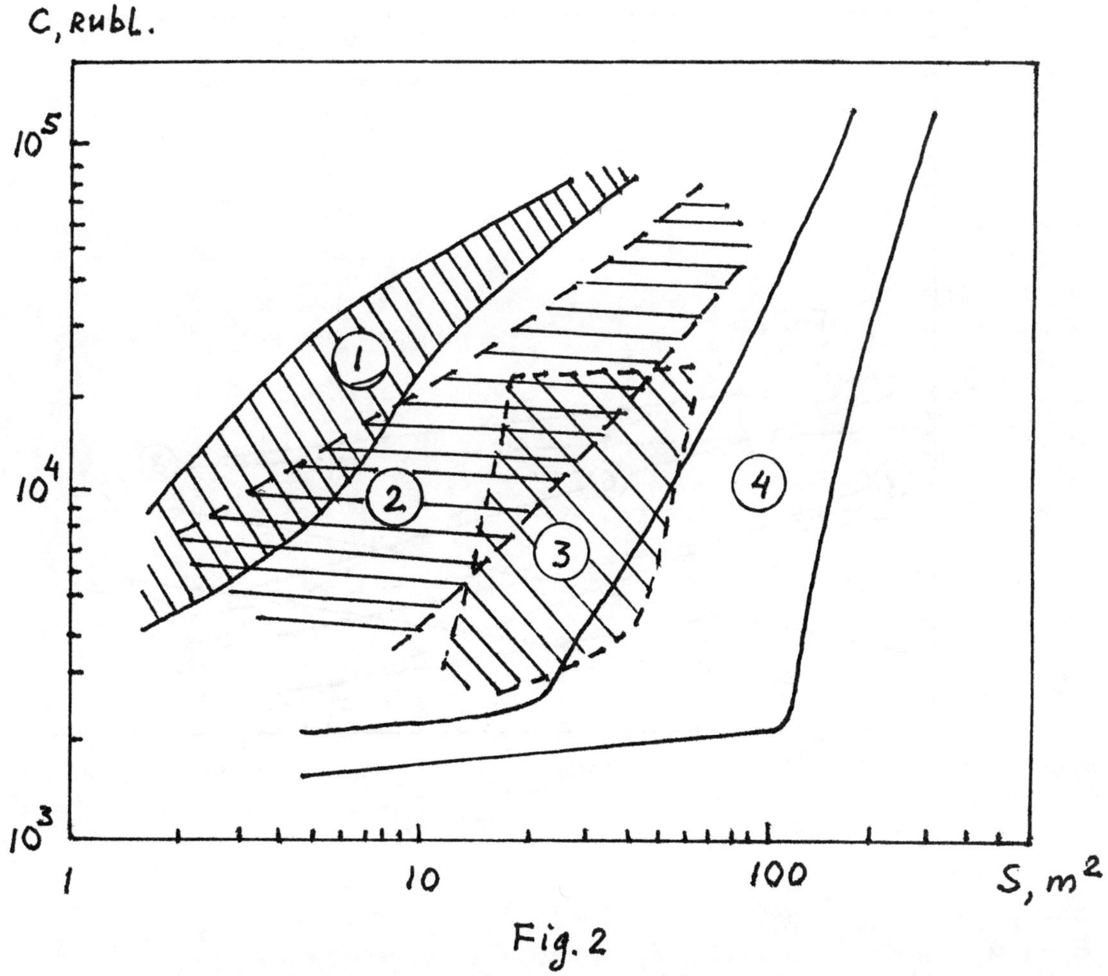
C, Rubl.
10^5
10^4
10^3
1
10
100
S, m^2
1
2
3
4

Fig. 2

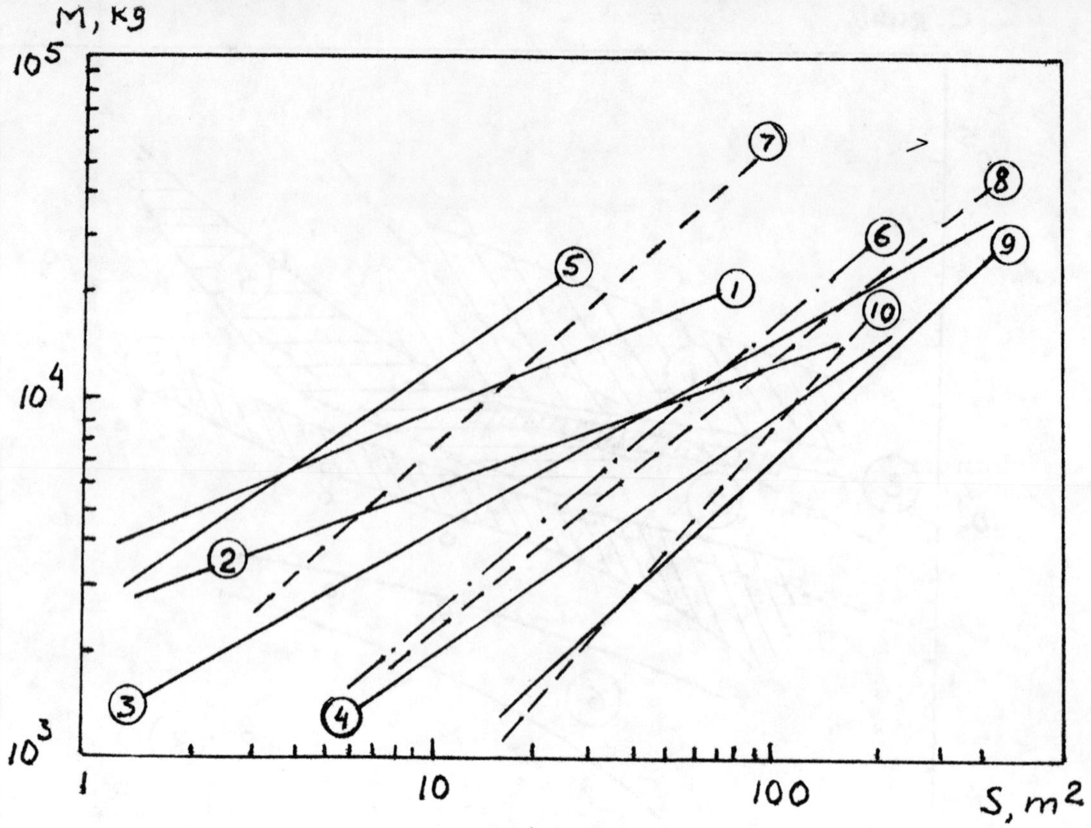
M, kg
10^5
10^4
10^3
1
10
100
S, m²
1
2
3
4
5
6
7
8
9
10

Fig. 3

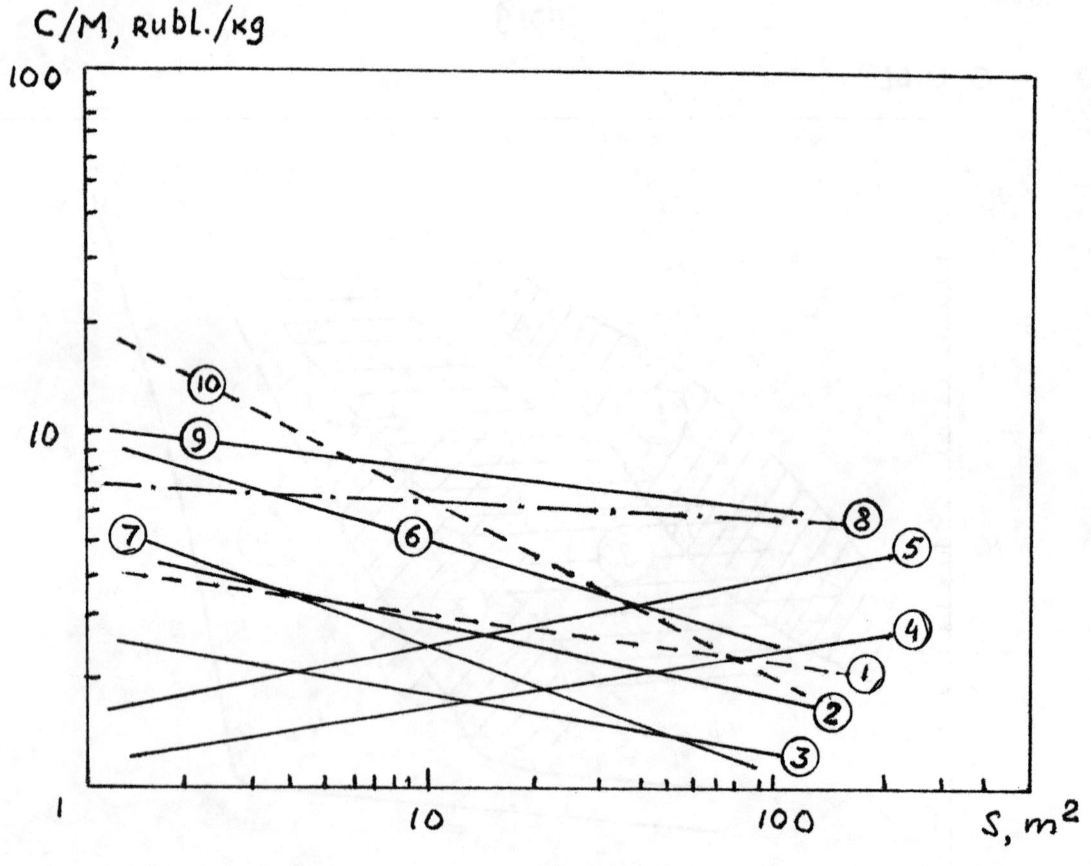
C/M, rubl./kg
100
10
1
10
100
S, m²
1
2
3
4
5
6
7
8
9
10

Fig. 4

Fig.1. The cost some types of filtering equipments C vs. S: 1- FPAKM; 2- drum vacuum filters; 3-chamber filter-presses with paper filtering ribbon; 4- disc vacuum filters; 5- mechanized filter-presses; 6- nonmechanized filter-presses; 7- special types of filter-presses.

Fiq.2. Dependence of C vs. S for: 1- belt vacuum filters; 2- catridqe filters; 3- horizontal leaf filters; 4- vertical leaf filters.

Fiq.3. Dependence of M vs. S. Numbers on Fig.3 are correspond with numbers in the Table 2.

Fig.4. Dependence of C/M vs. S. Numbers on Fig.4 are correspon with numbers in the Table 2.

USE AND APPLICATION CONCEPTS
OF GORE-TEX® MEMBRANE FILTER CLOTHS
IN MINERAL PROCESSING FILTRATION

Gordon R. S. Smith
Dwight R. Davis
Craig R. Rinschler
W. L. Gore & Associates, Inc.
P.O. Box 1100
Elkton, MD 21921

Expanded polytetrafluoroethylene (PTFE) membranes are laminated to strong felt support materials to make a membrane cloth capable of use with tubular and drum filters and filter presses. Used in combination with Back-Pulse Filtration it is possible to filter solids such as metal hydroxides and other compressible materials (as may be found in hydrometallurgical processing or waste treatment) producing filtrate solids as low as 0.05 ppm.

Product may also be recovered from dilute streams with equipment pay-back of less than one year.

The membrane filter cloths and back-pulse filtration with tubular filters will be described. Filter cycles are as short as 3 - 5 minutes and filter downtime between cycles may only be a few seconds. This method is compared with conventional tubular filter operation. Two examples are given: one of the recovery of a waste pigment (in which comparisons are made with a plate separator); and, second, the thickening of a waste bauxite mud.

Introduction

There are numerous streams of liquids containing small amounts of fine particles, many of which are very difficult to remove. Frequently, it is desired to separate these particles from the liquid either because they are considered pollutants or because they have value, or both. Separation is made difficult, however, by the very nature and origin of these suspensions. Frequently their source is thickener overflow or other gravity separations where all the large, easy filtering particles have already been removed. An additional source of dilute suspensions is precipitated solids such as found in plating wastes or various hydrometallurgical processors. All of these suspensions have in common dilute amounts of finely divided solids. Filtering by current methods is impractical as retention on existing filter cloths must be obtained by recirculation or by use of flocculants or precoat filter aids. When once retained and cake formation starts, filter pressure builds up and flow drops off rapidly. The filter must then be cleaned and clarity re-established. Cleaning may be difficult because the particles have penetrated into the filter cloth. This is a time-consuming process which when combined with filter cycles which may be only minutes long, results in uneconomic filtration.

There are also high density slurries which are then thickened, but for which it is desirable to remove additional liquids. Some of these slurries, despite their high densities may have solids which are very fine of a blinding nature resulting the same filtration problems as with dilute slurries.

This paper describes a method of filtration using an expanded PTFE membrane filter cloth combined with back-pulse filtration.

Back-pulse filtration is currently used for such applications as pulp mill white liquor, pigment recovery from drum filters, plating waste filtration, catalyst recovery and others.

This system and GORE-TEX® membrane filter cloths received the 1989 AFS new product award.

Expanded PTFE Membrane Filter Sleeves

GORE-TEX® membrane filter cloths for liquid filtration consist of expanded PTFE membranes bonded to a variety of felts and other fabrics (including fabric woven of expanded PTFE fiber). The expanded PTFE membrane/expanded PTFE fabric filter cloth is inert to practically all chemicals and can be used at temperatures up to 500°F (260°C). These materials combine the filtration efficiencies of membranes with the durability of high strength felts and woven cloths.

Porosity and particle retention of the membrane can be controlled to fit specific filtration requirements. Typically, membranes used in liquid filtration retain 90% of 3 μm down to submicrometer latex beads with corresponding air permeabilities of 8-10 cfm (4-5 l/s) to 1 cfm (0.5 l/s). One of the more open expanded PTFE membranes is compared with conventional filter media in Figure 1. This membrane (Figure 1a) has a nominal particle retention of 90% of 3 μm latex beads without cake formation. Woven polypropylene cloth and a heavily glazed polypropylene felt are shown in Figures 1b and 1c respectively. Note the very small proportion of open area and the large sizes of the openings in these media.

The result of this combination of the GORE-TEX® membrane with strong support materials is filter media with:

high permeability
sub-micron openings
high strength

resulting in benefits of:

high flow rates
high solids retention (filtrate clarity)
long media life
excellent cake release
chemical inertness (for the PTFE membrane/PTFE fabric filter media).

GORE Back-Pulse Filtration

The term "Back-Pulse Filtration" refers to the method of filtration discussed in this paper and the use of the expanded PTFE membrane filter cloth.

Using expanded PTFE membrane filter cloths, it is possible to operate tubular filters as "liquid baghouses" (Figure II). In this type of filtration, pressure differential is limited to 10-15 psi (70-100 kPa), and cake removal is by momentary flow reversal which takes only 1-3 seconds. Filtration starts immediately after pulsing and is, for all practical purposes, continuous. The filter vessel is not drained of liquid during or after pulsing. Suspended solids, which have been consolidated into a cake, settle rapidly to the bottom of the filter vessel where they are discharged as a paste-like slurry. The filter can be operated under vacuum (shown here) or pressure.

The objective of this mode of filtration is to operate at low filter pressures, and short cake removal times. Because cake removal time may be only a few seconds, filtration times may be as short as 2-3 minutes and filter cakes as thin as 0.1 mm.

Expanded PTFE membrane filter cloths for liquid filtration make this type of filtration possible because they give immediate filtrate clarity, high flow rates, and complete cake release.

For this filtration method to operate, it must be possible to form a filter cake on the surface of the membrane which will then release during the back-pulse. Throughput must be sufficient to allow for a back-pulse filtrate usage of approximately 0.05-0.1 gal/sq. ft. of filter area.

Testing

Testing for back-pulse filtration is very simple because of the short filter cycles (usually only a matter of minutes) and low pressures.

Bench scale testing is done with a leaf, Figure III, especially designed by Gore to prevent solids from by-passing the membrane. The leaf is connected to a vacuum receiver. A three-way valve is included in the line for filtration and back-pulsing. This arrangement is shown in Figure IV.

If the leaf tests so indicate, pilot tests can be carried

out using a 0.3 m^2 (3 ft^2) pilot filter which has been designed by W. L. Gore & Associates. It can operate completely unattended for many hours at a time.

Comparison of Back-Pulse Filtration with Conventional Tubular Filter Operation

To better understand back-pulse filtration, it is helpful to compare it with conventionally operated candle filters as shown in Table I. Operation can be divided into downtime, filtration time, and back-pulse (wash) time. Downtime includes time for cake removal plus time to obtain desired filtration efficiency. For back-pulse filtration, cake removal times of only 1.5 to 5 seconds are needed as the cake is simply dislodged from the filter tubes without draining the filter. As the cake is not redispersed, it settles rapidly to the bottom of the filter. For low density cakes, additional settling time of up to 30 seconds may be included. In conventional filtration air is compressed in the filter dome and used to "blow" liquid back through the filter tubes at pressures up to 50 psi. The filter vessel is entirely emptied and must, therefore, be refilled before a second air bump (if necessary) requiring substantial amounts of time and water and, in addition, diluting the slurry discharged from the filter. Cake solids are completely redispersed and, therefore, settle slowly.

Because the PTFE membranes used in the back-pulse system retain particles as low as 0.5 micrometers, filtration efficiency is obtained at once whereas time for desired filtration efficiency in conventional filtration may be 15-30 minutes if recirculation or precoating is required. (Note the substantial volumes of water added to the system for cleaning and precoating of filtrate is not used).

With back-pulse filtration, a high on-line factor is still possible with short filtration times resulting from the filtration of fine, compressible solids and because differential pressures are limited to 70-100 kPa. This minimizes cake adherence to the membrane and, therefore, facilitates cake release. These pressures are compared with 30-50 psi (200-400 kPa) for conventional filtrations.

Underflow concentration for back-pulse filtration can be as thick as can be handled. This will range from 10-12% for metal hydroxides to over 50% for other materials such as flue gas desulfurization scrubber sludge. These concentrations compare with underflow concentrations for conventionally operated tubular filters of 3-7% because the filter vessel is emptied at least once for each cleaning.

Back-pulse filtration can also be effective for the filtration of high density slurries. In pulp mills, a thirty percent caustic solution at approximately 170°F and containing 10-15% suspended solids is filtered in a combination filter settler. The unfiltered white liquor is fed into the side of a vessel 30 - 40 feet tall. Most of the solids settle directly to the bottom of the vessel which is equipped with a rake.

Those solids which do not settle out (making up a slurry approximately two percent by weight) are filtered through a series of tubes suspended from a tube sheet at the top of the vessel. These filters are made up to 1000 ft^2 (90 m^2) area and handle up to 1000 gpm flow. They are replacing gravity thickeners because of their greater reliability and better separation.

Waste Pigment Recovery

The following comparison between back-pulse and a plate separator was made by a Canadian pigment manufacturer. This comparison is supplied here, even though it is not a mining application, because exhaustive pilot tests have been run in both back-pulse filtration and gravity separation of a dilute suspension of compressible solids. As a result of this comparison, which was based on pilot tests, the company cancelled an order for a plate separator and placed an order for a back-pulse filtration system.

This company has a 250 US gpm (1 m^3/min) waste stream from their filter presses. Solids content ranges from several hundred to 6,000 ppm and averages 500 ppm.

Plate separator pilot tests indicated an overflow of 30 to 50 ppm which is above the Canadian environmental limit and would, therefore, require an additional polishing filter. A sand filter was considered, but rejected because it was only good down to 0.45 micrometer particles. Also, the sand filter, being a depth filter did not offer a positive separation.

Pilot tests were run, therefore, using a 3 ft^2 (0.3 m^2) Gore automated pilot filter designed and constructed by W. L. Gore & Associates. Test results and operating conditions are summarized below.

Cycle Time, min	
Filtration	30
Back-pulse and cake settling, min	0.5
Filter pressure, lb/in^2 (bar)	15 (1)
Filtrate solids, ppm	5
Cake solids, %	20
Feed slurry, average, ppm solids	500
Flow rate, gal/ft^2/min (m^3/m^2/hr)	0.6 (1.4)
Flow (total), gal/min (m^3/min)	250 (1)
Recovered pigment, lb/ft^2/yr (kg/m^2/yr)	1,000 (4800)
Total, lb/yr (kg/yr)	270,000 (120,000)

Recovered solids concentration is estimated at 10 to 20 percent depending on filter operation. This is a waste minimization factor of over 400 to one for the 20% solids cake and the 500 ppm feed. Because flocculants are not used, the pigment can be recovered and is considered to have a value of approximately $3.00 Canadian per kilogram.

Equipment costs for the plate separator and the filter are summarized below in Canadian dollars.

Equipment Costs	U.S. Dollars
Plate separator	83,000
Polymer addition	17,000
Structural improvements	42,000
floor reinforcement	
roof extension	
Polishing filter	80,000
Total	222,000
Back-pulse filter (skid mounted)	142,000
450 ft^2	
180 tubes	
Equipment savings	80,000

BACK-PULSE CROSS-CLEAN FILTRATION OF THICK SLURRIES

In the description of Back-Pulse Filtration at the beginning of this report, cake pulsed from the filter tubes settles to the bottom of the filter and is withdrawn as a dense underflow. For this to happen, there must be a significant difference in density between the liquid and the cake. Also, the liquid must be of low enough viscosity to allow the cake to settle. None of these conditions are present with high density slurries. Another technique is, therefore, proposed.

In the proposed technique a continuous flow past the surface of the filter tubes is maintained so that as the cake is pulsed from the surface of the tubes it is carried away, leaving the surface of the tube clean and ready for the next filter cycle. Because cake removal is by cross flow it is essentially instantaneous and no settling time is required. It should be possible, therefore, to operate on two - three minute cycles.

It is envisioned that filtration would take place in an open vessel with the filter tubes suspended from a series of subheaders and headers. Flow would be induced by pump suction on the inside of the tubes. Individual headers could be isolated and removed for replacement of the expanded PTFE filter sleeves.

Waste Bauxite Slurry Concentration

Following is a description and discussion of leaf tests run on waste bauxite slurry, the purpose of which was to determine the filter area required to increase the slurry's solids content from approximately 40% to a desired level of 50%. It is possible that this same technique could also be applied to some kaolin slurries.

Test Method

Tests were run on the waste bauxite slurry as received (containing 41.4% suspended solids) and diluted to 36.8% suspended solids. Two tests were run on the slurry as received with identical results and one run on the diluted slurry.

Because the filtrate volume was very low times were recorded when filtrate first appeared at the outlet of the leaf (35 ml volume) and at the 25 ml mark in the receiver. The leaf and vacuum lines were then drained into the receiver for a final reading of 70 ml. The cake was discharged onto a dish for wet and dry weights.

Test Results

Test conditions for the two slurries are summarized in the table below:

Test Conditions

	Slurry 1	No. 2
Feed slurry solids, %	41.4	36.83
Vac., in Hg	22	22
Temperature	ambient	ambient

Tests Results

	Slurry 1	No. 2
Filtrate times, sec		
at leaf exit (35 ml)	27	20
at end of test (70 ml)	160	115
Cake weight, wet, g	174	
Cake moisture, %	41.8	42.8

Discussion

Filtrate volume versus time for the 35 and 70 ml levels was plotted on log-log paper for the 41.4% slurry. Data for longer filter cycle times was then obtained by extrapolation. Due to the nature of the cake and the mode of operation, back-pulse volume is estimated at 0.05 gallons per square foot. This is based on our experience with these conditions. 0.05 gal/sq. ft. back-pulse volume is equal to 15 ml for the test leaf. This amount was deducted from the filtrate volumes for rate calculation. Time to recover the 15 ml was estimated by extropolation and was found to be less than two seconds. Pulse time is only a few seconds and is insignificant for these calculations. No time is required for settling. Filtration rates versus time are shown in the table below:

Filtration Rate Versus Time
(Undiluted Slurry)

Time Seconds	Rate gal/sq.ft./min.
90	0.08
120	0.07
160	0.067
240	0.052
300	0.047

A material balance can be obtained from the following information:

Cake weight, g	174
Cake moisture, %	41.8
Cake moisture, g	72
Cake solids, g	101

The filtrate weight of 70 g (70 ml) added to the wet cake weight equals the weight of the slurry filtered which, divided into the cake's solids weight of 101 g equals 41% which, in turn, was the measured solids value of the slurry.

Amount of Liquid to be Removed

The objective of increasing a slurry density can be phrased as the question: "How much liquid must be removed from the existing slurry to increase its solid content from its existing level to a desired level?" This can be calculated by subtracting the mass liquid in the desired slurry from the mass liquid in the original slurry using the following formula:

$$\Delta W_l = W_s/S - W_s - (W_s/S_d - W_s)$$

which can be simplified to:

$$\Delta W_l = W_s\ (1/S - 1/S_d)$$

where:

ΔW_l = mass liquid to be removed, lb/min

W_l = mass feed liquid, lb/min

W_s = mass feed solids, lb/min

S = mass fraction feed slurry solids

S_d = desired mass fraction slurry solids.

Comparison of Rates for Diluted and Undiluted Slurries

The diluted slurry resulted in a filtration time of 115 seconds for 70 ml filtrate versus 160 seconds for the undiluted slurry. ΔW_l was calculated for both slurry concentrations as shown below:

The amounts of liquid to be removed for a 50% slurry are:

for the 41.4% slurry = 0.42 W_s and for the 36.8% slurry ΔV_2 = 0.7 W_s.

The ratio for the two slurries (0.7 divided by 0.42) = 1.7. Thus, the flow rate for the diluted slurry would have to be 1.7 X that for the undiluted slurry for equal results. Since this was not the case, no calculations were made on the diluted slurry.

The following calculations are based on a 40% solids slurry and a solids rate of 1000 lb/min.

The amount of liquid to be removed to increase solids from 40% to 50% is calculated as follows from the above formula:

ΔW_1 = 1000 (1/0.4 - 1/0.5)
= 500 lb/min
= 60 gpm, the rate of liquid to be removed from the bauxite residue slurry at a solids rate of 1000 lb/min.

The filter area required for the above flow can be obtained by dividing this flow by the filtration rates as shown in the Table "Filtration Rates Versus Time" (above).

Cake volume is calculated below.

The mass of dry cake, W:

$$= \frac{\rho \Delta W_1 S}{1 - S/S_c}$$

$$= \frac{500 \times 0.41}{1 - 0.41/0.58}$$

= 720 lb/min

Cake specific gravity, ρ_c:

$= \rho\ (1 - S/\rho_s) + S$

where:

ρ = specific gravity of liquid
ρ_s = true specific gravity of solids

Thus:

ρ_c = (1 - 0.58/2.3) + 0.58

= 1.33 g/cc

$$\text{Cake volume} = \frac{720 \text{ lb/min}}{62.5 \text{ lb/cu.ft.} \times 1.33}$$

= 8.7 cu. ft./min

Cake volumes and areas are shown in the following table for 2, 4 and 5 minute filter cycles.

Filter Cycle, min	Filter Area sq. ft.	Number of Tubes*	Est. Equipment Capital Cost at $75/sq. ft.	Cake Volume
2	860	220	86,000	2.2
4	1200	306	120,000	3.1
5	1300	332	130,000	3.3

*6'L X 2.5"D, 3.92 ft^2

Feed volume is estimated at 40 cu. ft./min/1000 lb feed solids by the following method:

$$\frac{1000 \text{ lb/min solids}}{0.4 \text{ solids fraction} \times 62.5 \text{ lb/cu. ft.} \times 1.06}$$

The exit volume of 50% solids is 32 ft^3/1000 lb feed solids.

Equipment

With proper baffling, it should be possible, with a combination circular and downward flow, to sweep the cake off the tubes after each pulse and prevent reintrainment of cake on the tubes during subsequent pulses. Recirculation of the slurry may be considered for increased cleaning velocity.

A rough indication of costs to increase waste bauxite slurry from 40% to 50% solids is shown below.

1000 lb/min solids in a 40% slurry is equivalent to a flow of 300 gpm or 1.6×10^8 gallons per year.

A four minute cycle is assumed which results in a unit flow of 0.05 gal/sq. ft. filter area/min.

Power costs were taken from estimates for water filters of similar areas.

Operating Costs	
Capital $240,000 (equipment costs X2)	
Amortization (10 year straight line)	24,000
Expanded PTFE membrane filter sleeves	24,000
(two changes/year, 1200 ft^2 area at $10 ft^2)	24,000
Labor	
Two sleeve changes/year	
X 300 sleeves	
= 600 sleeves/year	
X 0.25 hours/sleeve	
= 150 hours*	
X $15/hours	
=	22,000
Operation labor (1/8 man @ $15/hour)	16,000
Power	
100,000 kwh/year	
X $0.08/kwh	
=	8,000
	74,000
divided by 1.6×10^5 k gal/year	
=	$0.46/k gal

*Although 150 hours are alloted for changing sleeves, no down time is required because the sleeves will be changed while the filter is running.

Conclusions

Expanded PTFE filter cloths can be used in conjunction with back-pulse filtration to filter dilute concentrations of finely divided solids.

Bench scale tests indicate that thick concentrated slurries can be further dewatered.

Sufficient solids can be recovered from some dilute streams and provide a payback of less than one year.

References

Smith, Gordon R. S., "Gore Back-Pulse Filtration Compared to Gravity Thickening." American Filtration Society Annual Meeting, March 27-29, 1989, Pittsburgh, Pennsylvania.

Smith, Gordon R. S., "Comparison with Gravity Separators of a New Filtration Method for Fine Particle Filtration." Society of Mining Engineers Annual Meeting, February 27 - March 2, 1989, Las Vega, Nevada.

Smith, Gordon R. S. & Rinschler, Craig R., "The Application of The New Gore Back-Pulse Filtration for Gold Recovery with High Silver/Gold Ratios." October 28, 1988, Perth International Gold and Silver Conference, Perth, Western Australia.

0683e

TABLE I
BACK-PULSE FILTRATION COMPARISON WITH CONVENTIONAL CANDLE FILTER FILTRATION

Operating Step	Back-Pulse Filtration	Conventional Filtration
Downtime		
Cake removal	1.5-30 sec.	5-10 min.
Start-up time for desired filtration efficiency	0-5 sec.	15-20 min.
Total downtime	1.5-35 sec.	20-30 min.
Water requirements Precoating	0	1500 gal* (5700 l)
Cleaning	0	1500 gal (5700 l)
Total		3000 gal (11000 l)
Minimum practical filtration time	2-5 min.	3-5 hrs.
Differential pressure to get above filtration time	(70-100 kPa) 10-15 psi	(200-400 kPa) 30-50 psi
Backwash pressure for cake removal	(20-35 kPa) 3-5 psi	(150-300 kPa) 20-40 psi
Underflow slurry concentration	10-50%	3-6%

*45m^2 (500 ft^2)

FIGURE I

GORE-TEX MEMBRANE VS. POLYPROPYLENE WOVEN AND FELT FILTER CLOTHS

FIGURE II

W.L.GORE & ASSOC., INC.
FILTRATION PRODUCTS

BACK-PULSE FILTRATION

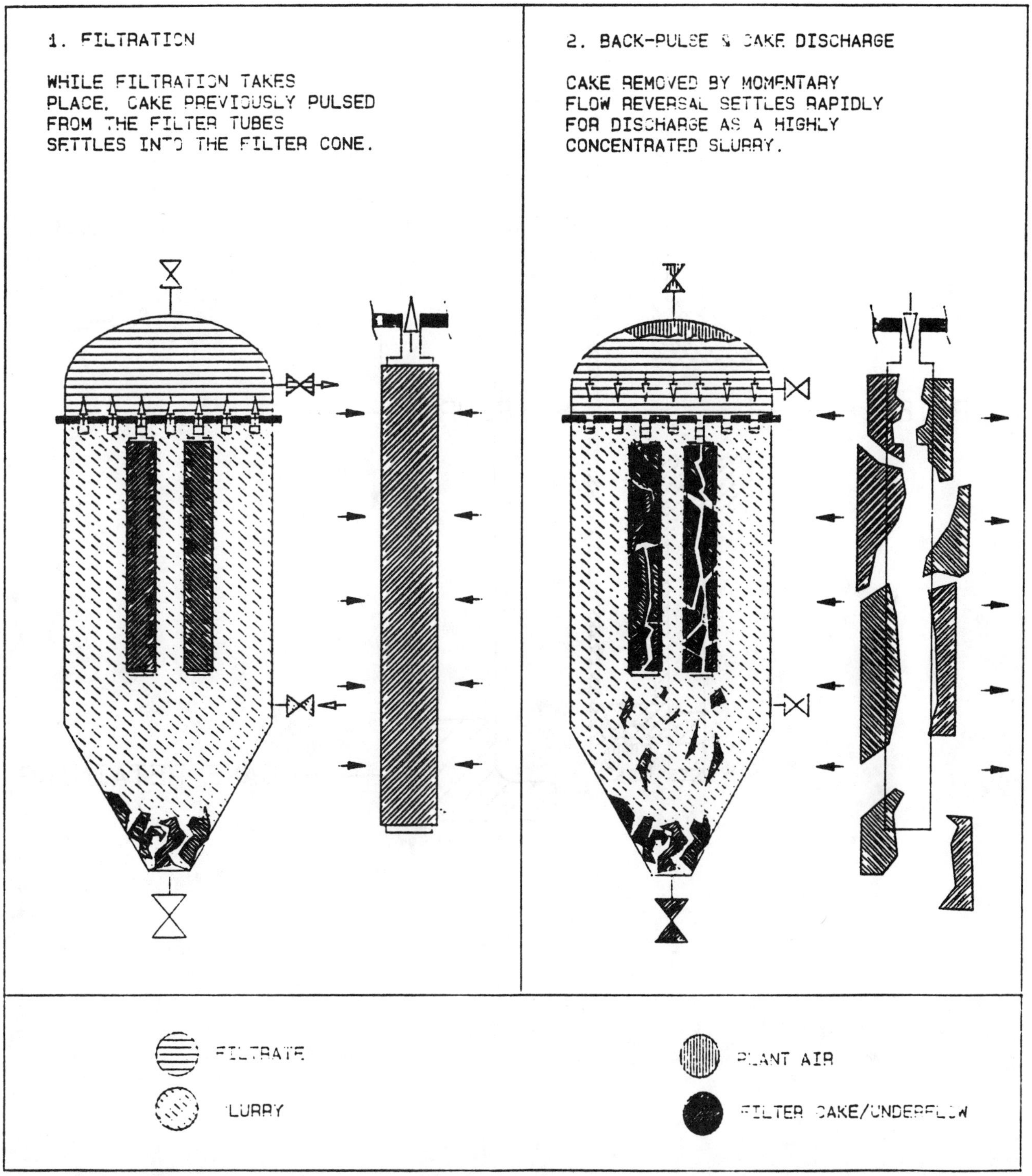

Figure III

TEST LEAF

DESIGNED TO PREVENT SOLIDS BY-PASSING THE MEMBRANE

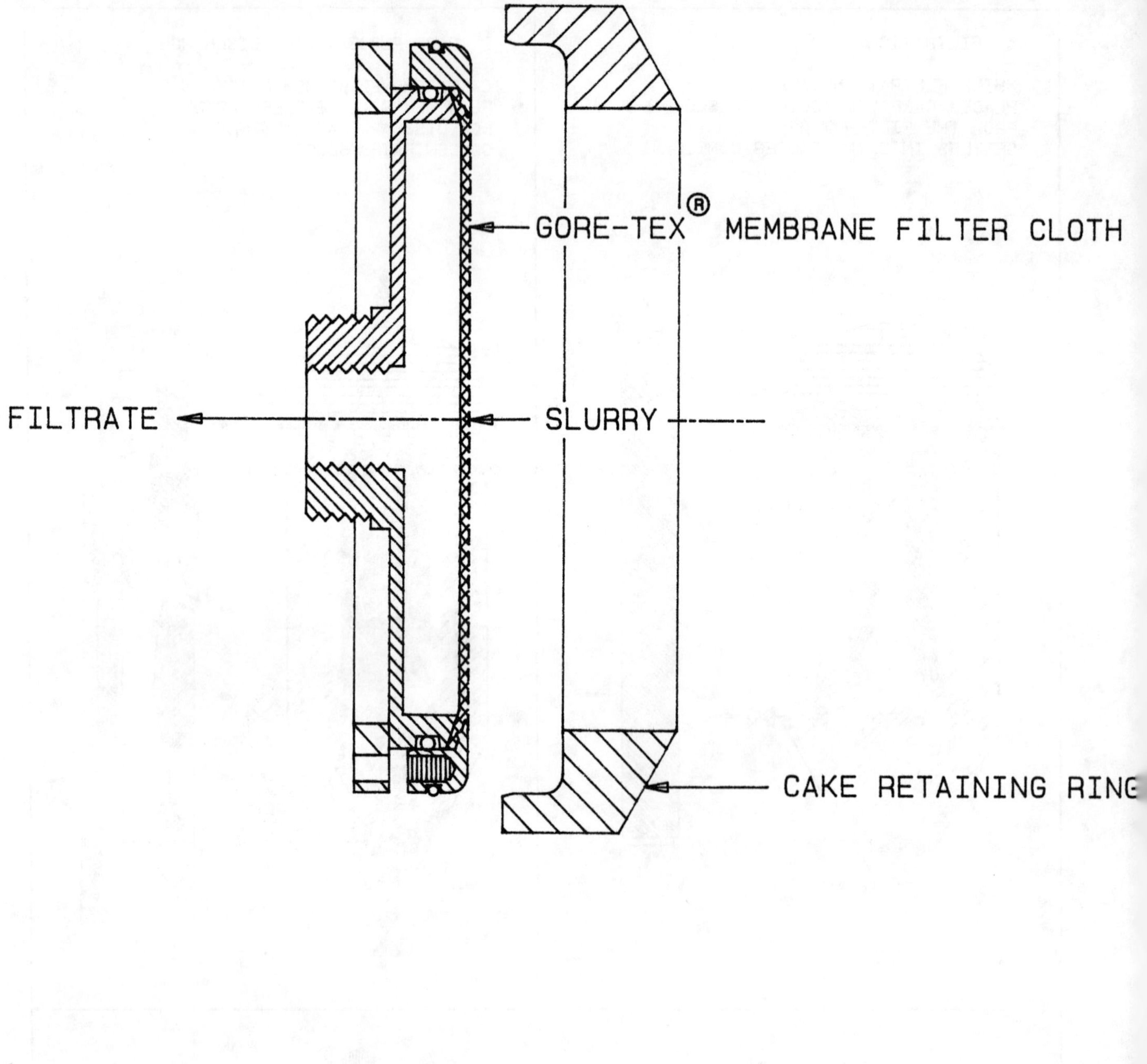

FIGURE IV

GORE W.L.GORE & ASSOC., INC.

FILTRATION PRODUCTS

SUGGESTED LAB INSTALLATION FOR LEAF TESTS

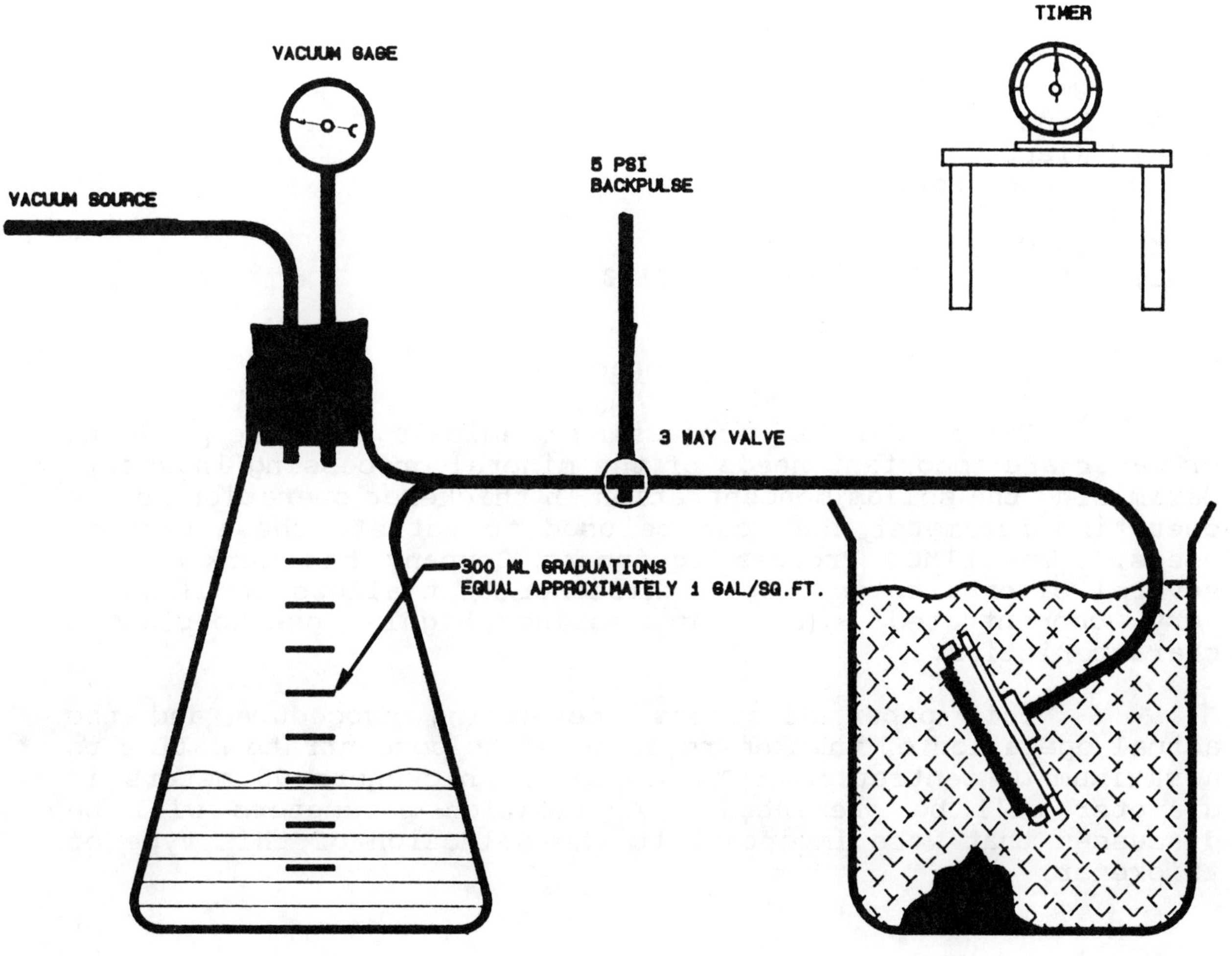

USE NON-RESTRICTIVE CONNECTIONS

HIGH TORQUE THICKENERS FOR PSEUDO-PLASTIC SLURRIES

Robert Emmett
Donald King
Ronald Klepper
EIMCO Process Equipment Co.
P.O. Box 300
Salt Lake City, UT 84110
(801) 526-2133, 526-2189, 526-2143

ABSTRACT

The recovery of processing chemicals, soluble products or water are important needs of the mineral processing industry. Maximizing the solids concentration in thickener operation is one operating parameter that can be used to satisfy these process needs. The EIMCO Process Equipment Company has worked with several clients to develop a thickener that allows handling of very concentrated solids processing highly pseudo-plastic characteristics.

This paper discusses the sizing procedure and the actual operation of thickeners designed to concentrate solids to a maximum concentration. Operation of thickeners 90 meters in diameter will be presented. Application parameters will be discussed that were important to the selection of this type of thickener.

Introduction

High solids thickeners have been proposed for more than 15 years by several groups of people working from different perspectives. One group proposed stacking thickened mineral tailings (Robinsky 1978). Much work was conducted by the British Coal Board using deep cone thickeners to dispose of coal refuse (Abbott 1979).

The greatest utilization of the technology has been in the alumina red mud processing (Chandler 1983, Want 1987). The deep cone thickener concept without rakes has had success on low tonnage throughput due to geometric requirements of the device.

EIMCO PEC combined process and mechanical technology to design and build a 90 meter diameter high solids thickener for ALCOA's Pinnjara project in Western Australia. The development work went through laboratory and pilot plant testing to generate the necessary data to design the equipment.

Now that this thickener has been operating for more than three years other opportunities are available for use of this technology.

Process Development

In order to determine the dimensions and mechanical strength of a thickener designed to produce an underflow of very high consistency, three separate functions must be studied:

1) Optimum conditions for flocculation;

2) Rate of compression or densification of the solids; and,

3) Torque requirements due to the drag on the mechanism produced by the thick slurry.

Approximate values for these functions can be estimated using laboratory techniques, but pilot plant test work must be performed to more accurately establish the depth, diameter, and torque rating of the mechanism.

Efficient flocculation is the key to producing the desired performance in the thickener. If the average particle size is smaller than 10 micrometers in diameter, the feed slurry apparent viscosity can increase to create a problem with flocculant mixing. Laboratory test work can be used to determine the optimum conditions for flocculation, including type of flocculant, dosage, and ideal feed solids concentration.

Initial settling rate is usually a good indicator of flocculation performance. However, some large flocs settle more rapidly than do others, but will not attain the same ultimate solids concentration. Both initial settling rate and final underflow density must be considered to select the best flocculant.

The flocculation performance can be evaluated alternatively by plotting the solids "flux" (the product of solids concentration and initial settling rate) versus the feed solids concentration. This is shown in Figure A. If the optimum solids concentration creating the maximum flux is lower than the expected thickener feed concentration, provision to dilute the feed slurry prior to flocculation must be included in the design.

Laboratory 2-liter graduated cylinder tests equipped with slow speed picket rakes) can be used to establish the attainable underflow concentration. The slurry must be flocculated under the optimum conditions and the final depth of thickened slurry must be greater than about 8 inches (20 cm), for most mineral pulps.

The amount of slurry to produce the final depth will almost always exceed the volume of the graduated cylinder. Continuous feeding and simultaneous flocculation, overflowing the clarified supernatant as the cylinder fills with partially thickened slurry, is an effective method to accumulate sufficient solids.

This small scale testing is subject to caution in measuring the ultimate solids concentration. Insufficient compressive force in the solids will yield lower than actual results. Classification of coarse particles (where 15 or 20 wt% are greater than +200 mesh) in the bottom of the graduated cylinder can yield higher than actual results.

A batch settling curve of the compressed solids can be used to calculate a densification rate expressed as the product of the average volume during compression and the time to reach the final density, divided by the weight of solids in the sample.

This "Unit Volume" (m^3 of compression zone per ton per day of solids), unlike unit area, is a function of depth of pulp. An increase in the compression zone depth by a factor of 10 can result in an increase in required detention time of 3-5. This means that a given thickener area will have increased capacity or an increase in ultimate solids concentration with greater pulp depth.

Scaling up from laboratory tests of 20 cm solids depth to full scale depth of several meters is subject to possible error. Therefore pilot plant data with pulp depths of 1 meter in compression provide operating data to more accurately predict the full scale performance at pulp depths of 3 to 4 meters. This approach is illustrated in Figure B.

Calculating the torque required for a thickener of this type can begin with bench-scale tests, based on rheological measurements of the pulp at its expected final density. If one assumes that the force exerted on the structure is uniform across the entire diameter and is independent of velocity, a torque requirement can be calculated as the sum of the forces acting on each component. Direct scale-up using this value is subject to error, and, therefore, pilot plant or large scale data are far more useful.

Torque scale-up can be based on a ratio of the diameters to the third power, (rather than the usual diameter squared), when data are available from a large pilot plant thickener or a full-scale unit. The procedure is illustrated in Table 1, showing projected and actual torques calculated by these methods.

Equipment Development

Through laboratory and pilot plant testing the diameter and depth of the prototype unit was established at 90 meters diameter with a side wall depth of 3 meters and a center depth of 9.5 meters.

Unlike the solids in most conventional thickeners, the settled solids required continuous raking to transport the viscous slurry to the discharge points at the center of the tank since the high solids concentration exhibited very little fluidity. Underflow pumps were located within a 6 meter "Caisson" at the center of the thickener due to the very viscous nature of the slurry exiting the thickener.

Minimal steel structure was required to create the least resistance of the rakes traveling through the viscous slurry as well as to contend with the inevitable deposition of scale on all submerged components. Even minimizing the resistance of the rake structure, a torque requirement of 13,600,000 Newton-meters (10,000,000 ft-lbs) was determined to be the continuous torque required to turn the rakes.

An EIMCO Swing Lift Thickener with cables pulling the rakes was modified to meet the design requirements. The modifications were mainly in the central support mechanism where structural mechanical strength was needed as well as minimal obstruction to flow so that the thickened solids could be removed. There were two long rake arms and two short rake arms to have adequate raking capacity to physically convey the viscous slurry to the four discharge points.

The drive mechanism was a tractor located at the tank perimeter that used a gear and pinion mechanism or a cog type traction device to develop the torque to move the rakes. A maximum speed of up to 3 rph was selected.

The feedwell mounted concentric to the Caisson was designed utilizing the difference in specific gravity of the feed slurry and the clear liquid at the surface of the thickener, to allow clear liquid to flow into the feedwell diluting the feed to a lower solids concentration, allowing optimum flocculation.

Start up of the 90 meter diameter unit had no major mechanical problems and the anticipated adjustments to solids concentration and flocculant dosage to obtain the optimum flux by proper flocculation were made.

A second unit 75 meter in diameter has been designed and installed using the same process and mechanical considerations. The process performance has been as predicted by the laboratory and pilot plant testing predicted.

Comparison of Various Types of Thickeners

Table II exhibits the relative comparison of conventional thickeners using flocculant at various amounts to the high rate thickener that uses flocculant very efficiently to high solids thickeners that use flocculant efficiently and use deep solids to create higher ultimate solids concentrations.

Deep cone thickeners developed by the British National Coal Board and ALCAN have exhibited higher solids concentrations than the conventional or high rate thickeners.

High solids thickener technology both in process and mechanical design aspects has been utilized to produce consistently high solids concentration in performance and has functioned on the largest scale in the world.

The alumina red mud thickening application has proven to have the utmost characteristics to which this technology applies. There certainly are other applications where water management is very important and where it is important to minimize the quantity of liquid that is lost with the solids.

The properties of the thickened solids must be such that the slurry remains fluid even though it exhibits pseudo-plastic in appearance. To have these properties the size distribution of the solids will typically be 100% minus 200 mesh and 80-90% minus 10 micron. This is typical of high clay content solids or materials where very fine grinding is used to liberate desired minerals.

TABLE I
TORQUE DETERMINATION

1) Based on $T=Ky_sD^3$
Measured operating torque for a 30 meter Ø thickener is 200,000 ft-lbs,
Scale-up to a 90 meter Ø thickener:
$T= [90/30]^3*200,000=5,130,000$ ft-lbs

Measured operating torque for a 38 meter Ø thickener is 328,000 ft-lbs,
Scale-up to a 90 meter Ø thickener:
$T= [90/38]^3*328,000=4,310,000$ ft-lbs

2) Based on summation of forces on submerged components:
Forces measured on a 38 meter Ø thickener averaged 70 lb/ft^2

Total area of the submerged components for the 90 meter Ø thickener multiplied by the radius equaled 80,600 ft^2*ft

The total torque=70 $lb/ft^2*80,600\ ft^2*ft=5,642,000$ ft-lbs

3) The torque installed into the 90 meter Ø thickener was selected to be 10,000,000 ft-lbs.
Actual normal operating torque = 4-5,000,000 ft-lbs
Actual maximum operating torque = 7,700,000 ft-lbs

TABLE 2

Comparison of Conventional, High Rate, and High Solids Thickeners

Equipment Type	Unit Area m2/mtpd	Underflow Concentration wt% Solids	Flocculant Dosage rams/tonn	Torque Factor k*
Base and Precious Metal Tailings				
Conventional Thickener	0.25-1.0	50-60	0-5	250
High Rate Thickener	0.02-0.1	45-55	10-55	400
High Solids Thickener	0.20-1.0	60-70	50-100	400
Laterite, Red Mud				
Conventional Thickener	0.8-1.2	20-40	15-35	200
High Rate Thickener	0.2-0.5	20-35	30-50	400
High Solids Thickener	0.8-1.2	40-50	40-80	1500-2500
Deep Cone Thickener (No mechanism)	0.5-1.0	30-45	80-100	-
Coal Refuse				
Conventional Thickener	0.4-0.8	35-40	50-100	250
High Rate Thickener	0.04-0.06	30-35	100-200	400
Deep Cone Thickener (with stirring mechanism)	0.15-0.25	60	150-350	3,500

* Based on formula: T=k D2; torque in Newton-meters, thickener diameter in meters.
Abbott, 1973; Abbott, 1979; Chandler, 1983; Robinsky, 1989)

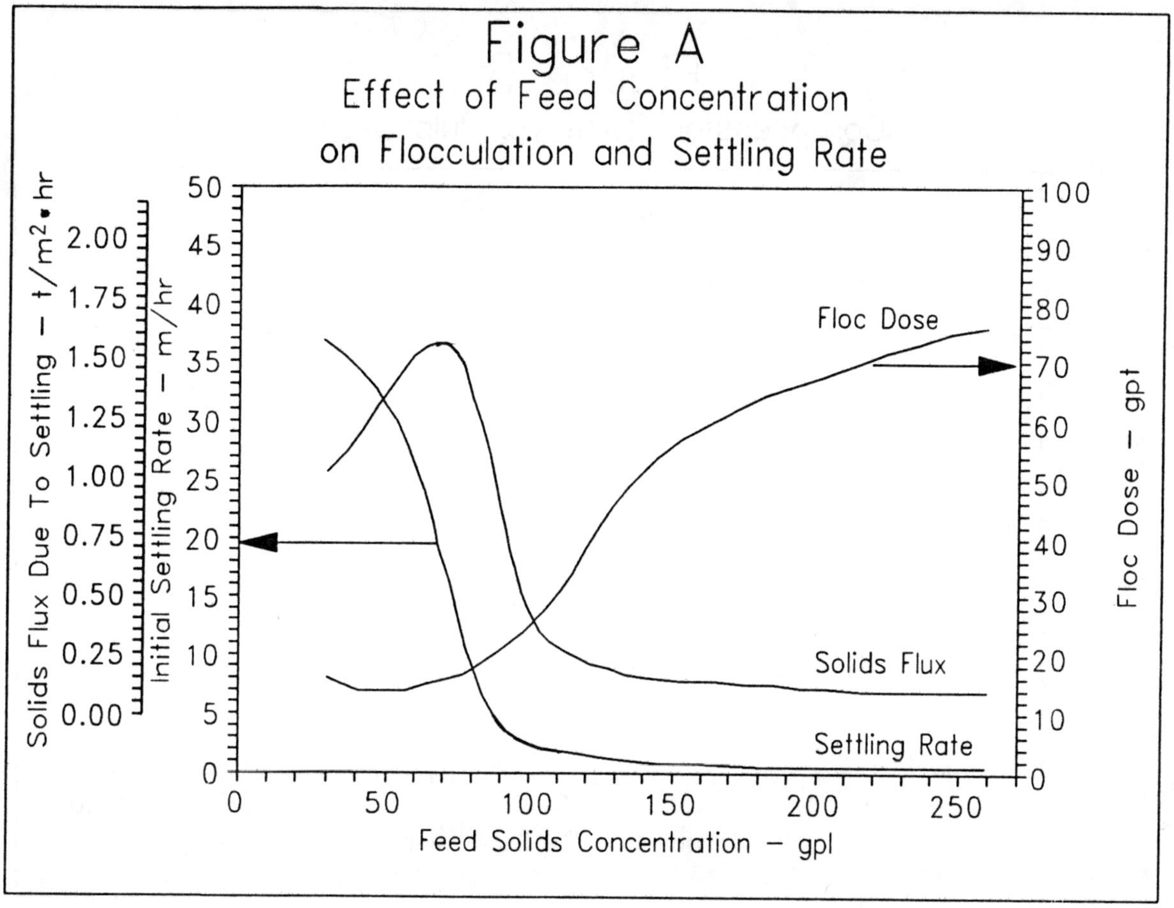
Figure A
Effect of Feed Concentration
on Flocculation and Settling Rate
Solids Flux Due To Settling – t/m²•hr
Initial Settling Rate – m/hr
Floc Dose – gpt
Feed Solids Concentration – gpl
Floc Dose
Solids Flux
Settling Rate
0.00
0.25
0.50
0.75
1.00
1.25
1.50
1.75
2.00
0
5
10
15
20
25
30
35
40
45
50
60
70
80
90
100
150
200
250

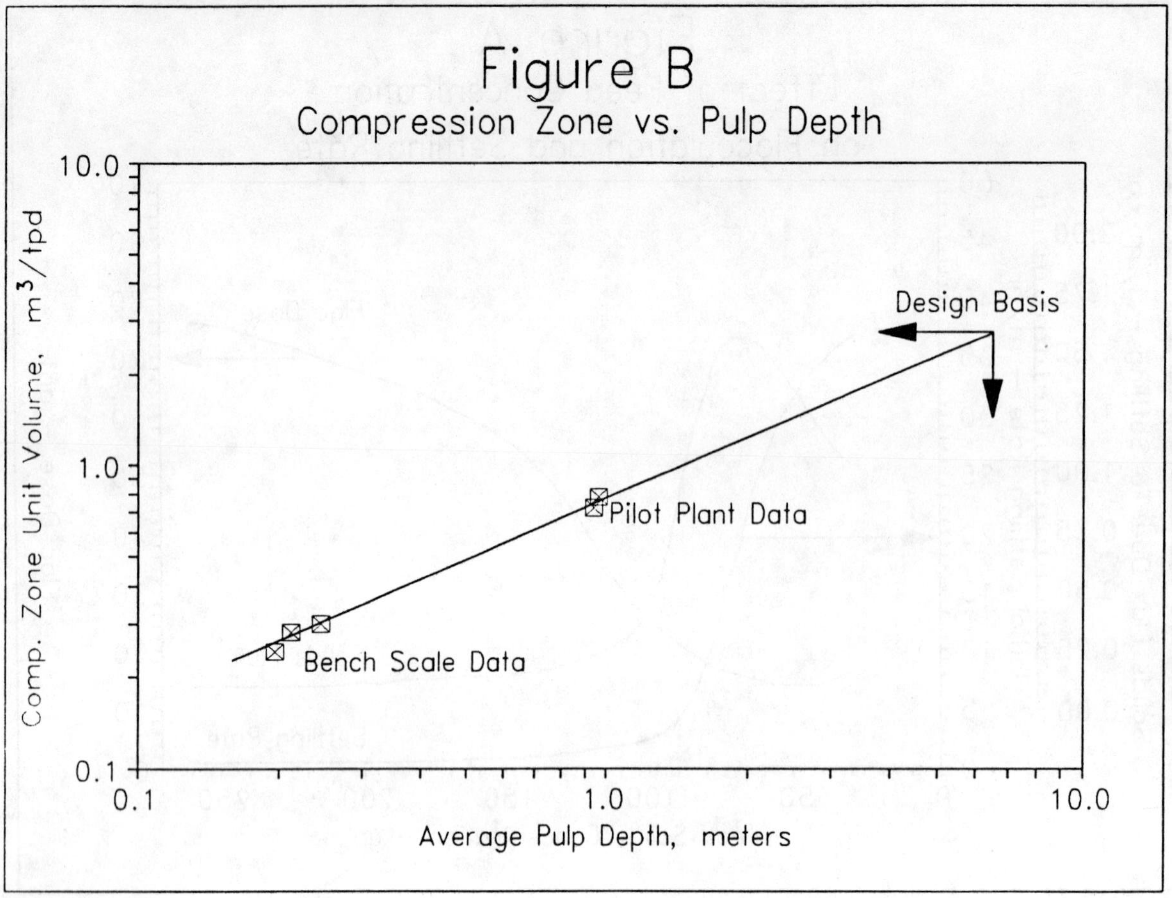

Figure B
Compression Zone vs. Pulp Depth
10.0
1.0
0.1
Comp. Zone Unit Volume, m3/tpd
Design Basis
Pilot Plant Data
Bench Scale Data
0.1
1.0
10.0
Average Pulp Depth, meters

SOLID-LIQUID SEPARATIONS OF SLURRIES OBTAINED FROM THE LEACHING OF PHOSPHATIC CLAY WASTES

J. G. Davis, Chemist, G. M.
Wilemon, Research Chemist, and
B. J. Scheiner, Research Supervisor,
U. S. Bureau of Mines
Tuscaloosa Research Center
P. O. Box L, University of Alabama Campus
Tuscaloosa, AL 35486

The Bureau of Mines has studied the extraction of phosphate values from fine-particle wastes generated during conventional mining and beneficiation of phosphate ore. Investigations have centered on leaching dried wastes using sulfuric acid (H_2SO_4) in the presence of methanol. Separation of insoluble gangue from desired leach liquor in these slurries is challenging because of the fine grain size of the unleached residues. Vacuum filtrations of the slurries yield filter cakes that are tight and difficult to wash, resulting in a loss of product and handling problems. A combination of flocculating agents has been discovered that enhances the settling properties of the solids in these slurries, facilitating solid-liquid separation. The effect of parameters such as polymer dosage and order of mixing on the efficiency of flocculation and settling are discussed.

Introduction

During hydrometallurgical processing of ores and concentrates, solid-liquid separation of aqueous slurries is commonly required. When these slurries contain clay particles, the solid-liquid separation is often difficult. For example, the beneficiation of Florida phosphate ore generates a waste stream containing mostly clay and having a solids content of approximately 2 to 6 pct[1-3]. This material, called phosphatic clay waste, is discarded into waste ponds. Because of the colloidal character of the waste, it settles very slowly, and after years it still may contain more than 70 pct water. Waste disposal using this method ties up large quantities of land and water and is also an environmental concern. Finding techniques to rapidly dewater these aqueous clay wastes has been the topic of several Bureau of Mines investigations[4-5].

In addition to clay, phosphate wastes contain up to one third of the phosphate present in the unbeneficiated ore. Recent efforts at the Bureau of Mines to recover the phosphate values from these wastes involve leaching with H_2SO_4 in the presence of methanol[6-8]. These leach slurries contain insoluble clay components that must be separated from the crude acid product. Separation of these very fine particulates from the leach liquor is challenging. Filtration of the slurries generates a tightly bound filter cake that is extremely difficult to wash. The inability to properly wash the cake results in low product recoveries. Therefore, filtration alone is not a facile method for separating the crude acids from the leach tails.

Previous research at the Bureau of Mines has shown that polyethylene oxide (PEO) enhances the filtration of leached clay slurries in both aqueous and alcohol media. Filtration rates based on filtration time alone exceeded 500 $lb/h{\bullet}ft^2$ in some cases, but again, the filter cakes obtained were very tight and difficult to wash[8]. However, if sufficient agglomeration and settling of the gangue could be achieved, the solids could be washed using a series of flocculation and decantation steps. Then, a single filtration step to remove the last vestiges of acid could be employed and no cake washing would be required. To determine if thickening or flocculation with counter-current decantation could replace the need for filtration to recover acid values, the effects of polymers on the settling rates of leach slurries were studied.

Materials and Procedures

Sample Description

The samples used in this study were obtained from reclaimed phosphatic clay waste ponds in central Florida. The solids content of the as-received ore was approximately 85 pct. The samples were air-dried for several days to remove surface water. Conventional grinding techniques were used to break up clay agglomerates so all the ore passed 28-mesh.

Settling Test Procedure

Leaching experiments were conducted in a 3-neck, 2,000-mL round-bottom flask that was equipped with an overhead stirrer, a reflux condenser, and an automated dispenser which was used to meter H_2SO_4 into the reaction vessel. In each experiment, 600 g of phosphatic clay waste were slurried in 1200 mL of methanol. The desired quantity of technical-grade H_2SO_4 (93 pct) was added dropwise via the automated dispenser to the stirring slurry over a 30-min period. The reaction mixture was allowed to stir for an additional 30 min following the acid addition. When the leaching step was completed, 200-mL samples of the slurry were siphoned into 250-mL mixing cylinders. Each cylinder received approximately 80 g of ore corresponding to a percent solids of 36 to 38. Polymer dosages were added to the cylinders using 5- or 10-mL syringes and the cylinders were inverted 5 times after each polymer addition to ensure even distribution of the flocculants. The tests were conducted at ambient temperatures and atmospheric pressure. Flocculation of the slurries was measured by evaluating the rate of settling. The tests were timed over a period of 30 min and the height of the interface as a function of time was recorded.

Filtration Test Procedure

For this series of tests, a 500-mL flask was charged with 100 g of ore and 200 mL of methanol. After the leaching step, the samples were transferred into 1000-mL beakers and gently agitated as the desired polymer dosages were added. The flocculated slurries were vacuum-filtered via a Buchner funnel that was equipped with an on/off valve. Tests at filtration times of 30, 60, and 120 s were conducted. The filtrate from each test was collected and its weight and volume were recorded. Filtration characteristics were evaluated as a function of the amount of filtrate collected per unit of time.

Results and Discussion

In an effort to determine if thickening could be used for the solid-liquid separation of acid-alcohol leached slurries, screening tests were conducted using the following methanol-soluble polymers: polyvinyl acetate, PEO, quaternized polyamines, hydroxypropyl cellulose (HPC), and polyethylene glycols. The polymers were added to the slurries singularly and in various mixtures in dosages ranging from 0.002 to 0.75 lb/ton. Different degrees of settling enhancement were observed, but the most significant results were noted when the slurries were treated with PEO or a mixture of PEO and HPC. Large strong flocs were formed that settled rapidly into an agglomerated mass; acid could then be easily decanted from the insoluble clay residues. The flocculated mass also continued to release fluid for several minutes. Although flocculation could be achieved using very small amounts of the two polymers, the most dramatic results were observed when the dosage of each polymer was at least 0.10 lb/ton or more, when the HPC was added prior to the PEO, and when the polymers were added in a 1:1 ratio. After decantation of the product acid, the percent solids of the leached tails ranged from 60 to 75 pct. The tails were repulped with methanol and reflocculated to recover additional phosphate values. Smaller polymer dosages were required to flocculate repulped slurries than were needed for initial slurries.

Since promising results were observed during the screening tests using PEO or a combination of PEO and HPC, tests to determine the settling rates of acid-alcohol leach slurries flocculated with these polymers were conducted. The polymers were added to the slurries as methanol or aqueous solutions. The results of these tests are shown in figure 1. The data indicate that overall, the polymer combination performed better than PEO alone. A slight increase in the settling rate and a greater degree of solids consolidation were also observed when the HPC was added as an aqueous solution instead of a methanol solution. These results were surprising because earlier attempts to use water-soluble polymers as filtration aids had been unsuccessful.

Tests to determine the effects of the HPC-PEO polymer combination on the filtration characteristics of the leach slurries were also conducted. Multiple tests using a polymer dosage of either 0.2 lb/ton PEO or 0.1 lb/ton HPC with 0.1 lb/ton of PEO were run. These experimental results are given in table I.

Collection of the product acid by filtration after flocculation with equal dosages of HPC-PEO or PEO alone indicated that the addition of the cellulose compound to the slurry prior to the addition of PEO appears to augment floc stability. The table shows that more filtrate was collected when the combination was used than when PEO was used alone, and that the amount of filtrate collected per unit of time varied over a larger range after PEO treatment than it did after treatment with HPC-PEO. These data seem to indicate that the PEO flocs are not as strong as those obtained using the HPC-PEO mixtures and that the PEO flocs will fall apart if too much shear is applied during mixing.

Based on the fact that HPC could be used as an aqueous solution to enhance slurry settling, the use of the analogous, water-soluble hydroxyethyl cellulose (HEC) was investigated. The use of HEC in conjunction with PEO as a settling or filtration aid exhibited behavior that was very similar to that obtained when the HPC-PEO mixture was used. However, figure 2 shows that the HEC-PEO combination produced better settling rates and consolidations when the HEC and the PEO polymers were added as aqueous solutions. This study demonstrates that both polymer combinations enhance the separation of the leach slurries better than PEO alone does. Also, since the cost of HEC (about $2.50/lb) is less than that of PEO or HPC (about $4.50/lb), the HEC-PEO polymer combination appears to be the best choice for agglomeration of the acid-alcohol leach slurries.

Conclusions

The solid-liquid separation of aqueous or alcohol slurries prepared by leaching phosphatic clay waste with H_2SO_4 was investigated. A variety of synthetic polymers was used in an attempt to enhance filtration and/or settling characteristics. None of the polymers tested improved the filtration of the slurries sufficiently to make filtration viable. However, flocculation of the slurry with PEO or a combination of HEC-PEO or HPC-PEO should enable solid-liquid separation to be achieved using a series of flocculation and decantation steps. Faster settling rates and better consolidation of the leach tails were obtained when the cellulose polymers were added as aqueous solutions. Use of the HEC-PEO combination for flocculation is the best choice for this application because the HEC is less expensive than PEO or HPC.

References

1. Stowasser, W. F., Phosphate Rock, Ch. in Mineral Facts and Problems, 1985 Edition. BuMines B 675, 1985, pp. 579-594.

2. Becker, P., Phosphates and Phosphoric Acid, Marcel Dekker, Inc., 1983, 585 pp.

3. U. S. Bureau of Mines, The Florida Phosphate Slimes Problem--A Review and Bibliography, BuMines IC 8668, 1975, 41 pp.

4. Scheiner, B. J., A. G. Smelley, and D. A. Stanley, Dewatering of Mineral Wastes Using the Flocculant Polyethylene Oxide. BuMines B 681, 1985, 18 pp.

5. Smelley, A. G., and I. L. Feld, Flocculation Dewatering of Florida Phosphatic Clay Wastes, BuMines RI 8349, 1979, 26 pp.

6. Wilemon, G. M., and B. J. Scheiner, Leaching of Phosphate Values From Two Central Florida Ores Using H_2SO_4-Methanol Mixtures, BuMines RI 9094, 1987, 9 pp.

7. Wilemon, G. M., and B. J. Scheiner, Extraction of Phosphate Values From Phosphatic Clay Wastes: An Acid-Alcohol Technique, Ch. in Challenges In Mineral Processing, ed. by K. V. S. Sastry and M. C. Fuerstenau. Soc. Min. Eng., AIME, Littleton, CO, 1989, pp. 421-436.

8. Wilemon, G. M., R. G. Swanton, J. G. Davis, and B. J. Scheiner, Leaching of Phosphate Values From Phosphate Wastes Using H_2SO_4-Methanol Mixtures, Metall. Soc. AIME Paper No. A89-4, 1989, 17 pp.

Table I.--Results of filtration tests using PEO and a combination of PEO and HPC

Filtration time, s	Grams of filtrate collected (polymer dosages in lb/ton)	
	HPC, 0.10 PEO, 0.10	PEO, 0.20
120	191	188
	189	153
	190	181
	186	180
		183
	avg: 189	177
60	175	151
	178	168
	178	177
	178	166
	177	177
	174	179
	178	136
	avg: 177	165
30	168	155
	169	172
	174	163
	173	158
	166	159
	172	171
	153	147
	173	167
	avg: 169	162

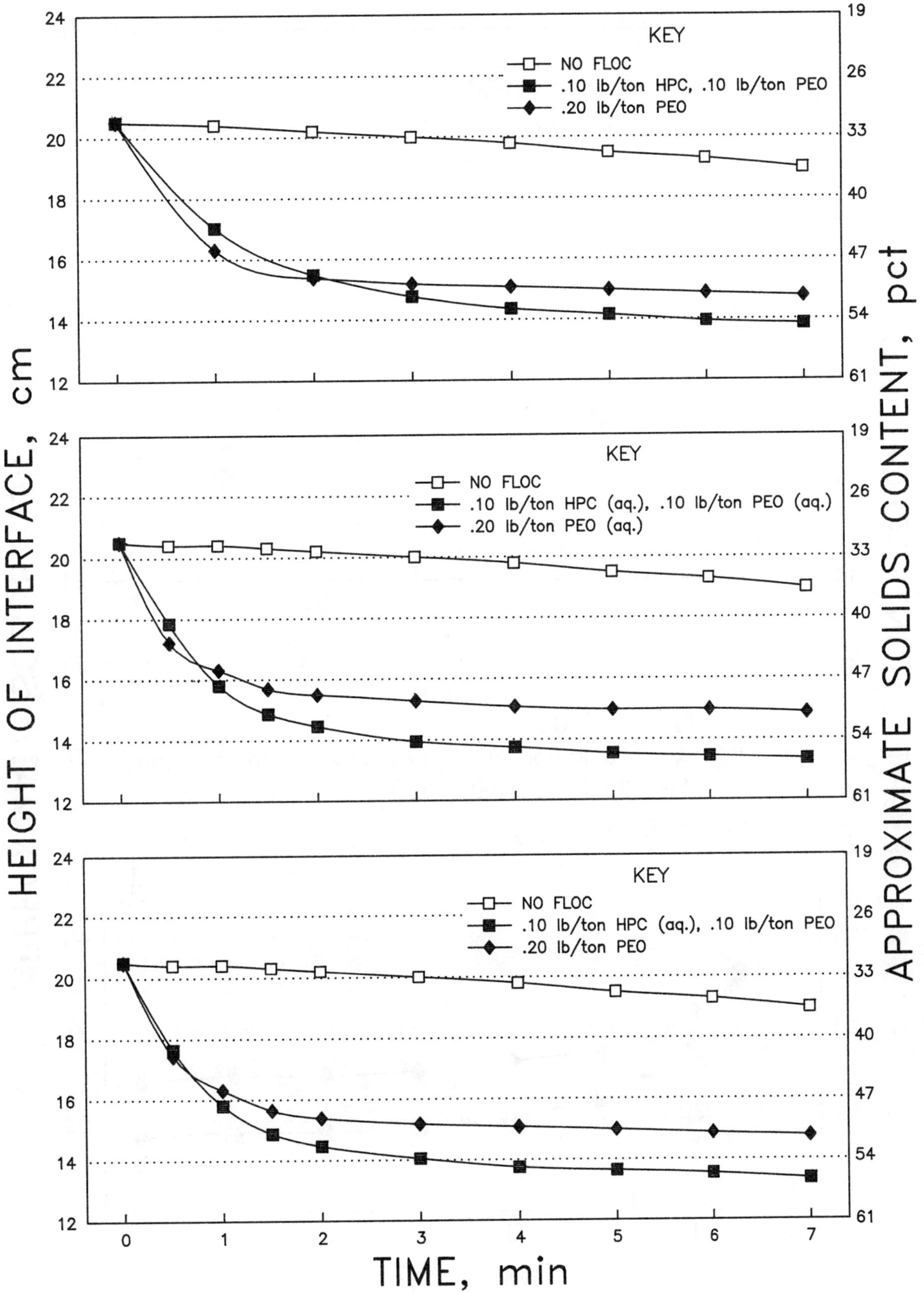

Figure 1 Comparison of the settling rates for slurries obtained by H_2SO_4-leaching of phosphatic clay waste in methanol when treated with PEO or HPC-PEO. (Polymers used as methanol solutions unless otherwise denoted.)

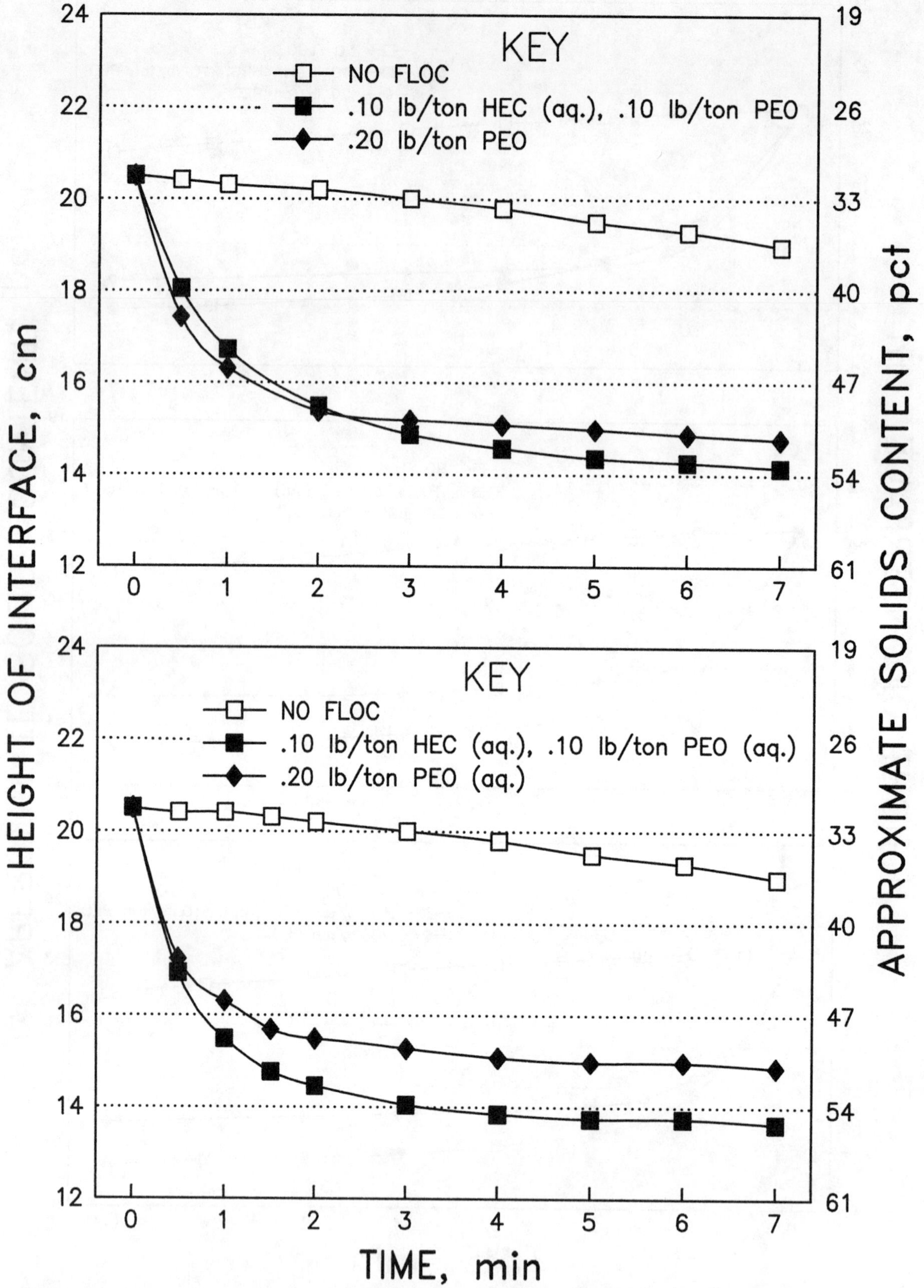

Figure 2 Comparison of the settling rates for slurries obtained by H_2SO_4-leaching of phosphatic clay waste in methanol when treated with PEO or HEC-PEO. (Polymers used as methanol solutions unless otherwise denoted.)

LIST OF KEYWORDS

Flocculation
Phosphate
Leaching
Solid-Liquid Separation
Fine-Particle Waste
Filtration
Settling
Hydroxypropyl Cellulose
Hydroxyethyl Cellulose
Polyethylene Oxide

THE USE OF HYDROCYCLONES FOR SOLIDS/LIQUID SEPARATIONS AN OVERVIEW

Mark P. Schmidt and Barry M. Buttler
KREBS ENGINEERS
1205 CHRYSLER DRIVE
MENLO PARK, CALIFORNIA 94025

(415) 325-0751

Hydrocyclones are centrifugal separators with no moving parts. They can be used to concentrate slurries, to classify solids in liquid suspension, to degrit liquids or suspensions, and to wash solids in suspension. They are frequently used as protection or pre-treatment devices to improve the performance or decrease the cost of downstream equipment such as centrifuges, filters, and screens. Hydrcyclones are also used in conjunction with thickeners, clarifiers, and strainers.

The hydrocyclone was first patented in 1891 for use as a water purifier; that is, a device to remove suspended solids from water. Commercial use of hydrocyclones was slow to develop, but in the late 1940's hydrocyclones were exploited for coal beneficiation by Dutch State Mines. Various mineral processing applications were soon to follow, and since that time, the use of hydrocyclones has crossed over into many other industries.

The application and commercial use of hydrocyclones for the following industries will be examined.

- Petroleum Production and Refining
- Plastics Recycling
- Chemical Processing
- Electronics
- Metal Working
- Power Generation
- Food & Beverage
- Pulp & Paper
- Environmental
- Iron & Steel

Petroleum Production And Refining

There are several applications within this field. During the drilling process, hydrocyclones are used in the oil patch in conjunction with vibrating screens ("shale shakers") to separate grit and cuttings from drilling muds. During production, hydrocyclones are used for removing sand and other forms of grit from raw crude oil and oil/water mixtures. At refineries, they are used to remove catalyst and coke solids from oil. Hydrocyclones (sometimes referred to as separators) are also used to remove grit, sediment, and scale from diesel and jet fuels.

Plastics Recycling

In this new but rapidly expanding industry, hydrocyclones are used to separate and recover discarded polymer based packaging materials for reuse. In the process of recycling plastic bottles, hydrocyclones are used for washing, dewatering, and separating different plastics; the primary application being the separation of plastics having different specific gravities. This is accomplished in a hydrocyclone by using a heavy medium (dense liquid or solution), allowing the unit to sort materials by specific gravity. In several stages, this process can be used to sort different plastics, metal fragments, and various particulate materials.

Chemical Processing

There are a variety of applications for hydrocyclones in the chemical process industries. Uses include classification of solids, thickening of slurries, clarification of solids, thickening of slurries, clarification of solutions, and washing of solids either in the process or for waste treatment.

Specifically, hydrocyclones are used extensively in crystallizer (evaporator) circuits for producing crystalline products. Hydrocyclones help maintain proper crystal size by recovering and thickening product-size crystals while returning finer crystals back to the crystallizer. Frequently, the stream containing the fine crystals is heated to dissolve the crystals in order to minimize the number of small nucleation sites in the crystallizer and promote growth of the coarser crystals.

Hydrocyclones are often used as pre-thickening devices ahead of centrifuges and filters, thus improving the capacity and performance of these pieces of process equipment. In certain applications, hydrocyclones can be used to remove only gritty, abrasive contaminants from a slurry ahead of a centrifuge.

Hydrocyclones can remove grit and scale from process water to prevent wear on piping and valves and fouling of heat exchangers. Similarly, hydrocyclones are used to remove solids from liquids ahead of pH meters and other on-stream analyzers.

Electronics

Hydrocyclones are used by printed circuit (PC) board manufacturers for recovery of photoresist (plastic) from stripping solutions, recovery of copper particulates from solutions, and recovery of fine pumice from rinse water in polishing operations.

Metal Working

In the metal working operations of machining, grinding, and honing, hydrocyclones are used to remove fine metal particulates, abrasives, and dirt from cutting oil coolants. The coolant, which also acts as a lubricant, is recycled for reuse. Coolant cleaning systems can incorporate a variety of equipment including screens, cyclone separators, and filters often hooked in series to recover very fine particulate material.

Power Generation

Hydrocyclones are used in flue gas desulfurization (FGD) systems and intake water systems at power generating plants. At coal fired power plants using flue gas desulfurization (FGD) systems, ground limestone ($CaCO^{3}$) is used to scrub sulfur dioxide (SO^{2}) out of the flue gas. In these systems, the reaction of the limestone with the sulfur dioxide produces inert gypsum particles as a byproduct. Hydrocyclones are used in the limestone grinding circuit to control the size of the limestone ($CaCO^{3}$) particles being sent to the scrubber. Hydrocyclones are used again to thicken the gypsum slurry ahead of belt filters. The final dewatered gypsum product is often sold for the production of wall board.

Coal-fired power plants are large users of water. Much of the water is used in ash handling systems, condenser cooling systems, and other equipment cooling operations. If the plant intake water comes from a nearby river, sand can erode pumps, valves, and nozzles. Cyclone separator systems have been installed in many power plants to eliminate and remove much of the sand material from the river water.

Food & Beverage

At food processing plants, wash water used to clean fruits and vegetables picks up a considerable amount of sand and grit. Hydrocyclones are used to separate this material from the wash water so it may be recirculated throughout the wash tank and spray nozzles.

To produce catsup, cooked tomatoes are pulverized using Urschel mills or homogenizers. Residual sand in the cooked tomatoes can be removed using cyclone separators to minimize or prevent damage to the high speed parts in the Urschel mills. In the preparation of juices and sauces, the cyclone separator is used to remove seeds, pits, and grit material.

Pulp & Paper

Hydrocyclones are used as cellulose fiber and glass, ceramic, or rock wool fiber pulp cleaners. In cellulose fiber applications, cyclones remove sand, grit and tramp oversize material from virgin wood pulp and a variety of other contaminants including baling wire, tramp metal, glass, and staples from recycled paper pulp. Removal of this material improves product quality and minimizes downtime and maintenance costs on downstream equipment such as pressure screens, refiners, calendar or press rolls, and paper and board machines.

Environmental

Along with the air pollution control (FGD systems) and solid waste management (plastics recycling) applications already mentioned, hydrocyclones are used in many other waste treatment applications.

Hydrocyclones can be used in countercurrent stages to wash hazardous wastes from contaminated soils. Cyclones are also used on industrial wastewater to lower the solids content before reuse of the water or discharge of the water to a river, lake, pond,or treatment plant. Municipal sewage treatment plants use hydrocyclones ahead of digesters to separate inorganic grit from the organic material.

Iron & Steel

Hydrocyclones (or separators) are used by steel rolling mills to remove grit and mill scale that accumulates in the cooling water pits. Cyclones or separators are also used in-line to remove particles ahead of spray nozzles. Both applications allow the use of recycled water, minimize the need for fresh makeup water, and reduce plant discharge.

In summary, hydrocyclones have considerable application in solving solids/liquid separation problems in many industries. It has been said "the most efficient and dependable machines are those with the smallest number of working parts". For this reason hydrocyclones have gained wide usage and the range of applications is rapidly expanding.

BAGHOUSE SYTEMS & COMPONENTS

Milton A. Buffington, P.E.
Buffington & Associates
3592 Oak Rim Way
Salt Lake City, UT 84109

A. Introduction

After the design or process parameters have been established defining the operating environment, the baghouse system and its components can be specified. This section describes the different types of baghouses and the principle components available. The function of the various components and some of the selection criteria is included.

Function

The function of a baghouse is to remove particulate from a dirty gas. This is done quite well by a baghouse. Efficiencies of 99.98% have been obtained with a baghouse, in utility service, the particulate emissions were in the order of 0.004 lb/MMBTU. The particulate size ranged from 0.02-2.0 microns. At the same time, the pressure drop was 4.5 in. water.

Classification

Baghouses are classified by the method of cleaning for which they are designed. Reverse-air; shake-deflate; and pulse-jet cleaning, are the most common methods of cleaning available. The utility field generally uses the reverse-air or shake-deflate system.

The majority of systems currently in service use the reverse-air method of cleaning, the bags are fiberglass with teflon coating. Test results indicate that the shake-deflate method of cleaning offers the power companies a substantial saving, an a/c ratio of from 2.5 to 3.5 acfm/sq.ft. vs. 1.5 to 2 acfm/sq.ft. with reverse air cleaning. If bags can give the same length of service in shake-deflate operation this type of unit would be more economical.

Using treated fiberglass bags which are presently the most common in power plants, a bag life of between three and five years is anticipated. The latest figure of 5 years is the expectancy of the bags in a unit in Utah. This application is a reverse air fabric filter.

The use of sonic horns together with both reverse-air cleaning and shake-deflate mode of cleaning is helpful in removing more cake than either method used alone. The horns are air powered and produce a number of frequencies. Presently, the most widely used range is from 100 to 500 hz. Very good results have been obtained using a horn of 200 hz operating on 100 psi compressed air.

A baghouse may also be defined as a positive or negative pressure unit. In general, the units used are negative pressure design. This is determined by the location of the draft fan in the system. The fan is either pulling the gas through the house or discharging the gas into the baghouse.

Appurtenances

The appurtenances available vary from one manufacturer to another. Valves, ash removal systems, and bag suspension systems are examples of componenets that vary from one design of baghouse to another. Maniflods, general arrangement, the flow of reverse air, by-pass arrangement, and hopper design are some of the items that are fairly common from one design to another.

B. Bag Cleaning Systems

1. Function and Description

Bag cleaning systems are required in baghouses in order for them to continue to function as filters. As the cake builds up on the fabric bags, the pressure drop across the bag, (that measured from one side of the tube sheet to the other), builds up and can reach a completely undesirable amount. The limitation is the differential pressure available from the Forced Draft Fan.

The cleaning systems periodically, on a predetermined cycle, causes the excess cake to fall from the surface of the bag. This reduces the pressure drop between the dirty and clean side of the bag, and keeps the unit in operation.

One point, when reverse air cleaning is refered to, actually the cleaning media is clean gas, not ambient air. The introduction of cold air from the atmosphere would have a detrimental effect upon the bags. Also, corrosion could be enhanced.

2. Types of Fabric Filter Cleaning System Designs

a. Reverse-air cleaning system

Reverse-air cleaning is the most common method of cleaning currently used in utility plants, and other large gas volume applications. In this system, the dirty gas enters the baghouse in the hopper area, it then passes through the bag, and exits at the top of the house through the exhaust manifold.

Cleaning is accomplished by the timed opening of valves in a separate manifold in which flows a steam of clean, hot gas. This gas is diverted from the main clean gas stream and its pressure is increased by a reverse air fan. This clean, hot gas moves through the reverse-air valve into the compartment calling for the cleaning cycle. At the same time, the flow of clean gas from the compartment is interrupted. The clean gas, moving in reverse to its normal direction of flow, passes through the bag in the opposite direction, causing most of the built-up cake to dislodge and fall into the hopper.

Upon completion of the bag cleaning cycle, the reverse-air valve closes, the exhaust or clean air outlet valve opens and the particular compartment resumes its normal operation. The valve at the inlet to the compartment is mainly for maintenance and repair work. It enables the plant operator to close off a compartment safely for men to enter and do work.

The bags used in a reverse air baghouse are usually 10 or 12 inches in diameter and 30 to 35 feet in length. Fiberglass is the basic material used here for bag material, a coating or "finish" is applied to the fabric. This additive, lubricates the fibers and prolongs the life of the bags. In addition these bags are provided with steel rings of approximately 1/8 in diameter wire spaced about 3 or 4 feet apart the length of the bag. These rings keep the bags from collapsing during the cleaning cycle. If this occured, the dust could not fall into the hopper.

Ventilating air is provided to each compartment. This is necessary to reduce the air temperature within the compartment in order for men to be able to work inside the baghouse while the unit is in operation.

b. Shake-Deflate System

A shake-deflate system is one in which the unit of bags to be cleaned are mechanically shaken from side to side in an amplitude of from 1 to 3 inches. This cleaning action is amplified by adding an air inlet valve after the clean gas valve. This valve opens when the clean gas valve is closed for the cleaning cycle. "Air", again actually clean, hot gas flows in a reverse direction, through the bags helping remove the built-up cake.

Mechanical shaking is harder on the filter cloth than the reverse air cleaning method. A baghouse of this design, however, does not require as low an air/cloth ratio and therefore costs less. As filter cloth materials improve, the attractivness of this cleaning method may surpass that of the reverse-air design.

Once again it is pointed out that the reverse-air is actually clean gas. It´s temperature is sufficient to prevent condensation and corrosion. This gas is re-introduced into the clean gas stream after it has caused the bags to deflate.

An example of the shaking device is shown. Also the position of the reverse air valve needed to deflate the bags is indicated.

c.Pulse-Jet Cleaning

Pulse-Jet Cleaned baghouses have only been used in a few power plants todate. The largest is on a 100MW plant. These units are also used as clean-up baghouses on such applications as conveyor transfer points, coal stackers, silo working areas.
This is a popular unit for relatively low volume of gases in industrial applications.

The biggest advantage of Pulse-Jet cleaning method is it high allowable A/C ratio. This can run as high as 12 or 15 to 1 in certain applications.

Pulse-jet cleaning is the use of compressed ambient air at a pressure of 80 to 100 psi in intermittent pulses to cause the cake to become dislodged and fall into the hopper. In this design, the dust is collected on the outside of the bag. The bag is pulled over a wire cage. At the top of the cage there is located a venturi (horn) through which the compressed air passes when discharged. The discharge time is very short, a few seconds, this sends a pulse of air down the bag, shaking the cloth and knocking off some of the cake. Larger volume pulse jet baghouses do use a lower pressure for the cleaning cycle. One unit used on a power plant uses about 40 psi for the pulsing air pressure.

At the present time, bag cloth of high temperature limits is not readily available for very hot service. Also the pressure drops experienced in these units is high for large volume applications. If sufficient cloth is provided to lower the differential pressure, it´s competative position with shake-deflate designs, for example, looses its advantage. With the small bags used, typically 6 inches in diameter and 10 to 12 feet in length, a size limitation exists regarding its use for high volume applications.

c. Selection Criteria

The most important design consideration influencing the selection of a baghouse is the pressure drop and air to cloth relationship (A/C). The pressure drop determines the steady energy requirment which comes from the power plant. The air to cloth ratio also effects the cost of a baghouse. This impacts the cost of the bags and also the cost of the baghouse structure.

Reverse-air baghouses require the lowest air to cloth ratio. A ratio of from 1.5 to 2.0 acfm per square foot of cloth is to be expected. The pressure drop will range from 4-8 inches of water. This is to be compared to the shake-deflate unit with an air to cloth ratio ranging from 2 to 3.0 acfm per sq.ft. of cloth. This is with a pressure drop of 5.2 to 6.5 inches of water.

Both of the cleaning methods mentioned above can be improved upon by the addition of sonic horns. With a plant using western, low sulfur coal a pressure drop reduction of as much as 50% can be realized. Using Eastern, high sulfur coal the amount reduction in pressure drop is less, in the order of 20 to 30 %.

About 90% of the current power plants, with baghouses, use reverse-air cleaning. From the standpoint of experience and guaranteed results, the reverse-air baghouse is presently the appropriate choice. Today this would, of course, include sonic horns.

One inch of pressure drop in a utility plant, results in an energy cost of $50,000 per year, (1984 costs). This is indicitive of the importance of pressure drop in the selection of a baghouse. Also a very important consideration would be the number of bags and the expected bag life.

Bag life on a reverse-air cleaned baghouse can be as long as 5 years. On a shake-deflate unit an average bag life of 2 1/2 years can be expected. No comparative data was available for pulse-jet cleaned bags when this was written.

SUMMARY

This is a brief over-view of the most common types of baghouses. There are others which in general have only application in the industrial field. Baghouses are used for a very wide range of gas volumes. Some as small as a few hundred acfm, up to the very large units which handle gas volumes of several million acfm. The temperature of the gases can range from ambient air temperatures, to very high temperatures found in power plants and metal processing facilities.

BEVERAGE FILTRATION - ALWAYS KNOW WHAT'S GOING ON

Herman C. Mouwen
BEAM Enterprises
Technical Applications Consultant
1290 Sunnycrest Avenue
Ventura, CA 93003

The Market

This article concerns the filtration of beverages: beer, wine, distilled spirits, soft drinks, wine coolers, bottled water and fruit beverages. Over 25 billion gallons of these beverages were consumed by the American public in 1989. However, not all of the 25 billion gallons were filtered. It is estimated that about 60% or 15 billion gallons of these beverages were filtered in one way or another.

Soft drinks are the number one beverage consumed and it is estimated that only 45% of these beverages are filtered. Many operators rely on the fact that the water supplied to their plant from the local municipal source is potable and therefore of an acceptable quality. The second most frequently consumed beverage is beer. It is well known that 100% of all beer is filtered. Next are fruit beverages; it is estimated that 20% of these are filtered. Bottled water is the fourth beverage on the consumption list with about 65% of this beverage being filtered.

What are the market trends? First of all, the bottled water industry will face increased government regulation through the FDA or EPA Safe Drinking Water Act. Higher standards of product testing and quality have been required by many states. Such testing is expensive. The expenses associated with production management, equipment, and procedures have increased dramatically due to filtration and sanitation requirements including such things as clean room type bottling and water treatment. The estimated figure of 65% of all bottled water being filtered will probably rise as high as 75 to 80% by the year 1994.

Beer filtration is heading toward final membrane filtration for the cold sterilization of this beverage. Although the beer industry has always been under tight regulations and product quality controls, the consumer appears to have reacted to the situation that cold filtered (sterilized) beer "tastes better."

Soft drinks and their filtration trends will follow the bottled water direction. As soon as the water sources for soft drink operations are subjected to the higher standards of product testing and quality control, the filtration of soft drinks will increase. The current estimate that only 45% of soft drinks are filtered will probably rise to 60 65% by the year 1994.

Government regulation has been the foundation for the new caution in the areas of bottled water and soft drinks. Health conscious consumers are also helping to "pave the way" for improved filtration in the beverage markets.

Clearly, the two areas of concern which dictate the filtration of beverages are:

1) Microbiological control
2) Cosmetic control

Microbiological control is self explanatory. Cosmetic control has to do with the appearance of the beverage in its final package or container. The consumer is usually impressed by the beverage's appearance.

With beer and wine, for example, there is no question that yeast cells and bacteria must be removed prior to final bottling. Even with filtration, certain preservatives are used in some of these beverages and their presence must be shown on the label according to new label laws.

Distilled spirits are a special case. Few bacteria can exist in these beverages, but they still must be cosmetically acceptable to the consumer. They must appear bright, sparkling and crystal clear in the bottle. So, they are filtered.

On the other hand, soft drinks and their filtration requirements are subject to several different considerations:

A) If the water source is up-to-date and highly controlled to conform to today's potable water standards per the FDA or EPA Safe Drinking Water Act, then the only real concern is producing a beverage which looks appealing in its container. If the container is a can, then the cosmetic appearance requirement may be relaxed.

B) If the bottling/canning facility is up-to-date, the filtration requirements may be less strict since modern bottling facilities tend to be very clean. This, of course, assumes that the water source is also very acceptable. Under such conditions, sanitization becomes paramount in order to keep the system very clean and under microbial control at all times. Good sanitization procedures apply to all these situations.

C) If the bottling facility is old and out-of-date and the water source is very acceptable, the product quality can still be challenged. Under these conditions, filtration may be required for both microbial and cosmetic controls.

D) The last consideration would be an old plant with a poor water source. This producer/operator faces a major situation. Filtration is probably not the most cost-effective way to keep this plant's beverage products under quality control. Under all these situations and conditions, the quality of the final packaged product will dictate its acceptance at the marketplace.

All of this means something very important. The beverage producer/operator must know, at all times, the microbiological and/or cosmetic characteristics of the beverage being produced. For example, the producer must know the yeast and bacteria levels of their beer or wine products. If cosmetic appearance is the real concern and under control according to consumer acceptance, just what are the particle count levels of the current product?

Nominal and Absolute Rated Filters

Filter ratings refer to the size of a specific particle or organism retained by the filter to a specific degree of efficiency.

Nominal - This rating describes the ability of a filter medium to retain the majority of particulates at a given rated pore size or greater, according to test conditions defined by the filter manufacturer. Nominal filters are usually filter media other than membranes.

Absolute - Under strictly defined test conditions, an absolute pore size filter medium retains 100% of the challenge organisms or particles. Among the test conditions that must be specified are: test organism (or particle), challenge pressure, and, concentration and detection method used to identify the contaminant. The HIMA procedures are an example of such defined test conditions which must be followed by all filter manufacturers who claim an absolute pore size rating. Absolute filters are usually membrane filter media.

Thus, the real difference between **nominal** and **absolute** pore size ratings is that the filter manufacturer establishes a nominal rating according to the filter manufacturer's test standards.

However, the filter manufacturer who claims an absolute pore size rating, for example a 0.2 micron sterilizing pharmaceutical grade filter cartridge, **must** follow industry test standards.

As discussed above, it is very important that the beverage producer/operator knows, at all times, the microbiological and/or cosmetic characteristics of the beverage being produced.

Let's take the case of a 0.45 micron absolute pore size rated membrane filter which has been HIMA tested and validated. Then let's suppose that a wine producer uses such a membrane filter from manufacturer "A" to control the wine's final state of sterilization and appearance. This producer, for whatever reason, could use a 0.45 micron absolute pore size rated filter (HIMA tested and validated) from manufacturer "B" and achieve the same state of sterilization and appearance controls. There may be other differences between the filters produced by manufacturers "A" and "B", such as initial pressure drops, throughput volumes, taste effects, filtration costs, etc., but the 0.45 micron absolute pore size rating will be the same.

On the other hand, let's discuss a situation in which the producer of a soft drink who uses a nominal rated filter could find him or herself. Suppose the soft drink producer uses a 5 micron nominal rated filter from manufacturer "A" to control the particle sizes and concentrations in the final beverage product. The bacteria levels, as verified by tests, are acceptable. A preservative may be used to provide the required shelf life. The beverage looks great in its container, a bottle: bright, sparkling and crystal clear.

Now, for whatever reason, suppose this producer changes from manufacturer "A" to manufacturer "B" and a 5 micron nominal rated filter is specified. Both filters are clearly identified as 5 micron nominal, the labels say so and the catalogs say so. By making this change, the quality and/or product could and usually do change. On one hand, the drink may no longer look bright, sparkling and crystal clear. The size and concentration of the particulate matter in the final beverage product has changed.

On the other hand, the drink may still be cosmetically acceptable, but whereas the producer used to get 60,000 gallons per cartridge with manufacturer "A", he is now getting 25,000 gallons per cartridge with manufacturer "B". This means that there will be a significant operating cost increase.

This is why it is so important for the beverage producers to know the microbiological and particulate characteristics of their final products. Then, quality control specifications can be generated and the current characteristics specified. If a change has to be made in the filtration system, tests can be conducted to show that the filter's performance is equivalent to that which was previously producing acceptable product.

Improved filtration is the direction in which the beverage industry is heading. The bottled water and soft drink markets will feel the impact of higher standards, more stringent government regulations, and improved product testing and quality control, more than the beer and wine markets.

The filter manufacturers also face the opportunity to improve their service to the beverage producers/operators. Filter manufacturers usually have particle counters and they can use this equipment to assist the producers in determining the product's particulate levels. Silt Density Index and Turbidity tests are other procedures which can be used to assist the producers. Major filter manufacturers may also have equipment which can be used to determine extraction levels (taste controls) and technical service personnel who can assist the producers in developing procedures for cartridge integrity testing, maximizing cartridge life and system sanitizing, for example.

The Future For Beverage Filtration

Higher standards of beverage testing and quality control will increase the use of filtration. The quality and performance of beverage filters will have to improve in order to offset the expenses resulting from increased filtration.

Absolute pore size rated filters will offer improved performance through greater throughput, increased effective filtration areas per cartridge, zero taste effects, and simplicity in integrity testing. Today, beverage producers associate absolute rated filters with high operating costs. The filter manufacturers will have to change this perception.

The use of sanitary style filter housings will increase. The higher standards and increased product quality control will necessitate improved sanitary operating conditions.

Filter manufacturers will increase their knowledge of their product's performance. They must work very closely with the beverage producers/operators so that the contamination characteristics of the final beverage products will always be known.

There are no nominal pore size rating industry standards or test procedures. Thus, nominal pore size rated filters, with the same numeric ratings, can have vastly different performances. These differences can play havoc with beverage producers who face filtration level changes due to improved standards and additional testing requirements.

In the past, there were attempts among the filter manufacturers to develop nominal pore size rating test procedures. Nothing ever happened, simply because the words "nominal pore size" could not be defined and agreed upon.

However, in the absence of standards for testing nominally pore size rated filters, the filter manufacturers must develop methods to assist the beverage producers/operators so that the specific performance of a given filter can be accurately defined in terms of the quality of the final beverage product.

SELECTED ENGINEERING PRINCIPLES OF CROSSFLOW MEMBRANE TECHNOLOGY

James L. Williamson, Senior Application Engineer
David J. Paulson, Director, Corporate R&D

OSMONICS, INC.
5951 Clearwater Drive
Minnetonka, Minnesota 55343

INTRODUCTION

Since the 1959 development by Sourirajan of the first viable RO membrane, crossflow membrane technology has been applied to an ever increasing range of liquid separation processes and markets. In food and beverage applications, crossflow membrane technology is used to purify the liquids we drink, concentrate and reconstitute juices, dewater and purify dairy products, and also helps keep our water supplies potable by treating wastes at all levels of production.

Over the years, the empirical study of the crossflow filtration process has led to several inter-related operational principles and engineering "rules of thumb." This paper discusses a few of these, supported by examples from actual food and beverage applications.

CROSSFLOW COMPARED TO NORMAL FLOW FILTRATION

Most people think of filtration in the "normal flow" mode; where the entire fluid volume passes through the filter media, producing two streams: a feed (influent) and filtrate (effluent). In the crossflow process, a feed stream flows across a membrane, with only a portion passing through the membrane pores to produce a "permeate." As the solvent or carrier liquid passes through the membrane, dissolved, colloidal and suspended solids retained by the membrane are concentrated, producing a "concentrate" stream. Thus, a single feed but two effluent streams define crossflow filtration.

As with normal flow filtration, two performance characteristics are used to measure the crossflow process; permeate **flux** (work rate) and **separation** (rejection of solids).

The flux is determined by several factors, including the feed composition, temperature and viscosity, driving pressure, the inherent membrane structure and its overall condition when in operation.

Factors which affect the separation achieved include pore volume and size distribution, solids concentration at the membrane/feed stream interface, chemical interaction between the membrane material and feed stream, and the net driving pressure.

DETERMINING FACTORS FOR CROSSFLOW MEMBRANE PERFORMANCE

Many factors influence flux and separation, so the engineer must understand how those factors affect one another to optimize the design and operational parameters of the overall process.

In crossflow membrane processes, the degree of solids concentration affects both the separation and flux of the system. The fluid dynamics of the system in turn, affect the degree of concentration possible. Thus, achieving the correct balance of fluid dynamics and fluid condition in the system ultimately determines the degree of success for an application.

A major concept the crossflow engineer must understand to account for the full impact on the system of solids concentration is the concept of **recovery:**

$$R = \frac{\text{permeate volume}}{\text{feed volume}}$$

As this ratio increases, the volume in which retained solids are concentrated decreases proportionately. Recovery has a mathematically corresponding **concentration factor,** which expresses the degree by which solutes are concentrated:

$$CF = \frac{1}{1\text{-Recovery}}$$

The engineer must consider what impact the solids concentration level will have on the process. In almost all cases, the feasible degree of solute concentration will be determined by one or more of the following:

° **Osmotic pressure** ($\Delta\pi$) - the natural tendency to equilibrate energy levels by solvent flow from a dilute to a more concentrated solution. Net osmotic pressure ($\Delta\pi$) is proportional to the concentration difference, and must be overcome by hydraulic energy (a pump) to achieve a suitable driving pressure for efficient separation.

- **Concentration polarization** (CP), or the natural phenomenon where a boundary layer near the membrane surface forms, with a substantially higher solute concentration than in the bulk stream. As a result, the flux and separation behavior of the system correspond to a substantially higher feed concentration than is measured in the bulk stream.

- **Saturation**, or the maximum concentration at which a solute or emulsoid exists without precipitation or agglomeration. If saturation is exceeded, the precipitate can coat the membrane. Colloidal and suspended solids also tend to settle on the membrane. This solids layer will restrict flux and affect the rejection performance of the membrane. This phenomenon is known as **fouling**.

In all cases, the effect of these parameters is proportional to the concentration factor. Therefore, selecting the correct recovery is critical in optimizing nearly all crossflow filtration applications, especially for success in high solids applications.

The following engineering parameters also directly affect crossflow membrane performance and, therefore, must be understood and considered in process design:

- **Temperature** and **viscosity**

- **Operating pressure** (or gauge pressure - psig) and **effective pressure**, (the net gauge pressure less net osmotic pressure - psi).

- **Crossflow velocity** and **turbulence**

- Membrane **pore structure**

- **Element configuration**

ENGINEERING RELATIONSHIPS

Temperature

Flux increases proportionally with temperature due to reduced kinematic viscosity. While this is well understood, the offsetting effects of increased temperature are often not considered. Increased flux leads to an increased rate of fouling. Also, temperature increases chemical activity and foulant deformation which can accelerate foulant layer densification, resulting in increased hydraulic resistance. Ultimately, flux in a fouling system is reduced at a higher rate from a temperature increase, and may actually stabilize at a lower value than would occur at lower temperatures.

Add to this fouling enhancement the fact that true membrane compaction (densification of the porous structure) is greatly increased above a critical temperature (which varies by membrane polymer and structure) and increased temperature may have further negative effects.[1]

In summary, although solvent flux independent of fouling increases with temperature, fouling effects increase as well and may offset that gain. Whether any net gain is achieved will depend on the nature of the foulants present, and on the operating time before shutdown and cleaning.

Pressure

Flux increases proportional to effective pressure. For ultrapure water, this increase is linearly proportional. In the real world, this relationship is never fully achieved.

As flux increases, concentration polarization and the $\Delta\pi$ effect at the boundary layer also increase and the curve of flux gained as pressure is increased gradually falls off the pure water ideal. This reflects the fact that as $\Delta\pi$ increases, effective pressure decreases at a constant operating pressure.

Where foulants are present (as opposed to solutes still in solution), the variation from ideal is even more dramatic. Not only will $\Delta\pi$ increase, but hydraulic resistance by the foulant layer will further reduce the realized flux. In this condition, flux is considered to be "foulant layer limited."

Figure 1 is an idealized representation of three aspects of the principle of flux versus increased pressure. A rule of thumb for predicting flux response to increased operating pressure in a system is that; fouling solutions will show the least benefit, and in non-fouling systems, the benefit will be inversely proportional to the solute concentration. Other factors influence that benefit as well, and chief among those are fluid velocity and turbulence.

Crossflow Velocity

Providing the required crossflow velocity (parallel to the membrane surface) is necessary for both work aspects of membrane; separation and rate. Although it is well known that velocity enhances flux by mitigating fouling and CP, a critical velocity is also necessary to achieve the separation potential, (particularly with solutes). If sufficient mixing of the boundary layer is not present, the CP layer will present a more difficult condition to the membrane than would be expected based on measurement of the bulk stream. Turbulence enhances mixing. The data in Figure 2 demonstrates that the critical velocity for 4-inch (10 cm) diameter "full-fit" style cepralators (with both standard and high rejection membrane) is not achieved until above 10 gpm feed flow. (For the more traditional concentrate seal/hard outer cover RO sepralator design, 4 to 5 gpm has been established as minimum feed rate.)

Pressure and Crossflow Velocity

Increasing the fluid velocity benefits flux as does increasing the operating pressure (up to a point of diminishing returns, as discussed above). In high fouling applications, they are also very much inter-related from an engineering perspective. Increasing pressure gains flux, until fouling and/or CP eclipse the gain. Crossflow velocity and turbulence reduce both CP and fouling, so for every system there will be an optimum combination of the two.

Cheryan's data in Figure 3 shows that if a 2-inch (5 cm) diameter RO sepralator is operated only at 1.1 gpm feed rate, the benefit achieved above 150 psig is not proportional to the added pressure and no benefit occurs above 280 psig. Yet, if the feed rate is doubled, substantial flux benefit is achieved up to at least 350 psig. Only a minor gain is realized, however, with an increased feed rate of another 50%.[2]

Membrane Pore Size

A variety of polymers and polymer blends have been formulated into membranes. Characteristics vary in regard to operating temperature, pressure, pH tolerance, solute rejection (organic and inorganic) and tolerances to oxidants.

Manufacturers usually assign a pore size or molecular weight cut-off (MWCO) rating to their membranes as well. This rating should not be used exclusively when selecting a membrane. For example, it seems reasonable to select a membrane with an MWCO rating of 20,000 over a smaller-pore membrane when concentrating an organic molecule such as pectin; since larger pore membranes yield higher pure water flux rates, this would maximize the flux-to-energy ratio required for the application. However, when removing pectin from a citrus waste, a membrane having a 2000 MWCO showed more favorable operation (Figure 4). Since all UF membranes have a fairly wide size distribution among their pores, this can be explained by larger pores of the 20,000 MWCO membrane which were similar in size to the solids being removed. Solvent flux was obstructed by deposition of pectin inside the pores. With the smaller pore membrane, the solids could only deposit above the plane of the membrane surface, and higher permeate flux resulted over time. Since interior pore-plugging is often worse than cake-layer deposition, for organic separations it is often preferable to select a membrane with an MWCO rating well below the MW of the target organic solid.

Membrane Configuration

As with membrane types, various configurations of crossflow membrane devices or elements are available. Table I lists the most common ones and the advantages and disadvantages of each. Discussion here is centered on the spiral-wound configuration (sepralator) which is the most widely used.

Membrane technologists are most familiar with the traditional hard-cover "brine seal" design which uses a mesh netting to define the feed stream channel. The first major improvement on this was the "full-fit" design, which eliminated the brine (concentrate) seal as well as the impervious outer cover and their attendant dead flow areas. It also provided better fluid dynamics for many fouling applications; probably due to the increased gap between the spacer and the membrane which results under operation.

A more recent development is the 'tubular spiral' design, which uses a clear-channel spacer instead of the usual mesh. The clear channel design lends itself to higher viscosity solutions due to lower pressure drops and also can handle higher levels and larger size suspended solids without plugging.

Applications which favor the full-fit and tubular designs include the concentration of second-press apple juice and gelatin concentration. Compared side-by-side for flux versus percent recovery at a constant pressure in the juice application, the turbulence provided by the mesh spacer resulted in higher flux than seen with the tubular spacer (which was run at feed rates below turbulence) (Figure 5). In contrast, the tubular design out-performed the full-fit for flux in concentration of food-grade gelatin, even below turbulence velocities (Figure 6).

Thus, the new designs are useful for applications unsuitable for the traditional design. The most effective design will vary from application to application, and often be determined by pilot testing.

Summary

Several inter-related process factors and their resulting relationships have been presented for crossflow membrane filtration processes. Correct application of these relationships will optimize the flux and separation through balancing the effects of temperature, pressure, crossflow velocity, pore size, and spiral-wound design.

(1) Rudie, B.J., et al "Reverse Osmosis and Ultrafiltration Membrane Compaction and Fouling Studies Using UF Pretreatment", Chapter 29, Reverse Osmosis and Ultrafiltration, Ed Sourirajan and Matsuura, ACS Symposium Series 281, Copyright 1985.

(2) Cheryan, M., et al "Reverse Osmosis of Milk with Thin-Film Composite Membranes", Journal of Membrane Science, Scheduled for 1990 Publication.

KEY WORDS

Flux

Separation

Crossflow

Recovery

Concentration Factor

Osmotic Pressure

Effective Pressure

Velocity

Temperature

Pore Size

Configuration

Relationship

Fouling

Turbulence

Engineering

Solids

FIGURE 1

Flux vs. Pressure

Idealized Behavior For Three Types of Feedstreams

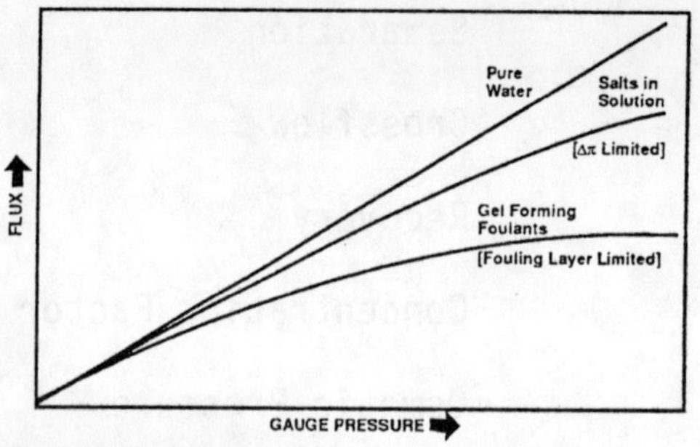

Assumes Crossflow Velocity Is Constant
(therefore fouling and concentration polarization increase proportional to flux increase)

©1990 OSMONICS

FIGURE 2

NaCl Passage vs. Crossflow

OSMO® 416 Model Full-fit Sepralators (4"D x 40"L)

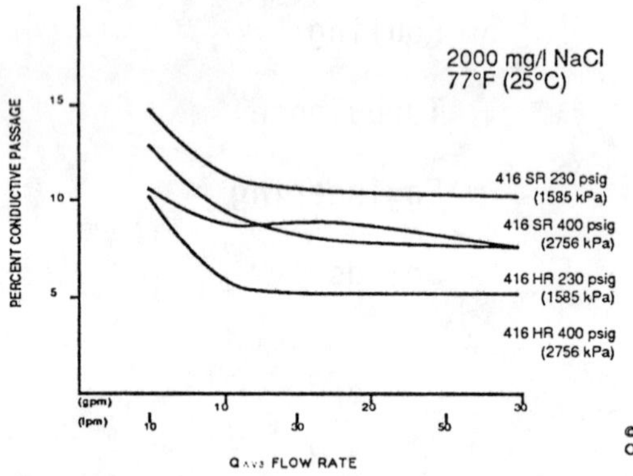

©1990 OSMONICS

FIGURE 3

Flux vs. Transmembrane Pressure

RO On Skim Milk At Various Crossflow Velocities

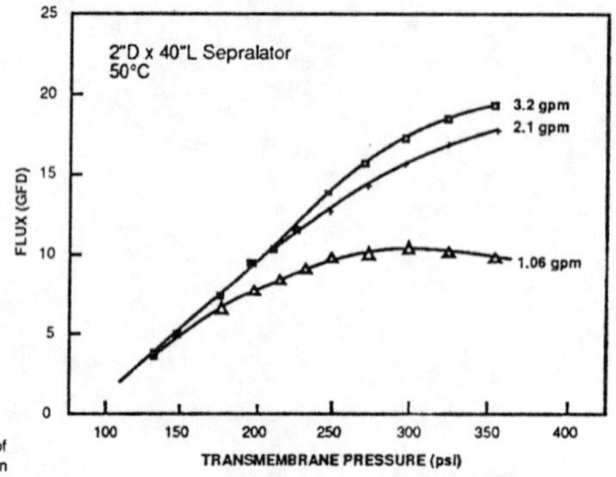

Data Courtesy of Prof. M. Cheryan

©1990 OSMONICS

FIGURE 4

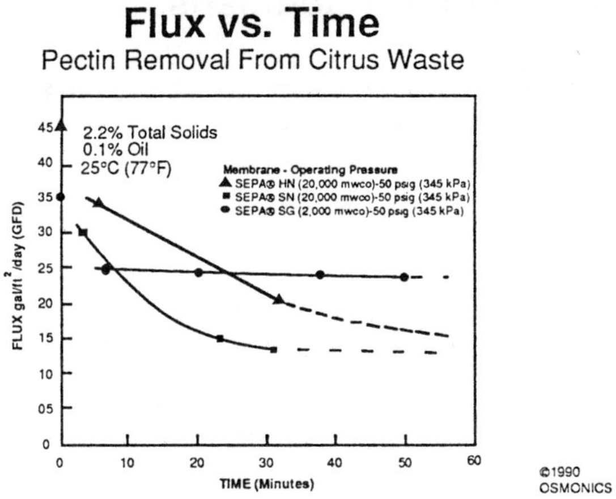

FIGURE 5

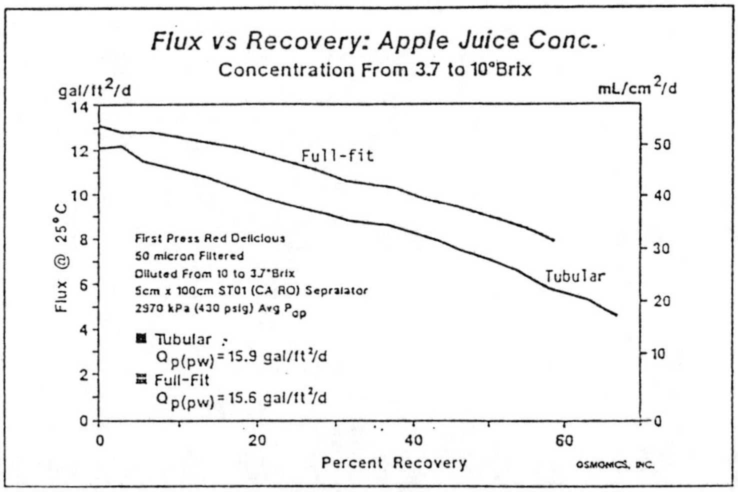

FIGURE 6

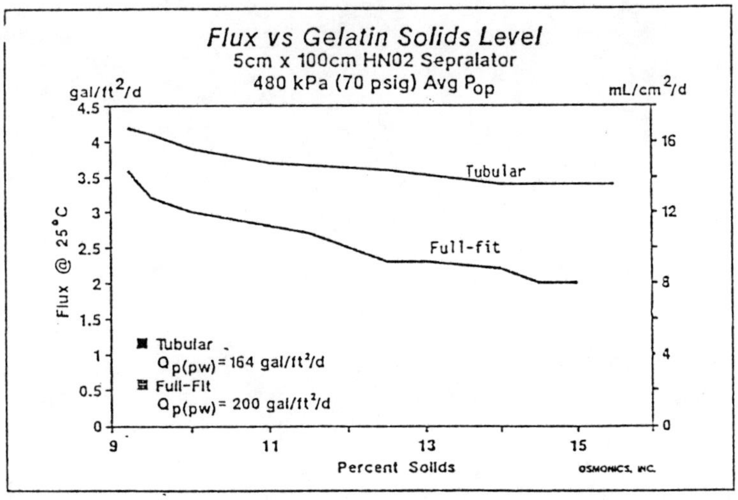

TABLE I

Crossflow Membrane Configuration Comparison

Characteristic	Traditional Spiral-Wound	Tubular Spiral	Fibers2	Tubular	Plate & Frame
Cost	Low	Moderate	Low	High	High
Packing Density	High	High	UF-High / RO-Very High	Low	Moderate
Operating Pressure Capability	High	High	UF-Low / RO-High	UF-Moderate	High
Membrane Polymer Choices (RO/UF)	Many	Many	Few	Few	Many
Fouling Resistance	Fair	Very Good	UF-Good / RO-Poor	Very Good	Fair
Cleanability	Fair	Very Good	UF-Good / RO-Poor	Very Good	Fair

1. Traditional spiral-wound sepralators use mesh spacer, concentrate (brine) seals impervious outer covers.
2. Fibers here mean self-supporting: lumen or tube-side feed for UF "fat fibers," shell-side feed for RO "fine fibers."

Crossflow Microfiltration Economics

Steve Valeriote
Memtec America Corporation
9690 Deereco Road, #700
Timonium, Maryland 21093
(301) 560-3029

Introduction

Crossflow microfiltration is a fast growing commercially accepted technology in the Beverage Industry. Full assessment on its value to the industry is now required

New technology must offer benefits to the end user without detracting from the other quality aspects of their business. CMF offers economical and operational advantages to the production of beverages and does not detract from the quality of the product compared to traditional technologies. To achieve this, it has been necessary to develop systems that give high flow rates for the capital investment required, utilize membranes that give clarity without removing desirable components and to allow automatic operation with the versatility to fit into a variety of processing conditions.

CMF makes no attempt to replacing present procedures such as depectinizing or fining, but reduces the use of disposable filtration aids and media. As cost of filtration media is increasing and disposal becomes more difficult, economic advantages result. These benefits are enhanced by the ability of CMF to eliminate multiple filtrations.

The utilization of PLC controlled backwashable systems presents the opportunity for automated operation. Correctly designed, a number of processing modes can be run giving considerable versatility.

Reduction of filter aids also eliminates the possibility of the off characters picked up during their use. With the use of inert materials, CMF does not contribute off characters to the beverage.

Economics

In the Beverage Industry, the traditional methods of clarification have been centrifuges, plate and frame, pressure leaf and rotary vacuum filters. All of these processes have relatively poor economic efficiencies due to high enery costs, high labor and operator content and expensive consumable costs.

Even though continuous microfiltration is still in its infancy, users are reporting substantial cost savings. Based on systems now in operation, we can use as an example a system sized to filter at a rate of 26,5000 gallons per day. A system of this size would cost less than $5,000 per year to operate, including the cost of replacement elements, gas, electricity and cleaning chemicals.

Surprisingly the cost of the system itself is very competitive when compared on available filtration area, to conventional systems. The initial reaction is that continuous microfiltration is slower than conventional D.E. or Pad filtration and this is a fact. We must, however, remember that conventional procedures are repeated several times during the clarification process, so total filtration time must be considered.

COST OF PRODUCT LOSS

(PER 1000 LITERS)

VALUE OF PRODUCT

%LOSSES	$0.50	$0.90	$1.30	$1.70	$2.10
0.1	$0.27	$0.67	$1.07	$1.47	$1.87
0.2	$0.54	$1.34	$2.14	$2.94	$3.74
0.5	$1.35	$3.35	$5.35	$7.35	$9.35
1.0	$2.70	$6.70	$10.70	$14.70	$18.70
2.0	$5.40	$13.40	$21.40	$29.40	$37.40

Flow Rates

To maintain high flow on continuous systems, a variety of techniques have been developed. These include very high crossflow rates, liquid backwashing, and most recently, gas backwashing.

High Crossflow Rates

By operating membrane systems at very high crossflow rates, the level of foulants deposited can be reduced, and the flow rates can be shown to increase with increase in crossflow velocities. Now backwash technologies are replacing this approach.

This technique requires high energy input and can result in significant temperature increases in the beverage. The high volume of material being pumped from tank to tank can create a great deal of surface agitation which can increase oxygen pickup in the product.

Liquid Backwash

High crossflow can be combined with liquid backwash to reduce foulant formation.

Liquid backwash involves the pumping of filtered product backwards, under pressure through the membrane. That does have a positive effect on the flow rate, but still suffers from the disadvantages of high crossflow. Since clean product is forced back through the membrane it must be re-filtered. This procedure has the effect of reducing overall throughput.

Gas Backwash

This method involves flowing compressed gas through the membrane from the filtrate side. The gas flows through the membrane at a high velocity and will expand in volume by 600% on its way through the membrane. The effectiveness of removing foulants is greatly superior to other techniques.

This process does not require high crossflow rates. Resulting in low energy input and negligible changes in product temperature.

Product Quality

Off Characters

Unlike traditional filter aids, such as Pads and D.E., CMF systems are constructed from inert materials and add no off characters to the beverage. By eliminating several filtration steps their benefits are enhanced.

Tests indicate that the reduction in color due to CMF is less than one half of the reduction found from a single D.E. filtration. This is due to the very small amount of filtration media (Hollow Fiber) that is in contact with the product, hence the amount of color component required to saturate the media is very small.

Clarity

CMF produces beverages of a turbidity less than 0.2 NTU, irrespective of the feed quality.

Even the haziest beverage can be clarified to this degree in a single filtration.

Colloid levels are reduced to a level similar to other sterile filtration media.

Organoleptic Properties

At 0.2 micron nominal pore size, there are no flavor materials removed by the CMF membrane. Small amounts of flavor compounds can be absorbed by the membrane, as with color components, but the effect is minimal.

Microfiltration For the Beverage Industry

Applications

- Sterile prebottling filtration
- Sterile juice
- Clarification of difficult to filter beverages
- Removal of fining agents
- Termination of fermentation

Filtration Rates

While many variables can affect rates, the following table summarizes typical rates for various types of products.

Application - Wine	Flux (gpm/m^2)	Application - Juice	Flux (gpm/m^2)
Prebottling White	1.8 - 2.4	Apple (Settled)	0.5 - 1.0
Prebottling Red	1.0 - 1.5	Cranberry (Pressed)	0.25 - 0.5
Post Fermentation	0.6 - 1.2	Lime	0.25 - 0.5
Cold Stabilized White	0.8 - 1.0	Pear	0.2 - 0.5
Bentonite Removal	0.5 - 1.0		
Botrytis Affected Wine	0.3 - 0.5		
Juice	0.4 - 0.7		

Filtrate Quality

Analytical

Microbiological counts consistently 0/100 ml from feeds as high as 8×10^8/100 ml.

Turbidity reduced to 0.5 NTU from feed streams over 500 NTU.

Complete removal of suspended solids.

"No significant difference" from controls after UV absorbance, gel filtration and HPLC for anthocyanin, protein, polysaccharide, and phenolic compounds.

Colloid Levels After Various Filtrations

	Test 1	Test 2
Unfiltered	3000	600
D.E. Filtration	920	470
Sterile Pad	360	270
0.45 Membrane	320	220
CMF	325	215
Ultrafiltration	62	20

(Colloid Levels Determined to DIN Standard)

Valuing The System

Cleaning Chemicals	________	Dollars/Cleaning	
Cleaning Frequency	________	Cleaning/YR	
		Cost of Cleaning	________
Operating Hours	________	HRS/YR	
Energy Consumption	________	KW	
Cost of Electricity	________	Dollars/KW	
Days of Operation	________	Days	
		Cost of Operation	________
Life of Modules	________	YRS	
# of Modules	________	Modules	
Cost of Modules	________	Dollars/Modules	
		Cost of Modules	________
Backwash Frequency	________	Times/HR	
# of Modules	________	Modules	
Hours of Operation	________	HRS/Day	
Days of Operation	________	Days	
Volume of Gas	________	CU/FT/MOD	
Cost of Gas	________	Dollars	
		Cost of Backwashing	________
Cost of Spare Parts	________	Pump Seals	
	________	Gas Filters	
	________	Pressure Gauges	
	________	________	
	________	________	
		Cost of Spare Parts	________
		Cost of Service	________
Cost of Labor	________	Dollars/HR	
	________	HRS/Day	
	________	Days/HR	
	________	Burden	
		Cost of Labor	________
		Total Cost	________
Volume Filtered	________	G.P.H.	
	________	HRS/Day	
	________	Days/YR	
		Total Gallons Filtered	________
	Cost Divided By Gallons Filtered		________

Perform the same analysis for the conventional filter.

Conclusions

Continuous Microfiltration avoids most of the inherent disadvantages of storage, handling, disposal, plugging and blinding which have become an accepted penalty for having to contend with such systems.

A technolofy is now available which will eliminate or enhance several stages of the filtration process. In many instances Continuous Microfiltration is becoming the best and most economical method of separation.

The benefits derived from Continuous Microfiltration are:

A. Lower Energy Costs

B. Decreased Labor

C. No Media Storage

D. Reduced Health and Safety Problems

E. Better Yeilds

F. Consistent Quality

G. Lower Maintenance Costs

H. No spent Media Disposal

We are now at a time where the industry is looking closely at the total costs involved in the filtration process. Continuous Microfiltration is not for everybody or every process, but it is the most versatile and economical piece of filtration equipment available. This is an efficient tool for producing more consistent products at lower costs.

I feel quite confident that when all the factors are considered, you will agree that this new Technology has tremendous potential.

DEVELOPMENT OF CERAMIC CROSS FLOW FILTERS FOR PARTICULATE CONTROL

T. E. Lippert
G. B. Haldipur
E. E. Smeltzer

Westinghouse Science & Technology Center
1310 Beulah Road
Pittsburgh, PA 15235

Introduction

The ceramic cross flow filter device is under development for particulate control in applications for high temperature, high pressure (HTHP) systems as Pressurized Fluidized Bed Combustion (PFBC), and Integrated Gasification Combined Cycle (IGCC) Power generation. The cross flow filter element is comprised of thin porous ceramic plates that contain channels formed by ribbed sections. The plates are stacked and fired to form a monolithic porous structure.

Process gas and particulates flow into the short, dirty side channels of the cross flow filter. Gas then passes through the porous ceramic plates that form the "roof" and "floor" of the channels and into the longer channels that form the clean gas side. One end of the clean side channels is sealed to force the filtered gas to flow to a central collection plenum where the filter is mounted. Dust collected in the short side channels is periodically removed by reverse pulse cleaning. The major attributes of this filter concept are its absolute particulate filtration, and its capability to be operated at relatively low pressure drop. Since each filter plate represents a filter surface, the cross flow configuration has a very high filter surface area to volume ratio, is compact, and potentially economical.

Approach

Since 1986 Westinghouse has focused its technology emphasis on developing the use of commercial-scale (12x12x4 in.) cross flow filters primarily for 1600-1700°F pressurized fluidized bed combustion (PFBC) cleanup applications. In-house filter qualification tests using the high temperature-high pressure (HTHP) Westinghouse single element simulator, have evaluated the influence of channel design and dimensions, as well as element mounting fixtures on filter performance and durability. Multielement durability testing has achieved over 750 cumulative hours of operational life at simulated PFBC conditions in the presence of re-entrained ash fines. Testing is scheduled to continue into 1990, tentatively accumulating 3000 hr of service life.

Westinghouse has also conducted in-field testing of multiple cross flow filter elements at the KRW PFB combustion simulator facility in Madison, PA, at the New York University PFB combustor, and at the Texaco entrained-bed gasifier in Montebello, CA. Future testing is planned at the Foster Wheeler second generation PFBC unit in Livingston, NJ, under both oxidizing and reducing conditions.

Results

Qualification and Long-Term Durability Testing

In-house HTHP testing of alumina/mullite cross flow filters demonstrated highest delamination resistance when the elements were manufactured with a mid-ribbed bond, one inch gas channel, and had a radiused flange. To date, long-term durability of this filter element configuration has been demonstrated during the >750 cumulative hrs (3000 scheduled hrs) test of a 2-4 element, double module array at the newly constructed Westinghouse HTHP facility. These filters which experience 1500-1600°F, 85 psi, 1000-2000 ppm re-entrained ash fines, 6-10 ft/min gas face velocities, permit only <1 ppm of dust to penetrate. After approximately 1000 pulse gas cleaning cycles, the filters continue to exhibit a consistent pressure drop recovery after each pulse.

Subpilot Pressurized Fluidized Bed Combustion Applications

KRW PFBC Simulator Facility. Westinghouse successfully completed 160 hours of testing of an 8 element, 4 module, mid-ribbed bond, cross flow filter array at KRW PFB combustion simulator test facility in 1987. The 12x12x4 inch cross flow filters used in these tests were exposed to temperatures between 1550-1625°F, pressures of 78-115 psia, and operated with average face velocities of 4.3-6.1 ft/min. Isokinetic sampling (total mass) of the filter outlet gas flow measured dust loadings of 2.2-5.2 ppm, in comparison to inlet dust loadings of 1000-2000 ppm.

During the course of continuous testing, the filters experienced 54 pulse cleaning cycles, and demonstrated an overall baseline pressure drop increase from 17 to 23 in. wg. The resulting minimal loss of filter permeability, suggested effective cleaning and long-term filter baseline pressure drop stability.

Following testing, the filter elements were disassembled and inspected. Of the 8 filters, cracks which permitted dust to breach the clean gas plenum were identified in 2 filter elements. The remaining 6 filter elements appeared virtually intact -- free of structural failures, and had no visible evidence of dust accumulation along their clean side channels. No visible evidence of dust leakage from the filter gasket seals was identified. Based on the very low outlet dust loadings that were measured, delamination or small cracks do not appear to lead to catastrophic cross flow filter failure.

New York University PFBC Facility. Westinghouse successfully completed 100 hours of testing of a 15 element, 5 module, mid-ribbed, cross flow filter array at the New York University PFBC facility. During the first 50 hours of testing, 12x12x4 in. cross flow filters were exposed to temperatures of <1500°F, pressures of 105 psig, nominal face velocities of 5.2 ft/min, and had a nominal baseline pressure drop of 35 in wg. (conditioned). Inlet PFBC dust loadings of 305-1056 ppm were reduced to outlet dust loadings of 2.9-8.9 ppm.

During the first 50 hours of operation a slight dust leak was identified, possibly indicating a filter delamination crack. Post test inspection of the filters with a borescope confirmed that one of the 15 elements had experienced a partial flange crack on the clean gas side. This element was removed from the filter module, and a replacement element was installed. All other filter elements appeared to be in excellent condition (no evidence of cracks or dusting on the clean side).

During the second 50 hours of testing, the filters experienced temperatures of 1300°F, pressures of 105 psig, and nominal face velocities of 10 ft/min. After 3 pulse solenoid valve failures, a hot gas leak in one of the facility knife valves, and a facility computer system malfunction which resulted in an inlet gas temperature excursion to 1880°F, dust loadings were observed to increase from 9.9 to 103 ppm. Following termination of the test, a preliminary borescope inspection of the filters showed significant dust in 2 of the 5 filter plenums, and some dusting in a third. After disassembly, 10 of the filters were shown to be undamaged, 5 had delamination cracks, and 9 had dust seal leaks. It was evident that the majority of dust penetrating the system occurred through the dust seal leaks rather than through the delamination cracks.

Subpilot Coal Gasification Applications

A 48 hour commissioning test was conducted at the Texaco entrained bed gasification plant in Montebello, CA, in April 1989. The gasifier was operated at 350 psig pressure in an oxygen blown mode with a Pittsburgh No. 8 seam coal, in the absence of in situ sulfur sorbents. Four cross flow filter elements were mounted on 2 stainless steel plenums attached to a metal tube sheet inside a 40 inch diameter refractory lined pressure vessel. The filters were successfully operated at temperatures of 876-1016°F, pressures of 330 psig, with face velocities of 0.9-3.3 fpm. During testing, the filters were periodically cleaned for a total of 270 pulses, and essentially exhibited consistent pressure drop recovery after each pulse. During filter operation inlet mass loadings were identified in the range of 2300 ppm, with outlet mass loadings of <5 ppm.

Future efforts will include testing of the cross flow filters in conjunction with Texaco's desulfurization program. This will include a 4 day air blown test using a Pittsburgh seam coal, and exposure to limestone and/or dolomitic sorbents. Westinghouse anticipates a minimum of 250 hours of successful operation of the full-scale ceramic cross flow filter system in the integrated gasification environment.

Conclusions

Significant technical advances have been made in the development of high performance cross flow filter elements. Long-term operational durability of the state-of-the-art cross flow filters is currently being exhibited at Westinghouse in the newly constructed HTHP test facility. The present filter element design minimizes crack formation, particularly at the base of the flange, and provides high bond strength between adjoining plate seams.

The relative chemical stability which was exhibited by the alumina/mullite cross flow filter material composition during exposure to KRW, NYU or Texaco process gases demonstrates the versatility of its use under either combustion or gasification process conditions.

CERAMIC BARRIER FILTRATION FOR DIRECT COAL FUELED GAS TURBINES

M. D. Stephenson, R. T. LeCren,
and P. B. Roberts
Solar Turbines Incorporated
2200 Pacific Highway
P.O. Box 85376
San Diego, CA 92138-5376

Introduction

Solar Turbines is developing a coal fueled version of its 3.8 MW Centaur H gas turbine for cogeneration applications. To protect gas turbine components from erosion and corrosion due to impacting particulates, and to meet New Source Performance Standards for particulate emissions, ceramic barrier filters are being employed. The test program includes evaluation of silicon carbide candle filters and Nextel 440 ceramic bag filters. Both the candle and bag filter tests are 50 h each using the particulate-laden gas stream from a two-stage slagging combustor system. This gas stream is at 1900°F. The testing includes determining collection efficiency and obtaining pressure drop versus time profiles for each device. Additionally, exposure testing of the candle filters is used to determine whether loss in strength, or changes in the chemical or mineralogical structure have occurred. This includes four-point bend testing, X-ray diffraction and scanning electron microscopy with energy dispersive X-rays.

The direct coal-fueled gas turbine (DCFT) system is comprised of a two-stage slagging coal combustor, a sorbent injection system for sulfur control, an impact separator for removal of molten slag and a ceramic barrier filter for particulate control. A one-tenth scale system, including all of the above components, was recently commissioned for the filter tests. In this paper, only the ceramic barrier filter system is discussed. Details of the overall system are described elsewhere[1].

Ceramic barrier filters

High temperature, high pressure (HTHP) particulate control technologies that are presently under development include barrier filters, granular bed filters, electrostatic precipitators, acoustic and electrostatic agglomerators, impact separators and nested fiber filters[2-3]. In Solar's slagging combustor system, particulate levels can reach 6000 ppmw due to injected dolomite or hydrated lime for sulfur control. It is this particulate that must be removed prior to gas entering the turbine. Also, environmental regulations mandate that over 99% of the particulate be removed. With the current state-of-the-art, these high levels of particulate removal are achievable with barrier filters which include candle filters, bag filters and cross flow filters. All function in similar ways, although differ markedly in the overall structure and materials. Periodic pulse cleaning is used to dislodge an accumulating dust cake.

The initial testing includes candle and bag filters only. Cross flow filters may be tested at a later date.

Collection efficiency is being determined by isokinetically sampling the inlet and outlet gas streams and passing the sampled gas stream through a Balston absolute filter. The face velocity (acfm/filtration area) is measured accurately by using a flow nozzle downstream of the filter vessel. In addition, the effective permeability (filter face velocity/filter pressure drop) will be measured after each cleaning cycle when the candles are effectively in a clean condition. This parameter indicates whether the filters are blinding over time due to particles imbedded in the interstices of the filter.

Secondary filter test facility

The refractory-lined filter vessel is 4 ft in diameter by 16 ft tall. The unfiltered gas enters at the bottom of the vessel and is directed downward towards an impaction plate. The gas sweeps upward and filters through the filtration device and exits from the downstream plenum into the exhaust piping. The filtered gas then enters a refractory-lined flow nozzle where the mass flow rate and hence the filter face velocity are accurately measured and displayed in real-time. The exhaust gases are quenched downstream of the nozzle before passing through a back-pressure valve. Two back-pressure valves are employed to control the flow rates of the filtered gas and the bypass. This permits the face velocity to be set accurately.

During filtration, the ceramic barrier filters are periodically cleaned by a pressurized pulse of filtered nitrogen. The duration of the pulse, generally less than one second, is set precisely using a programmable cycle timer. During a cycle, each manifold is purged once. The cycle time, i.e., the time interval between cleaning cycles, can be as short as 3 min or upwards of an hour depending on the dust-loading and other factors. This time is ultimately a function of the bulk density and mass flow rate of the dust and the properties of the filtration device. Typically, the onset of purging is driven by pressure drop or by elapsed time. Both methods will be employed in this testing.

The candle purge system is comprised of four manifolds (three for the eight bag filters), each of which contains a solenoid and a ball valve. In the event a solenoid valve sticks in an open position for longer than a second, a controller will close the air actuated ball valve. This design is intended to keep the candles from being thermally shocked by an inordinately long pulse of cold gas.

Because of the high temperature DCFT environment, certain components of the system, such as the tubesheet and nitrogen purge tubes, are either water-cooled or constructed of high temperature alloys. The candles or bags are suspended from water-cooled tubesheets. The tubesheet is basically a disk of type 316 stainless in which is embedded a serpentine channel of cooling water to maintain the tubesheet surface temperatures at 400°F. Twenty-two thermocouples are surface mounted on the tubesheet to monitor the upstream and downstream temperature profile. To minimize heat losses from the gas stream, Kaowool insulating boards (types 12-C and HS-45) are mounted on both upstream and downstream sides. The purge gas tubes are constructed of Haynes alloy 230. These tubes subjected to a very high temperature environment and experience stresses from the high pressure pulse gas used.

Ceramic candle filters

Ceramic candle filters are among the most promising HTHP devices presently available. These devices are typically about 4.5 ft in length and 2-4 in. in diameter. The ceramic is highly porous with an average pore size on the order of 40 μm.

For these tests, 14 Schumacher dia-Schumalith candle filters are being used to give a total filtration area of 42 sq ft. At a face velocity of 12 fpm, this corresponds to about one-third the maximum flow in this system or roughly 3% of the full scale system. Should a filter of these dimensions be used in the full-scale system, and at a face velocity of 12 fpm, nearly 400 candle filters would be necessary to filter the entire flow. These filters are noted for a low pressure drop at relatively high face velocities. For instance, at 12 fpm, these candles (without dust) exhibit about a 1 psi pressure drop. The candles are comprised of silicon carbide grains that are imbedded in a clay binder to impart strength. This particular candle design has two layers. The outermost layer has fine pores and is responsible for the filtration. It also contributes most to the pressure drop. The inner layer has much larger pores, a relatively low pressure drop, and provides structural support.

The tubesheet is a 43 in. diameter, 2 in. thick disk of type 316 stainless steel with an embedded serpentine water channel for cooling. The candles are mounted to the tubesheet using a patented refractory bonding method. Basically, the silicon carbide candle is cast into a Hastelloy-X cup. Tests have shown that this bond is quite strong and gas tight. A

Hastelloy-X sleeve is welded to the tubesheet and the cup and candle assemblies are edge-welded to the sleeve.

Ceramic bag filters

Ceramic bag filters are flexible filters comprised of ceramic fibers that are woven as a cloth. For these tests, Nextel 440 seamless bags are being used. These alumina silicate bags measure 4.25 in. in diameter and 66 in. long. Eight of these bags equate to 49 sq ft of filtration area and will be mounted from filter cages constructed from RA-330 expanded metal. The cages themselves are welded directly to the tubesheet. A maximum face velocity of 6 fpm will be employed for these tests which amounts to about a one-fourth slip stream.

Candle qualification testing

Acurex Corporation is under contract to examine unexposed and exposed candle filters. Scanning electron microscopy with energy dispersive X-rays (SEM/EDX), X-ray diffraction (XRD), and four-point bend testing are among the techniques being employed. Two samples are cut from each of three exposed and unexposed candles for SEM/EDX examination. Photographs are being taken from the SEM/EDX. Elemental compositions are determined qualitatively for species having molecular weights equal to or greater than sodium.

One XRD sample is taken from each candle. The XRD spectra enables comparison of the crystallinity and phases of the various components and determination of changes in crystalline size or composition which might occur. Four beams are also cut from each filter so that strength can be determined by four-point bend testing. The modulus of rupture determined from this test is essentially the same as the tensile break strength.

Additional testing to be done at Solar includes visual inspections and measurement of candle dimensions and curvature, C-ring or O-ring tests before and after exposure and if appropriate, additional SEM/EDX and XRD testing can be done on more candles.

References

1. P.B. Roberts, "Solar's advanced coal-fueled industrial cogeneration gas turbine system - a status report," <u>Proceedings of the Fifth Pittsburgh Coal Conference</u>, September, 1988.

2. R.C. Bedick, V.P. Kothari, "Gas stream cleanup papers from DOE/METC sponsored contractor's review meetings in 1988," DOE/METC-89/6099 (DE89000901). October, 1988.

2. S.C. Saxena, R.F. Henry, W.F. Podolski, "Particulate removal from high-temperature, high-pressure combustion gases," <u>Prog. Energy Combust. Sci.</u> 11: 193. (1985).

LIST OF KEYWORDS

bag filter
barrier filters
candle filter
coal
coal combustion
dia-Shumalith
direct coal fueled gas turbine
gas turbine
hot gas cleanup
Nextel 440
particulate removal
slag
slagging combustor

HOT GAS STREAM PARTICULATE REMOVAL WITH AN ACOUSTICALLY ENHANCED CYCLONE

Douglas C. Rawlins
Tammara K. Grimmett
Solar Turbines Incorporated
2200 Pacific Highway
San Diego, CA 92138-5376
(619) 544-5023

Abstract

Acoustically enhanced cyclone collectors offer the potential of achieving environmental particulate control standards under pressurized fluidized bed combustor (PFBC) conditions (1650 °F, 10 atm), without the need for post turbine particulate control. This paper reports on the development of a subpilot-scale test facility to study the effects of specified operating parameters on acoustic agglomeration and cyclone collection efficiency of fine fly ash. The major test facility components include a simulated PFBC effluent stream, a variable residence time acoustic agglomeration chamber, a natural gas fired pulse combustor sound source, and a two-stage high temperature, high pressure cyclone. The operating parameters to be studied include ash residence time within the sound field, sound frequency and intensity, gas stream particulate loading, and particle size distribution.

Introduction

The successful development of coal-fired pressurized fluidized-bed combined cycle combustion systems for electric power generation depends heavily on the development of an efficient and economically viable cleanup system for the high temperature, high pressure combustion gases. This paper presents work being performed on acoustically enhanced cyclone collectors.

Acoustic agglomeration can be described in simplified form as follows. As an acoustic wave passes through an aerosol, it generates an oscillating gas velocity. Particles suspended in the gas stream oscillate with the vibrating gas due to the drag induced by the gas viscosity. The magnitude of the particle motion depends upon the particle size. Very small particles oscillate exactly with the gas, but large particles remain motionless relative the oscillating gas due to a larger inertia-to-drag force ratio. The different amplitudes of motion of the various sized particles cause them to come into close contact, or collide with each other. Van der Waals molecular forces hold these agglomerates together once the collisions have occurred.

Facility Design

The goal of this program is to extend acoustic agglomeration technology from the laboratory-scale up to the subpilot-scale. The effects of high intensity sound on the agglomeration of fly ash particles will be studied in a simulated PFBC effluent stream. The sound generator for this system will be a natural gas fired pulse combustor developed by Manufacturing and Technology Conversion International, Inc.

The agglomeration chamber has been designed and built as a vertical, refractory lined, unit with upward flowing gases. The chamber has been designed for a fixed volumetric flow rate of 200 acfm with a maximum residence time of approximately 3 seconds. A natural gas preheat combustor is located at the bottom of the agglomeration chamber to heat the bulk air stream to an average temperature of 1650 ^{o}F. Ash is fed into the system through one of three ports located along the chamber length, using a high pressure rotary disk feeder. The location of the ash injection ports provides, respectively, 1, 2, and 3 second residence times for the ash prior to the ash sampling probe. A pulse combustor sound generator is located at the top of the chamber next to the 2-stage cyclone entrance. This position for the sound source generator provides for easy removal and replacement of the pulse combustors to achieve different sound frequencies. Pulse combustor sound generators have been designed to operate in the sound frequency range from 1000 to 3000 Hz at an intensity level between 140 and 165 db. Since the frequency of the pulse combustor is largely fixed by design, three different combustors have been provided to cover the desired frequency range for the test conditions. The 2-stage refractory-lined cyclone system has water cooled hoppers below each stage, with high temperature knife gate valves above and below each hopper for batchwise ash removal.

Agglomerated ash samples are removed from the agglomeration chamber through a water cooled isokinetic sampling probe located near the top of the chamber, and passed through a Cyclade multiple cyclone particle sizer for size distribution analysis. This probe has the capability of traversing across the diameter of the chamber to obtain particulate samples at different radial locations in the duct. Gas temperature measurements are made with thermocouples at the exit of the preheat combustor, and also at the entrance to the cyclones so that the temperature drop through the system can be determined. Wall temperature measurements are obtained from thermocouples embedded in the refractory insulation. A pressure tap is located on the chamber to measure system pressure. Sound intensity and frequency from the pulse combustor are measured with PCB Piezotronics sound pressure transducers. These water-cooled transducers are located next to the isokinetic sample probe and ash injection ports. Two optical view ports are located adjacent to the sample probe port at the top of the chamber. These optical ports allow visual observation within the agglomeration chamber and also provide capabilities for laser-based particle sizing instrumentation in possible future work in this facility.

Test Plans

The acoustically enhanced cyclone collector will be evaluated with three distinct goals in mind: 1) determine the effects of the major operating parameters on acoustic agglomeration efficiency, 2) optimize the agglomeration system performance for maximum particulate removal efficiency, and 3) demonstrate the system durability over a meaningful time frame. During the agglomeration tests, ash samples will be collected from the bulk gas stream at the exit of the agglomeration chamber, between the two cyclones, and also from the discharge pipe of the final cyclone. The main dependent variables of interest will be the mass mean particle size and size distribution of the ash before and after the agglomeration process, and the cyclone collection efficiency of the unagglomerated and agglomerated particulate streams.

Six independent variables have been identified for testing during the detailed parametric tests. These are: 1) the intensity (140-165 db), and 2) frequency (1000-3000 Hz) of the sound generator, 3) the ash residence time within the agglomeration chamber (1-3 seconds), 4) the ash loading in the bulk gas stream (1-12 gr/scf), 5) the operating pressure of the agglomeration chamber (50-150 psig), and 6) the particle size distribution of the fly ash (4-30 μm mass mean). These tests will be performed using a statistical experimental design approach.

The final test sequence will be a 200-hour durability assessment of the agglomerator system using the optimal test conditions for fly ash agglomeration. Periodic particulate sampling will be conducted during the durability testing to document any change in agglomeration rate and collection efficiency with time. Since the major impediment to long life with this system is likely to be acoustic fatigue rather than erosion by particulates, the particle feed system will be periodically turned off (keeping the pulse combustor sound generator operating) to allow for batch mode removal of the ash from the collection hoppers below the cyclones and refilling the ash feeder. The duration of time in which the feeder is on per sequence and the total number of sequences will be dictated by the ash loading level chosen for this test.

At the end of the durability test, the pulse combustor, agglomeration chamber, and cyclones will be inspected to identify any degradation in mechanical integrity. Additional inspections may be called for during the durability test if a significant decrease in collection efficiency occurs at some point in the test.

The data obtained during these tests will provide direct information on the effects of sound intensity and frequency, particle residence time, ash loading, pressure and durability on the enhancement of cyclone collection efficiency with acoustic agglomeration. These data, in conjunction with results from an acoustic agglomeration computer model will be used to evaluate the economic and engineering feasibility of an acoustically enhanced cyclone collector for a full-scale pressurized fluidized bed combustor. A major goal of this program will be to establish whether a full-scale agglomerating cyclone system can meet NSPS particulate emission standards and still provide a 15% reduction in the cleanup system fixed and operating costs relative to conventional hardware. If this cost reduction goal can be met, a commercialization plan for the scaled-up hardware will be prepared.

Current Status

As of the writing of this paper, the test facility installation is nearing completion. Acoustic agglomeration testing is expected to begin in early 1990, with final results obtained by mid-year 1990.

Keywords

Hot Gas Cleanup
Particulate Removal
Acoustic Agglomeration
Agglomeration
Pressurized Fluidized Bed Combustion (PFBC)
Ash
Fly Ash
Cyclone
Enhanced Cyclone
Pulse Combustion
High Intensity Sound
Coal
Coal Combustion
Experimental Testing
Subpilot-Scale Facility
Gas Turbine
Electric Power Generation
Particle Size
Submicron
Micron

HIGH TEMPERATURE FILTRATION
USING CERAMIC FILTERS

L. R. White and S. M. Sanocki
3M Co., 218-3S-03
St. Paul, Minnesota 55144-1000
(818)767-4888

INTRODUCTION

High temperature filters are needed for advanced coal conversion processes such as pressurized fluidized bed combustion (PFBC) and coal gasification/gas turbine combined cycle systems. Filters would remove particulates from hot gases before expanding them through turbines. Several approaches have been taken to hot gas filtration including baghouses, ceramic cross-flow filters, fixed and moving granular beds, etc. In this paper the focus is on baghouse filtration using both fabric and rigid filters with cleaning by pulse-jet. Two kinds of filters are discussed, ceramic fabrics and porous, rigid structures called candles.

Fabrics and finishes

In fabric filtration at temperature less than about 500 F, fabric finishes are used to minimize abrasion. At high temperatures such finishes cannot be used. However, if the fiber used in the filter is harder than the particulate being collected, then the particulate acts like a solid lubricant.

In weaving ceramic fabrics yarn must be selected carefully. Some yarns are too stiff to weave or are damaged during weaving. Key parameters are tensile modulus and diameter of the individual fibers in the yarn. The higher the modulus, the stiffer the yarn and the harder it will be to weave. But, the ability of fibers to tolerate sharp bends depends strongly on filament diameter and stiff yarns can be woven if filament diameter is small enough.

The ceramic fabric discussed here is woven from an alumina-boria-silica yarn (Nextel 312™-available from 3M Co.).

Seams

Conventional fabric filters are cylindrical in shape and prepared by sewing an axial seam and early versions of ceramic bags were made with seams. However, seam failures occurred, and woven seamless tubing was developed. Seamless tubing is flattened during weaving and takeup processes and the edges suffer some yarn damage, nevertheless, seamless tubing is stronger than tubes with seams.

Hardware

Ceramic fabrics are damaged during cleaning if they are supported over wire cages. On the other hand, hundreds of thousands of pulses have been applied to ceramic fabrics disposed over perforated metal cages with no failures observed. Most filter bags have a sewn-in bottom and the cage is completely covered by the bag. Although ceramic fabric bags can be produced with one end closed they are abraded by the bottom of the cage. As an alternative bags have been made which are open at both ends. Installation has been done by clamping to the cage at both top and bottom. Cages had to be made with solid bottoms of course. Good life has been observed with this method of installation. It has been necessary to use temperature-compensating clamps since ordinary clamps will not hold fabric in place after a few cycles in temperature.

Axial expansion of cages must be accounted for when bags are installed. Ceramic fabrics can be exposed to temperature swings in excess of 1000 degrees and the mismatch in coefficient of thermal expansion between the cage and the fabric can stretch the fabric. If cage growth is excessive then bags must be installed with longitudinal slack.

It might appear that use of metal cages and tube sheets leads to a restriction in filtration temperature. From the standpoint of filtration alone, the upper limit in temperature is determined by stickiness of the particulate. When particles begin to fuse filtration becomes difficult because the cake blinds and is hard to remove by pulse-cleaning. In filtering flue gas from coal this temperature depends on whether oxidizing or reducing conditions are present and the composition of the ash. It can be as low as 1700 F. In incineration where a variety of inorganic constituents are present and low-melting eutectics can be expected to form, the upper limit is much lower. In coal gasification, there is a temperature window defined on the low end by the dew point of tars and on the high end by the softening point of the ash. Therefore, the temperature at which a filter can be operated depends on the service more than on the metals used in the bag house.

In filtration of flue gas from combustion systems it is important to avoid collecting large glowing or burning particles. These "sparklers" can damage fabric if they fuse the dust cake. Molten ash can either react with the fabric and weaken it or simply embrittle it by rigidizing it. After all the fuel has burned out of a particle it will cool, solidify and if the ash has penetrated the fabric, rigidize it. Then when the fabric is pulse-cleaned the edge of this brittle region will tear. It is good practice to maintain a thick dust layer since the thicker the dust layer the larger the glowing particle that can be collected without damage to the filter. Thus overcleaning should be avoided. When high concentration of sparklers exists in the flue gas, mechanical separation of the largest of them is highly recommended and a variety of devices can be used for that purpose.

Chemical Attack

Ceramic fabrics used in filtration of flue gases are subject to chemical attack in two different ways, primarily. At temperatures greater than about 1400 F, alkalis have significant vapor pressure, will be present in the vapor phase, and will attack alumina-based ceramics. To avoid destruction of the ceramic fabric temperature must be low enough so that alkalis are present as solids only. If dust-covered fabrics are allowed to pass through the dew point then attack occurs after the fabric is reheated. The mechanism is not understood fully but this has been observed several times and there is no question that it occurs. A potential attack mechanism is reaction with low-melting eutectics which are leached from the ash and then concentrated on fiber surfaces.

PERFORMANCE OF CERAMIC FABRIC

Life test

A life test was run on ceramic fabric and test conditions along with results are in Table 1. Test conditions were set to simulate hot gas cleanup in the turbo-charged boiler concept of PFBC. The most important results are shown in Figure 1 which shows modest loss in fabric strength after 17000 hours.

Table 1 Typical operating conditions ceramic fabric filter test

Temperature, degrees C	400-450
Air/cloth ratio, m/min	1.8-2.0
Baghouse delta P, kPa	1.4
Flue gas rate, cu m/min	18-20
Collection efficiency	>99.95%

Application

In petroleum refining regenerated cat cracking catalyst is discharged to a storage hopper and the vent from that hopper must be filtered to capture catalyst dust. Because this stream is hot the simplest way to filter it involves a high temperature fabric. A baghouse with ceramic fabric filters was installed in the first quarter of 1988 and has been operating since.

CERAMIC CANDLES OF FIBER-REINFORCED SILICON CARBIDE

Candle filters are being developed which will be made by chemical vapor infiltration of silicon carbide on fibrous substrates.[1]

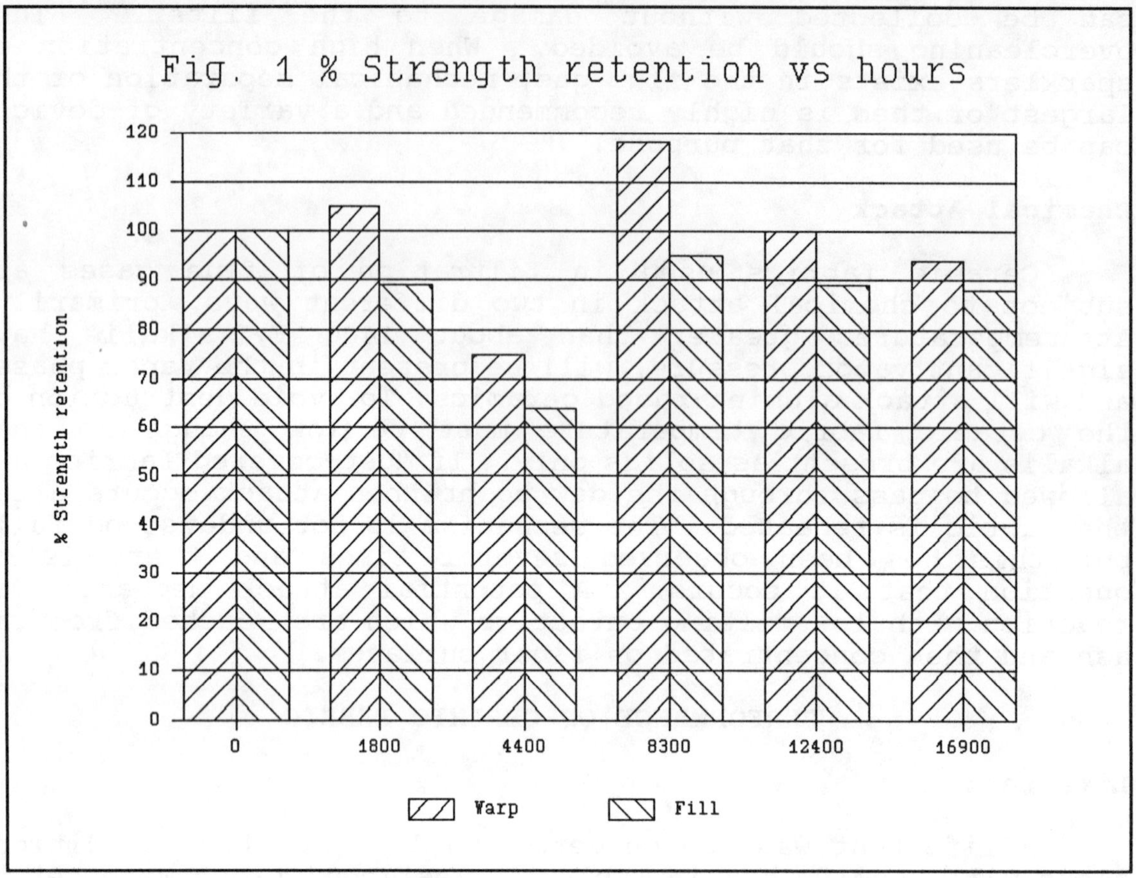

A candle is shaped like a large test tube, typical dimensions being five feet long by three inches in diameter. Wall thickness has been about 1/2 inch. A goal in making CVI candles is to reduce wall thickness thus reducing weight and also pressure drop. Less weight and pressure drop simplify tube sheet design and the thinner the wall, the shorter the filtration path and the more cleanable the filter.

Making CVI candles involves the following steps, making a ceramic preform, coating it with a carbon interface, and applying a ceramic coating by chemical vapor infiltration. The carbon barrier is needed to promote toughness.

At the time of writing this paper, preforms had been made and CVI had been scheduled.

---

[1] Stinton, D. P., R. A. Lowden, Ramsey Chang, Ceramic Engineering & Science Proceedings, vol. 9, no. 9-10, p.1233-1244, 1988

Keywords

Filtration
Hot gas filtration
Ceramic filter bags
Ceramic candle filters
CVI candles
Chemical vapor infiltration (CVI)
Ceramic/ceramic composite
Fiber-reinforced ceramic
Fiber-reinforced silicon carbide

THE GRANULAR BED FILTER FOR COAL FIRED GAS TURBINE PROTECTION

Keith Wilson
Combustion Power Company
1020 Marsh Road, Suite 100
Menlo Park, CA 94025

Introduction

Efforts to operate coal-fired gas turbines have been conducted in recent years. These attempts, like the one at Combustion Power Company in the early 1970's, have been plagued by turbine problems due to ash-ladened combustion gases. The U. S. Department of Energy is currently sponsoring programs to develop coal-fired, pressurized fluidized-bed combustors (PFBC's) to be used in combined-cycle power generating systems. These systems utilize a gas turbine driven by pressurized combustion gases to generate a portion of the electricity. It is recognized that a hot gas cleanup train must be used before the gas turbine to remove the major portion of the particulate. This is necessary to prevent erosion of turbine materials and deposition of particles within the turbine. The Granular Bed Filter (GBF) was developed to protect the gas turbine from particulate and in addition meets New Source Performance Standards for particulate emissions.

The design of this filter began in the early 1970's when Combustion Power was testing a gas turbine based power generator fired on coal and municipal solid waste. As part of this effort, work began on the development of the granular bed filter. The granules in these early tests were 2 mm alumina spheres arranged in various bed depths and configurations for removal of particulate from the hot flue gases. Gas cleaning is primarily by impaction and is analogous to filtering suspended solids from waste water through a sand bed. Tests of the pilot-scale filter element shown in Figure 1 at atmospheric pressure and at a temperature of about 1600F demonstrated successful operation over a 1500-hour test period. Collection efficiencies of 99% were obtained at a pressure drop at about 25 in H_2O, and degradation of the collection media did not occur.

Figure 1 shows the 5 ft. diameter, developmental GBF configuration, and Figure 2 shows the 5 ft. diameter filter essentials installed in the test pressure vessel. There are three zones of gas cleaning: a parallel-flow zone in which a high amount of ash is collected, a cross-flow zone, and a counterflow zone where the final ash collection/gas polishing takes place. Additional layers of media in the counterflow zone can increase gas cleanup effectiveness. A more compact and less costly filter will result by utilizing larger (3 mm or 4 mm) sized media.

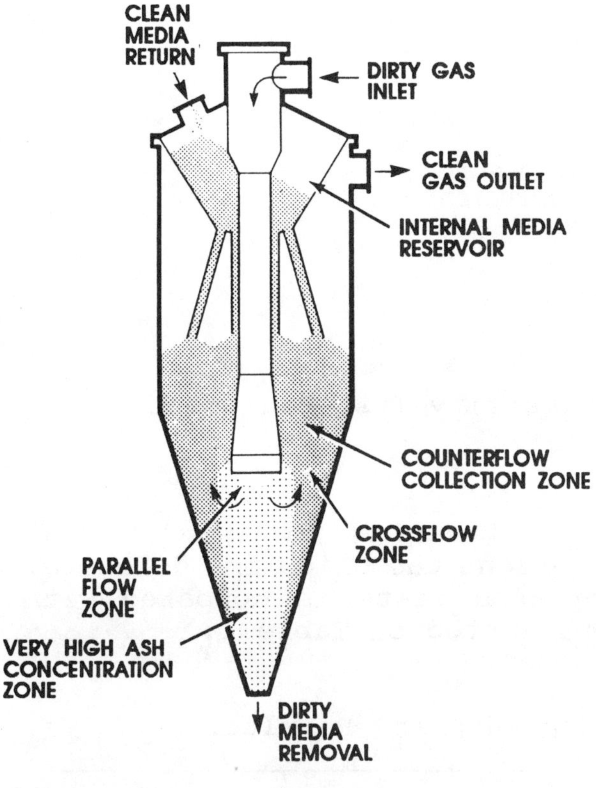

Figure 1. Screenless Counter-Flow Granular Filter Bed

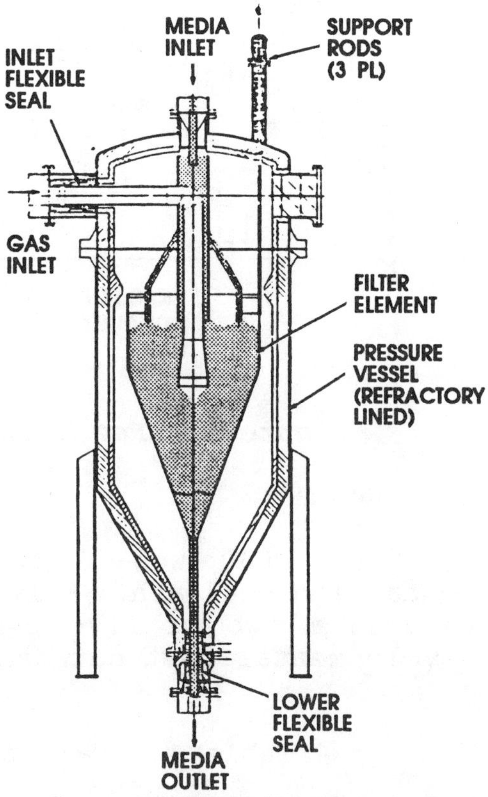

Figure 2. Granular Bed Filter Installation

Test Equipment

Figure 3 shows a schematic of a test unit recently utilized at New York University to demonstrate GBF effectiveness. Two to four diesel driven compressors operated against one or both sonic nozzles. Nozzle sizes were chosen prior to a test series to divide the PFBC gas flow between the filter and the bypass. Pressure changes during operation were normally made by adjusting the flow from the compressor bank. The unobstructed sonic nozzles were used to measure gas flow based on nozzle diameter, upstream pressure (above about 28 psia) and temperature. The GBF was complete with a pneumatic system to circulate media and remove ash from the filter on a continuous basis.[1] The filter is shown on Figure 2.

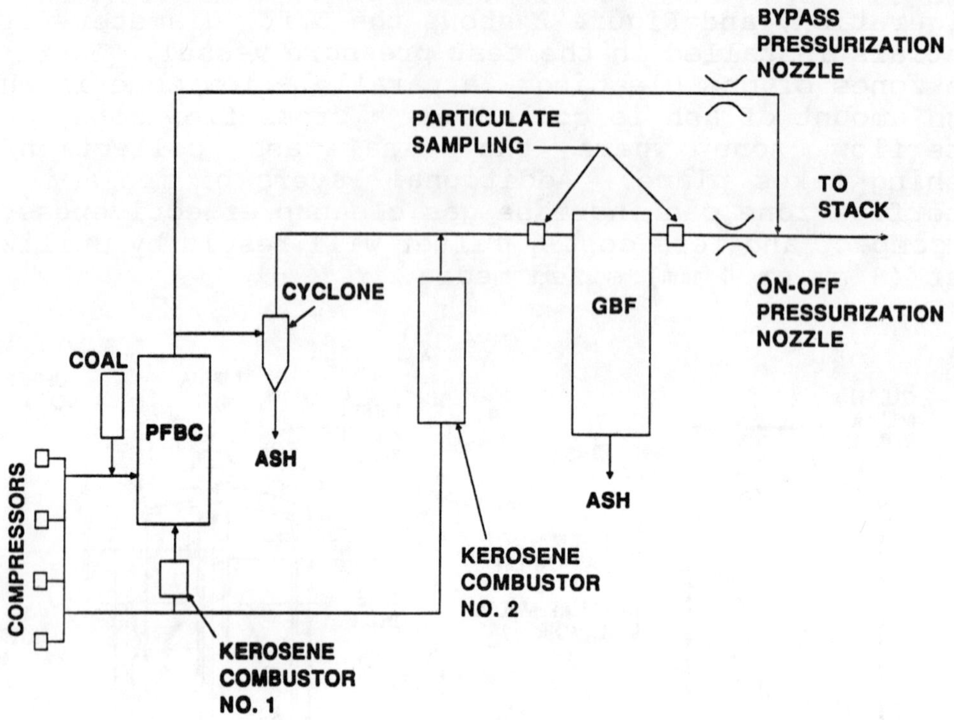

Figure 3. New York University Test Equipment

Test Results

The PFBC was fired on Kittanning bituminous coal containing 5-8% ash. Sulfur sorbent was Ohio lime dolomite. Data from two of five performance tests is compared with developmental test data and presented on Table 1.

Table 1. Particulate Sampling Results.

PARAMETER	LOW PRESSURE (DEVELOPMENTAL) GBF TESTS	CONTRACTURAL TARGET	REPRESENTATIVE NYU TESTS HG-204	HG-205
ASH CONCENTRATION				
@ GBF INLET, PPMW	1500-40,000	--	80-2800	160-1600
o AVG	--	≤ 1200	750	860
@ GBF OUTLET, PPMW	30-60	12	3-16	1-10
o AVG	--	~	6	4
EMISSIONS LB/10^6 BTU	--	< .03 LB/10^6 BTU	.003-.013	.001-.010
o AVG		(NSPS)	.005	.004
COLLECTION EFF %	98-99.2	99	91-99.9	98-99.3
o AVG	--	~	96.8	99.3

Inlet ash loadings varied over a wide range for many reasons but could be roughly controlled by adjusting sulfur sorbent feed (dolomite) to the PFBC. The averages shown for HG-204 are for 18 of 26 samples during 74 hours of operation where inlet loadings measured greater than 200 ppmw. For HG-205 there were 17 samples collected over 47 hours of operation.

Outlet loadings meet New Source Performance Standards (NSPS) of .03 lb/million Btu and meet turbine tolerance limits which actually can be more restrictive at large particulate sizes. With 3 mm media (HG-205) the outlet loadings were expected to increase over that measured at 2 mm (HG-204). One explanation for the higher efficiency of 3 mm media is that it was composed of alumina spheres ranging between 2.4 and 4.0 mm. The 2 mm media was more uniform at 1.9 to 2.0 mm. More opportunity for ash collection (by impaction, etc.) could exist with the wider size range of media than with an evenly sized media bed. Another explanation is that the higher gas velocity permitted by the larger media increased particulate collection by impaction.

Conclusion

The demonstration test at New York University was of short duration compared to the 1500 hours of developmental testing at low pressure. Nevertheless, it was a major step towards proving feasibility of hot gas cleanup utilizing a granular bed filter. Stable GBF operation was achieved downstream of the coal-fired PFBC simulating a gas turbine based power plant. Furthermore, particulate removal was achieved to below New Source Performance Standards and turbine tolerance limits. Design studies subsequent to testing show how the single filtration element can be packaged in a multi-element pressure vessel for commercial-scale use.

References

1. Wilson, K.B., Performance Evaluation of a Screenless Granular Bed Filter on a Subpilot-Scale PFBC, Proceedings of the Seventh Annual Gasification and Gas Stream Cleanup Systems Contractor's Review Meeting, DOE/METC 87/6079 (June, 1987)

A NEW AUTOMATED FILTER TESTER FOR LOW EFFICIENCY, HEPA GRADE, AND ABOVE FILTER MEDIA AND CARTRIDGES

Brian R. Johnson, Sandra K. Herweyer,
Edward M. Johnson, and Jugal K. Agarwal
TSI Incorporated
P.O. Box 64394
500 Cardigan Road
St. Paul, MN 55164

Abstract

An automated filter tester for production testing of filter media, filter cartridges, and respirator cartridges has been developed. Filter penetrations from 100 to 0.001% (efficiencies of 0 to 99.999%) can be measured with this tester using NaCl, DOP, or other oils as the challenge aerosol.

The automated filter tester's unique oil generator produces dioctyl phthalate aerosol with a 0.18 micrometer geometric number mean diameter with a geometric standard deviation of 1.6. This is accomplished by initial aerosol generation with Retec nebulizers and further conditioning with a coarse filter to narrow the size distribution.

A new photometer designed specifically for filter testing has been incorporated into this tester. This photometer features fast response and purge times, automatic gain selection, and a three position switching valve with a straight-through aerosol path on the inlet of the device. A solid-state laser diode and photodetector with 45 degree off-axis collection optics are used to measure aerosol concentrations from 1 microgram per cubic meter to over 100 milligrams per cubic meter.

Test data, including penetration, pressure drop, and flow rate, is output by the tester in three ways: via the digital display, through the front panel printer, and by the RS-232 serial port. The measurement ranges of the tester include:

filter penetrations down to 0.001%, filter pressure drop up to 150 mm of water, and challenge aerosol flow rates up to 100 liters per minute.

Comparison penetration data between the new automated filter tester and the commonly used Q-127 hot DOP machines show that the two methods have a one-to-one correlation.

Introduction

The demand for quality filters and filter media has seen enormous growth in the past decade. A few applications which have increased the demand for filters are: clean rooms in the semiconductor and pharmaceutical industries, filtering in Nuclear power facilities, and the increased use of particulate respirators in the workplace. Whether particulates are filtered to increase the yield in a semiconductor manufacturing situation, or to decrease the health risk for workers in hazardous environments, the ability of a filter to capture particulate matter is imperative for its user.

Efficiency and pressure drop testing of filters is required to determine filter quality. The efficiency of a filter is the ratio of the concentration of challenge aerosol retained in a filter during a test to the concentration which passes through. The pressure drop is a measure of the resistance of the filter to a given flow rate. Generally, the best filter for an application will have the lowest pressure drop at the required efficiency.

Traditionally, filter testing equipment falls into two categories. The first category consists of complete, commercially available testers which employ: flame photometry with NaCl, laser particle counting, or photometry with hot DOP. An example of this is the Q-127 which uses thermally generated DOP aerosol and a photometer. These machines were developed several years ago and lack the automation that is possible today. The other category is component systems that are put together using commercially available detection instrumentation and aerosol generators. While taking advantage of the latest technology, such systems require considerable technical expertise and expense to build and operate.

This paper describes a new automated filter tester (TSI Model 8110) that has been developed to meet the need for a complete turn-key tester based on current technology. It can operate with both solid (NaCl) and liquid (DOP, paraffin oil) aerosols. The importance of a complete automated tester is that it provides a test measurement rather than a group of instruments.

Data correlation of the results of new testing methods with traditional test methods is important for several reasons. With correlation, new data can be compared to past data to determine if there has been any long-term changes in production quality. In addition, data taken with a new method can be translated into a standard "language" for filter efficiency reporting, such as Q-

127 data in the United States. A correlation study to compare the new solid-state photometer based tester with the Q-127 hot DOP type tester is presented.

Technical Description of Tester

The new tester was designed to meet several needs, these include: ease of operation, accurate results, repeatable results--without operator variability, low maintenance, and fast test times.

A picture of the new automated filter tester is shown in figure 1. To operate the tester, the operator would simply insert a filter into the filter holder and press the "close" buttons to start a test. When the microprocessor controlled test is finished, the filter holder opens and the flow rate, pressure drop, and % penetration is given. A flow schematic of the tester is shown in figure 2.

Figure 1. TSI's Model 8110 Automated Filter Tester

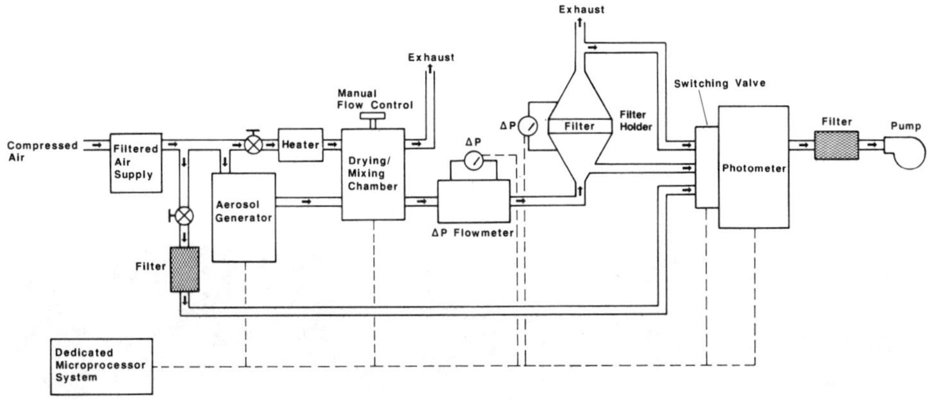

Figure 2. Flow schematic of TSI's Model 8110 Automated Filter Tester

The generated aerosol is mixed with dilution air in the drying/mixing chamber. This aerosol is then routed to the filter holder to challenge the test filter. The photometer measures the aerosol concentration before and after the filter. The flow rate is measured by monitoring the pressure drop across an orifice with an electronic pressure transducer. The pressure drop across the filter is also measured with an electronic pressure transducer which is connected to pressure taps before and after the test filter.

Some important design components of the new tester are: aerosol generation, aerosol detection, and the microprocessor controlled self-check diagnostics. These are described in detail in the following sections.

Aerosol Generation

The new tester includes two different aerosol generation systems. One is designed for liquid (oil) aerosol generation. The other is designed for solid (salt) aerosol generation. Both systems use Retec nebulizers which generate submicrometer particles by producing a fine spray of solution and impacting out the larger droplets. This paper will focus on use of the oil (DOP) generator for correlation with the Q-127 hot DOP machines.

The oil aerosol produced by the Retec nebulizers passes through a coarse filter before exiting the generator. This coarse filter narrows the oil size distribution. A diagram of the oil generator is shown in figure 3. The DOP used to challenge filters has a geometric mean diameter of 0.18 micrometers with a geometric standard deviation of 1.6, as measured by a Differential Mobility Particle Size Analyzer (DMPS) manufactured by TSI Incorporated, St. Paul, MN.

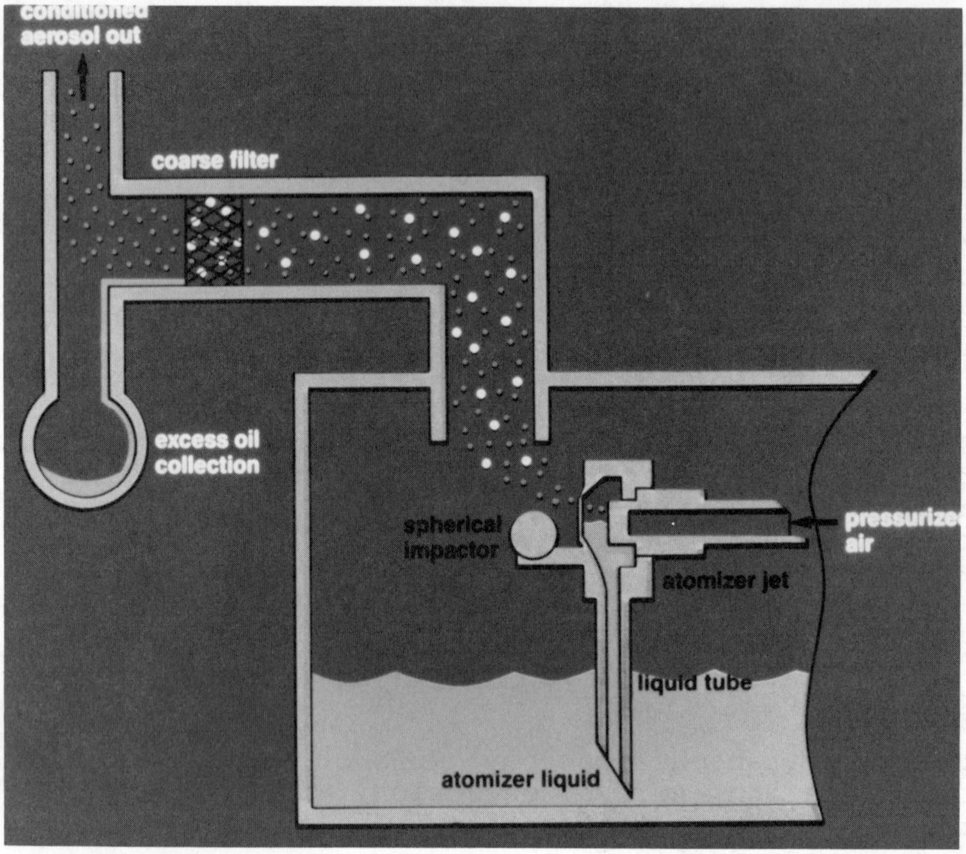

Figure 3. TSI's Model 8110 oil generator

Aerosol Detection

A light scattering photometer is used to make concentration measurements upstream and downstream of the test filter. A schematic of the photometer is shown in figure 4.

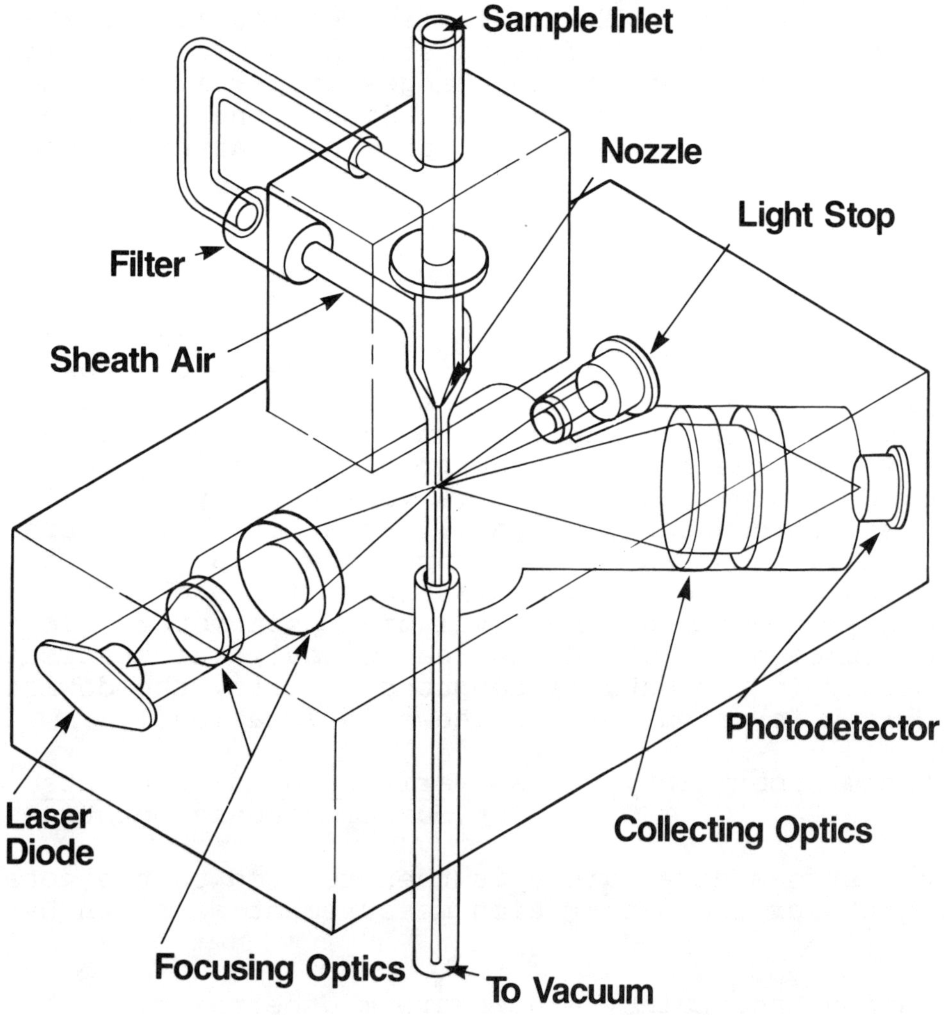

Figure 4. Photometer Schematic

The aerosol entering the photometer is split into two flow paths. One path is directed through a high efficiency filter and serves as clean sheath air around the aerosol sample being measured. The sheath air isolates the sample aerosol by encompassing it as it passes through the photometer, keeping the photometer optics clean as well as reducing the purge time between measurements.

The light scattering chamber contains a 5 milliwatt laser diode light source which is focused onto the aerosol stream. A laser diode light source was chosen because it is more stable than conventional white light sources which tend to change intensity over time. In addition, a laser diode light source can be collimated and focused more efficiently than a white light source.

The aerosol detection is accomplished by focusing the light scattered by the aerosol onto a photodiode. The total light scatter is converted to a voltage which is proportional to the aerosol concentration being measured. A photodiode offers stability over time. The collection optics are located 45 degrees off-axis from the light source to minimize light scattering variations due to changes in particle diameter. Forward scattering techniques accentuate the effect of larger particles which have a large portion of scatter in the forward direction.

Microprocessor Diagnostics

Statistical diagnostics. To ensure proper operation, the microprocessor based filter tester performs internal self-diagnostics during operation. One diagnostic feature is validation of the filter penetration data each time a filter is tested. The penetration data is validated by analyzing the downstream concentration signal. An accurate downstream signal is typically the area of most concern with a photometer-based filter tester. The signal downstream of a high efficiency filter can be near the background level of a photometer. To ensure a good measurement of the downstream signal, the statistical uncertainty in the data is computed. First, the downstream uncertainty is calculated as shown in equation 1.

$$\text{Downstream Uncertainty} = 2 \times \left(\frac{\text{Variance of data points}}{\text{Number of data points}}\right)^{1/2} \quad (1)$$

This downstream uncertainty is then related to the total uncertainty of the penetration measurement as shown by equation 2.

$$\text{Penetration Uncertainty} = \frac{\text{Downstream Uncertainty}}{\text{Upstream Signal}} \quad (2)$$

This penetration uncertainty is compared to the overall penetration measured. If the penetration uncertainty is greater than 10% of the total penetration measured, the test is invalidated. No data is output for an invalid test. This statistical check on the penetration verifies that the measured signal downstream of the filter is adequate and stable for a proper penetration measurement.

Other self-checks. In addition to the statistical analysis program, the tester monitors the photometer background level and upstream aerosol concentration level to alert the operator if a problem is diagnosed.

Data Correlation

As with all new technology, the problem arises in relating new test results with "standard" test results which have been collected over many years. To determine the correlation of the new tester with that of the Q-127 type hot DOP tester a correlation study was performed. This study included the testing of flat sheet glass fiber media produced by Lydall Incorporated, Technical Papers Division of Rochester, New Hampshire. Ten grades of filter media spanning the range of penetrations from 0.001% to 85% (ULPA to ASHRAE) were tested.

Both testers operated with dioctyl phthalate (DOP). The Q-127 tester uses a thermally generated DOP vapor. The new tester uses room temperature DOP aerosol produced by Retec nebulizers which is further conditioned by passing through a coarse filter. The flowrate through the media being tested was 32 liters per minute. The graph illustrates the correlation of the test results.

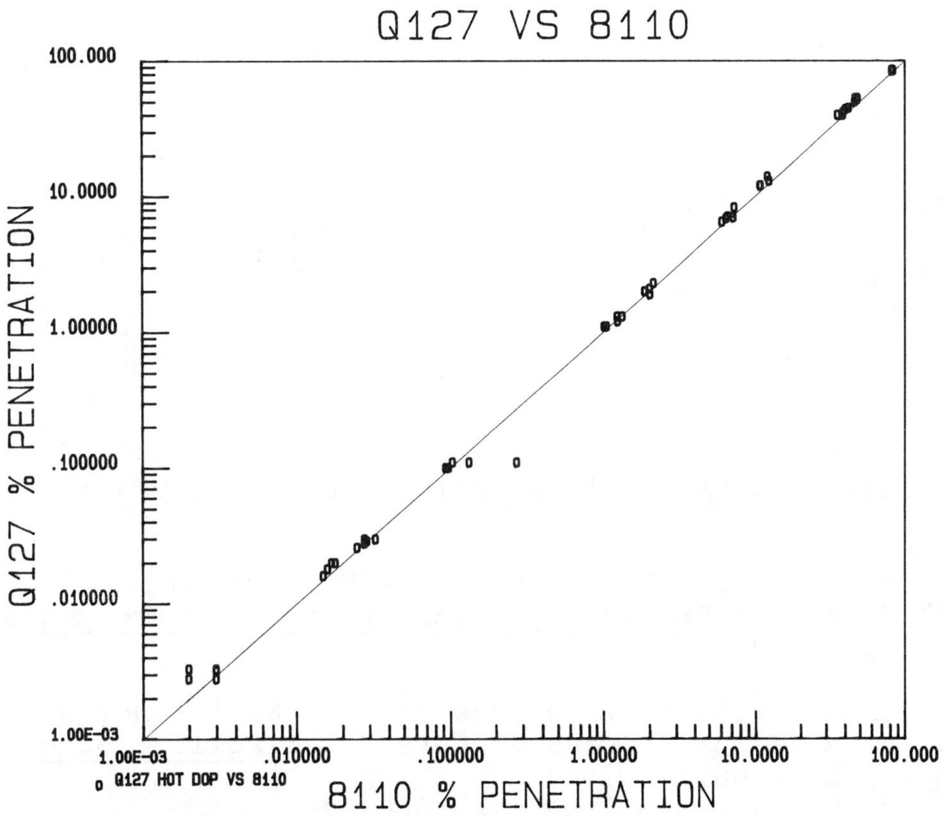

Figure 5. Q-127 versus TSI Model 8110 AFT

As can be seen from the graph, the line has a slope near one, with a y intercept of zero. This gives approximately a one-to-one correlation. A linear regression analysis performed on the test data resulted in a correlation coefficient of 0.9994. This indicates good agreement between the results of the two filter testers.

Conclusions

The new automated filter tester (TSI's Model 8110) can provide an automated filter test to measure penetrations from 0.001% to 100% (efficiencies of 99.999% to 0%) with statistical verification. The tester incorporates current technology to insure that operator variability and excessive maintenance is eliminated.

The new tester generates room temperature DOP, as well as other oils, by using Retec nebulizers for initial aerosol generation. The aerosol is then passed through a coarse filter to narrow the particle size distribution before exiting the generator.

The laser diode based, microprocessor controlled photometer offers an alternative to white light based photometers. The laser diode offers a stable light source which will not vary its intensity over time. In addition, the new microprocessor controlled photometer does not require rezeroing by the operator.

Clean sheath air which encompasses the aerosol entering the photometer eliminates downtime due to photometer maintenance requirements.

The correlation of the new tester with the Q-127 type hot DOP tester shows that the two methods have basically a one-to-one correlation. Therefore, the data taken with the new tester can be interpreted to Q-127 data without further data conversion.

References

Hinds, William C., Aerosol Technology, John Wiley & Sons, Inc., New York, 1982.

Lui, B.Y.H., Rubow, R.L. and Pui, D.Y.H., "Performance of HEPA and ULPA Filters," Proceedings, Institute of Environmental Sciences, pp 25-28 (1985).

Lee, K.W. and B.Y.H. Lui, "On the Minimum Efficiency and the Most Penetrating size for Fibrous Filters", Air Poll. Control Assoc. J. 30:377-381 (1980).

A NEW AUTOMATED FILTER TESTER FOR LOW EFFICIENCY, HEPA GRADE, AND ABOVE FILTER MEDIA AND CARTRIDGES

Brian R. Johnson, Sandra K. Herweyer,
Edward M. Johnson, and Jugal K. Agarwal
TSI Incorporated
P.O. Box 64394
500 Cardigan Road
St. Paul, MN 55164

KEYWORDS

Filter Testing, HEPA, Photometer, Efficiency, Penetration, Fibrous media, Automated, Media, Test

AMERICAN FILTRATION SOCIETY

P.O. Box 6269
Kingwood, Texas 77325
Telephone: (713) 359-1894
FAX: (713) 358-3939

ABSTRACT & COPYRIGHT

Meeting Location: Washington Conference **Meeting Date:** March 19-22, 1990

Session Chairman: Don Forester **Session Number:** 4D

SPEAKER (AUTHOR):

Name: Dr. Kenneth L. Rubow
Company: University of Minnesota
Address: 272 Mech. Mech. Dept.
City, State, Zip: Minneapolis, MN 55455
Telephone: 612/625-8354 **FAX:** 625-6069

CO-AUTHOR:

Name: Dr. Benjamin Y. H. Liu
Company: University of Minnesota
Address: 130 Mech. Eng. Dept.
City, State, Zip: Minneapolis, MN 55455
Telephone: 612/625-6574 **FAX:** 625-6069

CO-AUTHOR:

Name:
Company:
Address:
City, State, Zip:
Telephone: FAX:

CO-AUTHOR:

Name:
Company:
Address:
City, State, Zip:
Telephone: FAX:

TITLE OF PAPER: Experimental Evaluations of Highh Efficiency Fibrous and Membrane Filter Media

TEXT OF ABSTRACT: (200 Words in Space Below)

A new experimental system has been developed for measuring the pressure drop and efficiency characteristics of high efficiency air filter media. State-of-the-art pressure, flow and aerosol instrumentation are used for accurate filter performance evaluation. The system consists of a monodisperse aerosol generator to produce aerosols of a uniform, but adjustable particle size as challenge aerosols for filter testing and a condensation nucleus counter to measure the upstream and downstream aerosol concentration for determining the filter efficiency.

Pressure drop and particle penetration data for HEPA, ULPA and membrane filter media have been determined for face velocities ranging from 0.5 to 20 cm/s and particle diameters ranging from 0.03 to 0.3 μm, which includes the region of maximum penetration. The particle penetration and pressure drop data have been analyzed in terms of a single parameter called the "figure of merit". This parameter is used to determine the relative performance of these filter media.

****

PLEASE TYPE INFORMATION BECAUSE DATA WILL BE PUBLISHED IN THE ABSTRACT BOOKLET. RETURN FORM TO THE AMERICAN FILTRATION SOCIETY, P.O. Box 6269, Kingwood, Texas 77325, WITH A COPY TO YOUR SESSION CHAIRMAN.

AMERICAN FILTRATION SOCIETY

P.O. Box 6269
Kingwood, Texas 77325
Telephone: (713) 359-1894
FAX: (713) 358-3939

ABSTRACT & COPYRIGHT

Meeting Location: WASHINGTON,D.C.

Meeting Date: MARCH 19-22,1990

Session Chairman: DON FOSTER
EASTMAN KODAK
716/477-5674

Session Number:
FILTER TESTING 1&2

SPEAKER (AUTHOR): 1ST.

Name: ROBERT M. NICHOLSON
Company: DONALDSON CO. INC.
Address: P.O. BOX 1299, MPLS. MN.
City, State, Zip: 55440
Telephone: 612/887-3510
FAX:

CO-AUTHOR: 2ND.

Name: JOHN P. ORTIZ
Company: LOS ALAMOS NATL.LAB.
Address: P.O.BOX 1663
City, State, LOS ALAMOS,NM.87545
Telephone: 505/667-6135
FAX:

CO-AUTHOR:

Name:
Company:
Address:
City, State, Zip:
Telephone:
FAX:

CO-AUTHOR: 3RD.

Name: ARTHUR H. BIERMANN
Company: LAWRENCE LIVERMORE LAB. L386
Address: P.O. BOX 5505
City, LIVERMORE,CA.94550
Telephone: 415/422-8017
FAX:

TITLE OF PAPER: INTERLABORATORY RESULTS FROM TESTING A DOUBLE HEPA FILTER SYSTEM USING THE SINGLE AEROSOL SPECTROMETER METHOD.

TEXT OF ABSTRACT: (200 Words in Space Below)

TESTING RESULTS AND TECHNIQUES ARE PRESENTED FOR AEROSOL PENETRATION MEASUREMENTS OF A DOUBLE STAGED 1000CFM HEPA FILTRATION SYSTEM. USING A SINGLE LASER AEROSOL SPECTROMETER. THE RESULTS SUBSTATIATE FILTRATION THEORY ON THE MOST PENETRATING AEROSOL SIZE OF HEPA FLITRATION AND THE REPRODUCABILITY OF LOW PENETRATION TESTS WHEN THE TEST SYSTEM IS ADEQUATELY SPECIFIED. THE TEST METHOD COVERS THE AEROSOL RANGE OF 0.1 TO 1.0UM, THE THE PENTRATION RANGE OF 10(-5) TO 10(-8) AND THE CALIBRATION PROCEDURES REQUIRED FOR THE SYSTEM INSTRUMENTS.

THIS TESTING IS BEING CONDUCTED IN SUPPORT OF ASTM COMMITTEE F21.20 ON AIR/GAS FILTRATION. IT IS ROUND ROBIN TESTING FOR A PROPOSED STANDARD ON IN-PLACE TESTING OF HEPA FILTER SYSTEMS BY THE SINGLE PARTICLE SPECTROMETER METHOD.

****

PLEASE TYPE INFORMATION BECAUSE DATA WILL BE PUBLISHED IN THE ABSTRACT BOOKLET. RETURN FORM TO THE AMERICAN FILTRATION SOCIETY, P.O. Box 6269, Kingwood, Texas 77325, WITH A COPY TO YOUR SESSION CHAIRMAN.

Quantifying Particulate Contamination Control For Solutions Used For the Processing of Photographic Materials

Jane J. Janas, Ph.D.
Pall Corporation
30 Sea Cliff Avenue
Glen Cove, New York 11542

I. Introduction

To minimize the "sparkle" or white dirt on positive prints processed in photoprocessing fluid, improving filtration is necessary. Although there are numerous ways to determine the removal efficiency of a filter, the contaminants encountered in a typical motion picture film processing laboratory include fibers, gelatin, and particulates in amounts unique to photoprocessing fluids. Description of a filter grade is rarely well defined and not reproducible. Nominal terms, i.e., an arbitrary micron (μm) value assigned by a filter manufacturer, based upon the weight removal of a percentage of all particles of a given size or larger[1], is inadequate to predict the filter's ability to decrease the level of contaminant's which may adhere to film during processing. On the other hand, an absolute rating, the diameter of the largest hard spherical particle that will pass through a filter under specific conditions, is an indication of the largest pore/size in the filter element. For such a rating, specification is by use of a standard contaminant, such as well dispersed AC Fine Test Dust (AIChE), for automatic particle counting performed on-line both upstream and downstream of a test filter.[2,3] Due to the existence of unique interactions between materials of a given set of conditions, testing with the actual process fluid is more productionally specific. To correlate a depth filter efficiency measurement as an absolute rating to improved and consistent solution quality by removal of the contaminants found in the photoprocessing industry, was the purpose of this work. Since an efficient filtration scheme, produces cleaner solutions with a decreased level of contaminants which may adhere to the film during processing, the ultimate goal of the project was to minimize the dirt on processed photographic materials.

II. Processing Machine's Design

Typical photographic materials processing laboratories use continuous, deep tank machines, containing submerged rollers and rack-drive assemblies. Tables 1 and 2 describe the solutions and specifications for the current filters for the processing machines for color negative film and the process machine for color print film respectively[4,5]. The process steps indicated in the Tables were evaluated. Black and white negative and print film is processed with a similar set of solutions. The negative developer solution and the fixer solution were included in the test program.

III. Samples Chosen for Contaminant Level Evaluation

Ingression of materials from the atmosphere, from the replenishing solutions, from "drag-on" from previous tanks, and from the film itself, challenges the solution bath filters continuously. Test samples were taken from seven different solutions within the processing sequence for the development of color negative film, from eight solutions within the processing sequence for the development of color print film, and from two solutions within the sequence for the development of black and white negative and print film. Optical particle counting and photomicrographing the residual contaminants on an analysis membrane, were methods used to evaluate the contaminant levels of these solution sample.

Filterability studies were performed on each solution sampled to determine the optimum absolute rated filter grade which will provide good recirculating bath quality and contamination control[6]. One inch segments of polypropylene absolute rated depth filter elements of various grades (15μ, 30μ, 40μ) were utilized to filter each of the samples taken from the processing equipment. For better compatability, nylon absolute depth filter segments were used for the trichloroethylene used to clean the developed film. Filter effluents were collected in particle free bottles (i.e., bottles pre-rinsed with 0.2μm absolute filtered freon, and dried) and were evaluated utilizing the analytical techniques for the initial samples.

IV. Test Methods Used For Contamination Level Evaluation

(a) Optical Particle Counting (Modified Aerospace Recommended Practice[7])

Membrane preparation technique are discussed in Appendix A. For counting, two hundred and fifty milliliters of sample were drawn by vacuum through a 0.8μm absolute rated, cellulose acetate, gridded analysis membrane. The contaminants collected on the membrane are sized and counted using a microscope. In order to determine the total counts per milliliter in each particle size range, the counts per effective filtration area (EFA) are divided by two hundred and fifty.

On a 47mm analysis membrane, the EFA is the area where the sample fluid passes through is 9.6 cm^2 (100 boxes). Each gridded box on the membrane is 0.308 cm^2. Using a 100 division containing reticle, the size of each particle is measured. (The reticle is a scale in the microscope ocular.) Each division's length is dependent on the magnification used for observation and must be calibrated for each microscope. For this test, the lengths between each division were 19.9μm per division at 50X magnification, 9.9μm per division at 100X magnification, 4.9μm per division at 200X magnification, and 2.5μm per division at 400X magnification.

For each size range counted, the magnification for observation is chosen to optimize ease of correlating divisions to size ranges:

Correlation of Particle Size to Reticle Divisions

Size Range (μm)	Magnification	# Divisions
1-5	400X	0.5-2
5-15	200X	1-3
15-25	100X	1.5-2.5
25-50	50X	1.25-2.5
50-100	50X	2.5-5
>100	50X	>5
Fibers	50X	>5

where fibers have length at least 10X the size of the width

The counting unit of area is determined by the magnification of observation. A "box" is counted at 50X and 100X magnification. A box is defined by the area covered within form grid lines plus the area covered by two of the grid lines. Each box is selected at random. For this test, full boxes were counted until at least 100 particles in each size ranges were seen. To procure this amount, the entire EFA with particle boxes may have been counted.

A "diameter" (or the reticle width by the length of a box-including-one-grid-line) is the unit of area counted at 400X, 200X, and possibly 100X magnification. Random diameters were counted until at least 100 particles in each size range were seen. "Fractions of a diameter", i.e., a count of particles falling between 0 and 50 on the reticle by a box length equals 1/2 diameter, may have been counted if a sample was very contaminated.

To minimize error due to repetitious data, there was a limit set for the maximum area counted on very clean membranes. If 100 particles were not seen by this limit, raw count levels were reported for that area. In 1-5μm size range at 400X magnification and in the size range at 200X magnification, twenty diameters was the limit. In the 15-25μm size range at 100X magnification and in the >25μm ranges at 50X magnification, the entire effective filtration was the limit.

For the calculation of counts/ml, one must divide the raw counts by the volume used and multiply by a factor of the EFA counted to determine the total particle count in each size range per milliliter at 50X and 100X magnification when boxes were counted, the following equation is used:

$$\text{Counts/ml} = \frac{\text{Raw Counts}}{\text{\# of ml of sample}} \times \frac{\text{100 Boxes}}{\text{\# of boxes counted}}$$

To determine the total particle count in each size range per milliliter at 100X, 200X and 400X magnification when diameters were counted, the following equation is used:

$$\text{Counts/ml} = \frac{\text{Raw Counts}}{\text{\# of ml sample}} \times \frac{9.6\text{cm}^2}{\frac{(\mu\text{m})}{(\text{div})}\frac{(100\text{ div})\ (0.308\text{cm})}{(\text{diameter})}(10^4\text{cm}/\mu\text{m}\ (\text{\# diameters counted})}$$

As above, one must divide the raw counts by the volume used and multiply by a factor of the EFA counted.

b) Photomicrographing

After drawing two hundred and fifty milliliters of sample through the analysis membrane, photomicrographs of contaminants collected on a representative area of the analysis membrane were obtained at 100X magnification[8]. The reticle's divisions corresponded to 13.6μm/division on this microscope at this magnification. Use of the optical microscope with camera attachment allowed observation under polarized light and recording qualitatively the levels of contamination.

V. Discussion of Analytical Results

For all tested solutions, filtration through absolute rated depth filters provided improved solution quality over fluids filtered through nominally rated filters. Cleaner solutions reduced the level of contaminants which potentially adhere to processed film. Fewer contaminants on the processed film result in greater film quality when viewed on the screen.

Tables 3A, 3B, 3C, and 3D detail the results of optical particle counting of each solution. These qualitative data were confirmed by the photomicrographs which show the high contamination levels of the filtrates of existing nominally rated filters[8]. High contamination levels in solutions used for processing can cause defects on the processed film.

Based on testing with absolute rated filters on a single pass basis, significant reduction of contaminant level was achieved. Since solution baths are recirculated, absolute rated filters slightly coarser than those optimum for a single pass are expected to improve the current contamination level and provide consistent solution cleanliness. In order to maintain solution quality, the following levels of absolute filters will clarify the various solutions for processing photographic materials and help develop film of consistently high quality.

a) twenty micron absolute level of filtration for solutions of the continuous machine processing negative film;
b) forty micron absolute level of filtration for solutions of the continuous machine processing color print film;
c) seventy micron absolute level of filtration for the solutions of the continuous machine processing black and white negative and print film;
d) ten micron absolute level of single pass filtration for the rinse water;
e) thirty micron absolute level of single pass filtration for the trichloroethylene solution;

Final film quality when viewed on a screen is higher when cleaner solutions are used for processing.

Appendix A

Analysis Membrane

(1) Draw Down Funnel Set-Up:

Wash funnel with detergent. Rinse off soap with tap water. Rinse with deionized water, isopropyl alcohol, and freon, all of which were 0.2μm filtered. Cover funnel with aluminum foil which had been rinsed with filtered freon.

(2) Blank Membrane Preparation:

Rinse a 0.8μm gridded, cellulose acetate analysis membrane with filtered freon and place on funnel base. Rinse clean funnel with filtered freon and clamp on top of the membrane. Add 25 ml filtered freon and draw down with vacuum. Remove funnel and place membrane in clean petri slide holder. Count number of particles in all size ranges except the 1-5μ range in six random boxes at 100X magnification. Blanks should have <10% of the sample's number of particles in all ranges, no particles >25μm in diameter (although up to 2 is acceptable for the EFA), and have less than six particles in the 5-15μm and 15-25μ ranges respectively.

(3) Sample Membrane Preparation:

Shake two hundred and fifty milliliters of sample. Rinse the funnel and the analysis membrane with freon and set in place. Add a small amount (1-5ml) of filtered freon to funnel. Add sample and turn on vacuum. When fluid level reaches 0.25 inches, turn off and break vacuum. Rinse sides of funnel with filtered freon and draw downs as often as necessary to remove sample residue from the funnel. In order not to disturb the particle distribution giving an inaccurate count, do not let membrane go dry.

References

1. National Fluid Power Association, ANSI B93.31-1973 "Glossary of Terms for Fluid Power."

2. Fitch, F.C., "The Multi-Pass Filter Test - Now a Viable Tool", Paper No. P74-39, Eight Annual Fluid Power Research Conference, Stillwater, Oklahoma (1974)

3. National Fluid Power Association, ANSI B93.31-1973 "American National Standard Multi-Pass Method for Evaluation the Filtration Performance of a Fine Fluid Power Filter Element."

4. Kodak Publication No. H-26, "Abridged Specifications for Process ECN-2", Eastman Kodak Company, Rochester, N.Y. (1982)

5. Kodak Publication No. H-38, "Abridged Specifications for Process ECP-2A", Eastman Kodak Company, Rochester, N.Y. (1982)

6. Janas, J.J., "Improved Particulate Contamination Control for Solutions Used for the Processing of Photographic Materials", Journal of the Society of Motion Picture and Television Engineers, (in Press)

7. Aerospace Recommended Practice, ARP598B, "The Determination of Particulate Contamination in Liquids by the Particle Count Method", Society of Automotive Engineers, Inc., Warrendale, Pa (1986)

8. Janas, J.J., "Improved Contamination Control Using Pall Profile II Absolute Rated Filters for the Processing of Photographic Materials", FSR PRO4 and PRO4a, Pall Corporation, 1989.

TABLE 1

Present Filtration System of a Typical Continuous Processing Machine for Color Negative Film

Solution	Sample Analyzed	Filtration Mode	Present Filter Nominal Rating	Approximate Flow Rate
			(μm)	(gpm)
Prebath	yes	recirculation	10	4-8
Developer	yes	recirculation	5	4-8
Stop Bath	yes	recirculation	5	4-8
Persulfate Accelerator	yes	recirculation	5	4-8
Persulfate Bleach	yes	recirculation	10	4-8
Fixer	yes	recirculation	10	4-8
Stabilizer	yes	recirculation	10	4-8
Wash Water	yes	single pass	5	-
Silver Recovery	no	recirculation	25	4-8
Trichloroethylene	yes	-	-	-

TABLE 2

Present Filtration System of a Typical Machine Processing Color Print Film

Solution	Sample Analyzed	Filtration Mode	Present Filter Nominal Rating	Approximate Flow Rate
			(μm)	(gpm)
Prebath	yes	recirculation	20	10-15
Developer	yes	recirculation	20	30-45
Stop Bath	yes	recirculation	20	10-15
First Fixer	yes	recirculation	20	10-15
Accelerator	yes	recirculation	20	10-15
Persulfate Bleach	yes	recirculation	20	10-15
Second Fixer	yes	recirculation	20	10-15
Stabilizer	yes	recirculation	20	10-15
Sound Track Development	no	-	none	-

TABLE 3A:

RESULTS OF OPTICAL PARTICLE COUNT FOR SOLUTIONS OF CONTINUOUS MACHINE PROCESSING COLOR NEGATIVE FILM

Solution	Medium of Filtration	Number of particles/ml in size range (μm) 1-5	5-15	15-25	25-50	50-100	>100	fibers	[illegible]
Prebath	10μm Nominal		Too Numerous Count						
	15μm Absolute Depth	24	3	4	4	3	4	4	46
Developer	5μm Nominal	41	4	4	2	1	4	4	60
	15μm Absolute Depth	32	3	1	0	4	4	4	48
Stop Bath	5μm Nominal	277	46	81	3	1	1	4	413
	15μm Absolute Depth	71	6	1	1	4	4	4	91
Persulfate Accelerator	5μm Nominal		Too Numerous Count						
	15μm Absolute Depth	64	43	2	1	1	1	1	113
Persulfate Bleach	10μm Nominal	59	10	2	1	1	1	4	78
	15μm Absolute Depth	7	2	4	4	4	4	4	29
Fixer	10μm Nominal		Too Numerous Count						
	15μm Absolute Depth	11	3	4	6	3	4	4	35
Stablizer	10μm Nominal	12	7	6	6	3	3	2	39
	15μm Absolute Depth	15	5	1	4	4	4	4	37

* Because sample draw downs were not performed in a particle free environment, oversized particles found are most likely from background contamination.

TABLE 3B:

RESULTS OF OPTICAL PARTICLE COUNT FOR SOLUTIONS OF CONTINUOUS MACHINE PROCESSING COLOR PRINT FILM*

Solution	Medium of Filtration	Number of particles/ml in size range (μm) 1-5	5-15	15-25	25-50	50-100	>100	fibers	to[illegible]
Prebath	20μm Nominal		Too Numerous To Count						
	30μm Absolute Depth	145	12	1	1	4	4	4	171
Developer	20μm Nominal	146	66	20	1	4	4	4	245
	30μm Absolute Depth	40	19	8	1	4	4	4	80
Stop Bath	20μm Nominal	291	121	32	23	7	1	4	479
	30μm Absolute Depth	101	32	5	5	4	4	4	155
First Fixer	20μm Nominal	667	331	124	15	11	4	4	1,156
	30μm Absolute Depth	45	5	5	7	3	10	4	79
Accelerator	20μm Nominal	889	278	57	13	10	4	5	1,256
	30μm Absolute Depth	4	4	4	4	4	4	4	28
Persulfate Bleach	20μm Nominal	14	9	7	1	1	4	4	40
	30μm Absolute Depth	**	**	1	1	1	4	4	11
Second Fixer	20μm Nominal	315	172	9	1	1	4	4	506
	30μm Absolute Depth	19	3	1	6	5	1	4	39
Stabilizer	20μm Nominal	187	80	28	10	7	1	1	314
	30μm Absolute Depth	145	53	8	1	4	4	4	219

* Because sample draw downs were not performed in a particle free environment, oversized particles found are most likely from background contamination.

**Could not be counted due to salt agglomeration.

TABLE 3C:

RESULTS OF OPTICAL PARTICLE COUNT FOR SOLUTIONS OF CONTINUOUS MACHINE PROCESSING BLACK AND WHITE NEGATIVE AND PRINT FILM*

Solution	Medium of Filtration	Number of particles/ml in size range (μm) 1-5	5-15	15-25	25-50	50-100	>100	fibers	[illegible]
Fixer	10μm Nominal	259	71	15	3	2	1	4	355
	40μm Absolute Depth	1	1	1	1	4	4	4	16
Negative Developer	10μm Nominal		Too Numerous To Count						
	40μm Absolute	152	8	4	4	4	4	4	180

*Because sample draw downs were not performed in a particle free environment, oversized particles found are most likely from background contamination.

TABLE 3D:

RESULTS OF OPTICAL PARTICLE COUNT FOR SOLUTION COMMON TO CONTINUOUS MACHINES PROCESSING PHOTOGRAPHIC MATERIALS*

Solution	Medium of Filtration	Number of particles/ml in size range (μm) 1-5	5-15	15-25	25-50	50-100	>100	fibers	[illegible]
Trichloroethylene	Unfiltered	48	24	6	4	2	1	1	86
	30 Absolute Depth	20	17	9	3	2	1	4	56

*Because sample drawn downs were not performed in a particle free environment, oversized particles found are most likely from background contamination.

PROPOSED EVALUATION TEST
FOR POROUS METAL FILTER ELEMENTS

HARRY M. KENNARD
TECHNICAL CONSULTANT
MICHIGAN DYNAMICS, INC.
1876 MIDLAND ROAD
ROCK HILL, SOUTH CAROLINA 29731
(803) 366-8311 OR 800-553-5501

The most common micron rating test for porous metal filter elements is the bubble point test. This is a simple and genally useful production quality test. Unfortunately, it rejects what are potentially very acceptable elements based on the presence of a very few out-of-specification pores out of the hundreds of millions of pores in a typical element. Further, this effect is amplified for ratings under five micron where the ratio between the largest pore and the mean pore size is high.

Other tests which have been or are being used include: the mean pore size test and the one sigma test, both of which require tremendous air flows for typical elements; and the 10 LPM test which does not account for variations due to different filter areas.

This paper proposes a compromise test method which can be characterized as a multipoint Q/A test in which the following three measurements are made.

1. Bubble point
2. Pressure drop at .1 CFM/FT^2 (.003 LPM/CM^2)
3. Pressure drop at .3 to 3 CFM/FT^2 (.009 to .09 LPM/CM^2)

Since this method provides an improved characterization of a filter element, it should be possible to establish acceptable limits with minimum risk of accepting bad elements or rejecting good elements.

PROPOSED TEST FOR EVALUATING POROUS METAL FILTER ELEMENTS

The most common method for routine evaluation by fabricators and end users of porous metal filter elements is the bubble point test. This is a simple and generally useful method of testing in a production environment. The main weakness of the bubble point test is that it accepts or rejects a filter element based on the absence or presence of very few out-of-specification pores. In addition, since a bubble test rates an element on its largest pore, it is inconsistent with the growing use of the beta ratio as the criteria for filter rating.

Further, the use of the largest pore as the only criteria for evaluating a filter element can be misleading particularly at the finer filtration levels where the ratio between the largest pore and mean pore sizes is high. This factor is most important for filtration at or below the 5 micron level. Now more attention is being directed at media in the 3 micron and finer area. Hence, it is appropriate that efforts be made to establish a more meaningful filter rating criteria and test method.

It is important that the test method adopted be easy to incorporate into a production system and it is desirable that the data reduction be minimal.

Some methods in addition to the bubble point test that have been proposed and used are:

1. a mean flow pore size test
2. a 10 LPM test
3. a one-sigma test

These various methods are discussed in the following paragraphs. All use a wet flow curve to some degree; and all but the 10 LPM test use either a dry flow curve or knowledge of the media air permeability.

The mean flow pore size test (per ASTM F316 86 procedures or equivalent) is based on the reasoning that the air pressure where the wet air flow is half the dry air flow defines the mean flow pore size. This is a simple test to make provided the air flow rate required is reasonable. However, with all but small media samples the air flow required is quite large. Typically for a 3.0 sq. ft. element rated at 5 microns, the mean air flow will be 600 to 1000 CFM (17000 to 28000 LPM). Obviously, this flow exceeds practical limits.

The 10 LPM test consists of recording the wet air flow where the air flow rate is 10 liters per minute. This test has been recommended by Southern Metal Processing (SMP) to assist their customers in better evaluating the useful life of their elements. This test as currently performed gives a better evaluation of the useful life of the element than does the bubble test, however, it uses quite a low air flow and does not account for different filter areas. The use of a low flow rate means that much less than .1% of a typical elements pores have air flow through them during the test. And by not accounting for filter area the data represents a different point on a wet flow curve for a small element as compared to the data for a large element. As a result, large filters are penalized since the 10 LPM point for them corresponds to the area on the wet flow curve with predominately large pores, while the 10 LPM point for a small filter will include many small pores. In addition to these two problems, the method currently used to report the 10 LPM pressure drop is with the test stand tare (pressure drop) included. Since this tare value is typically 10 inches of water, it represents about one half of the reported pressure drop for a 20 micron element.

The one sigma test, first proposed by Fred Cole in 1975 when he was Manager of Technical Development for Michigan Dynamics, also uses the wet flow curve data. In this approach, the data is reduced to give not only the absolute rating from the bubble point and the mean flow rating but also the "sigma" rating. The sigma rating corresponds to the one sigma limit on a log-normal distribution curve. The main objection to this approach is that like the mean pore size approach it requires an excessive air flow capacity. In addition, it also requires data analysis that could be troublesome in a production environment.

In an effort to utilize the good points and overcome the shortcomings of the three methods discussed above, a method is proposed which appears to be a reasonable compromise between the mean pore size and sigma limit approaches which require excessive air flow rates; and the 10 LPM test which lacks good definition and uses too low an air flow rate. In the proposed test three measurements are made. They are:

1. the bubble point
2. the pressure drop at .1 CFM/FT^2 (.003 LPM/cm^2)
3. the pressure drop at .3 to 3 CFM/Ft^2 (.009 to .90 LPM/cm^2)

(The requirement for this third point's flow rate is that it be on the relatively flat part of the wet flow curve. The higher the flow, the better.)

These flow per unit area (Q/A) values correspond to air flow rates of .15CFM (4.2 LPM) and .45 to 4.5 CFM (12.7 to 127 LPM) for a 1.5 sq. ft. element; and 0.6 CFM (.17 LPM) and 11.8 to 18 CFM (51 to 510 LPM) for a 6 sq. ft. element. The data obtained in this test would be compared initially to wet flow data for the media under test to establish the acceptability of the filter element. Later, as a quantity of data is obtained, it should be possible to establish very good acceptance limits for production elements.

While this approach is a compromise made to keep the air flow requirements reasonable, it should permit an evaluation of the filter media in an area of the wet flow curve that has a micron rating near the mean value. This is expected to be true because even though the test still only covers a small part of the filter element, it would be in an area of the wet flow curve that is reasonably flat. As a example, for a 6 sq. ft., 15 micron absolute element this test would flow through only about .2% of the pores. However, the wet flow pressure at 3 CFM/Ft^2

appears to be about 70% of the pressure of the mean pore pressure even though the mean pore pressure occurs at a flow rate of about 700 CFM/Ft^2.

The specifying of the flow per unit area (Q/A) is of paramount importance. Examples of misleading results from the 10 LPM tests which did not specify Q/A can be found in test data. Table I shows 10 LPM values for a series of 1.5 sq. ft. 20 micron elements. If an acceptance limit of 8 inches (18 inches when the tare is included) is used, all but two of these elements would be acceptable. However, all these elements failed in their next use.

Additional evidence of the importance of specifying a Q/A value can be seen in Figure 1. This shows a theoretical curve for a 20 micron 5 sq. ft. element. Both the new condition and a damaged condition are shown. The damaged condition consists of adding 500 holes of 50 micron diameter to the element. (While 500 holes may seem to be a very large number of holes, a 5 sq. ft. element of this type would have approximately one billion pores. Hence, 500 oversize holes would represent a very small percentage of the total pores.) As the figure shows, at low values of Q/A, the effect of the 500 holes is easily detected. This is because they are flowing while few of the other billion pores have started to flow. As the Q/A value increases the effect of the 500 oversize holes is much less obvious.

When the proposed test is compared to either the mean flow pore size test or one-sigma test, its lower air flow requirements make it practical. When the test is compared to the 10 LPM test, its use of the flow per unit area makes it more consistant for elements of various areas. And the use of three points permits identification of the largest pore by the bubble point, the overall deterioration by the .3 to 3 CFM/ft^2 point and an indication of the number of oversize holes by the intermediate .1 CFM/ft^2 point.

Since the proposed method permits an improved characterization of the condition of a filter element, the producer and user of the elements should be able to establish acceptable limits with a minimum risk of accepting poor elements or rejecting good elements.

Table I

Filter Element 10 LPM Data and Subsequent Performance

Serial No.	10 LPM (w/o Tare) (Inches Of Water)	Performance
0005	9	Failed, Bypassed at 600 PSI
0007	7.7	Failed, Bypassed at 500 PSI
0009	9.8	Failed, Bypassed at 600 PSI
0021	11.5	Failed, Bypassed at 700 PSI
0022	13.5	Failed, Bypassed at 1300 PSI
0027	6.6	Failed, Bypassed at 500 PSI
0028	10	Failed, Bypassed at 600 PSI

Element area 1.5 sq. ft. Micron rating 20.

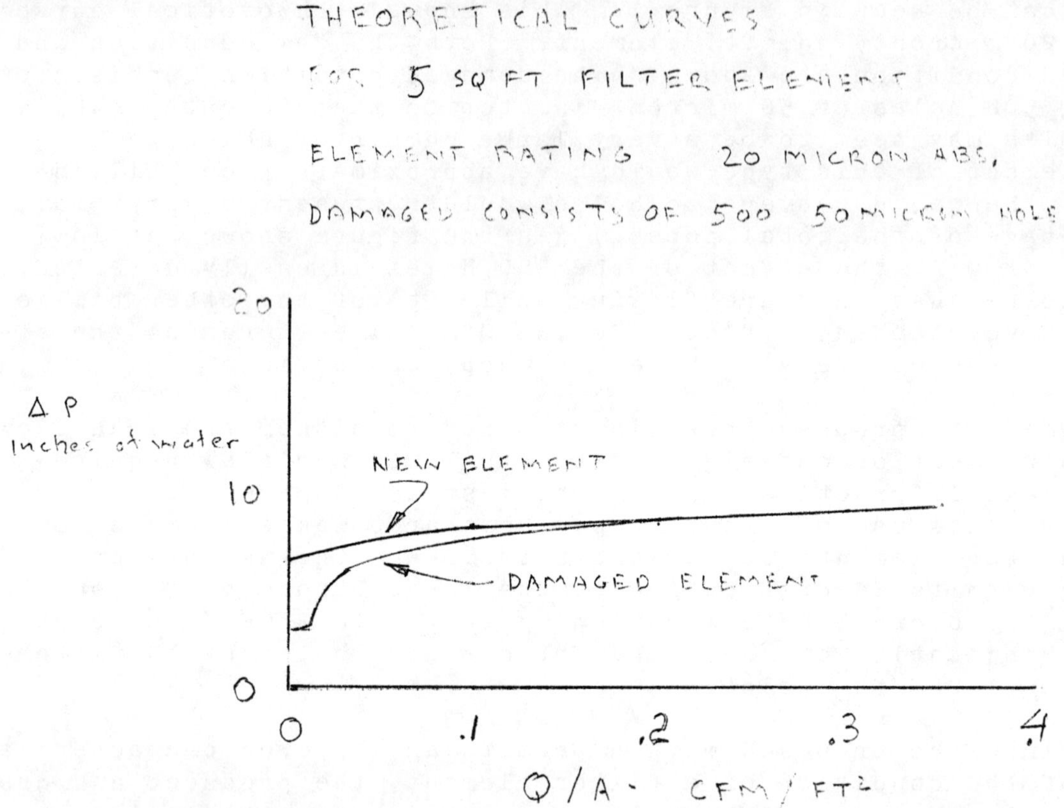

THEORETICAL CURVES
FOR 5 SQ FT FILTER ELEMENT
ELEMENT RATING 20 MICRON ABS,
DAMAGED CONSISTS OF 500 50 MICRON HOLES
ΔP
inches of water
20
10
0
NEW ELEMENT
DAMAGED ELEMENT
0
.1
.2
.3
.4
Q/A - CFM/FT²

FIGURE 1

PRECISION WOVEN FABRICS FOR FILTRATION

John R. Mollet
TETKO INC.
333 South Highland Avenue
Briarcliff Manor, NY 10510
914/941-7767

Weaving of engineered fabrics for filtration originated with the use of multifilament silk yarns. Today, a wide variety of polymers such as nylon, polyester, polypropylene, fluorocarbons and many more, are extruded into monofilament fibers, which are being used to produce precision woven fabrics. These filter media offer specific and controlled apertures, which determine filtration ratings and flow characteristics.

The choice of polymers allows for the selection of a product compatible with the chemical and physical environment of the filtration process. Weaving techniques, weave styles and finishing processes are discussed, in relationship to the filtration properties.

Several industrial filtration applications are reviewed, demonstrating how each requires its own specialized fabric. Furthermore, bio-medical filtration devices require fabrics of an ever higher degree of technology and process control, with the emphasis, shifting towards the post-weaving treatment, in view of biological requirements.

Well over 100 years ago, fine silk fibers were twisted into multifilament yarns, and woven on primitive looms. A cottage industry of individual home weavers produced fabrics for flour sifting. By varying mesh counts or number of yarns per inch width, and yarn diameters, fabrics could be engineered to specific apertures or opening sizes, in order to separate and grade various granular products. A few fluids were strained through such silk screens.

During World War II, silk supplies from the Orient were eliminated. At the same time the Dupont Co. introduced a new synthesized polymer fiber, called polyamide or nylon. An extruded plastic monofilament fiber was soon found to be much more uniform and reproducible in different diameters than a multifilament natural silk yarn. As nylon monofilament fibers could be woven into fabrics with precise apertures, at predictable tolerances, such woven screens began to find their way into filtration and separation processes. Technology improvements made the extrusion of finer monofilaments possible. This allowed to weave at higher mesh counts, resulting in fabrics with smaller but more openings. The result was finer filtration ratings combined with a high percentage of open area, therefore, a greater flow capacity. Today, besides nylon, a wide variety of polymer fibers is available. For instance, polyester, mainly in a PET, but also in a PBT formulation, polypropylene, polyethylene, poly-vinylidene fluoride and other fluorocarbons. Development keeps bringing new generations of plastics in extruded form to the market. This creates opportunities for precision fabrics in liquid, air and gas filtration applications, at ever higher temperatures, and in extremely corrosive environments, from which synthetics were previously excluded.

The choice of the polymer allows for tailoring a woven media to the chemical and physical requirements of the filtration process. For example, nylon is resistant to mild alkaline conditions, whereas polyester is suitable for mild acids. But nylon exhibits some creep under tension, which makes polyester, a better candidate for belt filter applications. Then again fluorocarbon fibers survive most chemical conditions. The nature of the polymer also influences the weaving feasibility, and puts limits on the number of products and filtration ratings which can be made available. The smallest openings or finest filtration ratings are typically obtained with nylon and polyester, because these are available in very small diameters or low denier (dtex). Typically, monofilament fabrics are produced with apertures ranging from 12500 microns (um) or a half inch, down to 5 microns (.0002 in). Hybrids of monofilament and multifilament and fabrics of all multifilament or spun yarns exist as well.

Originally, the weaving looms were of the shuttle type. The group of yarns or fibers, running parallel in the length direction of the fabric, are called the warp. The fibers are held at an equi-distance by the weaving reed, which consists of flat wires with controlled thicknesses, placed at precise distances from one another. A warp fiber runs through each opening in the reed. The threads in the width direction are called the weft or shute, and unwind from a spool placed in a shuttle, which is literally shuttling from one side to the other of the loom. Because the amount of weft fiber which can be placed in a shuttle is limited, and the speed of the shuttle, expressed in picks per minute, is considered relatively slow, the more productive shuttleless loom has taken over. The rapier system is one of the shuttleless methods, in which each weft fiber unwinds from a large spool, is taken through the warp fibers, and cut. As a result, a controlled, efficient weaving process, in a clean and climate-adjusted environment, delivers the precision products, we have come to rely on.

Besides the size and amount of fibers, the weave style is the other main factor which determines the filtration characteristics. The most basic style is called a plain square weave, in which the fibers go over one, and under the next fiber of the perpendicular direction. This pattern provides a straight-through flow path and the highest possible permeability of any woven media. The risk of blinding is minimal and cleaning or backwashing is easy. Such square weaves are used in high flow rate filters, such as blood transfusion, and water filters; in carrier belts, for instance used to form non-woven media, or to dry pasta products; in traveling water screens; in disc sector covers and many other applications.

A variation of the square weave is the twill weave, in which a fiber passes over two or more fibers. Some twill weaves are asymmetrical, such as a 3 over, one under (3:1); other are symmetrical, such as a 4:4. The symmetrical twill is of particular importance when assembling filters with stamped discs or other shaped pieces, and it is no longer possible or practical to identify the so-called cake-side. The cake side of the fabric normally refers to the smooth surface, for instance the "3 over" top side, compared to the knotty "one under" bottom. A smooth surface allows for easier cake release, and the ultimate smoothness is found on a "satin" weave, in for example a 7:1 or 6:2 weave pattern. Twill and satin cloths are used in filter belts for horizontal and rotary drum vacuum filters, filter presses, and bags for the chemical and food processing industries, and in belt filter presses for municipal and industrial sludge dewatering.

In the wire cloth industry, the most efficient filter product is a dutch weave, in which the wires in one direction are larger then in the other. Typically the smaller weft wires are driven up close to one another, which creates a dense, strong cloth. The fluid has to follow a tortuous path, making the filter effective in capturing solids. Synthetic weavers borrowed from this concept to design the PRD pattern, which is a reverse dutch weave in which a high number of small warp, and a lower number of larger weft fibers, are woven in a plain (1:1) pattern. The PRD's have a higher flow rate than twill or satin fabrics with a comparable filtration rating. They are, for example, mounted on rotary drum vacuum filters, but in totally different applications, they shape acoustical responses or function as sound attenuators.

Besides single layer fabrics, certain processes require double or multiple layer weaves. For instance, recently double layer products were developed with a top filtration layer in a twill pattern, interlocked with a heavy open square bottom layer. Applications vary from Nutsche filters to horizontal belt filters, for which this construction offers both improved durability and stability.

Contrary to wire cloth, where weaving alone virtually determines the properties, synthetic media are finished, after weaving. Finishing can mean heat setting, which is the fixing in place at a high temperature of the "plastic" fibers. It can include stretching or shrinking of the off-loom fabric, in one or both directions. Or calendaring in which the fabric is pressured between heated rollers. A calendered fabric exhibits a smoother surface, a finer pore size, hence improved particle retention, a lower permeability and a smaller thickness. All these parameters can be influenced, to create a product, which meets the needs of a specific application, such as dewatering of a mineral sludge on a vacuum filter, or the production of fruit juice on a belt filter press.

While finishing traditionally refers to the shaping or stabilizing of the "plastic" material, these days, it includes any of the post-treatment processes, which alter the surface characteristics of the fibers. A black dye reduces glare on filters for "light transmission", mounted on video display units. Anti-static agents prevent the build-up of static electricity, which when discharged can burn holes in the fabric. Also, sparks are to be avoided at all cost, when working with certain fine powders, for example in a fluid air dryer. Carbon and metallic coatings offer even higher electrical conductivity. Other agents change the surface tension or the hydrophylic/hydrophobic nature of the fibers, thereby influencing the flow and the solids capture rates. Specialized washing and treatment assures that precision fabrics meet the ever increasing biological requirements for medical filter devices. Fabrics have to pass in-vivo biological reactivity tests and in-vitro methods, such as hemolysis and LAL, are now performed in-house, as part of the process and quality control systems.

More and more markets demand such a high degree of technology. The modern weaver, who is positioned for the challenges of the 90's and looks ahead to the 21st century, can no longer rely on mechanical and textile engineering skills, even highly specialized ones, but is or has to become an expert in diverse disciplines, from chemistry and polymer engineering, to medical and pharmaceutical sciences, to name but a few. And a total quality assurance philosophy guiding a controlled process, ultimately yields the precision woven filtration products, which meet or exceed the customers' expectations.

Reference: E.G. Fugett, "Precision woven fabrics for filtration and other industrial applications", International Fabrics Association International, 1989 Annual Convention (Oct. 1989).

AIR/POLYMER INTERACTION EFFECTS ON THE STRUCTURAL FORMATION OF MELT BLOWN MICROFIBERS.

Hassan Bodaghi, Ph.D.
Biomaterials Section, Biosciences Laboratory
3M Company, St. Paul, MN 55144-1000

ABSTRACT

Superfine fibers produced by both melt and solution blowing have long been used for filtration applications. The smaller the fibers in a filter media, the higher the filtration efficiency. In melt blowing however, small fibers of uniform size are difficult to obtain. Any melt blown web often consists of: fibers of different sizes, sands or microshots, and other irregularities (e.g. bundles of molten filaments and polymer globules fused together). These irregularities, as a consequence of melt blowing process variables, are studied in this paper. The significance of resin melt rheology, process design, and development of molecular orientation and crystallinity during melt attenuation are discussed and also correlated to the properties of the final product.

INTRODUCTION

Melt blowing is a unique process for producing ultrafine fibers. In the early 1950's, scientists at the Naval Research Laboratory conducted extensive research and development in this area to produce superfine fibers for filtration media (1). To date, melt blown fabrics account for more than one-eighth of the total filtration market worldwide (2).

Over the past 30 years, during which melt blown fibers have come into wide commercial use, there has always been a recognition that the fibers processed by melt blowing have a broad size variation. This together with the lack of fiber strength and other non-uniformities (shots or resin blobs etc.), which are associated with the melt blowing process, have created numerous developments in the die design (nozzle etc.), process refinement, and resin modifications (3,4,5). Most, if not all of these efforts have been done in industry, with the main emphasis toward cost reduction and product development. Little information is published with regard to the technical aspects of the melt blowing process, and the fundamental concepts of this process are still not fully understood.

The lack of fiber uniformity and strength in melt blown fabrics limits their utility mainly to filtration and insulation applications. The objective of this paper is to investigate air/polymer interaction effects on fiber uniformity and structural morphology.

This paper is a continuation of earlier associated studies, in understanding the fundamental concepts of the melt blowing process, especially with regard to formation of fibers and their structural characterization (6).

EXPERIMENTAL

Materials and Conditions

Polymers used in this study were Foster Grant Nylon-6, 3M PET (Poyethylene terephthalate 379000), and Gulf 2253 Polypropylene (PP). Weight and number average molecular weights and polydispersities for all three resins were measured and are listed in Table I.

A schematic representation of the melt blown process in shown in Figure 1. All the resins were extruded through a multiorifice die and attenuated by high velocity air to produce fibers of different sizes. The fibers then were collected on a drum at the same die collector distance. The PET resin was dried for eight hours at 180°C and Nylon-6 was dried for 16 hours at 120°C before processing. Fibers were made from each resin at four different primary air pressures of 10, 25, 40, and 50 psi. The melt blown processing conditions are shown in Table II.

CHARACTERIZATION

Shear viscosities of each resin were measured by an Instron Capillary Rheometer at three different temperatures. These were 230, 285, 330 for nylon and polypropylene and 270, 285, and 330 for polyester.

The fiber size, uniformity, and distribution was investigated by a Baush & Lomb Omnicon 3500 image analyzer in conjunction with a Scanning Electron Microscope.

Crystallinity values were obtained from density measurements for Nylon-6 and PET and from heat of melting for PP. In the latter case, the heat of fusion of the fibers were determined from the area under peaks in the trace of heat content versus time using a Dupont model 1090 Differential Scanning Calorimeter. Heating was carried out at a rate of 10 degrees per minute. The heat of fusion of PP (100% crystalline) was taken to be 49.95 cal/g (7). The densities were determined using a density gradient column prepared with carbon tetrachloride and toluene. A technique to construct the density column is described elsewhere (8). The crystalline volume fraction was estimated by assuming that fibers are semicrystalline and have two phases (crystalline and amorphous).

Wide angle X-ray scattering patterns were obtained using a Philips ATD-3600 x-ray generator and CuK radiation (wave length 1.542 Å). Operating scattering were 45 Kv and 30 mA. From radial WAXS scans, average crystal sizes were measured using Scherrer equation (9). Contributions from crystal imperfections were ignored in the calculations. The (105) reflection for PET, (020) for nylon-6, and (110) for polypropylene were used.

Hermans-Stein orientation factors were computed from the azimuthal scans of (002),(200) reflection of nylon-6, (105) for PET, and (110),(040) for PP (10,11).

The optical retardation for the fibers was determined using a polarizing microscope and a Berek compensator. The bireferengence is the ratio of the optical retardation to the fiber diameter.

The mechanical properties were measured using a Table model Instron Tensile Tester.

RESULTS AND DISCUSSION

Factors affecting the final properties of a meltblown web are shown in Figure 2. Emphasis in this study has been toward the investigation of fiber structural formation as a function of process variables. The size, uniformity and the stability of fibers during melt attenuation is strongly affected by the geometry of the die nozzle, orifice size and spacing, and the position of the air knives relative to the tip of the nozzle (see Figure 1). In addition, the rheological properties of a polymer resin have to be controlled for steady state fiber formation to prevent any instability, shots or resin blobs etc. The physical properties of a polymer, as provided by the manufacturer, are often insufficient for prediction of flow behavior in processing. For example, rheological properties given by melt flow index represent flow at a single shear rate less than 10 sec -1. In the actual melt blowing process, a polymer melt is extruded under high shear rate flow through multiorifice dies. Therefore, to understand the resins characteristics better with regard both to shear and to thermal degradation, shear viscosities at different temperatures and shear rates were measured. Typical example for nylon-6 and PP is shown in Figure 3. The apparent shear rate for the process was calculated from the following equation (12).

Apparent shear rate: $\dot{\gamma} = 4Q/n x \pi R^3$

Where Q is Volume flow rate, R is the radius of the die orifice, and n is number of orifices. This value is also shown in Figure 3. Both nylon-6 and PET exhibit higher viscosities at any given temperature as compared to PP resin (see also Figure 4). PET and nylon-6 also show more newtonian behavior than PP resin. The curves for zero shear viscosities vs Temperature indicate that melt viscosities for all three resins decreases as temperature increases. PET and nylon-6 show much steeper slopes than PP. The PP melt viscosity decreases relatively slower than those of PET and nylon-6 as temperature increases. This means that the right spinning viscosity can be obtained for both PET and nylon-6 with a very small increase in temperature above their melting points. The rate of solidification for these resins would also be much faster than that of PP, during the melt attenuation and drawing. In contrast, the melt viscosity of PP decreases relatively slowly as temperature increases, meaning that higher temperature is required to achieve the right spinning viscosity. The rate of cooling is also slow and molecular orientation developed during melt attenuation can relax to some extent before solidification.

Figure 5 Shows typical SEM photomicrograghs for PP fibers processed using 10 and 50 psi primary air pressure. It can be seen that increase in air pressure from 10 to 50 psi results in breaking of the melt into droplets (microshots etc.). The same trend was also seen for PET and nylon-6 but at higher air pressures (above 70 psi data are not shown). Figure 6 shows SEM photomicrographs taken at the same magnification for PET, PP and nylon-6 fibers processed at the same primary air pressure. It is apparent from this Figure that both nylon-6 and PET fibers are less bonded and more uniform in size as compared to Polypropylene fibers. This can be attribute to rheological differences as seen in Figure 3 and 4. The formation of shots and other fiber irregularities seen here are mainly associated with the melt rheology (see also Table I) and air turbulence created during the melt blowing. The air turbulence increased as the primary air pressure and air gap increased (see Figure 7).

In order to eliminate the filament relaxation during the attenuation and increase molecular orientation, a secondary air chamber was placed 2 inch from the die exit. Cold air was applied through this chamber which increased both the fiber velocity and cooling rate. As a result, fibers with a narrow size distribution and high degree of molecular orientation were obtained. Typical SEM photomicrographs for PP, nylon-6, and PET fibers after passing through this chamber are shown in figure 8. An example of the size distribution histogram for PP fibers with and without using this chamber is shown in Figure 9.

Figure 10 shows the degree of crystallinity as a function of primary air pressure, for fibers made from all three resins. No significant increase in the degree of crystallinity is seen for these fibers, in the range of air pressure used. Similar results were also obtained for the average crystal size of these fibers, as a function of primary air pressure (Figure 11). These results clearly indicate that the nature of crystallization during melt attenuation is independent of the deformation rates at least in the range studied here. Since the spin line stress is not maintained on the filaments prior to complete solidification in the melt blowing, the crystallization rate is not expected to rise significantly. The molten filaments are drawn by, the hot high velocity compressible air, at the die nozzle. The tension developed on the filaments is released abruptly as the hot air dissipates into the ambient at a distance very close to the die tip (6).

The lack of sensitivity to deformation rate is also seen with regard to orientation (see Figure 12). However, a drastic increase in both crystallinity and orientation seen for the the fibers drawn by the secondary air chamber. The degree of crystallinity increased from an average 42 percent for undrawn PP microfibers to 51 percent for the drawn microfibers. These were 25, 31.5 percents for undrawn and 37, 49 percents for drawn PET and nylon-6 microfibers respectively.

Figure 13, 14, and 15 show typical WAXS film patterns for the fibers before and after passing through the secondary air chamber. These patterns do not indicate any increase in crystalline orientation for meltblown fibers processed from all three resins. Annealing of the PET fibers at 210 C for 1/2 hour under tension increase the crystallinity from 37 to 42 percent (compare Figure 14a with 14c). The fibers drawn by the secondary air chamber exhibit a high degree of crystalline orientation for all three polymers. The C-axis crystalline orientation factors for PP, nylon-6, and PET were 0.85, 0.71, and 0.82 respectively.

As one would readily infer from these data, the mechanical properties of the melt blown fibers are also essentially unaffected by increasing the primary air pressure(6). Increase of molecular orientation and crystallinity brought about by secondary air chamber increased the fibers tensile strength and modulus, while decreased the fibers elongation to break. A typical stress strain polts of the melt blown webs of the drawn and undrawn PP fibers are shown in Figure 16.

SUMMARY

This study indicates that the morphological features of the melt blown fibers, with regard to orientation and crystallinity, are relatively insensitive to the melt blowing rate of deformation used in this study. The reason for this is that the molten filaments are drawn by hot high velocity compressible air at the die nozzle. The tension developed on the filaments is released abruptly (as the hot air dissipates into the ambient) before the solidification is completed. Increasing the air pressure only determines the time of deformation-induced transformation. Increase in primary air pressure also increased fiber discontinuity, size distribution, and shots. These non-uniformities become more pronounced when the air becomes extremely turbulent. A drastic increase in both molecular orientation and crystallinity was only seen for the microfibers drawn by the secondary air chamber. These drawn fibers showed mechanical properties similar to those of conventional melt spun or spunbonded fibers. A substantial reduction in fiber size variation was also associated with drawing of the microfibers using the secondary air chamber.

REFERENCES

1. V.A. Wente, Tec. Rep. No. PB111437, Navel Res. Lab. 4364, 4/15/1954 and " Superfine Thermoplastic Fibers", Ind. Eng. Chem. 48, 1342(1956).
2. L. Bergmann, Nonwovens Industry, 19, 26(1988).
3. J.W. Harding , M.S. Patent No. 3,825,380.
4. L. Hartmann, U.S. Patent No. 3,502,763.
5. R.R. Buntin, U.S. Patents NO. 3,978,185 and No. 3,972,759.
6. H. Bodaghi, INDA Journal of Nonwonens Res., Vol. 1, No. 1. 14-27(1989).
7. J. Brandup, E.H. Immergut, Polymer Handbook, John Wiley, New York(1975).
8. H. Bodaghi, "A study and comparison of polymer films formed from isotropic and liquid crystalline solutions.",Univ. Microfilms Intern. Ann Arber, Michigan(1985).
9. L.E. Alexander, "X-ray Diffraction Methods in Polymer Science.", John Wiley New York(1979).
10. Z. Wilchinsky, J. Phy. 30,792(1959), ibid. 31 P.1969(1960)., Adv. X-ray Anal. 6, P.231(1963).
11. H. Bodaghi, J.E. Spruiell and J.L. White, Intern. Poly. Processing III/Issue 2, 100(1988).
12. M. Mooneny, J. Rheol., 2, 210(1931).

Figure Captions

Figure 1. Schematic representation of melt blown microfibers exiting the die.

Figure 2. Factors affecting the final properties of a melt blown web.

Figure 3. Viscosity vs Temperature for PP and Nylon-6 resins.

Figure 4. Zero shear viscosity for PET, Nylon-6, and PP vs Temperature.

Figure 5. SEM photomicrographs for PP microfibers processed at:
a) 10 psi primary air pressure
b) 50 psi primary air pressure

Figure 6. SEM photomicrographs of microfibers processed under the same condition for: a) Polypropylene, b) Nylon-6, c) PET

Figure 7. Re Number vs Primary air pressure for different die air gaps.

Figure 8. SEM photomicrographs for drawn microfibers using secondary chamber with 70 psi air pressure:
a) PP, b) Nylon-6, c) PET

Figure 9. Fiber size distribution histograms for PP microfibers before and after drawing.

Figure 10. Crystallinity vs primary air pressure

Figure 11. Average Crystal size vs primary air pressure

Figure 12. Average fiber Birefringence vs primary air pressure

Figure 13. Wide Angle X-ray film patterns for :
a) Undrawn PP microfibers
b) Drawn PP microfibers

Figure 14. Wide Angle X-ray film patterns for:
a) Undrawn PET microfibers
b) Drawn PET microfibers
c) Undrawn and annealed PET microfibers.

Figure 15. Wide Angle X-ray film patterns for:
a) Undrawn Nylon-6 microfibers
b) Drawn Nylon-6 microfibers

Figure 16. Stress-Strain curves for PP Melt blown webs of:
a) Undrawn microfibers
b) Drawn microfibers

Table I

Molecular Weights for Resins Before and After Extrusion Into Melt Blown Microfibers

		Mw	Mn	P
Nylon-6	Resin	5.40 e 4	1.00 e 4	5.37
Nylon-6	Fiber	2.80 e 4	3.05 e 3	9.16
3M PET	Resin	4.00 e 5	9.90 e 3	4.1
3M PET	Fiber	3.99 e 4	1.10 e 3	3.6
Gulf PP	Resin	9.98 e 4	2.07 e 4	4.8
Gulf PP	Fiber	2.67 e 4	8.40 e 3	3.2

Table II

Melt Blown Processing Variables

	Nylon-6	PP	PET
Poly. Flow Rate (lb/hr/in)	0.5	0.5	1.0
Melt Temp. (^{0}C)	260	220	300
Die Temp. (^{0}C)	275	250	316
Air Temp. (^{0}C)	275	245	300
Air Pressure (PSI)	10	10	10
	25	25	25
	40	40	40
	50	50	50
Orifice Size (Cm)	0.025	0.025	0.025
Air Gap (Cm)	0.038	0.038	0.038
	0.013	0.013	0.013
Secondary Chamber Air Pressure (PSI)	70	70	70
Secondary Chamber Air Temp. (^{0}C)	ambient	ambient	ambient

Schematic Representation of Melt Blown Fibers Exiting the Die

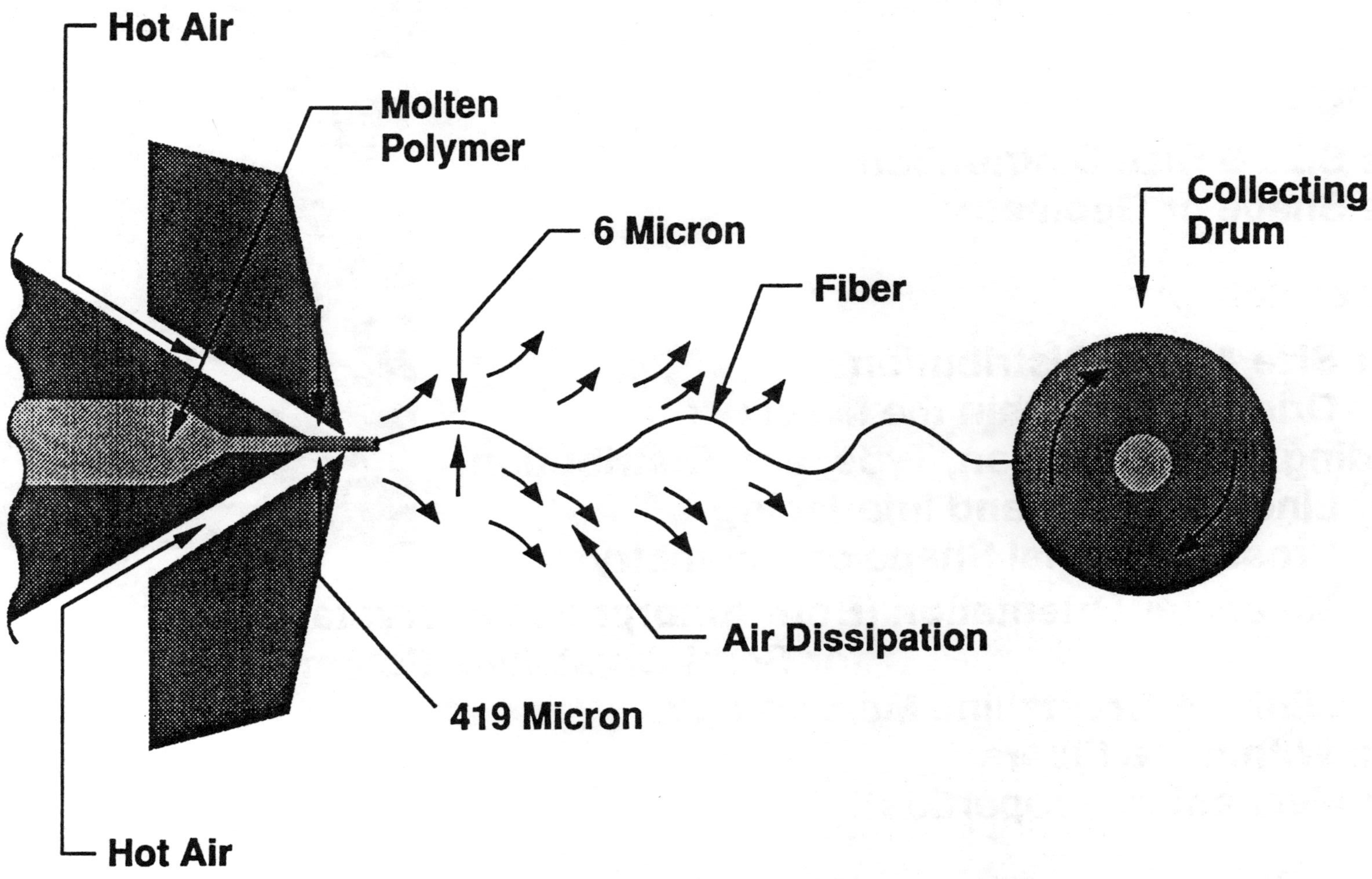

T510-18

Factors Affecting the Final Properties of a Melt Blown Web

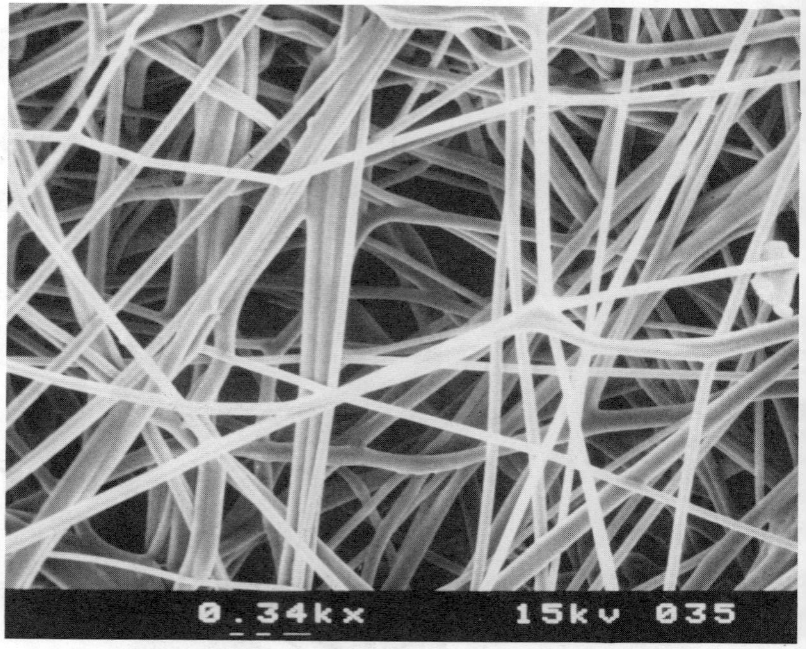

1. Porosity
   - Pore Size & Size Distribution
   - Pore Shape or Geometry

2. Polymer Network
   - Fiber Size & Size Distribution
   - Fiber Orientation Within the Matrix
   - Bonding (Concentration, Type, and Distribution)
   - Fiber Linear Density and Interlacing
   - Fiber Cross-Sectional Shape or Geometry
   - Fiber Molecular Orientation (Both Amorphous & Crystalline for Semi-Crystalline Polymers)
   - Crystallinity & Crystalline Morphology
   - Voids Within the Fibers
   - Fiber Mechanical Properties

3. Incorporation of Non-Fibrous Materials and Their Distributions

Figure 3

Comparative Viscosity vs. Shear Rate

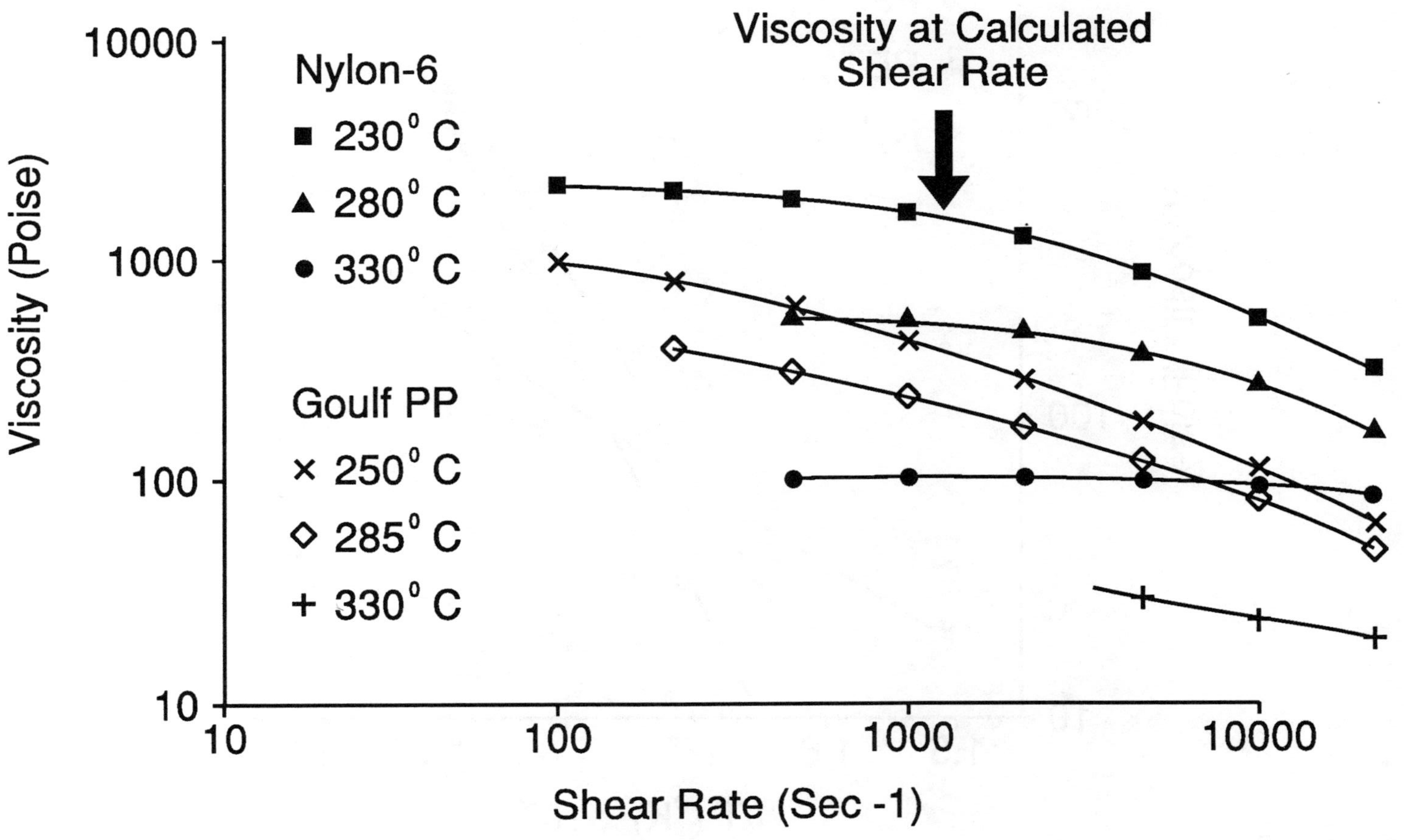

Figure 4

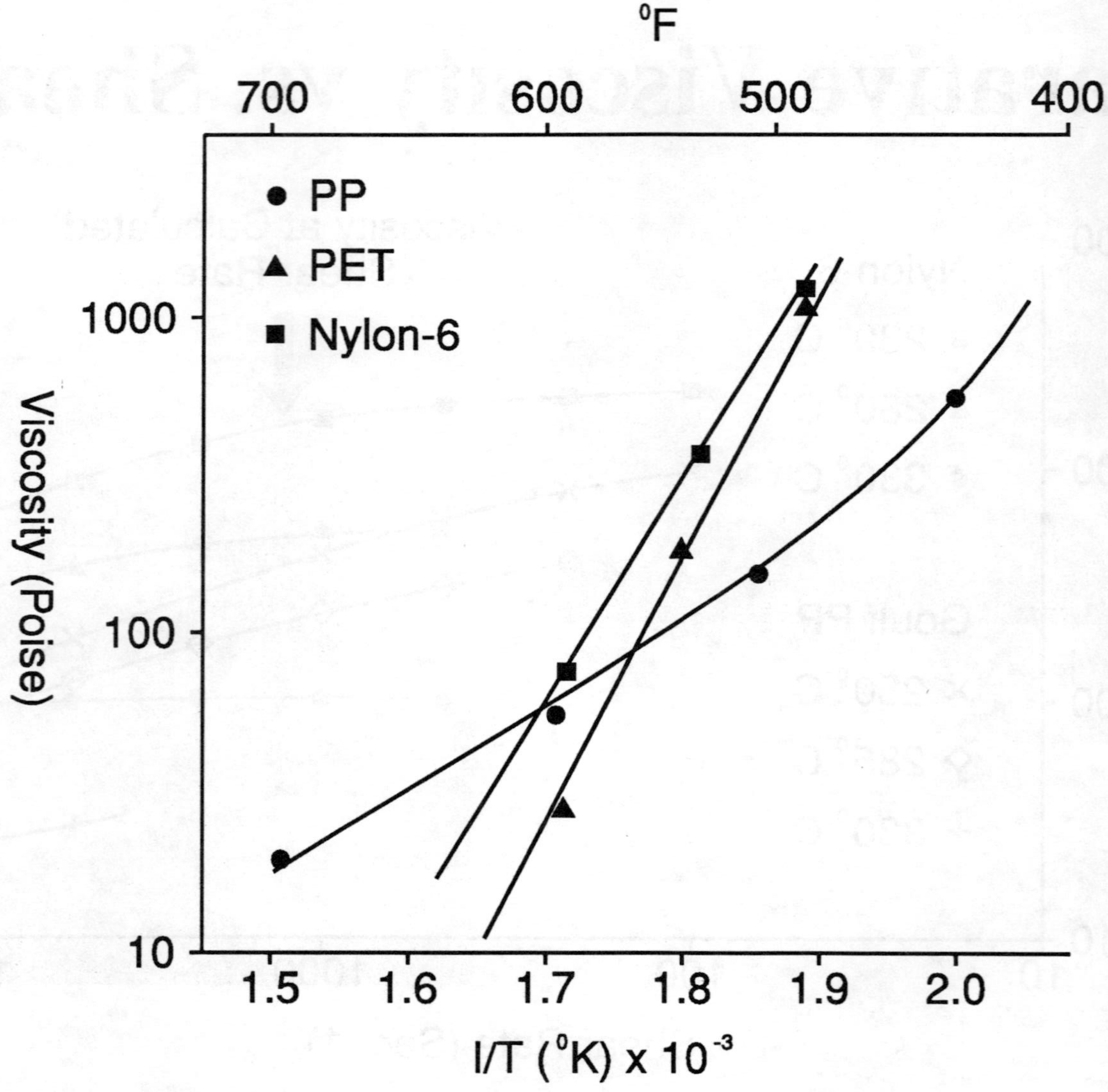
°F
700
600
500
400
PP
PET
Nylon-6
Viscosity (Poise)
1000
100
10
1.5
1.6
1.7
1.8
1.9
2.0
I/T (°K) x 10⁻³

Figure 5

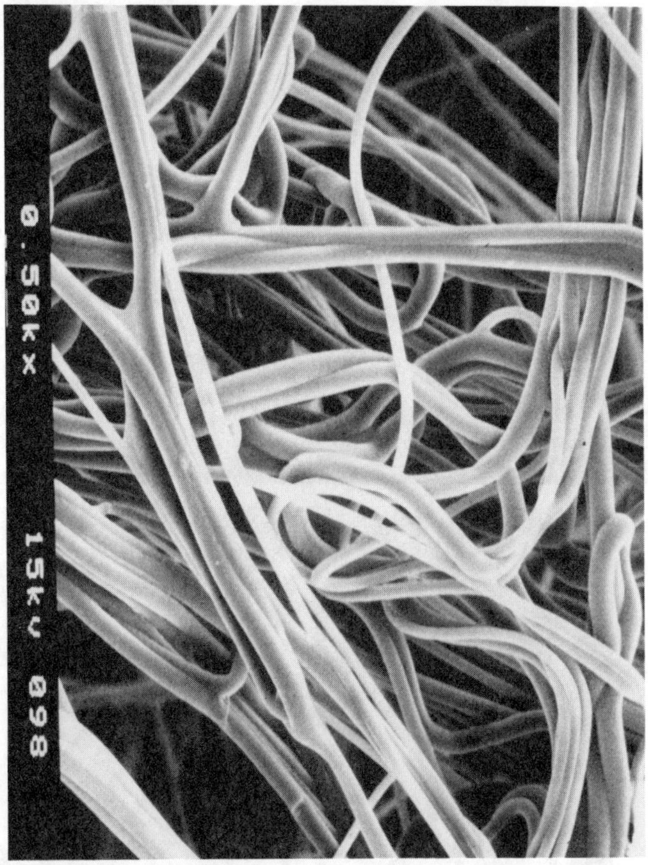

A

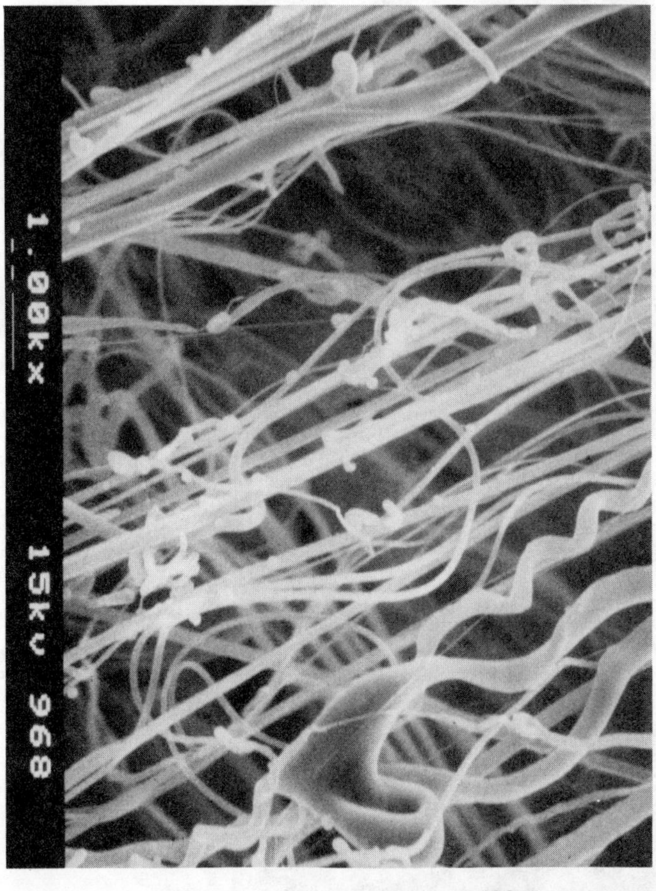

B

Figure 6

A

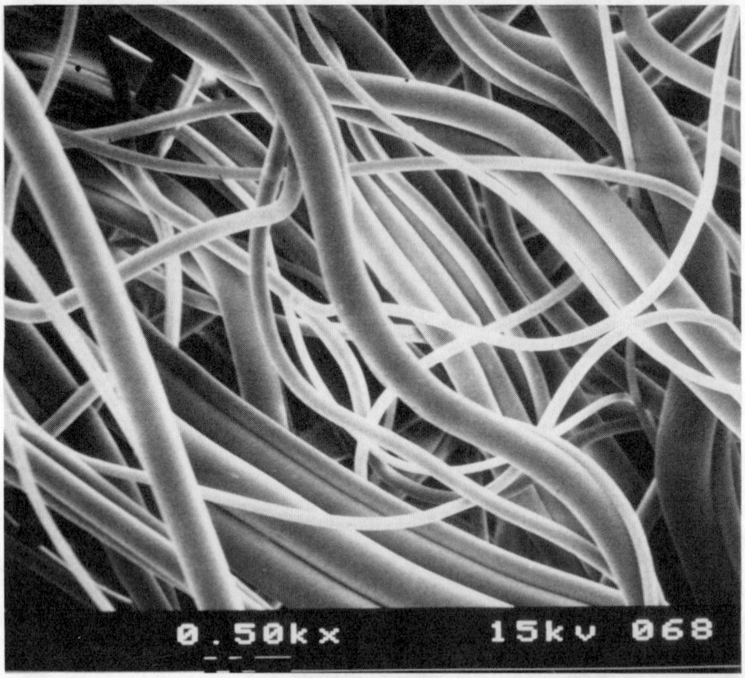

B

C

Figure 7

p vs Re

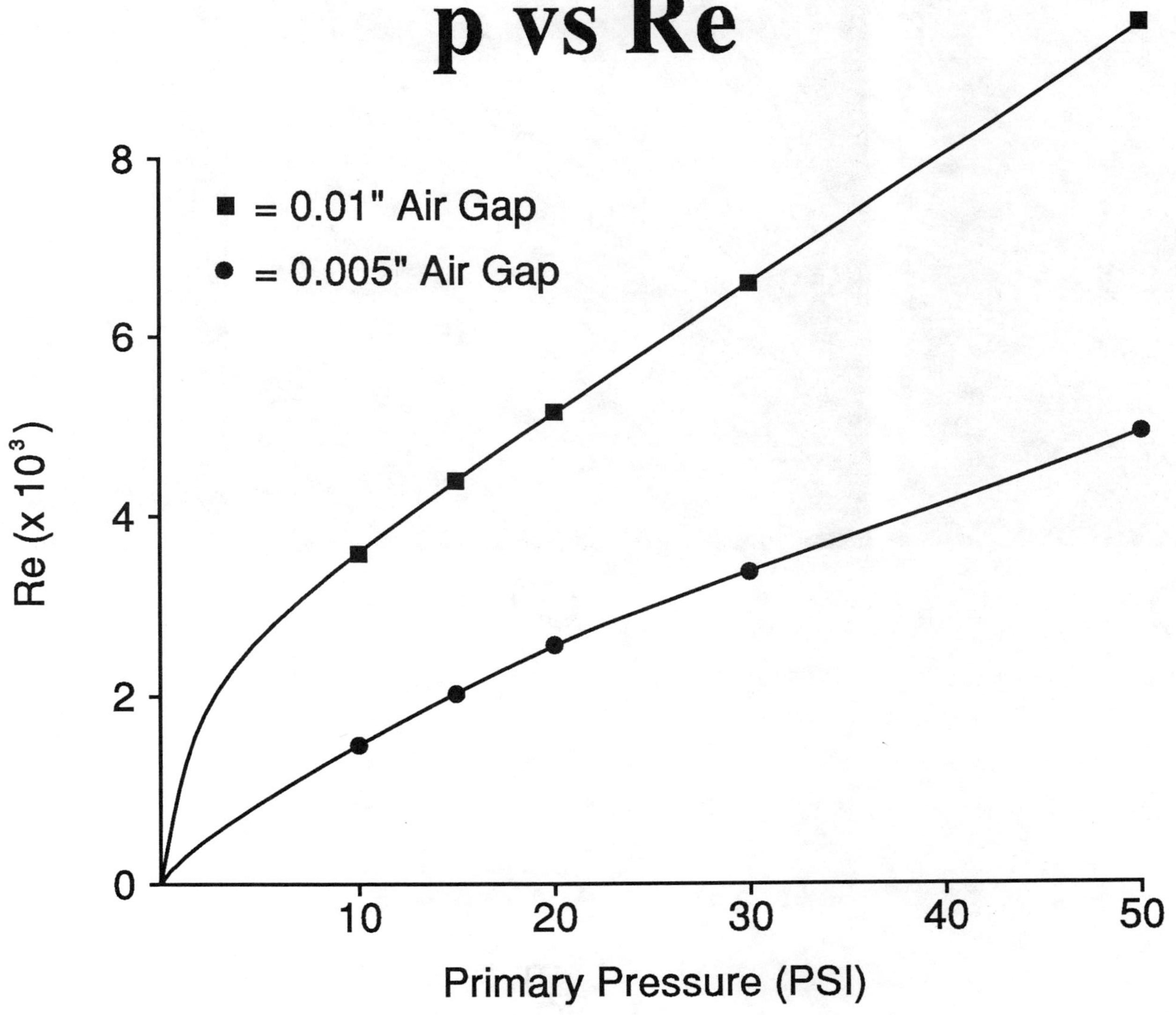

A

B

C

Figure 8

Figure 9

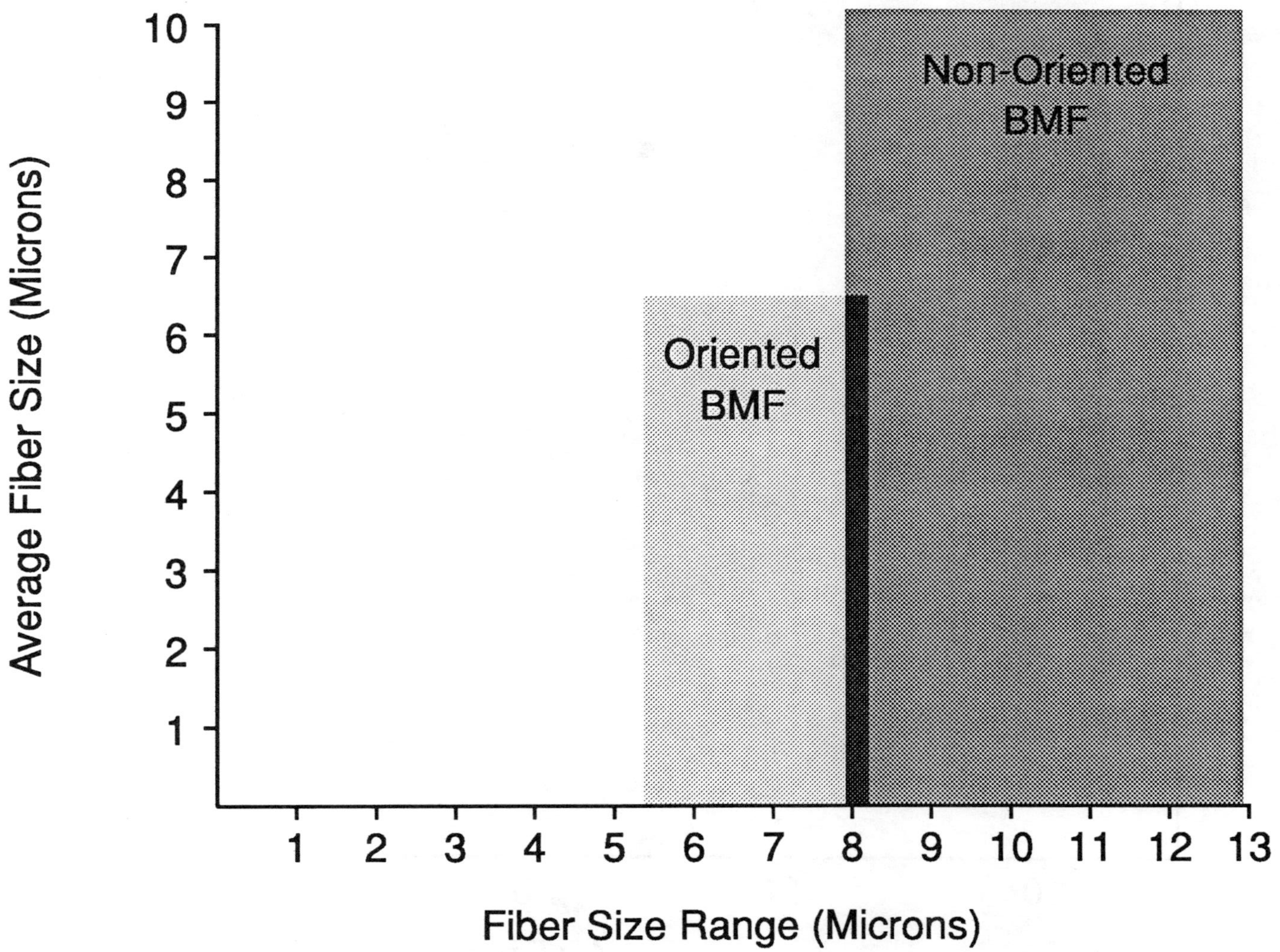

Average Fiber Size (Microns)
10
9
8
7
6
5
4
3
2
1
Oriented
BMF
Non-Oriented
BMF
1 2 3 4 5 6 7 8 9 10 11 12 13
Fiber Size Range (Microns)

Figure 10

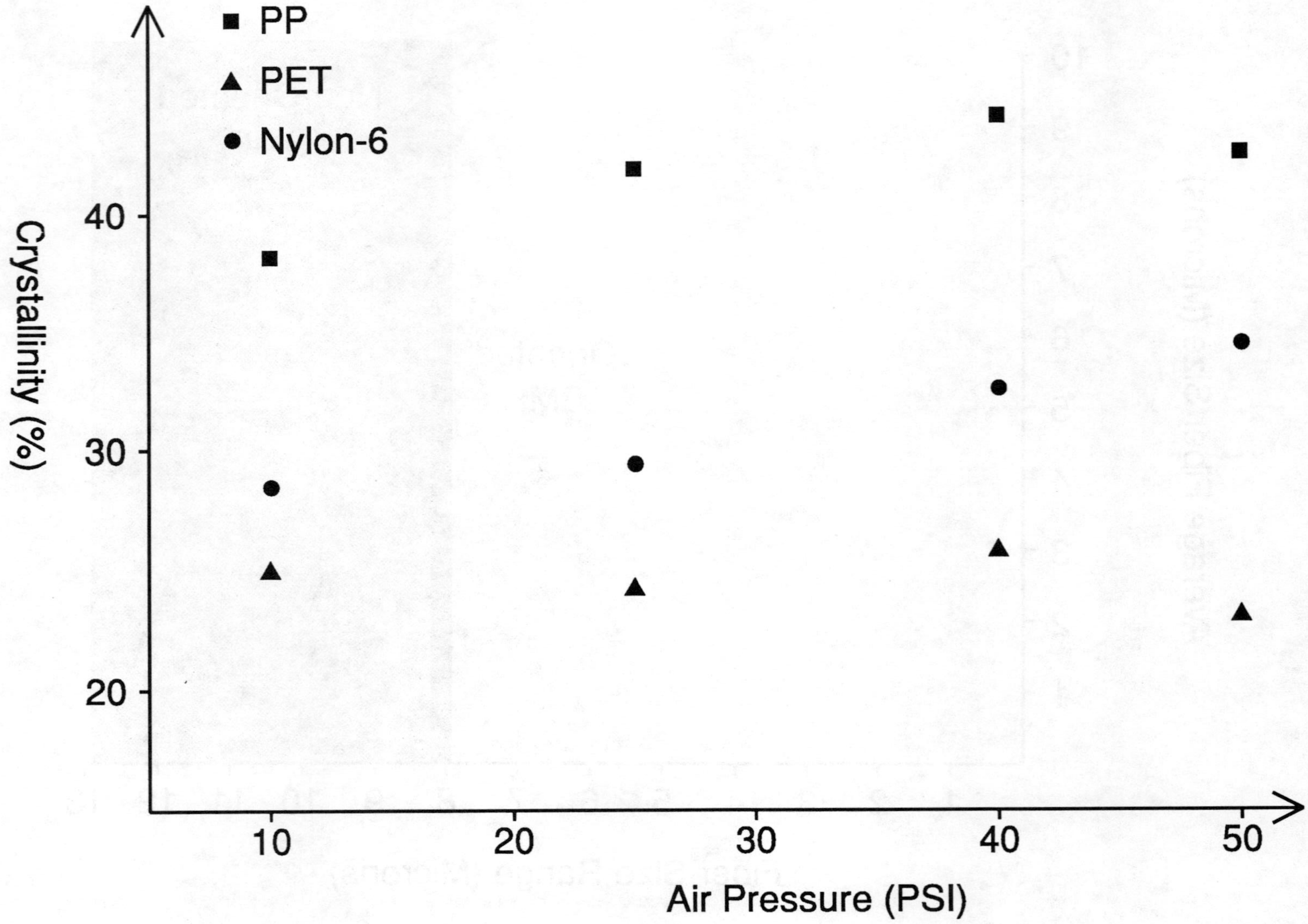
PP
PET
Nylon-6
Crystallinity (%)
40
30
20
10
20
30
40
50
Air Pressure (PSI)

Figure 11

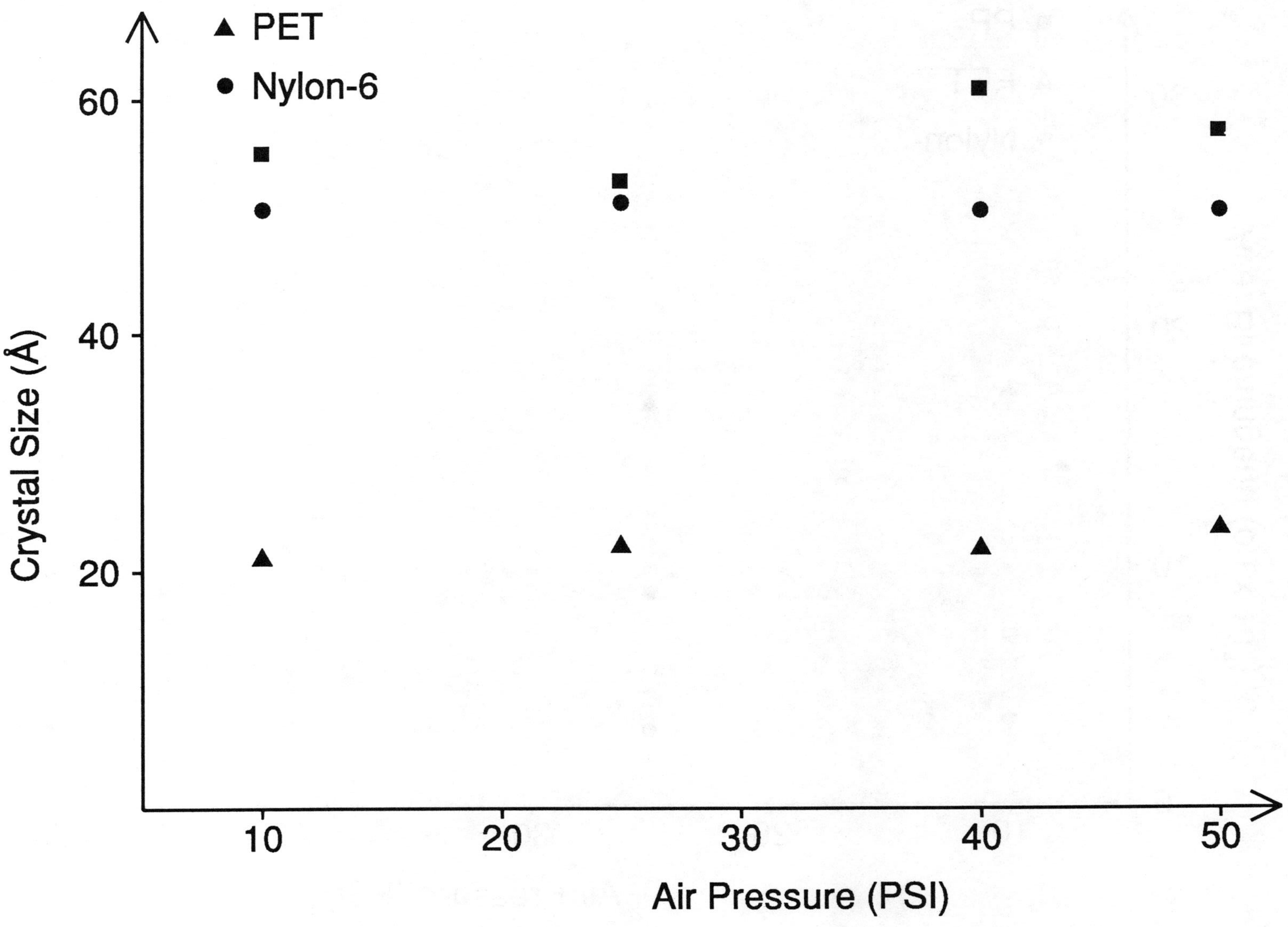
PP
PET
Nylon-6
Crystal Size (Å)
60
40
20
10
20
30
40
50
Air Pressure (PSI)

Figure 12

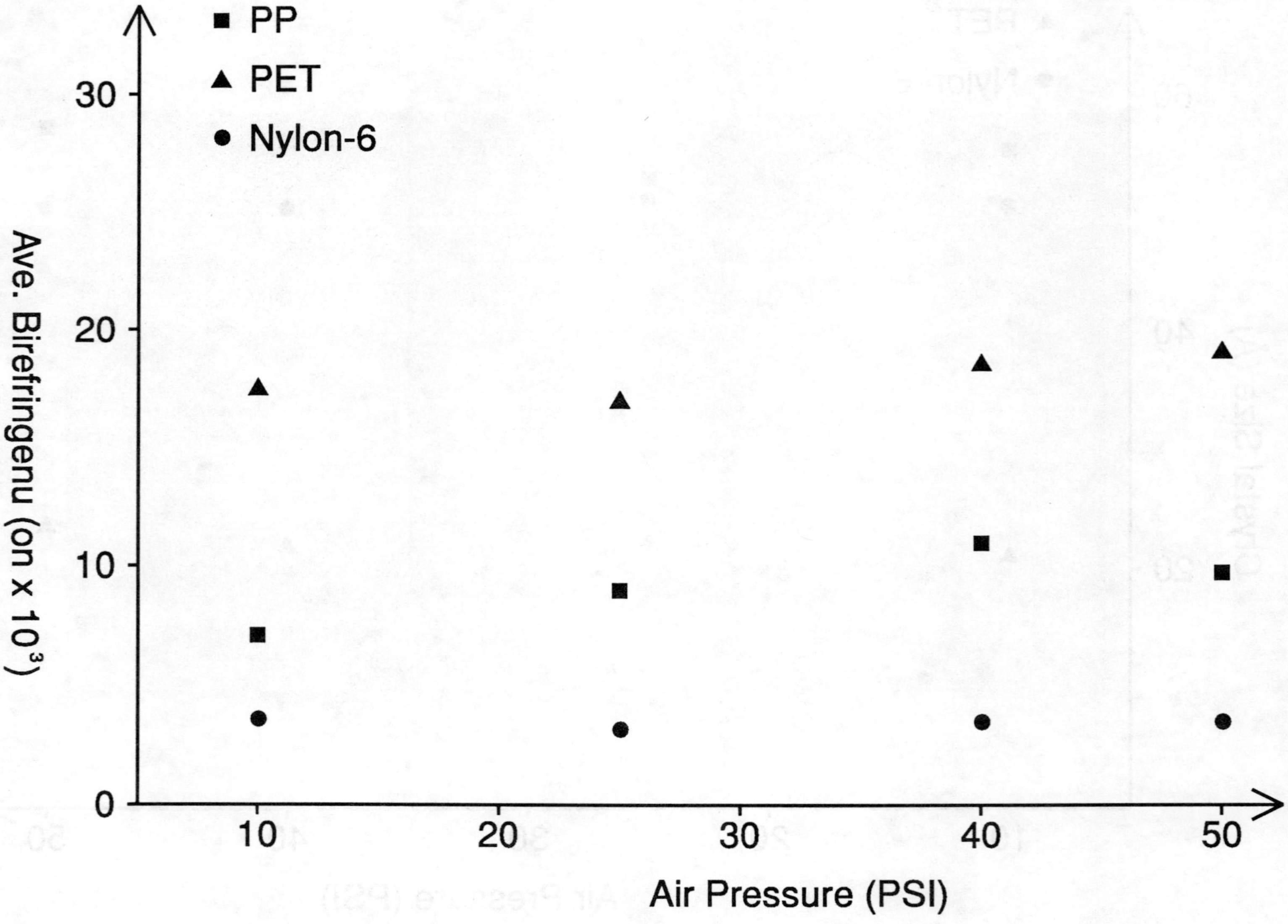
PP
PET
Nylon-6
Ave. Birefringenu (on x 10^3)
30
20
10
0
10
20
30
40
50
Air Pressure (PSI)

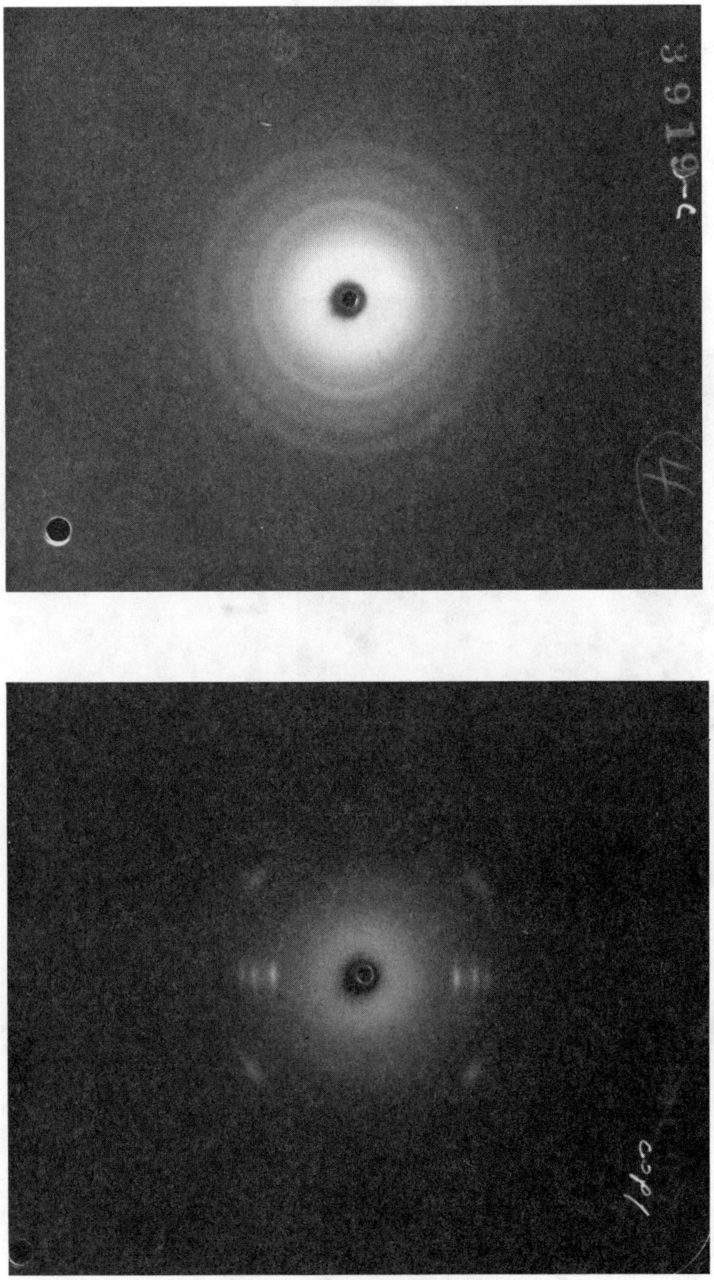

A

B

Figure 13

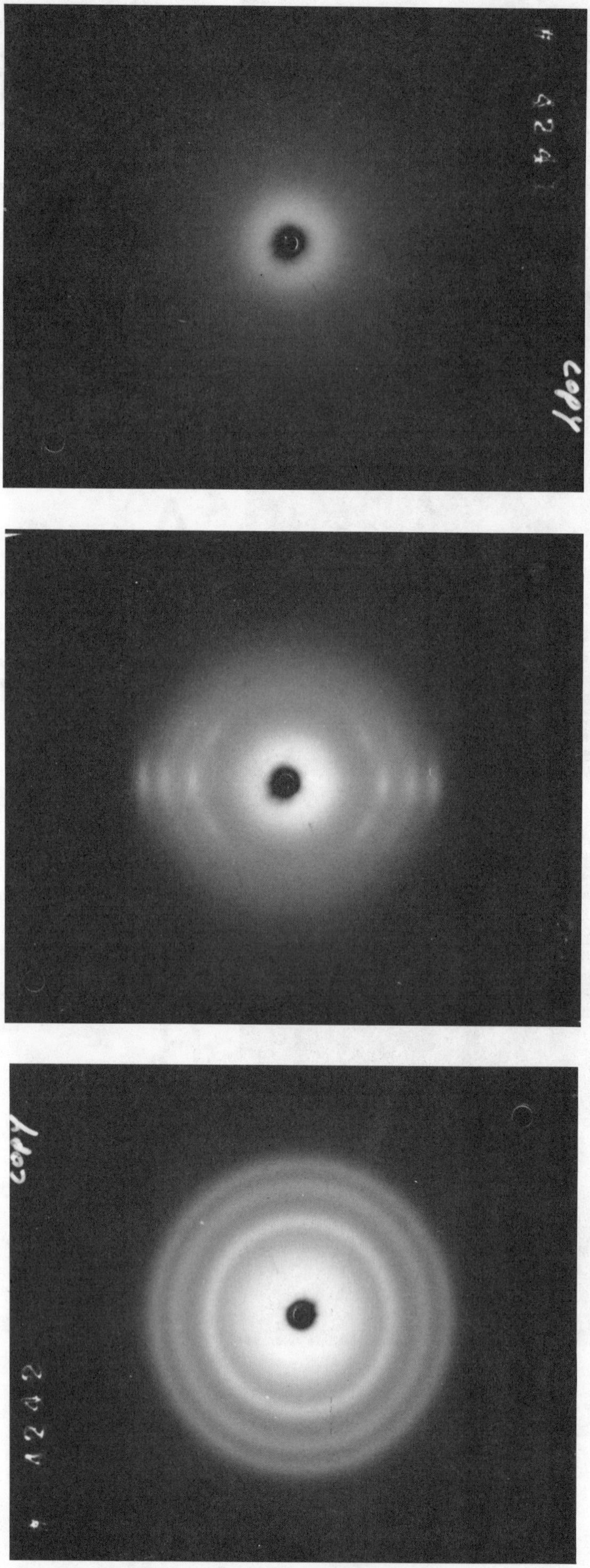

A

B

C

Figure 14

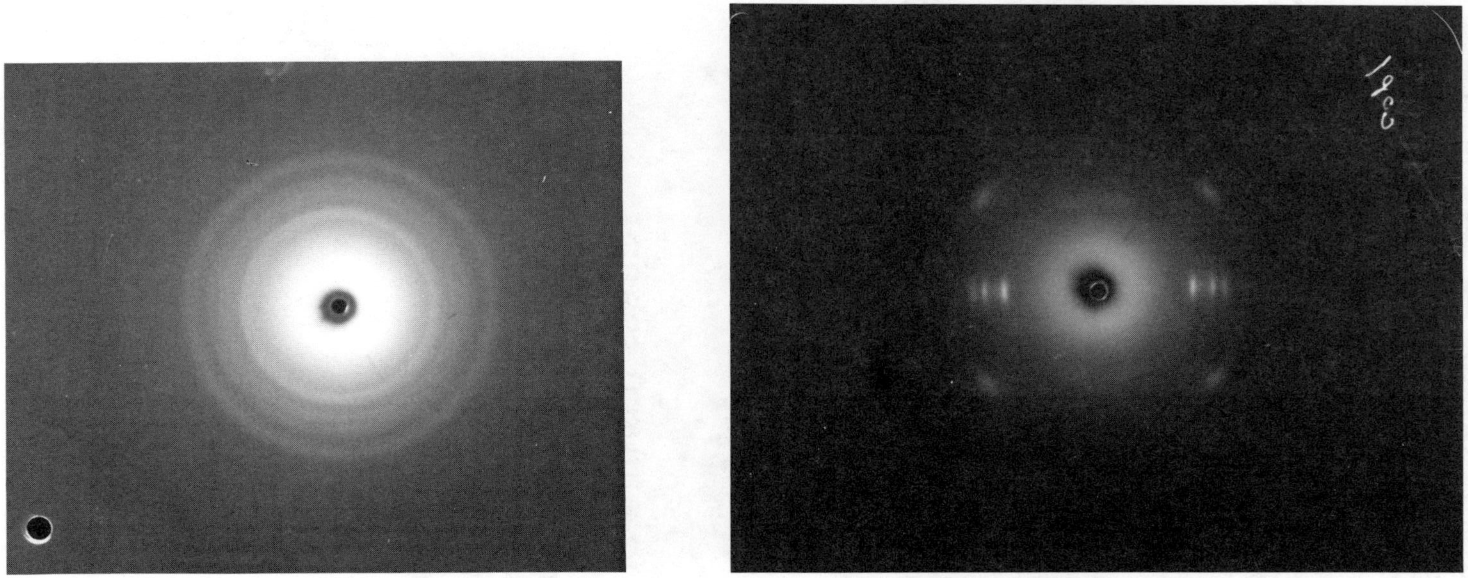

A B

Figure 15

Figure 16

Plastics Tensile Test

Specimen 3

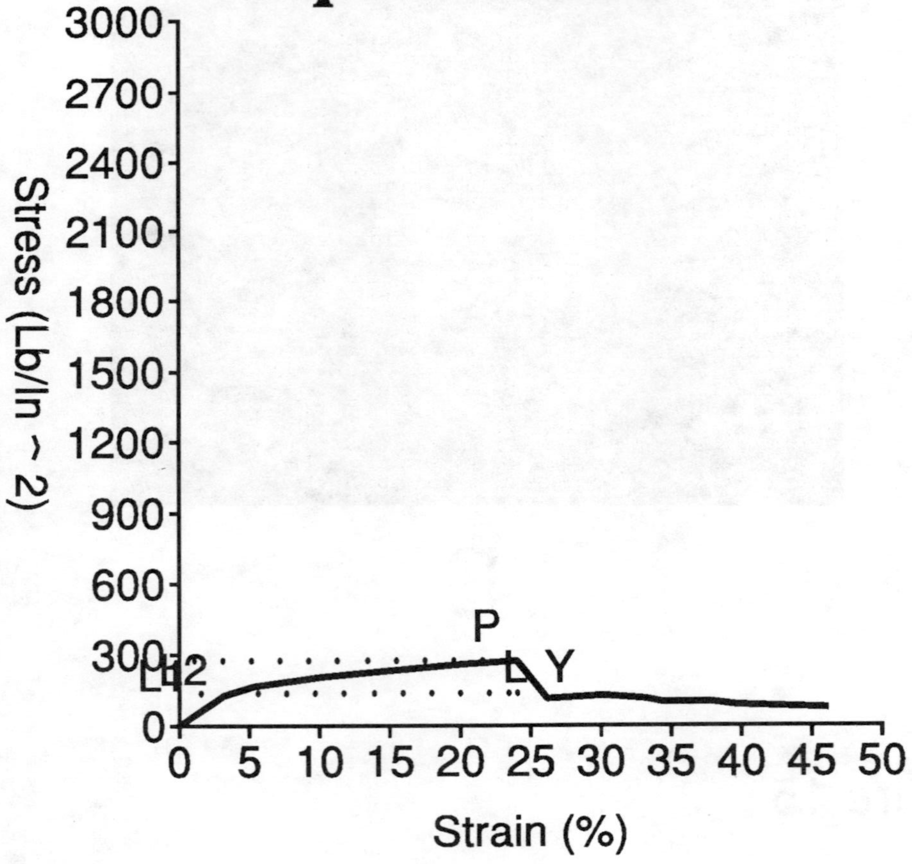

Specimen 7

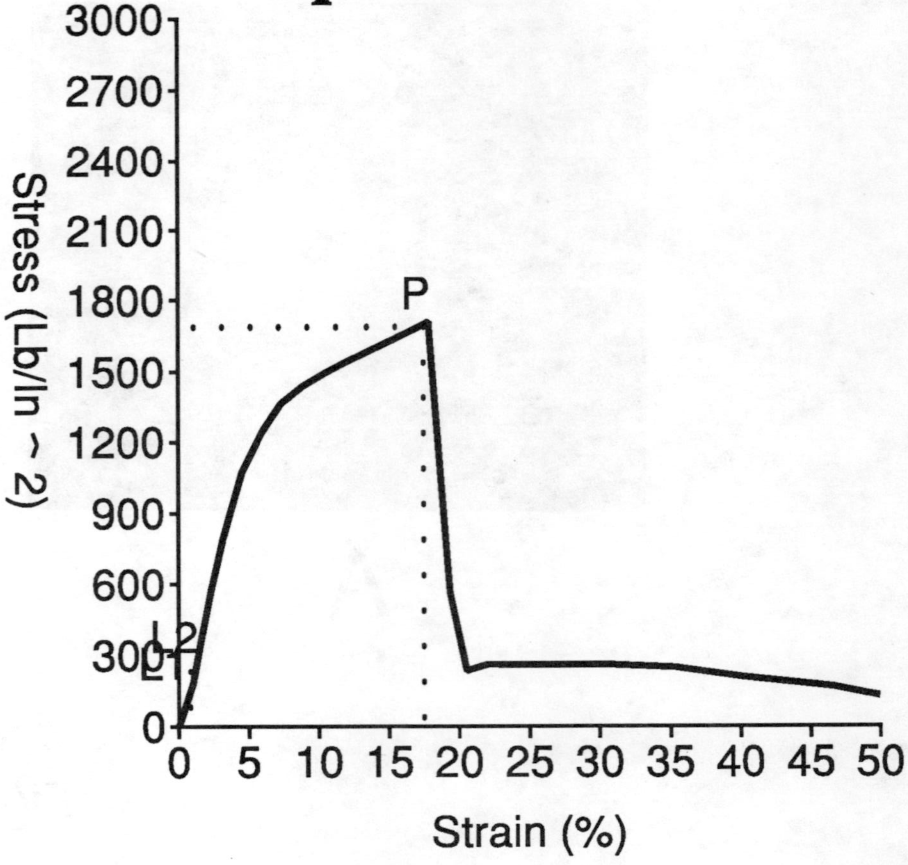

PARTICLE SEPARATION IN MICRO-GRAVITY

A.J. PALERMO
AIRFILTERS, INC.
5408 ASHBROOK
HOUSTON, TX 77081

A STUDY TO DETERMINE THE BEST POSSIBLE MEDIAS AVAILABLE TO SEPARATE MASS RESPIRABLE PARTICULATE IN THE CABIN AND COCKPIT OF THE SPACE SHUTTLES. ULTIMATELY TO BE USED IN SPACE STATION PROJECTS WHERE LONGEVITY OF MEDIA LIFE WAS AT AN OPTIMUM WITHOUT REPLACEMENT OR DETERIATION OF PERFORMANCE. MAJOR PERFORMANCE CRITERIA HAD TO BE ESTABLISHED AND APPROVED BY NASA ENGINEERING. SOME MAJOR CONCERNS WERE; OFF GASSING OF MEDIA IF BURNED, PORE SIZE, REUSABILITY, CLEANABILITY DURING A MISSION, AND MICRON RETENTION.

THE FIRST PRIORITY WAS TO DETERMINE THE CUBIC FOOTAGE OF SPACE TO BE CLEANED ON BOARD THE SPACE CRAFT. AFTER NUMEROUS MEETINGS WITH NASA ENGINEERS AND DESIGNERS WE DETERMINED THERE WAS APPROXIMATELY 1500 CUBIC FEET OF OVERALL CABIN AND COCKPIT AREA TO BE CLEANED.

THE NEXT PROCESS WAS TO ANALYZE THE SPACE SHUTTLE CABIN DEBRIS FOR PARTICLE SIZE DISTRIBUTION. THE MAJORITY OF PARTICLES FOUND IN THE SPACE SHUTTLE ARE FIBERS RANGING FROM 2-72 MICRONS IN DIAMETER. THESE FIBERS ARE CLOTHING FIBERS (COTTON, NOMEX, WOOL, ETC.), HAIR, PAPER, AND GLASS FIBERS. OTHER LARGER DEBRIS INCLUDED VISIBLE ITEMS SUCH AS PAINT CHIPS, FOOD, PENCIL LEAD, BUGS, AND TAPE. CHART A

BASED ON THESE FINDINGS THE ORIGINAL REQUEST FOR A .3 MICRON HEPA FILTER WHICH HAS A HIGHER PRESSURE DROP AND SHORTER LIFE CYCLE THAN OTHER RECOMMENDED FILTERS WAS NOT WARRANTED.

A LARGER PORE MEDIA 2-5 MICRON WOULD BE MORE EFFECTIVE, WHILE REQUIRING LESS ENERGY FOR ITS BLOWER, AND HAVE LONGER LIFE CYCLES.

MANY MEDIAS FIT THE QUALIFICATIONS OF REUSABLE, POLESTERS AND POLYPROPYLENE, COULD BE WASHED OR VACUUMED DURING ACTUAL MISSION TIME. THIS WOULD ENABLE THEM PRECIOUS SPACE SAVING REQUIREMENTS SO AS NOT TO STORE SEVERAL SPARE FILTERS. OBVIOUSLY TO BE ABLE TO CLEAN THE MEDIA DURING FLIGHT WOULD BE OF SIGNIFICANT ADVANTAGE.

THE PROCESS OF DETERMINING OPTIMUM AIR FLOW WITH MINIMUM PRESSURE DROP (DELTA P), MEASURED IN INCHES OF WATER, WAS OF VITAL IMPORTANCE. OPTIMUM PERFORMANCE FOR UTILIZATION OF ENERGY IS ALWAYS IMPORTANT. HERE IT WAS CRITICAL THAT WE COULD NOT USE MORE THAN WAS ABSOLUTELY NECESSARY TO DO THE JOB. PRECIOUS ENERGY FOR OPERATIONS OF ON BOARD COMPUTERS AND LIFE SUPPORT SYSTEMS ARE OF MORE IMPORTANCE.

OFF GASSING OF THE MEDIA DURING A 72 HOUR FLAME TEST WAS CONDUCTED AT THE WHITESANDS TEST FACILITY IN NEW MEXICO. NASA WANTED TO INSURE THAT THE FUMES GIVEN OFF BY THE MEDIA DURING SUCH A TEST WERE WITHIN THEIR TOXCISITY LEVELS FOR MATERIALS GOING INTO A SPACECRAFT FOR FLIGHT. ALL OF THE MEDIAS TESTED PASSED WITH VERY LOW LEVELS OF GAS BEING EMITTED. MOST WERE MEASURED IN HUNDREDTHS OF MICROGRAMS/GRAM WHERE A LIMIT OF 124 MICROGRAMS/GRAM ON XYLENE WERE THE MAXIMUM. REFER TO CHART B

BY THE TIME WE ARRIVED AT THIS PROCESS WE HAD NARROWED THE MATERIALS TO BE TESTED TO 3 MEDIAS WHICH ARE AS FOLLOWS:

1) NON WOVEN POLYESTER FIBER
2) POLYESTER PAPER
3) 400 MESH 316 S.S.

THE FOLLOWING IS HOW EACH MEDIA PERFORMED DURING THE TEST:
THE POLYESTER NON WOVEN DEVELOPED A .5" STATIC PRESSURE AT 825 CFM.
THE POLYESTER PAPER DEVELOPED A .5" STATIC PRESSURE AT 650 CFM. WHILE THE 400 MESH STAINLESS DEVELOPED A .5" STATIC PRESSURE AT 910 CFM. (THESE TEST WERE PERFORMED ONLY ON CLEAN UNITS) REFER TO CHART C

THE MICRON RETENTION WAS DETERMINED BY NASA, THEIR TEST PROCEDURES WERE NOT PUBLISHED OR RELEASED TO ME ONLY THE RESULTS. THE FOLLOWING ARE THE EFFICIENCY RATINGS AS DETERMINED BY NASA. THE POLYESTER NON WOVEN WAS 40% AT 2 MICRON. THE POLYESTER PAPER WAS DETERMINED TO BE 80% EFFICIENT AT 2 MICRON. THE 400 MESH S.S. WAS FOUND TO BE 98% EFFICIENT AT 38 MICRONS.

A PROCESS OF CLEANING DURING LONG TERM MISSIONS WAS THE NEXT CONSIDERATION. DURING A MISSION THEY MUST BE ABLE TO VACUUM OFF THE HEAVY OR LARGE PARTICULATE AS THE MEDIA BECOMES LOADED. AFTER SEVERAL VACUUMINGS THE MEDIA MAY NEED TO BE WASHED OR BACKFLUSHED WITH WATER. THIS PROCEDURE WOULD BE PERFORMED IN THE SHOWER WITH THE HAND HELD SHOWER HEAD. AFTER ARRIVING BACK ON EARTH THE FILTER COULD BE AUTOCLAVED TO BURN OFF ANY PARTICULATE AND KILL ANY BACTERIA OR

VIRUS THAT MAY ACCUMULATE ON THE MEDIA.

WE NOW ARRIVE AT THE HARDEST PART, THAT IS TO DECIDE WHERE THERE IS ENOUGH SPACE IN THE CRAMPED QUARTERS OF THE CABIN ON THE SPACE SHUTTLE TO PLACE THE FILTRATION UNIT. WE EXPLORED MANY POSSIBILITIES AS TO WHERE THE UNIT COULD FIT, OR COULD WE BUILD IT INTO THE CABINITRY THAT HOUSED MOST OF THE ELECTRONICS. FINALLY ALL AGREED ON THE HATCHWAY BETWEEN THE COCKPIT AREA AND THE LIVING QUARTERS. THERE WAS A NO TO THE LEFT SIDE HATCHWAY SINCE THAT IS WHERE THE SHOWER AND RESTROOM FACILITIES ARE. YES, THE RIGHT SIDE HATCHWAY IS DESIGNATED AS THE PLACE THAT WILL HOLD THE UNIT. BUT IT CAN NOT REMAIN THERE, IT IS ONLY TO BE USED DURING SLEEPING PERIODS.

TO OPERATE ONLY DURING SLEEPING PERIODS CREATES SOMEWHAT OF A HANDICAP, FOR NOW WE MUST HAVE A QUIET BLOWER. WE NOW REDESIGNED THE BLADE CONFIGURATION ON THE BLOWER SO THAT THE CORRECT PITCH IS USED TO MOVE THE AIR AT A GIVEN VELOCITY AND DELTA P YET BE QUIET. THE BLOWER BLADES AND DESIGN FINALLY BEHIND US THE QUESTION IS ASKED WHAT VOLTAGE AND HZ SHOULD THE MOTOR BE. TO CONSERVE PRECIOUS ENERGY THE MOTOR MUST OPERATE ON 24V AC., NO PROBLEM, BUT THE ON BOARD POWER GENERATOR PRODUCES IT AT 400 HZ. THIS CAUSES A MINOR PROBLEM BUT IS EASILY OVERCOME IN THE WINDINGS.

THE WHOLE UNIT MUST STOW INSIDE A 17" WIDE, BY 20" DEEP, AND 10" HIGH COMPARTMENT. THE HATCHWAY BEING APPROXIMATELY 24" BY 24" AND A COMPARTMENT 17" X 20" TO STOW IN DID NOT MATCH UP TO IDEAL CONDITIONS. THE FILTRATION UNIT WAS MADE SMALL ENOUGH TO GO INTO THE STORAGE COMPARTMENT WITH FOLD OUT ARMS AND CLOTH TO STOP AIR BYPASS. THIS WAS SECURED IN THE HATCHWAY WITH VELCRO CLOSURES.

THE NEED FOR A CABIN AIR PARTICULATE REMOVAL SYSTEM HAS ARISEN DUE TO THE NEW BREED OF ASTRONAUT WE HAVE GOING INTO SPACE ON TODAY'S MISSIONS. HIGHLY TRAINED IN THEIR SPECIALIZED FIELD BUT NOT IN PARTICLE CONTAINMENT. THE FOODS HAVE CHANGED AND SO HAS THE PACKAGING, IT IS NOT UNCOMMON TO FIND PEANUT BUTTER AND CRACKERS FOR SNACKS. THE NOMEX CLOTHING SHEDS CAUSING THE BLUE FUZZ SYNDROME ALL OVER THE ELECTRONIC CABINETS HOUSING THE ON BOARD COMPUTERS. THE HUMAN BODY CONTRIBUTES MUCH TO THE PARTICULATE CONTAMINATION ALSO. BUT THE HAIR THAT COMES OFF DURING BRUSHING CAUSES VISIBLE PROBLEMS. WHERE THE BUGS COME FROM IS QUITE A MYSTERY SEEING THIS AREA IS KEPT UNDER CLEAN ROOM CONDITIONS. THE LONGER THE MISSION THE MORE CONTAMINATION FROM THE MOST UNOBVIOUS PLACES. WHAT A LUXURY WE HAVE ON EARTH WHERE LARGE PARTICULATE SUCCUMBS TO GRAVITY AND IN GENERAL IS NOT INHALED.

PARTICLES	WEIGHT OR WEIGHT RANGE (GRAMS)
Metallics	.0003 - 1.7842
Paint chips	.0004 - .6320
Plastics	.0018 - .6188
Rubber	.0528 - .6188
Pencil Lead	Trace - .0490
Woven Tapes	.0235 - .3392
Wood	Trace - .0370
Paper	.0085 - .0520
Pieces of Velcro	Trace - .0024
Plastic Tubing	1.8258
Food	Trace - .0695
Braid/Tissue	.0485 - .1005
Banana/Peanuts	.8872
Cellophane Bag	2.7355
Waxed Paper	.0241
Goose Down/Rabbit Fur	Traces
Teflon	Traces
Wax	Traces
Glass	Traces
Acrilon/Kevlar	Traces
Miscellaneous: (nuts, insect parts, sand, yarn, finger nails, aluminum tape/masking tape/fiberglass tape, bugs, grass stem, toothpick, glass, metal-velcro loops, skin, aluminum tabs, silicones, glass tape)	Not measured

FIBERS	DIAMETER OR DIAMETER RANGE (MICRONS)
Cotton	12.9
Paper	22.0 - 23.1
Wool	20.0 - 23.0
Hair	52.0 - 72.0
Polyester	12.0 - 20.0
Glass	2.0 - 10.0
Acrylic	16.0 - 20.3
Nylon	16.0 - 34.0
Nomex	14.0
Cashmere	15.8 - 16.7
Rayon	15.0 - 18.0
Glass Spheres	8.0 - 30.0

OVERALL RANGES

PARTICLES WEIGHT (GRAMS)	FIBER DIAMETER (MICRONS)
Trace - 2.7355	2.0 - 72.0

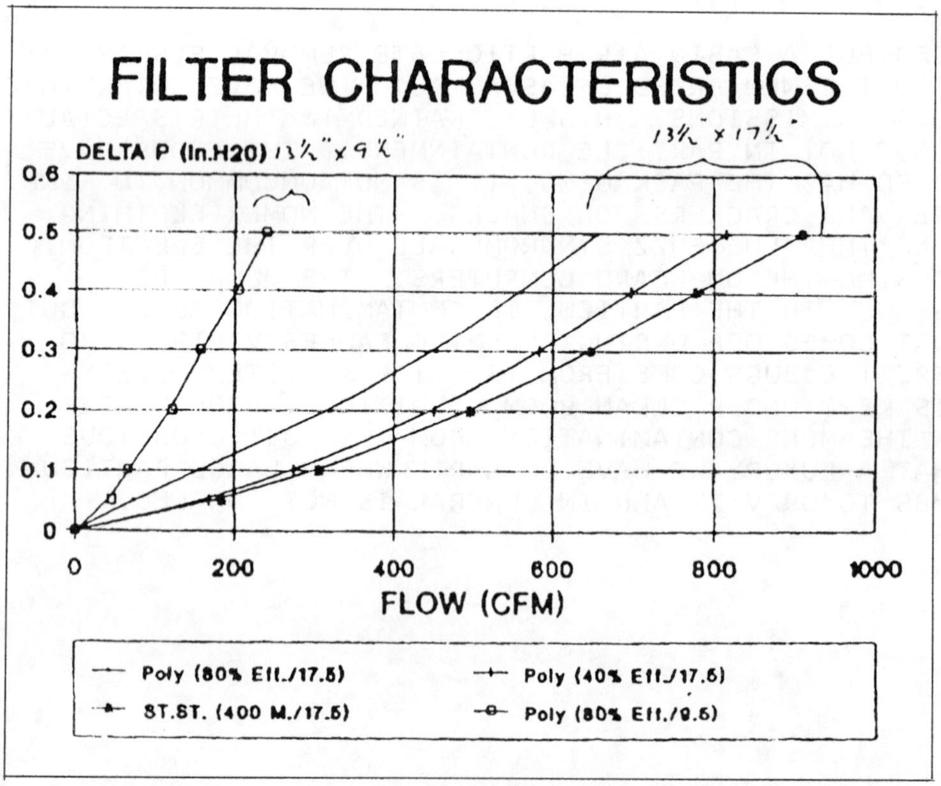

NASA HANDBOOK 8060.1B
TEST 16: DETERMINATION OF OFFGASSED PRODUCTS
FROM ASSEMBLED ARTICLES

TEST ARTICLE

Flame Shield

TEST SAMPLE DESCRIPTION

Weight: 315.79 g

TEST CONDITIONS

Test Chamber Free Volume: 42.8 liters

Test Atmosphere: 25.9% Oxygen
74.1% Nitrogen

Test Pressure: 81.4 kPa (11.8 psia)

Test Temperature: 322 K (120 °F)

Test Duration: 72 hr

TEST RESULTS, OBSERVATIONS, AND COMMENTS

TABLE 1. TEST RESULTS

Component	NASA Code	Toxic Limit (micrograms /gram	Quantity (micrograms /gram
Carbon monoxide	161000	40.90	0.047
Isopropyl alcohol	016400	140.00	0.027
Trichloroethylene	065700	0.77	0.035
Toluene	035200	108.00	0.019
Hexamethyl cyclotrisiloxane	164500	324.00	0.013
Xylene	039100	124.00	0.010
Dichlorobenzene	068800	42.99	0.013

The test atmosphere was established according to the directions in the NHB 8060.1B procedure.

NASA QUALITY ASSURANCE: [stamp]

DATE: [illegible]

NEW DEVELOPMENTS WITH SINTERED METAL FIBRE POROUS STRUCTURE AS FILTER MEDIA AND MEMBRANE SUPPORTS

Roger DE BRUYNE, Dr.ir.
BEKAERT CORPORATION
Bekaert Fibre Techn.
1395 S. Marietta Parkway
Building 500
Marietta, GA 30067 USA

Roland VERSCHAEVE,
Product Market Mgr Porous media
N.V. BEKAERT S.A.
Bekaert Fibre Technologies
B-8550 ZWEVEGEM
BELGIUM

The development of sintered metal fibre filter media is a continuing process driven through technological development and market requirements.

The development of purer stainless steels with lower inclusion content and especially smaller inclusion dimensions made it possible to develop finer metal fibres. Together with new developments in webbing, sintering and compaction this made it possible to create new filter media with finer absolute pore sizes and a much higher permeability than previously available.

This continuous development of metal fibre filter media made it possible to obtain very fine pore structures and smooth surfaces. This is the ideal surface to use as a membrane support structure due to the low pressure drop characteristics and high strength compared with sintered metal or ceramic powder layers. Both organic and inorganic membranes can be supported very efficiently.

It is also possible to incorporate the metallic support structure into the membrane or to form the membrane on this support layer. By this method renewable and sterilisable membranes can be supported in all types of materials : organic materials, sintered metal powders, ceramic structures, ...

1. METAL FIBRE FILTER MEDIA

1.1. Properties of standard Bekipor® media

1.1.1. Introduction

In order to meet the ever increasing requirements demanded by the market and thanks to improved technology the development of a new (third) series of Bekipor® (Bekipor® is a registered trade mark of N.V. Bekaert S.A. Zwevegem - Belgium. Patents granted and/or pending) sintered metal fibre filter media was becoming unavoidable. Higher permeability and more important higher dirt holding capacity were the main market demands and hence driving forces for the present developments. Breakthroughs in the improvements of the homogeneity of metal fibre webs of filter media provide the technological basis for the present development. A new series of higher performing filter media is the result.

1.1.2. Manufacturing of sintered metal fibre media

1.1.2.1. Metal fibres

Very fine wires can be produced by wire drawing. Production costs are extremely high, however, and increase exponentially as the diameter decreases. To overcome this problem bundle drawing was developed as early as 1936 (1). Instead of drawing wires singly, a number (in some cases several thousand) are bundled and drawn simultaneously (2). The sole difficulty remaining is to separate the individual fibres with a suitable material prior to bundling so that the final fibres can be individualized again. Whereas this method greatly reduces drawing and annealing costs, it only produces bundles of fibres and not reels of single wires.

Bundle drawn fibres can be produced in the form of continuous bundles, broken bundles (slivers), cut fibres, spun yarns, threads, cables, web, sintered web, needle felt, filter media, etc. All these morphologies are obtained by textile industry-like operations combined with metallurgical techniques, processes and methods. The fibre diameters depend to a large extent on the metallurgical and structural purity of the starting material (wire rod), itself a function of the alloying elements and the manufacturing and purification procedures during wire rod production. Standard diameters are 2, 4, 8, 12 and 22 μm (table 1).

1.1.2.2. Sintered metal fibre filter media

Starting from the broken metal fibre bundles highly porous webs are manufactured with standard weights and fibre diameters, by means of a proprietary process (table 2).
The webs are then combined and stacked to make up the "green" filter media. This stack is then compacted, sintered and calendered to obtain the required properties. An integrated quality control procedure measuring the permeability and the bubble point pressure is carried out on each panel.

1.1.3. Definition and measurement of relevant media properties, methodology

1.1.3.1. Pore size

The most controversial subjects in filtration undoubtedly are the definition, the determination and the measurement of the pore sizes.

- Maximum Pore Size, Bubble Point Pressure

The maximum pore size is derived from the bubble point test. This test was and still is primarily seen as a quality control test for filter cartridges.
Its definition : $\Delta p_{(BP)}$ equals the pressure required to pass the first dynamic bubble through an immersed filter (Pa) (3 and 4). The liquid has to wet completely the filter or filter medium. The maximum pore size d_m (3 and 4) is derived from the bubble point pressure as it is assumed that the first dynamic bubble is formed at the location of the "largest" pore, i.e. with the smallest capillary pressure. The definition is : d_m equals the calculated diameter of a capillary of circular cross section which is equivalent to the largest pore in the filter. This means with the same capillary pressure. This definition leads us to the relation between the bubble point pressure and the maximum pore size. As illustrated in ref. 3 one can calculate the maximum pore size from the pressure drop taking into account the capillary pressure theory

$$d_m\ (\mu m) = \frac{355}{\Delta p\ (\text{"WC})} \quad \text{or} \quad \frac{88.400}{\Delta p\ (Pa)}$$

when the liquid is isopropylic alcohol and the pressure measured in inches water column or Pa. This d_m value is a very good measure of the performance of the filter medium. However, it does not correspond to any physical dimension of a pore or a hole in the filter medium. As it is a rather straightforward and fast test, it can be used efficiently and economically as a quality control test for filter media. It is possible to relate this value to other characteristics e.g. absolute pore size, as will be shown later on.

- Absolute Pore Size

The next filter rating to be discussed is the absolute pore size or the absolute filter rating "a" (μm). This is the diameter of the largest hard spherical particle (glass bead, carbonyl iron, ...) which will pass through the filter element (5). The method to determine the absolute pore size is a very well described, but it is also a very tedious and time-consuming job, and should be performed on a large enough surface so that a reasonable representativeness will be reached. It was found that a good relationship between d_m and a exists for metal fibre media $d_m = 2.39$ a. Consequently a is calculated from the BPP : $a = \frac{37.000}{\Delta p\ (Pa)}$

The average of these values is the mean filter rating (see table 3).

- Flow Pore Size Distribution. Porometer

From the bubble point test in its extended form, it is possible to determine a flow pore size distribution curve as explained very brilliantly in Fred Cole's paper (6). By performing a continuing bubble point test (increasing Δp and flow rate) until the medium has been completely blown dry and simultaneously registering pressure difference and flow rate, it is possible to derive after some mathematics (6) a flow pore size distribution curve. From this it is possible to determine the mean, maximum and minimum flow rating. The mean flow rating (7) equals the diameter of the pore so selected that the pores in the filter larger than m μm carry a volume of flow equal to the pores in the filter element smaller than this diameter. The Coulter® Porometer, a precision analyzer for the automatic measurement of pore size distribution of filter media, textiles, paper, membranes and other porous products, can perform this test very conveniently. It uses an automated and extended version of the liquid displacement method (the bubble point test). The maximum, minimum and mean flow pore sizes are determined automatically. This technique, which used to be performed manually, eliminates the inevitable variations in the results due to differences in operators.

The mean flow pore size is used as an important characteristic for filter media evaluation. It is now possible to carry out a porometer test in a nondestructive way. The mean flow pore size is directly related to the absolute pore size (12) as d_m = 2.38 MFP and d_m = 2.39 a. It follows : a = MFP.

1.1.3.2. Permeability

- Definition

It is possible to define the permeability of a filter or a filter medium as the flow rate of a gas (e.g. air) at a specific pressure (e.g. atmospheric) under a certain pressure drop (e.g. 10 mm H_2O - 100 Pa ; 12.5 mm H_2O - 125 Pa or 20 mm H_2O - 200 Pa). The latter pressure drop is most commonly used.

- Permeability coefficient

In order to understand better filter media or filter medium material it is helpful to characterize the flow of gases through a porous material with the aid of the extended Darcy equation :

$$\frac{\Delta p}{H} = \frac{\mu . Q}{k . A} + \frac{\rho . Q^2}{\kappa . A^2}$$

where k (permeability coefficient) and κ (inertial permeability coefficient) actually are material constants. Standards are existing (8) which are very useful for characterizing filter media with relatively high pressure drop. Even nondestructive testing is possible and used for qualifying Bekipor® ST.

In the case of low flow rates or filters with a high inertial permeability coefficient, the second degree term becomes negligibly small and the Darcy equation applies :

$$\frac{\Delta p}{H} = \frac{\mu . Q}{k.A} \text{ or } \Delta p = \frac{Q}{A} \text{ or } \Delta p = \mu \frac{Q.H}{A.k}$$

Written in this form, the importance of k/H or H/k becomes clear : it is a controlling filter medium characteristic : higher k/H (i.e. high permeability coefficient k and low medium thickness H) or lower H/k values cause lower pressure drops for the same flow rate and filter surface. This leads to lower energy consumption or a smaller required filter surface.

1.1.3.3. Dirt holding capacity

- Definition

In general, the dirt holding capacity is the amount of artificial contaminant (e.g. ACFTD = air cleaner fine test dust) which can be added to the filter element before the pressure drop at constant flow becomes unacceptably high (e.g. 8 Δp initial) (9).

- Test method

The multipass test method is a standardized test method. The standard (10) exists only for filter elements. Consequently the test method has to be adapted to filter media for the present purposes (11). It is a destructive test in the sense that a sample has to be cut and will be contaminated with ACFTD afterwards. The best way to describe a multipass test stand is to state that it simulates a real hydraulic circuit.

1.1.3.4. Methodology

As a consequence of the continuous improvement of the web manufacturing methods it was possible to produce a much more homogeneous metal fibre web. This means a lower standard deviation, a lower hence better accuracy index Ca, a higher capability index Cp and a higher quality index Cpk. (Ca equals the difference between the specification average and the measured average divided by the tolerance times 100 %, Cp equals the tolerance divided by three times the standard deviation, Cpk equals the product of Cp with one minus Ca divided by 100 :

$$Ca = 100 \frac{\overline{Spec} - \bar{X}}{Tol} \qquad Cp = \frac{Tol}{3\sigma} \qquad Cpk = Cp\left(1 - \frac{Ca}{100}\right)$$

With these improved webs filter media were produced which show a very high degree of homogeneity. This has as a consequence that the pore size distribution will become much narrower and this finally will lead us to a higher porosity, permeability and dirt holding capacity for the same filter rating. For the end user this will mean longer cycle times, hence lower cleaning frequency, hence less down time and longer total filter cartridge life, lower pressure drops and thus lower energy consumption, ... all this for the same filtration quality.

Parallel with these developments a new instrument, the Coulter® Porometer, became available allowing us to characterize filter media in an objective way without operator interference. Especially the concept of the mean flow pore size is particularly interesting as it corresponds very closely with the absolute pore size as demonstrated elsewhere (12). Also the pore size distribution curves generate very useful information.

1.1.4. <u>Results of filter properties</u>

Table 3 gives the properties of the new series of filter media for the two types of media AL and BL.

Bekipor® ST-AL is a multi-layered metal fibre depth filter medium with very high dirt holding capacity and gel retention capability. Bekipor® ST-BL is a sintered metal fibre medium with a very high permeability and a low weight. Bekipor® ST-BL makes filtration of low viscosity fluids very economical : standard material : 316L. Supplied in standard panels 1.180x1.500 mm.

Figures I and II show typical cumulative pore size distribution curves obtained with the Coulter® Porometer. The most important improvements for these new media are the sometimes large increases in dirt holding capacities that could be realized especially for the AL media where the smallest increase is 27 % and the largest 140 %. Also in most cases the permeability coefficient was increased or kept constant. See table 3.

Figures III and IV give the relation between the important H/k and the product of the porosity and the absolute pore size. These figures illustrate clearly the underline logic structure for the designing of the standard series. They also show that the measurement of the absolute pore size, the thickness and the weight characterizes the filter medium very well. The permeability coefficient and hence the permeability can be determined directly from these values.
The same is true for the dirt holding capacity : DHC = $f(\epsilon . a . G^n)$ with $0.5 \leq n \leq 1$. (Figures V and VI)

1.2. <u>Finer media</u>

The manufacturing of media with absolute pore sizes smaller than three microns can be achieved along three different routes :

a = further compaction of existing media : classical approach
b = special compaction technique
c = finer fibres
d = combination of b and c

a) classical approach
By further compaction of 3AL3 samples it is possible to make a 2 and a 1 μm absolute filter medium with a rather low permeability.

b) special compaction
A special compaction technique has been developed applying isostatic pressure. This has the very beneficial effect of a very homogeneous porosity leading to a optimization of the permeability and dirt holding capacity.

THE USE OF ASPHALT EMULSION AS A DEWATERING AID FOR ULTRAFINE COAL

Wu-wey Wen
Pittsburgh Energy Technology Center
U.S. Department of Energy
P.O. Box 10940
Pittsburgh, PA 15236

Abstract

The Pittsburgh Energy Technology Center (PETC) is developing a vacuum filtration/in situ cake hardening process to deal with fine coal handling, transportation, storage, and dust problems. Asphalt emulsion has been used successfully as a filtration and reconstitution aid by mixing it into the coal slurry prior to filtration. The dewatered filter cake will then gradually harden into dust-free clumps after being discharged from the dewatering process.

This paper describes the effect of asphalt emulsion on fine coal filter cake moisture reduction and on filtration rate enhancement. Variables investigated were particle size distribution, type and dosage of asphalt emulsion, and the slurry mixing time. Test results indicated that with the addition of 2 wt% medium-setting asphalt emulsion into a minus-600-micron coal slurry, the cake moisture content decreased from 25 to 15 wt% and the filtrate removal rate was improved by an order of magnitude.

Introduction

A new concept of fine-coal dewatering and reconstitution (1,3), which was conceived by the author and is being researched at the Pittsburgh Energy Technology Center's Coal Preparation Division, focuses on the development of a combined vacuum filtration and in situ cake hardening process. The objective of this process is to (1) improve the effectiveness of mechanical dewatering and reduce cake moisture, (2) reduce the coal losses and dust emissions during transportation, handling, and storage, and (3) produce an economical, reconstituted, clean-coal product from advanced ultrafine-coal-cleaning processes. The process centers on the addition of a small amount of specially selected binding material by mixing it with the fine-coal slurry before filtration or by spraying it onto the filter cake during filtration in such a way that it becomes incorporated throughout the cake matrix. The dewatered clean-coal cake then gradually hardens into dust-free clumps after being discharged from the dewatering process. This paper describes the effect of asphalt emulsion on fine coal-cake moisture reduction and on filtration rate enhancement.

Experimental Methods

A Pittsburgh seam coal from Bruceton, Pennsylvania, was used in this study. It contained about 3.9 wt% ash, 1.2 wt% moisture and was crushed to three top sizes: 600 microns by 0, 74 microns by 0, and 37 microns by 0.

Three types of asphalt-in-water emulsions provided by Russell Standard Corporation were selected as binding agents for tests. They were cationic rapid-setting (CRS-2), cationic medium-setting (CMS-2), and cationic slow-setting (CSS-1h) according to ASTM D-2397. They were mixed into the coal slurry prior to vacuum filtration. An asphalt-in-water emulsion typically contains between 60% and 70% asphalt by weight. The dispersed droplets had an average diameter of approximately 30 microns.

The filtration system consists of an upper filtration unit and a lower vacuum chamber (2). The coal slurry is poured into the top of the filtration unit where the cake is formed on the filter medium (Whatman filter paper No. 41). The filtration unit's inner cylinder has a rubber sleeve that contracts around the cake, preventing channeling of filtrate flow near the cylinder wall. A vacuum-tight plexiglass cylinder encloses a load cell-transducer, a pressure-transducer, and a glass beaker. The signals from transducers are graphically recorded for rate and pressure data.

The coal slurry was prepared by mixing 100 grams of coal with deionized water to form a 20 wt% solids slurry. The binder was mixed into the slurry for 10 minutes prior to filtration. The total filtration and dewatering time was 7 minutes. The weight of the filtrate and the vacuum were graphically recorded. The total moisture of the wet cake was determined by LECO Proximate Analysis Determinator. The cake was then cured at 105°C for 4 hours. The dust index of the cake, defined as percent of minus-100-micron material, was measured and reported in another paper (3).

Results

Effect of Asphalt Emulsion on Cake Moisture. Figure 1 shows the trend of the effect of three different types of asphalt emulsion on filter cake moisture reduction. The cake moisture reduction is more effective on coarse coal than on fine coal. For example, the results indicate that at 2 wt% emulsion addition with 22 wt% minus-74-micron (600 micron top size) coal, the cake moisture was reduced from about 24 wt% without additive to about 20, 19, and 15 wt% with the addition of rapid-setting, slow-setting, and medium-setting emulsions, respectively. For the 95 wt% minus-74-micron (74 micron top size) coal, the cake moisture was reduced from about 28 wt% without additive to about 23-26 wt% for all three types of emulsion. These results suggest that excellent low cake moisture could be obtained by using CMS-2 emulsion. However, the rapid-setting (CRS-2) emulsion was selected for the remainder of testing in the in situ hardening process because it has a better binding property that provides better dust control, and it is 25% lower in cost.

Effect of Asphalt Emulsion Dosage on Cake Moisture. The effect of the CRS-2 emulsion dosages on filter cake moisture content was studied at three selected coal particle sizes. The results in Figure 2 indicate that, as expected, the higher the emulsion dosage, the lower the moisture content in the filter cake, and that the effect is roughly the same for all three coal sizes. The results also indicate that a moisture reduction of 6-8 percentage points was obtained with a 4 wt% dosage.

Effect of Asphalt Emulsion Dosage on Filtration Rate. The filtration rate was tested with minus-600-micron coal at 20 wt% solids concentration. Six different emulsion dosages were used: 0, 0.25, 0.50, 1.00, 2.00, and 4.00 wt%. Test results in Figure 3 indicate that the higher the asphalt emulsion dosage, the faster the filtration rate. For example, to reach a common cake moisture content of about 50 wt%, filtration times of 80, 40, 25, and 10 seconds were needed for emulsion dosages of 0, 0.25, 1.00, and 4.00 wt%, respectively.

Effect of Slurry Mixing on Cake Moisture. The slurry mixing for this series of tests was performed with a laboratory mixer at 500 rpm over different mixing times. The CRS-2 emulsion was used at 2 wt% with minus-600-micron coal. Test results in Figure 4 indicate that the best cake moisture contents of about 17-18 wt% were obtained at 30-60 second mixing times. The cake moisture content increased to 22 wt% for mixing times longer than 5 minutes. It seems that the best condition for dewatering occurred at the emulsion breaking point where the coal particles form micro-agglomerates. Further agitation might break the micro-agglomerates, thus resulting in a higher cake moisture content.

Conclusions

The advantages of using asphalt emulsion as a dewatering aid are as follows: (1) the asphalt emulsion significantly reduced the cake moisture by 6-8 percentage points by using up to 4 wt% asphalt dosage, (2) cake moisture reduction is more effective on coarse coal than on fine coal, (3) the filtration rate was improved by an order of magnitude, and (4) a short mixing time seems advantageous.

Acknowledgement

The author wishes to acknowledge K.J. Champagne of the Coal Preparation Division, Pittsburgh Energy Technology Center, Department of Energy for his experimentation.

References

1. W.W. Wen and A.W. Deurbrouck, "A New Strategy for Fine Coal Dewatering and Reconstitution," Fluid/Particle Separation Journal, V. 1, No. 2, (Dec. 1988).

2. H.B. Gala, "Use of Surfactants in Fine Coal Dewatering," Ph.D. Thesis, University of Pittsburgh, (1982).

3. W.W. Wen, R.P. Killmeyer, and A.W. Deurbrouck, "Fine Coal Reconstitution by an In Situ Hardening Process," Presented at the 21st Biennial Conf. of the Inst. for Briq. and Agglom., New Orleans, (Nov. 1989).

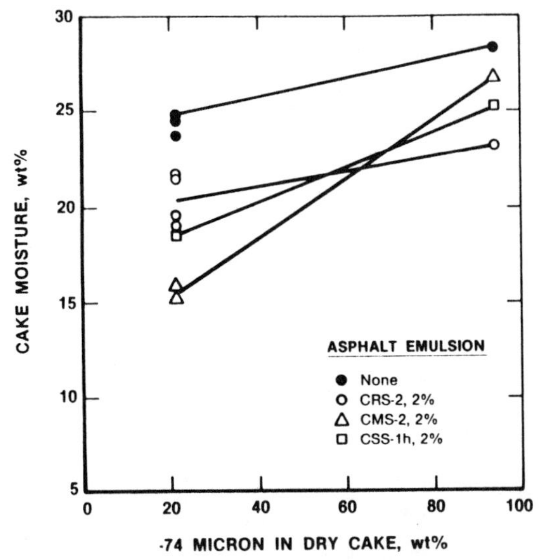

FIGURE 1. EFFECT OF EMULSION TYPES AND COAL PARTICLE SIZE ON CAKE MOISTURE CONTENT.

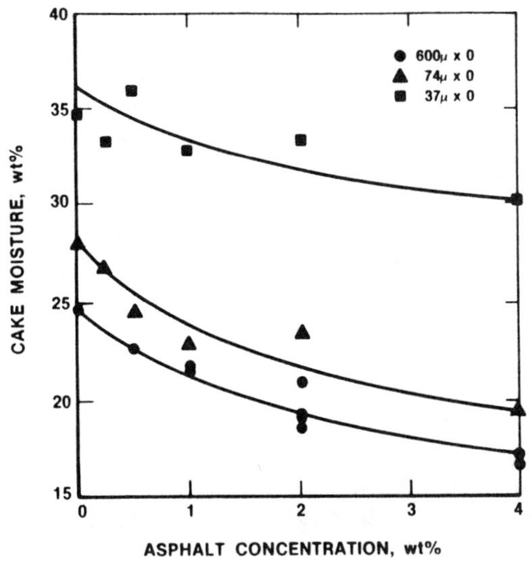

FIGURE 2. EFFECT OF ASPHALT (CRS-2) CONCENTRATION ON CAKE MOISTURE CONTENT FOR VARIOUS PARTICLE SIZES.

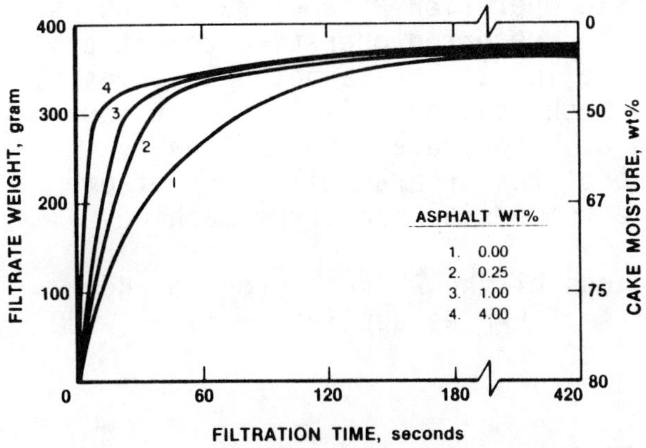

FIGURE 3. EFFECT OF ASPHALT (CRS-2) CONCENTRATION ON VACUUM FILTRATION RATE OF PITTSBURGH SEAM COAL (600 MICRONS X 0) AT 64 cm Hg AND 20% SOLIDS IN A LABORATORY VACUUM FILTER.

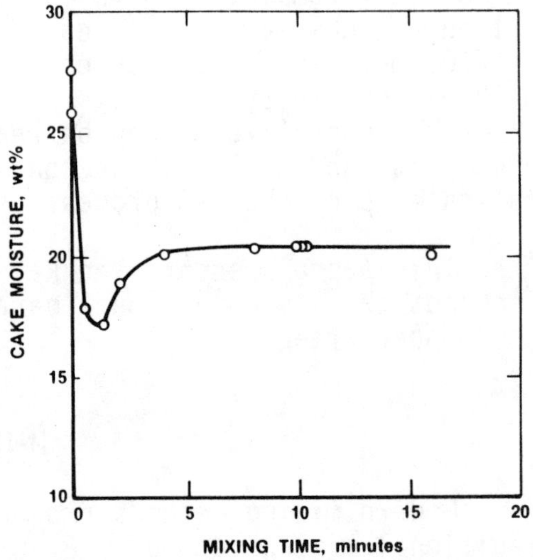

FIGURE 4. EFFECT OF SLURRY MIXING TIME ON CAKE MOISTURE CONTENT AT 2 WT% CRS-2 EMULSION ADDITION FOR 600 MICRON BRUCETON COAL.

BATCH ELECTROACOUSTIC DEWATERING (EAD) OF FINE COAL

B. Jirjis, H. S. Muralidhara, R. Menton,
N. Senapati, P. Hsieh, and S. P. Chauhan
Battelle Memorial Institute
505 King Avenue
Columbus, OH 43201 (614) 424-4374

ABSTRACT

Electroacoustic Dewatering (EAD) is a patented process being commercialized by Battelle and Ashbrook-Simon-Hartley for belt press applications. DOE/PETC has funded a program at Battelle to develop an advanced dewatering process for the preparation of fine (-100 mesh) and ultrafine (-325 mesh) coal. The program consists of two phases. Phase I was recently completed in which the basic operation parameters of EAD as applied to coal were investigated and the preferred operating conditions for -100 mesh and -325 mesh coals were defined. The target objectives of 15-20 percent moisture levels for -100 mesh coal and 25-30 percent for -325 mesh, respectively, can be reached at a moderate level of energy consumption and cost. The second phase of the program will consist of the scale-up of the EAD process using a commercial prototype machine.

This paper presents the key findings of the Phase I program and discusses the key operational parameters of EAD as applied to fine (-100 mesh) Upper Freeport coal.

INTRODUCTION

Modern mining methods produce more fine material (-28 mesh) than do traditional mixing methods. Even though fines may constitute 20 percent of a cleaning plant feed, they contain up to two-thirds of the product moisture. And with an increasing emphasis on cleaning these fines, the dewatering problems of conventional cleaning plants have increased tremendously. Traditional dewatering methods (vacuum disc filters and centrifuges) are proving inadequate for such fine coal dewatering resulting in an increased need for thermal (evaporative) drying, which is costly. The majority of the problem in fine (-28 mesh) coal can be attributed to the -100 mesh coal fraction, which has been the subject of field testing at EPRI's Homer City (now CQ, Inc.) facility. The goal for dewatering such fine coal is 15 to 20 percent moisture which is impossible to achieve with conventional techniques. Only the very expensive LAROX filter, which is a batch-type filter press, can achieve this target.

In addition to enhanced dewatering needs for conventional coal cleaning, there are needs associated with advanced coal cleaning processes. Here, the coal is often ground to ultrafine (-325 mesh) size to release the impurities. The U.S. Department of Energy (DOE) is supporting a number of such processes. Today, the best level of dewatering possible with such ultrafine coals is to about 45 percent moisture.

To improve mechanical dewatering of fine and ultrafine coal, Battelle is developing the electroacoustic dewatering (EAD) process. It is currently under contract from DOE's Pittsburgh Energy Technology Center to advance the EAD process through continuous dewatering.

EAD PROCESS BACKGROUND

The EAD process utilizes a synergistic combination of electric and acoustic (e.g., ultrasonic) fields in conjunction with conventional mechanical processes, such as belt presses, screw presses, plate and frame filters, and vacuum disc or drum filters[1]. The main mechanism of dewatering enhancement is electro-osmosis through the application of DC electric field. Electro-osmosis is characterized by the zeta potential[2]. In the case of coal slurry, the zeta potential at natural pH of slurry is negative, resulting in movement (pumping) of water to a cathode[3,4].

The liquid flow rate (Q) and energy consumption (E_{sp}) in idealized electroosmosis, at a given solids content of coal filter cake, can be expressed as follows[5,6]:

$$Q \ \alpha \ \frac{DZV}{\mu L} \qquad \text{Eq. (1)}$$

$$Q \ \alpha \ \frac{DZI}{\mu \lambda} \qquad \text{Eq. (1A)}$$

$$E_{sp} \ \alpha \ \frac{\mu \lambda V}{DZ} \qquad \text{Eq. (2)}$$

where

Q	=	flow rate of water per unit area
D	=	dielectric constant of water
Z	=	zeta potential
V	=	voltage across filter cake
μ	=	viscosity of water
L	=	cake thickness
I	=	current
E_{sp}	=	energy use per unit of filtrate removed
λ	=	electrical conductivity of cake.

The flow rate is thus proportional to zeta potential and voltage gradient (V/L). The energy use is directly proportional to voltage and conductivity and inversely proportional to zeta potential. High electrical conductivity results in high energy use and waste of energy by resistive heating. The practical range in our work with coal has been below 4 millimho/cm.

The electroosmosis in a filter cake quickly declines with loss of the liquid continuum from anode to cathode as dewatering proceeds. However, in the EAD process, the use of an ultrasonic field helps electroosmosis by consolidating the filter cake and releasing inaccessible liquid, which helps maintain a liquid continuum[2]. In addition, ultrasonics can aid electroosmotic and mechanical dewatering through other mechanisms[7].

The EAD process has been tested on over 75 suspensions in a variety of laboratory and bench-scale equipment designed to simulate a variety of commercial filters. The applications include dewatering of coal and minerals slurries, municipal and wastewater treatment sludges, process effluent sludges, food products, wood pulp, and clay suspensions[2,3,5,8-10].

The initial efforts on EAD of coal slurry were concentrated on vacuum-assisted filtration since it was envisioned that EAD could be easily retrofitted into vacuum filters. However, a prototype design effort with sewage sludges indicated some difficulties in coupling ultrasonics in such a system. It was considered easier to implement EAD in a belt filter press mode. Therefore, a developmental program for dewatering of fine (-100 mesh) and ultrafine (-325 mesh) coal utilizing a belt press configuration was initiated. The results of Phase I (laboratory studies) on -100 mesh coal are provided here.

EXPERIMENTAL DETAILS

Laboratory EAD tests were conducted on two coal particle sizes and two coal types to evaluate the EAD process on a bench scale. The objectives of the laboratory EAD tests were: (1) to determine the feasibility of achieving final cake moisture levels of 15 to 20 percent and 25 to 30 percent for -100 mesh and -325 mesh coals, respectively, and (2) to determine the effects of experimental factors on EAD performance, and (3) to establish the range of experimental conditions to be tested in Phase II using the PRU unit. The results of EAD tests with froth-flotation-cleaned, -100 mesh Upper Freeport coal are given in this paper.

EAD Variables

EAD bench scale tests were designed on the basis of operating conditions that exists in a normal belt press such as feed rate, belt speed, roll diameter and belt width, cake thickness, etc. The EAD bench scale parameters were as follows:

- Voltage gradient (volts/cm)
- Ultrasonic intensity (watts/cm^2)
- Solids loading (Kg(DS)/m^2)
- Dewatering time (min)
- Pressure (psi)

These parameters are related to EAD belt press parameters. For example, feed rate is a function of solids loading, dewatering time, speed of belt and the width of the belt.

The key dependent EAD variables were the final moisture content, specific energy consumption, and energy cost. The specific energy is defined as Kwh of electricity used to remove a pound of filtrate. And the energy cost is calculated as dollars per dry ton of coal to achieve a given moisture level assuming electricity cost at 5 cents per Kwh.

Coal Preparation

A froth-flotation-cleaned, -100 mesh, Upper Freeport coal was received from EPRI. The coal sample was prepared at EPRI's Homer City test facility. The as-received sample was centrifuged at Battelle using a Centra 7 centrifuge to dewater the coal slurry from 84 percent moisture to 40 percent moisture. The particle size distribution of the coal sample indicated that 50 percent of coal particles were smaller than 35 micron in diameter. The characteristics of feed coal are given in Table 1.

Experimental Setup

A 7.6 diameter pressurized batch EAD unit, shown in Figure 1, was used to obtain Phase I batch tests data. The -100 mesh coal sample at 40 percent moisture was placed between a polymetric filter, which was supported by a perforated ultrasonic plate (cathode), and a floating anode. A downward force was applied on the floating anode to press the coal sample at 3-7 psi. An appropriate electrical power input and ultrasonic power were applied for a given interval of time. The dewatering coal sample was then placed in a vacuum oven to determine the cake moisture.

Experimental Design

Approximately 100 tests were conducted on -100 mesh coal using statistically designed experiments. As shown in Table 2, the tests were divided into three series: preliminary tests, screening experiments, and higher-level designed experiments. At each stage, the results of the previous experiments were used to optimally design the next set of tests.

The experimental factors studied in the batch tests are shown in Table 3. The preliminary tests concentrated on the most important factors: feed solids content, dewatering time, DC voltage gradient, solids loading, pressure and ultrasonic intensity. Values for the other factors were held at fixed levels. The results from the preliminary tests were used to optimally design the screening experiments. The screening experiments were used to reduce the number of factors that needed to be studied further. A rational factorial design was used to minimize the number of tests to be performed.

Following the screening tests, higher-level experimental designs were used to study in more detail the effects and interactions of fewer factors. Employing response surface methods (composite and Box-Behnkin designs) we were able to estimate the linear, quadratic, and interaction effects in the multifactor experiments. These experiments were designed iteratively. After analyzing the data from previous experiments, a new

TABLE 1. CHARACTERIZATION OF -100 MESH UPPER FREEPORT (FEED)

PHYSICAL PROPERTIES

- moisture content of slurry as received: 84%
- moisture content of slurry before EAD: 38.05-40.55%
- viscosity for coal slurry before EAD: 300 Poise at 100 rad/S
- particle size: 50 percent smaller than 35 micron

PROXIMATE ANALYSIS

	Percent As Received	Percent Dry Basis
Moisture	1.07	xxxxx
Ash	5.67	5.73
Volatile	28.59	28.90
Fixed Carbon	64.67	65.37
	100.00	100.00

HEATING VALUE (BTU/LB)

- 14699 (as rec'd)
- 14858 (dry)
- 15761 (MAF)

FREE SWELLING

5.0

ELECTROKINETIC PROPERTIES

- Electrical conductivity: 3.25 mmho/cm
- pH of as-received slurry: 7.43
- zeta potential @ pH 7.2: -36 mv

ULTIMATE ANALYSIS

	Percent As Received	Percent Dry Basis
Moisture	1.07	xxxxx
Carbon	82.12	83.01
Hydrogen	5.01	5.06
Nitrogen	1.45	1.47
Sulfur	1.63	1.65
Ash	5.67	5.73
Oxygen (diff)	3.05	3.08
	100.00	100.00

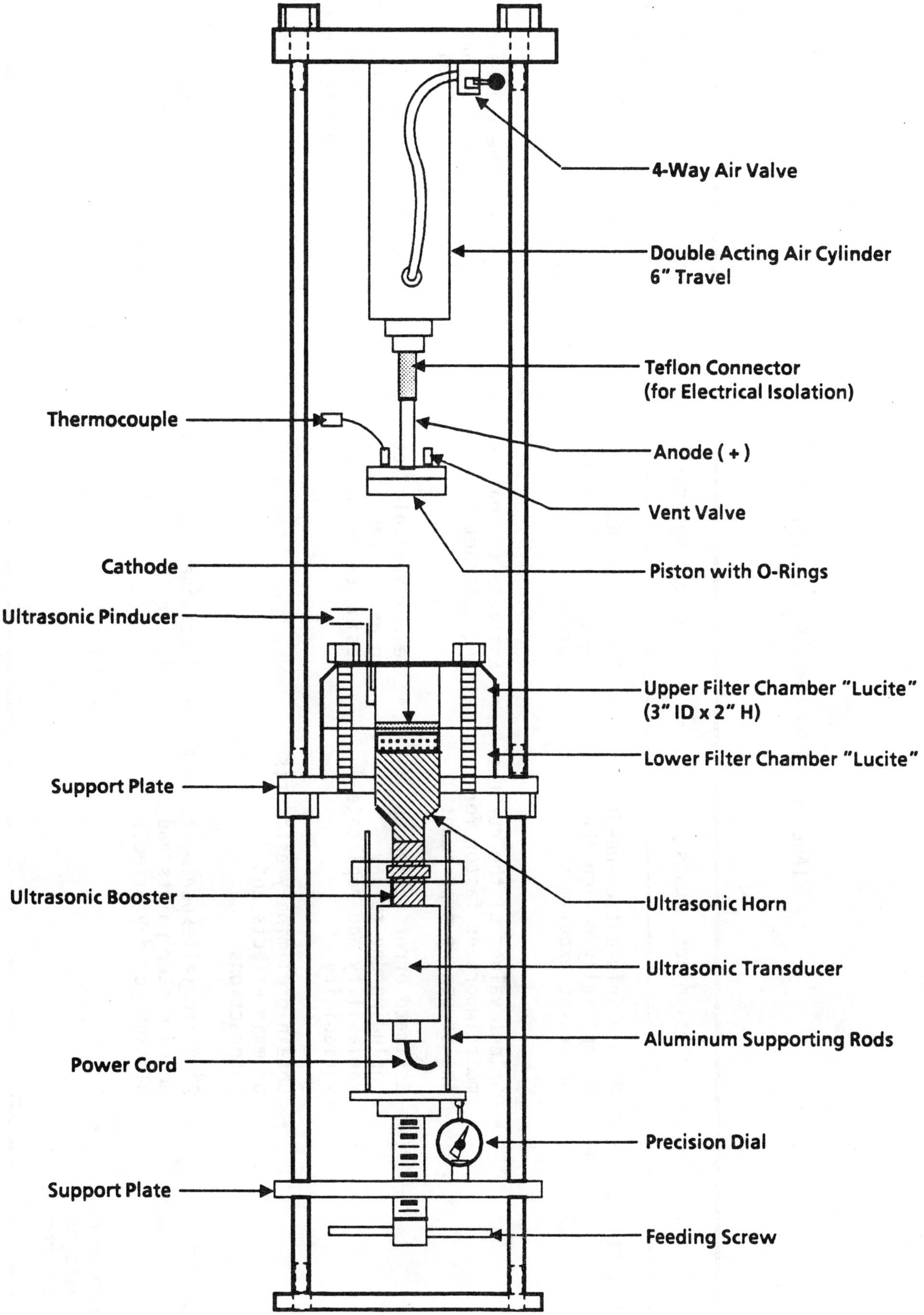

Figure 1. Schematic of Pressurized Batch EAD Unit

TABLE 2. EXPERIMENTAL TESTING STAGES

	Preliminary Tests	Screening Tests	Higher-Level Design Test
Objectives	• Obtain information needed to optimally design the screening experiments	• Reduce the number of factors to be studied further	• Estimate the linear, quadratic and interaction effects of importance
	• Select values of the current most important factor for the screening experiment	• Estimate the main efforts and interactions	• Determine the values of important factors that optimizes solids content
	• Estimate error variances including day to day variability and replicate variability	• Select variable settings to use in the higher-level design experiments	• Identify a range of conditions under which we expect to achieve near optimum performance
	• Obtain preliminary estimates of main effects and interactions		
Design	• 2^4 factorial experiments with added center points and blocked for day effects	• Fractional factorial design	• Sequentially designed response surface experiments
Number of Tests on -100 mesh coal	• 50	• 34	• 16

TABLE 3. EXPERIMENTAL FACTORS FOR BATCH TESTS

No.	Factor	Levels
1	Coal particle size and degree of cleaning	minus-100 mesh clean coal minus-325 mesh ultra clean coal
2	Source of coal	two sources
3	Feed moisture content	38% to 40%
4	Dewatering time	1-6 minutes
5	DC voltage gradient	0-300 volts/cm
6	Ultrasonic intensity	0-1.21 watts/cm^2
7	Pressure	3-10 psi
8	Cake thickness	0.6 to 2.6 cm
9	Polymer type	cationic, nonionic, mixture of the two
10	Polymer dosage	0.1 to 1.9 lb/dry ton

set of test conditions were generated for the unswayed set of tests. This approach enabled us to adjust the levels of important factors at each stage to obtain predicted responses (dewatering efficiency) in the optimal experimental region. This experimental design ensured that the effects of important factors or interactions could be estimated with sufficient precision and that these factors were not confounded with each other.

EAD RESULTS AND DISCUSSION

Effect of Flocculation

A number of exploratory tests were carried out to determine the effect of flocculation on solids capture and EAD performance. Percol 726 flocculent was used at a dosage of 0.11 lb/ton coal to flocculate the -100 mesh coal. EAD tests were performed on both flocculated, as well as unflocculated, coal with the pressure EAD batch unit. Results in Table 4 indicate that elimination of flocculation did not significantly decrease the percent solids capture. Also, flocculation seemed to have minor effects on the dewatering performance. Therefore, it was decided to eliminate flocculation and use unflocculated coal to evaluate the effect of EAD parameters on coal dewatering.

Parametric Tests

A series of 50 parametric tests were conducted on -100 mesh Upper Freeport coal (40 percent moisture) to determine the relevant ranges of key parameters (voltage gradient, ultrasonic intensity, solids loading, and dewatering time) on achieving lower cake moisture content and consumed energy. Two control tests (no EAD) were conducted by using pressure of 5 psi for 6 minutes. Control test results indicated a reduction in cake moisture content of 4 percent, thus, producing a cake of 36 percent.

Eight center point tests were conducted in order to estimate the error variability and the day to day variability. Results based on the final moisture content of the cake of the eight conducted tests indicate 0.77 percent experimental error. The experimental error based on energy values for the same tests indicates 0.024 percent experimental error. These data shows that there are very little experimental error associated with above 50 series of tests.

The final moisture content for the EAD tests varied from 19.77 to 32.06 percent. A final moisture content of 19.77 was achieved at 5 minutes residence time, 0.82 cm cake thickness, solids loading 6.1 kg/(DS)/m^2, 0.99 watts/cm^2 ultrasonic intensity, 145.6 volts/cm voltage gradient, a specific energy of 0.250 Kwh/lb filtrate, and at an energy cost of 10.07 dollars/ton (DS). The consumed specific energy for the 50 parametric tests varied from 0.067 - 0.496 Kwh/lb filtrate. Specific energy decreased with increased solid loading. An energy of 0.187 Kwh/lb filtrate was achieved at the highest solids loading of 19.1 Kg(DS)/m^2 and resulting in a cake final moisture content of 22 percent. The above tests (Test 46) was conducted at 5 psi, 6 min. residence time, 2.57 cm cake thickness, 0.99 watts/cm^2 ultrasonic intensity, 46.6 volts/cm voltage gradient, and at an energy cost of 6.59 dollars/ton (DS). It is believed that the energy cost can be reduced substantially by removing some of the flotation chemicals from coal by simple washing.

TABLE 4. EFFECT OF FLOCCULATION ON EAD PERFORMANCE (7 PSIG PRESSURE, 5 MIN. DEWATERING TIME, 155 VOLTS/CM VOLTAGE GRADIENT)

Flocculated/ Unflocculated	Initial Sample Moisture Percent	Cake Final Moisture Percent	Percent Solids Capture
Flocculated	40.1	23.17	97.38
Unflocculated	38.14	25.05	96.99
Flocculated	39.77	25.51	96.69

Effect of Dewatering Time on Moisture Content

Dewatering time is one of the critical parameters for establishing economic feasibility of the EAD process. Dewatering time effects the size of the EAD unit and the energy consumption. Figure 2 and Table 5 compare the final percent moisture obtained in the screening tests at dewatering times of 1, 4, 5, and 6 minutes. Results indicate that increasing the dewatering time decreases the moisture content of the cake. For example at 1 minute dewatering time, 121 volts/cm voltage gradient, 0.44 watts/cm^2 ultrasonic intensity, and 6.0 Kg(DS)/m^2 solids loading, the final moisture content was 27.9 percent, while at 6 minutes under the same conditions, the moisture content was 21.4 percent.

Figure 2 also indicates, that the final moisture content decreases with increased ultrasonic and voltage level at 6 minutes, but it is relatively unchanged with increased ultrasonic and voltage levels at 1 minute. Therefore, it was concluded that 1 minute of dewatering time may not be long enough for the EAD effect to take place.

Effect of Voltage Gradient and Ultrasonic Intensity

Electrokinetic potential across the cake is the electrical driving force for the electroacoustic dewatering process. Dewatering was observed to improve significantly with the application of ultrasonics in conjunction with the electric field, as shown in Figure 3. The final cake moisture decreased with increasing levels of voltage gradient and ultrasonic intensity, suggesting synergistic effects.

Table 6 displays the observed and predicted final cake moistures for selected tests conducted at 6 min, and 6 Kg(DS)/m^2 solid loadings. The measured and predicted final cake moistures are shown in the fourth and fifth columns of the table. The precision of the predicted values can be assessed by the 95 percent prediction bounds for final cake moisture given in the sixth, and seventh columns of the table; we are 95 percent confident that individual final cake moistures will fall between the lower and upper prediction bounds. The average final cake moisture for tests conducted at 6 min, 0 watts/cm^2, 73 volts/cm, and 6 Kg(DS)/m^2 was predicted to be 29.7 percent. When the ultrasonic intensity is increased from 0 to 0.99 watts/cm^2, final cake moistures are predicted to decrease to 26.1 percent from 29.7 percent. In the absence of ultrasonic intensity, final cake moistures are predicted to decrease by only 2.6 percent to 27.3 percent when voltage levels are increased from 73 to 121 volts/cm. However, in the presence of ultrasonic intensity at .99 watts/cm^2, final cake moistures are predicted to decrease by 4.4 percent from 26.1 percent to 21.7 percent when voltage levels are increased from 73 to 121 volts/cm. Figure 4 illustrates the above synergistic effect of voltage and ultrasonic intensity. Finally, final cake moistures are predicted to reach the target levels of 20 percent at 6 min, .99 watts/cm^2, 145 volts/cm, and 6 Kg(DS)/m^2.

The effects of voltage and ultrasonic watts are presented graphically in Figures 5 to 7. Final cake moistures are plotted against voltage in Figure 5 for tests conducted at 6 min dewatering time, .99 watts/cm^2, ultrasonic intensity, and approximately 6 Kg(DS)/m^2 solids loading. Average final cake moisture values predicted from the model fitted to all tests conducted at dewatering times of 4, 5, and 6 min are plotted as the solid line, and observed final cake moistures are represented by the asterisk symbols. Upper and lower 95 percent prediction bounds are shown by the dotted lines. Final cake moistures are plotted against ultrasonic intensity in Figure 6 for tests conducted at 6 min dewatering time, 121 volts/cm, and approximately 6 Kg(DS)/m^2.

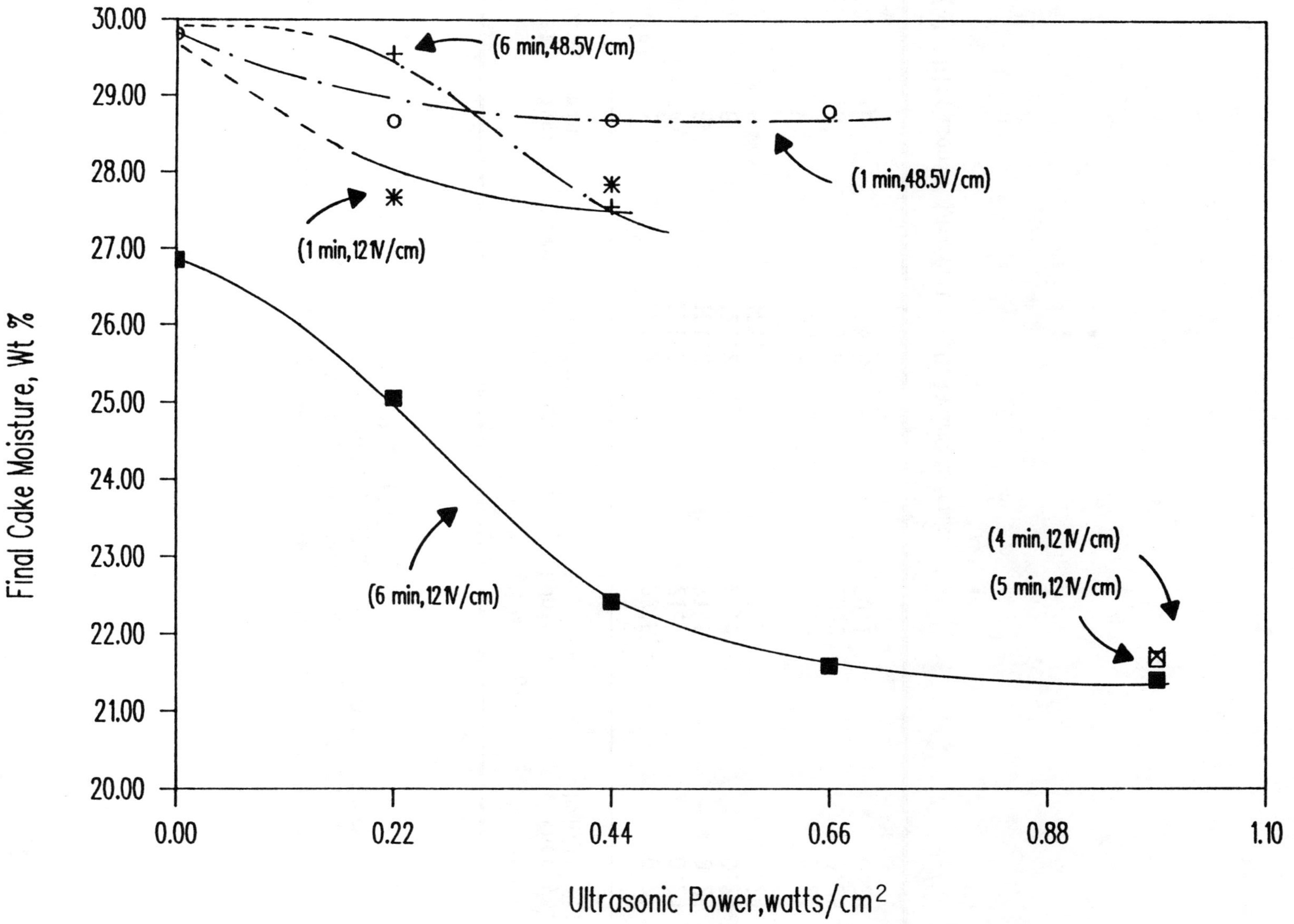

Figure 2. Variation of Final Cake Moistures with Ultrasonic Intensity for Selected Voltage Levels at Various Dewatering Times.

TABLE 5. COMPARISON OF FINAL CAKE MOISTURES AND SPECIFIC ENERGY LEVELS AT VARIOUS DEWATERING TIMES (6 KG(DS)/M^2 SOLIDS LOADING AND .99 WATTS/CM^2 ULTRASONIC INTENSITY)

Test No.	Time min	Voltage Gradient volts/cm	Final Moisture Content, Percent	Energy Use kwh/lb filtrate
16[a]	1	121.2	27.85[a]	0.22
48	4	121.2	21.7	0.24
47	5	121.2	21.7	0.25
41	6	121.2	21.6	0.24
45	6	121.1	21.2	0.24
49	4	145.6	19.9	0.24
50	5	145.6	19.8	0.25
42	6	145.5	20.1	0.28

(a) Ultrasonic intensity = 0.44 watts/cm^2

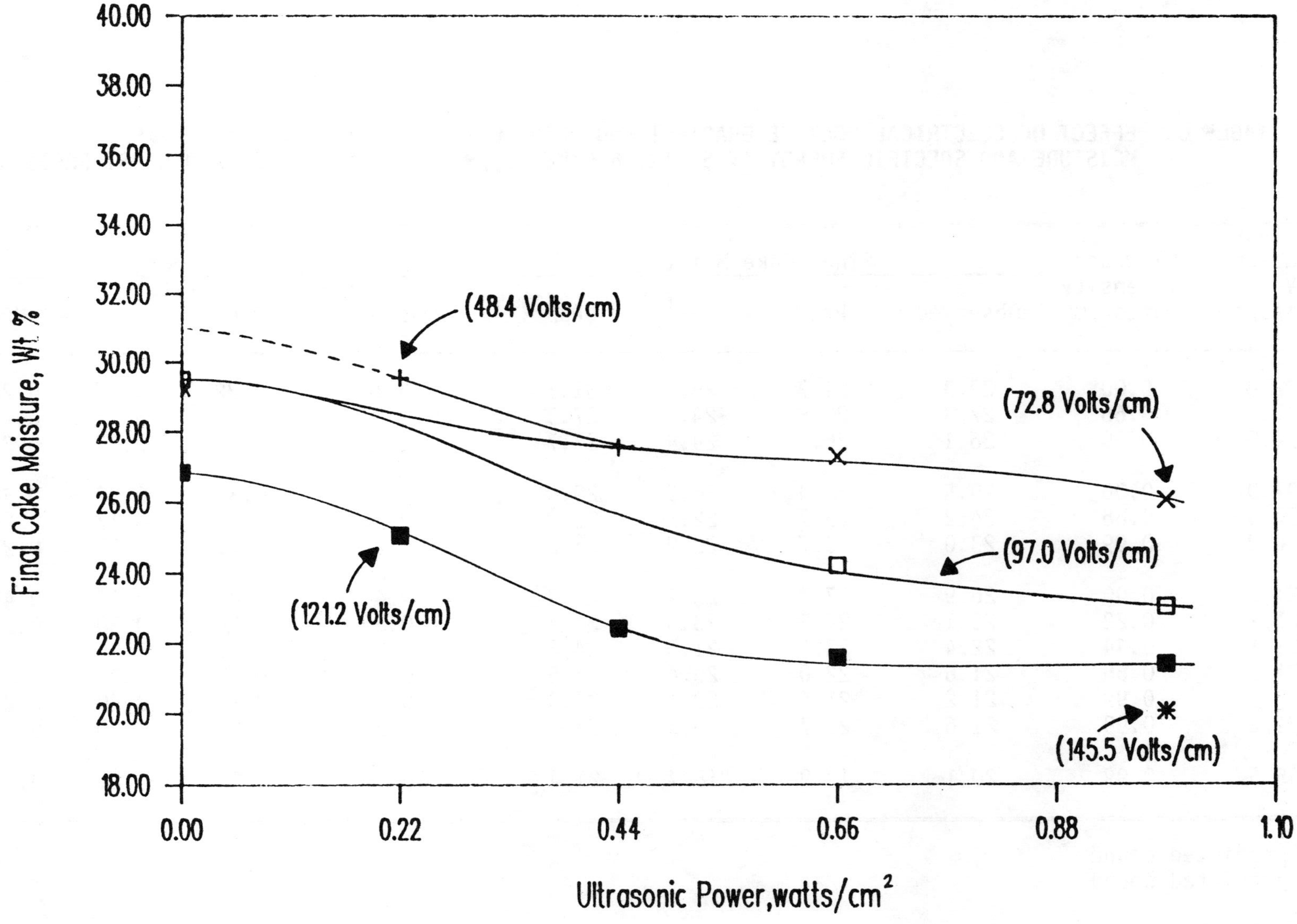

Figure 3. Effect of Ultrasonic Intensity on Final Cake Moisture at 6 Min. Dewatering Time and 6 Kg(DS)/m^2 Solid Loading

TABLE 6. EFFECT OF ELECTRICAL VOLTAGE GRADIENT AND ULTRASONIC INTENSITY ON FINAL CAKE MOISTURE AND SPECIFIC ENERGY AT 6 MIN, 6.0 KG(DS)/M^2 SOLID LOADINGS, AND 5 PSI PRESSURE

Test No.	Voltage Gradient Volt volts/cm	Ultrasonic Intensity watts/cm^2	Final Cake Moisture (%)				Specific Energy (Kwh/lb filtrate)			
			Observed	Pred.	LPB(1)	UPB(2)	Observed	Pred.	LPB	UPB
35	72.8	0.00	29.3	29.7	28.1	31.3	0.18	0.22	0.17	0.27
38	72.7	0.66	27.3	25.8	24.3	27.3	0.17	0.16	0.12	0.21
39	72.9	0.99	26.1	26.1	24.4	27.7	0.17	0.17	0.12	0.22
36	97.0	0.00	29.5	28.4	26.9	29.9	0.32	0.32	0.27	0.36
37	97.0	0.66	24.2	23.7	22.3	25.2	0.21	0.22	0.17	0.26
40	96.9	0.99	23.0	23.7	22.2	25.2	0.20	0.21	0.17	0.26
12	121.2	0.00	26.9	27.3	25.7	28.9	0.44	0.40	0.35	0.45
8	121.2	0.22	25.1	24.8	23.3	26.4	0.35	0.34	0.30	0.39
32	121.1	0.44	22.4	23.1	21.5	24.6	0.29	0.30	0.25	0.35
34	121.3	0.66	21.6	22.0	20.4	23.5	0.26	0.27	0.22	0.31
45	121.1	0.99	21.2	21.6	20.1	23.0	0.24	0.24	0.20	0.29
41	121.2	0.99	21.6	21.7	20.3	23.2	0.24	0.24	0.20	0.28
42	145.5	0.99	20.1	19.9	18.4	21.4	0.28	0.26	0.21	0.30

(1) Lower predicted bound
(2) Upper predicted bound

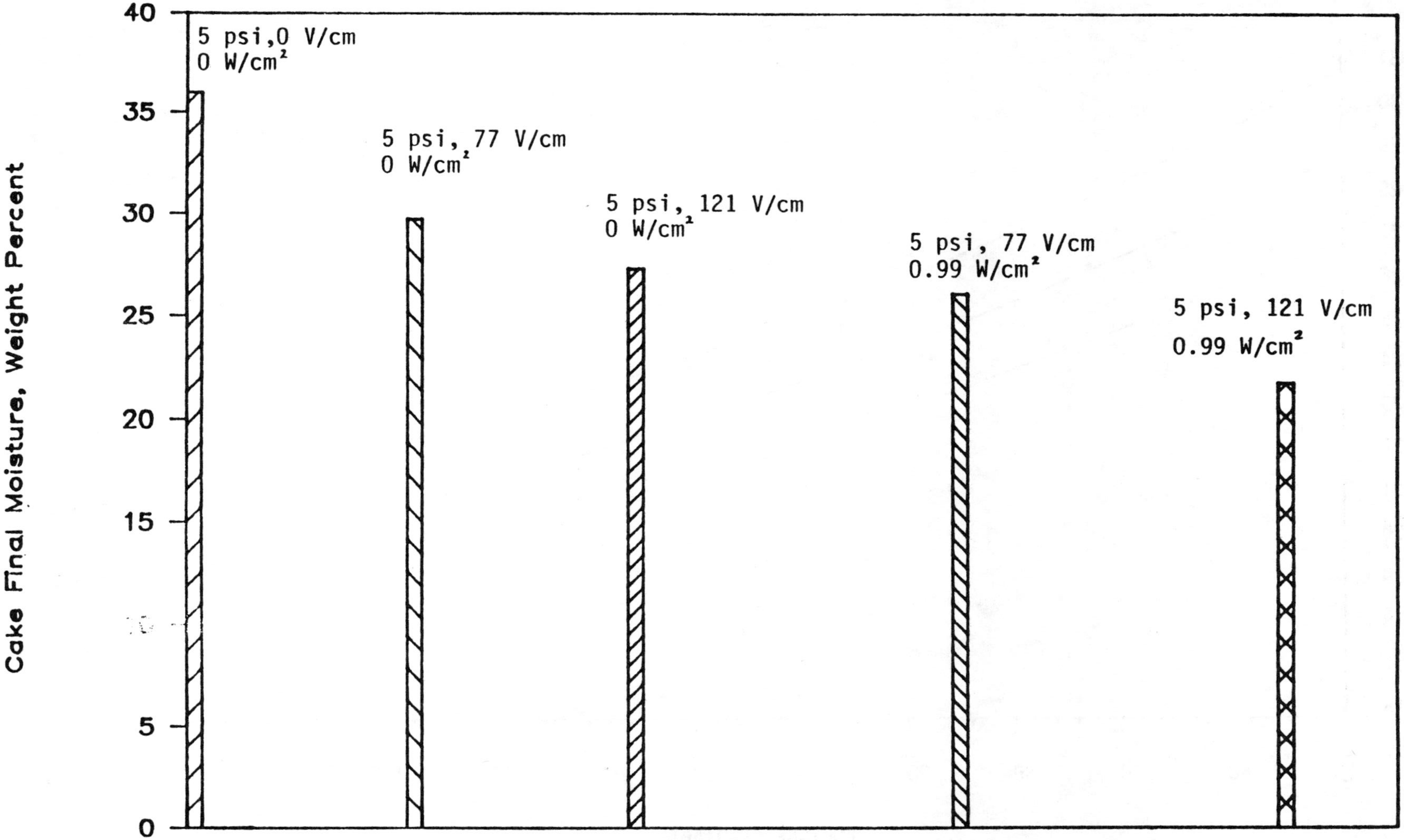

Figure 4. Synergistic Effect of Voltage and Ultrasonic Intensity on Final Cake Moisture at 6 Min. Dewatering Time, 6 Kg(DS)/m^2 Solids Loading

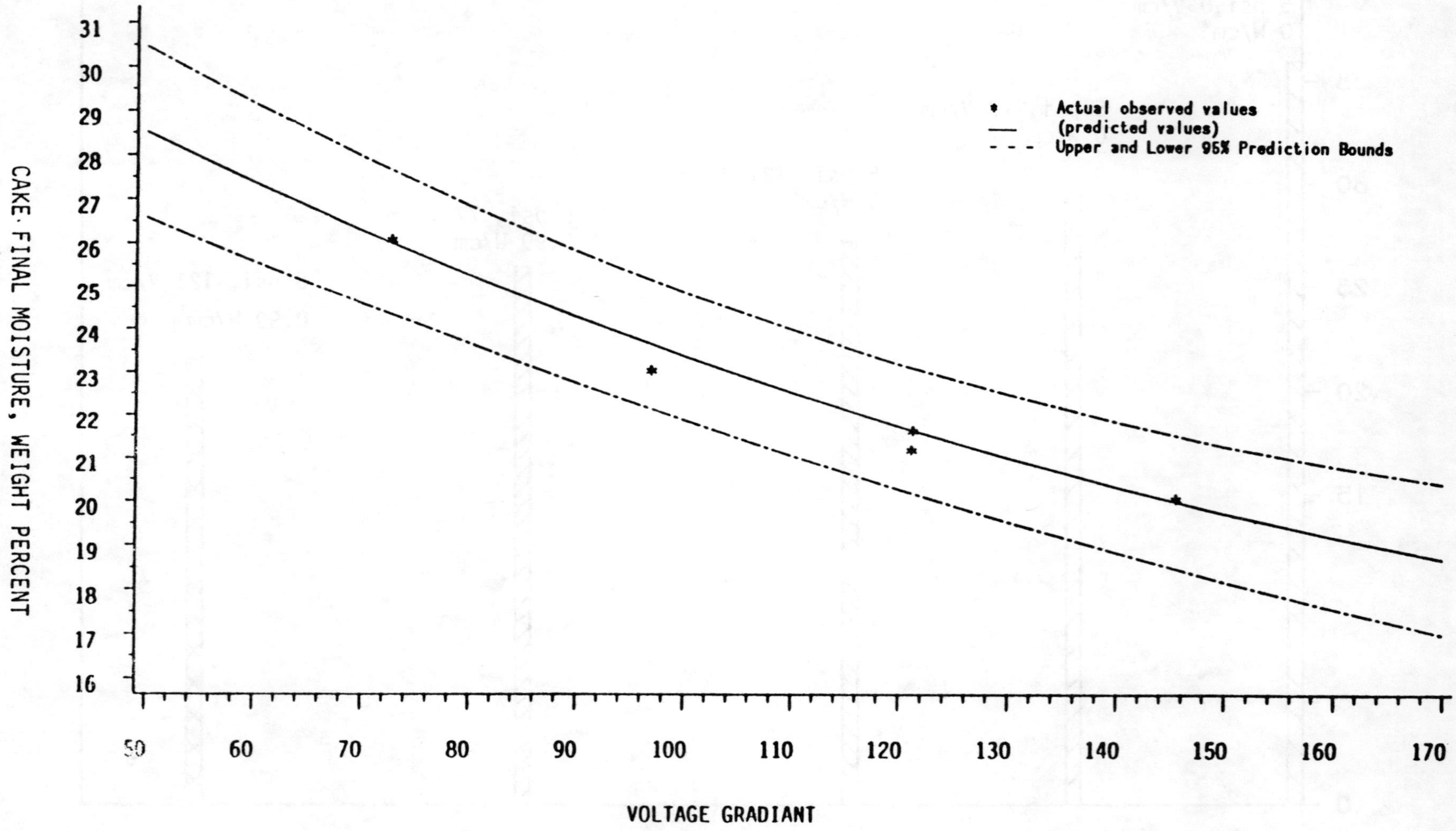

Figure 5. Predicted Final Cake Moisture Versus Voltage Gradient at 6 kg/(DS)/m^2, 0.99 watts/cm^2 and 6 Min. Dewatering Time

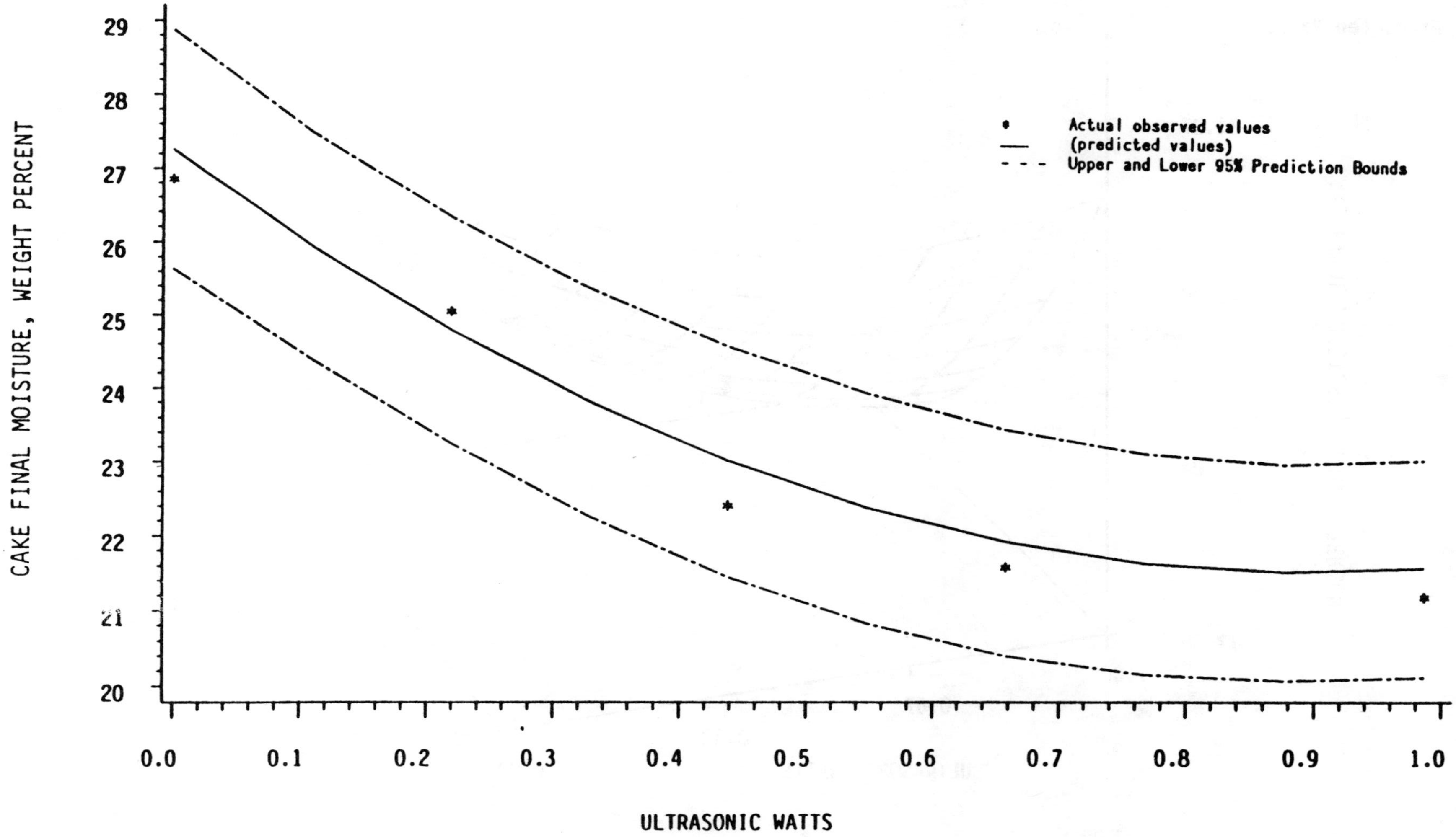

Figure 6. Predicted Final Cake Moisture Versus Voltage Gradient at 6 kg(DS)/m^2, 121 Volts/cm and 6 Min. Dewatering Time

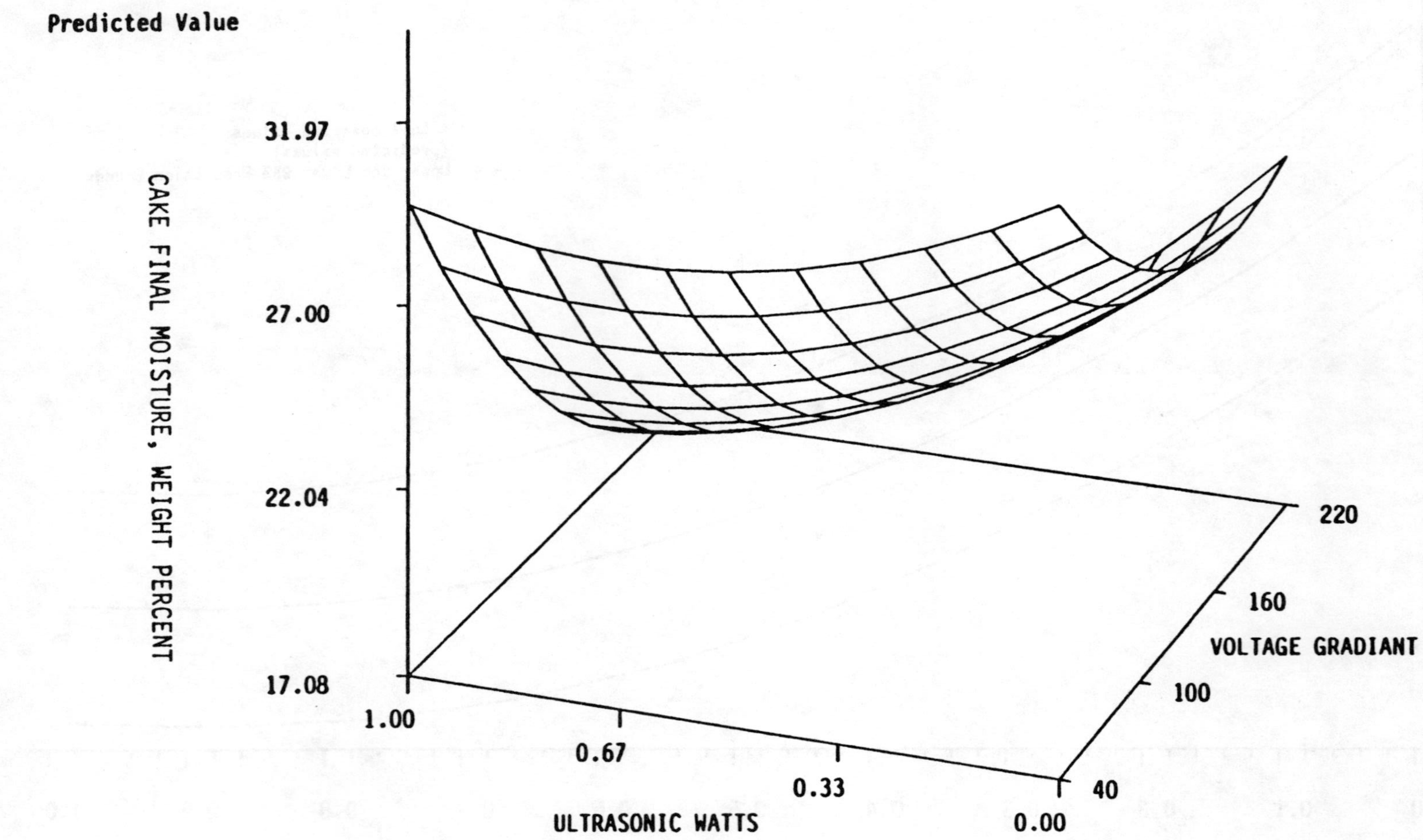

Figure 7. 3-D Plot of Predicted Final Cake Moisture Versus Ultrasonic Intensity and Voltage Gradient at 6 kg(DS)/m^2 and 6 Min. Dewatering Time

As seen in Figures 5 and 6, the predicted final cake moistures provide a good fit to the observed cake moistures. While final cake moisture levels are predicted to decrease almost linearly in Figure 5, the predicted trend flattens out at approximately 0.65 watts/cm^2 in Figure 6. Figure 7 displays a three dimensional (3-D) view of the relationship between final cake moisture and both ultrasonic power and voltage at 6 minutes dewatering time, and 6 Kg(DS)/m^2 solids loading.

Effect of Solids Loading on Moisture Content

The prior EAD experience with variety of materials such as corn gluten, corn fiber, and sludges indicated that the final moisture content of the dewatered material increased with increase in solids loading is the 3 to 12 kg(DS)/m^2 range. However, during the recent coal tests in the 3 to 12 kg(DS)/m^2 range, the final moisture content decreased with increase in solids loading. Results in Table 7 and Figure 8 show the effect of solid loading on final moisture content at different ultrasonic intensity levels and voltage gradients. For example, at a solids loading of 6 Kg(DS)/m^2, 73 volts/cm electric potential, 1 watt/cm^2 ultrasonic intensity and dewatering of 6 minutes, a final cake moisture time content of 26.06 percent was obtained. However, when the solids loading was doubled under the same above conditions, a final moisture content of 21.56 percent was achieved. This result is rather unusual because at higher solids loading (means higher cake thickness) the permeation rate through the cake should be longer (higher cake resistance) which means a lower cake solids. But, it is also possible that ultrasonic coupling efficiency increased with increase in cake thickness, thus, opening the channels for water movement through the cake resulting in a lower moisture cake. Pinducer data on ultrasonic penetration in coal samples support this hypothesis.

Further experimental investigation is warranted to confirm the above results specially at solids loading above 12 kg/(DS)/m^2. If this above phenomenon is true, it could make a significant economic impact on the use of EAD technology in the coal industry.

Effect on Energy Consumption

Figure 9 displays the effect of time, voltage, ultrasonic levels and solid loading on specific energy levels and cake moisture content. Results indicate that an increase in solids loading results in a lower consumed energy. Results also indicate that the ultrasonic power has a beneficial effect on the specific energy consumed. The higher the ultrasonic power, the lower the specific energy consumed. This is due to the better dewatering efficiency that takes place at higher ultrasonic power. Also, data indicate that the effect of dewatering time above 4 minutes is marginal in terms of increased in solids content even though a significant amount of EAD energy is consumed. This causes the specific energy consumption to increase with dewatering time at levels above 4 minutes.

It is believed that the energy consumption can be substantially reduced by washing the coal to remove some of the flotation chemicals and thus reducing the electrical conductivity of filter cake.

TABLE 7. EFFECT OF COAL SOLIDS LOADING ON MOISTURE CONTENT AT 5 PSI, 6 MIN. DEWATERING TIME.

Solids Loading (Kg(DS),^{2}	Voltage Gradient volts/cm	Ultrasonic Intensity watts/cm^2	Final Moisture Control Percent
6	73	0.99	26.06
12	73	0.99	21.56
6	121	0.99	21.39
8	109	0.99	20.59

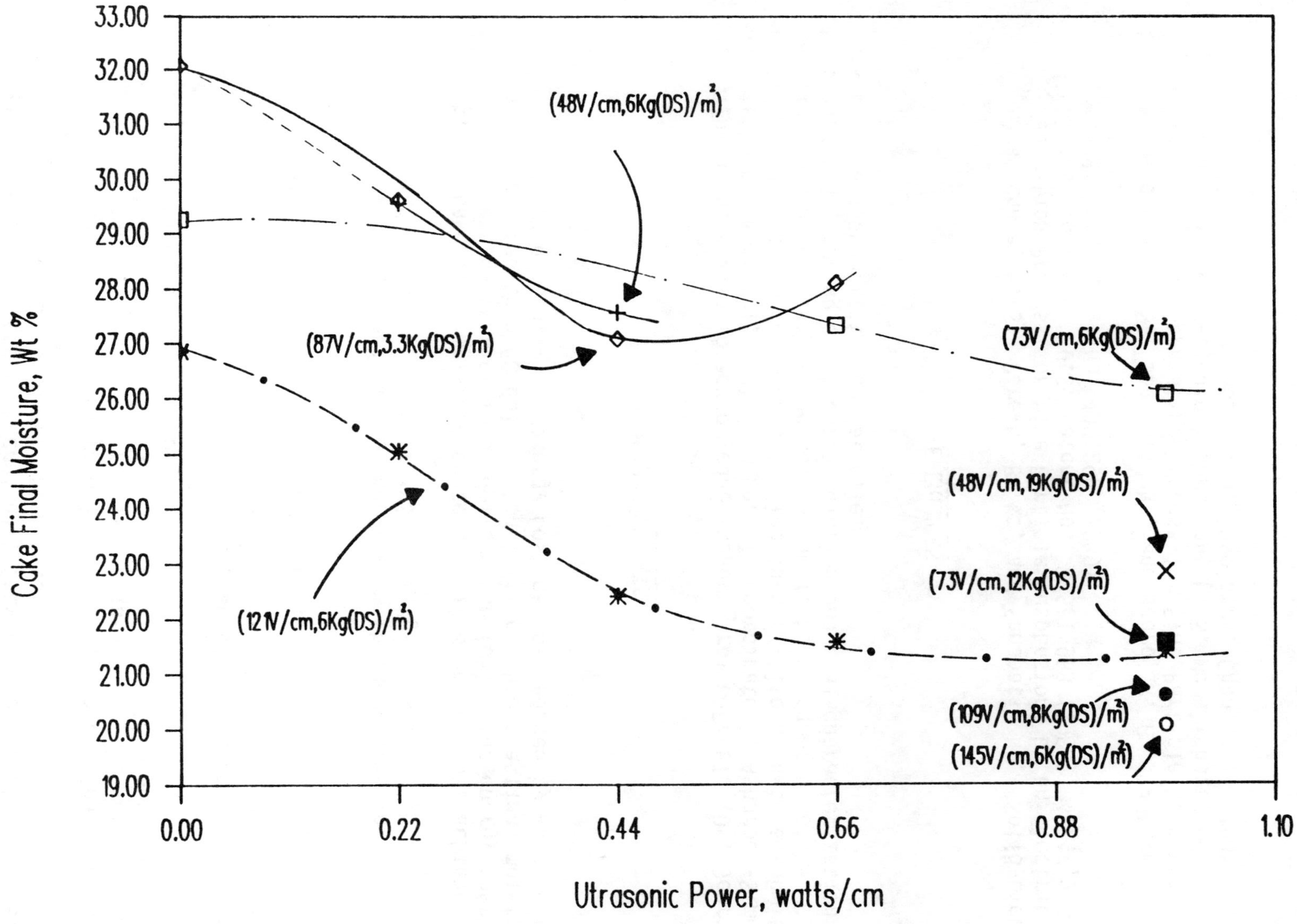

Figure 8. Effect of Solids Loading on Final Cake Moisture at 6 Min.

CONCLUSIONS

Based upon this study, the following conclusions are drawn regarding electroacoustic dewatering of fine (-100 mesh) Upper Freeport coal:

1. The EAD process is technically feasible for dewatering coal to less than about 20 percent moisture.

2. Voltage gradient, ultrasonic intensity, dewatering time and solids loading have strongest influence and the interactions between time and ultrasonic intensity and time and voltage gradient are very strong.

3. Specific energy consumption is predicted to decrease as the ultrasonic intensity is increased from 0 to 0.99 watts/cm.

4. Higher voltage gradient and a shorter time reduces specific energy.

5. Dewatering performance increases with increasing solids loading in the range of 3.3 to 12 kg(DS)/m^2.

6. A cake final moisture content of 19.8 percent was achieved at an energy cost of 10.07 dollars/ton (DS). The energy cost declined to 6.58 dollars/ton (DS) to achieve a final moisture of 22 percent. The energy cost can be reduced by washing the coal.

ACKNOWLEDGEMENTS

The authors wish to acknowledge support from DOE/PETC (Dr. George Wen, technical monitor), EPRI (Jim Hervol, coordinator), Ashbrook-Simon-Hartley (Bud Johnson, coordinator). The authors also wish to acknowledge the fine work by Dr. Shiao-Hung Chiang and Y. S. Cheng of U. of Pittsburgh in carrying out the flocculation and coal characterization work under a subcontract from Battelle.

REFERENCES

1. Muralidhara, H. S., Parekh, B. K., and Senapati, N., "Solid-Liquid Separation Process for Fine Particle Suspensions by an Electric and Ultrasonic field," U.S. Patent Nos. 4,561,953 (1985), 4,747,920 (1988), and Serial No. 400,296 (1989).

2. Muralidhara, H. S., Senapati, N., and Beard, R. E., "A Novel Electroacoustic Separation Process for Fine Particle Suspensions," Advances in Solid-Liquid Separation, H. S. Muralidhara, ed., Battelle Press/Royal Society of Chemistry, Chapter 14, 1986, 335-374.

3. Muralidhara, H. S., Senapati, N., Ensminger, D., and Chauhan, S. P., "A Novel Electroacoustic Separation Process for Fine Particle Suspensions", Proceedings of World Filtration Congress IV, held in Ostende, Belgium, April 1986.

4. Heath, L., and T. Demirel, "Pressurized Electro-Osmotic Dewatering", Paper presented at Engineering Foundation Conference on Flocculation, Sedimentation and Consolidation, Sea Island, GA, November 1984.

5. Chauhan, S. P., "Scale-up of Electroacoustic Dewatering of Sewage Sludges", Proceedings of Solid/Liquid Separation: Waste Management and Productivity Enhancement, held in Columbus, OH, December 1989.

6. Yukawa, H. Kobayashi, K. and Hakoda, H., "Study of the Performance of Electrokinetic Filtration Using Rotary Drum Vacuum Filter," J. Chem. Engr. Japan, 1980, 13 (5), 390-396.

7. Senapati, N., "Mechanisms of Ultrasonic Interaction During Electroacoustic Dewatering", Fifth International Drying Technical Symposium, Cambridge, MA, 1986.

8. Chauhan, S. P., Muralidhara, H. S. and Kim, B. C., "Electroacoustic Dewatering Process for POTWs (Sewage Sludges)," Proceedings of the National Conference on Municipal Treatment Plant Sludge Management, held in Orlando, FL, May 1986.

9. Chauhan, S. P., Muralidhara, H. S., Kim, B. C., Senapati, N., Beard, R. E., and Jirjis, B. F., "Electroacoustic Dewatering (EAD) - A Novel Process," paper presented at 1987 Summer AIChE Meeting, Minneapolis, MN, August 1987.

10. Chauhan, S. P., Kim, B. C., Muralidhara, H. S., Senapati, N., and Criner, C. L., "Scale-Up of Electroacoustic Dewatering (EAD) Process for Food Products," Paper No. 48a, presented at 1989 Summer AIChE Meeting, Philadelphia, PA, August 1989.

State of the Art Centrifugal Coal Dewatering

Mark J. Coholan
Bird Machine Company
100 Neponset Street
S. Walpole, MA 02181

The low speed solid bowl centrifuge, developed in the 1930's as a granular bulk solids classifier, was introduced to the coal industry in the 1940's. For nearly two decades, these first generation coal "filters" served the dual purpose of coarse coal dewatering and water polishing for recycle. In the early 1960's, with the advent of deeper coal cleaning circuits, the particle size consist reporting to the centrifuge became finer. In order to maintain product moistures and recovery, higher rotational speeds were required. Improved wear resistant materials were developed in response to the levels of abrasion associated with this increase in speed. During the mid 1960's, the screen bowl centrifuge was developed. Designed to operate at higher speeds and increased throughput, it yielded significant process improvements with product moistures reduced by 4-6 percentage points. These centrifuges were able to compete preferentially with disk filters. With the development of pipeline coal slurry systems in the late 1960's, a highly specialized machine design was required. To meet the specific needs of this application, design modifications were made improving machine performance and demonstrating favorable economics compared to alternative systems. In the 1980's, third generation machines were developed incorporating the best features of the earlier designs together with the most recent advances in tungsten carbide wear protection. This paper summarizes the development of high speed screen bowl centrifuge technology during our first 25 years.

LOW SPEED SOLID BOWL - PREPARATION PLANTS

The solid bowl centrifuge was developed in the 1930's to dewater/classify typically 600 micron (28 mesh) x 50 micron (300 mesh) solids. From the late 1940's through the mid-1960's, it found wide use in the coal industry for dewatering these "fines". The 54"x70" 600 rpm machine became the industry standard for dewatering $\frac{1}{4}$ inch x 0 Eastern Appalachian Coal from washing tables. The minus 45 micron (325 mesh) fraction of the feed was typically quite low consisting of up to 50% ash. These particles were classified out with the effluent. It was quickly realized that a significant cleaning or beneficiation occurred in the centrifuge yielding product ash levels consistently lower than feed. As the coal producers began to develop more mechanized mining techniques and apply cyclone and froth flotation technology in the preparation plants, greater amounts of 28 mesh x 0 fines were generated and cleaned. The fraction of minus 325 mesh particles reporting to the centrifuge increased to 10 - 15% with a resultant increase in product moisture. In order to maintain low moistures, rotational speeds of up to 1200 rpm were required. Machines were redesigned with stronger materials, greater torque capacity, and abrasion protection in key wear areas.

HIGH SPEED SOLID BOWL - PIPELINE SYSTEMS

Higher speed solid bowl machines also made it possible to meet the special requirements of slurry pipeline systems. In the late 1950's, for the Cadiz to Cleveland pipeline, a pumpable stabilized slurry was developed from Eastern Appalachian Coal consisting of 20 - 30% minus 325 mesh particles of which 20% were smaller than 10 microns. The slurry was typically 50%+ solids compared to froth concentrations of 18 - 25%. It became cost effective to heat this high solids slurry to 170°F in order to reduce the viscosity prior to dewatering as a means of lowering final product moisture. In order to meet the pipeline performance requirements, a 40"x60" machine was developed with improved hardfacing materials, rotational speeds of up to 1300 rpm, and a rotating solids gutter to improve cake discharge. With this increase in speed, abrasion resistance concerns were amplified. By the early 1960's, both the initial and operating costs were higher for solid bowl centrifuges than for disk filters, making filtration equipment the economical alternative.

However, in 1969, a specially designed 40"x80" solid bowl was chosen for the mechanical dewatering circuit of the Black Mesa Pipeline slurry system. It incorporated a rotating solids gutter, metallic hardfacing, a 600,000 inch pound gear unit, and components designed to withstand the 200°F feed temperature. Screen bowl centrifuges had only recently been commercialized and had not yet established a sufficient operational history and thus were not chosen for this installation. A total of 40 of these specialized solid bowl centrifuges were installed to dewater the pipeline slurry. They are still in operation today, 20 years later.

LOW SPEED SCREEN BOWL - PREPARATION PLANTS

In the mid-1960's, the Screen Bowl Centrifuge was developed. It combined the sedimentation principles of the solid bowl with the centrifugal filtration of the screen section. These machines were designed with large clarification volumes which increased liquid retention time and improved solids capture. The recovery of these fine solids resulted in higher cake moistures leaving the solid bowl section of the machine. However, the mechanical dewatering ability of the screen section was able to remove this additional moisture and yield cakes 4 - 6% drier than disk filters. As a result of these design improvements, including a 50% increase in capacity and metallic hardfacing protection, screen bowl centrifuges were first installed on fine coal circuits dewatering Eastern Appalachian Coal. The feed consist of the minus 325 mesh fraction was 10-20% with less than 1% minus 10 micron. The 54"x70" machine operated at 600 rpm and removed 90% of the bulk solids down to the minus 325 mesh range. The finer particles containing up to 40% ash were removed with the effluent and were discarded as refuse. With these early designs, high maintenance costs were major concerns. At this point in time, the centrifuge application on coal had been successful from a process standpoint but proved to be extremely expensive to operate.

SECOND GENERATION SCREEN BOWLS

With the recovery of additional minus 325 mesh material in the modern coal preparation plant, the original screen bowl design was modified and improved. Figure 1 shows the influence of the size consist (particle size distribution) on resulting product moisture for varying minus 45 micron (325 mesh) material in the product at different levels of the plus 600 micron (28 mesh) portion. The 36"x72" design was chosen to be the standardized coal centrifuge. This allowed interchangeability of parts and a machine size which could handle most coal applications. These second generation machines were designed with increased clarification capabilities through improvements in the geometrical design configuration coupled with speed increases that generated more than 1000 times the force of gravity. The correlation between product moisture and two different G levels can be seen in Figure 2. As a result, it became evident that the hardfacing technology was insufficient to address the levels of abrasion associated with the required rotational speeds. Improved wear resistant materials were needed to increase the life of components subject to abrasion.

A major development program was undertaken that proved to be a decisive factor in the acceptance of screen bowl centrifuges in the coal industry. High alumina content ceramic components were applied to key wear areas within the machine to extend operating life between major rebuilds. Previously, rebuilds were required as often as every 1000 hours at a maintenance cost of $.50 - $1.00 per ton. The use of ceramic wear components increased the time between rebuilds by a factor of five to ten times and reduced maintenance costs to $.40 - $.50 per ton.

By 1970, it was recognized that the lower product moisture resulting from the improved drainage capabilities of the screen bowl offset the higher price and higher operating costs as compared to disc filters. Screen Bowls were able to displace drum/disk filters by eliminating thermal dryers in many locations. Many preparation plants favored the screen bowl for all 28 mesh x 100 mesh or 28 mesh x 0 clean coal dewatering circuits. In the early 1970's, the 44"x132" machine was designed with increased clarification volume, screen area, and gearbox capacity. It was able to handle feed rates of 50+ TPH while improving recovery by an average of 2%. Table 1 presents the different sizes of centrifuges that have evolved from the coal industry. Also shown is the size of coal processed noting that the early machine capacity is high as a result of a coarse feed.

SCREEN BOWLS - PIPELINE SYSTEMS

When the opportunity arose to test actual pipeline slurry on a commercial size screen bowl, special design considerations had to be given to wear resistance, particle size degradation, and recovery of ultrafine particles. The slurry was also going to be heated to 200°F. The influence of feed temperature on the resulting product moisture can be seen in Figure

3. The parameters which govern screen bowl design are screen open area, machine speed, conveyor pitch, gearbox reduction ratio, pool depth, and the dimensions of the solid and screen portions of the machine. In addition, such factors as surface finishes, bowl angles, conveyor blade tilt, and materials of construction are contributing factors. Pilot scale tests were conducted confirming the need for more screen area in preference to clarification area for the full scale commercial design. They also demonstrated the need for additional torque capacity. Extra torque resulted in extra horsepower over that required on the standard 36"x72" machine. A third generation screen bowl was developed to address the specific requirements of this application. The design also permitted higher operating speeds of 1500 rpm. Actual tests at the Energy Transportation Systems Inc. Coal Evaluation Plant (ETSI CEP), produced pipeline slurry from Western sub-bituminous coal and dewatered the cake to 11 - 12% moisture. Mechanically, the machine was flawless thanks to the extra heavy duty components of the special pipeline design. Solids recovery averaged 90%, about 5% below expectation, confirming the fact that quantifying the minus 10 micron material takes on added importance when dealing with Western sub-bituminous seams.

ABRASION PROTECTION

When processing highly abrasive feeds, even the best ceramics were not adequate to insure maximum on-line availability. In the early 1980's, Bird Machine attained a major breakthrough with the initial installation of the tungsten carbide screen design. During this time, Bird worked closely with a tungsten carbide manufacturer to provide a material that was more than five times more wear resistant than the standard commercial carbides. Material costs were reduced, the design improved, quantity pricing became more attractive, and inherent toughness was significantly improved. The next step to increased conveyor life was the development of a tungsten carbide conveyor blade and a bonding technique that could withstand feed temperatures of 160°-200°F for slurry pipeline systems. With this design, expected machine life is well over 20,000 hours.

Currently, a third generation screen bowl incorporating the tungsten carbide conveyor and screen is being tested on Black Mesa Pipeline slurry. To meet the specifications for a low moisture product while increasing recovery above 95%, a modified third generation screen bowl was developed. It incorporated the high torque, high speed, and heavy duty components of the previous designs with an increase in screen area and a reduction of the clarification volume. Preliminary results have exceeded all expectations. Product total moistures of 19% - 21% (8% below current levels), including 10.7% inherent moisture, are the lowest recorded in the history of this pipeline operation while recovery has increased. Plant operators have noted an increase in the range of boiler feed control resulting from the low moisture cake.

CONCLUSIONS:

Today's screen bowl centrifuge represents a steady evolution from the early designs that were presented in the 1960's. With the development of improved abrasion protection, operation and maintenance costs have been reduced to a range from $.15 - $.25 per ton of product. Ongoing programs with new designs and materials continue to improve overall performance. Of primary importance, the development of the long bowl and short bowl designs has provided machines that can offer either extremely high recoveries of fine feed solids or the classification/beneficiation of high ash fine feed solids while maintaining a four to six percentage point decrease in product moisture compared to alternative systems.

References

1. N.D. Policow, J.S. Orphanos, "Development of the Screen Bowl Centrifuge for Dewatering Coal Fines"
2. N.D. Policow, G.A. Reierstad, "Development of the Screen Bowl Centrifuge for Coal Slurry Pipelines"

/mk

TABLE 1

SCREEN BOWL DESIGNS

COAL - MESH	28x100	28x0	28x0	28x0
FEED ORIGIN	CYCLONE UNDERFLOW	FROTH	FROTH	PIPELINE
MACHINE	54"x70"	36"x72"	44"x132"	36"x96"
THROUGHPUT T/HR	50	20	50	35
RECOVERY %	85 - 95	95 - 97	97 - 98	94 - 98
CLARIFICATION AREA SQ. IN.	4364	3058	5971	2131
SPEED	600	1400	1100	1500
SCREEN DESIGN	STAINLESS STEEL	CERAMIC	CERAMIC	TUNGSTEN CARBIDE
MAINTENANCE COSTS PER TON	\$.50 - \$1.00	\$.40 - \$.50	\$.40 - \$.50	\$.15 - \$.25

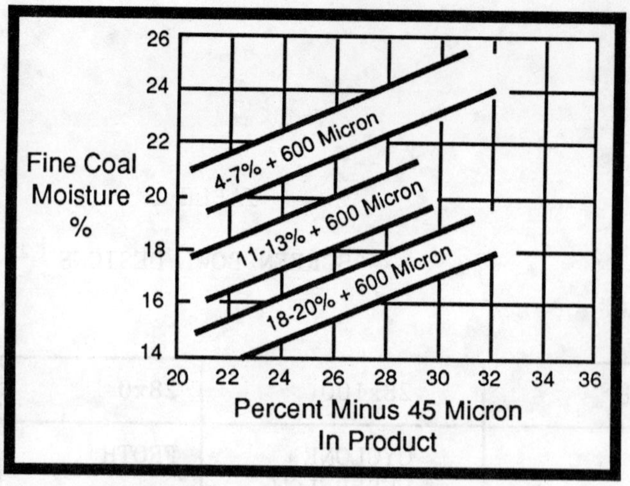

Figure 1 Screen bowl centrifuge influence of size consist on product moisture.

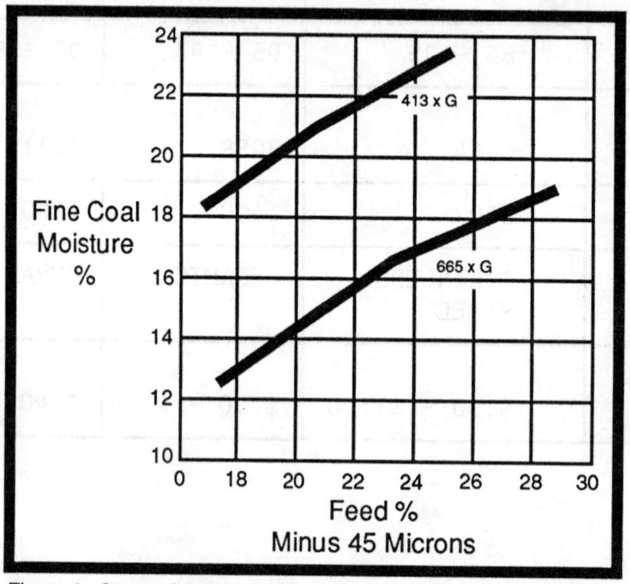

Figure 2 Screen bowl centrifugal force (xG) on product moisture nominal 600μm x 0 (28 mesh x 0) feed.

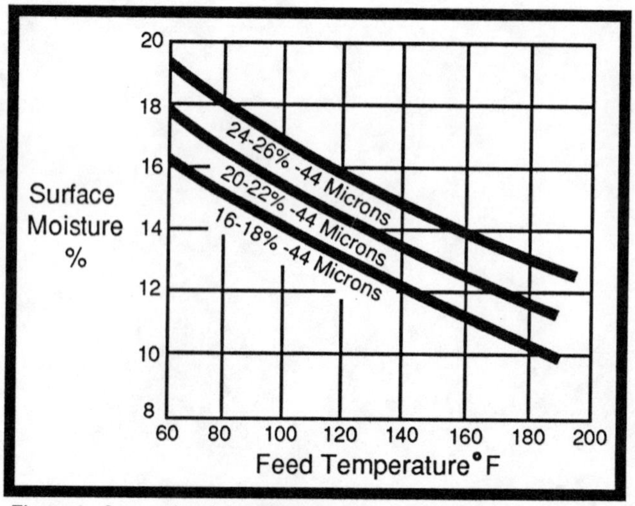

Figure 3 Screen bowl centrifuge tests on slurry pipeline.

AMERICAN FILTRATION SOCIETY
WASHINGTON CONFERENCE
ENERGY GENERATION III - FLUID/SOLID SEPARATION IN COAL PROCESSING
SESSION 5

ABSTRACT

"FILTRATION OF COAL FINES - STATE OF THE ART AND INNOVATIONS"

BY
RONALD KLEPPER
RAVINDER MENON

Although coal production has been in excess of demand for several years making research and development efforts minimal, there have been developments in vacuum filtration that have been important to other fluid/solid separation applications. This paper will discuss developments and their practicality in use for "Fluid/Solid Separation in Coal Processing."

There will be a review of work done in Europe as well as work done here in the United States using flocculants and surfactants and new filter media technology.

The principles of filtration have been affected by these developments. However, the acceptance of these novel ideas into coal processing of the future is uncertain.

THE ALTERNATING CURRENT ELECTRO-COAGULATION PROCESS

B.K. Parekh and J.G. Groppo
CAER, University of Kentucky
Lexington, KY 40511

J.H. Justice
Co-Ag Technology, Lexington, KY

Abstract

The Alternating Current Electro-Coagulation process has been successfully applied for effective coagulation of a stable ultra-fine solid suspension. The principles of electrostriction (charge neutralization) and electro-coagulation (polymeric hydroxy species of metal ions) facilitate rapid coagulation and improved dewatering of stable suspension.

The AC Electro-Coagulation (AC/EC) process was applied to a stable black water obtained from a sub-bituminous coal processing plant, containing particles with an average size of 2 μm. The process was successful in flocculating the particle using a slurry residence time of 20 to 60 secs. The estimated processing cost of the process ranged from \$0.09 to \$0.36/1000 gallons depending on mixing proportion of the AC/EC treated and untreated slurry.

Introduction

Removal of fine solid particulates from a mineral or coal processing plant waste discharge is important from environmental as well as water recycling point-of-view. Waste waters are of prime concern and warrant reassessment in terms of both volume reduction and pollutant removal. Most of the mineral and coal processing plants generally utilize organic polymers to promote flocculation of fine particles which are then separated from aqueous phase by sedimentation or filtration. However, when a substantial amount of ultra-fine clays is present then an effective flocculation of clays cannot be achieved efficiently and economically using the polymers.

As an alternative to chemical addition for coagulation, the AC Electro-Coagulator (AC/EC) process provide a technique which is efficient and cost effective. The AC/EC has been tested for coagulation and settling of fine solids and the process effectiveness has been demonstrated in various coal preparation plants for effectively removal of clays from coal (1,2,3).

This paper discusses results obtained on a stable black water suspension obtained from a sub-bituminous coal operation, a brief discussion on the theory of electro-coagulation and economics of the process are also given.

Background and Theory

Stability of fine solid suspension is attributed to the surface electrical charge carried by the particles. The presence of surface charge on solids is explained using the double layer theory. When the double layer collapses coagulation or destabilization of particles occurs. One of the ways to accomplish this is by adding metal ions.

The electro-coagulator process employs two main principles:

- Electrostriction - whereby the suspended particles are stripped of their charges by subjection to alternating current electrical field conditions in a turbulent steams, and
- Electro-coagulation - whereby minute quantities of metal hydroxy species generated from the dissolution of electrodes assist in coagulation of the suspended particles.

The current hypothesis of AC/EC operation is summarized as follows:

- Polar molecules adsorbed on the surface of small particles are neutralized by an equivalently charged diffuse layer of ions around the particle. A zero net change results.
- Non-spherical particles have non-uniformly distributed charges (dipoles) and elongated neutralizing charge clouds surrounding them.
- These dipoles come into play when the charge clouds are distorted by external forces or close proximity of other charged particles.
- External forces such as electric fields can: (a) cause dipolar particles to form chains; and (b) unbalance electrostatic forces resulting in dramatic phase changes (coagulation).

- AC electric fields do not cause electrophoretic transport of charged particles, but do induce dipolar chain-linking and may also tend to disrupt the stability of balanced dipolar structures.

Experimental

The laboratory setup used for the study is shown in Figure 1. The AC/EC cell consisted of two aluminum electrodes (1/8" x 5 1/2" x 60") placed inside a clear plastic cell. The distance between electrode was kept at 1/2", and were connected to a control box, which also monitors voltage and amperage applied to the electrodes. The slurry enters at bottom of the cell and exits at the top. The AC/EC treated (coagulated) slurry is collected in containers for settling studies. The three major operating parameters of the process are electrode spacing, applied current/voltage and retention time slurry.

Results

The black water slurry utilized in the present study had an average (50th percentile) particle size of 2 μm, with 90% of particle below 5 μm size. The electrophoretic mobility of particles was -2.3 μm/sec/volt/cm; which indicate a high magnitude of charge and hence a stable suspension.

Figures 2 and 3 show settling rate curves of the slurry at various voltages using 4 lit/min and 5 lit/min flow rates, respectively. These flow rates translate into 40 and 30 sec retention time, respectively. The data in Figure 2 shows that high (70 volts) voltages using 40 sec retention time provided the highest settling rate. The settling rates obtained at 55 and 60 volts (Figure 2) were similar. At high flow rate (Figure 3), 62 and 70 volts data provided a similar settling rate. The effect of amperage on floc size produced in the process is shown in Figure 4. It shows that for a given flow rate size of the flocs increases with increase in amperage.

A series of propagation tests were conducted on the slurry, where a portion of the slurry was treated with AC/EC and mixed with the untreated slurry. Figure 5 shows settling rate data obtained with 25%, 50%. 75% and 100% AC/EC coagulated slurry. Note, that all of them provided a similar settling rates, except 25% coagulated, which was slower than the others. These data indicate that only 50% of the slurry will need a treatment and by mixing the AC/EC treated and untreated slurry in 1:1 ratio provide an effective coagulation of solids. The final turbidity of suspension was 15 NTU compared 700 NTU of the 'as received' slurry.

Table 1 list the operating (power) cost data for treating 100 gallons of the slurry in various proportions. Note, that processing cost varies from \$0.09 to \$0.36/1000 gallons of slurry. Based on previous studies, it can be concluded that mixing of treated and untreated slurry in 1:1 should be optimum.

Results

Based on the laboratory studies, it can be concluded that:

- The AC Electro-coagulator is effective in coagulating the stable fine solids present in the black water providing a clear supernatant of low turbidity (~15 NTU).
- In laboratory unit, a residence time of 40 secs (flow rate 4 lit/min) and a current of 20 amp (60 volts) appears to be optimum for coagulation for the slurry.
- 1:1 mixture of AC/EC treated and untreated suspension provided similar settling characteristics as 100% treated slurry.
- Processing cost of the AC/EC process ranged from $0.09 to $0.36/1000 gallons of slurry depending on mixing ratio of treated and untreated slurry.

References

1. W.F. Berry, and J.H. Justice, "Electro-Coagulation: A Process for the Future", 4th International Coal Preparation Conference, 1987.

2. F.H. Nikerson, "Electrical Coagulation: A New Process for Prep Plant Water Treatment", Coal Mining & Processing, Sept. 1982.

3. R.E. Patrick, T.F. Stanczyk and B.K. Parekh, "Solid/Liquid Separation Using Alternating Current Electro-Coagulation", International Symposium on Solid/Liquid Separation, Battelle Columbus, 1989.

Power Costs

Percentage Exposed to Electrocoagulation	Dollars per 1000 Gallons
25	.09
50	.18
75	.27
100	.36

Table 1. Power Cost for AC Electro-Coagulator (Base: $0.05/KwH)

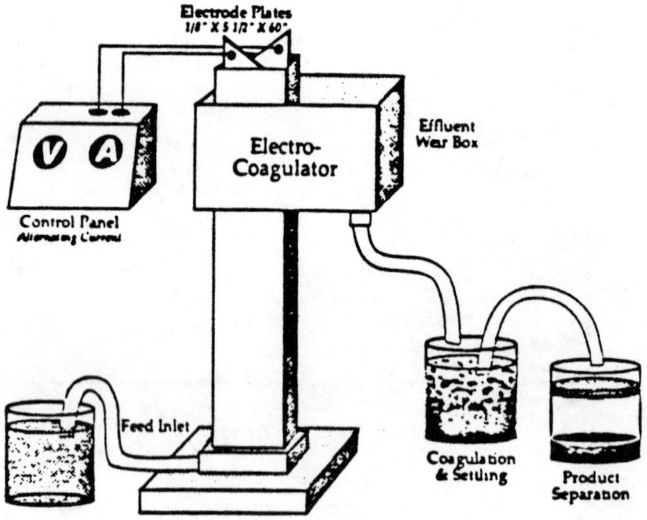

Figure 1. Laboratory AC Electro-Coagulator Setup

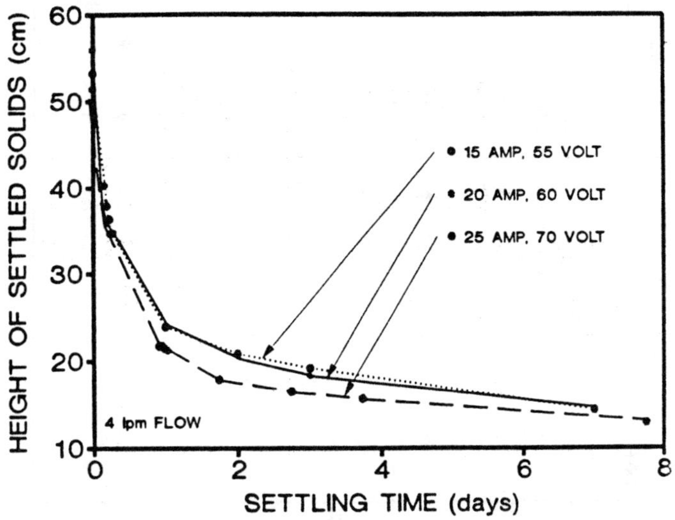

Figure 2. Settling Rate Curves at Various Current Densities Using 4 lit/min Flow Rate

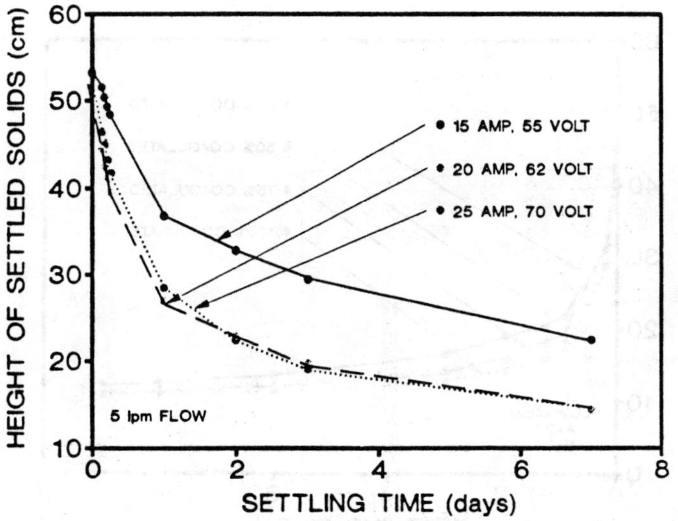

Figure 3. Settling Rate Curves at Various Current Densities Using 5 lit/min Flow Rate

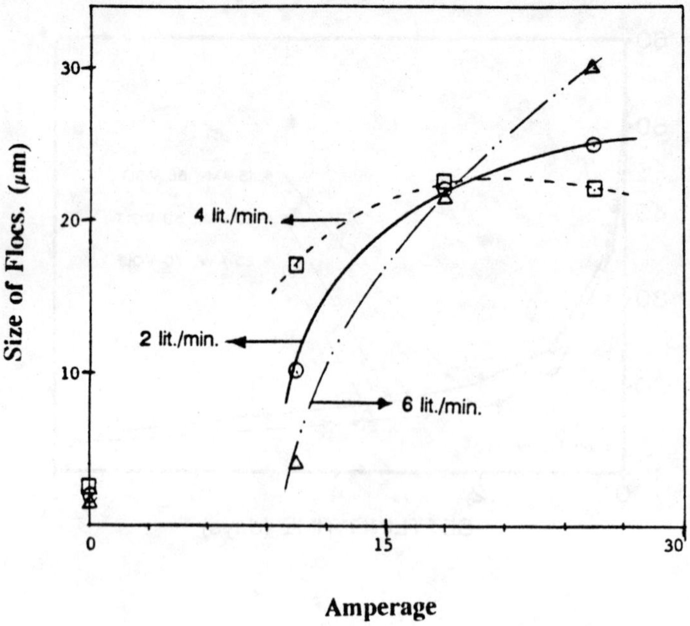

Figure 4. Effect of Flow Rates and Amperage on Size of Flocs

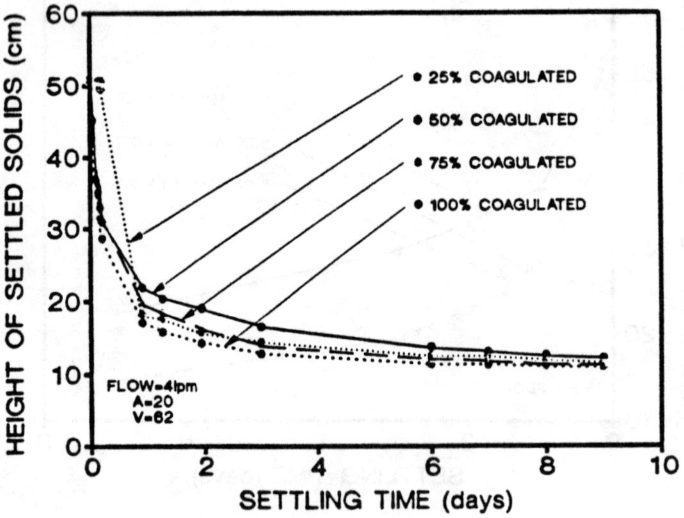

Figure 5. Propagation Tests Settling Rate Curves

Basic Review of Coolant Filtration

James J. Joseph
Joseph Marketing, Inc.
P.O. Box 232
East Syracuse, NY 13057
315-437-0217
FAX - 315-437-0487

The objective of this presentation is to provide a general overview of the coolant filtration industry and cover the salient points which make it different from other filtration arenas. The opening section describes typical approaches to coolant filtration. It will point out some of the fundamentals of closed loop and reveal the axioms that are involved. The second section will explain some of the unique features of coolant applications where liquid-solid separation is accomplished by many devices other than filters. The final section will report on the metalworking industry's attitude toward filtration. The user,machine builders and coolant suppliers play a role in the present trend to give filtration the recognition value it deserves.

TYPICAL APPROACHES

Figure 1 shows four general arrangements of a coolant filtration system. Each has a process which is a manufacturing function using a liquid. Liquid becomes contaminated during this operation and the device(s) are used to keep the fluid clean. The reservoir is the liquid holding tank. Some of these tanks go beyond just a holding chamber. They could be fitted with self-cleaning mechanisms to become contributors to the cleaning function. The schematic shows that liquid flows from components either by gravity or pump.

The four schematics are:

A. Closed loop full flow.
B. Closed loop bypass flow.
C. Batch arrangement, where a given volume of liquid in process reservoir is replaced with a just cleaned volume from a remote facility.

There are five axioms in the performance of these closed loop systems.

1. On a closed-loop, a system regardless of its capabilities, will attempt to reach a level of performance where the amount of contaminant it removes at any given interval will equal the amount of contaminant introduced at the process for the same interval. When this happens, the system is in a state of equilibrium with the process. In short, contaminant-in equals contaminant-out.

2. Bypass systems, although handling a smaller portion of the total liquid flow, will eventually see the total contaminant load of the process and attempt to remove an amount equal to the amount introduced at the process. In other words, a bypass filter, although handling 10 percent of the flow, will eventually see 100 percent of the contaminant produced by the process.

3. The amount of contaminant left in the fluid is dependent upon system performance and is a function of device efficiency.

4. A system applied on a bypass basis can be related to a full flow system when its efficiency level is multiplied by its bypass percentage. For example, a 90 percent efficiency system operating on a 50 percent bypass is equal to full flow system with 45 percent efficiency.

5. Time is an important factor in reaching equilibrium. A practical timetable for most installations is measured in weeks to months, not minutes to hours.

The first axiom is the most difficult to see but is the most important. Contaminant-in equals contaminant-out depicted in the equation 1+(100-performance/performance)= equilibrium. The explanation and proof is explained in the book Coolant Filtration(1). Once this phenomenon is understood then the selection process for a system and its hardware follow a format of logic, economics and trade-offs to satisfy the limits of time, space and money.

CLEANING DEVICES

There are a range of devices available in this industry. They include everything from a glorified shovel

2

to a very sophisticated membrane filter. We categorize them with two basic titles: Separator and Filter(2).

Separator

This is the device which capitalizes on physical characteristics of the contaminant and shows how it relates to the characteristics of the liquid. There is no barrier as such. A separator uses the physical characteristics of weight and magnetism. These include settlng tanks, magnetic separators, hydrocyclones, centrifuges and flotation.

Filter

These are the devices which actually provide a barrier for the liquid to pass through while the solids are being intercepted. In the coolant field, the type of filters used cover the range from cartridge, flat bed, pressure and suction devices. Since we cannot cover them all in this presentation, Figures 2 and 3 show some typical flat bed open and pressure filters.

Many times, the cleaning device (separator or filter), is picked for its contaminant holding capacity and its self-cleaning features. Remember the ratio of "dirt" load to liquid could be high for many installations. Therefore, the ease of removing solids from the system is as an important issue as the device's cleaning ability.

Also, unknowingly at times, the source of the contaminant makes an impact on the selection process.

The sources of contaminants fall into three groups.(3) These three groups establish a ranking order by which the liquid becomes contaminated and how the equipment is selected.

The first group is where the contaminant in the system is a direct result of the manufacturing operation. For example, a grinding facility generates contaminants in performing its function. If contaminants were not generated, then no work is accomplished. Therefore, the dirt level, load and characteristics offer some constant factors which must be evaluated.

The second group is where the contaminant generated during the operation is an indirect result of the process. For example, a rolling operation is not designed to remove stock but it does. These metallic fines are an indirect result of the operation. There is a difference in the amount and physical characteristics of the fines generated in an indirect-result manner such as rolling and in those generated in a direct-result manner such as grinding. The selection process must consider the volume generated and and the particle size range it will encounter. For example, fines from a rolling mill application will exhibit a narrow range of particle size while grinding yields a wider

3.

range because of the cutting action. Although there is more contaminant in a grinding process, it is easier to filter than the smaller, more dense rolling mill contaminant.

The third group is where the contaminant is foreign to the process. Actions extraneous to the operation result in the introduction of plant debris, floor cleaning material, and other foreign items because of the environment in which the coolant flows.

Most industrial processes become contaminated with foreign material as well as indirect and direct contaminants.

All this must be weighed in the decision for the best, most economical device to serve the operation.

Media

Since many filters use disposable media, it would be well to cover some general aspects of this section.

Disposable media used in this industry includes cartridges, bags, sleeves, sheets and rolls. Roll goods are used extensively with the flat bed filter family. Nonwovens are the most popular and they consist of materials ranging from rayon, polyester, polypropylene, nylon to cellulose blends. They are put up in various weights and densities: mainly between .5 to 2.5 ounces per square yard. Most filter equipment manufacturers design the hardware to accept standard widths which correspond to the paper mill's capabilities. The standard widths are 51 and 72 inches. Narrower widths are usually selected so the material can be slit from the standard widths with little waste.

Most metalworking applications with flat bed filters do not rely on the media as the actual filtering medium. It functions more like a septum upon which a cake or film of solids are accumulated. The cake in turn becomes the effective separating barrier. Media usage and performance depends upon the source of contaminants and the influence of other factors. Figure 4 shows some typical characteristics.

a. Permeability - good flow per square foot, good separation.
b. Particulate-tight and dense - good separation, lower flow per square foot.
c. Tramp oil: with water base coolants - coats the media and particulate to reduce flow rate.
d. Blended material - approaching depth filtration where particle size range is not inherent in the operaton.

ATTITUDE TOWARD COOLANT FILTRATION

<u>Reception</u>

In the past, coolant filtration equipment was looked upon as a necessary evil. It was not a contributing factor to productivity. It had only an indirect cost with no value added to the product. A system was treated as nothing more than the garbage can of the facility. Machine tool suppliers often ignored the need, saying it was the users responsibility. Or, if he was asked to include a filter, he would select anything as long as it was inexpensive and could be construed to be a filtration system. As the concern for better productivity, longer tool life and reduction of coolant disposal costs increase, so does the awareness of the value of good coolant filtration systems increase. Many facility planners are now considering the filtration needs at the same time they are selecting machine tools. It is no longer an afterthought. Also many industries are devoting resources to retrofit existing facilities with better filtration options and they find that this can be accomplished within their limits of time, space and money. They have found that even when they have to accept a trade-off and install a less than ideal system, the new system can still show economic justification. The irony is that, what was once considered a liability is proving to be an asset.

Shift from Oil

There is a general shift, in the coolant industry, away from oils toward water base solutions. The shift is mainly motivated by environmental concerns. Although some advocate that the political problems in the petroleum industry is a factor, the main force is the growing roster of regulations on mist, waste and fire control. Therefore, water base solutions are being introduced on applications heretofore were reserved for oil. The filtration concepts for the water base liquids are generally simpler than those for oil. There are many more options from which to choose and the liquid handling problems are fewer. However, stability, additives and bacteria control come into play when a water base fluid is selected.

Coolant Supplier Relationship

A positive relationship is growing between the coolant supplier and the filter system supplier. Each is more aware that the knowledge of the other will aid both in being successful in completing their respective tasks. Equipment and liquid compatibility is just as important as the ability of the products to perform their basic functions. The so called adversarial confrontation between the coolant supplier and filter supplier is rapidly disappearing and there is an element of team work and cooperative problem solving in its place. A classic case history which is now happening often, is a situation where a filter supplier had difficulty with the excess amounts of tramp oil in solution where it adversely affected the filter. A relationship developed with the coolant supplier, allowed him to discuss this openly and the coolant supplier reacted by providing a better oil shedding additive which did not change the coolant's performance but helped the filtration system cope with the oil. Or, the situation could have been reversed. The coolant may have an additive which is being stripped by a tight media and the filter supplier changes the media selection or alters other filter characteristics which still allows the filter to function properly but minimizes the additive loss.

Machine Tool Supplier Relationship

The relationship between the machine tool builder and the filter supplier is another positive change in this industry. Granted this change may be mainly customer driven since this awareness justifies his expenditures for good filtration and the machine tool builder is given more freedom to make good selections. However, there is more to it than just that. The machine tool supplier is experiencing greater demands on his equipment. He wants to offer longer tool life and less machine maintenance. Also, he is using more precision components, high pressure systems and sophisticated instrumentation. Many of his concepts use moving pallets, critical fixtures and transfer mechanisms. All these are flushed and lubricated with the same coolant. It has to be clean and in many cases the clarity levels demanded by the machine are greater than those needed at the tools. Therefore, he has a vested interest in good filtration and hence his growing respect for filtration know-how.

Media Supplier

Here is another industry which is serving the same market which recognizes the value of knowing more about filter hardware. There are greater demands on the media supplier to offer service beyond the application. He must help in considering disposal needs and reduction of environmental issues. Disposable media has a growing obstacle because of environmental issues. Many industries are having a difficult time dumping spent media. Also, in some cases, even incineration is a problem because of the impact on emission regulations.

Disposal

The growing restriction on disposal is not limited to media. Obviously a bigger impact is the liquid itself. Governmental requirements toward transporting spent liquid particularly oils, is a growing pressure on the industry to extend life, formulate with easily disposed materials and reduce cost.

CONCLUSION

The term coolant filtration in the past reflected an image of a large user keeping his liquid clean enough to save it as long as possible because of initial costs. Now coolant filtration takes on a different stature. It includes make-up requirements, disposal requirements, recycling and satisfying environmental issues. The awareness of this involvement is growing and the future for this industry is also growing. What may appear to be a mature industry to some is a growing potential for others.

Figure Captions

Figure 1 -Types of Clarification Systems
Figure 2 - Flat Bed Open Filters
Figure 3 - Flat Bed Pressure Filters
Figure 4 - Media/Cake Cross Sections

References

(1) Joseph, J., Coolant Filtration, Joseph Marketing, Inc., E. Syracuse, NY (1985) PP 13 - 19

(2) Ibid., PP 31 - 43

(3) Ibid., P 11

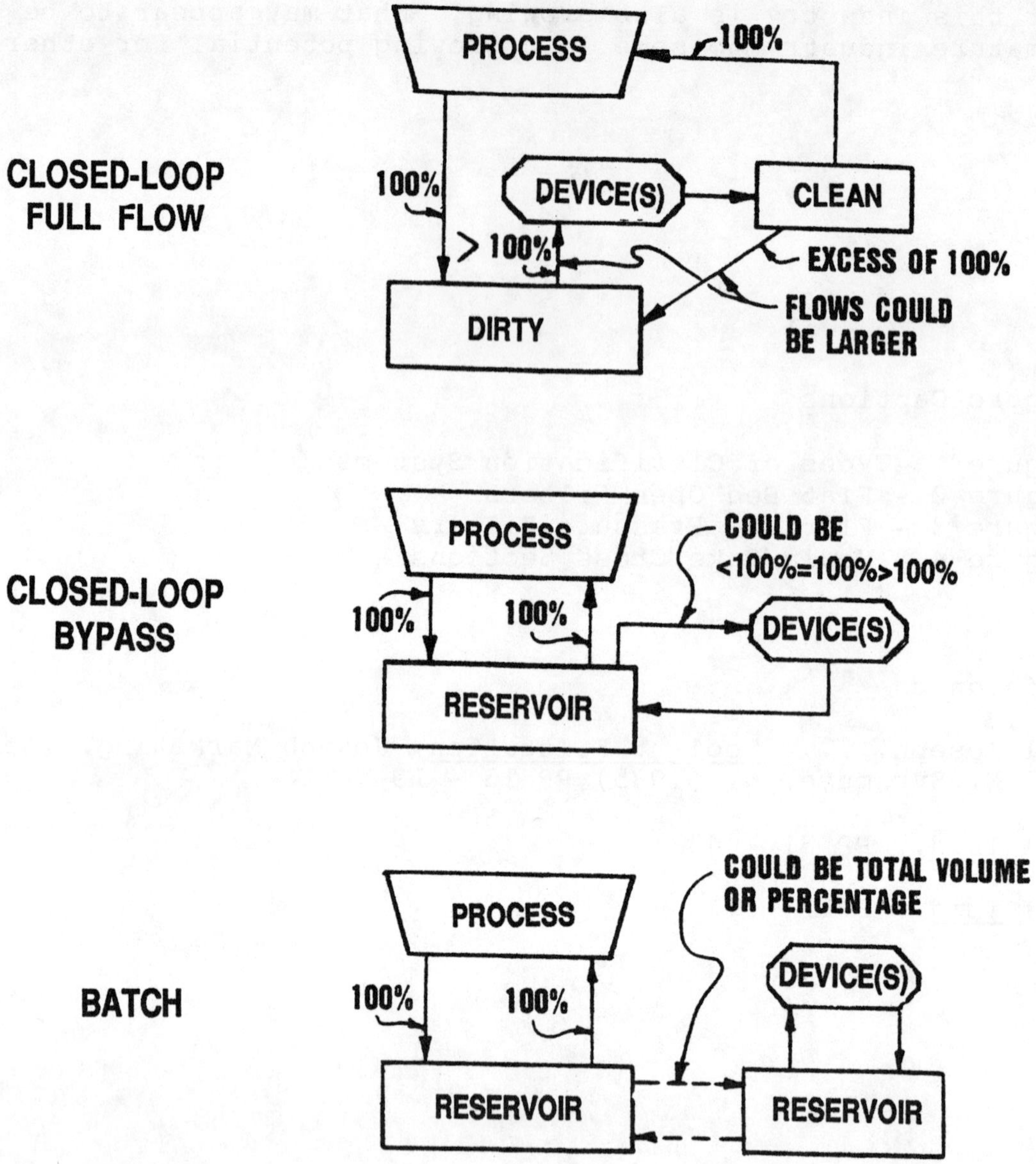

FIG. 1 TYPES OF CLARIFICATION SYSTEMS

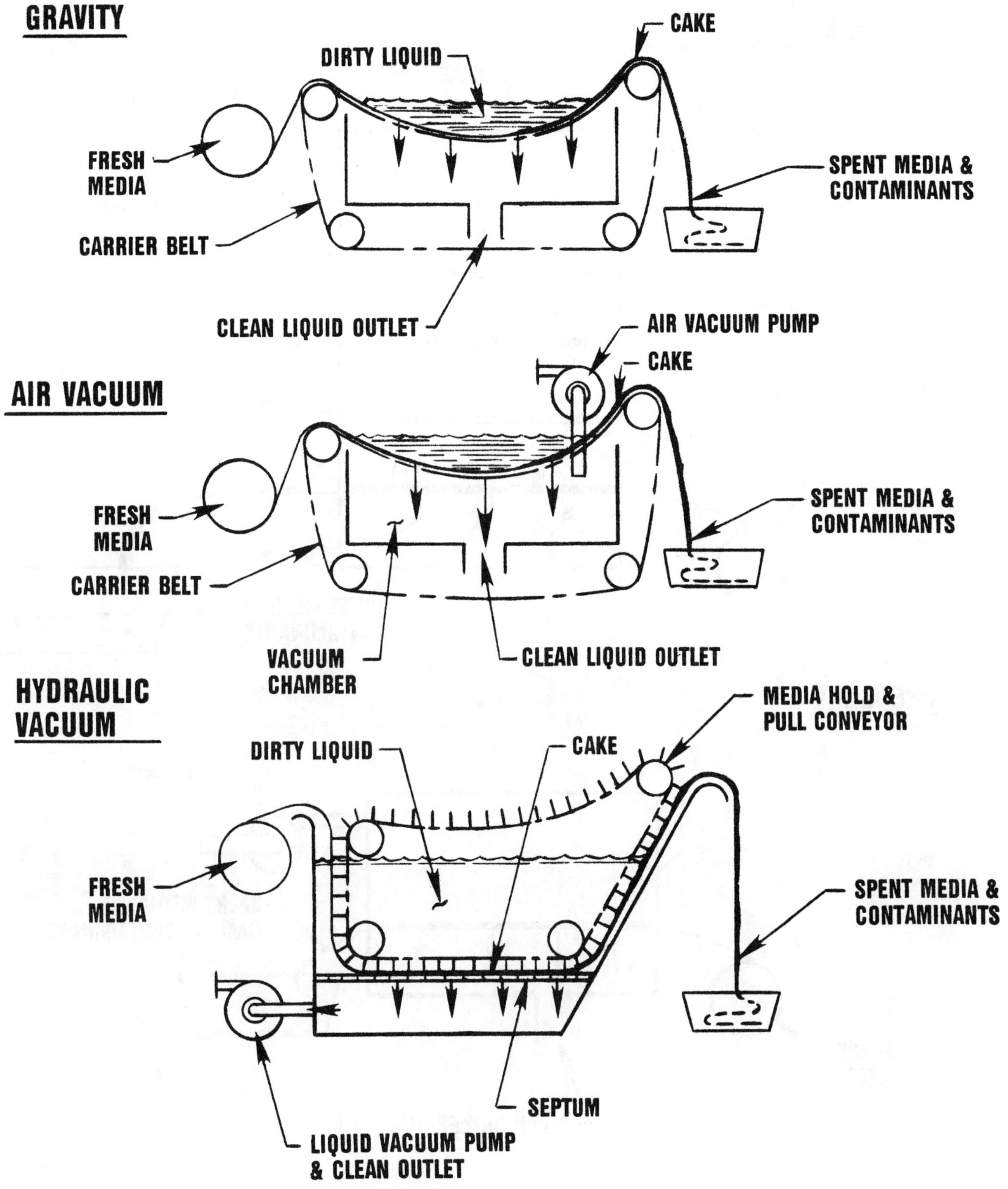

FIG. 2 FLAT BED OPEN FILTERS

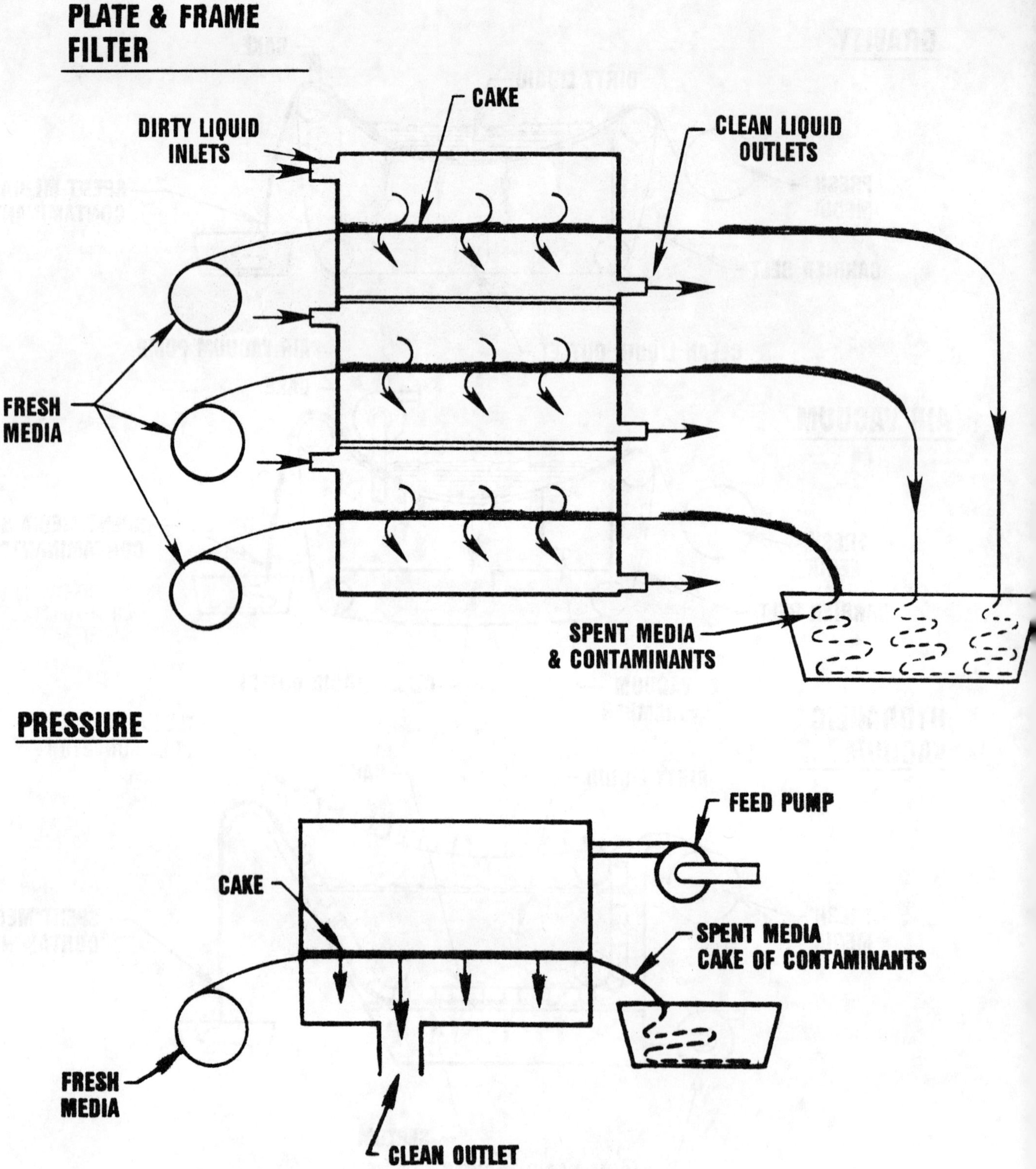

FIG. 3 FLAT BED PRESSURE FILTERS

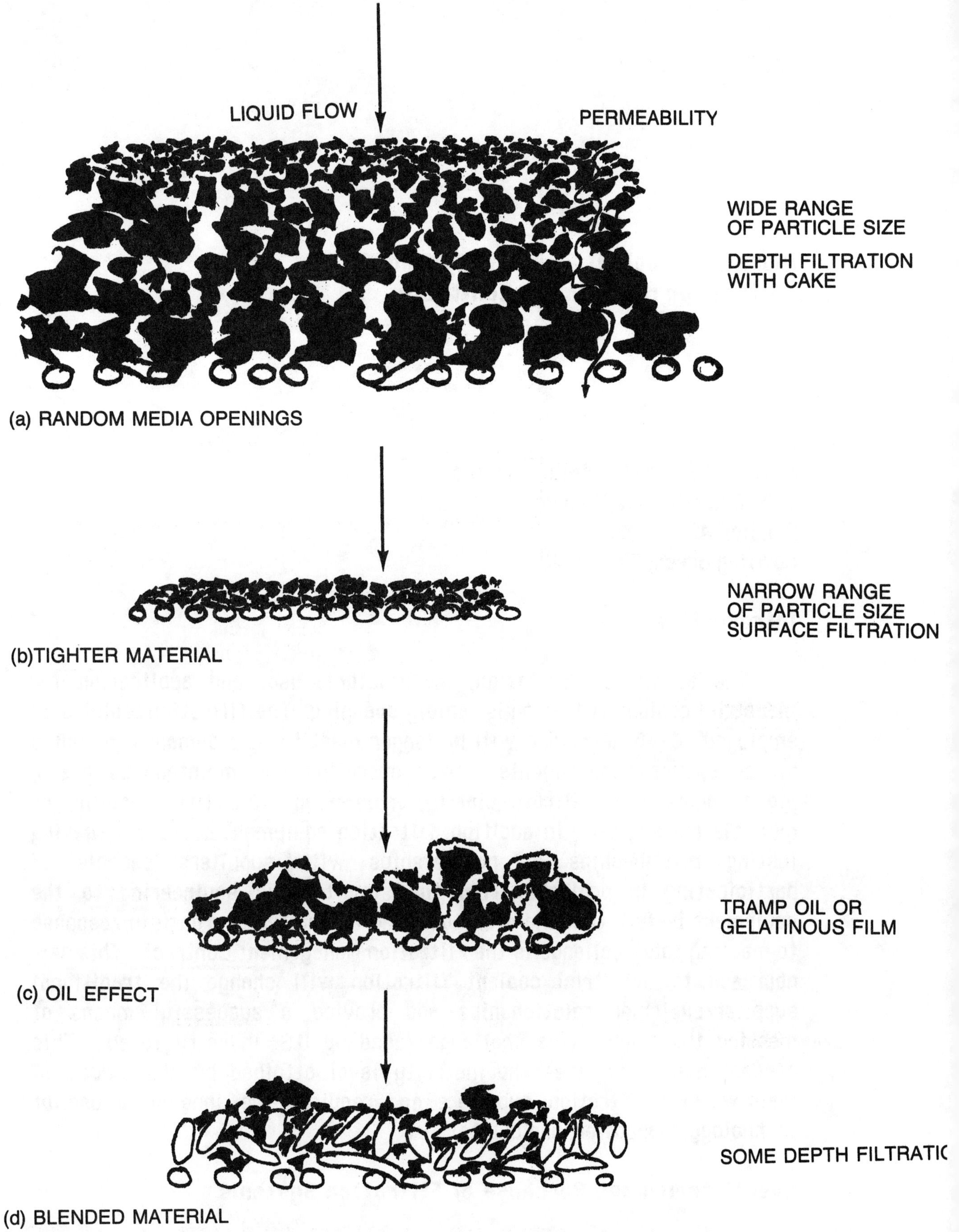

FIG. 4 MEDIA/CAKE CROSS SECTIONS

Industrial Coolant Filtration Trends: Filtration Management Contracts

Advanced Filtration Solutions, Inc.
19302 North Dixie Highway
P.O. Box 42
Bowling Green, Ohio 43402

Introduction

The specification, design, manufacture, use, and application for industrial coolant filtration is rapidly changing. The filtration "solutions" employed 10-15 years ago will no longer meet today's demands including higher uptime requirements, lower operating and maintenance costs, higher degrees of filtrate clarity, maximized fluid life, handling of multiple metals, etc. In addition, filtration equipment users are seeking lasting relationships or partnerships with suppliers capable of participating in projects from the simultaneous engineering to the equipment installation and start-up. One innovative approach in response to meeting this challenge is the Filtration Management Contract. This new approach to industrial coolant filtration will change the traditional supplier/customer relationships and provide a successful means of meeting the competitive challenge faced by U.S. industry today. This concept also recognizes the maturity level attained by the industrial metalworking filtration suppliers and requires the innovative use of technology, manpower, and capital.

Specification and Purchase of Filtration Systems

The demands on today's metalworking central coolant filtration system is greater than ever. The average size of a central system has grown as more machines are dedicated to the central filter concept.

It is therefore critical that the specification and design of the central filter system demand equipment which will offer "uptime" of 98% or more. Experienced filtration equipment buyers are now evaluating equipment bids which not only guarantee such performance, but the equipment design incorporates such features as primary work conveyors, dual filtration capabilities, multiple filtering elements, and sophisticated electrical controls with full diagnostic capabilities to insure performance. Significant penalties often accompany purchase orders for central filtration equipment not meeting the uptime and productivity goals specified, or the installation time table for the project.

In addition to central coolant system performance, often the following criteria are established which are equally as important:

1. Lower operating and maintenance costs.
2. Higher degrees of filtrate clarity.
3. Special coolant pressure requirements.
4. Temperature control and monitoring.
5. Built-in capabilities to expand or ramp-up.
6. Incorporating the ability to handle multiple metallic contaminates.
7. Additional processing of removed solids and filtration by-product.
8. Special equipment features which will prolong coolant life and enhance the work environment.

The task of preparing specifications and designs for central filtration equipment to accomplish all of these tasks can be a very formidable and expensive one. In addition, there is the bid evaluation process and finally the vendor selection process which can add weeks to the project not to mention substantially increase the overall project cost.

The metalworking manufacturers that are successfully competing in the global marketplace today are the ones that are the <u>low cost</u> producer of their respective components. In order to remain a low cost producer, every item which adds a cost to the final product is under constant scrutiny. Upon examination of the central filtration system portion of the total production costs, many plants find that in addition to the numerous costs mentioned earlier in the specification and capital purchase of the equipment, they find high maintenance, operating, energy consumption, and "down time" costs which threatens their low cost market position. While the need for quality filtration equipment and clean coolant is clearly justified, the costs of the present way of "doing business" seems to run counter to the requirements of the day. We need to question, "Is there a better way to conduct business such that we will gain all the benefits of working with a reputable filtration supplier, meet all the cleanliness and environmental goals, while reducing the expense burden upon the unit cost of the components we build?"

Supplier Relationships

Because of the complex nature of the metalworking processes, many have found working with or even becoming "partners" with filtration equipment suppliers has many advantages. First of all, forming partnerships in time to employ simultaneous engineering may prove cost effective as the machine tools, parts handling, utilities, and equipment layouts are developed along with the filtration equipment requirements. In some cases, having the ability to rent or lease filtration equipment is a desirable alternative to purchasing, especially for "short term" or soon-to-be phased out projects. In other cases, small pilot filters are needed to run-off particular "hot" or earlier machine requirements in the first stages of a major project. Having a working relationship or "partnership" with a qualified supplier can make these alternatives possible.

The Filtration Management Contract

No one is certain what the future will bring in the metalworking industry. Quality filtration will always be required where critical wet machining is performed. The Filtration Management Contract shifts the efforts of the filtration equipment user back to the business of building components and places the burden of providing the quality filtration where it belonged all the time: with the filtration equipment supplier. As we explore the Filtration Management Contract, we will examine all the advantages as well as the obstacles to employing such a contract at your facility. The concept will probably not be for everyone, but the Filtration Management Contract may be the best solution to remaining a low cost producer in the world market today.

Simply stated, the Filtration Management Contract places the entire responsibility for supplying filtered, clean coolant at the proper temperature, volume, pressure, and concentration to the respective machine tools to achieve the goals of the manufacturing production processes. As we will see, the actual details and structure for building a Filtration Management Contract can vary greatly. However, the underlying principle of every Filtration Management Contract is the shift of responsibility and equipment ownership from the user to the supplier, a radical departure from traditional buying and selling practices. Before plunging in to the contract itself, lets detail the anticipated benefits for both the filtration equipment user and supplier.

Filtration Management Contract : The Filtration Equipment User

In terms of returning to the business of building quality components, the filtration equipment user will find positive impact to several areas of his business including purchasing, plant engineering, maintenance, production, quality and inventory control. The incentives for the buyer of industrial filtration equipment to enter into a Filtration Management Contract easily outweigh any downside risks.

1. For the metalworking manufacturer, the Filtration Management Contract provides a release from peripheral business and allows full concentration on the core business. The manufacturer of automotive powertrain components is not an expert in industrial filtration. His expertise and efforts must be directed towards the production of the finest, lowest unit cost components as possible, i.e. the core business. A Filtration Management Contract allows the manufacturer to focus on process, production, and system design in lieu of filtration equipment design, specification, operation, maintenance, monitoring, and evaluation.

2. The Filtration Management Contract will assist the metalworking manufacturer in minimizing fixed assets from the business cost accounting. The manufacturer no longer owns the equipment. As we will discuss later as we look at the actual details of a typical contract, the equipment may be rented, leased, or even part of the management fees. In any event, with the ownership remaining with the filtration supplier, the fixed assets "disappears" from the cost accounting and may reduce the burden on various unit costs, not to mention the reduction in accounting time required for asset inventory, insurance, etc.

3. Since the filtration supplier owns the filtration equipment supplying the filtered coolant to the metalworking operation, any expendable or replacement parts required are no longer the responsibility of the metalworking manufacturer. This can greatly reduce or simplify the materials management and procurement manpower requirements and costs. Again the manufacturer can redirect these sources to the business of producing components, potentially lowering overall unit costs.

4. When the filtration supplier is fulfilling the Filtration Management Contract by supplying the quality coolant on a continuous basis as specified, a higher level of consistency of operation should be maintained. By greatly reducing or eliminating the "variableness" of the metalworking fluid quality, an application of S.P.C. to the metalworking operations can be accomplished with greater success than ever before.

Scrap rates and waste will no longer be at the mercy of fluctuating coolant cleanliness, temperature, concentration, and tramp oil levels.

5. With a Filtration Management Contract in place, a significant reduction in the number of non-value added personnel should be realized. All filter system maintenance, both routine and "breakdown" is the responsibility of the owner, the equipment supplier. Not only the overhead cost for the non-value added personnel, but also the training and retraining costs due to constant personnel turnover and varying types of filtration equipment will be eliminated.

6. While selecting a partner is perhaps the most critical part of implementing a Filtration Management Contract, a long term relationship of mutual trust and loyalty will be established with a single supplier. The cost and complexity of dealing with several suppliers for each type of central system equipment reduces any manufacturers competitiveness. Forming a "partnership" takes away any adversarial positioning by either the user or supplier, and places common goals as top priority for each.

7. The greatest gain for the metalworking manufacturer utilizing a Filtration Management Contract will be to lower and control unit cost. Consistency in high volume production is essential. The manufacturer is no longer "stuck" with equipment that does not perform, but now has a partner working for his best interests (and his own) long after the contract is put in place. Becoming a low cost producer and the ability to remain there is the best way to insure long term success, and the Filtration Management Contract provides the control to achieve it.

Filtration Management Contract : The Filtration Equip. Supplier

The Filtration Management Contract greatly expands the responsibilities of the filtration equipment supplier. No longer building and /or installing equipment to specifications as bid, selling the equipment and "moving on" to the next customer, the supplier is now a partner with long term commitment. Retaining ownership of the equipment he has built, the supplier's motivation for compensation is the consistent performance of his equipment. In addition, the success of the partnership may well rest on the response time and competency of the service personnel insuring uninterrupted performance of the central coolant system. The filtration equipment supplier is not simply a manufacturer of hardware, by a provider of a total service package. The challenge is a formidable one, but as we examine the benefits of such a relationship , the efforts to meet the demands are easily justified through the potential stabilization of the equipment suppliers business.

1. The Filtration Management Contract will permit the filtration equipment supplier to supply the most cost effective equipment solutions to the task. Under the pressures of competitive bidding and the historical purchasing habits of the U.S. automotive marketplace to buy "low dollars", the filtration equipment supplier often times was not in a position to offer the best, most cost effective equipment solutions. While sizing a particular filter system or using a particular type of system may have resulted in the highest quoted capital equipment costs in a competitive bidding situation, the supplier understands the "payback" or justification of such equipment in the long term partnership commitment. Since the supplier maintains ownership and is also responsible for the supply of disposable media and the maintenance of the equipment, he will install the very best equipment that he knows will maximize his success and therefore his compensation from the Filtration Management Contract.

2. Under the present system of receiving, interpreting, and complying with equipment specifications unique to almost every metalworking facility, the supplier will be able to significantly reduce his costs in the bidding procedure. The manpower and costs saved in this portion of his business can be redirected to the supply of better equipment to the Filtration Management Contract.

3. As a provider of capital goods to a market where capital spending is extremely variable from year to year, the Filtration Management Contract will assist the filtration equipment supplier in stabilizing his business activity and revenue generation. The long term commitment of a Filtration Management Contract permits the supplier to build his organization and resources to a level to maintain commitments; while greatly reducing the constant expansion and contraction of staff forced by the "whim" of capital spending. The supplier is able to maintain a trained staff to service the contract providing consistency to the partnership.

4. Because the incentive for the filtration equipment supplier is to maintain a continuous supply of quality coolant to the metalworking operations, he will give "constant" attention to the equipment. This daily monitoring will eliminate the "panic" emergency situations arising from equipment neglect. Also being responsible for all consumables and parts, the filtration supplier will have an incentive to keep such costs to a minimum to maximize the profitability of the contract. Because the filtration supplier is responsible for these items, it eliminates the spot purchases and/or inventory of such items by the metalworking manufacturer, reducing his costs.

5. As the result of a conservative buying attitude and the occasional

failure (or disaster) of a new filter product, it is often difficult for the filtration equipment supplier to introduce a new concept to the market. With the Filtration Management Contract, the supplier will be able to more rapidly bring new designs, concepts, and products to the field. Once a confidence level is established in the partnership, the metalworking manufacturer will be able to take advantage of the research and development efforts of his supplier/partner. With full access to the metalworking operations, the supplier will have the opportunity to take the ideas from the laboratory to the field more quickly, making such R & D efforts much more cost effective and justifiable.

6. For perhaps the first time ever, the supplier of the filtration equipment will now be forced to live with his own equipment. While not trying to be factitious, the supplier will now gain more factual data and information on the performance of operation of the equipment he has supplied. Quality filtration equipment suppliers desire this factual feedback on their equipment, but at present lack the total involvement necessary to ever receive all the facts. With the Filtration Management Contract, the supplier will be "operating" in a sense his own equipment, an experience which I am certain will be an educational one. Armed with this new bank of information, the supplier will be able to modify and enhance the design and subsequent performance of his equipment, better for the metalworking manufacturer and enhancing the supplier net compensation from the contract.

7. The saying among filtration equipment suppliers commonly is "there are no such thing as a customers, only purchase orders". Not to say that the equipment suppliers do not have well established customers and consistent users of their equipment, but even in the most "loyal" of metalworking manufacturing facilities, occasionally a new plant manager, engineer, or purchasing agent will decide to switch filtration suppliers. Lasting success can only be built upon mutual trust. The Filtration Management Contract will allow the filtration equipment supplier to demonstrate not only his best capabilities, but also loyalty and trust that a long-term relationship demands. The result is open, honest, and trusting communication between both the metalworking manufacturer and the filtration equipment supplier.

8. As a partner to the metalworking manufacturer, the filtration equipment supplier will increase his exposure to the production problems of the manufacturing processes. This increased exposure should provide the astute supplier the opportunity for potential integration into related peripheral equipment, expanding both his product line and market.

Developing a Filtration Management Contract

Having reviewed the potential advantages for both the metalworking manufacturer and the filtration equipment supplier, a closer look will give some insights into the development of the contract. Basically, the two types of Filtration Management Contracts to be examined are contracts for new program management and ones for existing equipment installation management. With over 6000 existing central coolant filtration systems presently in operation (500 GPM or greater), the latter Filtration Management Contract will be the most immediate area of savings to the metalworking manufacturer, and the place of opportunity for potential filtration equipment suppliers to prove their worth.

New Program Filtration Management Contracts

The most difficult task for the metalworking manufacturers in pursuing a Filtration Management Contract is selecting the right partner for the task. An evaluation of each potential supplier should be made with the assistance of an independent filtration consultant including:

1. The size and stability of the organization.
2. The reputation of performance of the filter equipment supplied.
3. The response of the supplier to problems.
4. The compatibility of each organization in terms of business philosophy, mission statement, and work ethics.
5. The manufacturing and research capabilities.
6. Supplier openness and willingness to communicate.

Selecting a partner is the most critical step in the process of setting a Filtration Management Contract in place. There are several fine suppliers of industrial filtration equipment in the marketplace. The metalworking manufacturer must select the one best suited to his needs. As a partnership, the filtration supplier must also select a metalworking facility that he believes fits with his organization and promises a successful contract for both.

The selection of the filtration supplier for the partnership should be made on the basis of the evaluation as outlined above, not upon a competitive bidding situation. The costs of the contract must be mutually beneficial and should therefore not become a deciding factor in the selection of a partner for the Filtration Management Contract.

Once the partnership has been formed, there should be a joint effort to build the Filtration Management Contract with the following activities:

1. Identification of the filtration requirements, including the system coolant flow, temperature, pressure, cleanliness level, chip loads and configuration, space limitations, and plans for future expansion.

2. The filtration equipment supplier should be prepared to supply all necessary equipment for the central coolant system, including the transport system interface, the central filtration equipment, and any auxiliary equipment such as temperature control, water treatment, tramp oil control, chip processing, high pressure coolant requirements, and polish filtration equipment.

3. A management program should be jointly developed to satisfy the "comfort level" of both partners. Such activities as on-site continuous management, periodic inspection, sampling and record keeping, meetings and progress reporting, and continuous improvement must be decided. The need for total and complete communication cannot be overemphasized.

4. The staffing requirements must be met by the filtration equipment supplier as agreed to by both partners. Union representation must be an integral part in developing the contract. In many cases, the union will be able to provide the on-site personnel to meet the staffing requirements for most activities.

5. The filtration equipment supplier is also responsible for all consumables and parts replacements required to maintain consistent performance. On site storage or just-in-time (JIT) delivery of these components must be part of the Filtration Management Contract.

6. Finally, there are several ways to structure the costs to the user (or compensation to the filtration supplier) depending on how the contract is written. The dirty coolant return system and the clean supply header system is normally the responsibility of the metalworking manufacturer. The filtration equipment supplier then is responsible for the delivery and installation of the filtration equipment, interfacing to the return systems, utilities, etc. Several options for remuneration are:

   a. Lease or rental of the equipment with a management fee based on the actual services provided.
   b. An annual fixed management fee.
   c. A variable annual management fee based on
      1. the amount of filtered coolant provided, gallons/shift
      2. the production rates, i.e. no. engines/hour
   d. Purchase of all equipment; with an annual service fee for parts and services provided.

Existing Central System Filtration Management Contracts

Unlike initiating a Filtration Management Contract for a new central system requirement, an existing central filter system has already demonstrated performance and conditions of operation which set the starting point for the contract. Regardless of the original manufacturer, the quality filtration equipment supplier should be able to enter into a Filtration Management Contract partnership on the existing central system to provide long term service to the customer. It is vital at this stage of the contract negotiations that open and honest communication on the actual performance and problems of an existing central system are clearly articulated to the potential filtration equipment supplier. Their should then be a joint effort to develop a Filtration Management Contract with the following sequence of events:

1. A joint study by the owner and the filtration equipment supplier perhaps utilizing and independent consultant should be conducted to verify the "facts" about performance, equipment condition, system requirements in terms of volume, pressure, etc. with complete documentation and agreement on the findings. As a base starting point, the filtration equipment supplier will then be in a position to make recommendations.
2. In order to improve the performance of the existing central system and meet the demands of the Filtration Management Contract, the filtration equipment supplier should then be able to recommend several options, including:
   a. retrofit and/or upgrade of the existing equipment
   b. rebuild, expansion, or replacement of the existing equipment
   c. the addition of auxiliary equipment such as by-pass filtration, tramp oil control, water treatment, coolant make-up, solids processing, and or modifications to controls and automatic monitoring.
3. As in the case of the Filtration Management Contract for a new central coolant system, the other details of the contract should be jointly agreed to as previously discussed.

Conclusion

In conclusion, the Filtration Management Contract is a concept whose time has come to meet the challenges of world competition. While the limitation of this discussion was of the central coolant filtration equipment, other capital equipment purchases such as industrial parts washers, parts handling, inspection, packaging, and even machine tool purchases could also benefit from the concept of the filtration management contract. Maintaining focus on what each respective company does best is a key ingredient to the success of the Filtration Management Contract. Working in an environment of mutual trust and openness is another. It is simply a smarter way to utilize technology, manpower, and capital to maintain a competitive edge in the metalworking marketplace.

If you are going to successfully introduce this new concept at your facility, be aware of the obstacles. Purchasing and plant engineering departments may feel threatened by the turning over of responsibilities and job activities to "outsiders", not to mention maintenance departments. Pride in achieving the companies goals of being the best low cost producer of quality components is needed in every employee; however it is also pride of job that could prevent a concept such as the Filtration Management Contract from becoming a reality. It will also take some filtration equipment suppliers the willingness to redefine themselves not as manufacturers and sellers of hardware but providers of service to the customer as never defined before. The challenge is there; the future potential tremendous to create a true "win-win" situation for two industries that need each other in order to survive.

GRAVITY SEPARATORS CUT THE COST OF CLEANING-UP LIQUIDS USED IN METALWORKING

Vincent Putiri
PhaSep, Inc.
1111 Jenkins Road
Gastonia, NC 28052

Introduction — Phase Separation by Gravity

A formal definition says, "gravity separators are passive devices that use natural gravity as the force to achieve separation of immiscible liquids, and solids and liquids". Gravity separation is what happens when muddy or oily water is allowed to stand. Eventually, the mud settles to the bottom and the oil floats to the top... and if there are other immiscible , or unmixable, liquids present, they will separate into layers also. (Figure 1)

This process also is called "phase separation", because we separate one or more phases of matter (liquid/solid/gas) from each other. So you can also compare gravity separators to a decanting system that simultaneously divides both lighter and heavier phases, either solid or liquid, from a process fluid. A properly configured separator acts as both a liquid/liquid and a solids/liquid separation system, and the separation takes place continuously-- not in batches, as in a decant vessel.

Gravity separators utilize the buoyancy (negative or positive) of the contaminants themselves. Differences in buoyancy remove these heavier or lighter particles or droplets from the process fluid-- such as the coolants and aqueous cleaning compounds used in the metalworking industry. The advantages of this type of separation over other forms of separation and filtration equipment include:

— Their relatively low cost to install and maintain.
— They do not need outside power, since they use gravity.
— They have no replaceable media.
— Maintenance normally consists of an occasional system cleanout.

Separation is Predictable-- There's a law!

The degree of separation of contaminants can be plotted by Stoke's Law, which predicts the terminal velocity of a rising or falling particle[1,5]. For example, the formula to predict the rise velocity of an oil droplet in water is:

$$V_r = g/18_\mu \ (S_w - S_o)\, D^2$$

where V_r = Separation Rate

g = Gravity Constant

μ = Viscosity of Continuous Phase

S_w = Specific gravity of water

S_o = Specific gravity of oil

D = Diameter of the oil droplet

From Stoke's Law it can be seen that relatively high rates of rise or fall can be achieved when :

- The viscosity of the process liquid is low, as is the case with water, water soluble coolants and aqueous cleaning solutions, and when--
- The diameter of the tramp oil droplet or particle is sufficiently large, and --
- The specific gravity differential is sufficient.

When To Use Gravity Separation...

In general, the gravity separation technique can be successfully employed when a specific gravity differential of .05 exists. As the differential increases, the diameter of the droplet that always can be removed will decrease. For example: With a specific gravity differential of .15 (water- 1.0 / oil- .85), oil droplets 20 microns and larger can always be removed economically with a properly designed system. Where very high specific gravity differentials exist, it is possible to remove particles down to below two microns in size.

It should be pointed out that, when the specific gravity differential and particle diameters are very small, the particles or droplets will remain in suspension. This factor can be used to advantage in applications such as coolant processing, because the coolant droplets pass through the system, while the rather coarse solids and free oil contaminants are removed.

The Benefits of Parallel Plate Separators

Gravity separators have significant advantages over straight decanting techniques. For one thing, the process takes place continuously as the liquid flows through the system... you don't have to rely on batch processing. This is possible since gravity separation will occur in a flowing liquid stream, provided laminar flow conditions are maintained. ("Laminar flow" is most easily described as smooth, non-turbulent flow.)

Most gravity separators include parallel plates in the separation chamber to further enhance the separation process. (Figure 2) The plates provide several benefits, including: A. Reduction of the distance the droplet or particle must rise or fall to reach an intercepting surface. B. Coalescence of tramp oil. C. Improved flow conditions.

Coolant Applications and Economics

The metalworking industry uses many water-based solutions whose effective life is shortened as contaminants build up while the liquids are in use. Water soluble oils used as coolants, as well as aqueous cleaning solutions, are particularly good candidates for cleanup using gravity separation techniques. In addition, removal of free and soluble oil from these solutions when they are spent is a necessary step prior to disposal.

Significant economic benefits from properly used gravity separators have been documented in both laboratory studies and in operating environments, reported in References which follow. The savings come from reduced disposal costs, extended chemical life, reduced tooling wear and improved quality of machined parts[2].

Aqueous Cleaning Systems: EPA Favored... Quick Pay Back

Recent studies by the Environmental Protection Agency specifically recommend that the less hazardous aqueous cleaning solutions be used in place of solvents, whenever possible. As a result, critical cleaning jobs once done by solvent cleaning now are performed by aqueous systems.

The gravity separator's unique ability to simultaneously remove oils and solid contaminants can be used to extend the effective life of these cleaning solutions. Improved cleaning efficiency results in reduced chemical costs, reduced disposal costs, improved surface finish on coated parts, and reduction in finish failure in the field.

Additional cost justifications for the use of separators in cleaning systems have come from users of immersion cleaning systems, multiple treatment cleaning and plating lines, and multiple stage coating operations.

Savings In Waste Pre-treatment

One final area where separators can effectively be used in the metalworking industry is in waste pre-treatment. Separators are being used in these sewage pre-treatment operations, either as stand-alone devices, or to remove oil and grease from water before membrane filtration or chemical treatment. The separators improve the efficiency of subsequent treatment systems by removing gross contaminants before they can foul membranes or consume chemicals used in downstream processing.

Summary...

Parallel plate separators do not answer every liquid cleaning need. But, for many purposes, separators provide a complete low-cost solution, and are often overlooked as a preliminary operation to reduce the load on the more expensive systems required for removal of soluble materials.

References:

1 — "Increased Machine Tool Productivity Resulting From Coalescing Plate Filtration", Turbo Conveyors, Inc., Gastonia, NC, 1982

2 — J. Joseph, Coolant Filtration, Joseph Marketing, 1985

3 — "Recycling cuts coolant costs 67%", Mod. Mach. Shop: (1989)

4 — E. Batutis, "Keep your cleaners clean", Prod. Fin.: 70-74 (Oct. 1989)

5 — C. Steiner, "A primer on separators and particle separation", Poll. Equip. News: 78-85 (June 1985)

Illustrations:

Figure 1 Typical separator.

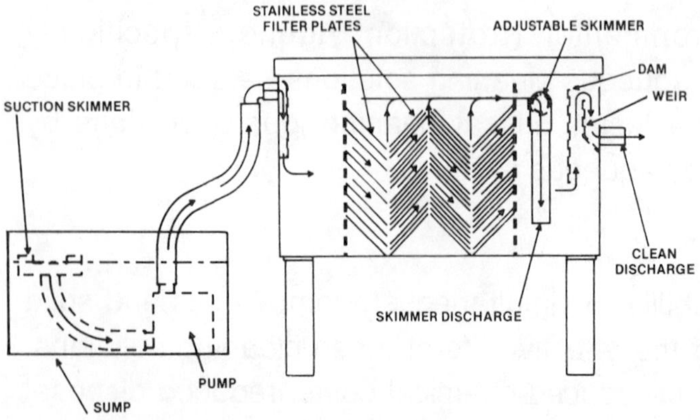

Figure 2 Coalescing plate stacks.

KEY WORDS ...

- separators
- gravity separators
-coolants
- metalworking
-Stoke's Law
- aqueous cleaning
- oil/water
- parallel plate
- coalescing
- liquid/solid

Filtration Awareness
At Saturn's New Automobile
Manufacturing Facility

Curt B. Stone
Powertrain Division
Saturn Corporation
P.O. Box 1500
Spring Hill, TN 37174

Introduction

In 1985, Saturn Corporation commenced a design and evaluation program for its manufacturing facilities to be located in Spring Hill, Tennessee. Taking a "clean sheet" approach to all aspects of this program, Saturn engineers attempted to take a comprehensive approach to the full spectrum of metal working operations. In the implementation of our plan, decisions were made on a number of operating parameters. This paper will attempt to outline actions taken,relative to providing adequate filtration systems, in bringing a quality powertrain on line.

LOST FOAM CASTING

Perhaps one of the biggest innovations in our Manufacturing Operations is the Lost Foam Casting Process. This process will be used on a high volume basis to produce various powertrain components. Patterns are made of expanded polystyrene beads and conform to very intricate shapes. A series of patterns are then assembled in a final form. Once immersed in a flask, molten aluminum displaces the foam (which vaporizes) and forms the desired shape.

This casting methodology changes conventional chip load calculations due to "near net" finish, the corresponding reduction in machining, and definitely influences decisions on filtration systems.

CHEMICAL MANAGEMENT

Metalcutting fluids, and their effect on the total machining/filtration process, have been a key consideration from the start. Through a chemical management program, all fluid applications have been evaluated on the basis of numerous criteria. The usual high proliferation of fluids has been kept to a minimum and the possibility of cross contamination has also been monitored in both the design and start-up phases.In the powertrain complex, the scope of this program ranges from coolants, to die release compounds, to floor cleaners.

With the program currently in place, on-site resources will be readily available to provide qualified technical involvement and in-house testing of fluids. Quick response, better accuracy, and isolation of filtration problems are anticipated with this arrangement.

The benefits of the chemical management program are numerous and definitely cost effective. Projected cost savings are estimated to be between 10 to 18%, when compared to tangible benchmarks. Of particular importance however, is the enhanced control we have of the M.S.D.S. (Material Safety Data Sheets) and M.A.R.S. (Material Assessment Request Sheet) process material <u>before</u> chemicals enter the building. Almost all of our metalcutting fluids are semi synthetic.

FILTRATION SYSTEMS

A comprehensive evaluation was made of coolant filtration equipment serving the automotive industry, with the eventual selection of one firm acting as "A Primary Supplier".

This is a departure from conventional sourcing in that an evaluation team of users, purchasing, and, financial people did a thorough analysis of each of the potential suppliers. The evaluation criteria included engineering capabilities, financial stability, delivery, service, and advances in technology. Cost estimates were prepared based on conceptual data, and then the final selection was made. The selected firm, Henry Corporation, had total responsibility in supplying filtration hardware, regardless of manufacturer. The final agreement covers a warranty period of 36 months and is contingent on 95% uptime (for a designated period of time).

Standardization of design and components have been the obvious benefits. Involvement in our process almost from the start, however, has been particularly helpful in the building design, sizing of systems requirements concurrent with machine line design changes, and having a consistent approach to problems that arise in the field installation/start-up-phases.

In total, there are over 30 filtration systems in place, ranging in size from 5 GPM to 12000 GPM, designed for clarity levels as clean as 5 parts per million. The types of filtration systems include wedgewire drum, precoat, cartridge, settling, and flatbed with disposable media.

WASHER FILTRATION

Part cleanliness is very critical to engine and transmission performance and the potential of warranty claims. In regard to parts washers, Saturn recognized the importance of minimizing sedimentation and spent many hours to properly evaluate design, specifications, and suppliers capabilities. Two suppliers were selected based on their ability to meet our specifications and long term viability to remain our partner. In total, we have 28 washers and their filtration was designed to protect and extend the life of washing fluid. The surface area has generally been increased, and by making the filter the main reservoir of the washer, we anticipate minimizing the settlement of solids. This has increased the floor space needs, but the operational/quality improvements should more than offset this penalty.

ENVIRONMENTAL PURSUIT

Saturn has been recognized for its continuing efforts to protect the environment, both above and below ground. Our aggressive environmental posture has been based on the opinion that just attaining compliance with current laws probably isn't satisfactory, and that we should anticipate resolution of some known historical problem areas. The establishment of a local citizens board for environmental concerns, developing baseline measurements of air/water quality, and double containment of all underground process piping have been some of the more notable accomplishments.

BUILDING LAYOUT

The plant layout process was driven by relationship diagrams that were established between all functions in the building. After basic relationship networks were configured, optimization of alternatives was driven by a comparison of total travel distances versus volume intensities. Normal modifying considerations include architectural or organizational parameters, although in our case, we were faced with the preference for gallery concepts instead of individual pits, and a combination of long transfer lines and smaller machine center groupings.

FILTRATION GALLERIES

The Spring Hill site presented a fairly difficult topography due to a limited amount of top soil above a solid limestone base. This condition existed across the site.

After considerable cut and fill work, a level platform was developed over one mile long, and a half mile in depth, on which all major buildings were constructed. In total, approximately 3.3 million cubic yards of rock and 2.8 million cubic yards of top soil were moved for site preparation. Major areas within Powertrain had to be blasted away, in a stepped fashion, to accommodate sloped flume runs (400 + feet). The longer flume runs did affect the depth of the gallery but this overall design was preferred due to uninterrupted machine configuration, structural design and enhanced mist control. Invert elevations are about 12 feet at the gallery wall. Shorter, but more complex flumes, were designed to use conventional drag line conveyor and shallower depth.

Two galleries were required, both of major proportions. The larger contains 14 filtration systems, chip processing, and pretreatment, waste water facilities. Its measurements are 55′ wide, 585′ long and 32′ deep. The second gallery, associated with rotational machining, has three filtration systems along with associated chip processing equipment. Its measurements are 55′ wide, 145′ long and 28′ deep.

All flumes and underground pits/galleries have been double contained using a 60 mil high density polyethylene liner, with geotextile material and polyethylene netting for fluid migration. Inspection wells are strategically located to monitor below grade seepage on a regular basis, with convenient accessibility for submersible pumps.

A number of operational and maintenance considerations led us to the gallery configurations. All major filtration systems are aligned together, allowing for immediate access to controls, equipment, and maintenance support. Chip collection and processing has been neatly packaged, and are located close to a large elevator in case of a major breakdown.

In spite of the longer, deeper flume runs, we chose to place the main gallery at the rough end of machining. The other option considered was placement of the gallery in the middle of the line. This could only be done by use of a cantilevered machine base, or adding more space to each line. Our manufacturing engineers were concerned over the structural integrity of the base design. Flumes entering both walls of the gallery would also constrain good maintenance accessibility. The merits of this decision have been reinforced by a considerable reduction in chip handling conveyor to the chip dock location. Bottom line, we anticipate that the operational benefits will more than offset the slightly higher costs of the gallery concept.

TRAINING

Training has been a critical part of our ramp up since we have new employees, in a new environment, using new equipment. Our basic approach has been to provide training on a "needs driven" basis. The support of our fluids and filtration suppliers has been exemplary. The preparation of maintenance manuals has been well thought out, logically formatted, and very well documented. In addition, the "hands-on" training being done in the field has raised the confidence of our whole team.

SUMMARY

We would like to think that the merits of good filtration have been recognized in our planning effort, and early enough to really impact our process development. This has been a rather broad overview, but it is difficult to discuss our project in small terms.

Will the ultimate customer ever appreciate what steps we've taken in engineering filtration equipment? Not directly, but it will be reflected in the purchase of a cost competitive, high quality vehicle.

Market introduction is planned for the fall of this year, and will migrate across the country from the west coast. We will offer a family of small, sporty, high value cars, designed to appeal to buyers in the compact and subcompact segments of the market. Saturn will compete directly against Toyota, Honda and Nissan in those market segments. First models will be 2-door and 4-door pricing will start at under $10,000 to about a $12,000 range.

We look forward to seeing you in the show room.

KEY WORDS

SATURN CORPORATION

LOST FOAM CASTING

CHEMICAL MANAGEMENT

PRIMARY SUPPLIER

DOUBLE CONTAINMENT

POWERTRAIN

MANUFACTURING FACILITY

METALCUTTING FLUIDS

SEDIMENTAION

GALLERY

MACHINE TOOL COOLANT MANAGEMENT SYSTEM OFFERS FILTRATION, FLEXIBILITY, AND CONTROL

Thomas P. Cassese, Executive Vice President
Hydroflow Incorporated
310 Tracy Road
Chelmsford, MA 01824

Coolant management systems for the machine tool industry can provide high levels of solids filtration and coolant conditioning functions that increase productivity and minimize costs. This paper offers specifications for the individual pieces of equipment that make up a typical coolant management system and provides explanations of their purpose and operation.

A fully automated coolant management system operating at a ball bearing manufacturer since 1987 is used as an example of how a coolant management system can be successfully designed, constructed, and installed. The system removes solid contaminants with self-cleaning barrier and diatomaceous filters from water soluble coolants, synthetic coolants, and oil. Tramp oils, which would otherwise cause the coolant to turn rancid, are removed from water soluble coolant using a high speed centrifuge. Coolant temperature is continuously controlled and the replenishment of fresh coolant is regulated by the coolant management system. Computerized controls integrate the requirements of eight distinct zones of machine tools into a single operation that provides higher production rates, reduced scrap rates, and minimized operation costs.

Coolant management systems in the machine tool industry offer more than the filtration of solid contaminants from liquids. Coolant systems can be designed to remove tramp oils that, when combined with coolant conditioning, can maximize coolant life and minimize operating costs. Automatic coolant temperature control can be integrated into a system to provide increased productivity and reduced scrap rate. Flexibility can be designed into a coolant management system, thereby providing the production flexibility necessary to keep pace with changing market forces.

Components of a Coolant Management System

A positive-barrier type filter (vacuum filter), in which liquid is drawn through filter medium weighing 1.5 to 2.5 oz per square ft, is recommended for the removal of solids from machine tool coolants. With a barrier-type filter, filtration with 99% of all solids removed to 5 microns (less than 10 ppm solids content) can be achieved without regard to the specific gravity of the solids being removed. A self-cleaning positive barrier-type filter offers the additional advantage of providing a continuous flow of clean coolant without having to stop production and almost dry discharge of solids for ease of disposal. A precoat filter may be required to remove solid contaminants down to 5 microns from grinding or cutting oils.

After filtration of solids from the machine tool coolant, tramp oil must be removed from the coolant to ensure that the maximum life of the coolant is obtained and to minimize the high cost of disposing spent coolant. Tramp oil reduces the life of coolant by promoting bacteria growth, thereby causing rancidity and dermatologic problems. A high speed centrifuge designed for liquid/liquid separation can be used for the removal of both free and mechanically emulsified tramp oils. Depending on the application requirements, a coalescing filter or one of several types of belt filters can be employed to remove tramp oils.

Fresh coolant must be added to a system to make up for coolant lost to evaporation, spillage, and coating of parts. Coolant concentrate must be mixed with deionized water to ensure the maximum coolant life. An automatic make-up proportioner is used to meter the proper amount of coolant into the supply water to maintain a fixed coolant concentration level.

Close tolerance work often demands that coolant be either heated or cooled prior to its being pumped back to the machine tool. A coolant management system should provide automatic monitoring and adjustment of coolant temperature to maintain it within 2°F of ambient conditions.

A Coolant Management System Case History

A Hydroflow designed and manufactured machine tool coolant management system has been operating at a leading manufacturer of precision bearings in New England since March 1987. The system offers complete filtration and coolant conditioning functions that are monitored and operated by a single computerized control system.

System Needs and Design

The ball bearing producer needed a centralized coolant management system that could handle the requirements of a variety grinders and several types of coolants, including water soluble coolant, synthetic coolant, and grinding oils. High filtration rates were required to obtain diameter specifications within 1 to 2 ten-thousandths inch, and surface smoothnesses to 2 CLA or better on a variety of tool steel and stainless steels. Careful coolant temperature control and adjustment was needed; coolant temperature had to be maintained within 2°F of the ambient temperature conditions while temperatures at the machine tools could vary up to 30°F in a single production shift.

A zone concept was used in the design of the coolant management system to obtain the advantages of a centralized coolant management system without sacrificing flexibility or jeopardizing continuous operation. Each of the eight zones was designed as a separate entity, each with its own needs for filtration and coolant conditioning. A master control system was used to tie the operation of each zone into the functioning of the system and to oversee coolant conditioning requirements, zone by zone, for coolant temperature control, tramp oil removal, and coolant make-up and mixing.

Filtration Equipment

Seven of the eight zones are served by self-cleaning barrier filters containing up to 24 square ft of filter media and capable of handling up to 250 gpm of water and soluble and synthetic coolants. Filtration down to 10 microns is achieved by drawing liquid from a contaminated coolant tank through a non-woven synthetic fabric filter medium supported by a perforated steel plate. As contaminants become trapped on the filter medium, a filter cake that traps additional solids is formed.

A precoat filter removes all solids down to one-half micron from oil at a flow of up to 150 gallons per min for the super finishing machine tools. Contaminated oil is pumped through diatomaceous earth, which forms a filter cake around a series of perforated stainless steel tubes within the precoat filter.

Each filter automatically enters a regeneration cycle as the pressure drop across the unit reaches a predetermined level. During regeneration of a vacuum filter, a dragout conveyor positioned between the medium and its supporting plate advances a short distance, bringing with it new fabric medium. Spent filter medium is automatically discharged from the filter and collected in a sludge hopper. When the precoat filter requires contaminated filter cake and a new cake of diatomaceous is formed. A continuous supply of clean coolant from each filter's clean coolant holding tank allows continuous production during regeneration.

As part of the zone concept, clean coolant flow can be directed from one zone filter to any desired coolant use zone to allow filter servicing without interrupting production.

Coolant Conditioning Equipment

The coolant management system is designed to monitor and control coolant temperature, remove tramp oils, and control coolant concentration levels. An air-cooled chiller is used for cooling requirements. Temperatures of ambient conditions and clean coolant are monitored individually at each zone and adjustments to the temperature of the coolant flowing to a particular zone are automatically performed.

A high-speed centrifuge is used to remove tramp oils, fine solids, and bacteria. Because tramp oil contamination levels vary from zone to zone, several zones need tramp oil removal processing only once a week, while other zones require tramp oil removal on a daily basis. Coolant concentration levels and addition of deionized make-up water in each zone are carefully monitored and adjustments are easily made.

System Control

The operation of the coolant management system is fully automated by a single computerized control that includes a video display screen offering system status reports in graphic form. Every facet of coolant flow and condition and equipment operation is monitored and controlled, zone by zone, as a integrated system.

Benefits of a Coolant Management System

The ball bearing producer has realized many benefits from using a coolant management system. Smoother finishes and reduced scrap rates have resulted from using clean coolant. Grinding to extremely close tolerances has become easier with less scrap since coolant temperature is now strictly maintained. Tramp oil removal has eliminated the cause of coolant rancidity and automatic coolant makeup and concentration level control have improved production. Disposal costs have been minimized because the nearly dry sludge from the filters is conveniently handled and more solids can be packed into 55-gal drums for disposal. The zone concept ensures that the coolant management system maintains the flexibility necessary to meet current production requirements and is prepared to take advantage of future opportunities.

Key words

MACHINE TOOL COOLANT MANAGEMENT SYSTEM
OFFERS FILTRATION, FLEXIBILITY, AND CONTROL

bacteria
barrier filter
coolant proportioner
coolant temperature
deionized water
machine tool coolant
precoat filter
tramp oil
vacuum filter

VELOCITY PROFILE MEASUREMENTS
WITHIN A LABORATORY FLOTATION CELL

C. E. Jordan, Supervisory Metallurgist
and B. J. Scheiner, Supervisory Metallurgist
U. S. Bureau of Mines
Tuscaloosa Research Center
P. O. Box L, University of Alabama Campus
Tuscaloosa, AL 35486

The Bureau of Mines used an electrochemical technique to measure the velocity profile within a laboratory flotation cell. A small dual electrode probe was inserted into a fluid containing ferric and ferrous cyanide. With a -300 mV potential at the current limiting cathode, the transport of ferric cyanide ions to this cathode controlled the electrode current. The fluid velocity moving past this small cathode surface (0.001 mm^2) was directly related to the ferric cyanide ion diffusion rate and the electrode current. After calibrating the electrode with known fluid velocities, the current measurements of the electrode probe instantaneously measured the fluid velocity at the cathode surface. Using a computer for automated data acquisition, the velocity profile was mapped throughout the laboratory flotation cell at different impeller rotation speeds. From the velocity profile, the distribution of the dissipation energy was computed and correlated with the flotation response of a copper ore.

Introduction

Froth flotation is an effective, low cost method of ore beneficiation. The process works well on mineral particles in the size range of 150 to 10 μm. To increase the effectiveness of conventional froth flotation, the mechanism for bubble-particle collision and bubble-particle attachment needs to be understood. In conventional flotation, the probability of collision between the particles and bubbles is fairly small[1-3]. To overcome the slow flotation kinetics and to obtain adequate recovery of the mineral product, the residence time in the flotation cell is increased. However, this always results in a lower grade concentrate. As part of the Bureau of Mines's program to improve the technology required for a strong domestic minerals industry, the hydrodynamics of flotation were investigated for a laboratory flotation cell to discover how to improve the flotation kinetics of fine size particles. Hydrodynamic theory indicated that increased agitation increased the bubble-particle collision rate which also improved the flotation recovery of fine size particles[1-3]. In a conventional flotation cell, the agitation is provided by the rotating impeller. The degree of agitation with the flotation cell is related to the dissipation energy of the turbulent fluid flow[4]. By measuring the fluid velocities and their fluctuations with time, the degree of agitation (dissipation energy) within the flotation cell can be determined.

Many methods, techniques, and instruments have been developed and used to measure fluid velocities in turbulent flow. Among these are hot wire anemometer, electrochemical, flow visualization, image analysis by photo-diode arrays, and laser-doppler measurements[3,5]. The electrochemical fluid velocity measurement technique provides a quick response with reasonable accuracy and should still be effective in the presence of air-bubbles and solids. For the electrochemical technique, the diffusion current is a function of the fluid velocity in the immediate vicinity of the cathode[5]. By limiting the exposed surface of the cathode, the electrochemical reactions at the cathode will control the current through the electrode. As the fluid velocity increases, the electrode current increases because the reactive ions diffuse faster to and from the cathode surface. With known fluid flows, the corresponding electrode current measurements can be calibrated. While the fluid velocity is actually a vector with magnitude and direction, this electrochemical measurement only measures the magnitude of velocity. The direction of the fluid is unknown. However, in a turbulent flow system, the fluid flow direction is fairly random due to the eddy currents. Using this technique, the magnitude of the fluid velocity in a laboratory flotation cell was measured and the dissipation energy was determined throughout the flotation cell.

Velocity Measurements

Apparatus

The dual electrode probe was constructed with a small platinum wire cathode sealed at the tip of a 3-mm-diam glass tube. Only the 0.001 mm^2 cross-section of the wire was exposed to the electrolyte solution. As shown in figure 1, a larger platinum anode was formed by wrapping platinum wire around the glass tube about 3 mm from the cathode, near the tip of the glass tube. With this configuration the cathode surface area was less than 1 pct of the anode surface. The electrolyte contained 0.001 molar potassium ferricyanide, 0.001 molar potassium ferrocyanide, and 0.2 molar NaOH. The electrochemical reactions are

$$Fe(CN)_6^{-3} + e^{-1} = Fe(CN)_6^{-4} \text{ at the cathode}$$

$$Fe(CN)_6^{-4} = Fe(CN)_6^{-3} + e^{-1} \text{ at the anode.}$$

When a -300 mV potential was applied to the cathode, the electrode current was limited by the diffusion rate of the ferricyanide ions to the small cathode surface. The electrode current was converted to voltage through a resistor and both the applied emf and current converted voltage were measured. The two measurements were linked to a computer through an analog to digital converter to record the data.

Procedure

The electrode was calibrated by pumping known volumes of the electrolyte through a 3-mm-diam tube that was directed at the small cathode surface of the probe. The fluid velocity was calculated from the flow rate and the cross-sectional area of the tube. The probe measured the limiting cathode current at each known fluid flow. To ensure accuracy, the electrode was calibrated before and after the experimental velocity measurements. Figure 2 shows a typical calibration curve. Changes in the electrolyte could be detected by comparing the pre- and post-measurement calibration curves. To minimize the changes in the electrolyte, velocity measurements were conducted only for about 1 h and then the electrolyte solution was replaced. The literature reports[5] that the electrode limiting current is a linear function of the square root of the fluid velocity. For the higher fluid velocities this was true; however, for the lower fluid velocities the electrode current proved to be less than what was expected from the linear relationship. Therefore, the calibration curve was used to determine the fluid velocity in all of the experimental measurements.

The electrolyte solution was placed in a standard Denver DR flotation cell. (Reference to specific products does not imply endorsement by the Bureau of Mines.) The electrode probe was clamped into each position as shown in figure 3. Measurements were taken at 2-cm intervals vertically and 1-cm intervals horizontally. At each position, 100 measurements were taken at 0.1-s intervals. The entire process, calibrations, measurements, calculations, and data recording, was supervised by computer.

Flotation

Flotation tests were conducted using a western porphyry copper ore from Arizona. The ore was ground below 38 μm size to study the flotation of fine particles. Virtually all of the particles were between 1 and 38 μm size. The ground ore was conditioned for 5 min at pH 12.4 with potassium amyl xanthate. After conditioning the slurry was diluted to 25 pct solids and floated for 5 min to recover the copper mineral.

Results and Discussion

Velocity measurements were made at 1,000, 1,500, 2,000, and 2,500 r/min in the laboratory flotation cell. Air was pulled into the flotation cell by the action of the impeller. Initially only the electrolyte and air were used in the flotation cell. Figure 4 show the typical fluid velocities observed over a 10-s monitoring period at 1,000 r/min and 4 in from the impeller. The mean fluid velocity with time was calculated for each position. The standard deviation of the velocity measurements at each position was a convenient measurement of the turbulent velocity fluctuations. The distribution of the fluid velocity fluctuations around the mean velocity was calculated as a function of the standard deviation from the mean velocity. Typical distributions of the fluid velocity fluctuations are shown in figure 5. Four different positions in the flotation cell are shown. Two of the positions had a mean velocity of 60 cm/s and the other two positions had a mean velocity of 10 cm/s. The fluid velocity fluctuations were distributed almost normally and 95 pct of the fluid velocity fluctuations were within two standard deviations of the mean. The standard deviation varied with each position, but ranged from 30 to 90 pct of the mean fluid velocity.

The velocity profile of the laboratory flotation cell was obtained by contouring the discrete mean velocity measurements. As shown in figure 6, the mean fluid velocity was roughly 10 times higher near the impeller than that of the lower velocity portions of the flotation cell. The effects of increasing the impeller speed are shown in figure 7. As the revolutions per minute of the impeller increased the fluid velocity increased proportionally, such that mean velocity squared was proportional to the impeller revolutions per minute cubed. This relationship was also reported in the literature[6-7].

Several measurements were made with ground quartz added to the electrolyte. Quartz did not react with the electrolyte or influence the electrochemical reaction. Both the calibration and experimental measurements were conducted with the quartz-electrolyte slurry. Figure 8 shows the velocity profile of a 12-pct solids quartz slurry in the laboratory flotation cell. The presence of the quartz solids dampened the fluid velocity about 15 pct throughout the flotation cell.

Near the impeller, the fluid velocity with the quartz was about 85 pct of the fluid velocity without quartz, but as the slurry moved farther away from the impeller the mean fluid velocity with quartz was nearly the same as without quartz.

Using Kolmogoroff's theory of uniform isotropic turbulence, Azbel and Liapis have shown that the dissipation energy within the laboratory flotation cell is proportional to the cube of the instantaneous fluid velocity[4]. As shown in figure 5, the fluctuating instantaneous fluid velocities were in a nearly normal distribution around the mean fluid velocity and could be approximated by the mean fluid velocity. Therefore, the dissipation energy at each position in the flotation cell was assumed to be proportional to the mean fluid velocity cubed. The total power consumption (dissipation energy) in the flotation cell was measured by linking a torque meter to the impeller shaft. To account for the bearing friction, the power consumption of the impeller spinning freely in air was subtracted from the power measurements with the flotation pulp. Table I shows the power consumption at the four impeller speeds. Dividing the power consumption by the volume of the flotation cell produced the mean dissipation energy. However, each unit volume dissipates a portion of the total power consumption that is proportional to its mean velocity cubed. Most of the dissipation energy occurred near the bottom of the flotation cell as shown in figures 9 and 10. The intensity of the dissipation energy in the upper portions of the flotation cell was much smaller than the lower regions. For effective bubble-particle attachment in froth flotation, the regions of high dissipation energy have a high probability of bubble-particle attachment. The capacity of this region determines how fast the flotation will take place. As shown in figure 10, at 2,500 r/min the region of high dissipation energy is much larger than the flotation cell at 1,000 r/min shown in figure 9. Increasing the volume of the effective dissipation energy will improve the flotation response. As shown in figure 11, the recovery of copper from a porphyry copper ore improved significantly by increasing the region of effective dissipation energy for these fine size particles. More effective agitation throughout the flotation cell will improve the attachment rate. However, this may hinder the formation of a quiescent region that's required at the top surface for bubble-pulp separation. One solution to this problem is discrete flotation unit operations, i.e., one unit for bubble-particle attachment with uniformly effective agitation followed by another unit operation for bubble-pulp separation with a quiescent region for bubble-pulp separation. Fluid velocity measurements by this electrochemical technique have been valuable in pointing out effectiveness of agitation and its limitations in the conventional laboratory flotation cell. This velocity measurement technique is an effective tool in the study of turbulent fluid systems.

References

1. Sutherland, K. L., and I. W. Wark, "Principles of Flotation," Australasian Institute of Mining and Metallurgy, Inc., 1955, pp. 347-378.

2. Trahar, W. J., and L. J. Warren, "The Floatability of Very Fine Particles - A Review," Int. Journal of Mineral Processing, Vol. 3, 1976, pp. 103-131.

3. Nonaka, M., T. Inoue, and T. Imaizumi, "A Micro-Hydrodynamic Flotation Model and its Application to the Flotation Process," XIV Intern. Miner. Proc. Congr., Toronto, Canada, 1982, Vol. III, pp. 9.1-9.19.

4. Azbel, D., and A. I. Liapis, "Motion of Solid Particles in a Liquid Medium," Handbook of Fluids in Motion, 1983, pp. 895-926.

5. Mizushina, T., "The Electrochemical Method in Transport Phenomena," Advance Heat Transfer, Vol. 7, 1971, pp. 87-161.

6. Bennett, C. O., and J. E. Myers, "Momentum, Heat, and Mass Transfer," McGraw-Hill Book Co., Inc., 1962.

7. Jones, W. P., and B. E. Launder, "International Journal of Heat and Mass Transfer," Vol. 16, 1973, 119 pp.

TABLE I.--Power consumption in the laboratory flotation cell

Impeller speed, r/min	Power, watts
1,000	9
1,500	28
2,000	64
2,500	124

Destabilization of Fine Particle Suspensions by Polymeric Flocculants

R. Hogg
Mineral Processing Section
115 Mineral Sciences Building
University Park, PA 16802

D. T. Ray
Department of Mineral Engineering
National Cheng-kung University
Taiwain, 700, Republic of China

L. Lundberg
Tin Mill Quality Assurance Department
Bethlehem Steel
Sparrows Point Plant
Sparrows Point, MD 21219

ABSTRACT

High molecular weight polymers can be extremely effective in promoting floc growth in relatively unstable fine-particle suspensions, but perform rather poorly when the suspension is stable and well dispersed. At sufficiently high dosage, however, these polymers can serve to destabilize suspensions, leading eventually to good flocculation. The role of high molecular weight polymers in the destabilization of suspensions has been investigated; the results provide some useful insights into the mechanisms and dynamics of bridging flocculation.

Experimentally, it is observed that as polymer is added to stable dispersions some large flocs appear but the suspension remains highly turbid. Floc size measurements reveal that bimodal distributions are established and that, as more polymer is added, the coarser mode is subject to rapid growth and shifts to progressively larger sizes while the finer mode remains at about the same size but is gradually depleted by the disappearance of fine particles. The experimental results are interpreted using an approximate model for simultaneous polymer adsorption and floc growth.

AC ELECTRO-COAGULATION PROCESS

B.K. Parekh, J.G. Groppo and J. Justice
Center for Applied Energy Research
University of Kentucky
3572 Iron Works Pike
Lexington, KY 40511

This paper describes application of a novel process where alternating current is utilized for effective coagulation of stable ultra-fine solids suspension. The principles of electrostriction (charge neutralization) and flocculation (polymeric species of metal ions) facilitate rapid coagulation and improved dewatering of stable suspension.

The process is very simple and can be augmented in existing plants with ease. It can also be used as a stand-alone process for "on-site" treatment step. The coagulator has been operated in both continuous and batch flow models in flow rates ranging from 1 to 750 gpm.

This paper provides overview of the process, its application to coal mine and preparation plant discharge streams and provides preliminary economic data.

"ABOVE-GROUND HAZARDOUS FILTRATION FOR DEEP-WELL INJECTION"

Dr. Ernest Mayer
E. I. du Pont de Nemours & Co.
Engineering Department
P.O. Box 6090
Newark, DE 19714-6090

and

Dr. Reza Hashemi
Pall Corporation
2200 Northern Blvd
East Hills, NY 11548

ABSTRACT

One of the key disposal techniques for hazardous liquid wastes is deep-well injection. Many injection wells are into tight strata so fine, micron filtration is needed to prevent down-hole plugging. Cartridge filters are typically used just prior to injection, to provide absolute particle removal. However, they are quite costly, especially when solids loading is high. As a consequence, coarse prefilters are used upstream. This paper describes various prefilter types and their general performance. Particular emphasis will be placed on renewable surface filters that are fully enclosed to prevent exposure and dry cake filters that generate dry cakes for secure landfilling or incineration.

INTRODUCTION

Deep-well injection for disposal of hazardous wastes is in selected cases increasingly more cost-effective than landfilling or incineration. The Environmental Protection Agency (EPA) has adopted strict regulations governing the use of this technology because of the potential for contamination of usable groundwater aquifers. These regulations require demonstrated containment and no migration outside the injection zone (Ref 1). The EPA as well as various state agencies are increasingly requiring that industry demonstrate no threat to underground water sources (Ref 2). Within Du Pont, extensive research has been conducted to assure safe injection practices.

Existing hazardous wells, as well as those used for oil industry enhanced oil recovery (EOR) processes (Refs 3,4), encounter serious fouling problems which result in expensive workovers. The primary foulant is usually solids which are inefficiently removed by the above-ground filtration system. This inefficiency occurs because cartridge filters are primarily used and production personnel usually balance life and costs and sometimes overlook solids carry through to the well. To further complicate this situation, cartridge filter vendors aggressively market their products with little regard to what type cartridge is most suitable, or if cartridges are appropriate for the application, eg, Refs 4,5,6,7. As a consequence, cartridge filters are almost exclusively used for well protection and typically they are unreliable because of high solids loading and poor seals. Furthermore, cartridges are becoming more difficult to dispose of due to recent RCRA hazardous waste regulations. This paper examines new techniques to improve upstream solids removal so final cartridge life is extended, and the use of renewable surface filters to replace cartridges (Ref 8).

TRADITIONAL TREATMENT

The traditional (or conventional) above-ground treatment filtration system usually consists of a bank of string-wound cartridges in parallel for continuous injection. These cartridges are at best 60-70% efficient (Ref 9), are difficult to seal so bypassing routinely occurs, have low dirt holding capacity, and can unload dirt due to pressure or flow disturbances. As a first attempt at upgrading, these string-wound cartridges should be replaced with a quality, absolute-rated cartridge that can be sealed properly. The examples cited here will show where this approach has been successfully applied.

If the solids loading in the waste being injected is too high for cartridges, then precoated pressure leaf filters are conventionally used upstream to reduce the dirt load. These filters suffer from two distinct disadvantages: solids bleed-through due to pressure/flow upsets and a sluicing type backwash which can no longer be easily disposed of due to RCRA land-ban regulations.

CASE HISTORIES

Acidic Waste Injection

This waste contained about 20 ppm total suspended solids (TSS) in a dilute acid stream. This TSS load is typically too high for cartridges so dual pressure leaf filters were used upstream of a standard bank of 1 micron string-wound cartridges. Costly well reworks were normally done because this above-ground filtration system was unable to protect the nominal 2 micron formation (as determined by core sampling). The first upgrade done was to replace the string-wounds with Pall 3 micron absolute Ultipor cartridges with soft-sponge seals so the existing housing cups would seal better. As a result, well reworks were practically eliminated; as an added benefit, cartridge life was extended almost two weeks. In addition, an Oberlin automatic pressure filter (APF), (Ref 8) was installed to produce dry cakes from the pressure leaf sluicings that could be readily incinerated. Pall well-guard filters are also being installed for final well protection.

Organic Waste

This waste stream contained typically about 300 ppm TSS (with excursions to about 1500 ppm) and about 200-1000 ppm oil. Primary pressure leaf filters followed by precoated tubular-bag secondary filters reduced TSS levels to only about 150 ppm. Excessive well rework costs exceeded $1 Million annually. Sluicings from both these filters were returned to an upstream clarifier that required annual cleanout at a cost of about $0.5 Million.

The first upgrade occurred when the precoat recipe was changed from 50/50 diatomaceous earth (DE)/coarse cellulose (for oil absorption) to 93/7. This change tightened the precoat so TSS were reduced to about 90 ppm. This still was not low enough for adequate well protection so further upgrades were necessary. The primary pressure leaves were outfitted with 3 micron Goretex bags which reduced TSS to less than 20 ppm. The secondary filters were eliminated and well rework costs were reduced dramatically. An added benefit was lower maintenance costs because the Goretex bags prevented the leaves from fouling with oil. In addition, another Oberlin APF system was installed to dewater all sludges/sluicings and produce dry cakes quite suitable for incineration. This improvement eliminated the annual cleanouts at cost savings of $0.5 million.

Cyanidic Waste

This waste stream contained about 100 ppm TSS so combined pressure-leaf filters and cartridges were used. Well core sampling revealed a 10 micron requirement. The plant used 10 micron string-wound cartridges which are really about 50 micron absolute. As a consequence, the first upgrade involved replacement with Pall Profile absolute-rated cartridges of similar rating.

The Pall cartridges exhibited 5X longer life at similar effluent quality but they plugged in about 15 minutes during plant upsets. Costs became prohibitive. Commercial's new 1 micron XTL cartridges were substituted as a compromise between cost and well protection until a renewable surface filter could be installed. XTL cartridge costs still exceeded $200 Thousand annually and disposal was troublesome. Thus, a Mott 2 micron IPHP sintered metal filter (Ref 8) was installed to replace the cartridges. This dual Mott filter system resulted in excellent filtrate quality (<10 ppm TSS) and four-day on-stream life. However, as soon as an upset occurred, the Mott tubes blinded within an hour requiring shutdown and insitu cleaning. A solution to this problem is being developed. A large membrane filter press (Ref 10) has been installed to dewater all sluicings/sludges from this entire waste area. This press can easily produce dry cakes that pass the new 50 psig unconfined EPA "Paint Filter Test" (Ref 10).

Low-Solids Content Waste-Well Stream

This waste stream required 1 micron absolute filtration for adequate well protection. Despite only 4 ppm TSS loading, cartridge replacement for the three-stage 1 micron string-wound/10 micron melt-blown PP bags/ 1 micron absoluted pleated cartridge system cost about a $1 Million. Incineration disposal costs were about $600/single cartridge change. Thus, significant economic stake existed to upgrade this filtration system.

Initial attempts to replace the costly 1 micron absolute final filters were unsuccessful because of well injectivity losses. It was concluded that the inefficient, 1 micron string-wounds should be replaced with a low-cost, absolute-rated cartridge to increase the life of the absolute final filters. Extensive long-term testing demonstrated reasonable success with 5 micron absolute pleated cellulose cartridges which provide about 10X more area than the string-wounds. As a result, the 1 micron absolute cartridge changes were reduced almost eight-fold and costs decreased by $800 Thousand. Further testing showed that the 10 micron bags were ineffective because of high flux (30 gpm/ft^2) so they were removed from service.

Subsequent development work demonstrated a Verti-Press (Refs 8,11)/Tyvek® (Ref 12) APF process that could produce excellent filtrate quality (about 0.2 ppm TSS) and that will significantly increase cartridge life about ten-fold. A project has been authorized to install this technology.

Acidic/Oily Waste

This acidic waste contains very little TSS (about 10 ppm) but about 300 ppm of oil. Pressure leaf filtration produces excellent quality filtrate that can be directly injected into the open 10 micron formation, ie, the DE precoat removes practically all oil by absorption. However, these pressure leaf filters were sluiced and the sluicings were lagooned. Recent RCRA regulations required dry cakes that would pass the EPA "Paint Filter Test."

An Oberlin APF (Ref 8) system was installed that produced acceptable cakes for landfilling. A polymer flocculant is added in-line for oil deemulsification so it can be removed by the APF filter cake.

SUMMARY

The case histories highlighted here show how above-ground, deep-well filtration facilities can be upgraded to prolong well life as well as produce dry cakes for RCRA-approved landfilling or incineration. Final cartridge filter improvements were highlighted, especially with respect to replacement of inefficient string-wound cartridges. Renewable-surface cartridge filter replacement units have demonstrated success in several treatment applications.

CARTRIDGE FILTERS IN WELL DISPOSED APPLICATIONS

In virtually every disposal fluid treatment system, cartridge filters are used as the necessary polishing stage. They are needed to provide reproducible and high quality effluent to protect the well formation pores from further plugging.

Cartridge Types

1. **Non-Fixed Random Pore Depth Type Media**

   Non-fixed random pore depth type media depend principally on the filtration mechanisms of inertial impaction and or diffusional interception to trap particles within the interstices or spaces of their internal structure. Examples of this type of media are felts, woven yarns, asbestos pads, and packed fiberglass. Such filters are constructed of non-fixed media of a thickness sufficient to trap particles in a given size range on a finite statistical basis.

   The release of collected particles is likely to occur if the structure of the medium is such that pore dimensions can enlarge as the pressure drop increases. Also, every type of filter will collect particles finer than its pore size but under impulse conditions these finer particles are more likely to be released by a filter whose pores can enlarge.

   This is always a problem with non-fixed random pore depth type media. Such media contain many tortuous passages, and there are many paths through which fluid can flow. Naturally, the small passages become blocked first, resulting in more and more flow being taken by the large passages. Since the structure of the medium is not an integral one, increased pressure on the large passages can cause the medium to separate, thereby widening the passages. It should be obvious that channeling of this nature adversely affects the performance of the filter.

Non-fixed random pore depth type filters depend not only on trapping but also on adsorption to retain particles. As long as the dislodging force exerted by the fluid is less than the force retaining the particle, the particle will remain attached to the medium. However, when such a filter has been on-stream for a length of time and has collected a certain amount of particulate matter, a sudden increase in flow and or pressure can overcome these retentive forces and cause the release downstream of some of the particles. This unloading will frequently occur after the filter has been in use for some time and can give the false impression of long service life for the filter.

Furthermore, most non-fixed random pore depth type filters are subject to media migration. This means that parts of the filter medium become detached and continue to pass downstream contaminating the effluent (fluid that has passed through the filter). Media migration is also sometimes incorrectly taken to include release of "built-in" contamination - for example, dust and fibers picked up by the filter during its manufacture.

2. **Fixed Random Pore Depth Type Media**

Fixed random pore depth type filters consist of either layers of medium or a single layer of medium having depth, depend heavily on the mechanism of direct interception to do their job, and are so constructed that the structural portions of the medium can not distort and that the flow path through the medium is tortuous. It is true that such filters retain some particles by adsorption as a result of inertial impaction and diffusional interception. It is also true that they contain pores larger than their removal rating. However, pore size is controlled in manufacture so that quantitative removal of particles larger than a given size can be assured. Moreover, by providing a medium with sufficient thickness, release of particles collected that are smaller than the removal rating can be minimized, even under impulse conditions.

3. **Surface Type Media**

In the strict definition of the term, a surface or screen filter is one in which all pores rest on a single plane, which therefore depends largely upon direct interception to separate particles from a fluid. Only a few filters on the market today, for example woven wire mesh, woven cloth and a membrane filter qualify as surface filters.

Naturally, a surface or screen filter will stop all particles larger than the largest pore (provided, of course, the structure of the medium is an integral one). While particles smaller than the largest pore may be stopped because of factors previously discussed (bridging, etc.), there is no guarantee that such particles will not pass downstream. Woven wire mesh filters are currently available with openings down to 8 micrometers.

REMOVAL RATINGS

Various rating systems have been evolved to describe the filtration capabilities of filter elements. Unfortunately, there is at the moment no generally accepted rating system and this tends to confuse the filter user. Several of the systems now in use are described below.

1. **Nominal Rating**

Many filter manufacturers, rely on a Nominal Filter Rating which has been defined thusly by the National Fluid Power Association (NFPA): "An arbitrary micron value assigned by the filter manufacturer, based upon removal of some percentage of all particles of a given size or larger. It is rarely well defined and not reproducible." In practice, a "contaminant" is introduced upstream of the filter element and subsequently the effluent flow (flow downstream of the filter) is analyzed microscopically. A given Nominal Rating of a filter means that 98% by weight of the contaminant above the specified size has been removed; 2% by weight of the contaminant has passed downstream.

Note that this is a gravimetric test rather than a particle count test. Counting particles upstream and downstream is a more meaningful way to measure filter effectiveness.

The various tests used to give non-fixed random pore depth type filters a Nominal Rating yield results that are nebulous if not misleading. Typical problems are as follows:

A. The 98% contaminant removal by weight is determined by using a specific contaminant at a given concentration and flow. If any one of the test conditions are changed, the test results could be altered significantly.

B. The 2% of the contaminant passing through the filter is not defined by the test. It is not uncommon for a filter with a Nominal Rating of 10 um to pass particles downstream ranging in size from 30 to over 100 microns.

C. Test data are often not reproducible, particularly among different testing laboratories.

D. Some manufacturers do not base their Nominal Rating on 98 percent contamination removal by weight but instead a contamination removal efficiency of 95%, 90%, or even lower. Thus it often happens that a filter with an Absolute Rating of 10 um is actually finer than a competitive filter with a Nominal Rating of 5 um. Always check the criteria upon which a competitor's Nominal Rating is based.

E. The very high upstream contaminant concentrations used for such tests are not typical of nominal system conditions and produce misleading high efficiency values. It is common for a wire mesh filter medium with a mean (average) pore size of 17 um to pass a 10 um nominal specification. However, at normal system contamination concentrations, this same filter medium will pass almost all 10 um size particles.

One cannot therefore assume that a filter with a Nominal Rating of 10 um will retain all or most particles 10 um or larger. Yet some filter manufacturers continue to use only a Nominal Rating both because it makes their filters seem finer than they actually are, and because it is impossible to place an Absolute Rating on a non-fixed random pore depth type filter.

2. **Absolute Rating**

The NFPA defines Absolute Rating as follows: "The diameter of the largest hard spherical particles that will pass through a filter under specified test conditions. It is an indication of the largest opening in the filter element. "Such a rating can be assigned only to an integrally bonded medium (such as fixed pore depth media or sintered metal media).

The original test and term for the Absolute Filtration Rating was proposed in the mid-1950's. It was considered by the Filter Panel of SAE Committee A-6 and with minor changes subsequently adopted.

One point that confused users of the absolute rated filters is that when measuring downstream contamination, contaminants larger than the pore size indicated by the Absolute Rating are invariably found. At first glance, this would seem to cast doubt on the very concept of an "absolute" rating. However, one must realize that it is impossible to take effluent samples, transfer them, run the test, or wash out a newly manufactured filter without adding a quantity of contaminants. Even a new filter is contaminated when it is removed from its packing! All of this is called "background contaminant" and before running any test an experienced laboratory will have determined the amount of such contamination in this test set-up. A test will be invalidated if the background contamination count is above established limits.

There are several recognized tests for establishing the Absolute Rating of a filter. What test is used will depend on the manufacturer, on the type of medium to be tested, or sometimes on the processing industry. In all cases the filters have been rated by a "challenge" system. A filter is challenged by pumping through it a suspension of a readily recognized contaminant (e.g. glass beads or a bacterial suspension), and both the influent and effluent examined for the presence of the test contaminant.

3 Beta (B) Rating System

While absolute ratings are clearly more useful than nominal ratings, a more recent system for expressing filtration rating is the assignment of Beta ratio values. Beta ratios are determined using the Oklahoma State university, "OSU F-2 Filter Performance Test." The test, originally developed for use on hydraulic and lubricating oil filters, has been adapted by some filter manufacturers for rapid semi-automated testing of filters for service with aquerous liquids, oils, or other fluids.

The Beta rating system is simple in concept and can be used to measure and predict the performance of a wide variety of filter cartridges under specified test conditions.

The basis of the rating system is the concept of titer reduction. If you measure the total particle counts at several different particle sizes, in both the influent stream and effluent stream, a profile of removal efficiency emerges for any given filter.

CHOOSING THE PROPER FILTER

Among the more important factors that must be taken into consideration when choosing a filter for deep well disposal system are the size, shape, and hardness of the particles to be removed, the quantity of those particles, the nature and volume of fluid to be filtered, the rate at which the fluid flows, whether flow is steady, variable and or intermittent, system pressure and whether that pressure is steady or variable, available differential pressure, compatibility of the medium with the fluid, fluid temperature, properties of the fluid, space available for particle collection, and the degree of filtration required.

REFERENCES

1. W. D. Constant and J. L. Alixant, Paper presented at First Regional Meeting of the American Filtration Society, Houston, TX, October 30-November 1, 1989: "Correlation Between Gulf Coast Shale Permeability and Resistivity-Application to Disposal Well Integrity Tests."

2. D. D. Reible, S. Rhee, and W. D. Constant, Ibid., "Modeling of Confining Layer Integrity in Deep-well Disposal."

3. Society of Petroleum Engineers 1988 Forum Series on "Water Quality Issues-Handling and Treatment of Produced and Injected Water," Mt. Crested Butte, Colorado, July 31-August 5, 1988.

4. J. E. Schmitz & C. H. Likens, Paper presented at First Regional Meeting of the American Filtration Society, Houston, TX, October 30-November 1, 1989; "Filtration Specifications for Oilfield Applications"; as well as other papers in this same symposium.

5. W. J. Fiero, A. Shucosky, V. Geraci, Ibid., "Characterization of Absolute Rated Filter Media for Oil and Gas Production Filtration."

6. J. Hampton, Ibid., "A cost Effective Procedure for Evaluating Cartridge Performance on Site."

7. T. W. Johnson, Ibid., "A Systems Approach to Filtration of Completion Fluids."

8. E. Mayer, Filtration News, 6, 24-27 (May/June, 1988).

9. E. Ostreicher, "Fluid Filtration: Liquid," Vol. II, ASTM STP 975, Johnston, P. R., and H. G. Schroeder, eds., American Society for Testing Materials, Philadelphia (1986)

10. E. Mayer, Paper presented at Battelle International Symposium on Solid/Liquid Separations, Columbus, Ohio, December 5-7, 1989.

11. R. A. Mau, Paper presented at American Filtration Society National Meeting, Pittsburg, PA, March 27-30, 1989.

12. H. S. Lim and E. Mayer, Fluid/Particle Separation J., 2 (2), 17-21 (March, 1989)

NEW CONCEPTS FOR THE RATING OF MEMBRANE FILTERS FOR LIQUID APPLICATIONS

K.L. Roberts, D.J. Velazquez and D.M. Stofer
Gelman Sciences Inc.
600 S. Wagner Road
Ann Arbor, MI 48106

Absolute rated membrane filters are used extensively to remove submicron particles from chemicals and ultrapure water used throughout the manufacturing process of semiconductors and other integrated devices. As device line widths approach 1 μm and smaller, the "rule of ten" dictates the need to remove all particles greater than 0.1 μm from process fluids. Therefore, absolute membrane filters rated 0.1 μm and smaller are often employed. It is assumed that these filters remove all particles greater than their labeled ratings. However, due to the lack of relevant standards the pore size ratings of these filters are inaccurate and cannot even be compared from one to another. The term "absolute" can only be interpreted on a relative basis.

Historical methods for rating membrane filters are inadequate for many current filtration applications. The one and only standard for rating membrane filters was established ten years ago for the pharmaceutical industry and is only meaningful for the cold sterilization of pharmaceutical drugs.

Tests conducted using uniform latex spherical particles provide more accurate information on the true pore size and absolute retention capabilities of membrane filters in liquid applications. Latex sphere challenge tests could provide the rating standardization needed to make definitive choices when selecting a filter for semiconductor and other related applications.

INTRODUCTION

One of the most important characteristics considered when choosing a microporous membrane filter, or at least that characteristic which is usually first considered, is the pore-size rating. Yet, this very important criterion is probably the most misunderstood property of filters.

The measurement of pore sizes of microporous membrane filters by direct methods is impractical; therefore, indirect rating methods are employed. These ratings do not refer to the actual size of the filter pore, but to the size of the challenge particle retained by the filter. This is often conducted under a very specific set of operating conditions.

The label on a 0.2 μm absolute rated filter implies that the filter retains 100% of all challenge particles greater than 0.2 μm. In an industry which lacks sufficient standards to rate microporous membrane filters, this implication continues unchallenged.

Historically the designing and rating of membrane filters was dictated by the needs of the pharmaceutical industry. Membrane filters were rated or categorized according to their ability to retain unwanted organisms and thereby cold sterilize critical fluids. Filters which produced sterile effluent when challenged with 10^7 *Pseudomonas diminuta* per square centimeter of filter surface area under conditions specified by the Health Industry Manufacturers Association (HIMA) were accepted as 0.2 μm absolute rated sterilizing filters. Indeed, they are sterilizing grade filters under the conditions tested.

However, the *P. diminuta* organism is known to be much larger than 0.2 μm in size. It is characterized at a minimum 0.3 μm in diameter and 1 μm in length. Thus, the filter designation of 0.2 μm becomes inaccurate. Filters with actual pore sizes of 0.3 μm and even larger could retain this organism.

THE INACCURACY CONTINUES

The organism challenge is a destructive test and cannot be conducted on each filter manufactured. Instead the organism retention level is conveniently correlated to a non-destructive test, such as the bubble point integrity test. The bubble point test is a simple, reproducible test which confirms the functional integrity of membrane filters. When a sufficiently high pressure is applied to the pores of a thoroughly wetted membrane, the liquid within those pores is ejected. The pressure is high enough to overcome the capillary force of the liquid within the pores. Liquid from the largest pores will evacuate first. This is according to the bubble point equation, where the pressure necessary to evacuate the liquid from the pore is inversely proportional to the diameter of the pore. Therefore, the bubble point value of the membrane is indicative of the size of the largest pore.

Filters which retain 100% of the challenge organism P. diminuta normally have water bubble point values of greater than or equal to 45 psi. As stated earlier, these filters were inaccurately labelled 0.2 μm. When finer pore-sized membrane filters were produced with bubble point values two times those of the 0.2 μm rated membrane filters or 90 psi, these were deemed 0.1 μm rated, based on the bubble point equation. This would be a correct assumption if a 0.2 μm rated filter were really a 0.2 μm pore sized filter. Instead, the inaccuracy expands.

The pharmaceutical standard for rating filters is perfectly suited for the complete removal of P. diminuta under HIMA conditions, but it does not accurately address the retention capabilities of filters when challenged with particles other than this specific organism. It also does not address conditions other than those specified in the HIMA document.

WHY LATEX SPHERES

True absolute retention can only be possible when sieving is the capture mechanism. Absolute retention ratings should ideally be independent of the filtration conditions. Conditions such as flow rate, differential pressure, and pH can drastically affect retention capabilities of small particles by membrane filters. Charge interactions between the particle and the membrane surface can cause adsorption of the particle to the membrane which can distort true sieving retention.

Latex particles are spherical and nondeformable. They can be monodispersed in a solution containing a surfactant to eliminate or reduce adsorption caused by particle and membrane interactions.

The effects of flow rate, pressure, and pH cannot be completely eliminated but can be controlled to simulate "worst case conditions" in order to maximize penetration. Uniform latex spherical particles are readily available in a variety of diameters which have very narrow size distributions.

THE CHALLENGE PROCEDURE

Uniform latex microspheres were used to challenge the test filters. Each 47 mm filter disc was challenged with 50 ppm of monosized latex spheres which were single dispersed in 0.1% Triton X-100 and ultrapure water. SEM's were used as verification of single dispersion. The addition of surfactant produced maximum penetration of the spheres through the membrane. In the absence of surfactant much higher retention efficiencies were observed for all membranes tested, in agreement with Johnston and Meltzer (1987).

Spheres were detected downstream via absorbance using a dual beam spectrophotometer. Each sphere diameter has a characteristic wavelength at which it has maximum absorbance sensitivity. For each sphere size a unique calibration curve was prepared by measuring the absorbance of serial dilutions of the challenge. All the curves were linear within the concentration range tested.

Bubble point integrity tests were conducted on each membrane type tested. The filters were challenged at 10 psid with 150 mL of sphere test solution, collecting 15 successive 10 mL aliquots and measuring their absorbance. The decay in flow rate at each sample collection point was measured. An overview of the test matrix can be seen in Table I

SUMMARY AND INTERPRETATION OF RESULTS

0.1 μm rated hydrophobic PTFE membrane, double layer, was challenged with 0.106 μm spheres and found to only be 33% retentive (Figure 1). This same configuration was challenged with increasingly larger spheres until 100% retention was achieved with 0.222 μm spheres.

Did the two layers of PTFE membrane benefit retention efficiency? Both single and double layers of 0.1 and 0.2 μm rated PTFE were challenged with 0.215 μm spheres to demonstrate the effect of membrane layering (Figure 2). The data show that layering does increase the probability of retention of spheres that are smaller than the actual pore size. Using a double layer filter could be advantageous over choosing the next tighter pore size.

Hydrophilic polysulfone rated 0.2 μm and having water bubble point values from 54-56 psi is nearly 100% retentive of 0.415 μm spheres. Two different runs of the tests are represented on the graph by the light and dark bars (Figure 3). The retention efficiency is plotted on the y-axis and the volume filtered in aliquots is plotted on the x-axis. Flow rate data are plotted on the y′-axis and superimposed over the retention bars. Each hydrophilic membrane type was evaluated in this manner using a wide range of sphere diameters.

Hydrophilic nylon 66 rated 0.2 μm and having a water bubble point value of 43 psi was challenged using the same procedure and sphere size as was used to challenge the polysulfone (Figure 4). Although rated the same as the polysulfone in Figure 4, the retention efficiencies of the two membranes differ drastically. The nylon 66 is only 30-40% retentive of the 0.415 μm sphere. This difference in retention efficiency would be predicted from its lower bubble point value as compared to the polysulfone.

Note that the retention efficiency was higher within the first 30 mL of the challenge volume after which the retention levels off. This effect occurred with all pore sizes and types of membrane when challenged with a sphere size which was smaller than the true pore size of the membrane. Some of the studies conducted in the past were only carried out to volumes which were too small to reveal this phenomena. Thus the membrane could appear more retentive than it actually is over prolonged time or higher throughput volumes.

The mean retention values representing a full range of sphere sizes was plotted for the three 0.2 μm rated hydrophilic membranes tested (Figure 5). Their corresponding bubble point values are referenced. <u>None</u> of the 0.2 μm rated membranes are retentive at 0.2 μm. Only two of the membranes retain even 0.3 μm spheres. The polysulfone and PVDF membranes attain 100% efficiency in the 0.43 μm range. The nylon 66 reaches 100% retention at a range greater than 0.55 μm.

The mean retention profiles are plotted for three 0.1 μm rated hydrophilic membranes (Figure 6). They were challenged with spheres ranging from 0.110 to 0.300 μm in diameter. All the membranes tested show less than 10% retention at 0.110 μm. 0.1 μm rated membranes are far from absolute at 0.1 μm. Even the polysulfone with a bubble point of 132 psi does not attain 100% retention below the 0.173 μm range. The two PVDF membranes have identical bubble point values, and their corresponding profiles mimic one another achieving 100% retention in the 0.300 μm range.

CONCLUSION

Studies conducted using uniform latex particles to investigate true retention capabilities of microporous membrane filters have been ongoing for the past decade. These studies, originally reported by Pall et al (1980), indicate that although 0.2 μm absolute rated membrane filters may retain 100% of the <u>P. diminuta</u> organism, they fall far short of 100% retention when challenged with 0.5 μm and even larger diameter latex spherical particles. Additional studies have been published by Wrasidlo et al (1983, 1984), Schroeder (1984), Krygier (1986), and many others.

This is the most comprehensive study conducted to date to characterize membrane pore sizes through the absolute retention of latex spheres. Challenging the membranes with a wide range of sphere sizes provides a detailed profile of the membrane's smallest to the largest pores. It provides a full characterization of the membrane's retention capability under worse case conditions.

This study and each of the other studies confirm the original work reported by Pall (1980); membrane filters which are rated 0.2 μm do not retain latex spheres which approach their published pore size ratings. This even includes filters rated 0.2 μm with <u>P. diminuta</u> organism retention.

The latex sphere challenge provides a reproducible test

method which can absolutely discriminate between different pore sizes of membrane filters. The use of latex spheres provides more accurate retention efficiency and pore size rating information on membrane filters than the current methods employed. The spheres offer a more rigorous test, since a method for testing membrane filters which retains by size exclusion of a singly dispersed contaminant will be consistently more reliable than those methods which retain by other means.

Bubble point testing would retain its value of practicality for routine, nondestructive testing. However, the bubble point values should be correlated to latex sphere-derived retention efficiencies and pore size ratings.

The rating of membrane filter pore sizes with latex spherical particles deserves consideration as an industry standard. When a reliable method of characterizing retention efficiencies of membrane filters is standardized, then other important factors that affect filtration processes such as flow rate can be evaluated. There is no substitute for testing a filter in its actual application, but a standardized rating system would allow for valid comparisons between membrane filters.

Latex Sphere Retention

Comprehensive Overview

Membrane Types	Rated Pore Size	Challenge Profile Sphere Diameter (um)
Polysulfone (PS) PVDF-1 N_{66}	0.2um	0.300, 0.320, 0.337 0.360, 0.378, 0.397 0.415, 0.426, 0.451 0.497, 0.523
Polysulfone (PS) PVDF-1 PVDF-2	0.1um	0.110, 0.135, 0.155 0.162, 0.173, 0.198 0.246, 0.300
PTFE	0.1um 0.2um	0.106, 0.173, 0.215 0.222

Table I

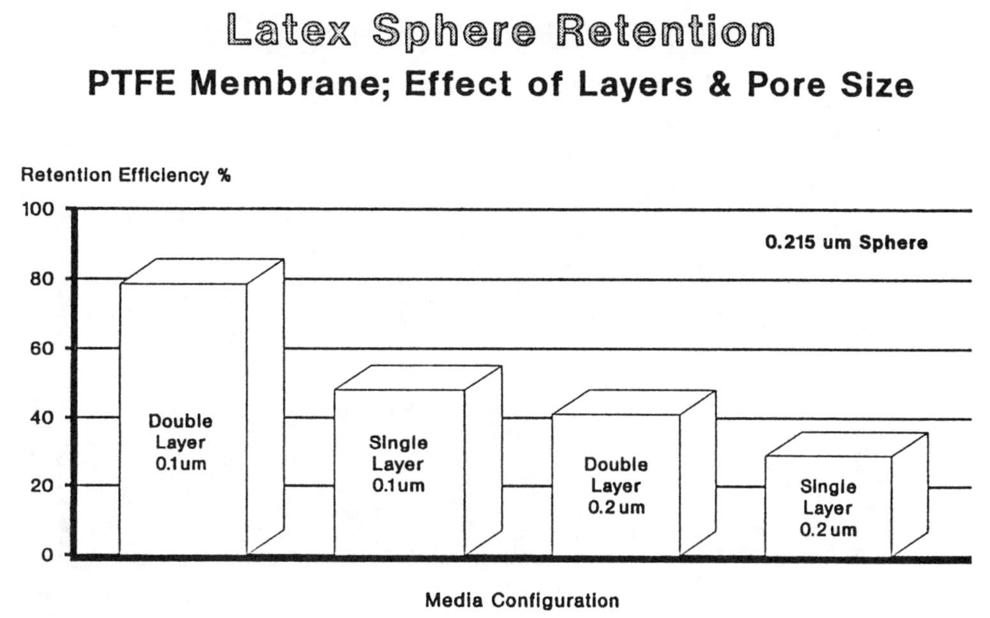

Figure 1

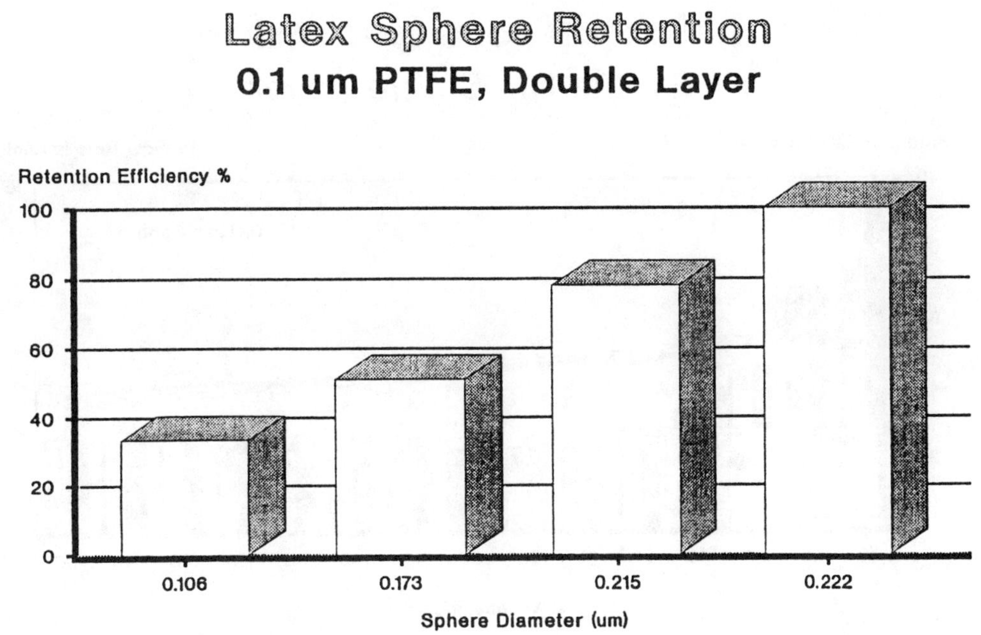

Figure 2

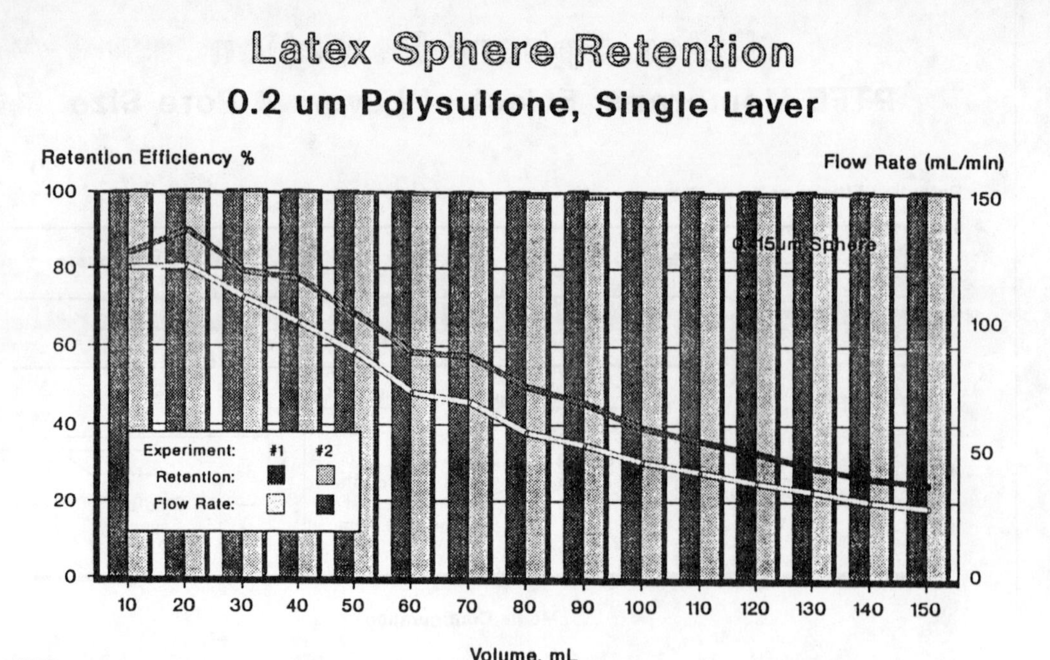

Figure 3

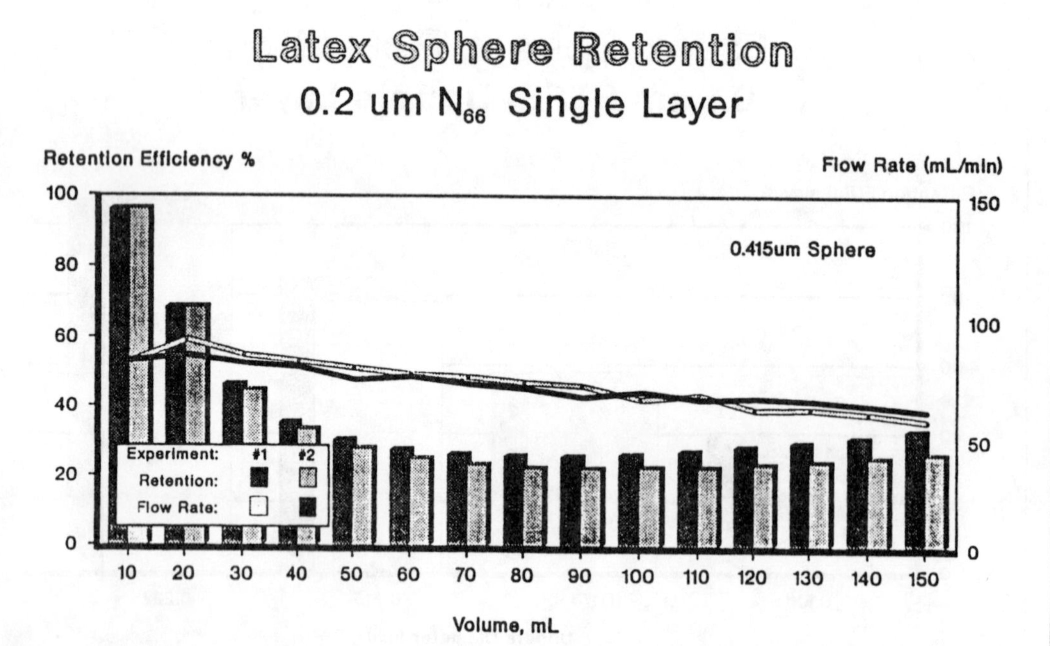

Figure 4

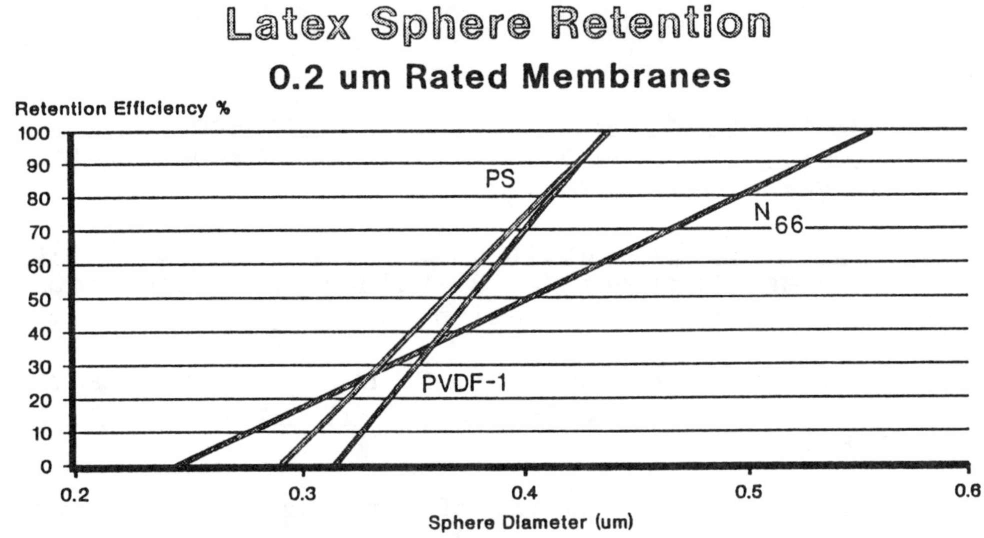

Membrane Type:	PS	PVDF-1	N_{66}
Actual Water Bpt. (psi):	54-56	48-49	43

Figure 5

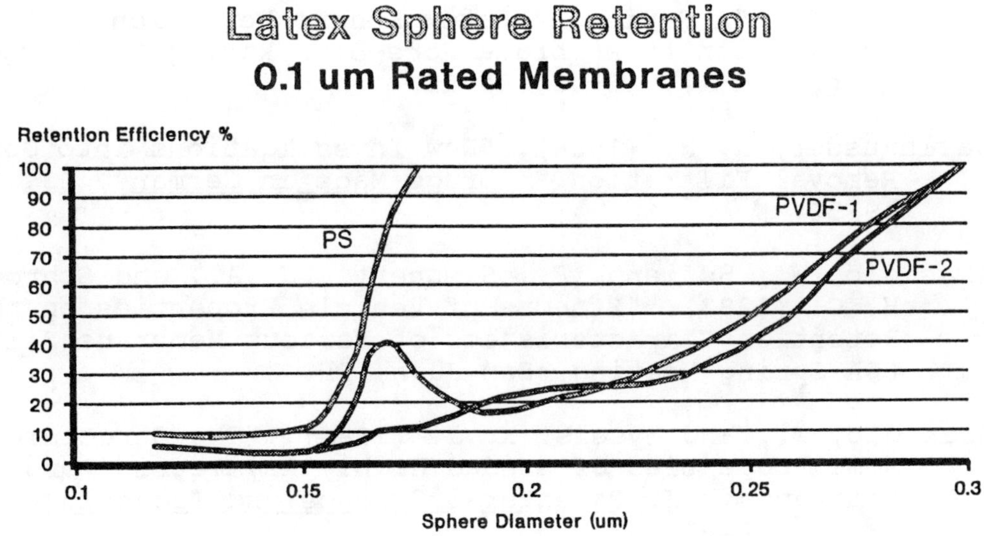

Membrane Type:	PS	PVDF-1	PVDF-2
Actual Water Bpt. (psi):	132	100	100

Figure 6

BIBLIOGRAPHY

FDA, Guideline on Sterile Drug Products Produced by Aseptic Processing, Food and Drug Administraton, Rockville, Maryland, July 1987.

HIMA, Microbial Evaluation of Filters for Sterilizing Liquids, Document No. 3, Volume 4, Health Industry Manufacturers Association, Washington, D.C. (1982).

Johnston, P. R., and Meltzer, T. H. (1987) "The Rating of Membrane Filter Pore Sizes." Ultrapure Water, November 1987.

Krygier, V. (1986). "Rating of Fine Membrane Filters Used in the Semiconductor Industry." Proceedings of the Fifth Annual Semiconductor Pure Water Conference, San Francisco, CA, January 1986.

Pall, D. B., Kirnbauer, E. A., and Allen, B. T. (1980). "Particulate Retention by Bacteria Retention Membrane Filters." Colloids and Surfaces, 1, 235-256.

Schroeder, H. G. (1984). "Selection Criteria for Sterilizing Grade Filters." Society of Manfuacturing Engineers, Philadelphia, PA.

Simonetti, J. A., Schroeder, H. G., and Meltzer, T. H. (1986), "A Review of Latex Sphere Retention Work: Its Application to Membrane Pore-Size Rating." Ultrapure Water, 3 (4), 46-48, 50-51.

Wallhausser, K. H. (1986). "New Investigations into Germ Removal Filtration." Drugs Made in Germany, 23, 10-15, 49-52.

Wrasidlo, W., Hofmann, F., Simonetti, J. A., and Schroeder, H. G. (1983). "Effects of Vehicle Properties on the Retention Characteristics of Various Membrane Filters." PDA Spring Meeting, San Juan, PR.

Wrasidlo, W., and Mysels, K. J. (1984). "The Structure and Some Properties of Graded Highly Asymmetric Porous Membranes." J. Parenteral Science and Technology, 38 (1), 24-31.

A Superior Biocide for Disinfecting Reverse Osmosis Systems

Jo-Ann B. Maltais,[1] Ph.D., and Thomas Stern[2]
Minntech Corporation[1] and FilmTec Corporation,[2]
Minneapolis, Minnesota

In an effort to determine the effectiveness of Minncare™ as a disinfectant compared to the biocides traditionally recommended for use on reverse osmosis (RO) membranes and their associated water distribution systems, Minncare, formaldehyde, bleach, and hydrogen peroxide were tested for efficacy by an independent testing laboratory. The concentrations chosen for the test were those normally recommended for cleaning and sanitizing RO membranes and water systems. Survival curves were generated for each biocide against Bacillus subtilis var. niger (ATCC 9372) spores. These provided data for calculating individual D-values. A D-value is the time required to reduce the microbial population by 1 log. A 6 D-value is generally accepted as the time required for disinfection. Results show that Minncare exhibited a 6 D-value of 36 minutes, compared to 11 hours for formaldehyde, 7 hours for bleach, and 1 to 25 hours for hydrogen peroxide.

Normal maintenance of RO membranes and water distribution systems requires routine cleaning and disinfection to prevent bacterial growth. Microbial contaminants commonly found in feedwater will decrease the operating efficiency of RO membranes, lower membrane flux and thereby increase the cost of producing high purity water. Fouling microorganisms may actually destroy some RO membranes by promoting decomposition of the membrane polymer. Bacterial growth acceleration, particularly during periods of shutdown, requires a microbiocidal agent to control growth and protect the water system.

Organic and inorganic fouling of the RO membrane can be removed with cleaning agents. Sanitization, which must achieve a 3 log reduction of bacteria,[1] is usually accomplished post-cleaning to eliminate fouling microorganisms. Since naturally occurring microbial contaminants on the membrane can easily reach levels of 10^4 colony forming units per square centimeter (CFU/cm^2),[2] sanitization may not completely destroy all the microbial bioburden present. Disinfection, requiring a 6 log reduction of bacteria[3], may be more appropriate. The biocides historically used for maintenance of RO membranes and their water distribution systems include formaldehyde, bleach, and hydrogen peroxide.[4] Hydrogen peroxide requires either extended soak times at low concentrations or high concentrations which may be destructive to the RO membrane. The high concentrations, however, are necessary for adequate disinfection of the water distribution system. Bleach, while suitable for disinfection of cellulose acetate membranes, is contraindicated for thin-film composite membranes. Formaldehyde, while compatible with both thin-film composite membranes and water distribution systems, presents toxicity and environmental hazards. Additionally, chemicals of the formaldehyde family, although effective for destroying microbial contaminants, do not remove the extracellular material associated with biofilms. This can result in a more rapid re-fouling of the water system.[5]

Minncare, a proprietary, stabilized mixture of hydrogen peroxide and peracetic acid, is an oxidizer compatible with thin-film composite membranes. Oxidizers are known to not only kill bacteria but also effectively remove biofilms[5]. In order to evaluate the efficacy of Minncare versus other biocides historically used for RO and water system maintenance, the following study was undertaken.

Materials and Methods

Test Organism

A B. subtilis var niger (ATCC 9372) aqueous spore suspension (North American Science Associates, Inc.) was used as the test bioburden. While spore-forming bacteria are not the predominant microorganisms associated with RO systems, they are highly resistant to disinfectants and were therefore chosen as the challenge organism for this study. A sufficient concentration of spores, in 0.1 milliliter (ml) volumes, was added to the following biocides to achieve a final concentration of 2 x 10^6 spores/ml: Minncare at 1% concentration, formaldehyde at 2% concentration, sodium hypochlorite (bleach) at 0.001% concentration, and hydrogen peroxide at 0.2%, 5%, and 10% concentrations. Biocide/spore mixtures were equilibrated to 20°C.

Aliquots from each biocide/spore mixture were removed for analysis of surviving spores at 0, 15, 30, and 60 minutes, and 2, 4, 6, 12, and 24 hours. Each sample was neutralized, as described below, and then diluted and plated (0.1 ml of each neutralized sample and dilutions as necessary) on Tryptic Soy Agar using the pour plate technique. Plates were incubated at 35°C for 7 days. Colonies derived from surviving spores were counted and reported as CFU/ml. **It should be noted that since 0.1 ml samples were analyzed, the minimum number of survivors detectable per ml was 10 CFU.**

Neutralization

The activity of each biocide was neutralized at the time each sample was drawn to ensure that survivors detected at each time interval accurately represented the killing power of the biocide for that exposure. The neutralizing agents used are shown in Table I.

Table I
Neutralizing Agents and Concentrations

Biocide	Neutralizing Agent	Final Concentration[1]
Minncare	D/E Neutralizing broth[2]	100X dilution
Formaldehyde	Sodium bisulfite	1% Aqueous
Sodium hypochlorite	Sodium thiosulfate	Equimolar
Hydrogen peroxide	D/E Neutralizing broth[2]	100X dilution

[1]Final concentration of neutralizing agent when combined with biocide/spore sample.

[2]DIFCO, Cat #0819-17-2

Validating the Neutralization

A separate test was done to validate the effectiveness of the neutralizers chosen. Test biocides, at use-concentration and equilibrated to 20°C, were sampled at the time intervals selected for efficacy testing, and each sample was mixed with the appropriate volume and type of neutralizing agent to achieve the final concentrations given in Table I. A 0.1 ml sample of each biocide/neutralizer mixture was plated using the pour plate technique and Tryptic Soy Agar. Ten B. subtilis spores were then added to each plate. Controls were set up without biocide, and without biocide or neutralizer, to ensure that neutralizers were not toxic to the spores. Plates were incubated at 35°C for 7 days. At the end of 7 days, all plates showed recovery of all B. subtilis spores added, demonstrating in each case that neutralization was achieved.

Generating D-Values

A D-value was generated for each biocide tested. The log of the number of surviving spores at each time frame tested was plotted and a "best fit" line generated by log regression analysis. The slope of the regression line in each case was determined. The negative reciprocal of the slope of each line is the D-value in minutes. A 6 D-value is determined by multiplying the D-value by 6 to give the time required to achieve a 6 log reduction of the spore population. Since the 5% and 10% hydrogen peroxide efficacy data showed a 1 log reduction in 60 and 30 minutes respectively, but <10 CFU/ml thereafter, use of the data only up to the 60 and 30 minute time frames respectively would give a D-value far larger than that actually reflected by the data. A value of "0" for the <10 CFU/ml could not be used, since the log of 0 is undefined. Therefore, a value of 10 CFU/ml was assigned at the 60 minute time frame for 5% hydrogen peroxide and at the 30 minute time frame for 10% hydrogen peroxide. This was not necessary for the other biocides tested since there was sufficient data covering at least 3 log reductions of the spore population to generate a reliable D-value.

Results

Disinfectant Efficacy

The results of efficacy testing are shown in Table II. Minncare, at a 1% concentration, showed a 5 log reduction of bacterial spores in 30 minutes. By contrast, 2% formaldehyde required 12 hours and 0.001% sodium hypochlorite 6 hours to achieve a 5 log reduction. Hydrogen peroxide, at 10%, 5%, and 0.2% concentrations, required approximately twice the amount of time as Minncare and as much as 24 hours to exhibit the same 5 log reduction.

Table II
Biocide Efficacy Results

Exposure Time	Minncare (1%)	Formaldehyde (2%)	Sodium hypochlorite (0.001%)	Hydrogen peroxide (0.2%)	Hydrogen peroxide (5%)	Hydrogen peroxide (10%)
0 minutes	2.3×10^6	2.3×10^6	2.1×10^6	2.2×10^6	2.1×10^6	2.0×10^6
15 minutes	1.1×10^6	2.3×10^6	2.2×10^6	2.3×10^6	2.0×10^6	2.0×10^6
30 minutes	3.0×10^1	2.1×10^6	1.1×10^6	2.3×10^6	2.0×10^6	2.0×10^5
60 minutes	<10	2.0×10^6	4.0×10^4	2.0×10^6	1.0×10^5	<10
2 hours	<10	1.5×10^5	1.0×10^4	2.0×10^6	<10	<10
4 hours	<10	1.2×10^5	1.0×10^3	1.5×10^6	<10	<10
6 hours	<10	1.0×10^3	<10	1.0×10^6	<10	<10
12 hours	<10	<10	<10	1.0×10^3	<10	<10
24 hours	<10	<10	<10	<10	<10	<10
D-values	6 minutes	113 minutes	69 minutes	250 minutes	22 minutes	11 minutes
6D	36 minutes	678 minutes	414 minutes	1500 minutes	132 minutes	66 minutes

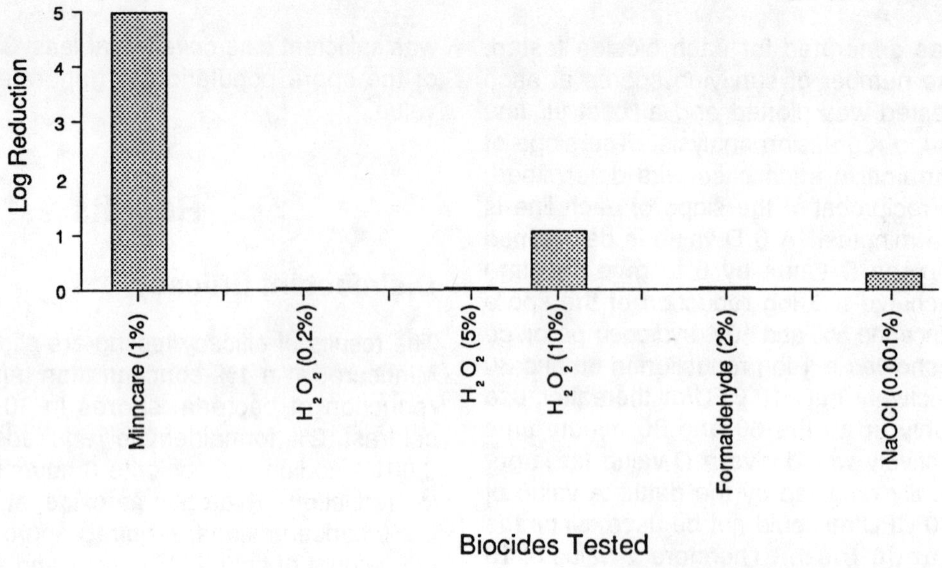

Figure 1: Log Reduction of B. subtilis var niger (ATCC 9372) Spores after 30 minutes Exposure to Various Biocides.

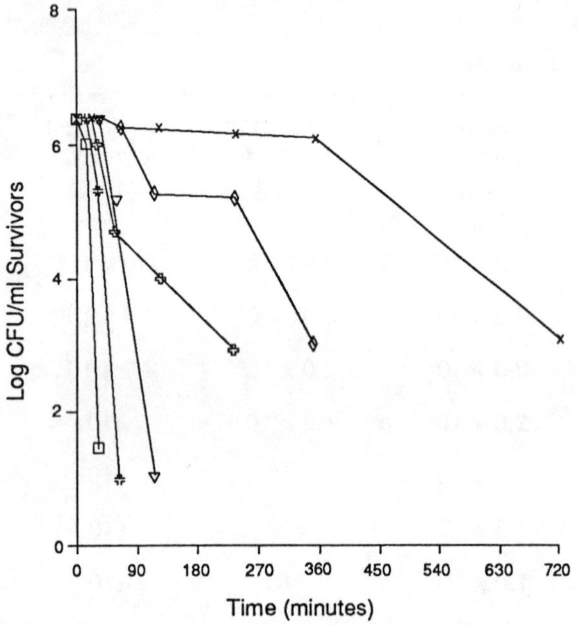

Figure 2: Survival Curves for Biocides Tested for Efficacy by Suspension Test against B. subtillis var niger (ATCC 9372) Spores. □ = Minncare (1%); X = Hydrogen peroxide (0.2%); ▽ = Hydrogen peroxide (5%); ⁜ = Hydrogen peroxide (10%); ◇ = Formaldehyde (2%); ✙ = Sodium hypochlorite (0.001%).

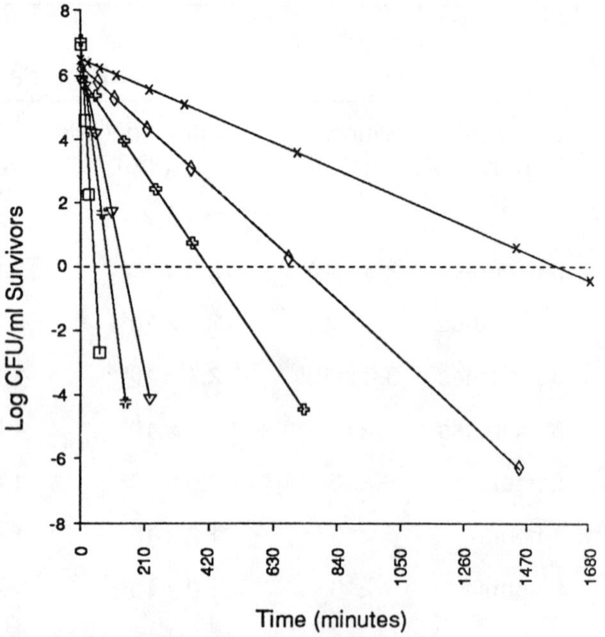

Figure 3: Log Regression Lines from Analysis of Survival Curves for Biocides Tested for Efficacy by Suspension Test against B. subtillis var niger (ATCC 9372) Spores. □ = Minncare (1%); X = Hydrogen peroxide (0.2%); ▽ = Hydrogen peroxide (5%); ⁜ = Hydrogen peroxide (10%); ◇ = Formaldehyde (2%); ✙ = Sodium hypochlorite (0.001%).

Survival Curves, Regression Analysis, and D-Values

Survival Curves for each biocide, based on the log of the number of CFU/ml of spores surviving at various exposure times, is presented in Figure I. The regression lines generated from each survival curve are also plotted on the same axes for each biocide. The D-values, resulting from the regression analyses as described in the Materials and Methods section, reflect the diversity of effectiveness seen among the biocides tested. The D-values are listed at the bottom of Table II.

Minncare had the smallest D-value of 6 minutes, while the other test biocides gave D-values ranging from 11 minutes for 10% hydrogen peroxide to 69 minutes for sodium hypochlorite, and 113 minutes for formaldehyde.

The 6 D-value is an indicator of the time necessary for a 6 log reduction in the bacterial population, giving a 1 in 6 log probability of having an organism survive the disinfection cycle.

Minncare, with a 6 D-value of 36 minutes, achieved disinfection in the shortest period of time. The other biocides required between 66 minutes and 25 hours (1500 minutes) to achieve the same 6 log reduction.

Conclusions

Minncare, at a 1% concentration, was the most effective biocide tested, as compared to all the commonly used biocides, for disinfecting reverse osmosis membranes and their associated water distribution systems. A 6 log reduction of B. subtilis spores was achieved in less than one hour only with Minncare. Formaldehyde, hydrogen peroxide, and bleach all required longer exposure times. Minncare can be used to disinfect both the water distribution system and the RO membranes at the same 1% concentration with excellent results. Both thin-film composite and cellulose acetate membranes are compatible with Minncare. Its use eliminates the problems commonly associated with both hydrogen peroxide and formaldehyde.

References

1. U.S. Environmental Protection Agency, Office of Pesticide Programs. 1976. Efficacy Data Requirements: Sanitizer Test (for inanimate, non-food contact surfaces). Washington, D.C.: Government Printing Office.

2. Enkowitz, J., and Maltais, J. 1989. MINNCARE,™ A Disinfectant Compatible with Reverse Osmosis Systems. Cypress, CA: G.E.M. Water Systems International.

3. U.S. Food and Drug Administration, Center for Devices and Radiological Health. 1987. Guidelines for the Premarket Testing and Labeling of Antimicrobial Agents for Medical Devices. Washington, D.C.: Government Printing Office.

4. Parekh, B. S. (ed) 1988. *Reverse Osmosis Technology: Applications for High-Purity-Water Production.* Marcel Dekker, Inc., New York, NY, p. 384.

5. Costerton, J. W. 1989. Personal communication. Department of Biology, University of Calgary, Alberta, Canada.

EVALUATING FILTRATION MEDIA
FOR USE WITH PHOTORESIST

Michael S. Schmidt, Michael A. Sakillaris,
Guiseppe Forcucci, Ronald P. Sienkowski
Shipley Company, Inc.
455 Forest Street
Marlborough, MA 01752

Photoresists are used to produce semiconductor devices with circuit geometries smaller than 0.5µm. Particle contamination can reduce yields and the industry considers any level of microcontamination undesirable. Made of phenolic resins, photoactive diazoquinone compounds, and trace amounts of performance additives, all dissolved in highly polar solvent mixtures, photoresists are processed through sub-micron membrane filters.

A variety of filter media were challenged with a formulation similar to commercial photoresists, using ethyl lactate as the solvent. Initial levels of microcontamination were varied, and filter media performance was evaluated in terms of filterability, flow rate, bulk liquid laser particle detection, and surface scanning of coated substrates.

This study highlighted the relative insensitivity of some accepted evaluation techniques to subtle differences in filter performance. The study also showed that material of construction as well as the porosity have a detectable effect on performance; polypropylene delivered higher flow rates than PTFE at comparable particle retention levels, while in some cases smaller pores reduced flow rates with no apparent improvement in particle retention.

Introduction

Photoresist is used to create the microscopic patterns that become circuits in semiconductor devices. As circuit densities increase, circuit geometries get smaller; semiconductor device manufacturers are using geometries as small as 0.7µm in production now and expect to transfer current laboratory processes of 0.5µm into production by 1992. Particle contamination, as shown in Figure 1, can reduce yields by producing either gaps or bridges in the patterns, and the industry considers any level of microcontamination undesirable. The semiconductor industry's "one tenth rule" indicates that particle control will have to be at 0.05µm by 1992 and even smaller after that.

Photoresists are made from phenolic resins, photoactive diazoquinone compounds, and trace amounts of performance additives such as dyes, leveling agents, contrast enhancers, or cross-linking agents, all dissolved in highly polar solvent mixtures. Photoresist makers process their products through sub-micron membrane filters that are part of proprietary filter trains set up to optimize particle elimination, flow rate, and cost. Most semiconductor manufacturers then filter the photoresist again at point of use.

Filtration systems for photoresist manufacture must balance the needs for high flow rates and low particle levels, without adding contaminants that might affect functional performance, and while preserving batch integrity. While flow rates can be measured simply enough, there is no generally accepted method in the semiconductor industry for monitoring particle levels. Techniques range from the venerable filterability method, which some users accept as a surrogate for particle levels, to bulk liquid laser particle counting, to surface scanning, where photoresist is coated onto substrates and laser scanned for defects.

Experimental Methods

The study included several different membrane filter media. All of the media were in the form of 47 mm flat, unsupported discs. The discs were made from PTFE, nylon, and polypropylene and had pore ratings of 0.2µm, 0.1µm, and 0.04-0.05µm. Only polypropylene was not available in the very small pore size.

The photoresist in the study was similar to commercial photoresists, being made from typical solid raw materials. The solvent was straight ethyl lactate, popular because of its relatively low hazard. A single solvent was used instead of a blend to avoid confounding the study. The test photoresist was 27% solids, with a viscosity of 45 cSt, a specific gravity of 1.096, and a surface tension of 31.2 dyne/cm. A single batch of this photoresist was prepared; part of it was used as prepared and was labeled "dirty", while the part labeled "clean" was filtered through a filter train that ended with a 0.2µm pleated membrane cartridge.

The photoresist was filtered using a constant pressure of 15 psig supplied by filtered nitrogen. Flow rate data were collected for each test filter by recording cumulative flow as a function of time. Instantaneous flow rates were calculated from this data and flow rate as a function of cumulative flow for the test filters is shown in Figure 2.

The filtrate from each run was collected in "micro-clean" glass bottles and tested for cleaness. Each sample was evaluated by four tests. The sample was first tested with a Tencor 5500 Surfscan Analyzer. The photoresist was coated and baked onto three virgin 4" flat, scanned, silicon wafers and scanned for particles. The Surfscan detects surface defects as small as 0.13µm, so scanned virgin wafers were used to assure that background levels were insignificant prior to coating.

Filtrate samples were next tested for particles with a modified PMS Model LLPS-X Liquid Laser Particle Spectrometer. At least four test sets were run for each sample to insure reproducibility.

The third and fourth tests were variations of a flow decay test. The filtrate samples were tested for filterability constant using a test based on ASTM F583-82, with the filtration through 0.2µm silver membranes. The test was extended beyond the usual 15 minutes required for determining filterability constants so that cumulative flow as a function of time could be collected in much the same manner as described above.

Results

The flow rate data shown in Figure 2 demonstrated, not surprisingly, that flow rate decreased as pore rating got smaller. Pore ratings indicated the general grouping of performance. Generally, the reduction in flow rate seemed to be proportional to the square of the pore rating, as would be predicted by Poiseulle's Law. What was very interesting was the deviation by the 0.2µm polypropylene membrane from the general trend. The flow rates were significantly greater than flow rates through comparable 0.2µm PTFE membranes. It was not apparent from the data shown in Figure 2 whether or not the higher flow rates were simply due to poorer particle retention.

One of the original intents of this study was to explore the utility of the beta rating, or filter efficiency, in determining the optimum filtration. Two levels of particulate loading, "clean" and "dirty", were filtered, but the flow decay of the "dirty" material through the 47 mm discs was so rapid as to make analysis of the filtrate impractical. A typical decay is shown in Figure 3.

The comparison in Figure 4 of filterability constants and the comparison in Figure 5 of filtrate flow through 0.2μm silver membranes both showed essentially no differences between any of the filter media tested. While the difference between the worst flow (0.1μm polypropylene) and the best flow (0.05μm PTFE) may arguably be significant, neither the filterability constants nor the filtrate flows followed any discernable trends and were best considered as part of a normal distribution. Unfiltered material, however, was distinctly worse. These tests did detect the difference between obviously dirty material and clean material, but when comparing filter media, the results implied that either there was no difference in the performance of the filter membranes, or that the tests, long established in the semiconductor industry, were insensitive to differences in particle levels below a certain threshold.

Analysis of the filtrates by surface scanning of coated substrates was a little more discriminating. The size distribution of particles on 4" wafers is shown in Figure 6; all eight of the filter media yielded clean filtrate. The distribution of particles greater than 0.4μm were indistinguishable among the eight. At very small particle sizes (less than 0.4μm), however, two populations emerged. The cleaner materials were produced by the polypropylene membranes and the 0.2μm rated nylon membrane. The PTFE membranes and the other nylon membranes were all in the other group.

Bulk liquid laser detection also yielded results which allowed the performance of the filter media to be differentiated. Reported as counts (greater than 0.5μm) per unit volume, the results indicated that filter media rated as 0.1μm generally produced a filtrate with fewer counts than filter media rated as 0.2μm. The comparison shown in Table I of the total counts by surface scanning and the total counts by bulk liquid laser detection demonstrate very little correlation (multiple r-squared = 0.036), at least when comparing these filtrates, which were all relatively clean.

Conclusions

Conclusions can be drawn from this study in two broad areas: the utility of pore ratings for predicting filter media performance, and the utility of various test methods for evaluating filter media.

Flow rate analysis gave unambiguous evaluations of the cost of using various filter media. That cost, expressed in terms of resistance to flow, was related to the pore size rating in a generally predictable way. Nonetheless, deviations were large enough to preclude pore rating from being used as a surrogate for actual testing of the application.

Pore ratings were of little use in predicting the cleanness of filtered photoresist. Some of the cleanest material produced in the study was filtered through a 0.2μm polypropylene membrane, the filter media which also gave the highest flow rates. PTFE, which has long had a reputation for producing the cleanest possible filtrates, performed no better than the other materials of construction.

The study also revealed a great deal about the test methods that are used by the semiconductor industry. Filterability constants and flow decay through a known membrane distinguished between filtered and unfiltered photoresist. To that extent, they were useful as "turkey-catchers". However, they were unable to distinguish between levels of quality, and should not be used to monitor improvements in filtration.

The two more sophisticated tests were better at distinguishing between levels of quality. Surface scanning detects surface defects in a coating due to the presence of contaminants, while bulk liquid laser detection detects particles in the liquid itself. Both methods depend on "seeing" the contaminants, and the lack of correlation between the results indicates that the two methods did not see the same thing. Given the high optical density of liquid photoresist and given the similarity of optical composition of many of the particles in the photoresist with the photoresist itself, this was not surprising. The two methods are both being used in the semiconductor industry; it was apparent from this study that one method cannot be used as a substitute for the other.

Regarding the filter media themselves, it was tempting to conclude that the 0.2μm polypropylene membrane was the superior filter for use with photoresist. However, these media were evaluated as unsupported, small diameter discs. In a production setting, the pleated cartridges used would be composed of multiple membrane layers supported both upstream and downstream. How these filter products would perform would depend not only on the combination of membranes used but on the quality of manufacture. No conclusions regarding the specific media tested would be appropriate.

Acknowledgement

The authors gratefully acknowledge the support given this study by Gary Parks for the SEM photomicrograph and by Tom Penniman and Jianping Chen for the data generation and collection.

Table I. Particle Count Results

	Surface Scanning		Bulk Liquid Laser Detection	
Filter Media	Total Counts *	Standard Deviation (n=3)	Total Counts **	Standard Deviation (n=5)
0.2μm polypropylene	147	22	51	0.4
0.2μm nylon	155	9	63	0.3 (n=4)
0.2μm PTFE	249	115	62	0.6
0.1μm polypropylene	168	29	42	0.4
0.1μm nylon	258	13	73	0.3
0.1μm PTFE	248	185	35	0.5
0.04μm nylon	339	60		
0.05μm PTFE	229	20		

* 0.13μm - 0.70 μm per unit area
** greater than 0.5 μm per unit area

Figure 1. An SEM photograph shows a commercial photoresist, Shipley's SPR2, imaged to create 0.5µm geometries and then salted with particles. The 0.5µm particles could destroy the device by creating bridges across the circuit.

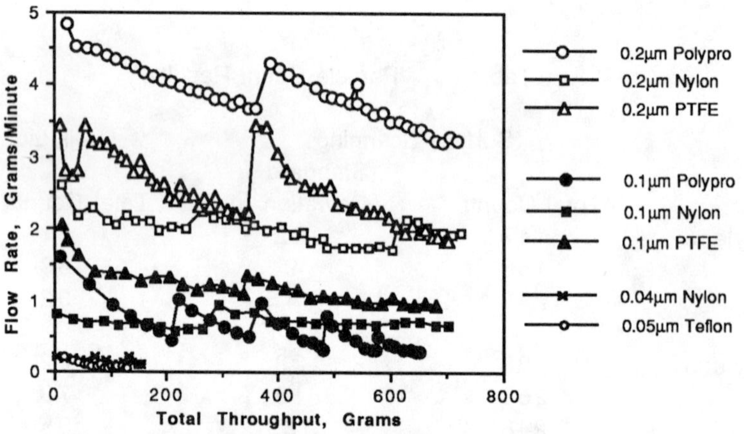

Figure 2. The flow rate of common starting material decreased as the pore rating of the filter media got smaller. While the relationship roughly followed that predicted by Poiseuille's Law, the 0.2µm polypropylene membrane performed far better than expected.

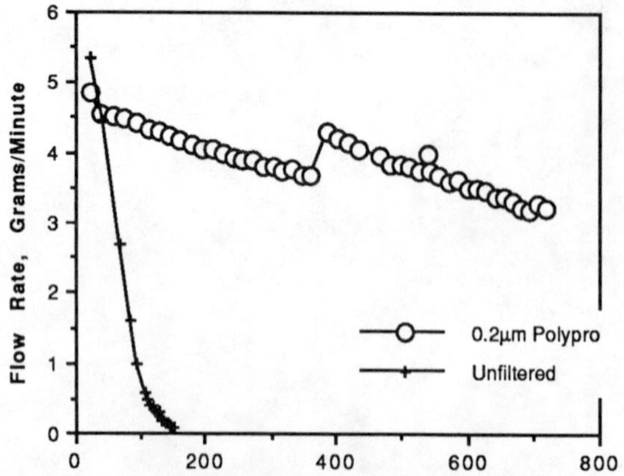

Figure 3. A typical comparison of unfiltered and pre-filtered photoresist through a membrane filter: the flow decay of unfiltered photoresist through a 0.2µm polypropylene membrane was so much worse than the flow decay of pre-filtered photoresist through the same media that further testing was precluded.

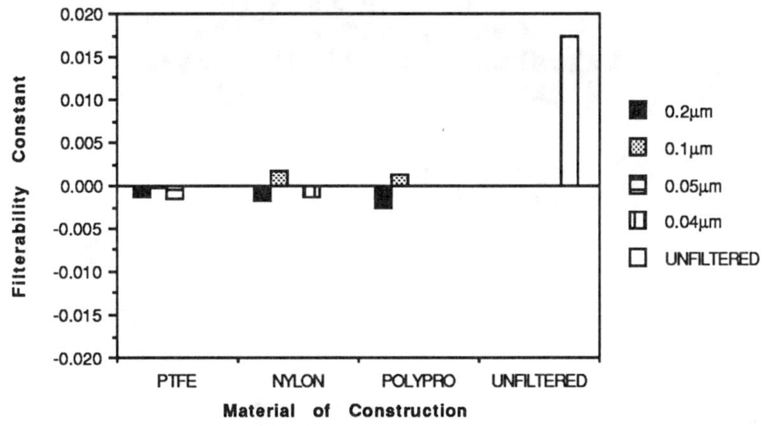

Figure 4. The filterability constants of the various filtrates were essentially identical, while unfiltered material was clearly worse.

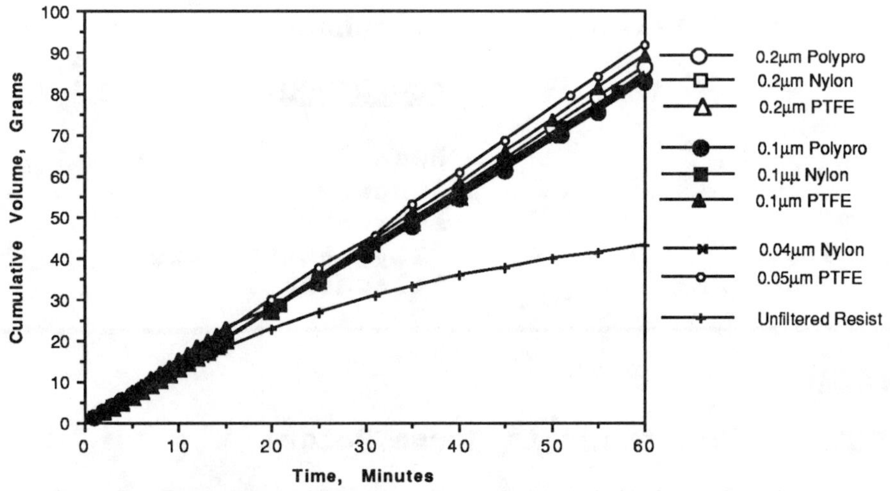

Figure 5. The flow rates of the various filtrates through a double 0.2μm sintered silver membrane were also essentially the same. They were clearly better than the flow rate of unfiltered material.

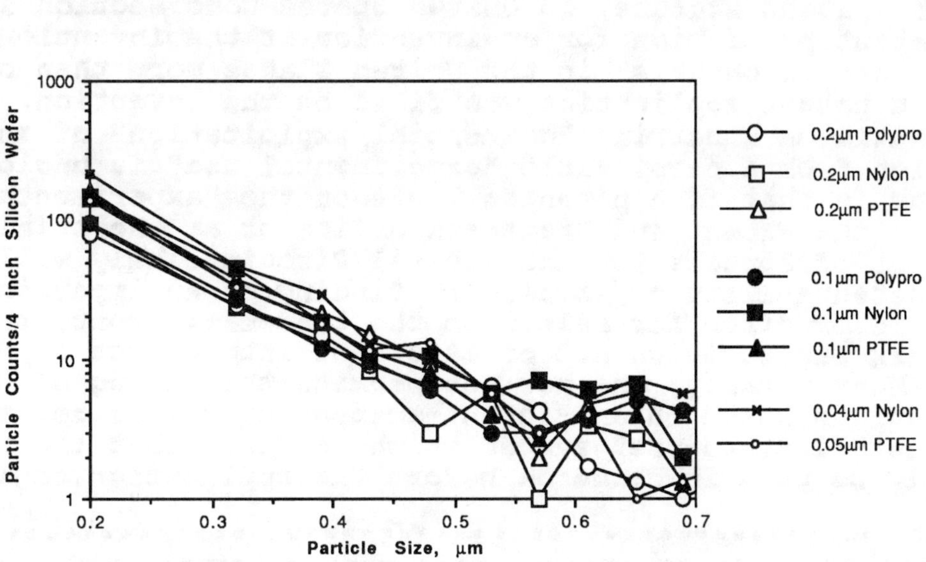

Figure 6. The analysis of particle size distribution by surface scanning showed that all eight media were indistinguishable at particle sizes greater than 0.4μm. Two distinct populatons emerged at the lower particle sizes.

AMERICAN FILTRATION SOCIETY

P.O. Box 6269
Kingwood, Texas 77325
Telephone: (713) 359-1894
FAX: (713) 358-3939

ABSTRACT & COPYRIGHT

Meeting Location: **Meeting Date:**

Session Chairman: **Session Number:**

SPEAKER (AUTHOR):

Name: GLENN RHODES
Company: ARNOLD, WHITE & DURKEE
Address: 400 ONE BERING PARK
City, State, Zip: HOUSTON, TX 77057
Telephone: 713-781-1427 FAX: 713-789-2679

CO-AUTHOR:

Name:
Company:
Address:
City, State, Zip:
Telephone: FAX:

CO-AUTHOR:

Name:
Company:
Address:
City, State, Zip:
Telephone: FAX:

CO-AUTHOR:

Name:
Company:
Address:
City, State, Zip:
Telephone: FAX:

TITLE OF PAPER:

TEXT OF ABSTRACT: (200 Words in Space Below)

Recent Developments in Patent and Trademark Law

The patent statute, 35 United States Code Section 102(b), bars patent protection for an invention if the invention was in "public use or on sale" in the United States more than one year before a patent application was filed on the invention. What differentiates a barring "commercial exploitation" of an invention from a permissible "experimental use" is unclear. What is clear is that if a patentee loses on the "experimental use" issue in the Patent and Trademark Office or at the trial court, the Court of Appeals for the Federal Circuit likely will affirm the unpatentability or invalidity finding of an impermissible use, sale or offer for sale. On the trademark front, the Trademark Law Revision Act of 1988 went into effect in November 1989. Most notably, the new act permits the filing of trademark applications based on an "intent-to-use" system rather than the previous "first-to-use" system which required that the mark actually be used in commerce before the application could be filed.

****

PLEASE TYPE INFORMATION BECAUSE DATA WILL BE PUBLISHED IN THE ABSTRACT BOOKLET. RETURN FORM TO THE AMERICAN FILTRATION SOCIETY, P.O. Box 6269, Kingwood, Texas 77325, WITH A COPY TO YOUR SESSION CHAIRMAN.

SELLING TO EUROPE IN 1992 AND BEYOND

Robert Feather
Filtration & Separation
Uplands Press Ltd
28 Centre Point House
St Giles High Street,
London WC2H 8LW

The Single Act of the European Community (EC) enacted in July 1987 sets out objectives for the Community members to achieve in the realms of the internal market, economic and social cohesion, research and development, the environment, and monetary union (Ref 1). Those objectives which must be achieved by 1992 include:

- Free movement of people,goods, services, capital and transport
- Removal of tax and customs frontiers
- Harmonisation of European standards
- Free competitive market in goods and services
- Unified standards on foodstuffs, animals, etc
- Unified health and safety legislation
- Common fisheries policy
- Merger and takeover codes

Adding impetus to these policies will be completion of the Channel Tunnel in 1993linking the UK to Europe and the consequent heavy investment in high speed rail links across the EC.

For companies in America and outside the EC the significance of some of these developments is:

It will solidify a zone of riches from a population of 350million which generates one quarter of the world's gross domestic product

($5300billion compared to a US GDP of $5800billion and Japanese of $3000billion) and at present rates of growth GDP of the EC will exceed that of the USA by the year 2000

Takeovers and mergers which would result in worldwide sales of over $5.95billion (5billion Ecu) or EC-wide sales of more than $300million (250million Ecu) must be vetted by Brussels to ensure they don't impair competition

Central Government procurement contracts over $155,000 and Local Government contracts over $240,000 must be open to bidders from all member states

Internal trading in Europe will be easier than in between some parts of America. (For example the Glass Steagal Act which prohibits US banks from engaging in securities operations)

The relevance of EC harmonisation standards on air and water pollution have considerable importance for the filtration and separation industries (Refs 2,3,4).

- In July 1980 an EC Directive set levels of sulphur dioxide and suspended particulates in the atmosphere to be attained by 1983 (with exmptions to 1993 for certain special circumstances). Similar Directives followed for nitrogen oxides and lead.

- In 1983 limits were set on emissions from road vehicle exhausts of 27% carbon monoxide, 17.5% hydrocarbons, 10.5% nitrogen oxides

- In December 1984 a limit came into force for lead in the air of 2 micrograms Pb per cum as an annual mean concentration

-In January 1987 a limit was set of 200 micrograms per cum of nitrogen dioxide in the air

- In 1988 the "Large Combustion Plants Directive" committed member states to reductions in sulphur dioxide and nitrogen oxides emissions from fossil fuel power plant. For the UK a ceiling of 3106 ktonnes/year for sulphur dioxide emissions was set,falling to 2330 ktonnes in 1998 and 1553 ktonnes in 2003. For nitrogen oxides the targets are 864 ktonnes by 1993 and 711 ktonnes by 1998, resulting in an overall reduction in sulphur dioxide of 60% and nitrogen oxides of 30%

- In June 1988 a decision was taken to ratify the Montreal Protocol of 11th September 1987 to the Vienna Convention for the protection of the ozone layer, emanating from the United Nations Environemntal Programme. This came into force in January 1989 and aims to reduce chlorofluorocarbon emissions by 50% by 1999.

- In June 1989 strict emission controls for cars under 1400cc were laid down, requiring limits on emissions of 19g per test of carbon monoxide and 5g per test of HC and NOx. Other measures limit the lead content of petrol, the sulphur content of gas oils

The principles of control of the environment from hazardous wastes and materials are based on the following precepts:

a) Best Practicable Environmental Option
b) Best Available Technology Not Entailing Cost
c) Polluter Pays Principle

In the UK , for example, principles a) and b) tend to be dealt with on a Best Practible Means basis. This requires firstly -:

- prevention of all escapes of potential pollutants irrespective of whether they may prove harmful or not

and secondly

- where prevention is impractible using modern technology such as absorbers, bag filters, electrostatic precipitators, etc then the discharge must be such that discharges are relatively harmless and inoffensive for public health purpose (ie high Chimneys in the case of gases)

These various legislations are generating enormous opportunities for filtration and separation equipmnent on power stations, water purification, sewage treatment, automative vehicles, chemical plant, clean rooms, fume cupboards, asbestos removal.
Some companies , like Pall and American Air Filter, have been established in Europe for over 20 years. Pall's UK manufacturing base is at Portsmouth and they have some 1,000 employees, as their entree to the EC and beyond. Mainly in high- tech disposable filters -cartridge, capsule, assemblies - Pall are gearing up with new factory in Cornwall to serve the European pharmaceuticals and biotechnology industries and have opened a sales office in Spain , in addition to already established wholly owned subsidiaries in Germany , France, and Italy.

American Air Filter manufacture in the UK at Cramlington, County Durham, but use Amsterdam as their sales headquarters into the EC, with manufacturing also in Holland and sales offices in other EC countries. Mainly in pollution and air filtration they find it necessary to "do their own thing" in Europe as the parent company's products need modifiying for the European market.

Johnson Filtration Systems Inc, St Paul, exploits lower the labour costs of Ireland and France for manufacture, with sales offices in Feltham, Middlesex, in the UK and in France. Whilst many other US companies are busy appointing European agents. Recent examples being Womak appointing Cedar Technology , of Wimbourne, Dorset, UK, to sell their pressure plate filters, and Osmonics Membrane Systems Inc, Minnetonka, appointing A.J.G.Waters, Croydon, UK, sales agents

There have been two main commercial effects on EC businesses of the realisation that trade and tariff barriers are due to dissolve by 1992.
1. Companies have been rushing about trying to tie up strategic, defensive, and where possible dominant market position within the EC.
2. Companies have been doing deals outside the EC to try and strengthen 3

their competitive position within the EC.

Examples of the second phenomena are a number of major acquisitions in the US by French companies, invariably backed by State funding. Rhone-Poulenc has successively acquired Union Carbide's agrochemicals business, Stauffer's minerals operation, Monsanto's aspirin outfit, the speciality chemicals business of RTZ and GAF, Canada's Connaught vaccines and a majority holding in US pharmaceuticals firm Rorer, whilst Bull also in America has purchased Venus computers and Pechiney has taken over American Can.

Factors to consider when choosing a country to set up in

1. Currency stability - W.Germany, Holland , Belgium have proved the most stable in recent years
2. Language compatibility - The UK and Eire are natural choices
3. Labour costs - Highest in Germany, Holland, Belgium , lowest in Portugal, Spain, Eire.
4. Subsidies available - All EC countries, but particularly favourable in Portugal, Greece, Spain, Corsica, Eire
5. Resistance to foreign companies - High in Germany and France, low in the UK, Northern Italy, Holland , Belgium

On balance the progressively more difficult ways of entering the market are through appointing agents, joint ventures, mergers, acquisitions, setting up independently.

A good way of assessing the market is to join an inward trade mission (Ref 5,6), or visit an exhibition , the most relevant being Achema, IFAT, Envitec, Filtech, Elmia. (Ref 7)

It is worth noting that Europe does not just comprise the 12 EC countries of W.Germany, France, Italy, UK, Spain, Portugal, Greece, Eire, Belgium, Netherlands, Denmark, Luxembourg; there is also another strong economic block namely EFTA which comprises the Nordic countries together with Austria and Switzerland.

Eastern Europe

At the time this paper was first considered it was intended to be exclusively about opportunities in Western Europe. However political events have moved so dramatically fast that it is now necessary to also consider the interaction of ex-Communist countries and even their possible integration into the EC.

Not only are West Germany, France, and the UK leading the vanguard of investment eastwards, but the EC itself is proposing financial support for Central and Eastern Europe to the tune of $240million in 1990, $1,000million in 1991, and $1225million in 1992.

Eastern Europe is a major polluter of the atmosphere, inland and coastal waterways, and over the next generation we are likely to see vast expenditure and effort put into bringing emission standards nearer to those in Western Europe. There will be extensive opportunities in Poland,

Czechoslovakia, East Germany, Romania, Hungary,and Bulgaria in the energy, water, metals refining, chemicals and carmaking industries.

The six countries just mentioned on average use more than twice as much energy and water as Western European countries for the same output of product. Coal is the main energy source , with Poland for example 78% coal dependent, and as good coal deposits are worked out lignite, with its high sulphur and ash content, is increasingly being used. Desulphurisation equipment in power plants is unknown, sewage is rarely treated, and toxic waste is improperly handled. In 1984 sulphur dioxide emission in East Germany amounted to 253 kg per person, and east European countries averaged 150 kg per person, compared to an EC average of 60 kg per person. Clearly there is lots to be done.

There will also be golden opportunites to utilise the relatively low cost but skilled labour force of Eastern Europe for manufacturing and as an entree both into Western Europe and further east - to Russia. This is an aspect the Japanese have not been slow to recognise, and they are "homing in " on Poland and its 38million population with Daihatsu propositioning the Polish State carmaker FSO, Mitsubishi lining up a food processing project, and Asahi Glass selling licences and equipment. The American company, Pall, however have had an operation in Poland for many years and are therfore well placed to take on this expanding market. Suzuki has also recently gone into a $140million joint venture with Hungary's main carmaker. South Korea has rapidly established trade delegations in Warsaw and Budapest. Nevertheless Japan's main European investment thrust has been into the UK with one in six workers in manufacturing industry expected to be employed by Japanese companies by 2010.

American industry, like the Japanese, has one great advantage in this "race" when competing with Germany. America has never invaded Poland or Czechoslovakia, and these countries are still nervous of a German reunification.

References

1. The Single Market- The Facts. Department of Trade and Industry, 1-19 Victoria Street, London SW1H OET.
2. Selling in the Single Market. Department of Trade and Industry, Room378, Victoria Street, London SW1H OET.
3. Testing and Certification. Department of trade and Industry, 1-19 Victoria Street,London SW1H OET.
4. Eurobases. 200 Rue de la Loi, B-1049,Bruxelles.
5. Task Forces. Department of Trade and Industry, 1-19 Victoria Street, London, SW1H OET.
6. Trade Promotions Guide. DTI, 88-89 Eccleston Square, London, SW1V 1PT.
7. Exhibition Bulletin. 266-272 Kirkdale, London SE26 4RZ.

SOURCING EMERGING TECHNOLOGIES FOR STRATEGIC ALLIANCES

J. Woody Stanley
Technology Catalysts Inc.
605 Park Avenue
Falls Church, VA 22046

Introduction

In the 1990s, companies operating in the filtration and separation industry face a rapidly changing business environment due to the increasing globalization of the market and the rapid pace of new technology developments. Despite the rising costs of developing and commercializing new technologies in this environment, many growth opportunities exist in this billion dollar marketplace as sales of equipment worldwide are expected to increase at annual rates of between 12 and 14 percent over the decade[1]. Those innovative companies that base their marketing and product development strategies on emerging technologies should have the advantage by capitalizing on these new opportunities while maintaining existing market share.

One strategy that is being increasingly adopted by companies is to look outside their own organizations to create and nurture alliances with other companies that allow them to access new technology developments and new markets. These strategic alliances can take a variety of forms: cooperative research agreements, marketing arrangements, technology licensing agreements, or the creation of joint ventures to develop new products or market existing products in new territories. Two of the more visible examples of this approach are the recent marketing agreement between Daicel Chemical Industries

Ltd. and Hoechst Celanese Corporation and the creation of a U.S. joint venture (TosoHaas Inc.) between Rohm & Haas and Tosoh Corporation. Strategic alliances offer many advantages to participating companies such as an expanded R&D base with greater capabilities and shared costs, access to new markets by geographic territory or industry, and a window on developments in unfamiliar markets.

Once a commitment is made to pursue strategic alliances as a corporate objective, a key task for many companies is accessing information about companies and emerging technologies in their immediate and related markets. In many companies, this process of searching for new technology developments has been established in-house using full time technology scouts. One of the first companies in the separation and filtration industry to establish such a program began their efforts in the early 1980s and now has an annual budget of approximately $250 thousand dollars for such activities. An alternative approach is to use independent technology licensing consultants. In the past decade, a few firms have specialized in this field and now provide this service to several companies in the filtration and separation industry. As one of these consulting groups, we believe there are many more situations in which companies can benefit from this approach. The remainder of this paper describes some of the sources we monitor for these new developments and the techniques used by in-house technology scouts and technology licensing consultants to identify emerging technologies.

Sources of Emerging Technology

New filtration and separation technologies are being developed by a variety of innovative organizations including:

- private and public companies with ongoing R&D programs;
- universities and private research institutions;
- government research laboratories;
- small research-oriented (SBIR) companies;
- private inventors.

In our efforts to monitor new technology developments and identify licensing or cooperative research opportunities for our client companies, we have found that the major source of emerging technologies are private and public companies. In the past four years, our firm has identified over 400 licensing opportunities for a variety of devices and processes related to industrial separation and filtration practices. From this group, approximately 55% were found to be offered by small and medium-sized companies (private and public) with ongoing research programs. The second largest source of new technologies are available from universities and private research institutions which account for 25% of the total. Approximately 14% are available from government research laboratories and the remaining 6% have been developed by private inventors and small research-oriented (SBIR) companies. A brief description of each source group is provided below.

Since the 1950s, the business of technology licensing has grown to include an estimated 50,000 private and publicly-owned companies worldwide[2]. Companies have traditionally looked to their suppliers and customer networks as sources of new ideas and technology. However, many companies have expanded this effort and now have dedicated staff members involved in locating technologies from many different sources which can be incorporated into the company's product line. At the same time, research programs produce new technologies which are not immediately adopted but still have commercial potential. Today, many companies are promoting these technologies developed within their organizations for out-licensing to potential licensees. In the filtration and separation industry, companies such as Dow Chemical, Rohm & Haas, and Mitsubishi are very active in this effort.

A more recent development is the emergence of universities and government research laboratories as a source of new technologies. This process began with the passage of federal legislation designed to stimulate the transfer of technology from the public to the private sector. The Stevenson-Wydler and Bayh-Dole Acts of 1980 (PL 96-480) and the Federal Technology Transfer Act of 1986 provided incentives to encourage scientists and engineers working on federally sponsored research programs to participate in arrangements with industry to commercialize their technologies[3]. As a result of this impetus, nearly 8000 inventions from federal laboratories are now available for licensing, according to the U.S. Department of Commerce Center for Utilization of Federal Technology. In the past year alone, over 1200 licensing opportunities and 300 patent filings were made available through this program. Research programs sponsored by agencies such as the Department of Energy and the National Institute of Health are leading to new filtration and separation products and processes that are available for commercialization.

At the same time, universities have become increasingly active in out-licensing activities. It is now estimated that at least 125 universities have active licensing programs. In 1988, approximately 400 licensing agreements were made between universities and private industry[4]. The most visible university programs are at MIT and Stanford where at least 160 agreements were concluded. One example of a university with a large filtration and separation research program and an active licensing program is the University of Cincinnati. A number of devices and processes related to membrane filtration and separation developed at the university's Center of Excellence in Membrane Technology are currently available for licensing.

University and government research laboratories in other countries are also excellent sources of technology for licensing. Examples of organizations active in developing filtration and separation technologies and out-licensing include the TNO Centre for Polymeric Materials (The Netherlands) and the Industrial Products Research Institute (Japan). Recent initiatives within the European Economic Community (Basic Research in Industrial Technologies) and Japan (MITI-Next Generation Base Technologies) have established specific research programs for membrane filtration and separation processes which could be commercialized in the coming decade.

Another federal program established in the 1980's presents new opportunities for sourcing technology. In 1982, the Small Business Innovation Development Act (PL 97-219 and PL 99-443) was enacted to shift some of the federal R&D from government research laboratories to small research-oriented companies. At least 11 federal agencies now participate in this program which has awarded over 12,000 research contracts in the period from 1983 to 1988[5]. In 1988 alone, 1400 companies received over 2700 awards with an estimated value of $389 Million dollars. A company which has used SBIR contract funding to develop new membrane products and processes is Bend Research Inc. Many more companies are nearing completion of research programs and are seeking a commercialization partner to finance the continuing development of these new technologies. The recent agreement between Ceramem Corp., an SBIR awardee that has developed a new ceramic membrane microfilter, and Exxon Chemical is an example of how a company can benefit from research initially funded by the SBIR program.

Techniques for Sourcing Technology

A variety of techniques are being employed by companies to monitor and identify new technologies[6]. Some of the more commonly used methods are listed below:

- participation in industry conferences and trade associations;
- monitoring company patent activities;

- scanning technical literature and information databases;
- attending technology transfer fairs and meetings;
- using specialized consultants for technology licensing information and services.

The first three methods listed above have been effectively used for many years by companies to monitor industry trends and locate new technology developments. Professional and trade organizations such as the American Chemical Society continue to meet these needs through a variety of publications, industry conferences, and the creation of specialized database and information services. Other private companies provide a variety of specialized services to the filtration and separation industry such as an annual membrane planning conference and newsletters with information about business trends and technology developments.

The last two approaches are more recent developments which are less widely appreciated. A variety of technology transfer conferences and fairs oriented towards strategic partnering through technology licensing arrangements are now held worldwide. Some of the more important meetings for companies involved in filtration and separation are Tech Ex, Chem Spec, Tech Mart, Innova, Techno Tokyo, and the High Tech R&D Trade Fair. Unlike conferences which are industry-oriented, these meetings are forums for innovators and potential licensees from a broad spectrum of industries to interact and focus on building strategic alliances.

Another recent trend is the use of consultants specialized in identifying technology licensing opportunities for companies. Unlike technology brokers, these consultants serve needs that are defined by client objectives and maintain an expertise in certain technical areas that can aid in determining the applicability of a given technology. They offer the added advantage of maintaining the confidentiality of the client in the initial search and inquiry process when various technologies are being considered. The Licensing Executive Society, an organization devoted exclusively to the technology licensing industry, is an excellent source for identifying consultants offering these services.

Summary

Strategic alliances are becoming more widespread as companies seek to access new markets and emerging technologies. In addition to other companies, organizations such as universities, federal laboratories, and small research-oriented companies are also candidates for alliances to commercialize new technologies. A number of filtration and separation companies have established in-house technology scouts or are using technology licensing consultants to identify these opportunities.

References

1. H. Strathmann, "Economical Evaluation of the Membrane Industry", Future Industrial Prospects of Membrane Processes, Elsevier Science Publishers Ltd., England, 1989, pp. 41-55).

2. E. Dougherty, "Tech Scouts: R&D's Globetrotters", Research & Development, October 1989, pp. 45-50.

3. J.D. Bagur and A.S. Guissinger, "U.S. Technology Transfer Incentives", les Nouvelles, 23(2) 90-97, (June 1988).

4. J. Preston, "The MIT Technology Licensing Office", presented at a meeting of the MIT Forum, Washington, DC, February 12, 1990.

5. U.S. SBA Office of Innovation, Research, and Technology, Washington, DC, Small Business Innovation Development Act: Sixth Year Results, June 1989.

6. R.L. DiCicco and W. Manfroy, "Sourcing Technology From Small Firms in the Chemical Field", les Nouvelles, 22(4) 196-199, (December 1988).

LONG TERM BUSINESS PLANNING: A SIMPLE AND PRACTICAL APPROACH THAT WORKS.

Don E. Smith, President
American Consulting Group, Inc.
35 Wedgewood Drive
Ithaca, New York 14850
607-257-4991

WHERE IS YOUR BUSINESS GOING?

Most companies simply move from one year to the next as best they can. The goal is to "grow sales and profits by 10%." On this basis, each department manager submits a budget. The controller consolidates the "plan" and presents the summary to the general manager...who fine tunes the plan by increasing the sales forecast by x% and cutting expense budgets by y%. During the year, the business plans (reacts) and functions on a daily level. Long term planning is focused on next month.

Unfortunately, this simplified exaggeration is descriptive of too many companies. Thus, they move from one year to the next and do "ok" but never "break out of the pack."

"WHERE ARE WE GOING?" "HOW ARE WE GOING TO GET THERE?"

Good planning is critical to real success...but few companies do it well. <u>One of the most common problems is that companies spend huge time on their one year plan without a specific long term goal</u>. This is equivalent to simply "driving west" when your goal is to go to San Diego.

The first rule of planning is:

<u>BEFORE</u> DEVELOPING YOUR ONE YEAR PLAN,
DEVELOP A <u>LONG TERM</u> PLAN

In this paper, I will discuss how you can develop, on ONE PAGE, a plan worksheet that really works. This plan worksheet will answer the following questions about your business:

. How large is your market?
. What is your market share?
. Where are your profits coming from?
. What are your primary strengths? Weaknesses? Opportunities? Threats?
. Where will you be in 3 (5) years based on your current programs? Sales? Profits? Market share?
. Is this where you want to be? If not, where?
. How do you make your plan work?

To start your plan, set aside one day. Personnel involved should include sales, marketing, engineering, finance and executive management. If possible, have your meeting away from the office.

Your long term planning program consists of five steps:

I COMPLETE THE FOLLOWING MATRIX FOR EITHER THE CURRENT OR PRIOR YEAR.

*************CURRENT OR PRIOR YEAR************

PRODUCT	MKT SIZE	GRWTH %/YR	SALES	% SHARE	% GROSS MGN	GROSS PROFIT	% OF PROFIT
COURSE STRAINERS	20000	5	10000	50	45	4500	51%
DUPLEX VALVES	20000	15	2000	10	60	1200	14%
DISPOSABLE FILTERS	100000	5	5000	5	50	2500	29%
CLEANABLE FILTERS	2000	15	1000	50	55	550	6%
TOTAL	142000		18000	13%	49%	8750	100%

Completion sequence:

1 Group your products into 5 to 10 "families" that make sense for your business. In grouping, be certain that you can easily identify your sales, costs and profits.

2 For each group, record your sales, gross margin $ and gross margin % of sales.

3 For each group, estimate your market and market share. In most cases, this information is not known and must be estimated. Initially, one option is to ask "For this product group, what share of the market do you think you have?" On this basis, your first pass market size is: sales/share = market.

4 For each group, estimate the growth rate of your market. Exclude inflation. This information is also typically unknown and must be estimated.

5 Calculate totals for your market, market growth/year, sales, % share, % gross margin and gross margin $.

II DEVELOP A PRELIMINARY FORECAST OF YOUR BUSINESS IN THREE (FIVE) YEARS.

	*********CURRENT OR PRIOR YEAR********						*********THREE OR FIVE YEAR FORECAST**************						
PRODUCT	MKT SIZE	GRWTH %/YR	SALES	% SHARE	% GROSS MGN	GROSS PROFIT	MKT SIZE	% SHARE	SALES	% GROSS MGN	GROSS PROFIT	CHANGED PROFIT	% OF CHANGED PROFIT
COURSE STRAINERS	20000	5	10000	50	45	4500	25526	45	11487	50	5743	1243	19%
DUPLEX VALVES	20000	15	2000	10	60	1200	40227	20	8045	60	4827	3627	56%
DISPOSABLE FILTERS	100000	5	5000	5	50	2500	127628	5	6381	55	3510	1010	16%
CLEANABLE FILTERS	2000	15	1000	50	55	550	4023	55	2212	50	1106	556	9%
TOTAL	142000		18000	13%	49%	8750	197403	14%	28126	54%	15187	6437	100%

Completion sequence:

1 Choose your timeframe. Common choices: 3 or 5 years.

2 Project the market based on the growth rate defined.

3 For each group, estimate your market share and % gross margins. Make these estimates based on discussions of your marketing-sales programs, cost reduction programs, new products, competition,...

Encourage in-depth discussions on your programs, problems, opportunities, competition, trends,.... Concisely, record these points on an easel pad for everyone to see and for later reference.

Based on this discussion, record your first pass estimates of market share and % gross margin.

4 Calculate your sales and gross margins.

5 Calculate the "changed profit." This is the difference between your long term forecasted profits and your base year profits. Then calculate the % of these additional profits coming from each product group.

6 For each product, define the 2-3 keys to success. These are those critical programs or events that spell success or failure.

III SUMMARIZE YOUR PLAN AND, AS A TEAM, DISCUSS THE OUTCOME. Good discussion questions include:

. Where are your profits coming from?
. Long term, what are your sales? Market share? Profits?
. Is this where you want to be?
. Are your goals realistic?
. What key programs are required to make this happen?
. What assumptions are questionable? Where do you need better information? Who is going to get this info?

Prior to concluding the meeting, make assignments as appropriate. Agree to meet again in 1 to weeks.

IV MODIFY THE PLAN. AS A TEAM, DEFINE THE MAJOR PROGRAMS REQUIRED TO PERMIT YOU TO MEET YOUR OBJECTIVES.

V IMPLEMENT YOUR PLAN.

In my consulting practice, I have worked with companies to develop excellent five year plans in as little as two days. Common impact from this type of planning includes:

. Major changes in program directions and priorities.

. Increased communication and teamwork.

. Better focus, at all levels, to the programs that will really make a difference.

In summary, once you have defined your long term goals, you will find that your short term plans automatically work better...and everyone will "get to San Diego"...on time and under budget.

Note: Options to increase the effectiveness of your planning include:

1 Add columns for "Gen'l and Administrative" and "Profit before Taxes." This will require some tough decisions on the allocation of variable and fixed costs. Even with the limitations of such estimates, this information can be crucial to better understanding your business and in developing your plans.

2 Place your worksheet on a computer spreadsheet. This permits you to quickly and easily modify your assumptions.

GETTING GOOD MANAGERS -- DEVELOPING THEM -- AND KEEPING THEM

Guy E. Weismantel, President
Weismantel International
P. O. Box 6269
Kingwood, Texas 77325

During a flight from New York City returning home to Houston, I picked up the latest issue of Inc magazine. Like many frequent flyers I often find that an airplane ride gives me a block of reading time that I don't schedule during my normal daily routines -- yet should. The book review section of Inc carried a blurb on Peter Drucker's latest business book with a comment that: "Drucker talks a lot about business, but obviously hasn't been there." My reaction was: "I have!"

Immediately I picked up my ballpoint and began to write a book that was titled: What Drucker Doesn't Tell you*. Since then, neither my life nor my consulting practice has quite been the same. Why? Because I began to compare some of the major differences between big business vs. small entrepreneurship vs emerging companies and found many startling similarities and a host of differences. The findings have led my own company into adopting a business "system" that is applied to client problems as well as that of Weismantel International. From the standpoint of an employed engineer, however, you may find some of the findings distasteful -- others helpful. Before discussing either let's examine recent history including some of the effects of buyouts (BO).

(*) This book was published in January, 1990 under the title: Managing Growth: Keys to Success in Expanding Companies.

Subtitle: Commodity Engineers

It is no secret what happened to petroleum, refining and chemical engineers during the last business downturn and throughout the 1980's, during a rash of acquisitions and mergers. When Chevron rescued Gulf from the jaws of an unfriendly takeover, and over 20,000 people lost their jobs in the friendly transition -- many of them engineers who took their place at the door of the Social Security office alongside exploration geologists and secretaries. Research and development teams from Getty were left pounding the pavement when Texaco made that acquisition. And, one can list a half dozen more stories that led to engineers and scientists -- especially on the Gulf Coast of the United States -- wondering how they were going to make the house payment. In Kingwood, Texas, where housing prices plunged, some engineers simply walked away from their homes because equity turned to zero.

I think everyone recognizes that during this last LBO season there have been a lot of inequities. Over the long haul this will dramatically affect the development of engineering managers, and keeping them. In the last several years, some engineers got a golden parachute or a golden handshake and/or a retirement bonus that left them well-heeled. Others got virtually nothing. Young engineers with lower salaries faired fairly well, however many of the engineers who were "on the cusp" so to speak, (that is, close to retirement but not with enough time on the job to qualify for the gold, came out losers depending upon where and how hard the ax fell.

We can blame some of the problem on the economy.
We can blame some of the problem on the LBO specialists.
We can blame some of the problem on poor management.
We can blame some of the problem on overstaffing and poor hiring and employment practices to begin with.
We can blame some of the problem on bad luck.
We can blame some of the problem on timing.
We can blame some of the problem on Reagan's hands-off policy in respect to acquisitions.
We can blame some of the problem on politics, and
We can blame some of the problem on engineers themselves.

Let's dig into that last "blame" more thoroughly.

Subtitle: The Fast Track

Did you ever wonder why some of your peers climb the job ladder so quickly while others -- even though technically talented -- tread water? There are lots of reasons, but at least one of them is that some engineers are hired to be on the fast track to begin with. The Exxons and Conocos of the world, for example, search out places to find these fast trackers and go after them. In the late 50's one such pool of talent was the staff of Admiral Rickover. Rickover was a hard taskmaster and many companies recognized. They successfully sought out his proteges and moved them onto a fast track lane within the firm. You can be sure that only a select few get into this clique and almost always they end up in key management positions.

Just as important to the fast track is "having the right mentor." This choice can keep you moving up in a company because managers want to be surrounded with people they know and trust. That's not bad in itself, however, an engineer can come in record best and even lose their job if they hook their hanger to the wrong star. This is born out in reorganizations ranging from PPG to duPont.

Obviously, then, the majority of engineers are treated like commodity items. This isn't - or may not be bad when companies are treating employees fairly, but the last decade of employment in the oil and gas industries certainly raises questions about fairness. More than one parent is now suggesting that their child rethink their choice of engineering as a profession. Even employed engineers should constantly evaluate their job condition and the future keeping in mind that no tenure is guaranteed. Middle managers, for example, have to comply to the big business plan of a big company, and a decision made by one who is higher up can entirely negate plans and promises made at lower levels. So -- should you always choose to work for a smaller company?

There are no peaches and cream promises in that petroleum patch either. In a simple proprietorship or closely held family firm, for example, management can bleed the profits and wipe out potential bonuses by everything from philandering to acting like a fruitcake.

Subtitle: What's the Answer?

To begin with, be part of a company with a mission statement you agree with and that practices what they preach in the statement. The statement should include a policy of advancement and fairness to employees.

For example: is it fair to ask a 25-year old piping engineer with a wife and two small children to move from Baton Rouge to Midland, Texas at the same salary, with additional responsibilities, and with no discussion about what that young man's next assignment might be? It happened. Do you believe this engineer is on the fast track?

I think not; but he does take on the label of a commodity item. And --- this "commodity" label probably fits best for almost all low-level engineers working for Kellogg, Fluor, Bechtel, McDermott and others. The same curve hiring and firing policy of engineering design and construction (ED&C) firms is well documented since construction of the Hoover Dam. In this area, anyone without a work contract should have an alternate work strategy in preparation for the downturn likely in 1991.

A fair question to your boss is: What are you doing to assure I am going to be gainfully employed throughout my career? If you get a lukewarm answer, consider the alternative. We all recognize that no job is absolutely secure (this is a point in favor of portable pensions) however, every engineer should have a written out success plan for himself or herself that should be understood and accepted at the very minimum by the boss, and by the spouse.

Subtitle: Analyzing What's Happening To You

In my recent book (Managing Growth: Keys to Success In Expanding Companies), there is some advice to chief executive officers in the area of: Selecting Managers, Developing Good Managers and Keeping Competent Managers. For readers of Hydrocarbon Processing, (particularly those who may not be in key management positions) some of the advice can be looked at in reverse. That is to say: "If the CEO is implementing a management development plan, do I see this implementation being performed on me? And to what extent?

In that light, let's review what the CEO may be doing. Often the best way to view the team and its assignment is with an organizational chart. Although there is some truth to the theory that organizational charts polarize the job and deter conservative effort, it is still a good management-hiring planning tool. This is particularly true when the chart is used to determine the makeup of the team needed to achieve the company's strategic and operational objectives.

Where do you fit into the chart?

Where will you fit into the chart 2 years from now, 5 years from now and 15 years from now?

How credible are these answers?

Keep in mind that emerging and expanding companies become large companies. At some point a salary structure is established for the organization. The managers don't negotiate compensation. Although any theory on compensation is certain to create arguments, here are some facets of the problem of the real key to building an emerging organization. This key is to motivate people and to develop teamwork coupled to an incentive plan that includes strategic goals, operational goals, and individual goals. The secret to making an incentive plan like this work is making sure that people understand how it operates. At the same time, design the plan to be discretionary to the extent where it is not an entitlement program where the individual's performance in the eyes of management is a significant part of the reward. The best plan is one where the people share in the achievement of the company's profit goals. It is best to tie this achievement factor or some other facet of the job that measures accomplishment.

It takes quite a bit of work initially to establish the wage rate for various jobs. Each of the jobs in your company has to be evaluated on the basis of its requirements and its responsibilities. Based on the analysis, the jobs are put into different classifications. The classifications are compared and analyzed for balance and equity. The salary range is established for each job classification. Once the program is established it can be maintained with a minimum of effort. Larger companies actually establish a point system when doing job evaluations. What becomes important to the individual is his or her analysis of stability,

Subtitle: Self-Analysis

In an emerging company the phenomena of "magnetic polar separation" occurs. Employees gravitate toward one pole or another depending upon their attitudes, desires, willingness, and ability to accept change. On one side you will find a group that likes the fast-track, responds to challenge, and aggressively pursues the new opportunities that constantly occur. On the other hand is a group that is complacent and happy to work a simple 8 hour shift and/or finds themselves asked to do things beyond their abilities.

It is not unusual, especially in large companies, for managers to keep the latter employee well into a decade of service before taking action that helps both the employee and the company. This action could include training, retraining, reassigning, or even firing. Pigeon-holing is not a good answer for an emerging company because everyone has to be able to make a contribution toward growth.

A more important worry for some engineers is the seriousness of company commitment to jobs in R&D. Many HPI companies pay lip service to growth in salary responsibilities in this area while others make a time commitment.

In closing, many psychologists believe each person has a hierarchy of needs beginning with physiological (needs of the body). This continues from the lowest level and progresses through safety (both physical and emotional), social (to belong), esteem (recognition by others), to growth (or self-actualization) at the highest level. The reason for really knowing your people is to understand their needs and nourish them, therefore creating a climate for motivation and releasing the energy that would have been spent trying to satisfy that need. An emerging company should point to satisfying those needs by achieving company-related goals.

As a start, motivation, and its sibling known as satisfaction, often come by way of achievement. But it is not just achievement that motivates or satisfies but achievement that is formally recognized. Responsibility and growth, which are related directly to the job you give a person, should provide the road to achievement and the destination policies that appropriately reward this achievement (as a formal company policy). Without this, you have the potential to cause dissatisfaction.

Keep in mind that managerial attitudes become employee attitudes. If troubles develop, you should go through a process of introspection. If people problems keep reappearing, it is possible that the only common denominator is: you! In that case, the introspection process should include a close look at those relationships that exist between people problems and performance.

Improved Paint Filtration

Howard L. Andrus
Ronningen-Petter
9151 Shaver Road
Portage, MI 49081
(616) 323-1313

Foreign competition and EPA have forced changes in paint manufacture and application. Yesteryear's auto paints were thick shiny colors. Today's auto paints are hybrid-like formulations, and contain enhancers of aluminum and mica.

Aluminum and mica must be allowed through a filter media. A feat not always done with woven or felt media. Ronningen-Petter has successfully marketed a vibrating paint filter to accomplish this task.

The design resembles a compressed coil spring with stringers welded to the I.D. Instead of using a round wire for the coil configuration, the wire is triangular like a prism, with the flat side out. A pneumatic vibrator is added above the media to give the media a soft and even vibration.

This newly adapted design, a product of 30 years of technology, allows the flat aluminum or mica through the filter media while keeping extraneous materials from downstream. The paint enters from the bottom and utilizes suspension as filtration takes place.

Successful applications have been made in the U.S., Canada, Mexico and Europe. Possibilities include duals, triples and quads. (The triples and quads are on 3" headers).

SELECTING THE CORRECT INLINE STRAINER FOR PAINTS AND OTHER HIGH SOLIDS PRODUCTS

ABSTRACT

Controlling costs, increasing productivity and achieving better quality products are now more than ever the prime objectives of U.S. manufacturers. To become more competitive we require better front end work in the design of our plants. For years pipeline strainers were considered just another pipe fitting. Todays process engineers realize that by choosing the correct pipeline strainer, the systems they design run longer, require less maintenance and produce a better quality product. In no process is this more important than in the production of paints and coatings. It does not take an expert to notice imperfections in a coating systems appearance. Ask anyone buying their first car. Depending on the process, straining equipment is used in all phases and at multiple locations throughout the coating industry. Whether it is protecting sprayheads, prescreening for filters or ensuring that all the "Fish Eyes" are removed prior to loading product containers, the pipeline strainer helps keep the system clean.

The selection of the correct strainer for the correct application is not extremely difficult, but it does require knowledge of the system and the type of contaminants that must be retained. The strainer itself is a simple device to operate and with minimal maintenance will last for many years. Strainers are macro filters. They range in particle retention from as large as .50 inch down to 400 mesh or 30 micron. Typically we see a range from .25 inch to 100 mesh or 165 microns. The degree of straining required is usually determined by the manufacturer of the equipment to be protected. The strainer is to retain all particles greater than those acceptable by this downstream equipment. Straining too fine will cause operational and maintenance problems due to premature fouling of the straining media. This can unnecessarily increase the frequency of cleaning and cause flow obstructions to the downstream equipment. A good rule of thumb is to size the strainer opening to 1/2" the size of the opening to be protected.

This article is a guide to the selection of pipeline strainers. It should prove helpful to the novice as well as the veteran process engineer.

Gerard J. Lynch
Engineering Manager

FILTERING
THE KEY TO HIGH QUALITY COATING

Alan Ponchick
March 22, 1990

A DEFECT FREE FINISH

All automobile manufacturers share a common objective: Provide as close to a defect free finish as is possible. Auto dealerships should never have to alter the "factory paint job" based on a car buyers justified complaint. The newest and most efficient assembly plants have developed programs to limit the generations of dirt and dust throughout the painting process. The painting operations are in some cases approaching "clean room" standards. The air handling systems have been upgraded with all make up air filtered to insure cleanliness. Lint free clothing, shoe coverings, air baths, and robotics are being used. All other people and environment generated dirt and dust sources are minimized or eliminated. All these efforts are directed toward achieving a defect free finish.

PAINT QUALITY SPECIFICATIONS

In order to achieve a defect free finish automakers have tightened paint quality specifications on the OEM coatings suppliers. OEM coatings suppliers usually Q.C. test at various manufacturing stages and after final filtration. The paint or coating must pass Q.C. before the batch can be filled. Test procedures vary, but in general, a "screen test" is done. A sample of filtered product is poured through a piece of solvent washed nylon mesh of a specific pore size or opening. For example, a clear topcoat, after final filtration could be tested using a 400 mesh (36 micron) screen. A gallon sample of the clear topcoat would be gravity fed through the nylon screen and the Q.C. standard would be zero counts. This means zero visible dirt or fiber contaminants left on the nylon mesh screen. If too many defects are visible, the batch would be refiltered.

CLEAN PAINT

Clearly, OEM coatings manufacturers must supply clean paint to the automaker if a zero defect finish is to be achieved. The paint manufacturer must begin with the highest quality raw materials - solvents, pigments, wetting agents, metallics etc. The raw materials must perform per their specifications. Cleanliness, and housekeeping standards have to insure that all paint mixing and transporting equipment, including all piping and lines, are free of contamination. Closed systems are used to the greatest extent possible and all containers are tightly shut during all transportation. The containers themselves must be super clean. Purchase specifications for drums or totes are to medical standards. Cleaning procedures for totes and tank wagons are explicit and demanding. It does absolutely no good to fill the cleanest paint made into dirty containers.

DIRT, FIBERS, AND THEIR ELIMINATION

OEM coatings manufacturers employ filtration to remove contaminants at different processing points. Filter selection must be based on the nature and volume of the contaminant to be removed. (SEE CHART)

Paint manufacturers have found that their efforts have led to significant improvements in Q.C. screen test results. Use of wound and resinated filter cartridges have essentially eliminated dirt. However, the removal of fibers which cause defects on a car's finish has become another challenge toward achieving a defect free finish. Fibers can be generated from many external sources. But fibers can also originate from filtration products. See photo's of dirt and fibers.

The optimal filter design will insure removal of the undesired contaminants, removal of fibers, and will not allow any media migration. This means no release of the filter media itself throughout its filtration life. To date excellent filtration performance has been achieved through the use of a bi-component fiber technology product. This type of filter cartridge is manufactured using long fibers having an inner core of polypropylene and an outer sheath of polyethylene. These fibers are processed onto a teflon coated bar while computer controlled heating of the fibers causes rigid bonding within the filter cartridge.

The resultant depth filter cartridge yields interesting filtration characteristics:

1. A uniform pore size dictated by the fiber diameter.

2. A rigid non-compressible mass offering little resistance (ΔP) to flow.

3. Uniform classification or sieve type filtration.

4. No media migration.

For example, in filtration of metallic paint only oversize particles or metallic agglomerates are removed. A filter that compresses or flexes under pressure can dislodge previously captured contaminant and pass it downstream of the filter. Additionally, a filter cartridge that provides good shear yields a lower thixotropic index which optimizes tension, distinction of image, and color.

Coatings shipped from the manufacturing site to the satellite plant or to the assembly plant may reform agglomerates in transit. Also, despite all efforts, some contamination of the coating may occur. A final trap filter is advisable. A rigid, non media releasing filter is again recommended here to capture and retain defect causing contaminants that would prevent attainment of a defect free finish.

In summary, filtration must provide the final insurance to both the OEM coatings manufacturer as well as the assembly plant. Quality, however, is the driving force. A defect free finish can only be achieved by the elimination of dirt, lint, and fibers up to the point of the coatings use - the spray nozzles and the undercoating tank.

A bi-component filter (Betapure) is loaded into a housing at an OEM coatings manufacturer.

FILTER SELECTION CHART

PRODUCT	CONTAMINANT	TYPE OF FILTRATION
CLEAR COAT	DIRT, SOLID MATERIALS	CLARIFICATION
BASE COAT (SOLID COLORS)	PIGMENT AGGLOMERATES, GELS, FIBERS	CLASSIFICATION
METALLICS	OVERSIZE AND AGGLOMERATED ALUMINA OR MICA, FIBERS	CLASSIFICATION

DEFINITIONS:

CLARIFICATION: Filtration of liquids containing small quantities of suspended solids; filtration takes out most of these solids and increases the clarity of the liquids.

CLASSIFICATION: (Sieving) To arrange or separate larger particles from smaller particles.

DEPTH FILTRATION: Filtration accomplished by flowing a fluid through a mass filter media providing a tortuous path with many entrapments to capture the contaminants.

GRADED DENSITY: Having an increasing weight per unit volume from the upstream side to the downstream side of the filter. Decreasing size pore structure from outside to inside diameter.

MEDIA MIGRATION: Carryover of fibers from filter, separator elements, or other filter material into the effluent.

FILTER CARTRIDGE TYPES

STRING WOUND

MATERIAL - YARN OR ROVING (COTTON, POLYPROPYLENE, FIBERGLASS)

MANUFACTURING - PRECISE WINDING WITH A RATIO CHANGE TO BUILD PRODUCT GRADE STRUCTURE. SOME INCLUDE A MEDIA BLANKET WOUND WITHIN.

RESIN BONDED

MATERIALS - MEDIA - POLYESTER, RAYON, ACRYLIC, CELLULOSE, GLASS.

MANUFACTURING - WINDING A STRAND OR BLANKET OF MEDIA THAT HAS OR WILL BE IMPREGNATED WITH A BONDING RESIN AROUND A CORE OR MANDREL. SOME ARE HEAVILY GRADED DENSITY.

BI-COMPONENT POLYOLEFIN DEPTH

MATERIALS - BI-COMPONENT POLYPROPYLENE AND POLYETHYLENE FIBER.

MANUFACTURING - MECHANICALLY PRODUCE A FINE WEB OF FIBER THEN THERMALLY BONDING THE FIBER TOGETHER WHILE PROCESSING.

POLYOLEFIN DEPTH

MATERIAL - POLYPROPYLENE

MANUFACTURING - MELT BLOWING MOLTEN POLYPROPYLENE EXTRUDED THROUGH A SMALL DIE OPENING AT A GIVEN PRESSURE AND FLOW RATE AND CONCENTRICALLY WINDING ON A CORE.

POLYOLEFIN SURFACE - BAG TYPE

MATERIAL - POLYPROPYLENE, NYLON, POLYESTER ETC.

MANUFACTURING - SEAM SINGLE & MULTIPLE LAYERS OF MEDIA INTO A RINGED, OR TIE ON, BAG CONFIGURATION.

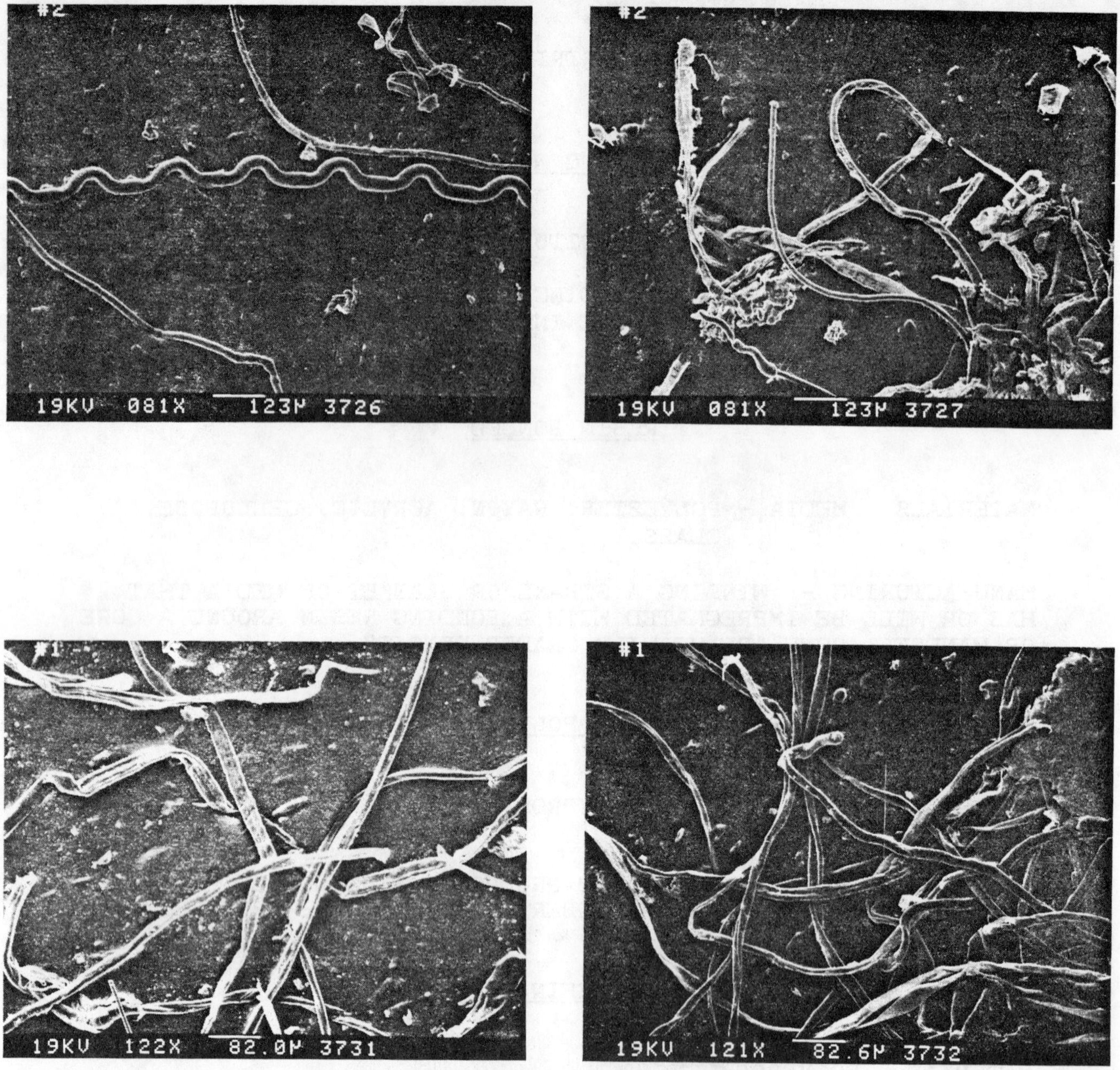

These photos show fibers that turned up in a screen test involving a clear topcoat.

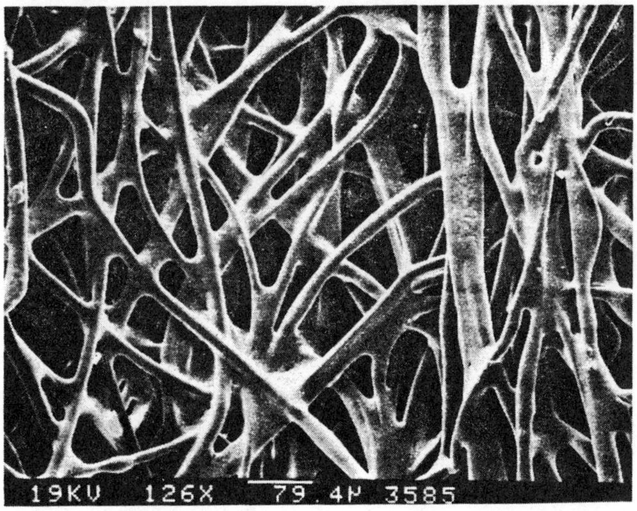

1 Micron Betapure

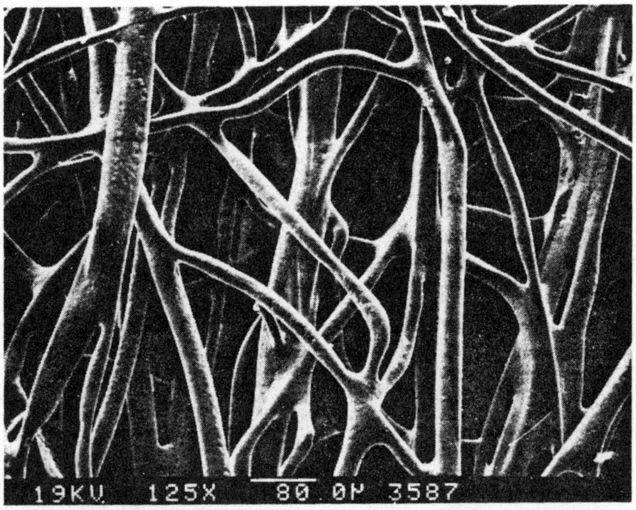

10 Micron Betapure

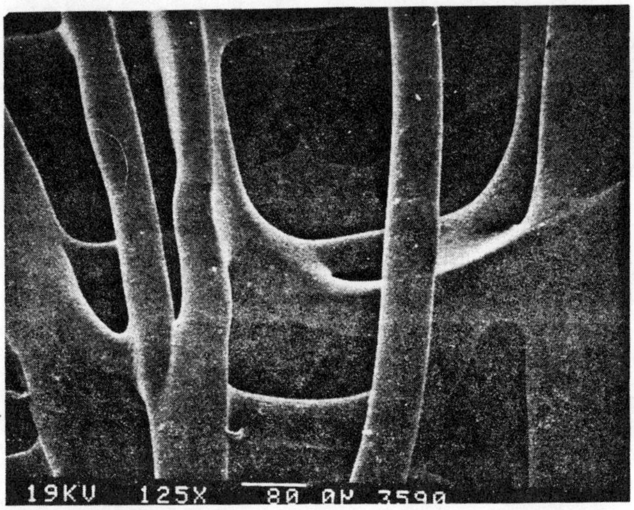

75 Micron Betapure

A Comparitive Analysis of Fabrics used in Liquid/Solid Separation

C S & S Filtration
P.O. Box 2400
2901 Long Street
Chattanooga, TN 37409

Phone number (615) 756-7067

In this paper, I will endeavor to expound on the physical variations of filter fabrics in making enlightened approximations as to liquid and solid separation. The word filtration will be used in describing the process of passing a flow of liquid through this porous medium termed fabric. Correspondingly, I will discuss the principles of resistance exhibited by fabrics on the flow and the subsequent collection of suspended solids by the fabric. I will also touch on the factors relevant in projecting the functional period of the media.

Before proceeding into a technical discussion of fabrics and their separation abilities, I must first make a disclosure that will be wise in keeping in regard to fabric filtration technology. In fabric media there is a variable degree of inexactness. No established standards apply due to the mechanical and physical variations that can occur in the manufacturing process. We must carefully compare 2 or more materials on the basis of simple resistance characteristics inherent in all porous media and consider recognized behavioral characteristics in generally broad application and also to what degree the physical characteristics of the system becomes involved in the process.

To gain a more comprehensive knowledge in supporting our decision making process, we must first become familiar with what we are to filter, and identify the variables that will affect the filtration process. In separation processes there are 3 particle properties that are important;(1) shape (2) size (3) density. The combination of these properties will determine what porous filter fabrics can be used and the ones that can not. But in selecting a fabric to perform a specific filtration task, size warrants the primary consideration. Particles are measured in several ways depending on what industry you may be associated with, although the scientific measurement 'micron' is the most widely used and the term I will be referring to in this discussion. A micron is defined as one millionth of a meter. To give you some perspective as to actual size for visualization, examples are; (1) table salt-100 microns (2) sand-50 and up (3) human hair-50-80 (4) smallest visible to naked eye-25 to 30 (5) talcum powder-10 (6) iron rust-1 to 10 (7) clay-0.1 to 1.0 (8) pigments-0.2 to 0.4. A liquid or slurry, as commonly referred, will have a combination of sizes with a specific distribution across a particular range. Dealing with the upper range of sizes becomes of no apparent difficulty, but the smaller range will be the area of most concern in selecting the most effective media. A fabric must be able to retard passage of not only the larger particles but must also facilitate deposition of even the smallest group within the distribution.

The properties of the fluid that are most important in industrial filtration are viscosity and density. Density,or the amount of dissolved solids,are relevant in terms of projecting the necessary capacity of filtering system. Viscosity deserves more attention in that it can affect filtration in many ways by altering conditions already established and identified. Temperature for example can change the viscosity of a fluid and subsequently impede valid predictions of expected results. Temperature can also affect the density of a fluid but not to the degree as it affects viscosity. One should analyze these properties and be aware how they may become altered by such influences as temperature, suspended solids, concentrations or dilutions of their slurry and formulate corrective measures in pre-treating to overcome undesirable situations. In touching on the area of pre-treatment, I will define the process as physically altering the slurry to facilitate more acceptable filtration results. This can be done by an additive such as a filter aid to enhance the development of a filter cake. In many applications, the deposited particulate forms a dense layer on the filter surface in which the fluid must pass. At this point I need to mention a mechanism that occurs in many cases when a new or clean fabric is introduced into the filtration process. Initially, the new or clean fabric will allow passage of some particles for a period of time, until at such time that a substantial cake develops to aid in stopping the smaller particles. Filter aids expedite this process and also increase the media's inherent efficiency. Many other methods are at your disposal. Flocculation and coagulation procedures can be implemented into a pre-treatment process. Some others would be freezing and thawing, ultrasonic radiation, thermal and batching. Pre-treatment should be viewed as a definite alternative and in many cases the economic rewards can be good.

Next on our list of prerequisites is to determine the objective of the particular filtration process. There are 3 primary categories; (1) refining a fluid or slurry (2) recovery of one or more of the components (3) clarifying a liquid or filtrate. Most all filtration tasks will fall into one of these categories, and awareness of the objective must be incorporated into the selection process. In identifying this factor, we will be led into another important consideration in selecting a filter material. That being cost. Many decisions, if not all, must be tempered with economic feasibility. The huge majority of filtration situations are relatively inexpensive in comparison to many of the cost incurred in an industrial environment. But there are situation that exist that require more than normal or standard treatment. In these situations, the objective will in most cases dictate the allowable expense in terms relative to the function's integral association with the overall process. To give an example, say there is a particular filtration function which purpose is to clarify a liquid coming from a vital production function. This liquid must be cleaned up to a degree to allow expulsion into an adjoining waterway. If this expulsion does not occur, a back-up will occur and the process must be halted until corrective measures are done. In this hypothetical situation the filtration becomes of extreme importance to the overall process, and the allowable cost of choosing components of the filtration function increase. Of course, many other variables may exist in establishing such a criteria, and must be taken into account. But I hope by this example I have conveyed the thought process necessary in this particular area.

The next area to consider in building the foundation for the decision making process is to analyze the physical means in which the filtration function will occur; the equipment. There are basically 5 forces involved in facilitating filtration by any piece of equipment. These are: (1) gravity (2) vacuum (3) pressure (4) centrifugal (5) electrical. Approximately 95% of all equipment manufacturers utilize one or more of these forces in designing their equipment. Each one has it's own spectrum of considerations in determining ultimate results in any given process. Each one performs quite differently in any given situation. Consequently, awareness and incorporation of this factor is a very important detail. Most operations have equipment in place and need only to identify and recognize the significance of the function characteristics. The force utilized and equipment implementing that force must be used in making a decision on the correct filter media. This brief summary of the many variables involved in selecting the optimum filter fabric is completed. I will now move into a discussion on the fabrics and how some of these variables come into play to produce desirable filtration results.

In an effort to select the proper filter fabric, we find 4 things that are important:

(1) The ability to prevent passage of particles beyond acceptable levels.

(2) The ability to allow an even flow.
(3) The ability to resist plugging or blinding.
(4) Ease of cake removal or cleaning.

The characteristics of a filter fabric depends in practice partly on the intrinsic properties of the material from which it is made, and partly on the fabricating techniques employed. As I mentioned earlier, there exists a degree of inexactness in fabrics due to this fact. No fabric can remain truly constant even being made on the same equipment and using the same raw materials, and considerable variation can occur from one manufacturer to another on the same fabric. But usually enough consistency is available to formulate a decision based on a particular fabric in 9 out of 10 situations.

There are 2 main types of mechanisms that should be distinguished. These are: (1) surface filtration (2) depth filtration. As I mentioned in earlier discussion, a new or clean filter cloth will allow passage of an amount of small particulate until it stabilizes. What occurs to provide this stabilization is that solid particles embed themselves between and within the individual yarns. Once this initial depth filtration occurs a cake is formed and the filter is now able to provide a clear filtrate. Depth filtration, as the name implies, is not restricted to just the surface, but continuous throughout part or all of its thickness. Surface filtration on the otherhand, is done on the outer most plane of the media and resist infiltration of particulate.

To initiate my review, the fabrics we will be discussing can be divided into 3 categories; (1) woven (2) non-woven (3) knit. Any of the 3 can be further divided into (1) natural fibers or (2) synthetic fibers. Woven fabric will be comprised of those materials that incorporate yarn mechanically laid into the length and width in a predetermined alignment. In the construction of a woven fabric, all of the parameters are established. The yarns are laid 90 degrees to each other, and the size of the yarn is a factor in determining the characteristics of the fabric. There are 3 forms of yarn; (1) spun staple (2) multifilament (3) monofilament. A woven fabric can be made up of any one of or a combination thereof. Multifilament yarn is made up of a number of continuously extruded very fine fibers, bound or twisted together. The filament count can vary and will have an affect on the fabric in that the higher the count the better the fabric will be able to retain small particles. Another factor of multifilament yarn is the amount of twist applied to the yarn. Twisting reduces the surface area allowing yarns to be packed more tightly thus adding weight to the fabric without using larger yarn, and increasing the number of intersects. This process is usually incorporated in situations where high efficiency is needed in conjunction with a more substantial piece of material. Filament yarn can only be made from synthetic fibers since natural fibers such as cotton, wool, cellulose, or ramie come from nature in variable lengths well short of being considered continuous. Spun yarn, however, is made of short sections of fiber, bound or twisted into a particular size. Synthetic fibers as well as natural fibers can be made into spun yarns. Spun yarn is usually identified as having a hairy like surface which is caused by the many ends protruding from the fabric surface. Multifilament yarn on the other hand has a very smooth and uniform surface. Monofilament yarn is the other form of fibrous yarn. As multifilament it too is extruded from synthetic raw material, but instead of being multi-stranded, it is a monocord. This means that it is one, single yarn with no plied or twisted configurations. Fishing line is a monofilament yarn. Usually it is clear or opaque, very resistant to abrasion, and has a very slick nature. Fabrics can be constructed of a combination of any 2 of the yarn types to gain characteristics not present in just one. This is done in situations where special demands must be made of a filter fabric.

The next area of discussion will involve the fiber types that are usually found in filtration processes. The most common found in liquid applications are:
(1) Polypropylene
(2) Polyester
(3) Nylon (4) Polyethylene
(5) Cotton
There are a number of other fibers that are used in filtration such as teflon, saran, acrylics, fiberglass, and some newer more specialized fibers. These fibers yield themselves to specialized filtration tasks and are not routinely selected for the vast number of applications. To determine the correct fiber to use, certain questions must be asked in regard to the physical conditions found in the filtration environment. Some of the conditions that must be considered are temperature, chemistry of liquid, and if abrasion a factor. In selecting a fiber to operate in these particular conditions, referral to a listing of the physical characteristics of each individual fiber is in order. By comparing the resistive features of the fiber with the existing conditions, it is possible to determine what will work best in a given situation. When extremes in either chemistry or temperature exist, it becomes necessary to use special considerations in making a selection. Certain fibers will function in extreme conditions when other will not. But in a relatively neutral environment, several fibers may perform equally well, and there are less variables present to complicate the decision making process.

Since I have identified the type fiber available and discussed the form of this fiber to be yarn, the next step in the sequence is the actual weaving. There are basically 3 weave patterns to become familiar with. These are:
(1) Plain
(2) Twill
(3) Satin (sateen)
There are of course many other weaves, such as basket, duck, chain, crowsfoot, oxford, and a host of minor variations. But they are essentially variations of the basic trio of plain, twill, and satin. The influence of these weaves on the properties of a filter medium is significant. Later in the discussion I will elaborate on these properties.

Continuing with the description of the weaving process we come to the matter of the number of yarns per inch. As mentioned earlier, yarn is woven in 2 directions on a section of fabric; the length termed warp direction and the width termed the fill direction. The construction is measured by counting the number of yarns in each direction from 1 square inch of fabric. In most cases the yarn count in the warp direction will be the higher figure. As in the case of the weave pattern, the yarn or thread count, as it is also called, will influence the properties of the filter fabric greatly. It should be mentioned here that the construction, that being the weave pattern in conjunction with the thread count, and given a particular size yarn, will dictate the weight of the filter medium. Weight is measured in terms of 1 square yard of the fabric. Weight comes into consideration when such factors as mechanical wear may be of concern, abrasion, duty requirements, gasketing, or a number of conditions that may exist that would require a filter medium with substantial integrity. It should be pointed out that even though these factors are separate, they cannot be varied independently at will, but are dictated to a large extent by the problems and techniques of the manufacturing process.

After the weaving process is completed, the fabric is in an unfinished state. This state is termed 'in the greige'. The woven fabric is subjected to a variety of finishing processes. The 3 most common are:
(1) Calendering
(2) Napping
(3) Heat treatment
The finishing process changes the physical characteristics of the fabric to provide a more effective filter medium. One of these characteristics that is vital in determining the optimum filter fabric is the permeability. This is defined as the amount of liquid or gas that will pass through a filter medium. The method of determining this flow is to measure the amount of cubic feet of air that will pass through 1 square foot of fabric at 1/2" of water pressure. Calendering, which is actually crushing the yarn then heatsetting it into place, and heat treating, which actually shrink the fabric, are the 2 main methods of adjusting the permeability.

Woven fabrics have traditionally been the bulk of filter mediums in years past. However, In recent years non-woven have reached a technological level to make them suited for filtration applications. The term non-woven refers to a process by which staple yarn is made into batts and bonded mechanically by means of a network of needles. This process yields what is commonly termed felt. Felt is essentially a pad of short fibers arranged at random. They have excellent mechanical properties, such as dimensional stability, resistance to tearing, and the absence of fraying after cutting. They provide rigidity which in many filter applications is very desirable. As in the case of woven fabrics, the physical and chemical properties of felt are ultimately limited by the properties of the fibers from which they are made. But, an aspect of felt that has made it more applicable in filtration, is the versatility of the manufacturing process. In the felting process, it is possible to change the physical properties with greater ease and with a greater range than in the case of wovens. This allows felts to be tailored to specific situations. Felts are also subject to finishing techniques similar to wovens. Calendering and heat-treatment are the primary means of adjusting properties. Felts are typically characterized by depth filtration properties, but new technology has transformed them into having surface filtration properties. This is attained by means of physically treating the surface with heat or applying a chemical coating. These surface treatments and coatings are broadening the application of their use. Some of these treatments include teflon, acrylic, urethane, acid resistant finishes, fire retardant finishes, and many other to address specific situations. Felts have been experiencing a growing acceptance in replacing many woven fabrics and this trend appears to be continuing. Non-wovens encompass numerous other types of materials such as spun bonded fabrics, paper, cellulose, and some chemically bonded materials, but for reasons of time and practical application, I have chosen not to detail.

The final category I will be discussing is knitted fabric. Knitted fabrics have seen only limited use in filtration in years past. This was due partly to the fact that the technology did not exist to provide our industry with the necessary materials needed to perform in industrial filtration processes. But in recent years there has been developments in knitting technology that has provided new and interesting materials. Knitting a fabric involves mechanically constructing yarn in a similar arrangement as wovens with the depth and versatility of felts. One strong advantage of knitted fabrics over woven and felt fabrics, is the consistency of the manufacturing process. There are not as many mechanical variables to interfere with the construction. The manufacturing process is extremely versatile and adaptable, which mean fabrics can be constructed to address specific filtration situations more readily. By the process providing greater uniformity and continuity, more satisfying results can be achieved from situations that may have been limited with woven or felt fabrics. Since present technology has provided limited fabrics, the applications being serviced by knitted media is also limited. But research is continuing, and as long as there is interest and demand for the fabrics, this research should intensify. Knitted fabrics are subject to similar finishing techniques as woven and felt fabrics, and properties of function are also similar.

This concludes the format involved in establishing a process in selecting filter media. By following this sequence, referring to the data provided on fabrics, and paying close attention to include relevant variables that may be unique in your particular system, you are now prepared to embark upon the task at hand. There is one final thing I should mention that can be of aid in the decision making process, and that is preliminary,small scale testing. This is usually performed in a laboratory setting, and can gives insight into the actual performance of a media. This is extremely advisable in situations where very sensitive conditions exist, and an erroneous selection could cause expensive outcomes. The bench-scale model is very basic and inexpensive to construct. It requires only a lab beaker, some tubing, a very small lab type vacuum pump and meter, and a few other readily available items. Equations and formulas are available to duplicate the conditions in the actual filtration system and apply them to the model. The leaf test set-up, as it is called, will generate data that can be analyzed to validate the selection process. Since the details of this testing procedure is quite involved, and cannot be described in this discussion due to time restraints, anyone interested is encouraged to let me know at the close of the presentation or later at CS&S, and I will be happy to provide information and/or assistance.

In conclusion, I want to issue a word of caution in regard to selecting filter fabric. What I have outlined today is a basic format in establishing a decision making process in better enabling you to chose a filter medium that will address the particular needs of your respective system. Filtration technology has not yet developed to the level of an exact science. There exists too many related and unrelated variables in a given process to accurately predict the ultimate performance of a filter media. As a student of filtration for many years, I have found many known and predictable areas, but I also been baffled in areas that available information was very gray or almost nonexistent. To an extent one could venture to say that accurate selection is due partly to science and partly to an acquired art. But using the format I have outlined will enable the layman to be well prepared in efforts to chose more effective filter fabrics. Data sheets outlining fabric characteristics and properties have been provided to aid you in your selection process, as well as a format for using the leaf test procedure. Since there exists some degree of complexity in the analysis of these findings, it is advisable to consult a professional before committing too deeply to media selection that may have profound influences in a given environment.

Thank you for your time and attention.

PHYSICAL PROPERTIES OF FIBRES

	Maximum safe temperature °F	Specific gravity	Absorbency of water %	Wet breaking tenacity gm./den.	Elongation at breaking %	Resistance to wear
Acetate	210	1.30	9-14	0.8-1.2	30-50	Poor
Acrylic	275-300	1.14-1.17	3-5	1.8-3	25-70	Good
Cotton	200	1.55	16-22	3.3-6.4	5-10	Fair
Fluorocarbon	400	2.3	nil	1-2	13-27	Fair
Glass	550-600	2.50-2.55	up to 0.3	3-6	2-5	Poor
Modacrylic	160-180	1.31	0.04-4	2-4	14-34	Fair
Nomex	400-450	1.38		4.1	14	Excellent
Nylon	225-250	1.14	6.5-8.3	3-8	30-70	Excellent
Polyester	300	1.38	0.04-0.08	3-8	10-50	Excellent
Polyethylene						
low density	150-165	0.92	0.01	1-3	20-80	Good
high density	200-230	0.92	0.01	3.5-7	10-45	Good
Polypropylene	250	0.91	0.01-0.1	4-8	15-35	Good
PVC	150-160	1.38		1-3		Fair
Rayon	210	1.50-1.54	20-27	0.7-4	6-40	Poor
Saran	160-180	1.7	0.1-1.0	1.2-2.3	15-30	Fair
Wool	180-200	1.3	16-18	0.76-1.6	25-35	Fair

EFFECT* OF FABRIC GEOMETRY ON ITS PERFORMANCE

Variable	Maximum filtrate clarity	Minimum resistance to flow	Minimum moisture in cake	Easiest cake discharge	Maximum cloth life	Least tendency to blind
Yarn size	Large	Small	Small	Small	Large	Small
	Medium	Medium	Medium	Medium	Medium	Medium
	Small	Large	Large	Large	Small	Large
Twists/in.	Low	High	High	High	Medium	High
	Medium	Medium	Medium	Medium	Low	Medium
	High	Low	Low	Low	High	Low
Threads/in.	High	Low	Low	High	Medium	Low
	Medium	Medium	Medium	Medium	High	Medium
	Low	High	High	Low	Low	High

* In decreasing order of preference

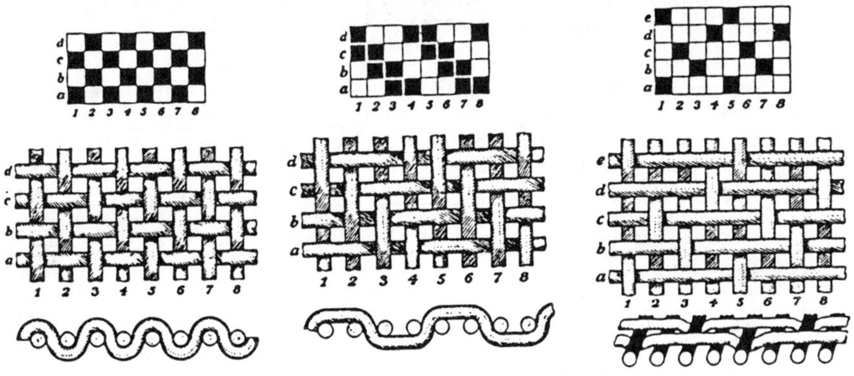

(i) PLAIN | (ii) TWILL | (iii) SATIN

THE THREE BASIC WEAVES

EFFECT* OF WEAVE PATTERN ON CLOTH PERFORMANCE

Maximum filtrate clarity	Minimum resistance to flow	Minimum moisture in cake	Easiest cake discharge	Maximum cloth life	Least tendency to blind
Plain	Satin	Satin	Satin	Twill	Satin
Twill	Twill	Twill	Twill	Plain	Twill
Satin	Plain	Plain	Plain	Satin	Plain

* In decreasing order of preference

EFFECT* OF TYPE OF YARN ON CLOTH PERFORMANCE

Maximum filtrate clarity	Minimum Resistance to flow	Minimum moisture in cake	Easiest cake discharge	Maximum cloth life	Least tendency to blind
Staple	Monofil	Monofil	Monofil	Staple	Monofil
Multifil	Multifil	Multifil	Multifil	Multifil	Multifil
Monofil	Staple	Staple	Staple	Monofil	Staple

* In decreasing order of preference

PROPERTIES OF TEXTILE FIBERS* FOR DRY FILTRATION

**All Information is based on data of the different fiber manufacturers, verified and acknowledged.*

FIBER — GENERIC NAME	COTTON	WOOL	POLYAMID	POLYPROPYLENE	POLYESTER	ACRYLIC COPOLYMER	HOMOPOLYMER ACRYLIC	AROMATIC ARAMID		POLYTETRAFLUORETHYLENE	
TRADE NAME			NYLON 66	HERCULON®	DACRON®	ORLON®	DRALON T	NOMEX®	TEIJINCONEX®	TEFLON®	TOYOFLON®
Recommended continuous operation temperature (dry heat)	180° F 82° C	200° F 94° C	200° F 94° C	200° F* 94° C	270° F 132° C	248° F 120° C	257° F 125° C	400° F 204° C	392° F 200° C	500° F* 260° C	500° F 260° C
Water vapor saturated condition (moist heat)	180° F 82° C	190° F 88° C	200° F 94° C	200° F 94° C	200° F* 94° C	230° F 110° C	260° C 125° C	350° F 177° C	356° F 180° C	500° F* 260° C	500° F 260° C
Maximum (short time) operation temperature dry heat	200° F 94° C	230° F 110° C	250° F 121° C	225° F 107° C	300° F 150° C	248° F 120° C	302° F 150° C	465° F 240° C	482° F 250° C	550° F 290° C	550° F 290° C
Specific density	1.50	1.31	1.14	0.9	1.38	1.16	1.17	1.38	1.37-1.38	2.3	2.3
Relative moisture regain in % (at 68° F & 65% relative moisture)	8.5	15	4.-4.5	0.1	0.4	1.0	1.0	4.5	4.5	0	0
Supports combustion	Yes	No	Yes	Yes	Yes	No	Yes	No	No	No	No
Biological resistance (bacteria, mildew)	No, if not treated	No, if not treated	No Effect	Excellent	No Effect	Very Good	Very Good	No Effect	No Effect	No Effect	No Effect
Resistance to alkalies	Good	Poor	Good	Excellent	Fair	Fair	Fair	Good	Good	Excellent	Excellent
Resistance to mineral acids	Poor	Good	Poor	Excellent	Fair+	Good	Very Good	Fair	Fair	Excellent	Excellent
Resistance to organic acids	Poor	Good	Poor	Excellent	Fair	Good	Excellent	Fair+	Fair	Excellent	Excellent
Resistance to oxidizing agents	Fair	Fair	Fair	Good	Good	Good	Good	Poor	Poor	Excellent	Excellent
Resistance to organic solvents	Very Good	Very Good	Very Good	Excellent	Good	Very Good	Very Good	Very Good	Very Good	Excellent	Excellent
Comments:				*250°F for Type 154	*Not Recommended						

PROPERTIES OF TEXTILE FIBERS FOR DRY FILTRATION*
Continued

**All Information is based on data of the different fiber manufactures, verified and acknowledged.*

FIBER — GENERIC NAME	EXPANDED PTFE	POLYETHERIMIDE	SULFAR (PPS)			POLYKETONE	POLYIMIDE	GLASS	METAL *	CERAMIC	
TRADE NAME	RASTEX®	AKZO® PEI	RYTON®	TEIJIN PPS	BAYER PPS	ZYEX®	P 84®	FIBERGLASS®	BEKINOX®	NEXTEL 312®	FIBERFAX®
Recommended continuous operation temperature (dry heat)	500° F 260° C	338° F 170° C	375° F 190° C	375° F 190° C	375° F 190° C	460° F 240° C	500° F 260° C	500° F 260° C	1020° F 550° C	2100° F 1150° C	2300° F 1260° C
Water vapor saturated condition (moist heat)	500° F 260° C	338° F 170° C	375° F 190° C	375° F 190° C	n.y.e.*	460° F 240° C	383° F 195° C	500° F 260° C	1020° F 550° C	2100° F 1150° C	2300° F 1260° C
Maximum (short time) operation temperature dry heat	550° F 290° C	392° F 200° C	450° F 232° C	446° F 230° C	446° F 230° C	570° F 300° C	580° F 300° C	550° F 290° C	1110° F 600° C	2600° F 1427° C	3260° F 1790° C
Specific density	1.6	1.28	1.38	1.34-1.35	1.37	1.30	1.41	2.54	7.9	2.7	2.7
Relative moisture regain in % (at 68° F & 65% relative moisture)	0	1.25	0.6	0.24-0.25	<0.6%	0.1	3.0	0	0	0	0
Supports combustion	No	No	No	No	Self Quenching LOI 39-41%	No	No	No	No	No	No
Biological resistance (bacteria, mildew)	No Effect	No Effect	No Effect	No Effect	n.y.e.*	No Effect	No Effect	No Effect	No Effect	No Effect	No Effect
Resistance to alkalies	Excellent	Good ph<9	Excellent	Excellent	Excellent	Excellent	Fair	Fair	Excellent	Good	No Effect
Resistance to mineral acids	Excellent	Good	Excellent	Excellent	Excellent	Very Good	Very Good	Very Good	Good	Very Good	Very Good
Resistance to organic acids	Excellent	Excellent	Excellent	Excellent	Excellent	Excellent	Very Good	Very Good	Very Good	Very Good	Excellent
Resistance to oxidizing agents	Excellent	Good	Fair*	Fair	Fair**	Good	Very Good	Excellent	Excellent	Excellent	Excellent
Resistance to organic solvents	Excellent	Good*	Excellent	Excellent	Excellent	Excellent	Excellent*	Very Good	Excellent	Excellent	Excellent
Comments:		*PEI fiber is dissolved by partially chlorinated hydrocarbons	*PPS fiber is attacked by strong oxidizing agents. For example at 200°F for 7 days.		*n.y.e. Not Yet Examined **Depending on Concentration		*Soluble only in strong polar solvents (DMF, DMAC, DMSO, NMP)		*INCONEL 601™		

A NOVEL FILTER DESIGN

Yuan-Ming Tang
3M Company
St. Paul, MN 55144
Tel : 612-733-0093

INTRODUCTION

Cartridge filters may be classified according to their use of "depth" or "surface" filter media. The depth filters primarily trap particles within the interstices of their internal structure. Examples are string wound and resin bonded cartridges. The surface filters act as sieves, trapping solids on their surface forming a filter cake.[1] Examples are pleated cartridges. All filters are designed to maximize their dirt loading capability. Quite simply the more dirt held per unit of cartridge cost, the lower the cost of filtering. It has long been known that traditional pleated cartridges provide more surface area and loading capacity as compared with filters having no pleats with the same dimensions. However there are a number of practical drawbacks inherent in the commercial use of such filters. These problems have until recently received little attention. 3M company has investigated the fundamental drawbacks and has demonstrated the viability of a novel design to overcome these drawbacks.

THE CHARACTERISTICS AND DRAWBACKS OF CONVENTIONAL PLEATED CARTRIDGES

Center core diameter restricts the number of pleats

Inside the conventional pleated cartridge, the filter medium consists of a filter layer or multiple filter layers, sandwiched between two protective layers . These outer layers protect the filter media from damage during the pleating process and serve as pleat separation and drainage layers in the finished cartridge. As a consequence of the sandwich construction, each pleat, naturally, has a certain thickness. The number of pleats inside the cartridge can be defined as:

$$N = \frac{2\pi r}{2t} \qquad (1)$$

N : number of pleats
r : core radius
t : thickness of filter medium

Pleating increases the surface available for filtration within the limited volume of a filter housing. Thus pleating increases the dirt holding capacity of a cartridge. The surface area is a function of pleat height, length, and number of pleats as shown in equation 2 :

$$A = N \frac{h}{\cos(360\ t / \pi r)} * 2 * l \qquad (2)$$

l : cartridge length
r : core radius
h : pleat height
A : surface area

Most cartridge manufacturers supply cartridges 25.5 cm long x 7.0 cm in diameter. The core diameters range from 3.3 to 4.45 cm. From equation 1 we know that if the core diameter and the filter medium thickness is fixed, then the number of pleats is fixed. Normally, the cartridge manufacturers compress the media near the inside core in order to allow more pleats than what is shown in equation 1. Under this compression, the cartridge can produce end effects which increase the element's pressure drop. Also bunching or pinching of the medium occurs. The particles being removed may be large enough to bridge across a pleat, blocking the internal channel between two adjacent pleats. Small particles may, after their individual deposition on the filter, agglomerate and grow large enough to cause bridging. Whatever the mechanism, the bridging serves to deny the liquid being processed to useful flow channels bordered by filter media.[2,3,4]

Scale up problems

Michael J. Matteson indicated that for a conventional pleated cartridge, the optimum pleat height is about one-fouth of the cartridge diameter.[2] By using equation 2 and assuming a filter medium thickness 0.075 cm and cartridge length of 25.4 cm, the equation becomes :

$$A = \frac{0.2128\ r\ (R-r)}{\cos(8.6 / r)}\ m^2 \qquad (3)$$

R : cartridge radius

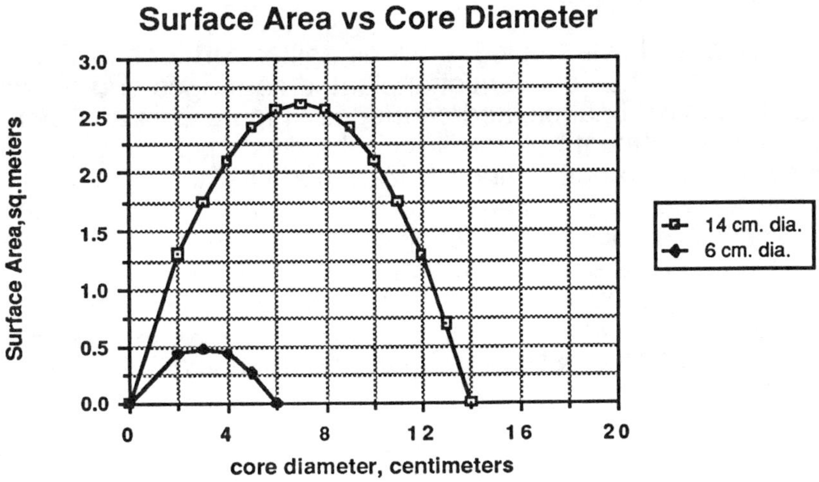

Figure 1

Figure 1 shows the results of equation 3. As shown in Figure 1, the maximum surface area for 14 and 6 cm cartridge diameter is at core diameters of 7 and 3 cm, respectively. There are three practical problems inherent here. First, if the cartridge diameter is large, a great amount of surface area is lost because of the large core.

The second problem is that conventional pleated cartridges used in high pressure fluid systems must of necessity possess sufficient strength to withstand the fluid pressure to which they are subject to during use. With larger diameters and longer pleat heights, the space between the pleats is bigger and the pleats have a strong tendency to collapse, or layover, arising from the folding together of pleats in groups, as contaminants collect on the surface of the filter and differential pressure builds up. Layover, of course, reduces the available surface for flow, further reducing the dirt capacity of the filter and giving a rapid raise in differential pressure across the filter.

Lastly, the factor governing effective filtration area in a cartridge, (in addition to cartridge diameter and core diameter) is the pleat height. Obviously, for any given pleat, the longer its height, the greater its surface area. In liquid filtration, present pleating machines do not make pleat heights much beyond 2.54 cm. The filter media in liquid filtration cartridges are made of very thin filter materials of small pore size. It is very difficult to pleat thin materials unless the filter media is mixed with a high resin content or supported by a thick support screen. The resin or thick support screen reduce the number of pleats.[3] Even if it might be desirable to make a 20 cm diameter cartridge with a 10 cm center core to maximize the surface area for liquid filter applications, conventional pleating machines do not easily make pleats with a 5 cm height. Because of the problems discussed above, the designing of a cartridge usually begins with a defining of its overall outside diameter. Given a maximum pleat height of 2.54 cm., the maximum size of the center core becomes determined. As the cartridge diameter becomes larger, maximizing the surface area becomes more difficult. This is one of the reasons liquid cartridges with diameters over 7.5 cm are rarely seen in the market.

One commonly used method to obtain larger surface areas in a single filter housing is to use multiple cartridges in a kettle style housing. This method involves a large void volume between individual cartridges. Space is taken by multiple cores and many individual seals must occur simultaneously when the housing is closed.

<u>Depth filter in pleat form</u>

It has been impossible to provide a unitized depth filter in pleated form due to the high compression near the inside core.

THE NEW DESIGN AND ADVANTAGES

Description of the new design

The novel design we have studied provides a filter cartridge having a higher surface area per unit volume than conventional pleated cartridges and overcomes the drawbacks of conventional pleated cartridges. Figure 2 is an illustration of a small section of the filter element before packing the disk shaped layers. The filter comprises a nested arrangement of disk shaped filter layers having a regular radial pleated pattern. The surface area of the filter can be controlled over a wide range simply by varying the packing or nesting density of the disk shaped layers. The filter is prepared from tubular elements which are embossed with a series of transverse raised and recessed shapes. The disk shaped layers are generated by folding the tubular elements to produce composite radially pleated disk shaped layers. The embossed wavy pattern creates flow channels when layers are folded into a complete cartridge.

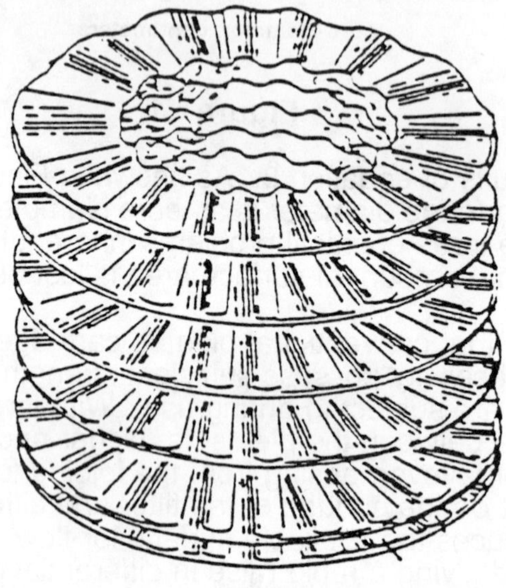

Figure 2 Illustration of a section of longitudinally collapsed filter element of the new design

Typical construction components of the new design are similar to conventional cartridges including endcaps, outer support screen, outer sleeve, drainage member and sealing materials.

Advantages of the new design

1. The number of pleats is not limited by the center core and no compression on the filter media near the center core is present as with conventional pleated cartridges.

2. The surface area of the new design can be controlled over a wide range simply by varying the packing density of the disk shaped layers along the length of the center core. For example a new filter design with the structure described above will have a surface area from 0.79 m^2 to 0.99 m^2 for 2 to 20 micron rated cartridges. Conventional pleated cartridges having the same filter media and external dimensions, have surface areas ranging from 0.4 to 0.71 m^2. Thus the new design provides a higher surface area per unit volume than is possible with conventional pleated cartridges.

3. The cartridge diameter can be increased more easily in the new filter design. This allows the addition of more useful filtering area than it is possible to add by increasing

the outside diameter of a traditional pleated cartridge. Assuming the cartridge diameter is 16 cm. Figure 3 shows the surface area comparison of the new design with a conventional pleated cartridge (25.4 cm long, 0.075 cm in medium thickness). The data shows that the new design can have a surface area of 6.4 m^2 with a 4 cm core diameter compared to the conventional cartridge with a maximum surface area of 3.4 m^2. The 4 cm core was chosen for the new design instead of a smaller diameter to allow for flowrates up to 22.70 m^3/hr.

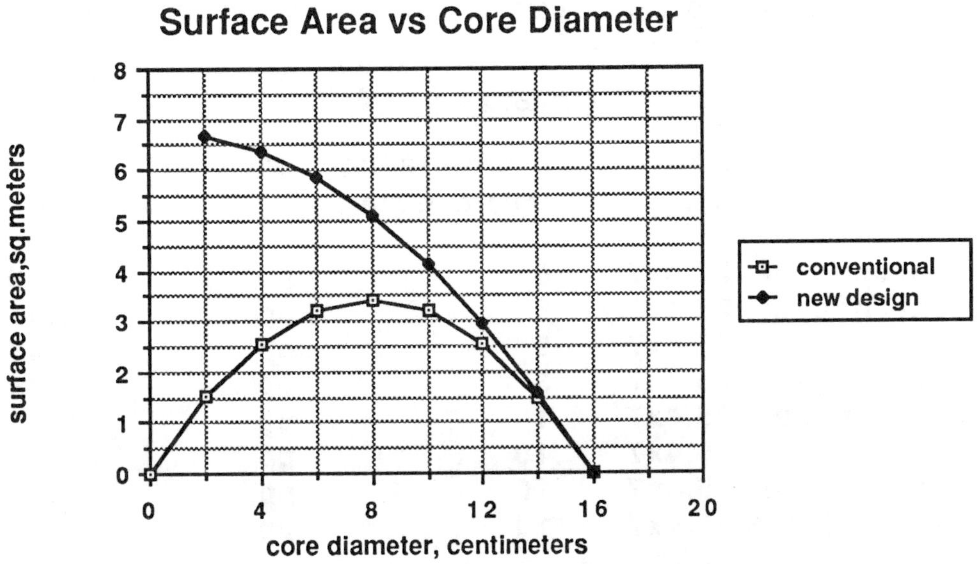

Figure 3

4. The embossed pattern and the channels formed by the embossed pattern keep the pleats from closing off on one another. The pleats are self supported by the pleats on the top and the bottom layers.

COMPARATIVE TEST RESULTS

This section will present and interpret laboratory testing of the new design and it will compare those results with commercially available filter cartridges. The objectives of the tests were to compare the dirt loading capacity and the particle removal efficiency.

Test method and equipment

The test equipment consisted of [4,5,6]:

1. Test stand : delivers a feed stream to the test cartridge at constant flowrate along with a constant concentration of test particles, uses pressure drop monitoring sensors.
2. Analytical instruments: including in-line Hiac Particle Size Analyzers (series 4300, 346 and HR 120 sensors), in line turbidimeters, multichannel analyzers and remote control computer systems.
3. Test particles: A.C. fine test dust at a constant concentration of 0.25 grams per liter for loading test and 0.01 grams per liter for efficiency test.
4. Test liquid : clean water (18 megohm water).
5. Test flowrate : 11.4 liter per minute for loading test and 15.2 liter per minute for efficiency test. The tests were stopped at a final pressure drop of 2.41 bars.

The performance characteristics used to evaluate the cartridge filters in this study were :

1. Pressure drop vs dirt added.
2. Turbidimeteric efficiency vs dirt added.
3. Particle removal efficiency vs particle size.

For functional examination, conventional pleated cartridges were compared with the new design (normally 25.4 cm long x 7.0 cm OD). The new design with scaled up diameter (25.4 cm long x 16.5 cm OD) was also tested to compare with conventional cartridges. To simplify reporting the cartridges will be referred to by a designated code letter as indicated in Table 1.

Table 1

code	construction	size (cm)	micron rating (micro meter)	material	surface area (m^2)
P5	conventional	25.4x7	5	pp**	0.46
G5	conventional	25.4x7	5	pp	0.54
M5	conventional	25.4x7	5	pp	0.46
N5	new design	25.4x7	5	pp	0.79
P10	conventional	25.4x7	10	pp	0.71
G10	conventional	25.4x7	10	pp	0.55
M10	conventional	25.4x7	10	pp	0.65
N10	new design	25.4x7	10	pp	0.79
NS5*	new design	25.4x16.5	5	pp	4.36

* tested at 45.6 liter per minute
** polypropylene filter material

RESULTS AND DISCUSSION

In Figure 4, the pressure drop and turbidimetric efficiency are plotted as a function of the dirt added. All of the pleated cartridges showed the same pattern of performance behavior. All showed an early period of turbidimetric efficiency of 70 to 72 % and very low initial pressure drop. The length of this early period, the level of efficiency and the onset of the middle period appears to be a direct function of the surface area. The middle period exhibited a faster increase in efficiency to 98+% accompanied by a corresponding increase in pressure drop. The final phase is characterized by a rapid increase in the pressure drop. The increase in efficiency gradually leveled off to a constant rate of 99+% which appeared to be similar for all cartridges. During the early period, the filter media served as simple sieves of different pore sizes. When the dirt started to cover the filter surface, cake filtration became the dominant mechanism. As pressure drop continued to increase due to the buildup of the dirt, the cake was compressed by pressure and became less permeable as shown by the rapid increase of pressure drop during the final period.

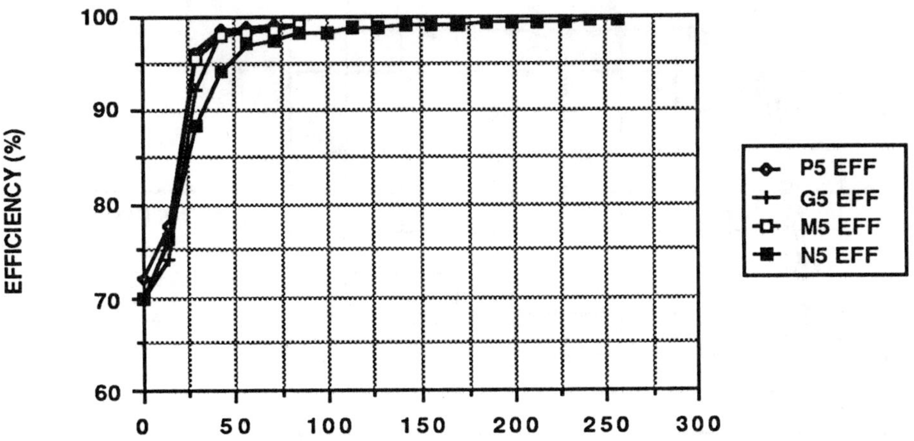

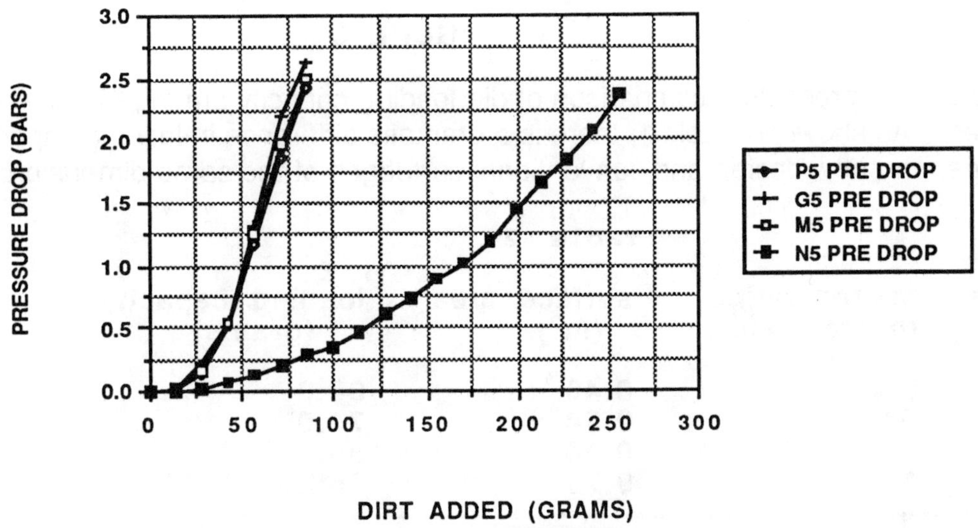

FIGURE 4

The superior dirt loading capacity of the new design is apparent in Figure 4. In this figure the dirt loading capacity for the new design is 256 grams. All of the other cartridges are below 99 grams. The development of the pressure drop with increasing dirt loading is of special interest . Although the pressure drop shows similar pattern in both the new and conventional cartridges , there are significant differences in the horizontal scale as the dirt was added. The turbidimetric removal efficiency (Figure 4) and initial removal efficiency (Figure 5) shows that the efficiency of the cartridges are similar.

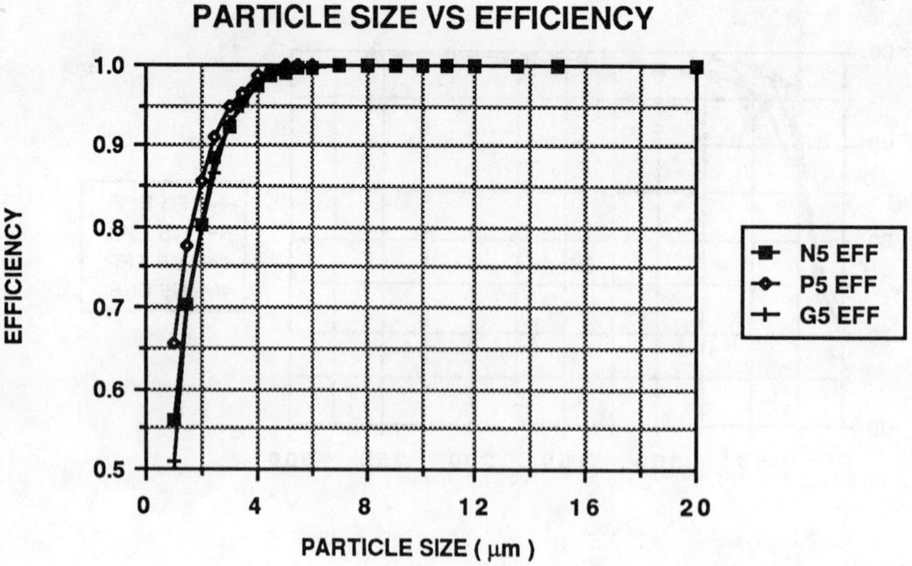

FIGURE 5

Table 2 also presents a comparison of dirt loading capacity among the different types of filters. As shown in Table 1, there is a dramatic difference in loading capacity between the new design and conventional cartridges of the same dimension.

Table 2

code	micron rating (micro meter)	surface area (m^2)	loading capacity (grams)
P5	5	0.46	88.0
G5	5	0.54	74.0
M5	5	0.46	85.2
N5	5	0.79	255.6
P10	10	0.71	161.9
G10	10	0.55	105.0
M10	10	0.55	130.0
N10	10	0.79	270.0
NS5	5	4.36	3010.0

What factors account for the superiority of the new design? Since the challenge concentration, temperature and flow rate were the same, it appears that the improved performance of the new design must stem from the higher surface area and the structure of the new design. The increase in the surface area is only 50 to 70 % for the 5 micron cartridge. The rest of the improvement should come from the new structure (less compression near the center core, and channels created from the embossed pattern).

NS5 is a scaled up new design 16.5 cm in diameter and 25.4 cm in length. The loading capacity is 3010 grams which is approximately 30 times greater than the conventional pleated cartridge. If the length of the new design is increased from 25 to 75 cm , the loading capacity will be more than 9500 grams.

CONCLUSIONS

Studies on the dirt loading capacity and efficiency were conducted. The study yields the following conclusions:

1. The new design is shown to have significantly higher loading capacity with similar efficiency.
2. The superior performance of the new design is believed to result from both the higher surface area and the structure of the design.
3. The scaled up new design dramatically increases the loading capacity compared to the conventional cartridges.

REFERENCES

1. Walter F. Lorch. " Handbook of water Purification." McGRAW-HILL, 1981. p131-140.
2. Michael J. Matterson. " Filtration, Principles and Practices". Marcel Dekker, 1987. P540-550.
3. Theodore H. Meltzer. "Filtration in the Pharmaceutical Industry", Marcel Dekker, 1987. P772-781.
4. H.N. Sandstedt. , and J.J. Weisenberger. "Cartridge Filter Performance and Micron Rating", Filtration & Separation, Vol. 22, No. 2, March/April 1985, p101-105
5. ASTM F795-82. "Determining the Performance of A Filter Medium Employing A Single-pass, Constant Rate Liquid Test". 1983.
6. E.A. Ostreicher, "Performance Evaluation of Industrial Filter Cartridges", Fluid Filtration: Liquid, Volume II, ASTM , STP 975, 1986.